Naval Education and Training Command	NAVEDTRA 12960 October 1992 0502-LP-213-9400	Training Manual (TRAMAN)

Principles of Naval Engineering

ALSO AVAILABLE:
Principles of Naval Engineering Addendum -Color Diagrams
ISBN: 978-0-9825854-4-3

ISBN: 978-0-9910923-6-9

Digitally Reproduced in 2015 by:
CONVERPAGE
23 Acorn Street
Scituate, MA 02066
781-378-1996

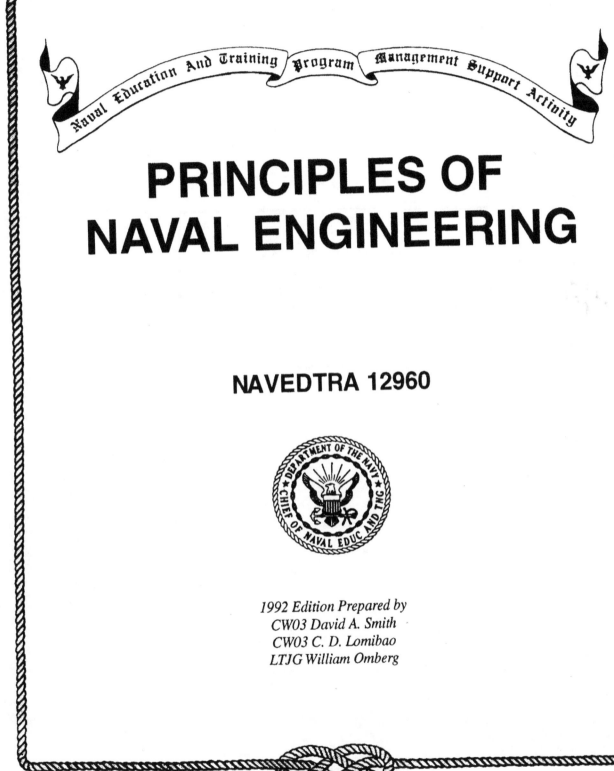

Naval Education And Training Program Management Support Activity

PRINCIPLES OF NAVAL ENGINEERING

NAVEDTRA 12960

1992 Edition Prepared by
CW03 David A. Smith
CW03 C. D. Lomibao
LTJG William Omberg

PREFACE

This training manual (TRAMAN) and nonresident training course (NRTC) form a self-study package that is designed for individual study rather than formal classroom instruction. This package provides an introduction to the theory and design of engineering machinery and equipment aboard ship. Primary emphasis is placed on helping the student acquire an overall view of shipboard engineering plants and an understanding of basic theoretical considerations that underlie the design of machinery and equipment.

This self-study package was prepared by the Naval Education and Training Program Management Support Activity, Pensacola, Florida, for the Chief of Naval Education and Training. Technical assistance was provided by the Naval Sea Systems Command (Code 08), Washington, D. C.

1992 Edition

Published by
NAVAL EDUCATION AND TRAINING PROGRAM
MANAGEMENT SUPPORT ACTIVITY

UNITED STATES
GOVERNMENT PRINTING OFFICE
WASHINGTON, D.C.: 1992

THE UNITED STATES NAVY

GUARDIAN OF OUR COUNTRY

The United States Navy is responsible for maintaining control of the sea and is a ready force on watch at home and overseas, capable of strong action to preserve the peace or of instant offensive action to win in war.

It is upon the maintenance of this control that our country's glorious future depends; the United States Navy exists to make it so.

WE SERVE WITH HONOR

Tradition, valor, and victory are the Navy's heritage from the past. To these may be added dedication, discipline, and vigilance as the watchwords of the present and the future.

At home or on distant stations as we serve with pride, confident in the respect of our country, our shipmates, and our families.

Our responsibilities sober us; our adversities strengthen us.

Service to God and Country is our special privilege. We serve with honor.

THE FUTURE OF THE NAVY

The Navy will always employ new weapons, new techniques, and greater power to protect and defend the United States on the sea, under the sea, and in the air.

Now and in the future, control of the sea gives the United States her greatest advantage for the maintenance of peace and for victory in war.

Mobility, surprise, dispersal, and offensive power are the keynotes of the new Navy. The roots of the Navy lie in a strong belief in the future, in continued dedication to our tasks, and in reflection on our heritage from the past.

Never have our opportunities and our responsibilities been greater.

CONTENTS

CREDITS

The illustration listed below is included in this edition through the courtesy of the designated source. Permission to use this ilustration is gratefully acknowledge. Permission to reproduce this illustration and other materials in this publication must be obtained from the source.

SOURCE	FIGURE
John Wiley & Sons, Inc.	8-11

CHAPTER 1

DEVELOPMENT OF NAVAL SHIPS

LEARNING OBJECTIVES

Upon completion of this chapter, you should be able to do the following:

1. Identify the pioneers in the development of steam machinery.

2. Identify the first naval warship to use screw propellers.

3. Explain the advantages of turboelectric drive.

4. Explain the advantages of nuclear propulsion.

5. Describe the development of gas turbine ships and their advantages and disadvantages.

6. Identify the pioneers in submarine development.

7. Describe the reason that submarines require both diesel engines and batteries.

8. Describe how the merging of nuclear propulsion and a submarine produced the first "true submarine."

INTRODUCTION

The story of the development of naval ships is the story of prime movers: oars, wind-filled sails, reciprocating steam engines, steam turbines, internal combustion engines, and gas turbine engines. It is also the story of the conversion and use of energy: mechanical energy, thermal energy, chemical energy, electrical energy, and nuclear energy. Seen in broader context, the development of naval ships is merely one fascinating aspect of our long struggle to control and use energy and thereby release ourselves from the limiting slavery of physical labor.

We have come a long way in the search for a better use of energy, from the muscle power required to propel an ancient Mediterranean galley to the vast reserves of power available in a shipboard nuclear reactor. Today's naval ships sail on the seas, under the sea, and skims across the top of the sea (hydrofoils). No part of this search has been easy; progress has been slow, difficult, and often beset with frustrations. And the search is far from over. Even within the next few years, new developments may drastically change our present concepts of energy use.

This chapter touches briefly on some of the highlights in the development of naval ships. In any historical survey, a few names will stand out and a few discoveries or inventions will appear to be of crucial significance. We may note, however, that our present complex and efficient fighting ships are the result not only of brilliant work by a relatively small number of well know people but also of the steady, continuing work of thousands of lesser known or anonymous contributors who have devised small but important improvements in existing machinery and equipment. The primitive man who invented the wheel is often cited as an unknown genius; we might do well to remember also the unknown genius who discovered that wheels work better when they turn in bearings. Similarly, the basic concepts involved in the design of steam turbines, internal combustion engines, and gas turbine engines may be attributed to a few men; but the innumerable small improvements that have resulted in our present efficient machines are very largely anonymous.

THE DEVELOPMENT OF STEAM MACHINERY

One of the earliest steam machines of record is the aeolipile developed about 2,000 years ago by the Greek mathematician Hero. This machine, which was actually considered more of a toy or novelty than a machine, consisted of a hollow sphere that carried four bent

Figure 1-1.--Hero's steam turbine.

nozzles. The sphere was free to rotate on the tubes that carried steam from the boiler, below, to the sphere. As the steam flowed out through the nozzles, the sphere rotated rapidly in a direction opposite to the direction of steam flow. Thus, Hero's aeolipile may be considered as the world's first reaction turbine. Hero's aeolipile (turbine) is shown in figure 1-1.

In 1698, Thomas Savery patented a condensing steam engine that was designed to raise water. This machine consisted of two displacement chambers (or one, in some models), a main boiler, a supplementary boiler, and appropriate piping and valves. The operating principles were simple, though most ingenious for the time. When steam is admitted to one of the displacement vessels, it displaces the water and forces it upward through a check valve. When the displacement vessel has been emptied of water by this method, the supply of steam is cut off. The steam already in the displacement vessel is condensed as cold water is sprayed on the outside surface of the vessel. The condensation of the steam creates a vacuum in the displacement vessel, and

the vacuum causes more water to be drawn up through suction piping and a check valve. When the displacement vessel is again full of water, steam is again admitted to the vessel and the cycle is repeated. In a model with two displacement chambers, the cycles are alternated so that one vessel is discharging water upward while the other is being filled with water drawn up through the suction pipe.

Although technological difficulties prevented Savery's engine from being used as widely as its inventor would have liked, it was successfully used for pumping water into buildings, for filling fountains, and for other applications that required a relatively low steam pressure. The machine was originally designed as a device for removing water from mines, and Savery was convinced that it would be suitable for this purpose. It was never widely used in mines, however, because very high steam pressures would have been required to lift the water the required distance. The metalworking skills of the time were simply not up to producing suitable pressure vessels for containing steam at high pressures.[1]

Although Savery's machine was used throughout the eighteenth century and well into the nineteenth century, two new steam engines had meanwhile made their appearance. The first of these, Newcomen's "atmospheric engine" represents a real breakthrough in steam machinery. Like Savery's device, the Newcomen engine was originally designed for removing water from mines, and in DOING this, it was highly successful. However, the significance of the Newcomen engine goes far beyond mere pumping. The second was the Watt engine, which brought the reciprocating steam engine to the point where it could be used as a prime mover on land and at sea.

The Newcomen engine was the first workable steam engine to use the piston and cylinder. As early as 1690, Denis Papin (who is also credited with the invention of the safety valve) had suggested a piston and cylinder arrangement for a steam engine. The piston was to be raised by steam pressure from steam generated in the bottom of the cylinder. After the piston was raised, the heat would be removed and condensation of steam in the bottom of the cylinder would create a vacuum. The downward stroke of the piston would be caused by atmospheric pressure acting on top of the piston. Papin's

---

1 It is reported that Savery attempted to use steam pressures as high as 8 or 10 atmospheres. When one considers the weakness of his pressure vessels and the total lack of safety values, it appears somewhat remarkable that he survived.

theory was good but his engine turned out to be unworkable, chiefly because he attempted to generate the steam in the bottom of the cylinder. When Papin heard of Savery's engine, he stopped working on his own piston and cylinder device and devoted himself to improving the Savery engine.

The Newcomen engine (fig. 1-2) was built by Thomas Newcomen and his assistant, John Cawley, in the early part of the eighteenth century. (The year 1712 is frequently given as the date of the Newcomen engine, and it is probably the year in which the engine was first demonstrated to a large public. It is likely, however, that previous versions of the engine were built at a considerably earlier date, and some authorities give the year 1705 as the date of the Newcomen engine.) The Newcomen engine differs from the engine suggested by Papin in several important respects. Most important, perhaps, is that Newcomen separated the boiler from the cylinder of the engine. As seen in figure 1-2, the boiler is located directly under the cylinder. Steam is admitted through a valve to the bottom of the cylinder, forcing the piston up. The piston is connected by a chain to the arch on one side of a large, pivoted, working beam. The arch

on the other side of the beam is connected by a chain to the rod of a vertical lift pump.

After the steam has forced the piston to its top position, the steam valve is shut and a jet of cold water enters the cylinder, condensing the steam and creating a partial vacuum. Atmospheric pressure then causes the down stroke (work stroke) of the piston.

As the piston comes down, the working beam is pulled down on the cylinder side. As the beam rises on the pump side, the pump rod also rises and water is lifted upward. As soon as the pressure in the cylinder equals atmospheric pressure, an escape valve in the bottom of the cylinder opens and the condensate is discharged through a drain line into a sump.

The use of automatic valve gear to control the admission of steam and the admission of cold water made the Newcomen engine the first self-acting mechanism since the invention of the clock. In the earliest versions of the Newcomen engine, the admission of steam and cold water was controlled by the manual operation of taps rather than by automatic gear. The origin of the automatic gear is a matter of some

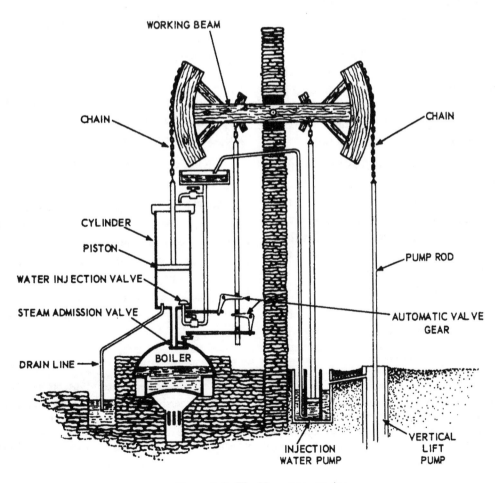

Figure 1-2.–The Newcomen engine.

dispute. One story has it that a young boy named Humphrey Potter, who was hired to turn the taps, invented the valve gear so that he could go fishing while the engine tended itself. This story, although persistent, is considered absurd by some serious historians of the steam engine.[2]

James Watt, although often credited for inventing the steam engine, did not even begin working on steam engines until some 50 years or so after the Newcomen engine was operational. However, Watt's brilliant and original contributions were ultimately responsible for the use of steam engines in a wide variety of applications beyond the simple pumping of water.

In 1799 Watt was granted a patent for certain improvements to "fire-engines" (Newcomen engines). Since some of these improvements represented major contributions to steam engineering, we should see how Watt described the improvements in a specification:

"My method of lessening the consumption of steam, and consequently fuel, in fire-engines, consists of the following principles:

"First, That vessel in which the powers of steam are to be employed to work the engine, which is called the cylinder in common fire-engines, and which I call the steam vessel, must, during the whole time the engine is at work, be kept as hot as the steam that enters it; first by inclosing it in a case of wood, or any other materials that transmit heat slowly; secondly, by surrounding it with steam or other heated bodies; and thirdly, by suffering neither water nor any other substance colder than the steam to enter or touch it during that time.

"Secondly, In engines that are to be worked wholly or partially by condensation of steam, the steam is to be condensed in vessels distinct from the steam-vessels or cylinders, although occasionally communicating with them; these vessels I call condensers, and whilst the engines are working, these condensers ought at least to be kept as cold as the air in the neighbourhood of the engines, by application of water or other cold bodies.

"Thirdly, Whatever air or other elastic vapour is not condensed by the cold of the condenser, and may impede the working of the engine, is to be drawn out of the steam-vessels or condensers by means of pumps, wrought by the engines themselves or otherwise.

"Fourthly, I intend in many cases to employ the expansive force of steam to press on the pistons, or whatever may be used instead of them, in the same manner in which the pressure of the atmosphere is now employed in common fire-engines. In cases where cold water cannot be had in plenty, the engines may be wrought by this force of steam only, by discharging the steam into the air after it has done its office."

As a result of these and other improvements, the Watt engine achieved an efficiency (in terms of fuel consumption) that was twice that of the Newcomen engine at its best. Among the other major contributions made by Watt, the following were particularly significant in the development of the steam engine:

1. The development of devices for translating reciprocating motion into rotary motion. Although Watt was not the first to devise such arrangements, he was the first to apply them to the task of making a steam engine drive a revolving shaft. This one improvement alone opened the way for the application of steam engines to many uses other than the pumping of water; in particular, it paved the way for the use of steam engines as propulsive devices.

2. The use of a double-acting piston–that is, one that is moved first in one direction and then in the opposite direction, as steam is admitted first to one end of the cylinder and then to the other.

3. The development of parallel motion linkages to keep a piston rod vertical as the beam moved in an arc.

4. The use of a centrifugal "flyball" governor to control the speed of the steam engine. Although the centrifugal governor had been used before, Watt brought to it the completely new and very significant concept of feedback. In previous use, the centrifugal governor had been capable of making a machine automatic; by adding the feedback principle, Watt made his machines self-regulating.[3]

---

2 See, for example, Eugene S. Ferguson, "The Origins of the Steam Engine," *Scientific American,* January 1964, pp. 98-107.

3 The distinction between automatic machines and self-regulating machines is of considerable significance. An automatic pump can operate without a human attendant, but it cannot change its mode of operation to fit changing requirements. A self-regulating pump, on the other hand, operates automatically and can change its speed (or some other characteristic) to meet increased or decreased demands for the fluid being pumped. To be self-regulating, a machine must have some type of feedback information from the output side of the machine to the operating mechanism.

Neither Newcomen nor Watt were able to use the advantages of high-pressure steam, largely because a copper pot was about the best that could be done in the way of a boiler. The first high-pressure steam engines were built by Oliver Evans, in the United States, and Richard Trevethick, in England. The Evans engine, which was built in 1804, had a vertical cylinder and a double-acting piston. A boiler, made of copper but reinforced with iron bands, provided steam at pressures of several atmospheres. The boiler was one of the first "fire tube" boilers; the "tubes" were actually flues that were installed in such a way as to carry the combustion gases several times through the vessel in which the water was being heated. This type of boiler, with many refinements and variations, became the basic boiler design of the nineteenth century. Trevethick, using a similar type of boiler, built a successful steam carriage in 1801; in 1804, he built what was probably the first modern type of steam locomotive.

Continuing efforts by many people led to steady improvements in the steam engine and to its eventual application as a prime mover for ships. For many years, the major effort was to improve the reciprocating steam engine. However, the latter half of the nineteenth century saw the introduction of the first practicable steam turbines. Sir Charles Parsons, in 1884, and Dr. Gustaf de Laval, in 1889, made major contributions to the development of the steam turbine. The earliest application of a steam turbine for ship propulsion was made in 1897, when a 100-ton vessel was fitted with a steam turbine that was directly coupled to the propeller shaft. After the installation of the steam turbine, the vessel broke all existing speed records for ships of any size. In 1910, Parsons introduced the reduction gear, which allowed both the steam turbine and the screw propeller to operate at their most efficient speeds, the turbine at very high speeds, and the propeller at much lower speeds. With this improvement, the steam turbine became the most significant development in steam engineering since the development of the Watt engine. With further refinements and improvements, the steam turbine is today the primary device for using the motive power of steam.

THE DEVELOPMENT OF MODERN NAVAL SURFACE SHIPS

The nineteenth century saw the application of steam power to naval ships. The first steam-driven warship in the world was the *Demologos* (voice of the people) that was later renamed the *Fulton* in honor of its builder, Robert Fulton. The ship, which is shown in figure 1-3, was built in the United States in 1815. The ship had a displacement of 2,475 tons. A paddle wheel, 16 feet in diameter, was mounted in a trough or tunnel inside the ship, for protection from gunfire. The paddle wheel was driven by a one-cylinder steam engine with a 48-inch cylinder and a 60-inch stroke.

The next large steam-driven warship to be built in the United States was the *Fulton 2nd*. This ship was built

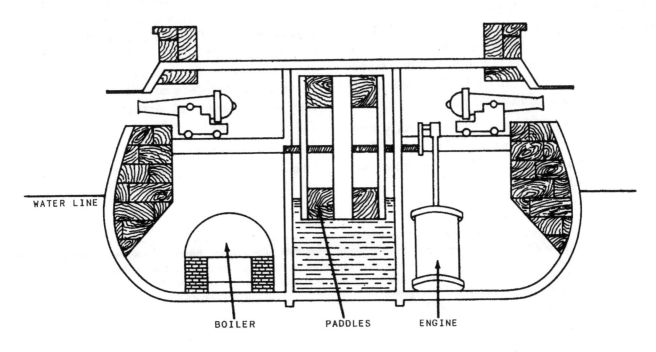

WATER LINE

BOILER   PADDLES   ENGINE

Figure 1-3.–The *Demologos:* the first steam-driven warship.

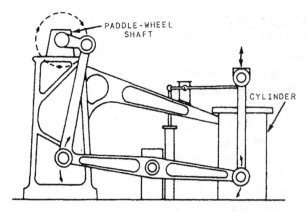

Figure 1-4.–Side-lever engine, USS *Mississippi* (1842).

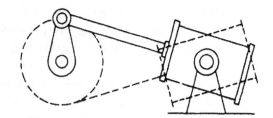

Figure 1-5.–Oscillating engine, USS *Princeton* (1844).

in 1837 at the Brooklyn Navy Yard. The *Fulton 2nd*, like the *Fulton 1* (or *Demologos*) before it, was fitted with sails as well as with a steam engine. The plant efficiency of the *Fulton 2nd* has been calculated to be about 3 percent. Its maximum speed was about 15 knots with a shaft horsepower (hp) of approximately 625.

The *Fulton 2nd* was rebuilt in 1852 and named the *Fulton 3rd.* The *Fulton 3rd* had a somewhat different kind of steam engine, and its operating steam pressure was 30 psi, rather than the 11 psi of the *Fulton 2nd*. Several other significant changes were incorporated in the *Fulton 3rd*, but the ship still had sails as well as a steam engine. The Navy was still a long way away from abandoning sails in favor of steam. The *Mississippi* and the *Missouri,* built in 1842, are sometimes regarded as marking the beginning of the steam Navy even though they, too, still had sails. The two ships were very much alike except for their engines. The *Missouri* had two inclined engines. The *Mississippi* had two side-lever engines of the type shown in figure 1-4. Three copper boilers were used on each ship. Operating steam pressures were approximately 15 psi.

The *Michigan,* which joined the steam Navy about 1843, had iron boilers rather than copper ones. These boilers lasted for 50 years. The *Michigan* operated with a steam pressure of 29 psi.

The *Princeton,* which joined the steam Navy in 1844, was remarkable for a number of reasons. It was the first warship in the world to use screw propellers, although they had been tried out more than 40 years before.[4] The *Princeton* had an unusual oscillating, rectangular piston type of engine (fig. 1-5). The piston rod was connected directly to the crankshaft, and the cylinder oscillated in trunnions. This ship was also noted for being the first warship to have all machinery located below the waterline, the first to burn hard coal, and the first to supply extra air for combustion by having blowers discharge to the fireroom. But even the *Princeton* still had sails.

Almost twenty steam-driven warships joined the steam Navy between 1854 and 1860. One of these was the *Merrimac.* Another was the *Pensacola,* which was somewhat ahead of its time in several ways. The *Pensacola* had the first surface condenser (as opposed to a jet condenser) to be used on a ship of the U.S. Navy. It also had the first pressurized firerooms.

It was not until 1867 that the U.S. Navy obtained a completely steam-driven ship. In the Navy's newly created Bureau of Steam Engineering, a brilliant designer, Benjamin Isherwood, conceived the idea for a fast cruiser. One of Isherwood's ships, the Wampanog, attained the remarkable speed of 17.75 knots during trial runs, and maintained an average of 16.6 knots for a period of 38 hours in rough seas.

The *Wampanog* had a displacement of 4,215 tons, a length of 335 feet, and a beam of 45 feet. The engines consisted of two 100-inch, single-expansion cylinders

4 In 1802, Colonel John Stevens applied Archimedes' screw as a means of ship propulsion. The first ship that Stevens tried the screw on was a single-screw ship, which unfortunately ran in circles. The second application, a twin-screw ship with the screws revolving in opposite directions, was more successful. John Ericsson, who designed the *Princeton*, developed the forerunner of the modern screw propeller in 1837. Note that the original problem with screw propellers was that they were inefficient at the slow speeds provided by the large, slow engines of the time. This is just the opposite of the present-day problem. Both then and now, the solution is gears: step-up gears, in the old days, and reduction gears, at the present time.

turning one shaft. The engine shaft was geared to the propeller shaft, driving the propeller at slightly more than twice the speed of the engines. Steam was generated by four boilers at a pressure of 35 psi and was superheated by four more boilers. The *Wampanog* propulsion plant, shown in figure 1-6, was a remarkable power plant for its time.

Sad to relate, the *Wampanog* came to an ignominious end. A board of admirals concluded that the ship was unfit for the Navy, that the four-bladed propeller was an interference to good sailing, and that the four superheater boilers were merely an unnecessary refinement. As a result of this expert opinion, two of the four propeller blades and all four of the superheater boilers were removed. The *Wampanog* was reduced from a superior steam-driven ship to an inferior sailing vessel, with steam used merely as an auxiliary source of power.

The modern U.S. Navy may be thought of as dating from 1883, the year in which congress appropriated funds for the construction of the first steel warships. The major type of engine was still the reciprocating steam engine; however, the latter part of the nineteenth century saw increasing interest in the development of internal combustion engines and steam turbines.

Ship designers approached the close of the nineteenth century with an intense regard for speed. Shipbuilders were awarded contracts with bonus and penalty clauses based on speed performance. In the construction of the cruisers *Columbia* and *Minneapolis*, a speed of 21 knots was specified. The contract stipulated a bonus of $50,000 per each quarter knot above 21 knots and a penalty of $25,000 for each quarter knot below 21 knots. The *Columbia* maintained a trial speed of 22.8 knots for 4 hours, and thereby earned for her builders a bonus of $350,000. Her sister ship, the *Minneapolis*, made 23.07 knots on her trials, earning $414,600 for that performance. Other shipbuilders profited in similar fashion from the speed race. And some, of course, were penalized for failure. The builders of the *Monterey* lost $33,000 when the ship failed to meet the specified speed.

By the early part of the twentieth century, steam was here to stay; the ships of all navies of the world were now propelled by reciprocating steam engines or by steam turbines. Coal was still the standard fuel, although it had certain disadvantages that were becoming increasingly apparent. One of the problems was the disposal of ashes. The only practicable way to get rid of them was to dump them overboard, but this left a telltale floating line on the surface of the sea, easily seen and followed by the enemy. Furthermore, the smoke from the smokestacks was enough to reveal the presence of a steam-driven ship even when it was far beyond the horizon. The military disadvantages of coal were further emphasized by the fact that it took at least one day to coal the ship, another day to clean up–a minimum of two

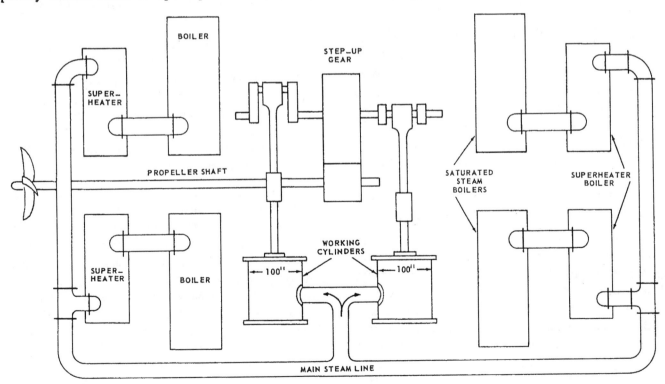

Figure 1-6.–Propulsion plant of the *Wampanog*.

days lost–and the coal would only last for another two weeks or so of steaming.

Then came oil. The means for burning oil were not developed until the early part of the twentieth century. Once the techniques and equipment were perfected, the change from coal to oil took place quite rapidly. Our first oil-burning battleships were the *Oklahoma* and the *Nevada*, which were laid down in 1911. All coal-burning ships were later altered to burn oil.

While the coal-to-oil conversion was in progress, a tug-of-war was going on in another area. The reciprocating steam engine and the steam turbine each had its proponents. To settle the matter, the Bureau of Engineering made the decision to install reciprocating engines in the *Oklahoma* and steam turbines in the *Nevada*. Although there were still many problems to be solved, the steam turbine was well on its way to becoming the major prime mover for naval ships.

With the advent of the steam turbine, the problem of reconciling the speed of the prime mover and the speed of the propeller became critical. The turbine operates most efficiently at high speed, and the propeller operates most efficiently at low speed. The obvious solution was to use reduction gears between the shaft of the prime mover and the shaft of the propeller; and basically, this is the solution that was adopted and that is still in use on naval ships today. However, other solutions were possible; and one–the use of turboelectric drive–was tried out on a fairly large scale.

During World War I, the collier *Jupiter* (later converted to the aircraft carrier *Langley*) was fitted with turboelectric drive. The high-speed turbines drove generators that were electrically connected to low-speed motors. The "big five" battleships–the *Maryland*, the *Colorado*, the *West Virginia*, the *California*, and the *Tennessee*–were all built with turboelectric drive. Ultimately, however, starting with the modernization of the Navy in 1934, the turboelectric drive gave way to the geared turbine drive; and today there are relatively few ships of the Navy that have turboelectric drive.

The period just before, during, and after World War II saw increasing improvement and refinement of the geared turbine propulsion plant. One of the most notable developments of this period was the increase in operating steam pressures–from 400 psi to 600 psi and finally, on some ships, to 1200 psi. Other improvements included reduction in the size and weight of machinery and the use of a variety of new alloys for high-pressure and high-temperature service. Although the development of naval surface ships, unlike the development of submarines, has been largely dependent upon the development of steam machinery, we should not overlook the importance of an alternate line of work–namely, the development of internal-combustion engines. In the application of diesel engines to ship propulsion, Europe was considerably more advanced than the United States; as late as 1932, in fact, the United States was in the embarrassing position of having to buy German plans for diesel submarine engines. A concerted effort was made during the 1930s to develop an American diesel industry, and the U.S. Naval Engineering Experiment Station (now the Marine Engineering Laboratory) at Annapolis undertook the testing and evaluation of prototype diesel engines developed by American manufacturers. The success of this effort may be seen in the fact that by the end of World War II the diesel hp installed in naval vessels exceeded the total hp of naval steam plants.

Since World War II the gas turbine engine has become more and more prominent as prime movers of our naval combatants of today. Throughout the course of history exists examples of other scientists using the principle of expanding gases to perform work. Among these were inventions of Leonardo da Vinci and Giovanni Branca.

The first patent for a design that used the thermodynamic cycle of the modern gas turbine was submitted in 1791. John Barber, an Englishman, submitted the patent. It was also suggested as a means of jet propulsion.

The patented application for the gas turbine as we know it today was submitted in 1930 by another Englishman, Sir Frank Whittle. His patent was for a jet aircraft engine. Whittle used his own ideas along with the contributions of scientists, such as Coley and Moss. After several failures, he came up with a working GTE. Up to this time the early pioneers in the gas turbine field were European born or oriented.

The concept of using a gas turbine to propel a ship goes back to 1937. At that time a Pescara free-piston gas engine was used experimentally with a gas turbine. The free-piston engine, or gasifier is a form of diesel engine. It uses air cushions instead of a crankshaft to return the pistons. It was an effective producer of pressurized gases. The German navy used it in their submarines during World War II as an air compressor. In 1953 the French placed in service two small vessels powered by a free-piston engine-gas turbine combination. In 1957 the liberty ship *William Patterson* went into service on a transatlantic run. It had six free-piston engines driving two turbines.

At that time applications of the use of a rotary gasifier to drive a main propulsion turbine were used. The gasifier, or compressor, was usually an aircraft jet engine or turboprop front end. In 1947 the Motor Gun Boat 2009 of the British navy used a 2500-hp gas turbine. It was used to drive the center of three shafts. In 1951 the tanker *Auris*, in an experimental application, replaced one of four diesel engines with a 1200-hp gas turbine. In 1956 the *John Sergeant* had a very efficient installation. It gave a fuel consumption rate of 0.523 pounds per hp/hr. The efficiency was largely due to use of a regenerator that recovered heat from the exhaust gases.

By the late 1950s the gas turbine marine engine was becoming widely used, mostly by European navies. All the applications combined the gas turbine plant with another conventional form of propulsion machinery. The gas turbine was used for high-speed operation. The conventional plant was used for cruising. The most common arrangements were the combined diesel or gas (CODOG) or the combined diesel and gas (CODAG) systems. Diesel engines give good cruising range and reliability. But they have a disadvantage when used in antisubmarine warfare. Their low-frequency sounds travel great distances through water. This makes them easily detected by passive sonar. Steam turbines have been combined to reduce low-frequency sound in the combined steam and gas (COSAG) configuration like those used on the British *County* class destroyers. However, these require more personnel to operate. Also they do not have the long range of the diesel combinations. Another configuration that has been successful is the combined gas or gas (COGOG) such as used on the British-type 42 DDG. These ships use the 4500-hp Tyne GTE for cruising. They use the Rolls Royce Olympus, a 28,000-hp engine, for high speed.

The U.S. Navy entered the marine gas turbine field with the *Asheville* class patrol gunboats. These ships have the CODOG configuration with two diesel engines for cruising. A General Electric LM1500 gas turbine operates at high speed. The Navy has now designed and is building destroyers, frigates, cruisers, and patrol hydrofoils that are entirely propelled by GTEs. The first frigate propelled by GTE was of the *Oliver Hazard Perry* class (fig. 1-7) and the first cruiser was of the

24.80

Figure 1-7.–USS *Oliver Hazard Perry* (FFG-7).

Ticonderoga class (fig. 1-8). The newest addition to the fleet is the gas turbine destroyer USS *Arleigh Burke* (DDG-51) (fig. 1-9). In the future, fast combat support ships will also be gas turbine. This is a result of the reliability and efficiency of the new gas turbine designs.

The gas turbine, when compared to other types of engines, offers many advantages. Its greatest asset is its high power-to-weight ratio. Compared to the gasoline piston engine, the gas turbine operates on cheaper and safer fuel. In a warship, the lack of low-frequency vibration of gas turbines makes them preferable to diesel engines. There is less noise for a submarine to pick up at long range. Modern production techniques have made gas turbines economical in terms of horsepower-per-dollar on initial installation. Their increasing reliability makes them a cost-effective alternative to steam turbine or diesel engine installation. In terms of fuel economy, modern marine gas turbines can compete with diesel engines. They may be superior to boiler/steam turbine plants, when these are operating on distillate fuel.

However, there are some disadvantages to gas turbines. Since they are high-performance engines, many parts are under high stress. Improper maintenance and lack of attention to details of procedure will impair engine performance. This may ultimately lead to engine failure. A pencil mark on a compressor turbine blade or a fingerprint can cause failure of the part. The turbine takes in large quantities of air that may contain substances or objects that can harm. Most gas turbine propulsion control systems are very complex because you have to control several factors. You have to monitor numerous operating conditions and parameters. The control systems must react quickly to turbine operating conditions to avoid casualties to the equipment. Gas turbines produce high-pitched loud noises that can damage the human ear. In shipboard installations, special soundproofing is necessary. This adds to the complexity of the installation and makes access for maintenance more difficult. Also, the large amount of air used by a GTE requires large intake and exhaust

24.83

Figure 1-8.–USS *Ticonderoga* (CG-47).

147.195

Figure 1-9.–USS *Arleigh Burke* (DDG-51).

ducting. This takes up much valuable space on a small ship.

From a tactical standpoint, there are two major drawbacks to the GTE. The first is the large amount of exhaust heat produced by the engines. Most current antiship missiles are heat-seekers. The infrared (IR) signature of a gas turbine makes an easy target. Countermeasures, such as exhaust gas cooling and IR decoys, have been developed to reduce this problem.

The second tactical disadvantage is the requirement for depot maintenance and repair of major casualties. The turbines are not overhauled in place on the ship. They must be removed and replaced by rebuilt engines if any major casualties occur. Here too, design has

reduced the problem. An engine can be changed wherever crane service and the replacement engine are available.

As improved materials and designs permit operation at higher combustion temperatures and pressures, gas turbine efficiency will increase. Even now, gas turbine main propulsion plants offer fuel economy and installation costs no greater than diesel engines. Initial costs are lower than equivalent steam plants that burn distillate fuels. Future improvements have made gas turbines the best choice for nonnuclear propulsion of ships up to cruiser size.

One of the most dramatic events in the entire history of naval ships is the application of nuclear power, first to submarines and then to surface ships. The first three ships of our nuclear surface fleet are shown in figure 1-10. They are from top to bottom: the guided missile destroyer, USS *Long Beach* (CGN-9); the aircraft carrier, USS *Enterprise* (CVN-65), on which crew members are shown on the flight deck forming Einstein's famous equation, which is the basis of controlled nuclear power; and the guided missile cruiser, USS *Bainbridge* (CGN-25). The fourth nuclear surface ship to join the fleet was USS *Truxton* (CGN-35). The USS *Nimitz* (CVN-68) was the first of her class of nuclear-powered aircraft carriers. The sixth of the *Nimitz* class, USS *George Washinton* (CVN-73), will be added to the fleet in the near future. Although the full implications of nuclear propulsive power may not yet be fully realized, one thing is already clear: the nuclear-powered ship is virtually free of the limitations on steaming radius that apply to ships using other forms of fuel. Because of this one fact alone, the future of propulsive power seems assured.

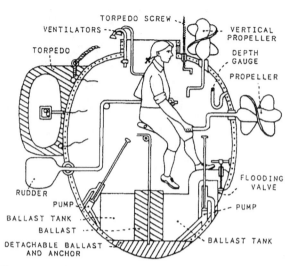

Figure 1-11.–The *Turtle*–the first submersible used in naval warfare.

THE DEVELOPMENT OF SUBMARINES

Although ancient history records numerous attempts of varying degrees of success to build underwater craft and devices, the first successful submersible craft and

147.6

Figure 1-10.–The nuclear surface fleet, 1965: USS *Long Beach* (CGN-9) (top); USS *Enterprise* (CVN-65) (middle); and USS *Bainbridge* (CGN-25) (bottom).

certainly the first to be used as an offensive weapon in naval warfare was the *Turtle* (fig. 1-11), a one-man submersible invented by David Bushnell during the American Revolutonary War. The *Turtle*, which was propelled by a hand-operated screw propeller, attempted to sink a British man-of-war in New York Harbor. The plan was to attach a charge of gunpowder to the ship's bottom with screws and to explode it with a time fuse. After repeated failures to force the screws through the copper sheathing of the hull of the British ship, the submarine gave up, released the charge, and withdrew. The powder exploded without any result except to cause the British man-of-war to shift to a berth farther out to sea.

The *Turtle* looked somewhat like a lemon standing on end. The vessel had a water ballast system with hand-operated pumps, as well as the hand-operated propeller. It also had a crude arrangement for drawing in fresh air from the surface. The vent pipes even closed automatically when the water reached a certain level.

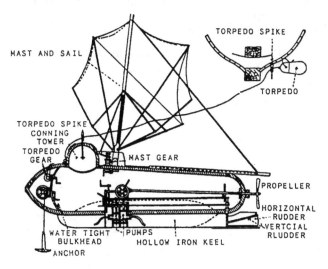

Figure 1-12.–The first *Nautilus*.

In 1798, Robert Fulton built a small submersible that he called the *Nautilus*. This vessel, which is shown in figure 1-12, had an overall length of 20 feet and a beam of 5 feet. The craft was designed to carry three people and to stay submerged for about an hour. The first *Nautilus* carried sails for surface propulsion and a hand-driven screw propeller for submerged propulsion. The periscope had not yet been invented, but Fulton's craft had a modified form of a conning tower that had a porthole for underwater observation. In 1801, Fulton tried to interest France, Britain, and America in his idea, but no nation was willing to sponsor the development of the craft, even though this was the best submarine that had yet been designed.

Interest in the development of the submarine was great during the period of the Civil War, but progress was limited by the lack of a suitable means of propulsion. Steam propulsion was attempted, but it had many drawbacks, and hand propulsion was obviously of limited value. The first successful steam-driven submarine was built in 1880 in England. The submarine had a coal-fired boiler and a retractable smokestack.

In 1886, an all electric submarine was built by two Englishmen, Campbell and Ash. Their boat was propelled at a surface speed of 6 knots by two 50-hp electric motors operated from a 100-cell storage battery. However, this craft suffered from one major defect: its batteries had to be recharged and overhauled at such short intervals that its effective range never exceeded 80 miles.

In 1875, in New Jersey, John P. Holland built his first submarine. Twenty-five years and nine boats later, Holland finally built the U.S. Navy's first submarine, USS *Holland* (fig. 1-13). Although Holland's early models had features that were later discontinued, many of his initial ideas, perfected in practice, are still in use

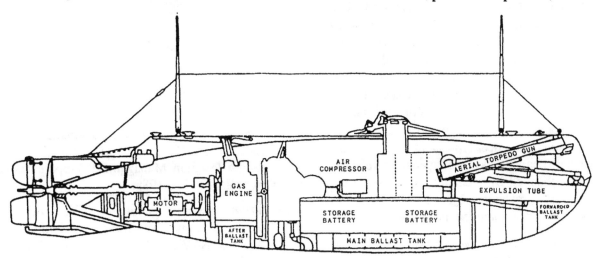

Figure 1-13.–The USS *Holland*; the first U.S. Navy submarine.

today. The *Holland* had a length of 54 feet and a displacement of 75 tons. A 50-hp gasoline engine provided power for surface propulsion and for battery charging; electric motors run from the storage batteries provided power for underwater running.

Just before Holland delivered his first submarine to the Navy, his company was reorganized into the Electric Boat Company, which continued to be the chief supplier of U.S. Navy submarines until 1917. After the acceptance of the *Holland*, new contracts for submarines came rapidly. The A-boats, of which there were seven, were completed in 1903. These were improved versions of the Holland; they were 67 feet long and were equipped with gasoline engines and electric motors. This propulsion combination persisted through a series of B-boats, C-boats, and D-boats turned out by the Electric Boat Company.

The E-boat type of submarine was the first to use diesel engines. Diesel engines eliminated much of the physical discomfort that had been caused by fumes and exhaust gases of the old gasoline engines. The K-boats, L-boats, and O-boats of World War I were all driven by diesel engines.

There was little that was spectacular about submarine development in the United States between 1918 and 1941. The submarines built just before and during World War II ranged from 300 to 320 feet in length and displaced approximately 1500 tons on the surface. These included such famous classes as *Balao*, *Gato*, *Tambor*, *Sargo*, *Salmon*, *Perch*, and *Pike*. In the latter part of World War II, the Germans adopted a radical change in submarine design known as the "schnorkel." The spelling was reduced to "snorkel" by the Americans and to "snort" by the British.

The snorkel is a breathing tube that is raised while the submarine is at periscope depth. With the snorkel in the raised position, air for the diesel engines can be obtained from the surface. The snorkel was developed and improved by the U.S. Navy at the end of World War II and was installed on a number of submarines. Another post-war development was the Guppy (Great Underwater Propulsion Power) submarine. The Guppy was a conversion of the fleet-type submarine of World War II. The main change was in the superstructure of the hull; this was changed by reducing the surface area, streamlining every protruding object, and enclosing the periscope shears in a streamlined metal fairing.

With the advent of nuclear power, a new era of submarine development began. The first nuclear submarine was the USS *Nautilus* (SSN-571), which was commissioned on 30 September 1954. At 1100 on 17 January 1955, the *Nautilus* sent its historic message: "Underway on nuclear power." The *Nautilus* broke all existing records for speed and submerged endurance, but even these records were soon broken by subsequent generations of nuclear submarines. The modern nuclear-powered submarine is sometimes considered the first "true submarine" because it is capable of staying submerged almost indefinitely.

With the development of the modern missile-firing submarine–a nuclear-powered submarine that is capable of submerged firing of the Polaris or Poseidon missile–the submarine has become one of the most vital links in our national defense. As time passes, newer classes of submarines are being developed. Historically, the mission of a submarine has been to seek out and destroy enemy surface ships, both combatant and noncombatant. In the recent past, the basic mission changed and now the primary mission of submarines is to seek out and destroy enemy submarines. The first fleet ballistic missile (FBM) submarine, USS *George Washington* (SSBN-598) was launched in June 1959. The Trident submarines, such as the USS *Ohio* (SSBN-726) (fig. 1-14), are replacing the aging FBM submarines built during a short period in the 1960s. All Trident submarines have exceeded their performance design specifications in speed and quietness and have successfully launched Trident test missiles. Poseidon submarines will be retired and replaced by Trident submarines by the late 1990s. The Seawolf submarine is under development now.

The modern nuclear-powered missile-firing submarine is a far cry from David Bushnell's hand-propelled *Turtle*, but an identical need led to the development of both types of vessels: the need for better and more effective fighting ships.

REFERENCES

Fireman, NAVEDTRA 10520-H, Naval Education and Training Program Management Support Activity, Pensacola, Florida, 1987.

Gas Turbine System Technician (Mechanical) 3 & 2, NAVEDTRA 10548-2, Naval Education and Training Program Management Support Activity, Pensacola, Florida, 1988.

Gas Turbine Systems Technician (Electrical) 3/Gas Turbine Systems Technician (Mechanical) 3, Volume 1, NAVEDTRA 10563, Naval Education and Training Program Management Support Activity, Pensacola, Florida, 1989.

3.347

Figure 1-14.--USS *Ohio* (SSBN-726).

CHAPTER 2

SHIP DESIGN AND CONSTRUCTION

LEARNING OBJECTIVES

Upon completion of this chapter, you should be able to do the following:

1. Identify the basic considerations involved in the design of naval ships.

2. State Archimedes' law.

3. Describe the three types of stress to which a ship may be subjected.

4. Identify the structural parts of a ship and discuss the purpose of the parts.

5. Describe the compartment symbols for ships.

6. Describe the three plans of the ship's hull and their associated planes.

INTRODUCTION

As technology has increased, plans for building ships have become more detailed and more numerous. Today only meticulously detailed plans and well-conceived organization, from the naval architects to the shipyard workers, can produce the ships required for the Navy.

After intensive research, many technical advances have been adopted in the design and construction of naval ships. These changes were brought about by the development of welding techniques, by the rapid development in aircraft, submarines, and weapons, and by developments in electronics and in propulsion plants.

This chapter presents information concerning basic ship design considerations, basic ship structure, ship compartmentation, and the geometry of the ship.

BASIC DESIGN CONSIDERATIONS

Basic considerations involved in the construction of any naval vessel include the following:

• Cost–The initial cost is important in ship design, but it is not the only cost consideration. The cost of maintenance and operation, as well as the cost and availability of the required manning, are equally important considerations.

• Life expectancy–The life expectancy of a ship is limited by ordinary deterioration in service and also by the possibility of obsolescence due to the design of more efficient ships.

• Service (mission)–The mission of a naval ship is the job it is designed to perform, and that mission is a prime determinant in the planning and construction of the ship. As an example, the basic concerns during hull construction and design for an antisubmarine-warfare destroyer will be the sonar gear, weapons, embarked helicopters, detection and information processing gear, and required personnel. But a destroyer with an antiair-warfare mission will need a different set of design parameters. An even greater difference is evident between ships used in amphibious operations and fleet support.

• Port facilities–The port facilities available in the normal operating zone of the ship affect the design to some extent. Dockyard facilities available for drydocking and maintenance work must be taken into consideration.

• Prime mover–The type of propelling machinery to be used must be considered from the point of view of the required speed of the ship, the location in the ship, space and weight requirements, and the effect of the machinery on the center of gravity of the ship.

• Special considerations–Special considerations, such as the fuel required, the crew to be carried, and special weapons, are factors that restrict the designer of a naval ship.

Naval ships are designed for maximum simplicity that is compatible with the requirements of service. Naval ships are designed as simply as possible to lower building and operating costs and to ensure greater availability of shipyard facilities.

The following operating considerations affect the size of a naval ship:

• Effect of speed–For large ships, speed may be maintained with a smaller fraction of displacement devoted to propulsion machinery than is the case for smaller ships. Also, large ships lose proportionately less speed through adverse sea conditions.

• Effect of radius of action–An increasing cruising radius may be obtained by increasing the fraction of displacement allotted to fuel and stores. If the fraction of displacement set aside for fuel and stores is increased, some other weight must be decreased. Since the hull weight is a constant percentage of the displacement, an increase in the fraction of the displacement assigned to one military characteristic involves the reduction of other fractions of displacement. By increasing the displacement of the ship as a whole, the speed and the radius of action may be increased without adversely affecting the other required characteristics.

• Dock facilities–These considerations obviously have an effect on the size of ships that must use canals or dock facilities.

• Effect of seagoing capabilities–Larger ships are more seaworthy than smaller ships. However, smaller ships are more maneuverable because they have smaller turning circle radii than larger ships, where all other factors are proportional. The maneuverability of large ships may be increased somewhat by the use of improved steering gear and large rudders.

In general, larger ships have the advantage of greater protection because of their greater displacement. From the point of view of underwater attack, larger ships also have an advantage. If compartments are of the same size, the number of compartments increases linearly with the displacement. It is apparent, then, that protection against both surface and subsurface attacks may be more effective on larger ships without impairing other military characteristics.

Many compromises must be made in designing any ship, since action that improves one feature may degrade another. For example, in the design of a conventionally powered ship there is the problem of choosing the hull line for optimum performance at a cruising speed of 20 knots and at a trial speed of 30 knots or more. One may select a hull type that would minimize resistance at top speed, and thus keep the weight of propulsion machinery to a minimum. When this is done, however, resistance at cruising speed may be high and the fuel load for a given endurance may be relatively great, thus nullifying some of the gain from a light machinery plant. On the other hand, one may choose a hull type favoring cruising power. In this case, fuel load will be lighter but the shaft horsepower required to make trial speed may be greater than before. Now the machinery plant is heavier, canceling some of the weight gain realized from the lighter fuel load. A compromise based on the interrelationship of these considerations must usually be adopted. The need for compromise is always present, and the manner in which it is made has an important bearing on the final design of any naval vessel.

It is increasingly difficult to design and build a ship without regard to the systems engineering approach. This has led to the need for a way to deal with complex assemblies made up of many specialized components. If ships are to be capable of optimum performance, complex assemblies, such as the nuclear-powered aircraft carrier or fleet support ships, must be designed and built in an orderly manner. The designer of a ship should try to incorporate as many desirable characteristics as possible, keeping in mind the purpose of the ship.

BASIC SHIP STRUCTURE

In considering the structure of a ship, it is common practice to liken the ship to a box girder. Like a box girder, a ship may be subjected to tremendous stresses. The magnitude of stress is usually expressed in pounds per square inch (psi).

When a pull is exerted on each end of a bar, as in view A of figure 2-1, the bar is under the type of stress called tension. When a pressure is exerted on each end of a bar, as in view B of figure 2-1, the bar is under the type of stress called compression. When a shaft, bar, or other material is subjected to a twisting motion, the resulting stress is known as torsional stress.

When a material is compressed, it is shortened. When it is subjected to tension, it is lengthened. This change in shape is called strain. The change of shape (strain) may be regarded as an effect of stress.

If a simple beam is supported at its two ends and various loads are applied over the center of the span, the beam will bend (fig. 2-2). As the beam bends, the upper section of the beam compresses and the lower part stretches. Somewhere between the top and bottom of the beam, there is a section that is neither in compression

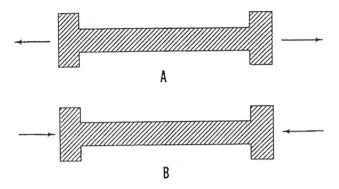

Figure 2-1.–Stresses in metal: (A) tension; (B) compression.

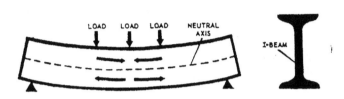

Figure 2-2.–I-beam with the load placed over the center.

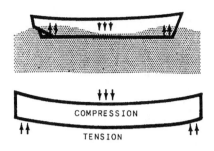

Figure 2-3.–Sagging.

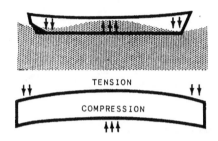

Figure 2-4.–Hogging.

nor in tension; this is known as the neutral axis. The greatest stresses in tension and compression occur near the middle of the length of the beam, where the loads are applied.

LONGITUDINAL BENDING AND STRESSES

In an I-beam, the greater mass of structural material is placed in the upper and lower flanges to resist compression and tension. Relatively little material is placed in the web that holds the two flanges so that they can work together; the web, being near the neutral axis, is less subject to tension and compression stresses than are the flanges. The web does take care of shearing stresses, which are sizeable near the supports.

A ship in a seaway can be considered similar to this I-beam (or, it can be likened to a box girder) with supports and distributed loads. The supports are the buoyant forces of the waves; the loads are the weight of the ship's structure and the weight of everything contained within the ship.

The ship shown in figure 2-3 is supported by waves, with the bow and stern each riding a crest and the midship region in the trough. This ship will bend with compression at the top and tension at the bottom. A ship in this condition is said to be sagging. In a sagging ship, the weather deck tends to buckle under compressive stress and the bottom plating tends to stretch under

tensile stress. A sagging ship is undergoing longitudinal bending–that is, it is bending in a fore-and-aft direction.

When the ship advances half a wave length, so that the crest is amidships and the bow and stern are over troughs, as shown in figure 2-4, the stresses are reversed. The weather deck is now in tension and the bottom plating is in compression. A ship in this condition is said to be hogging. Hogging, like sagging, is a form of longitudinal bending. The effects of longitudinal bending must be considered in the design of the ship, with particular reference to the overall strength that the ship must have. In structural design, the terms *hull girder* and *ship girder* are used to designate the structural parts of the hull. The structural parts of the hull are those parts that contribute to its strength as a girder and provide what is known as longitudinal strength. Structural parts include the framing (transverse and longitudinal), the shell plating, the decks, and the longitudinal bulkheads. These major strength members enable the ship girder to resist the various stresses to which it is subjected.

The ship girder is subjected to rapid reversal of stresses when the ship is in a seaway and is changing from a hogging condition to a sagging condition (and vice versa), since these changes occur in the short time required for the wave to advance half a wave length. Other dynamic stresses are caused by pressure loads forward due to the ship's motion ahead, by panting (a

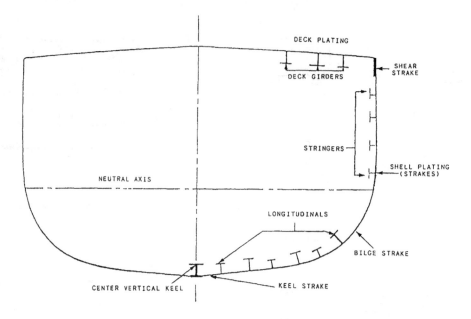

Figure 2-5.–Shell section of a destroyer.

small in and out working of the plating at the bow) of forward plating due to variations of pressure, by the thrust of the propeller, and by the rolling of the ship.

HULL MEMBERS

The principal strength members of the ship's structure are located at the top and bottom of the hull where the greatest stresses occur. The top section includes the main deck plating, the deck girders, and the sheer strakes of the side plating. The bottom section includes the keel, the outer bottom plating, the inner bottom plating, and any continuous longitudinals in way of the bottom. The side webs of the ship girder are composed of the side plating, aided to some extent by any long, continuous fore-and-aft bulkheads. Some of the strength members of a destroyer hull girder are indicated in figure 2-5.

Keel

The keel is the most important structural member of a ship. It is considered to be the backbone of the ship. The keel is built up of plates and angles into an I-beam shape (fig. 2-6). The lower flange of this I-beam structure is the flat keel plate, which forms the center strake of the bottom plating. (On large ships, an additional member is attached to this flange to serve as the center strake.)

The web of the I-beam is a solid plate that is called the vertical keel. The upper flange is called the rider plate; this forms the center strake of the inner bottom plating. An inner vertical keel of two or more sections,

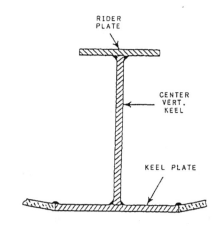

Figure 2-6.–Typical welded keel section.

consisting of I-beams arranged one on top of the other, is found on many large combatant ships.

Framing

Frames used in ship construction may be of various shapes. Frames are strength members. They act as integral parts of the ship girder when the ship is exposed to longitudinal or transverse stresses. Frames stiffen the plating and keep it from bulging or buckling. They act as girders between bulkheads, decks, and double bottoms, and transmit forces exerted by load weights and water pressures. The frames also support the inner and outer shell locally and protect against unusual forces, such as those caused by underwater explosions. Frames are called upon to perform a variety of functions, depending upon the location of the frames in the ship. Figure 2-7 shows various types of frames used on board ship.

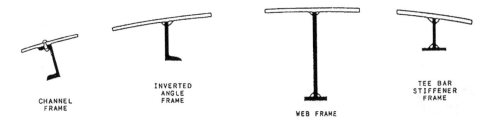

CHANNEL FRAME · INVERTED ANGLE FRAME · WEB FRAME · TEE BAR STIFFENER FRAME

Figure 2-7.–Types of frames used on board ship.

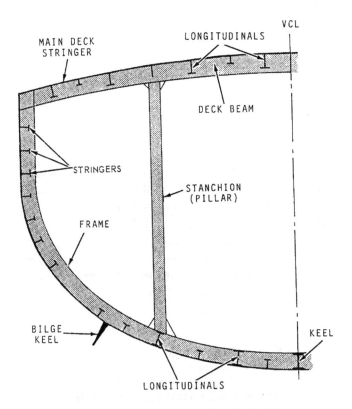

Figure 2-8.–Basic frame section (longitudinal framing).

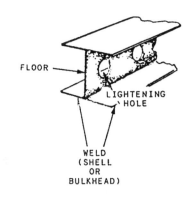

Figure 2-9.–Built-up longitudinal.

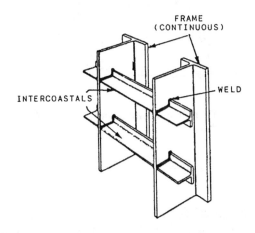

Figure 2-10.–Intercostal and continuous frames.

There are two important systems of framing in current use: the transverse framing system and the longitudinal framing system. The transverse system provides for continuous transverse frames with the widely spaced longitudinals intercostal between them. Transverse frames are closely spaced and a small number of longitudinals are used. The longitudinal system of framing consists of closely spaced longitudinals, which are continuous along the length of the ship, with transverse frames intercostal between the longitudinals.

Transverse frames are attached to the keel and extend from the keel outward around the turn of the bilge and up to the edge of the main deck. They are closely spaced along the length of the ship, and they define the form of the ship.

Longitudinals (fig. 2-8) run parallel to the keel along the bottom, bilge, and side plating. The longitudinals provide longitudinal strength, stiffen the shell plating, and tie the transverse frames and the bulkheads together. The longitudinals in the bottom (called side keelsons) are of the built-up type (fig. 2-9).

Where two sets of frames intersect, one set must be cut to allow the other to pass through. The frames, which are cut and thereby weakened, are known as intercostal frames; those that continue through are called continuous frames. Both intercostal and continuous frames are shown in figure 2-10.

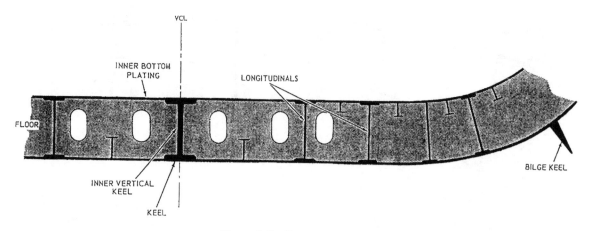

Figure 2-11.–Bottom structure.

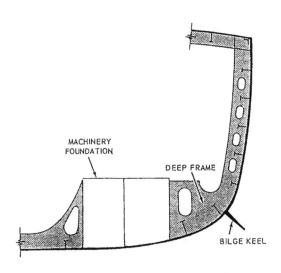

Figure 2-12.–Deep floor assembly for machinery foundations.

A cellular form of framing results from a combination of longitudinal and transverse framing systems using closely spaced deep framing. Cellular framing is used on most naval ships.

In the bottom framing, which is normally the strongest portion of the ship's structure, the floors and keelsons are integrated into a rigid cellular construction (fig. 2-11). Heavy loads, such as the ship's propulsion machinery, are bolted to foundations that are built directly on top of the bottom framing (fig. 2-12). (This method is outdated and is being replaced by block assembly technology.)

In many ships, the top of this cellular region is covered with shell plating, which forms many tanks or voids in the bottom of the ship. The plating over the intersection of the frames and longitudinals is known as the inner bottom plating. The inner bottom plating is a watertight covering laid on top of the bottom framing. It is a second skin inside the bottom of the ship. It prevents flooding in the event of damage to the outer bottom, and it also acts as a strength member. The tanks and voids may be used for stowage of fresh water or fuel oil or they can be used for ballasting. This type of bottom structure, with inner bottom plating, is called double-bottom construction.

Bow and Stem Construction

The ship's bow, which is the front of the ship, varies in form from one type of ship to another as the requirements of resistance and seakeeping dictate the shape. The external shape is shown in figure 2-13 and is commonly used on combatant ships. This form is essentially bulbous at the forefoot, tapering to a sharp entrance near the waterline and again widening above the waterline. Internally, the stem assembly has a heavy centerline member that is called the stem post. The stem post is recessed along its after edge to receive the shell plating, so that the outside presents a smooth surface to cut through the water. The keel structure is securely fastened to the lower end of the stem by welding. The stem maintains the continuity of the keel strength up to the main deck. The decks support the stem at various intermediate points along the stem structure between the keel and the decks.

At various levels and at regular intervals along the stem structure between the keel and the decks are horizontal members called breast hooks. Breast hooks rigidly fasten together the peak frames, the stem, and the

Figure 2-13.–Bulbous-bow configuration.

outside plating. Breast hooks are made of heavy plate and are basically triangular in shape.

Deep transverse framing and transverse bulkheads complete the stem assembly. The stem itself is fabricated from castings, forgings, and heavy plate, or in the case of smaller ships, heavy, precut structural steel plate.

Stern Structure

The aftermost section of the ship's structure is the stern post, which is rigidly secured to the keel, shell plating, and decks. On single-screw ships, the stern post is constructed to accommodate the propeller shaft and rudder stock bosses. The stern post as such is difficult

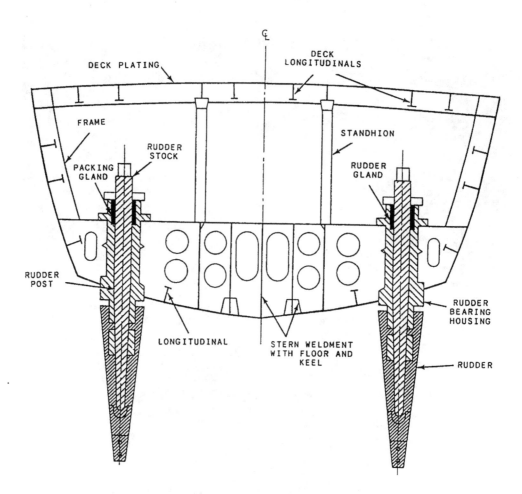

Figure 2-14.–Stern structure.

to define, since it has been replaced by an equivalent structure of deep framing. This structure (fig. 2-14) consists of both longitudinal and transverse framing that extends throughout the width of the bottom in the vicinity of the stern. To withstand the static and dynamic loads imposed by the rudders, the stern structure is strengthened in the vicinity of the rudder post by a structure known as the rudder bearing housing.

Plating

The outer bottom and side plating forms a strong, watertight shell. Shell plating consists of approximately rectangular steel plates arranged longitudinally in rows or courses called strakes. The strakes are lettered, beginning with the A strake, which is just outboard of the keel, and working up to the uppermost side strake.

The end joint formed by adjoining plates in a strake is called a butt. The joint between the edges of two adjoining strakes is called a seam. Seams are also welded flush. Butts and seams are illustrated in figure 2-15.

In general, seams and butts are located so that they do not interfere with longitudinals, bulkheads, decks, and other structural members. Since the hull structure is composed of a great many individual pieces, the strength and tightness of the ship as a whole depend very much upon the strength and tightness of the connections between the individual pieces. In today's modern naval vessels, joints are welded flush together to form a smooth surface.

Bilge Keels

Bilge keels are fitted in practically all ships at the turn of the bilge. Bilge keels extend 50 to 75 percent of the length of the hull. Bilge keels consist of two plates forming a "V" shape welded to the hull, and on large ships may extend out from the hull nearly 3 feet. Bilge keels serve to reduce the extent of the ship's rolling.

Decks

Decks provide both longitudinal and transverse strength to the ship. Deck plates, which are similar to

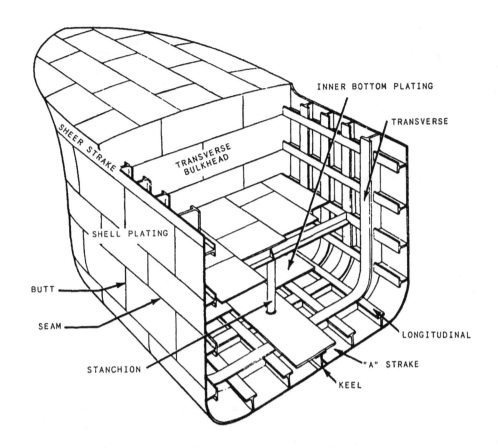

Figure 2-15.–Section of a ship, showing plating and framing.

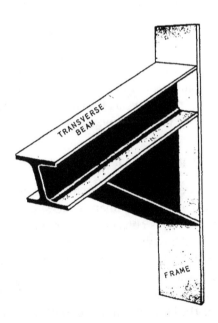

Figure 2-16.–Transverse beam and frame.

the plates used in side and bottom shell plating, are supported by deck beams and deck longitudinals.

The term *uppermost strength deck* is applied to the deck that completes the enclosure of the box girder and

the continuity of the ship's structure. It is the highest continuous deck–usually the main or weather deck. The term *strength deck* also applies to any continuous deck that carries some of the longitudinal load. On destroyers, frigates, and similar ships in which the main deck is the only continuous high deck, the main deck is the strength deck. The flight deck is the uppermost strength deck on aircraft carriers (CVs and LHDs) that carry helicopters, but the main or hangar deck is the strength deck on older types of carriers.

The main deck is supported by deck transverses and deck longitudinals. Deck transverses are the transverse members of the framing structure. The transverse beams are attached to and supported by the frames at the sides, as shown in figure 2-16. Deck girders are similar to longitudinals in the bottom structure in that they run fore and aft and intersect the transverse beams at right angles.

The outboard strake of deck plating that connects with the shell plating is called the stringer strake. The stringer strake is usually heavier than the other deck strakes, and it serves as a continuous longitudinal stringer, providing longitudinal strength to the ship's structure.

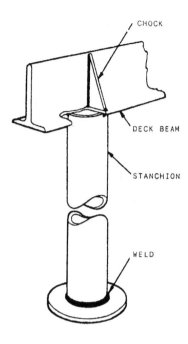

Figure 2-17.–Pipe stanchion.

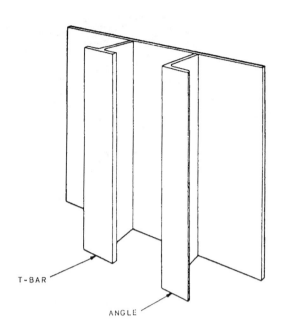

Figure 2-18.–Bulkhead stiffeners.

Stanchions

To reinforce the deck transverses and to keep the deck transverse brackets and side frames from carrying the total load, vertical stanchions or columns are fitted between decks. Stanchions are constructed in various ways of various materials. Some are made of pipe or rolled shapes. The stanchion shown in figure 2-17 is in fairly common use; this pipe stanchion consists of a steel tube that is fitted with special pieces for securing it at the upper end (head) and at the lower end (heel).

Bulkheads

Bulkheads are the vertical partitions that, extending athwartships and fore and aft, provide compartmentation to the interior of the ship. Bulkheads may be either structural or nonstructural. Structural bulkheads, which tie the shell plating, framing, and decks together, are capable of withstanding fluid pressure; these bulkheads usually provide watertight compartmentation. Nonstructural bulkheads are lighter; they are used chiefly for separating activities aboard ship.

Bulkheads consist of plating and reinforcing beams. The reinforcing beams are known as bulkhead stiffeners (fig. 2-18). Bulkhead stiffeners are usually placed in the vertical plane and aligned with deck longitudinals; the stiffeners are secured at top and bottom to any intermediate deck by brackets attached to deck plating. The size of the stiffeners depends upon their spacing, the height of the bulkhead, and the hydrostatic pressure that the bulkhead is designed to withstand.

Bulkheads and bulkhead stiffeners must be strong enough to resist excessive bending or buckling in case of flooding in the compartments that they bound. To form watertight boundaries, structural bulkheads must be joined to all decks, shell plating, bulkheads, and other structural members with which they come in contact. Main subdivision bulkheads extend through the watertight volume of the ship, from the keel to the bulkhead deck, and serve as flooding boundaries in the event of damage below the waterline.

SHIP COMPARTMENTATION

Every space in a naval ship (except for minor spaces such as peacoat lockers, linen lockers, cleaning gear lockers, and so on) is considered a compartment and is assigned an identifying compartment number. The compartment number is stenciled on the bulkhead in a "bull's eye." The bull's eye identifies the space, setting forth the frames that bound the space and the division responsible. The bull's eye has a yellow background that is 12 inches by 15 inches in size. The first line of the bull's eye identifies the compartment number. The second line identifies the frame number at the forward and aft bulkhead of the compartment. The third line identifies the division assigned and responsible for the compartment.

For naval ships, the compartment number consists of a deck number, a frame number, a number indicating

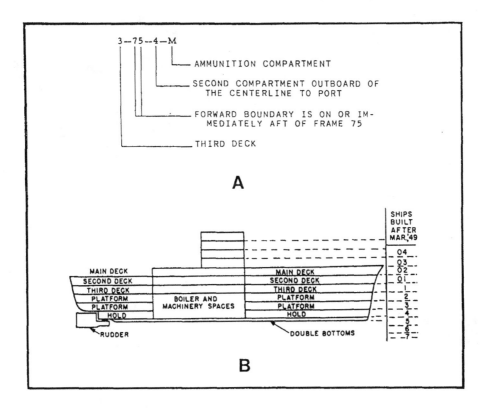

Figure 2-19.–Examples of compartment and deck numbering symbols.

the relationship of the compartment to the centerline of the ship, and a letter showing the primary use of the compartment. These are separated by dashes.

The main deck is always numbered 1. The first deck or horizontal division below the main deck is numbered 2, the second below is numbered 3, and so on, consecutively for subsequent lower division boundaries. Where a compartment extends down to the bottom of the ship, the number assigned the bottom compartments is used. The first horizontal division above the main deck is numbered 01, the second above is numbered 02, and so on, consecutively for subsequent upper divisions. The deck number becomes the first part of the compartment number and indicates the vertical position within the ship.

The frame number at the foremost bulkhead of the enclosing boundary of a compartment is its frame location number. Where these forward boundaries are between frames, the frame number forward is used. Fractional numbers are not used. The frame number becomes the second part of the compartment number.

Compartments located so that the centerline of the ship passes through them carry the number 0. Compartments located completely to starboard of the centerline are given odd numbers, and those located completely to port of the centerline are given even

numbers. Where two or more compartments have the same deck and frame number and are entirely to port or entirely to starboard of the centerline, they have consecutively higher odd or even numbers, as applicable, numbering from the centerline outboard. In this case, the first compartment outboard of the centerline to starboard is 1, the second is 3, and so on. Similarly, the first compartment outboard of the centerline to port is 2, the second is 4, and so on. When the centerline passes through more than one compartment, the compartment having that portion of the forward bulkhead through which the centerline passes carries the number 0, and the others carry the numbers 01, 02, 03, and so on, in any sequence found desirable. These numbers indicate the relationship of the compartment to the centerline and are the third part of the compartment number.

The fourth and last part of the compartment number is the capital letter that identifies the assigned primary use of the compartment. A single capital letter is used, except that on dry- and liquid-cargo ships a double letter designation is used to identify compartments assigned to cargo carrying. An example of a compartment number is shown in figure 2-19, view A, and an example of deck symbols is shown in view B.

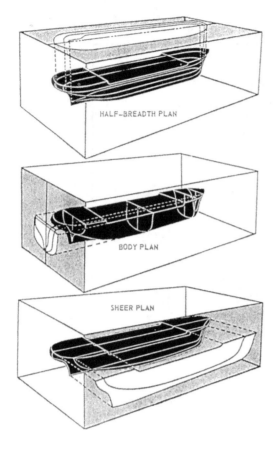

Figure 2-20.–Half-breadth plan, body plan, and sheer plan.

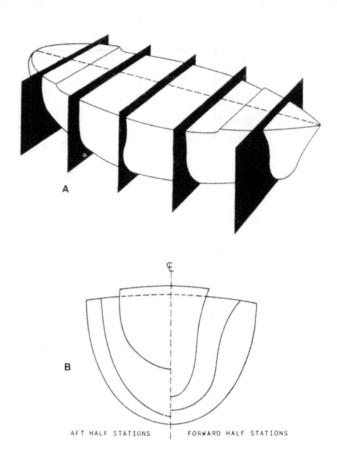

Figure 2-22.–Transverse planes and body plan.

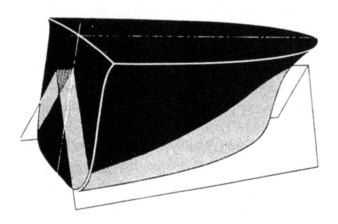

Figure 2-21.–Diagonal plane.

GEOMETRY OF THE SHIP

Since a ship's hull is a three-dimensional object having length, breadth, and depth, and since the hull has curved surfaces in each dimension, no single drawing of a ship can give an accurate and complete representation of the lines of the hull. In naval architecture, a hull shape is shown by means of a lines drawing (sometimes referred to merely as the lines of the ship). The lines drawing consists of three views or projections–a body plan, a half-breadth plan, and a sheer plan–that are obtained by cutting the hull by transverse, horizontal, and vertical planes. The use of these three planes to produce the three projections is illustrated in figure 2-20.

In addition to using transverse, horizontal, and vertical planes, ship designers frequently use a set of planes known as diagonals. A diagonal plane is illustrated in figure 2-21. As a rule, three diagonals are used; these are identified as diagonal A, diagonal B, and diagonal C. Diagonals are frequently shown as projections on the body plan and on the half-breadth plan.

BODY PLAN

To visualize the projection known as the body plan, we must imagine the ship's hull cut transversely in several places, as shown in figure 2-22, view A. The shape of a transverse plane intersection of the hull is obtained at each cut; when the resulting curves are projected onto the body plan (fig. 2-22, view B), they show the changing shape of transverse sections of the hull.

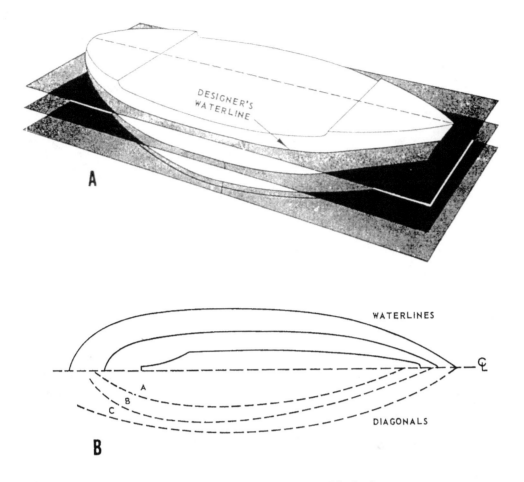

DESIGNER'S WATERLINE

WATERLINES

DIAGONALS

A

B

Figure 2-23.–Horizontal planes and body plan.

Since the hull is symmetrical about the centerline of the ship, only one-half of each curve obtained by a transverse cut is shown on the body plan. Each half curve is called a half station. The right-hand side of the body plan shows the forward half stations–that is, the half stations resulting from transverse cuts between the bow and the middle of the ship. The left-hand side of the body plan shows the aft half stations–that is, the half stations resulting from transverse cuts between the stern and the middle of the ship.

As may be inferred, a station is a complete curve such as would be obtained if each transverse cut were projected completely, rather than as half a curve, onto the body plan. The stations are numbered from forward to aft, dividing the hull into equally spaced transverse sections. The station where the forward end of the designer waterline (the waterline at which the ship is designed to float) and the stem contour intersect is known as the forward perpendicular, or station O. The station at the intersection of the stern contour and the designer waterline is known as the after perpendicular. The station midway between the forward and after perpendiculars is known as the midship section.

HALF-BREADTH PLAN

To visualize the half-breadth plan, we must imagine the ship's hull cut horizontally in several places, as shown in figure 2-23, view A. The cuts are designated as waterlines, although the ship could not possibly float at many of these lines. The base plane that serves as the point of origin for waterlines is usually the horizontal plane that coincides with the top of the flat keel. Waterlines are designated according to their distance above the base plane; for example, we may have a 6-foot waterline, an 8-foot waterline, a 10-foot waterline, and so forth.

The waterlines are projected onto the half-breadth plan, as shown in figure 2-23, view B. Since the hull is symmetrical, only half of the waterlines are shown in the half-breadth plan. Diagonals are frequently shown on the other half of the half-breadth plan.

SHEER PLAN

To visualize the sheer plan, we must imagine the ship's hull cut vertically in several places, as shown in

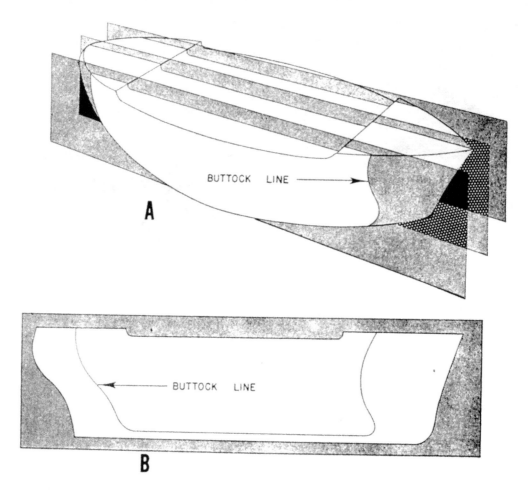

Figure 2-24.–Vertical planes and sheer plan.

figure 2-24, view A. The resultant curves, known as buttocks or as bow and buttock lines, are projected onto the sheer plan, as shown in figure 2-24, view B. The centerline plane is designated as zero buttock. The other buttocks are designated according to their distance from zero buttock. The spacing of the vertical cuts is chosen to show the contours of the forward and after quarters of the ship.

GROUP TECHNOLOGY SHIPBUILDING

The shipbuilding industry is centuries old. Shipbuilding techniques have changed in response to changes in vessel design, materials, markets, and construction methods.

The organization of shipbuilding companies has also changed to match this progression. Throughout its early history, shipbuilding, like most early industries, was craft oriented. It was almost exclusively dependent on the skills of the craftsmen doing the work. Little planning was performed before construction began. As owners became more specific in defining the desired characteristics of a new ship, shipbuilders were required

to do more planning. Nevertheless, before the use of iron and steel for ships, little more than a scale model or a simple drawing of a proposed ship was used to guide construction.

As industrial processes became more complex and efficient, shipbuilders kept pace with changing technology. Shipbuilding began to be subdivided into specialties such as hull construction, machinery outfitting, and painting. More recently, the development of mass production techniques and welding have had profound impacts on shipbuilding. As late as the 1960's and 1970's, shipbuilders continued to try to employ mass production or assembly line approaches. Since then a different approach to shipbuilding has emerged and has proven to be better suited to the economic and technical condition of the industry. This approach is based on the application of *group technology* to shipbuilding.

Currently, there is a new ship construction project titled GROUP TECHNOLOGY SHIPBUILDING (Block Construction Method). The application of these principles requires an alteration in the ship design and engineering process. The outputs of the design cycle required by a group technology shipbuilder are different from those of conventional shipbuilders. Rather than a set of detailed plans suitable for use by any shipyard, the group technology shipbuilder requires work instruction packages that provide specific information for construction organized by the problem areas defined for the individual shipyard. The stages of design and engineering, while similar to the traditional stages of concept, preliminary, contract, and detail design, produce considerably different outputs. Once the Navy has identified the need for a new ship, the operational requirements, and the basic characteristics of the vessel, the preliminary or concept design stage can be done internally by the owner's staff, by a design agent hired by the owner, or by the staffs of one or more shipyards. Common practice in the United States has been to use a design agent for preliminary design. An exception to this is the U.S. Navy, which has a large internal preliminary design section. Owners with considerable experience with particular ship types may, to satisfy specific operational requirements, approach a shipyard directly.

The aim is to develop a design that will meet the requirements while taking advantage of the building experience and capability of a particular shipyard to minimize construction time and cost. The end product of the design stage is a general definition of the ship including dimensions, hull form general arrangement, powering, machinery arrangement, mission, combat systems, habitability, capacities of variable weights (fuel, water, crew, stores), and preliminary definition of major systems (structural, piping, electrical, machinery).

Shipbuilding involves the purchasing of tons of raw materials and many thousands of components, the manufacturing of thousands of parts from the raw materials, and the assembly of these parts and components. Therefore, complex and very detailed planning is required. Detail design and planning must answer the questions of what, where, how, when, and whom. Determining what parts, assemblies, and systems are to be built and what components are to be purchased is primarily detail design. Determining the location within the shipyard and the construction tools and techniques to be used answers the questions of where and how. Considerations of subcontracting and in-house manufacture versus purchasing are also answers to these questions. These questions are resolved as part of planning. Determining the sequencing of all operations, including purchasing and manufacturing as well as need times for information, answers the question of when. This is the scheduling function. Finally, the question of by whom relates to the use of the shipyard workforce. The success of any shipyard or shipbuilding project is directly related to the answers to the questions or to the detail design and planning processes. The final stage of the shipbuilding process is the actual construction of the vessel.

Ship construction can be considered to occur in four manufacturing levels:

- Level 1–Parts manufacturing using raw materials, such as steel plate, sections of pipe, sheet metal, and cable, to manufacture individual parts. The purchasing and handling of components also can be considered to be a part of this lowest manufacturing level.

- Level 2–Involves the joining of parts and components to form subassemblies or units.

- Level 3–The small collections of joined parts are combined to form hull blocks. Hull blocks are commonly the largest sections of ships built away from the final building site.

- Level 4–The final manufacturing level, erection, involves the landing and joining of blocks at the building site (launching ways, graving dock, or drydock). The actual construction phase of shipbuilding is primarily involved with the assembly of parts, subassemblies, or blocks to form a completed vessel. An important part of the construction phase is verification that the ship complies with the contractual requirements. Consequently, the vessel is subjected to a series of tests and trials before delivery to the Navy.

REFERENCES

Hull Maintenance Technician 3 & 2, NAVEDTRA 10571-1, Naval Education and Training Program Management Support Activity, Pensacola, Florida, 1987.

CHAPTER 3

STABILITY AND BUOYANCY

LEARNING OBJECTIVES

Upon completion of this chapter, you should be able to do the following:

1. Be familiar with the basic trigonometric functions.

2. Explain the basic principles of physics

3. Identify the three types of stability.

4. Explain the relationship between draft, displacement, underwater volume, and the height of the center of buoyancy from the keel (KB).

5. Describe the information displayed on stability curve graphs.

6. Describe how the movements of weights affect a ship's stability.

7. Identify the three different states of loose water in a ship.

8. Discuss trim and how a ship's trim may be changed.

9. Describe the changes in stability and the changes present when a ship runs aground or is in drydock.

INTRODUCTION

This chapter deals with the principles of stability, stability curves, the inclining experiment, effects of weight shifts and weight changes, effects of loose water, longitudinal stability and effects of trim, and causes of impaired stability. The damage control aspects of stability are discussed in chapter 4 of this manual.

PRINCIPLES OF STABILITY

From Archimede's law, we know that an object floating on or submerged in a fluid is buoyed up by a force equal to the weight of the fluid it displaces. The weight (displacement) of a ship depends upon the weight of all parts, equipment, stores, and personnel. This total weight represents the effect of gravitational force. When a ship is floated, it sinks into the water until the weight of the fluid displaced by its underwater volume is equal to the weight of the ship. At this point, the ship is in equilibrium–that is, the forces of gravity (G) and the forces of buoyancy (B) are equal, and the algebraic sum of all forces acting upon the ship is equal to zero. If the underwater volume of the ship is not sufficient to displace an amount of fluid equal to the weight of the ship, the ship will sink because the forces of gravity are greater than the forces of buoyancy.

The depth to which a ship will sink when floated in water depends upon the density of the water, since the density affects the weight per unit volume of a fluid. Thus, we may expect a ship to have a deeper draft in fresh water than in salt water, since fresh water is less dense (and therefore less buoyant) than salt water.

Although gravitational forces act everywhere upon the ship, it is not necessary to attempt to consider these forces separately. Instead, we may regard the force of gravity as a single resultant or composite force that acts vertically downward though the ship's center of gravity. Similarly, the force of buoyancy may be regarded as a single resultant force that acts vertically upward through the center of buoyancy located at the geometric center of the ship's underwater body. When a ship is at rest in calm water, the center of gravity and the center of buoyancy lie on the same vertical line.

TRIGONOMETRY

Trigonometry is the study of triangles and the interrelationship of the sides and the angles of a triangle. In determining ship stability, only a small part of the subject of trigonometry is used–the part dealing with right triangles. There is a fixed relationship between the angles of a right triangle and the ratios of the lengths of

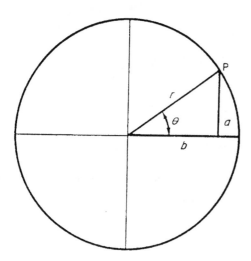

Figure 3-1.–Trigonometric relations.

the sides. These ratios, which are known as trigonometric functions, have been given the names of sine, cosine, tangent, cotangent, secant, and cosecant.

The three ratios required for ship stability work are the sine, cosine, and tangent. Angles are often represented by the Greek letter theta, θ. The sine of an angle θ, abbreviated as sin θ, is the ratio expressed when the side of a right triangle opposite the angle is divided by the hypotenuse. Figure 3-1 shows the trigonometric relations.

For example $sin\ \theta = a/r$, or the altitude, a, divided by the hypotenuse, r. If the hypotenuse, r, is also the radius of a circle, point P moves along the circumference as the angle changes in size. As angle θ increases, side a grows longer. The length of the hypotenuse or radius remains the same. Therefore, the value of the sine increases as the angle increases. Changes in the value of the sine corresponding to changes in the size of the angle are shown on the curve in figure 3-2. This curve is called a sine curve. The size of the angle is plotted horizontally and the value of the sine vertically.

At any angle, the vertical height between the baseline and the curve is the value of the sine of the angle. This curve shows that the value of the sine at 30° is half of the value of the sine at 90°. At 0°, sin θ equals 1.

The cosine is the ratio expressed by dividing the side adjacent to the angle by the hypotenuse. Therefore,

$$cos\ \theta = \frac{b}{r}$$

In contrast to the sine, the cosine decreases as the angle θ becomes larger. This relationship between the value of the cosine and the size of the angle is shown by the curve in figure 3-3. At 0° the cosine equals 1; at 90° the cosine equals zero; and at 60° the cosine is half the value of the cosine at 0°.

The tangent of the angle θ is the ratio of the side opposite the angle θ to the side adjacent. Therefore,

$$tan\ i = \frac{a}{b}$$

DISPLACMENT

Since weight (W) is equal to the displacement, it is possible to measure the volume of the underwater body (V) in cubic feet and multiply this volume by the weight

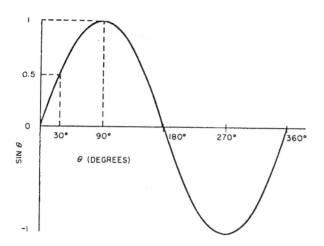

Figure 3-2.–A sine curve.

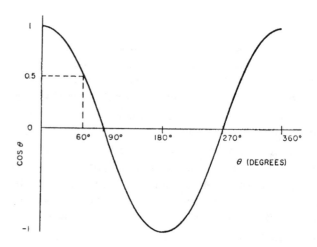

Figure 3-3.–A cosine curve.

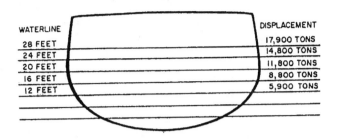

WATERLINE		DISPLACEMENT
28 FEET		17,900 TONS
24 FEET		14,800 TONS
20 FEET		11,800 TONS
16 FEET		8,800 TONS
12 FEET		5,900 TONS

Figure 3-4.–Displacement data.

of a cubic foot of seawater, to find what the ship weighs. This relationship may be written as:

(1) $W = V \times \dfrac{1}{35}$

(2) $V = 35W$

where

V = volume of displaced seawater, in cubic feet

W = weight, in tons

35 = cubic feet of seawater per ton (When dealing with ships, it is customary to use the long ton of 2,240 pounds.)

It is also obvious, then, that displacement will vary with draft. As the draft increases, the displacement increases. This is indicated in figure 3-4 by a series of displacements shown for successive draft lines on the midship section of a cruiser.

The volume of an underwater body for a given draft line can be measured in the drafting room by using graphic or mathematical means. This is done for a series of drafts throughout the probable range of displacements in which a ship is likely to operate. The values obtained are plotted on a grid on which feet of draft are measured vertically and tons of displacement horizontally.

A smooth line is faired through the points plotted, providing a curve of displacement versus draft, or a displacement curve as it is generally called. The result is shown in figure 3-5 for a cruiser.

To use the curve in figure 3-5 for finding the displacement when the draft is given, locate the value of the mean draft on the left side of the draft scale and proceed horizontally across the diagram to the curve. Then drop vertically downward and read the displacement from the scale. For example, if the mean draft is 24 feet, the displacement found from the curve is approximately 14,700 tons.

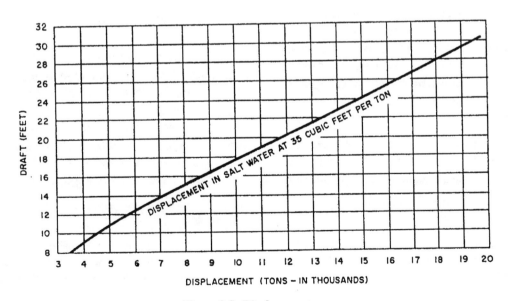

Figure 3-5.–Displacement curve.

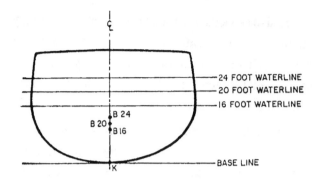

Figure 3-6.—Successive centers of buoyancy for different drafts.

KB VERSUS DRAFT

As the draft increases, the center of buoyancy (B) rises with respect to the keel (K). Figure 3-6 shows how different drafts result in different values of KB, the height of the center of buoyancy from the keel. A series of values for KB is obtained and these values are plotted on a curve to show KB versus draft. Figure 3-7 shows a typical KB curve.

To read KB when the draft is known, start at the proper value of draft on the left side of the scale and proceed horizontally to the curve. Then drop vertically downward to the base line (KB). Thus, if a ship were floating at a mean draft of 19 feet, the KB found from the chart would be approximately 10.5 feet.

RESERVE BUOYANCY

The volume of the watertight portion of the ship above the waterline is known as the ship's reserve buoyancy. Expressed as a percentage, reserve bouyancy is the ratio of the volume of the above-water body to the volume of the underwater body. Thus, reserve buoyancy may be stated as a volume in cubic feet, as a ratio or percentage, or as an equivalent weight of seawater in tons. (In tons it is 1/35 of the volume in cubic feet of the above-water body.)

Freeboard, a rough measure of the reserve buoyancy, is the distance in feet from the waterline to the main deck. Freeboard is calculated at the midship

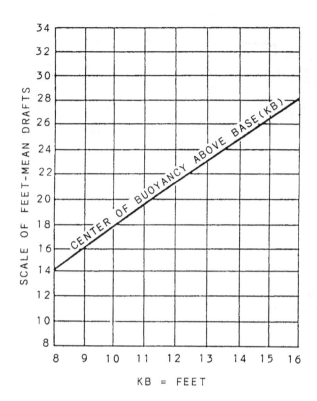

Figure 3-7.—KB curve.

section. As indicated in figure 3-8, freeboard plus draft is equal to the depth of the hull in feet.

When weight is added to a ship, draft and displacement increase in the same amount that freeboard and reserve buoyancy decrease. Reserve buoyancy is an important factor in a ship's ability to survive flooding due to damage. It also contributes to the seaworthiness of the ship in very rough weather.

INCLINING MOMENTS

The moment of a force is the tendency of the force to produce rotation or to move an object about an axis. The distance between the point at which the force is acting and the axis of rotation is called the moment arm or the lever arm of moment[1]. To find the value of a moment, multiply the magnitude of the force by the distance between the force and the axis of rotation. The magnitude of the force is expressed in some unit of

---

1 The significance of the distance between the force and the axis of rotation may be seen if we consider a simple seesaw. If two persons of equal weight sit on opposite ends, equally distant from the center support, the seesaw balances. But if one person moves closer to or farther away from the center, the person farthest away from the support moves downward because the effect of his or her weight is greater.

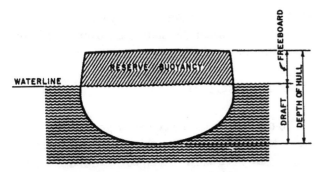

Figure 3-8.–Reserve buoyancy, freeboard, draft, and depth of hull.

weight (pounds, tons, and so on) and the distance is expressed in some unit of length (inches, feet, and so on); hence, the unit of the moment is the foot-pound, the foot-ton, or some similar unit.

When two forces of equal magnitude act in opposite and parallel directions and are separated by a perpendicular distance, they form a couple. The moment of a couple is found by multiplying the magnitude of one of the forces by the perpendicular distance between the lines of action of the two forces.

When a disturbing force exerts an inclining moment on a ship, causing the ship to heel over to some angle, there is a change in the shape of the ship's underwater body and a consequent relocation of the center of buoyancy as the center of buoyancy seeks the new geometric center of the underwater portion of the hull. Because of this shift in the location of B, B and G no longer act in the same vertical line. Instead of acting as separate equal and opposite forces, B and G now form a couple.

The newly formed couple produces either a righting moment or an upsetting moment, depending upon the relative locations of B and G. The ship shown in figure 3-9 develops a righting moment, the magnitude of which is equal to the magnitude of one of the forces (B or G) times the perpendicular distance (GZ), which separates the lines of action of the forces. The distance GZ is known as the righting arm of the ship. Mathematically,

$$RM = W \times GZ$$

where

RM = righting moment, in foot-tons

W = displacement, in tons

GZ = righting arm, in feet

For example, a ship that displaces 10,000 tons and has a 2-foot righting arm at a certain angle of inclination

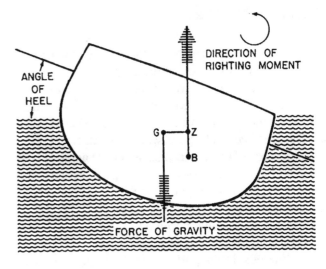

Figure 3-9.–Development of a righting moment when a stable ship inclines.

has a righting moment of 10,000 tons times 2 feet, or 20,000 foot-tons. This 20,000 foot-tons represents the moment, which in this instance tends to return the ship to an upright position.

Figure 3-10 shows the development of an upsetting moment resulting from the inclination of an unstable ship. In this case, the high location of G and the new location of B contribute to the development of an upsetting moment rather than a righting moment.

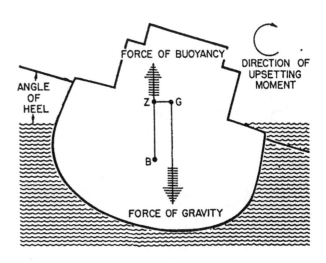

Figure 3-10.–Development of an upsetting moment when an unstable ship inclines.

THE METACENTER

A ship's metacenter (M) is the intersection of two successive lines of action of the force of buoyancy as the ship heels through a very small angle. Figure 3-11 shows two lines of buoyant force. One of these represents the ship on an even keel, the other is for a small angle of heel. The point where they intersect is the initial position of the metacenter. When the angle of heel is greater than 7°, M moves off the centerline (due to the shape of the hull and the path of the center of buoyancy) and the path of the movement is a curve (see fig. 3-10). However, it is the initial position of the metacenter that is most useful in the study of stability. In the discussion that follows, the initial position is referred to as M. The distance from the center of buoyancy (B) to the metacenter (M) when the ship is on even keel is the metacentric radius.

METACENTRIC HEIGHT

The distance from the center of gravity (G) to the metacenter is known as the ship's metacentric height (GM). Figure 3-12 shows a ship heeled through a small angle (the angle is exaggerated in the drawing), establishing a metacenter at M. The ship's righting arm, GZ, is one side of triangle GZM. In triangle GZM, the angle of heel is at M. Side GM is perpendicular to the

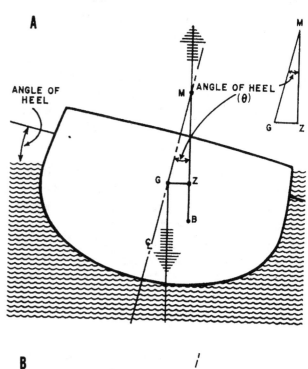

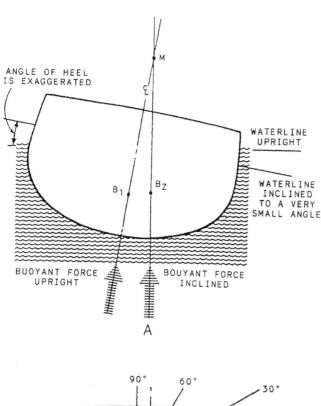

Figure 3-11.–The metacenter.

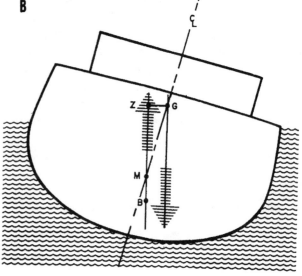

Figure 3-12.–(A) Stable condition, G is below M. (B) Unstable condition, G is above M.

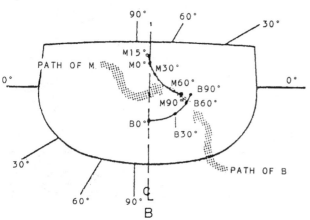

waterline at even keel, and ZM is perpendicular to the waterline when the ship is inclined.

It is evident that for any angle of heel not greater than 7°, there will be a definite relationship between GM and GZ because GZ = GM sin θ. Thus, GM acts as a measure of GZ, the righting arm.

GM is also an indication of whether the ship is stable or unstable at small angles of inclination. If M is above G, the metacentric height is positive, the moments that develop when the ship is inclined are righting moments, and the ship is stable (fig. 3-12, view A). But if M is below G, the metacentric height is negative, the moments that develop are upsetting moments, and the ship is unstable (fig. 3-12, view B).

INFLUENCE OF METACENTRIC HEIGHT

When the metacentric height of a ship is large, the righting arms that develop at small angles of heel are also large. Such a ship resists roll and is said to be stiff. When the metacentric height is small, the righting arms are also small. Such a ship rolls slowly and is said to be tender. Some GM values for various naval ships are FFs, 3 to 5 feet; DDs, 4 to 6 feet; CGs, 4 to 7 feet; and CVs, 9 to 13 feet.

Large GM and large righting arms are desirable for resistance to the flooding effects of damage. However, a smaller GM is sometimes desirable for the slow, easy roll, which makes for more accurate gunfire. Thus, the GM value for a naval ship is the result of compromise.

STABILITY CURVES

When a series of values for GZ at successive angles of heel are plotted on a graph, the result is a stability curve. The stability curve shown in figure 3-13 is called a curve of static stability. The word static indicates that it is not necessary for the ship to be in motion for the curve to apply; if the ship is momentarily stopped at any

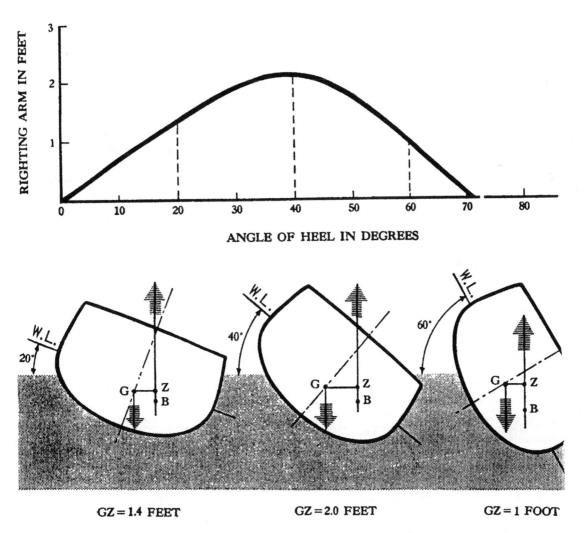

Figure 3-13.–Righting arms of a ship inclined at successively larger angles of heel.

angle during its roll, the value of GZ given by the curve will still apply. Design engineers usually use GM values as a measure of stability up to about 7° heel. For angles beyond 7°, a stability curve is used.

To understand the stability curve, you should consider the following facts:

— The ship's center of gravity does not change position as the angle of heel is changed.

— The ship's center of buoyancy is always at the center of the ship's underwater hull.

— The shape of the ship's underwater hull changes as the angle of heel changes.

Putting these facts together, we see that the position of G remains constant as the ship heels through various angles, but the position of B changes according to the angle of inclination. Initial stability increases with increasing angle of heel at an almost constant rate; but at large angles the increase in GZ begins to level off and gradually diminishes, becoming zero at very large angles of heel.

EFFECT OF DRAFT ON A RIGHTING ARM

A change in displacement will result in a change of draft and freeboard; and B will shift to the geometric center of the new underwater body. At any angle of inclination, a change in draft causes B to shift both horizontally and vertically with respect to the waterline. The horizontal shift in B changes the distance between B and G, and thereby changes the length of the righting arm, GZ. Thus, when draft is increased, the righting arms are reduced throughout the entire range of stability. Figure 3-14 shows how the righting arm is reduced when the draft is increased from 18 feet to 26 feet, when the ship is inclined at an angle of 20°. At smaller angles up

to 30°, certain hull types show flat or slightly increasing righting arm values with an increase in displacement. A reduction in the size of the righting arm usually means a decrease in stability. When the reduction in GZ is caused by increased displacement, however, the total effect on stability is more difficult to evaluate. Since the righting moment is equal to W times GZ, the righting moment will be increased by the gain in W at the same time that it is decreased by the reduction in GZ. The gain in the righting moment, caused by the gain in W, does not necessarily compensate for the reduction in GZ. In brief, an increase in displacement affects the stability of a ship in several ways. Although these effects occur at the same time, we should consider them separately. The effects of increased displacement are as follows:

— Righting arms (GZ) are decreased as a result of increased draft.

— Righting moments (foot-tons) are decreased as a result of decreased GZ (for a given displacement).

— Righting moments may be increased as a result of the increased displacement (W), if GZ times W is increased.

CROSS-CURVES OF STABILITY

To facilitate stability calculations, the design activity inclines a lines drawing of the ship at a given angle, and then lays off on it a series of waterlines. These waterlines are chosen at evenly spaced drafts throughout the probable range of displacements. For each waterline the value of the righting arm is calculated, using an assumed center of gravity rather than the true center of gravity. A series of such calculations is made for various angles of heel–usually 10°, 20°, 30°, 40°, 50°, 60°, 70°, 80°, and 90°–and the results are plotted on a grid to form

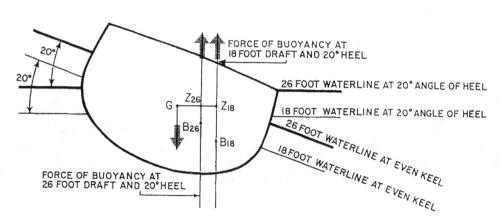

Figure 3-14.–Effect of draft on a righting arm.

a series of curves known as the cross-curves of stability (fig. 3-15). Note that, as draft and displacement increase, the curves all slope downward, indicating increasingly smaller righting arms.

The cross-curves are used in the preparation of stability curves. To take a stability curve from the cross-curves, a vertical line (such as line MN in fig. 3-15) is drawn on the cross-curve sheet at the displacement that corresponds to the mean draft of the ship. At the intersection of this vertical line with each cross-curve, the corresponding value of the righting arm on the vertical scale at the left can be read. Then this value of the righting arm at the corresponding angle of heel is plotted on the grid for the stability curve. When a series of such values of the righting arms from 10° through 90° of heel have been plotted, a smooth line is drawn through them and the uncorrected stability curve for the ship at that particular displacement is obtained. The curve is not corrected for the actual height of the ship's center of gravity, since the cross-curves are based on an assumed height of G. However, the stability curve

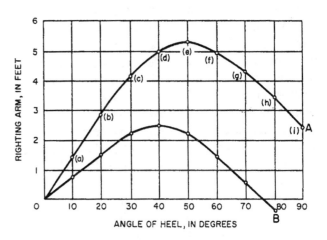

Figure 3-16.–(A) Uncorrected stability curve taken from cross-curves. (B) Corrected stability curve.

does embody the effect on the righting arm of the freeboard for a given position of the center of gravity.

Figure 3-16 shows an uncorrected stability curve (A) for a ship operating at 11,500 tons displacement, taken from the cross-curves shown in figure 3-15. This

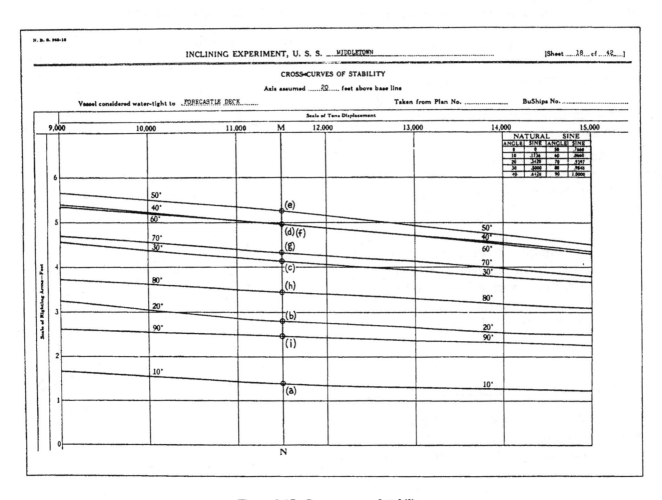

Figure 3-15.–Cross-curves of stability.

3-9

stability curve cannot be used in its present form, since the cross-curves are made up on the basis of an assumed center of gravity. In actual operation, the ship's condition of loading will affect its displacement and, therefore, the location of G. To use a curve taken from the cross-curves, therefore, you must correct the curve for the actual height of G above the keel (K)–that is, you must use the distance KG. As far as the new center of gravity is concerned, when a weight is added to a system of weights, the center of gravity can be found by taking moments of the old system plus that of the new weight and dividing this total moment by the total final weight. Detailed information concerning changes in the center of gravity of a ship can be found in *NSTM*, chapter 079, volume 1.

Assume that the cross-curves are made up on the basis of an assumed KG of 20 feet, and the actual KG, which includes the added effects of free surface, for the particular condition of loading is 24 feet. This means that the true G is 4 feet higher than the assumed G and that the righting arm (GZ) at each angle of inclination will be smaller than the righting arm shown in figure 3-16 (curve A) for the same angle. To find the new value of GZ for each angle of inclination, multiply the increase in KG (4 feet) by the sine of the angle of inclination, and subtract the product from the value of GZ shown on the cross-curves or on the uncorrected stability curve. To facilitate the correction of the stability curves, a table showing the necessary sines of the angles of inclination is included on the cross-curves form (fig. 3-15).

Next, find the corrected values of GZ for the various angles of heel shown on the stability curve (A) in figure 3-16 and plot them on the same grid to make the corrected stability curve (B) shown in figure 3-16.

When the values from 10° through 80° are plotted on the grid and joined with a smooth curve, the corrected stability curve (B) shown in figure 3-16 results. The corrected curve shows maximum stability to be at 40°; it also shows that an upsetting arm, rather than a righting arm, generally exists at angles of heel in excess of 75°.

THE INCLINING EXPERIMENT

The vertical location of the center of gravity must be known to determine the stability characteristics of a ship. Although the position of the center of gravity as estimated by calculation is sufficient for design purposes, an accurate determination is required to establish the ship's stability. Therefore, an inclining

experiment is performed to obtain a precise measurement of KG, the vertical height of G above the keel (base line), when the ship is completed. An inclining experiment consists of moving one or more large weights across the ship and measuring the angle of list produced (see fig. 3-17). This angle of list, produced by the weight movement and measured by means of a pendulum and a horizontal batten or an inclinometer device designed for this purpose, usually does not exceed 2°. The metacentric height is calculated from the following formula:

$$GM = \frac{wd}{W \tan \theta}$$

where

w = inclining weight, in tons

d = distance weight is moved athwartships, in feet

W = displacement of ship, including the inclining weight, w, in tons

$\tan \theta$ = tangent of angle of list

The results of this experiment are calculated and tabulated in the Inclining Experiment Data Report, which consists of two parts. Part 1, Report of Inclining Experiment, contains the observations and calculations that determine the displacement and location of the center of gravity of the ship in the light condition. Part 2,

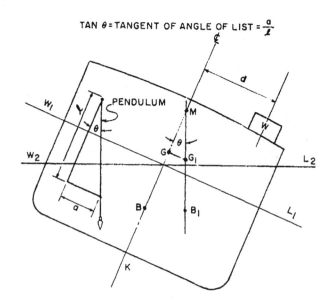

Figure 3-17.–Measuring the angle of list produced in performing the inclining experiment.

Stability Data For Surface Ships and Stability and Equilibrium Data For Submarines, contains data relative to the characteristics of the ship in operating condition. These booklets are prepared by the inclining activity, and Part 2 is issued to the ships for their information.

The KG obtained from the inclining experiment is accurate for the particular condition of loading in which the ship was inclined. This is known as condition A, or the "as-inclined" condition. The ship may have been in any condition of loading at the time of the experiment, and this may not have been in operating condition. To convert the data obtained to practical use, KG must be determined for various operating conditions.

INCLINING EXPERIMENT DATA

It is customary to perform an inclining experiment (section 4) only on one or two of any class ship. The results of this experiment are contained in the *Booklet of Inclining Experiment Data* and include the following information for forces afloat:

— Complete stability information for certain prescribed conditions of loading. These loadings are determined by NAVSEA and include the extreme light and heavy loadings to be expected in operations. In many cases, intermediate loading conditions are tabulated.

— A detailed statement indicating the weight and location of boats, ordnance equipment, planes, and permanent ballast.

— A summary of the consumable loads such as fuel water, ammunition, and stores included in each condition. This summary includes information on displacement KG, GM, trim, and drafts for each loading.

— A table of approximate changes in metacentric height due to added weights. The effect on GM of adding or removing weights at different deck levels as indicated is accurate only for weights approximately twice the unit weight specified. A change in specified displacement or a greater ratio of unit weight employed will reduce the accuracy of this table.

— Graphic illustrations of curves of displacement, stability, statistical, and other curves. Stability curves are drawn for the ship in the standard conditions of loading. Dotted line stability curves represent a situation where all liquids in the ship are assumed to be solid. Solid lines represent the ship under normal operation with free surface in fuel and water tanks. The overall stability features of each curve are indicated in tabular form.

— A detailed statement of consumables in standard loading, which includes those items composing the standard loading in various conditions.

— Although a ship has not been inclined, data from another ship of the class are usally available and applicable until alterations involving major weight changes are made, in which case NAVSEA will authorize the performance of another experiment.

STANDARD LOADING CONDITIONS

Standard conditions of loadings are as follows:

Condition A - Light

Condition A1 - Light, without permanent ballast (this condition is listed only when ships have permanent ballast)

Condition B - Minimum operating condition

Condition C - Optimum battle condition

Condition D - Full load

After obtaining the displacement and locating the center of gravity for a ship in condition A, corresponding values may be computed for other standard conditions of loading. The weights and vertical moments of all consumables to go aboard are determined and, starting with the displacement and KG for condition A, a new displacement, KG, and GM are calculated for each of the other conditions of loading. The GM obtained is in each case corrected for the free surface assumed to exist in the ship's tanks for that particular condition of loading.

Having determined displacement and KG, it is possible to draw a curve of stability for each condition of load. Additional information concerning inclining experiment data can be found in *NSTM*, chapter 079, volume 1, and chapter 096.

EFFECTS OF WEIGHT SHIFTS

If one weight in a system of weights is moved, the center of gravity of the whole system moves along a path parallel to the path of the component weight. The distance that the center of gravity of the system moves may be calculated from the formula

$$GG_1 = \frac{ws}{W}$$

where

    w = component weight, in tons

    s = distance component weight is moved, in feet

    W = weight of entire system, in tons

    GG_1 = shift in center of gravity, in feet

Weight movements in a ship can take place in three possible directions–athwartships, fore and aft, and vertically (perpendicular to the decks). The most general type of movement is inclined with respect to all three of these. Such a diagonal movement can be divided into components in each of the three directions, and one component can be studied at a time without reference to the others. For example, if a weight is moved from the main deck, stardboard side, aft, to a storeroom on the fourth deck, port side, forward, this movement may be regarded as taking place in three steps, as follows:

1. From main deck to fourth deck (down)

2. From starboard side to port side (across)

3. From stern to bow (forward)

VERTICAL WEIGHT SHIFT

If a weight is moved straight up a vertical distance on a ship, the ship's center of gravity will move straight up on the centerline (fig. 3-18). The vertical rise in G (explained later in the chapter) can be computed from the formula mentioned previously.

Example: A ship is operating with a displacement of 11,500 tons. Its ammunition, totaling 670 tons, is to be moved from the magazines to the main deck, a distance of 36 feet. Find the rise in G.

$$GG_1 = \frac{670 \times 36}{11,500} = 2.1 \text{ feet}$$

Since moving a weight which is already aboard will cause no change in displacement, there can be no change in M, the metacenter. If M remains fixed, then the upward movement of the center of gravity results in a loss of metacentric height:

$$G_1M = GM - GG_1$$

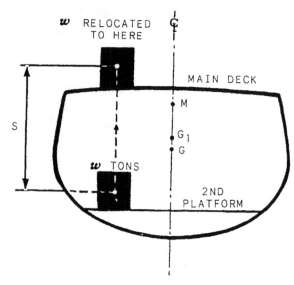

Figure 3-18.–Shift in G due to a vertical weight shift.

where

    G_1M = new metacentric height (after weight movement), infeet

    GM = old metacentric height (before weight movement), in feet

    GG_1 = rise in center of gravity, in feet

If the ammunition on the main deck is moved down to the sixth deck, the positions of G and G_1 will be reversed. The shift in G can be found from the same formula as before, the only difference being that GG_1 now becomes a gain in metacentric height instead of a loss (fig. 3-19).

If a weight is moved vertically downward, the ship's center of gravity, G, will move straight down on the centerline and the correction is additive. In this case, the sine curve is plotted below the abscissa. The final stability curve is that portion of the curve above the sine correction curve.

A vertical shift in the ship's center of gravity changes every righting arm throughout the entire range of stability. If the ship is at any angle of heel, such as in figure 3-19, the righting arm is GZ with the center of gravity at G. But if the center of gravity shifts to G_1 as the result of a vertical weight shift upward, the righting arm becomes G_1Z_1, which is smaller than GZ by the amount of GR. In the right triangle GRG_1, the angle of heel is at G_1; hence, the loss of the righting arm may be found from the formula

$$GR + GG_1 \times \sin \theta$$

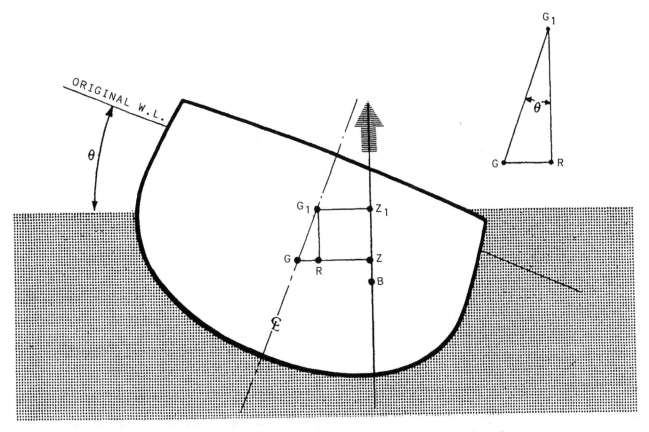

Figure 3-19.–Loss of a righting arm due to a rise in the center of gravity.

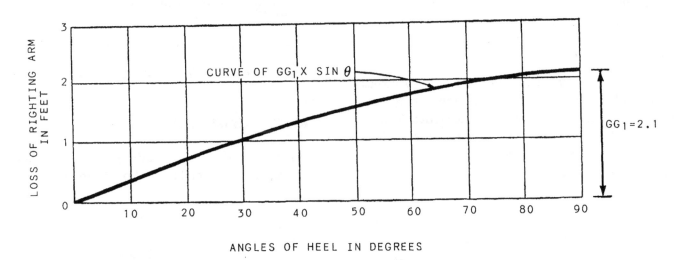

Figure 3-20.–Sine curve showing the loss of a righting arm at various angles of heel.

This equation may be stated in words as The loss of righting arm equals the rise in the center of gravity times the sine of the angle of heel. The sine of the angle of heel is a ratio, which can be found by consulting a table of sines.

If the loss of GZ is found for 10°, 20°, 30°, and so forth by multiplying GG_1 by the sine of the proper angle,

a curve of loss of righting arms can be obtained by plotting values of $GG_1 \times sin\ \theta$ vertically against angles of heel horizontally, which results in a sine curve. When plotted, the curve is as illustrated in figure 3-20.

The sine curve may be superimposed on the original stability curve to show the effect on stability characteristics of moving the weight up in a ship. Since

3-13

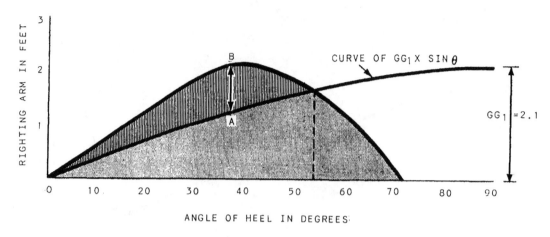

Figure 3-21.–Sine curve superimposed on an original stability curve.

displacement is unchanged, the righting arms of the old curve need to be corrected for the change of G only, and no other variation occurs. Consequently, if $GG_1 \sin \theta$ is deducted from each GZ on the old stability curve, the result will be a correct righting arm curve for the ship after the weight movement.

In figure 3-21 a sine curve has been superimposed on an original stability curve. The dotted area is that portion of the curve that was lost due to moving the weight up, whereas the lined area is the remaining or residual portion of the curve. The residual maximum righting arm is AB and occurs at an angle of about 37°. The new range of stability is from 0° to 53°.

The reduced stability of the new curve becomes more evident if the intercepted distances between the old GZ curve and the sine curve are transferred down to the base, thus forming a new curve of static stability (fig. 3-22). Where the old righting arm at 30° was AB, the new one has a value of CB, which is plotted up from

the base to locate point D (CB = AD) and thus a point is established at 30° on the new curve. A series of points thus obtained by transferring intercepted distances down to the base line delineates the new curve, which may be analyzed as follows:

— GM is now the quantity represented by EF.

— Maximum righting arm is now the quantity represented by HI.

— Angle at which maximum righting arm occurs is 37°.

— Range of stability is from 0° to 53°.

Total dynamic stability is represented by the shaded area.

HORIZONTAL WEIGHT SHIFT

When the ship is upright, G lies in the fore-and-aft centerline, and all weights on board are balanced.

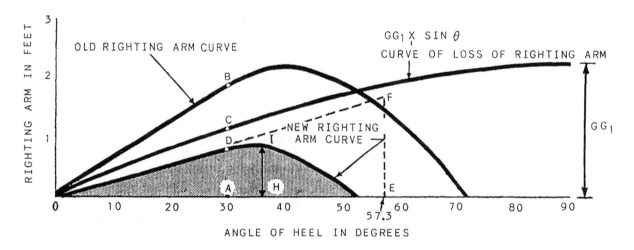

Figure 3-22.–Curve of static stability as corrected for loss of stability due to a vertical weight shift.

Moving any weight horizontally will result in a shift in G in an athwartship direction, parallel to the weight movement. B and G are no longer in the same vertical line and an upsetting moment exists at 0° inclination, which will cause the ship to heel until B moves under the new position of G. In calm water the ship will remain at this angle, and in a seaway it will roll about this angle of permanent list. This shift of G can be computed from the following formula:

$$G_1G_2 = \frac{ws}{W}$$

where

G_1G_2 = athwartship shift of G, in feet

w = weight moved over, in tons

s = distance w moved, in feet

W = displacement of ship, in tons

Let us further assume that ship's stores totaling 185 tons are shifted from port storerooms to starboard storerooms, a horizontal distance of 56 feet. Using the formula:

$$G_1G_2 = \frac{185 \times 56}{11,500} = 0.90 \text{ foot}$$

In figure 3-23, the righting arm has been reduced from G_1Z_1 to G_2Z_2 by this weight shift. G_2Z_2 is smaller than G_1Z_1. However, the distance G_1T is equal to $G_1G_2 \times cos\ \theta$. Thus, loss of a righting arm involved in an athwartship movement of G is equal at any angle of heel to $G_1G_2 \times cos\ \theta$. This variable distance ($G_1G_2 \times cos\ \theta$) is called the ship's inclining arm; when this value is multiplied by the displacement, W, the product is the ship's inclining moment.

Just as the sine curve was superimposed on the GZ curve, so may the cosine curve be superimposed on the stability curve to show the effect on stability of moving a weight athwartship. The cosine curve has been placed on the original stability curve, corrected for the actual

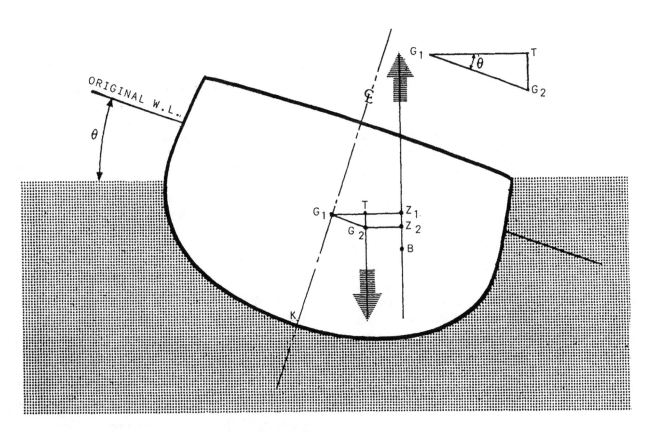

Figure 3-23.–Loss of a righting arm when the center of gravity is moved off the centerline.

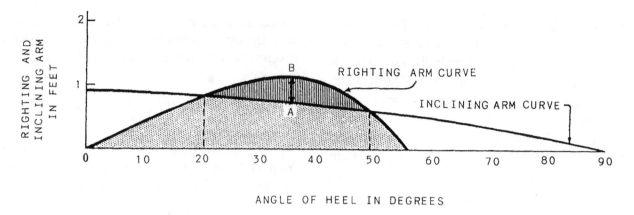

Figure 3-24.–Cosine curve superimposed on an original stability curve.

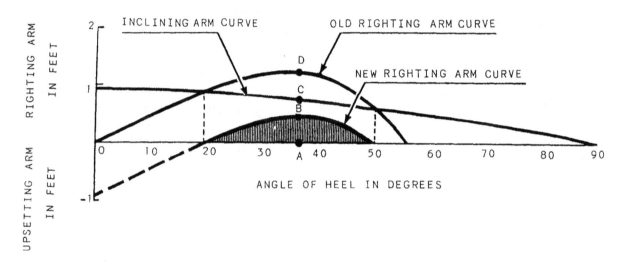

Figure 3-25.–New curve of static stability after correction for horizontal weight movement.

height of the center of gravity. The dotted area (fig. 3-24) is that portion of the curve that was lost due to the weight shift, and the lined area is the remaining or residual portion of the curve. The residual maximum righting arm is AB, which develops at an angle of about 37°. The new range of stability is from 20° to 50°.

The new curve of static stability can be plotted on the base by transferring down the intercepted distances between the cosine curve and the old GZ curve. For example, in figure 3-25 the old righting arm at 37° was AD, the loss of the righting arm (inclining arm) at this angle is AC, leaving a residual GZ of CD. This value has been plotted up from the base as AB to provide one point on the final curve. The residual stability may be analyzed on the new curve as follows:

– Maximum righting arm AB.

– Angle of maximum righting arm at A.

– Range of stability 20° to 50°.

– Total dynamic stability is represented by the lined area.

The ship will have a permanent list at 20°, which is the angle where B is under G, the inclining arm equals the original righting arm, the cosine curve crosses the original GZ curve, and the residual righting arm is zero. In a seaway the ship will roll about this angle of list. If it rolls farther to the listed side, a righting moment develops, which tends to return it toward the angle of list. If it rolls back towards the upright, an upsetting moment develops, which tends to return it toward the angle of list. The upsetting moment (between 0° and the angle of list) is the difference between the inclining and righting moments.

DIAGONAL WEIGHT SHIFT

A weight may be shifted diagonally, so that it moves up or down and athwartship at the same time, or by moving one weight up or down and another athwartship. A diagonal shift should be treated in two steps; first, by finding the effect on GM and stability of the vertical shift, and second, by finding the effect of the horizontal movement. The corrections are applied as previously described.

EFFECTS OF WEIGHT CHANGES

The addition or removal of any weight in a ship may affect list, trim, draft, displacement, and stability. Regardless of where the weight is added (or removed), when determining the various effects it should be considered first to be placed in the center of the ship, then moved up (or down) to its final height, next moved outboard to its final off-center location, and finally shifted to its fore or aft position.

Assume that a weight is added to a ship so that the list or trim is not changed, and G will not shift. The first thing to do is find the new displacement, which is the old displacement plus the added weight:

$$\text{New displacement} = W + w \text{ tons}$$

where

$$W = \text{old displacement, in tons}$$
$$w = \text{added weight, in tons}$$

With the new value of displacement, enter the curves of form and on the displacement curve find the corresponding draft, which is the new mean draft. Figure 3-26 shows typical displacement and other curves generally referred to as curves of form.

If the change in draft is not over 1 foot, the procedure can be reversed. Find the tons-per-inch immersion for the old mean draft from the curves of form, divide the added weight (in tons) by the tons-per-inch immersion to get the bodily sinkage in inches, and add this bodily sinkage to the old mean draft to get the new mean draft. Using the new mean draft, enter the curves of form and find the new displacement.

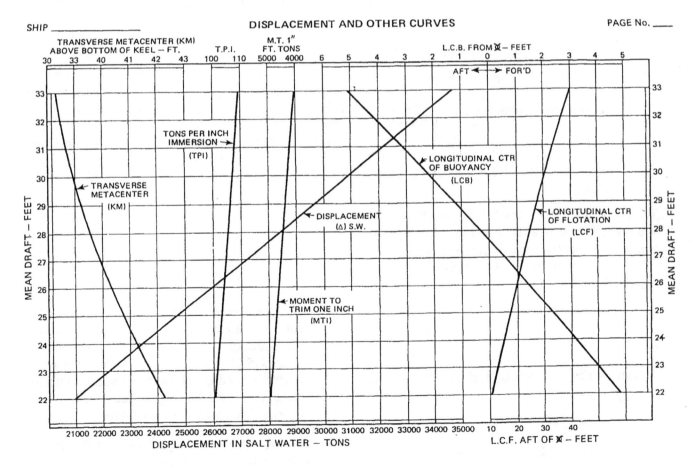

Figure 3-26.—Curves of form.

Assume that the weight added above is shifted vertically on the ship's centerline to its final height above the keel. This movement will cause G to shift up or down. To compute the vertical shift of G use the following formula:

$$GG_1 = \frac{wp}{W + w}$$

where

GG_1 = shift of G up or down, in feet

w = added weight, in tons

p = vertical distance, w, is added above or below original location of G, in feet

W = old displacement, in tons

$(W + w)$ = new displacement, in tons

This vertical shift must be added to or subtracted from the original height of the center of gravity above the keel. To do this, the original height, KG, must be known:

$$KG_1 = KG + GG_1$$

where

KG_1 = new height of G above keel, in feet

KG = old height of G above keel, in feet

GG_1 = shift of G from formula $GG_1 = \dfrac{wp}{(W + w)}$

If the final position of the added weight is below the original position of G, then GG_1 is minus; if it is above, then GG_1 is plus.

To find the new metacentric height, enter the curves of form with the new mean draft and find the height of the transverse metacenter above the base line. This is KM_1. The new metacentric height is determined by the following formula:

$$G_1M_1 = KM_1 - KG_1$$

where

G_1M_1 = new metacentric height, in feet

KM_1 = new KM

KG_1 = new KG

With the new displacement $(W + w)$, enter the cross-curves and pick out a new, uncorrected curve of stability. Correct this curve for the new height of the

ship's center of gravity above the base line. This is acomplished by finding AG_1 (which is $KG_1 - KA$) and subtracting $AG_1 \times \sin\theta$ from every vertical on the stability curve, provided G_1 is above A. If G_1 is below A, the values of $AG_1 \times \sin\theta$ must be added to the curve, as previously explained. The resulting curve of righting arms is now correct for the loss of freeboard due to the added weight and for the final height of the ship's center of gravity resulting from weight addition.

EFFECTS OF LOOSE WATER

When a tank or a compartment in a ship is partially full of liquid that is free to move as the ship heels, the surface of the liquid tends to remain level. The surface of the free liquid is referred to as free surface. The tendency of the liquid to remain level as the ship heels is referred to as free surface effect. The term *loose water* is used to describe liquid that has a free surface; it is not used to describe water or other liquid that completely fills a tank or compartment and has no free surface. (Treat any tank or compartment that is solidly flooded as a weight addition.)

FREE SURFACE EFFECT

Free surface in a ship always causes a reduction in GM, with a consequent reduction of stability, superimposed on any additional weight that would be caused by flooding. The flow of the liquid is an athwartship shift of weight that varies with the angle of inclination. Wherever free surface exists, a free surface correction must be applied to any stability calculation. This effect may be considered to cause a reduction in a ship's static stability curve in the amount of

$$\frac{i}{V} \times \sin\theta$$

due to a virtual rise in G

where

i = the moment of inertia of the surface of water in the tank about a longitudinal axis through the center of the area of that surface (or other liquid in ratio of its specific gravity to that of the liquid in which the ship is floating. It is usual to assume all liquids are salt water, and thus neglect density, unless very accurate determinations are required.

V = existing volume of displacement of the ship in cubic feet. For a rectangular compartment, i may be found from

$$i = \frac{b^3 l}{12}$$

where

b = athwartship breadth of the free surface (with the ship upright), in feet

l = fore-and-aft length of the free surface, in feet

To understand what is meant by a virtual rise in G, refer to figure 3-27. This figure shows a compartment in a ship partially filled with water, which has a free surface (fs) with the ship upright. When the ship heels to any small angle, such as , the free surface shifts to $f_1 s_1$, remaining parallel to the waterline. The result of the inclination is the movement of a wedge of water from $f_0 f_1$ to $s_0 s_1$. Calling g_1 the center of gravity of this wedge when the ship was upright, and g_2 its center of gravity with the ship inclined, it is evident that a small weight has been moved from g_1 to g_2.

Point G is the center of gravity of the ship when upright, and G would remain at this position if the compartment contained solids rather than a liquid. As the ship heels, however, the shift of a wedge of water along the path $g_1 g_2$ causes the center of gravity of the ship to shift from G to G_2. This reduces the righting arm, at this angle, from GZ to $G_2 Z_2$.

To compute GG_2 and the loss of GZ for each angle of heel is a laborious and complicated task. However, an equivalent righting arm, $G_3 Z_3$ (which equals $G_2 Z_2$), can be obtained by extending the line of action of the force of gravity up to intersect the ship's centerline at point

G_3. Raising the ship's center of gravity from G to G_3 would have the same effect on stability at this angle as shifting it from G to G_2.

The distance $G_3 Z_3$ is the righting arm the ship would have if the center of gravity had risen from G to G_3, and this virtual rise of G may be computed from the formula:

$$GG_3 = \frac{i}{V}$$

Referring to the formula, loss is

$$GZ = \frac{i}{V} \times \sin \theta.$$

This formula is accurate for small angles of heel only, due to the pocketing effect as the angle increases. In case several compartments or tanks have free surface, their surface moments of inertia are calculated individually and their sum used in the correction for free surface. The effect of a given area of loose liquid at a given angle of heel is entirely independent of the depth of the liquid in the compartment, as is apparent in the formula

$$i = \frac{b^3 l}{12}$$

where the only factors are the dimensions of the surface and the displacement of the ship. The free surface effect is also independent of the free surface location in the ship, whether it is high or low, forward or aft, on the centerline or off, as long as the boundaries remain intact.

The loss of metacentric height can obviously be reduced by reducing the breadth of the free surface, by installing longitudinal bulkheads. However, off-center flooding after damage then becomes possible, causing the ship to take on a permanent list and usually bringing about a greater loss in stability than if the bulkhead were not present.

The loss of righting arm due to free surface is always lessened to some extent by pocketing. This is the contact of the liquid with the top of the compartment or the exposure of the bottom surface of the compartment, either of which takes place at some definite angle and reduces the breadth of the free surface area. To understand how pocketing of the free surface reduces the free surface effect, study

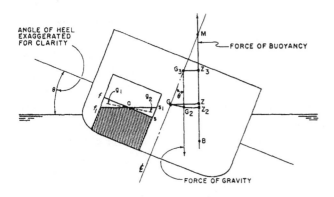

Figure 3-27.–Diagram showing virtual rise in G.

figure 3-28. View A shows a compartment in which the free surface effect is not influenced by the depth of the loose water. The compartment shown in view B, however, contains only a small amount of water; when the ship heels sufficiently to reduce the waterline in the compartment from $w1$ to $w111$, the breadth of the free surface is reduced and the free surface effect is thereby reduced. A similar reduction in surface effect occurs in the almost full compartment shown in view C, again because of the reduction in the breadth of the free surface. As figure 3-28 shows, the beneficial effect of pocketing is greater at larger angles of heel.

The effect of pocketing in reducing the overall free surface effect is extremely variable and not easily determined. In practice, therefore, it is usually ignored and tends to provide a margin of safety when computing stability.

Most compartments of a ship contain some solid objects, such as machinery and stores, which would

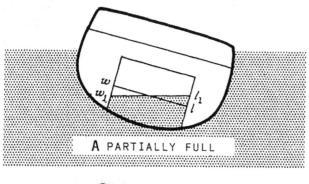

A PARTIALLY FULL

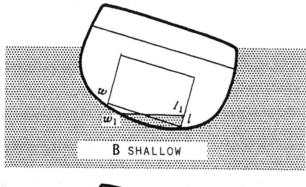

B SHALLOW

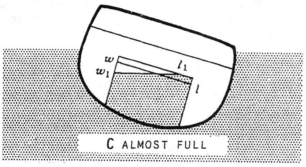

C ALMOST FULL

Figure 3-28.—Pocketing of free surface.

project through and above the surface of any loose water. If these objects are secured so that they do not float or move about, and if they are not permeable, then the free surface area and the free surface effect is reduced by their presence. The actual value of the reduction (surface permeability effect) is difficult to calculate and, like the value of pocketing, if ignored when calculating stability, will provide a further margin of safety.

Swash bulkheads (nontight bulkheads pierced by drain holes) are fitted in deep tanks and double bottoms to hinder the flow of liquid in its attempt to remain continuously parallel to the waterline as the ship rolls. They diminish the free surface effect if the roll is quick, but they have much less effect when the roll is slow. A ship taking on a permanent list will incline just as far as if the swash plate were not there. When a fore-and-aft bulkhead separating two adjacent compartments is holed (ruptured) so that any flooding water present in one is free to flow athwartship from one compartment to the other, a casualty duplicating the effect of a swash bulkhead has occurred. In this case, it is incorrect to add the free surface effects of the two compartments together; an entirely new figure for the flooding effect must be computed, regarding the two as one large compartment.

In summary, the addition of a liquid weight with a free surface has two effects on the metacentric height of a ship. First, there is the effect on GM and GZ of the weight addition (considered as a solid), which influences the vertical position of the ship's center of gravity, and the location of the transverse metacenter, M. Secondly, there is a reduction in GM and GZ due to the free surface effect.

FREE COMMUNICATION EFFECT

If one or more of the boundaries of an off-center compartment are ruptured so that the sea may flow freely in and out with a minimum of restriction as the ship rolls, a condition of partial flooding with free communication with the sea exists. The added weight of the flooding water and the virtual rise in G due to the free surface effect cause what is known as free communication effect. With an off-center space flooded, a ship will assume a list that will be further aggravated by the free surface effect. As the ship lists, more water will flow into the compartment from the sea and will tend to level off at the height of the external waterline. The additional weight causes the ship to sink further, allowing more water to enter, causing more list, until some final list is reach. The reduction of GM due to free communication effect is approximately equal in magnitude to

$$\frac{ay^2}{V}$$

where

a = area of the free surface, in square feet

y = perpendicular distance from the geometric center of the free surface area to the fore-and-aft centerline of the intact waterline plane, in feet

V = new volume of ship's displacement after flooding, in cubic feet. Reduction in GM is additional to and separate from the free surface effect.

The approximate reduction in GZ may be computed from the following formula:

$$GZ = \frac{ay^2}{V} \sin \theta$$

This may be considered as a virtual rise in G, superimposed upon the virtual rise in G due to the free surface effect.

If two partially filled tanks on opposite sides of an intact ship are connected by an open pipe at or near their bottoms allowing a free flow of liquid between them, the effect on GM is the same as if both tanks were in free communication with the sea. Hence, valves in cross-connections between such tanks should never be left open without anticipating the accompanying decrease in stability. Such free flow is known as sluicing.

SUMMARY OF EFFECTS OF LOOSE WATER

The addition of loose water to a ship alters the stability characteristics by means of three effects that must be considered separately:

- The effect of added weight
- The effect of free surface
- The effect of free communication

Figure 3-29 shows the development of a stability curve with corrections for added weight, free surface, and free communication. Curve A is the ship's original stability curve before flooding. Curve B represents the situation after flooding; this curve shows the effect of added weight (increased stability) but it does not show the effects of free surface or of free communication. Curve C is curve B corrected for free surface effect only. Curve D is curve B corrected for both free surface effect and free communication effect. Curve D, therefore, is the final stability curve; it incorporates corrections for all three effects of loose water.

LONGITUDINAL STABILITY AND EFFECTS OF TRIM

The important phases of longitudinal inclination are changes in trim and longitudinal stability. A ship pitches longitudinally in contrast to rolling transversely, and it trims forward and aft, whereas it lists transversely. The difference in forward and after draft is defined as trim.

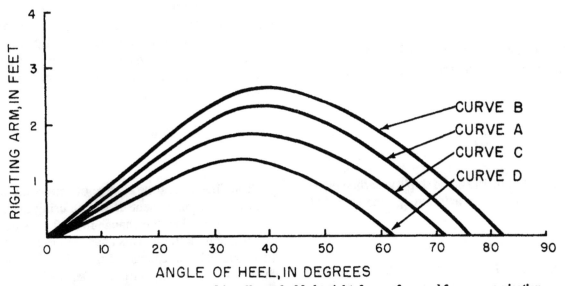

Figure 3-29.—Stability curve corrected for effects of added weight, free surface, and free communication.

CENTER OF FLOTATION

When a ship trims, it inclines about an axis through the geometric center of the waterline plane. This point is known as the center of flotation. The position for the center of flotation aft of the midperpendicular for various drafts may be found from a curve on the curves of form (fig. 3-26). When a center of flotation curve is not available, or when precise calculations are not required, the midperpendicular may be used in lieu of the center of flotation.

CHANGE OF TRIM

Change of trim may be defined as the change in the difference between the drafts forward and aft. If in changing the trim, the draft forward becomes greater, then the change is said to be by the bow. Conversely, if the draft aft becomes greater, the change of trim is by the stern.

Changes of trim are produced by shifting weights forward or aft or by adding or subtracting weights forward of or abaft of the center of flotation.

LONGITUDINAL STABILITY

Longitudinal stability is the tendency of a ship to resist a change in trim. For small angles of inclination, the longitudinal metacentric height multiplied by the displacement is a measure of initial longitudinal stability. The longitudinal metacentric height is designated GM and is found from

$$GM' = KB + BM' - KG$$

where

KB and KG are the same as for transverse stability BM'. The longitudinal metacentric radius is equal to

$$BM' = \frac{I'}{V}$$

where

I' = the moment of inertia of the ship's waterline plane about an athwartship axis through the center of flotation

V = the ship's volume of displacement

The value of BM' is very large–sometimes more than a hundred times that of BM. The values of BM' for

various draft may be found from the curves of form (fig. 3-26).

MOMENT TO CHANGE TRIM ONE INCH

The measure of a ship's ability to resist a change of trim is the moment required to produce a change of trim of a definite amount, such as 1 inch. The value of the moment to change trim 1 inch is obtained from

$$MTI = \frac{GM' \times W}{12L}$$

where

GM' = longitudinal metacentric height, in feet

W = displacement in tons

L = length between forward and after perpendiculars, in feet

For practical work, BM' is usually substituted for GM' since they are both large and the difference between them is relatively small. When this is done, however, MTI is called the approximate moment to change trim 1 inch. This value is often found as a curve in the curves of form (fig. 3-26). When such curve is not available, the moment to change trim 1 inch may be calculated from the formula

$$MTI = \frac{BM' \times W}{12L}$$

CALCULATION OF CHANGE OF TRIM

The movement of weight aboard ship in a fore-and-aft direction produces a trimming moment. This moment is equal to the weight multiplied by the distance moved. The change of trim in inches may be calculated by dividing the trimming moment by the moment to change trim 1 inch:

$$\text{Change of trim} = \frac{w \times t}{MTI}$$

The direction of the change of trim is the same as that of weight movement. If we are using midships as our axis of rotation, the change in draft forward equals the change in draft aft. This change of draft forward or aft is one-half the change of trim; for example, for a change of trim by the stern, the after draft increases the same amount the forward draft decreases–that is, one-half the change of trim. The reverse holds true for a change of trim by the bow.

Example: If 50 tons of ammunition is moved from approximately 150 feet forward of the center of flotation to approximately 150 feet aft of the center of flotation (300 feet), what are the new drafts?

Draft fwd = 19 feet 9 inches

Draft aft = 20 feet 3 inches

Mean draft = 20 feet

Trimming moment = $50 \times 300 = 15{,}000$ foot-tons

Moment to change trim 1 inch = 1,940 foot-tons, from the following calculation:

$$\text{MTI} = \frac{\text{BM}' \times \text{W}}{12L} = \frac{1{,}150 \times 11{,}800}{12 \times 582} = 1{,}940 \text{ foot-tons}$$

(BM' from curve in fig. 3-26)

Change in trim = $\frac{15{,}000}{1{,}940} = 8$ inches by the stern

Change of draft - 4 inches fwd, 4 inches aft

New draft fwd = 19 feet 5 inches

New draft aft = 20 feet 7 inches

LONGITUDINAL WEIGHT ADDITION

The addition of a weight either directly above or below the center of flotation will cause an increase in mean draft but will not change trim. All drafts will change by the same amount as the mean draft. The reverse is true when a weight is removed at the center of flotation.

To determine the change in drafts forward and aft due to adding a weight on the ship, the computation is in two steps. First, the weight is assumed to be added at the center of flotation. This increases the mean draft and all the drafts by the same amount. The increase is equal to the weight added, divided by the tons-per-inch immersion. With the ship at its new drafts, the weight is assumed to be moved to its ultimate location. Moving the weight fore and aft produces a trimming moment and therefore a change in trim, which is calculated as previously described.

FLOODING EFFECT DIAGRAM

From the flooding effect diagram in the ship's *Damage Control Book*, you can obtain a change in draft fore and aft due to solid flooding of a compartment. The weight of water to flood specific compartments is given and the trimming moment produced may be computed, as well as list in degrees which may be caused by the additional weight. Additional information on the flooding effect diagram can be obtained from *NSTM*, chapter 079, volume 1.

DRAFT DIAGRAM AND FUNCTIONS OF FORM

The draft diagram and functions of form of the ship's *Damage Control Book* indicates the location of draft marks and a nomograph for easy determination of displacement from drafts, vertical heights above the keel, and longitudinal information of numerous reference points discussed earlier and found on the curves of form. The draft diagram provides a simple means of obtaining a ship's displacement when the forward and after drafts are known. The drafts forward and aft are spotted on their respective scales, and a straight line connects these points. Displacement and KM are read directly at the point where this line crosses these scales; all other functions are read on a horizontal line drawn through this point (fig. 3-30).

EFFECT OF TRIM ON TRANSVERSE STABILITY

The curves of form prepared for a ship are based on the design conditions–that is, with no trim. For most types of ships, so long as trim does not become excessive, the curves are still applicable and may be used without adjustment.

When a ship trims by the stern, the transverse metacenter is slightly higher than indicated by the KM curve because both KB and BM increase. The center of buoyancy rises because of the movement of a wedge of buoyancy upward. The increased BM is the result of an enlarged waterplane as the ship trims by the stern.

Trim by bow usually means a decreased KM. The center of buoyancy will rise slightly, but this is usually counteracted by the decreased BM caused by the lower moment of inertia of the trimmed waterplane.

CAUSES OF IMPAIRED STABILITY

The stability of a ship may be impaired by several causes, resulting from mistakes or from enemy action. A summary of these causes and their effects are given in the following paragraphs.

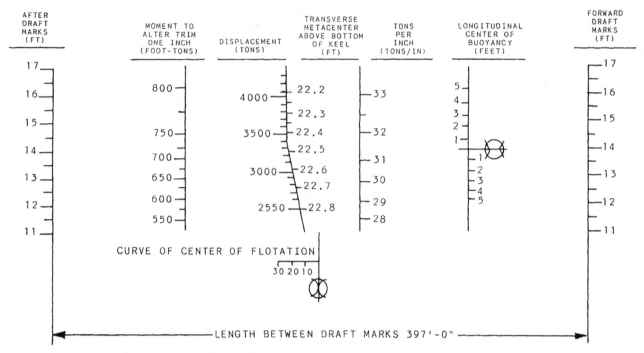

Figure 3-30.–Draft diagram and functions form.

ADDITION OF TOPSIDE WEIGHT

The addition of appreciable amounts of topside weight may be occasioned by unauthorized alterations; icing conditions; provisions, ammunition, or stores not struck down; deck cargo; and other conditions of load. Whenever a weight of considerable magnitude is added above the ship's existing center of gravity, the effects are as follows:

- Reduction of reserve buoyancy

- Reduction of GM and righting arms due to raising G

- Reduction of GM and righting arms due to loss of freeboard (change of waterplane)

- Reduction of righting arms if G is pulled away from the centerline

- Increase in righting moment due to increased displacement

The net effect of added high weight is always a reduction in stability. The reserve buoyancy loss is added weight in tons. The new metacentric height can be obtained from the formula

$$G_1M_1 = KM_1 - KG_1$$

Stability is determined by selecting a new stability curve from the cross-curves and correcting it for $AG_1 \sin \theta$ and $G_1G_2 \cos \theta$.

LOSS OF RESERVE BUOYANCY

Reserve buoyancy may be lost due to errors, such as poor maintenance, failure to close fittings properly, improper classification of fittings, and overloading the ship; or it may be lost as a result of enemy action, such as fragment or missile holes in boundaries, a blast that carries away boundaries or blows open or warps fittings, and flooding that overloads the ship. When the above-water body is holed, some reserve buoyancy is lost. The immersion of buoyant volume is necessary to the development of a righting arm as the ship rolls; if the hull is riddled it can no longer do this on the damaged side, toward which it will roll. In effect, the riddling of the above-water hull is analogous to losing a part of the freeboard, thus reducing stability. When this happens, if the ship takes water aboard on the roll, the combined effects of high added weight and free surface operate to cut down the righting moment. Therefore, the underwater hull and body should be plugged and patched, and every effort should be made to restore the watertightness of external and internal boundaries in the above-water body.

FLOODING

Flooding may take place because of underwater damage, mines or missile hits below decks, collision, missile hits near the waterline, running aground, fire-fighting water, ruptured piping, sprinkling of magazines, counterflooding, or leakage. Regardless of how it takes place, it can be classified in three general categories, each of which can be further broken down, as follows:

1. With respect to boundaries

   a. Solid flooding

   b. Partial flooding

   c. Partial flooding in free communication with the sea

2. With respect to height in the ship

   a. Center of gravity of the flooding water is above G

   b. Center of gravity of the flooding water is below G

3. With respect to the ship's centerline

   a. Symmetrical flooding

   b. Off-center flooding

Solid Flooding

The term *solid flooding* designates the situation in which a compartment is completely filled from deck to overhead. For this to occur, the compartment must be vented by an air escape, an open scuttle or vent fitting, or through fragment holes in the overhead. Solid flooding water behaves exactly like an added weight and has the effect of so many tons placed exactly at the center of gravity of the flooding water. It is more likely to occur below the waterline, where it has the effect of any added low weight. Since as G is usually a little above the waterline in warships, the net effect of solid flooding below the waterline is most frequently a gain in stability, unless a sizeable list or a serious loss of freeboard results in a net reduction of stability. The reserve buoyancy consumed is the weight of flooding water in tons, and the new GM and stability characteristics are found as previously explained.

Partial Flooding with Boundaries Intact

The term *partial flooding* refers to a condition in which the surface of the flooding water lies somewhere between the deck and the overhead of a compartment. The boundaries of the compartment remain watertight and the compartment remains partially but not completely filled. Partial flooding can be brought about by leakage from other damaged compartments or through defective fittings, seepage, shipping water on the roll, downward drainage of water, loose water from fire fighting, sprinklers, ruptured piping, and other damage.

Partial flooding of a compartment that has intact flooding boundaries affects the stability of the ship because of the effect of added weight and the effect of free surface. The effect of the added weight will depend upon whether the weight is high in the ship or low, and whether it is symmetrical about the centerline or is off-center. The effect of free surface will depend primarily upon the athwartship breadth of the free surface. Unless the free surface is relatively narrow and the weight is added low in the ship, the net effect of partial flooding in a compartment with intact boundaries is likely to be a very definite loss in overall stability.

Partial Flooding in Free Communication with the Sea

Free communication can exist only in partially flooded compartments in which it is possible for the sea to flow in and out as the ship rolls. Partial flooding with free communication is most likely to occur when there is a large hole that extends above and below the waterline. It may also occur in a waterline compartment when there is a large hole in the shell below the waterline, if the compartment is vented as the ship rolls. Where free communication does exist, the water level in the compartment remains at sea level as the ship rolls. When a compartment is partially flooded and in free communication with the sea, the ship's stability is affected by (1) added weight, (2) free surface effect, and (3) free communication effect. In general, the net effect of partial flooding with free communication is a decided loss in stability.

GROUNDING (STRANDING)

Running aground, whether deliberate or unintentional, involves three major problems:

- Ability to get off again

- Structural strength

- Stability

If the ship runs aground unintentionally, there is a strong temptation to use the engines in an effort to get off again. However, ship's screws become less effective in shallow water and attempting to back down may do more harm than good.

When the ship is aground, the beach exerts an upward force on the hull equal to that portion of the ship's weight that is not supported by the buoyancy of the water. Its effect on stress is the reverse of flooding.

The magnitude of the ground pressure or number of tons aground is the reduction in displacement caused by beaching. It may be found as the difference between the displacement when floating free and the displacement corresponding to the mean draft after grounding. (To be very accurate, these displacements must be corrected for trim, list, and draft.) The ground pressure is applied at the keel or other point of contact and has all the effects on draft list, trim, and stability of removing that many tons of weight from the location of the point of contact. If the point of contact is at the keel, it constitutes a low weight removal. The virtual center of gravity goes up by an amount computed with the following formula:

$$GG_1 = \frac{P\,(KG)}{W - P}$$

where

GG_1 = rise of G, in feet

P = tons aground

KG = height of G above keel, in feet, if floating free

$W - P$ = displacement in tons, corresponding to mean draft after grounding

The shift in the metacenter (MM_1) may be found in the curves of form as the difference in KM for the drafts floating free and after grounding. The metacentric height when aground is then as follows:

$$G_1M_1 = GM \pm MM_1 - GG_1$$

If the point of contact is at one end, the ship will trim by an amount equivalent to removing the tons aground from that end. If the point of contact is off-center, the ship will list to an angle corresponding to the off-center weight removal. If the GM calculated is negative, the effect of this condition on list must be considered. The result may be capsizing if the ship is aground only at the bow where it is narrow athwartship. However, if grounded throughout their length on a level surface, ships do not usually capsize even if high and dry.

REFERENCES

Hull Maintenance Technician 1 & C, NAVEDTRA 10574, Naval Education and Training Program Management Support Activity, Pensacola, Florida, 1972.

Damage Controlman 3 & 2, NAVEDTRA 10572, Naval Education and Training Program Management Support Activity, Pensacola, Florida, 1986.

Naval Ships' Technical Manual, Chapter 079, Volume 1, "Damage Control Stability and Buoyancy," Naval Sea Systems Command, Washington, D.C., 1976.

Naval Ships' Technical Manual, Chapter 096, "Stability and Buoyancy," Naval Sea Systems Command, Washington, D.C., 1976.

CHAPTER 4

PREVENTIVE AND CORRECTIVE DAMAGE CONTROL

LEARNING OBJECTIVES

Upon completion of this chapter, you should be able to do the following:

1. Describe the importance of watertight integrity in damage control.

2. Describe a shipboard damage control organization.

3. Describe the three material conditions of readiness, the degree of protection each offers, and the fittings that are closed for each condition.

4. Describe the means that damage control central (DCC) has to correct for list, trim, buoyancy, stability, and hull strength.

5. Identify the immediate local measures that repair locker personnel can take to correct damage.

6. Identify the three components of the fire triangle.

7. Identify the four types of fires in terms of chemical reactions and by-products, and the most effective extinguishing agent for each type.

8. Describe the function, operating principles, and applications of damage control systems and equipment.

9. Describe the organization of a fire-fighting team and the responsibilities of its members.

10. Identify the local measures that repair parties can take to control damage.

11. Describe the methods used to prevent progressive flooding and the importance of controlling flooding.

12. Identify the methods of shoring.

13. Identify the basic principles of defense against chemical warfare, biological warfare, or a nuclear attack.

14. Identify the preparatory and active measures of chemical, biological, and radiological (CBR) defense.

15. Describe CBR protective equipment.

16. Identify the methods and equipment used for detecting CBR agents.

17. Identify the different CBR contamination markers.

18. State the reason for decontamination, and identify the methods used for decontamination.

19. Explain the reasons investigators must constantly check for damage and the importance of reporting all damage to DCC.

INTRODUCTION

Aboard ship, the overall damage and casualty control function is composed of two separate but related phases: the engineering casualty control phase and the damage control phase. The engineer officer is responsible for both phases. Equally important to both phases is a prevention program to ensure the reliability of all equipment.

The engineering casualty control phase is concerned with the operational and battle casualties to the propulsion plant and associated equipment. Engineering casualty control is handled by engineering department personnel.

The damage control phase, on the other hand, involves practically every person aboard ship. The damage control phase is concerned with such things as the preservation of stability and watertight integrity, the control of fires and flooding, the repair of structural damage, and the control of CBR contamination. Damage control is the responsibility of all hands. All personnel must know their assignments within the damage control organization and understand the importance of those assignments. Damage control cannot be overemphasized. The necessary state of readiness can only be achieved through a reliable program.

This chapter presents some basic information on the principles of the damage control phase of the damage and casualty control function.

PREPARATIONS TO RESIST DAMAGE

Naval ships are designed to resist accidental and battle damage. Damage resistant features include structural strength, watertight compartmentation, stability, and buoyancy. Maintaining these damage-resistant features and maintaining a high state of material and personnel readiness before damage occurs is far more important for survival than are any damage control measures that can be taken after the ship has been damaged. It has been said that 90 percent of the damage control needed to save a ship takes place before the ship is damaged and only 10 percent can be done after the damage has occurred. In spite of all preparatory measures, however, the survival of a ship sometimes depends upon prompt and effective damage control measures taken after damage has occurred. It is essential, therefore, that all shipboard personnel be trained in damage control procedures.

The maintenance of watertight integrity is a vital part of any ship's preparations to resist damage. Each undamaged tank or compartment aboard ship must be kept watertight if progressive flooding is to be prevented after damage. Watertight integrity can be lost in a number of ways. Failure to secure access closures and improper maintenance of watertight fittings and compartment boundaries, as well as external damage to the ship, can cause loss of watertight integrity. The condition of watertight boundaries, compartments, and fittings is determined by visual observation and by various tests. All defects discovered by any test or inspection must be remedied immediately.

The proper setting of material conditions of readiness is very important for maintaining a ship's watertight fittings and compartment boundaries. The compartment checkoff list (CCOL) (fig. 4-1) is an itemized list of all classified fittings and all equipment or facilities useful for damage control within a compartment or an area. The CCOL shows the compartment name, number, location, purpose, and classification of each fitting within the compartment or area. It also states who is responsible for its proper closure. A CCOL is permanently posted in each compartment. Duplicate CCOLs shall be prepared, marked "Duplicate," and posted at each entrance of compartments with two or more entrances. A master copy of each CCOL must be kept by the damage control assistant (DCA). Further guidance on CCOLs can be found in NSTM, chapter 079, volume 2.

DAMAGE CONTROL ORGANIZATION

To ensure damage control training and to provide prompt control of casualties, a damage control organization must be set up and kept active on all ships.

As previously noted, the engineer officer is responsible for damage control. The DCA, who is under the engineer officer, is responsible for establishing and maintaining an effective damage control organization. Specifically, the DCA is responsible for the prevention and control of damage, the training of ship's personnel in damage control, and the operation, care, and maintenance of certain auxiliary, machinery, piping, and drainage systems not assigned to other departments or divisions.

Although naval ships may be large or small, and although they differ in type, the basic

COMPARTMENT CHECKOFF LIST

NAVSHIPS 9880/2 (REV. 2-67) (Formerly NAVSHIPS 184) S/N 0105-530-2000

COMP'T. NO. 2-108-1-L   NAME Crew's Berthing

ITEM	FITTING	NUMBER	LOCATION & PURPOSE	CLASS	DIV RESP
	ACCESS				
1.	QAWTD	2-108-1	Access to 2-96-1-L	Z	REP III
2.	WTD	2-108-3	Access to 2-120-1	Z	REP III
3.	WTH	2-120-3	Access to 3-108-1-L	X	S
	MISCELLANEOUS CLOSURES				
4.	ATC	2-109-1	in WTD 2-108	X	E
	DRAINAGE				
5.	DDV	2-112-1		Z	E
	FIREMAIN, SPRINKLING, AND WASHDOWN				
6.	FMCOV	2-109-1	Cut out to FP 2-	W	REP III
7.	FMCOV	2-109-1	Cut out to	W	REP III
	FUEL OIL				
8.	STC	2-118-1	Sound FO Tank	X	B
	JP-5				
9.	STC	2-118-1	Sound JP-5	X	B
	VENTILATION				
10.	EVC	2-113-1	For Exhaust	Z	REP III
	MISCELLANEOUS				
11.	Loud Speaker		1MC system		
12.	151b CO_2		Portable		

Figure 4-1.—A CCOL.

principles of the damage control organization are more or less standardized. Some organizations are larger and more elaborate than others, but they all function on the same basic principles.

A standard damage control organization, suitable for large ships but followed by all ships as closely as practicable, includes DCC and repair stations. On most naval ships DCC is a separate station. On all ships with gas turbine propulsion, DCC is integrated with propulsion and electrical control in the central control station (CCS). Repair parties are assigned to specifically located repair stations. Repair stations are further subdivided into unit patrols to permit dispersal of personnel and a wide coverage of the assigned areas.

DAMAGE CONTROL CENTRAL

The primary purpose of DCC is to collect and compare reports from the various repair stations to determine the condition of the ship and the corrective action to be taken. The commanding officer (CO) is kept posted on the condition of the ship and on important corrective actions taken. The DCA, at his battle station in DCC, is the nerve center and directing force of the entire damage control organization. The DCA is assisted in DCC by a stability officer, a casualty board operator, and a damage analyst. In addition, representatives of the various divisions of the engineering department are assigned to DCC.

In DCC, repair party reports are carefully checked so that immediate action can be taken to isolate damage and to make emergency repairs in the most effective manner. Graphic records of the damage are made on various damage control diagrams and status boards as the reports are received. For example, reports concerning flooding are marked up, as they come in, on a status board which indicates liquid distribution before damage. With this information, the stability and buoyancy of the ship can be estimated and the necessary corrective measures can be determined.

If DCC is destroyed or is for other reasons unable to retain control, the repair stations, in designated order, take over these same functions. Provisions are also made for passing the control of each repair station down through the officers, petty officers, and nonrated personnel, so that no group will ever be without a leader.

REPAIR PARTIES

A standard damage control organization on large ships includes the following repair stations:

Repair 1 (deck or topside repair)
Repair 2 (forward repair)
Repair 3 (after repair)
Repair 4 (amidship repair)
Repair 5 (propulsion repair)
Repair 8 (electronics repair)

Aircraft carriers have two additional repair stations—repair 7 (gallery deck and island structure repair) and repair 6 (ordnance repair). Carriers also have an aviation fuel repair and an ordnance disposal team. All ships capable of manned aircraft operations have a crash and salvage team to handle all flight deck mishaps. On smaller ships, there are usually three repair stations—repair 2, repair 3, and repair 5.

The organization of repair stations is basically the same on all ships; however, more personnel are available for manning repair stations on large ships than on smaller ships. The number and rating of personnel assigned to a repair station, as specified in the battle bill, are determined by the location of the station, the portion of the ship assigned to the station, and the total number of personnel available.

Each repair party has an officer in charge, who may in some cases be a chief petty officer. The second in charge is usually a chief petty officer who is qualified in damage control and who is capable of taking over the supervision of the repair party.

Many repair stations have unit patrol stations at key locations in their assigned areas to supplement the repair station. Operating instructions should be posted at each repair station. In general, instructions should include the purpose of the repair party; the specific assignment of space for which that station is responsible; instructions for assigning and stationing personnel; methods and procedures for damage control communications; instructions for handling machinery and equipment located in that area; procedures for CBR defense; sequence and procedure for passing control from one station to another; a list of current damage control bills; and a list of all damage control equipment and gear provided for the repair station.

MATERIAL CONDITIONS OF READINESS

Material conditions of readiness refers to the degree of access and system closure to limit the extent of damage to the ship. Maximum closure is not maintained at all times because it would interfere with normal operation of the ship. For damage control purposes, naval ships have three material conditions of readiness, each condition representing a different degree of tightness and protection. The three material conditions of readiness are called X-RAY, YOKE, and ZEBRA. These titles, which have no connection with the phonetic alphabet, are used in all spoken and written communications concerning material conditions.

Condition X-RAY, which provides the least protection, is set when the ship is in no danger from attack, such as when it is at anchor in a well-protected harbor or secured at a home base during regular working hours.

Condition YOKE, which provides somewhat more protection than condition X-RAY, is set and maintained at sea. It is also maintained in port during wartime and at other times in port outside of regular working hours.

Condition ZEBRA is set before going to sea or entering port, during wartime. It is also set immediately, without further orders, when manning general quarters stations. Condition ZEBRA is also set to localize and control fire and flooding when not at general quarters.

The closures involved in setting the material conditions of readiness are labeled as follows:

- X-RAY, marked with a black *X*. These closures are secured during conditions X-RAY, YOKE, and ZEBRA.

- YOKE, marked with a black *Y*. These closures are secured during conditions YOKE and ZEBRA.

- ZEBRA, marked with a red *Z*. These closures are secured during conditions ZEBRA.

Permission to break a material condition of readiness must come from an authorizing officer. The DCC watch supervisor, when assigned, may authorize a fitting to be opened. If a DCC watch supervisor is not assigned, the officer of the deck (OOD) authorizes the opening of fittings. During general quarters, permission must come from the DCA or repair party officer. The repair party officer controls the opening and closing of all fittings in the assigned area during general quarters.

Other fitting markings that are modifications of the three basic conditions are as follows:

CIRCLE X-RAY fitting, marked with a black *X* in a black circle, are secured during conditions X-RAY, YOKE, and ZEBRA. CIRCLE YOKE fitting, marked with a black *Y* in a black circle, are secured during conditions YOKE and ZEBRA. Both CIRCLE X-RAY and CIRCLE YOKE fittings may be opened without special permission when going to and securing from general quarters, when transferring ammunition, or when operating vital systems during general quarters; but the fittings must be secured when not in use.

CIRCLE ZEBRA fittings, marked with a red *Z* in a red circle, are secured during condition ZEBRA. CIRCLE ZEBRA fittings may be opened during prolonged periods of general quarters, when the condition may be modified. Opening these fittings enables personnel to prepare and distribute battle rations, open limited sanitary facilities, ventilate battle stations, and provide access from ready rooms to flight deck. When open, CIRCLE ZEBRA fittings must be guarded for immediate closure if necessary.

DOG ZEBRA fittings, marked with a red *Z* in a black *D*, are secured during darken ship condition. The DOG ZEBRA classification applies to weather accesses not equipped with light switches or light traps.

WILLIAM fittings, marked with a black *W*, are kept open during all material conditions. This classification applies to vital sea suction valves supplying main and auxiliary condensers, fire pumps, and spaces that are manned during condition X-RAY, YOKE, and ZEBRA; it would also apply to vital valves that, if secured, would impair the mobility and fire protection of the ship. These items are secured only as necessary to control damage or contamination and to effect repairs to the units served.

CIRCLE WILLIAM fittings, marked with a black *W* in a black circle, are normally kept open (as WILLIAM fittings are), but must be secured as defense against a CBR attack.

INVESTIGATION OF DAMAGE

All available information concerning the nature and extent of damage must be given to the DCA so that damage may be analyzed and decisions made on appropriate measures of control. The repair parties that are investigating the damage at the scene are normally in the best position to give dependable information on the nature and extent of the damage. All repair party personnel should be trained to make prompt, accurate, and complete reports to DCC. Items that should normally be reported to DCC include the following:

● Description of important things seen, heard, or felt by personnel

● Location and nature of fires, smoke, and toxic gases

● Location and nature of progressive flooding

● Overall extent and nature of flooding

● Structural damage to longitudinal strength members

● Location and nature of damage to vital piping and electrical systems

● Local progress made in controlling fire; halting flooding; isolating damaged systems; patching or bypassing ruptured piping; and rigging casualty power and emergency communications

● Compartment-by-compartment information on flooding, including depth of liquid in each flooded compartment

● Conditions of boundaries (decks, bulkheads, and closures) surrounding each flooded compartment

● Local progress made in reclaiming compartments by plugging, patching, shoring, and removing loose water

● Areas in which damage is suspected but cannot be reached or verified

The DCA must find out just what information the CO desires concerning the extent of the damage incurred and the corrective measures taken. The DCA must also find out how detailed the information to the CO should be and when it is to be furnished. With these guidelines in mind, the DCA must sift all information coming into DCC and pass along to the bridge only the type of information that the CO desires.

CORRECTIVE MEASURES

Measures for the control of damage may be divided into two general categories: (1) overall ship survival measures, and (2) immediate local measures.

OVERALL SHIP SURVIVAL MEASURES

Overall ship survival measures are those actions initiated by DCC for the handling of list, trim, buoyancy, stability, and hull strength. Operations in this category have five general objectives: improving GM and overall stability, correcting for off-center weight, restoring lost freeboard and reserve buoyancy, correcting for trim, and relieving stress in longitudinal strength members.

Improving GM and Overall Stability

The measures used to improve GM and overall stability in a damaged ship include (1) suppressing free surface, (2) jettisoning topside weight, (3) ballasting, (4) lowering liquid or solid weights, and (5) restoring boundaries.

Correcting for Off-center Weight

Off-center weight may occur as the result of unsymmetrical flooding or as the result of an athwartship movement of weight. Correcting for off-center weight may be accomplished by (1) pumping out off-center flooding water, (2) pumping liquids across the ship, (3) counterflooding, (4) jettisoning topside weights from the low side of the ship, (5) shifting solid weights athwartships, and (6) pumping liquids overboard from intact wing tanks on the low side.

Restoring Lost Freeboard and Reserve Buoyancy

Restoring lost freeboard and reserve buoyancy requires the removal of large quantities of weight.

In general, the most practicable way of accomplishing this is to restore watertight boundaries and to reclaim compartments by pumping them out. Any corrective measure that removes weight from the ship contributes to the restoration of freeboard.

Correcting for Trim

The method used to correct for trim after damage includes (1) pumping out flood water, (2) pumping liquids forward and aft, (3) counterflooding the high end, (4) jettisoning topside weights from the low end, (5) shifting solid weights from the low end to the high end, and (6) pumping liquids over the side from intact tanks at the low end. The first of these methods—that is, pumping out flood water—is in most cases the only truly effective means of correcting a severe trim.

The correction of trim is usually secondary to the correction of list, unless the trim is so great that there is danger of submerging the weather deck at the low end.

**Relieving Stress in Longitudinal
Strength Members**

When a ship is partially flooded, the longitudinal strength members are subject to great stress. In cases where damage has carried away or buckled the strength members amidships, the additional stress imposed by the weight of the flooding water may be enough to cause the ship to break up. The only effective way of relieving stress caused by flooding is to remove the water. Other measures, such as removing or shifting weight, may be helpful but cannot be completely effective. In some instances, damaged longitudinals may be strengthened by welding.

IMMEDIATE LOCAL MEASURES

Immediate local measures are those actions taken by repair parties at the scene of the damage. In general, these measures include all on-scene efforts to investigate the damage, to report to DCC, and to accomplish the following:

1. Establish flooding boundaries by selecting a line of intact bulkheads and decks to which the flooding may be held and by rapid plugging, patching, and shoring to make these boundaries watertight and dependable.

2. Control and extinguish fires.
3. Establish secondary flooding boundaries by selecting a second line of bulkheads and decks to which the flooding may be held if the first flooding boundaries fail.
4. Advance flooding boundaries by moving in toward the scene of the damage, plugging, patching, shoring, and removing loose water.
5. Isolate damage to machinery, piping, and electrical systems.
6. Restore piping systems to service by the use of patches, jumpers, clamps, couplings, and so on.
7. Rig casualty power.
8. Rig emergency communications and lighting.
9. Rescue personnel and care for the wounded.
10. Remove wreckage and debris.
11. Cover or barricade dangerous areas.
12. Ventilate compartments that are filled with smoke or toxic gases.
13. Take measures to counteract the effects of CBR contamination or weapons.

Immediate local measures for the control of damage are of vital importance. It is not necessary for DCC to decide on these measures; rather, they should be carried out automatically and rapidly by repair parties. However, DCC should be continuously and accurately advised of the progress made by each party so that the efforts of all repair parties may be coordinated to the best advantage.

PRACTICAL DAMAGE CONTROL

Both the immediate local measures and the overall ship survival measures have, of course, the common aim of saving the ship and restoring it to service. The following subsections deal with the practical methods used to achieve this aim: controlling fires, controlling flooding, repairing structural damage, and restoring vital services.

It should be noted that controlling the effects of CBR warfare weapons or agents may in some situations take precedence over other damage control measures. Because of the complex nature of CBR defense, this subject is treated separately in a later section of this chapter.

CONTROL OF FIRES

Fire is a constant potential hazard aboard ship. All possible measures must be taken to prevent its occurrence or to bring about its rapid control and extinguishment. In many cases, fire occurs in conjunction with other damage, as a result of enemy action, weather, or mishap. Unless fire is rapidly and effectively extinguished, it may cause more damage than the initial casualty and it may, in fact, cause the loss of a ship even after other damage has been repaired or minimized.

Fire Components

Three components are required for a fire. These are a combustible material, a sufficiently high temperature, and a supply of oxygen. You will hear these components referred to as the fire triangle, consisting of fuel, heat, and oxygen (fig. 4-2). Fires are generally controlled and extinguished by eliminating one side of the fire triangle. That is, if you remove either fuel, heat, or oxygen, you can prevent or extinguish a fire. We will discuss the extinguishment of fires later in this chapter.

HEAT.—Fire is also called burning or combustion. This is a rapid chemical reaction that releases energy in the form of light and noticeable heat. Most combustion involves rapid oxidation. Oxidation is the chemical reaction by which oxygen combines chemically with the elements of the burning substance.

Even when oxidation proceeds slowly, such as a piece of iron rusting, a small amount of heat is generated. However, this heat usually dissipates before there is any noticeable rise in the temperature of the material being oxidized. With certain types of materials, slow oxidation can turn into fast oxidation (fire) if the heat is not dissipated. These materials are normally stowed in a confined space where the heat of oxidation cannot be dissipated rapidly enough. This is known as spontaneous combustion. Materials, such as rags or papers, that are soaked with either animal fats, vegetable fats, paints, or solvents are particularly subject to spontaneous combustion.

For a combustible fuel or substance to catch on fire, it must have an ignition source and be hot enough to burn. The lowest temperature at which a flammable substance gives off vapors that will burn when a flame or spark is applied is known as the flash point. The fire point is the temperature at which the fuel will continue to burn after it has been ignited. The fire point is usually a few degrees higher than the flash point. The auto-ignition or self-ignition point is the lowest temperature to which a substance must be heated to give off vapors that will burn without the application of a spark or flame. In other words, the auto-ignition point is the temperature at which spontaneous combustion occurs. The auto-ignition point is usually at a much higher temperature than the fire point.

The range between the smallest and the largest amounts of vapor in a given quantity of air that will burn or explode when ignited is called the flammable range or the explosive range. For example, let us say that a substance has a flammable or explosive range of 1 to 12 percent. This means that either a fire or an explosion can occur if the atmosphere contains more than 1 percent but less than 12 percent of the vapor of this substance. In general, the percentages referred to in connection with flammable or explosive ranges are percentages by volume.

FUEL.—Another component of the fire triangle is fuel. Fuels take on a wide variety of characteristics. A fuel may be a solid, a liquid, or even a vapor. Some of the fuels you will come into contact with are rags, paper, wood, oil, paint, solvents, and magnesium metals. This is by no means a complete list, but only examples.

OXYGEN.—The air you breathe contains 20.8 percent oxygen. All fires need oxygen to continue to burn. Some fires will burn with only 6 percent oxygen. However, there are some fires that will produce their own oxygen.

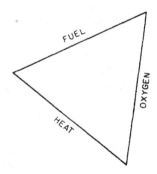

Figure 4-2.—The fire triangle.

Types of Fires

Fires are classified according to the nature of the combustible material involved. The classification of any particular fire is of great importance since it determines the manner the fire must be extinguished. Fires are classified as being either class A, class B, class C, or class D. Class A fires are those that involve ordinary solid combustible material, such as wood, paper, mattresses, and canvas. Class B fires are those that involve flammable liquids, such as fuel, oils, and greases. Class C fires are those that occur in electrical equipment. Class D fires are those of a chemical nature that involve metals, such as magnesium, potassium, powdered aluminum zinc, sodium, titanium, and zirconium.

Fire-fighting Agents

The agents commonly used by Navy fire fighters include water, aqueous film forming foam (AFFF), dry chemicals, carbon dioxide (CO_2), and Halon. The agent or agents that you will use in any particular case will depend upon the classification of the fire and the general circumstances.

WATER.—Cooling is the most common method of fire extinguishment, and water, which is readily available, is the most effective cooling agent. Water is most commonly used on class A fires. Water can also be used on class B fires, but you must remember the possibility of reflash exists until the fuel is cooled down below the flash point. Water is not recommended for use on class C fires. A solid stream of water will conduct electricity, and the possibility of electrical shock to the fire fighter exists. However, water in the form of a fine spray or fog does not conduct electricity, and can be used on a class C fire with extreme care.

AQUEOUS FILM FORMING FOAM (AFFF).—Foam is a highly effective extinguishing agent for smothering large fires, particularly those in oil, gasoline, and jet fuels. AFFF, also known as "light water," is a synthetic, film forming foam. AFFF is provided in a concentrate form and is mixed in a proportion of 6 parts concentrate and 94 parts water. When mixed it is a "light water" film that will float on flammable fuels. This film prevents the escape of vapors and forms a vapor-tight film. It enhances extinguishment and prevents reflash. AFFF is often used in combination with a dry chemical known as purple-K-powder (PKP).

DRY CHEMICALS.—Dry chemical powders extinguish a fire by a rather complicated chemical process. Instead of smothering or cooling the fire, they interrupt the fire. Dry chemicals prevent the three elements of fire from combining to form fire. However, dry chemicals do not provide protection against reflash after the fire is out. When used, PKP should be followed with an application of AFFF to prevent reflash. The combined use of PKP and AFFF together is three times more effective and faster than the individual use of either fire-fighting agent alone.

The most common dry chemical used is potassium bicarbonate, also known as PKP. It is normally blown into the fire by a charge of pressurized inert gas (usually CO_2). PKP is best suited for use on class B fires. PKP leaves a residue that is a corrosive and abrasive chemical, which is difficult to clean up or remove after use. Therefore, it is not recommended for use on class C fires, on internal parts of engines, on electronic gear, or on equipment with difficult to access surfaces. PKP is available in 18- or 27-pound portable extinguishers. It is not charged until it is to be used. The extinguisher (fig. 4-3) has a

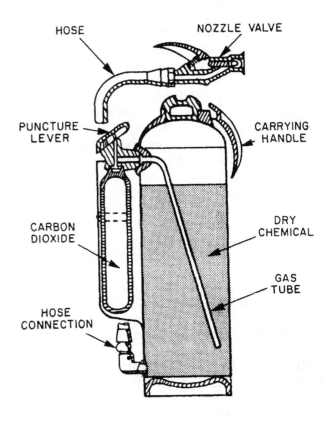

Figure 4-3.—A portable dry chemical extinguisher.

range of approximately 20 feet. To use, remove the locking pin from the puncture lever, then press sharply on the puncture lever to puncture the CO_2 charge. Use the PKP in short bursts, by squeezing the nozzle valve. Aim at the base of the fire. A short burst of PKP in the air will provide a heat shield. PKP is excellent at extinguishing a fire, but it has limitations. It provides no reflash protection and no cooling effect. When the fire is out, a backup agent should be quickly applied to prevent reflash. When you are finished with the PKP extinguisher, invert the extinguisher, squeeze the nozzle valve, and tap the nozzle on the deck. This action will release the charge and clear the hose and nozzle of powder. PKP powder can cake and become very difficult to remove, or it may block the hose.

CARBON DIOXIDE (CO_2).—Carbon dioxide is a dry, noncorrosive, dense, inert gas that is effective for extinguishing fires by smothering them. It is nonconductive and ideal for use on class C fires. It leaves no residue and does no damage to equipment. CO_2 is dangerous in that it displaces oxygen and can cause asphyxiation. When used, the discharge of CO_2 produces a frost that conducts electricity. Therefore, you should be sure that the extinguisher is not allowed to come in contact with electrical components. The CO_2 extinguisher is ideal for electrical fires, and is normally of a 15-pound size. Maximum range is 5 feet from the horn. It should be applied to the base of fire in a sweeping motion. To use, remove the locking pin from the valve, grab the handle of the horn with one hand, and squeeze the grip with the other hand. In continuous operation, a 15-pound extinguisher will last 30 to 40 seconds. When using the extinguisher, you should keep your hand on the insulated handle of the horn. When released, the CO_2 expands rapidly from under pressure. The rapid expansion and pressure drop cause the temperature of the CO_2 to drop significantly. The horn will freeze with "snow" from the temperature drop. Also, the rapid change may cause a buildup of static charge. To prevent the buildup of static charge, keep the base of the extinguisher on the deck to ground it.

HALON.—Halon is a halogenated carbon gas that is colorless, odorless, and five times as dense as air. This extinguishing agent is effective against class A, class B, and class C fires. There are some burring materials that Halon is not effective on, most notable are those associated with class D

fires. Halon interrupts the chemical reaction of the fire, just as PKP does, and does not provide reflash protection. For Halon to function effectively as an extinguishing agent, it must decompose. However, as it decomposes, several other products, such as hydrogen fluoride and hydrogen bromide, are formed. These gases are dangerous and irritating to the eyes, skin, and upper respiratory tract. They can also produce chemical burns. Halon is stored in gas cylinders or is available in portable extinguishers. When discharged, a 5 to 7 percent concentration can extinguish a fire.

There are two types of Halon used on board ship, Halon 1301 (bromotrifluoromethane) and Halon 1211. Halon 1301 is used for fixed compartment flooding systems, and Halon 1211 is used in portable extinguishers.

Halon 1211 portable extinguishers are for use primarily on class B fires. Halon 1211 is stored in a liquid state and retained under pressure with a booster charge of nitrogen. Halon 1211 should never be used on a class D fire, because if it reaches the fire in a liquid state, it could result in an explosion. When actuated, the discharge stream consists of a mixture of liquid droplets and vapor. Halon 1211 extinguishers are marked with a reflective silver band 6 inches in width around the extinguisher body. When you are using the extinguisher, the extinguisher should be aimed at the base of the fire and swept back and forth. This extinguisher has a range of 10 to 30 feet, and will last for 15 to 40 seconds of operation.

Halon 1301 flooding installations are provided in major machinery spaces. These installations have visual and audio alarms to warn personnel that Halon has been released. Because machinery spaces are normally occupied, there is a delay mechanism to prevent release of the Halon for 60 seconds. This system will always have controls to secure ventilation fans and dampers. If Halon appears to have extinguished the fire, you should wait 15 minutes to allow the space to cool and minimize the possibility of reflash before re-entry. Otherwise, the fresh air admitted to the space during re-entry may cause the fire to reflash. Halon flooding systems will differ some with manufacturer's design. A typical system is illustrated in figure 4-4.

You should not stay in a space where Halon has been released unless you are wearing an oxygen breathing apparatus (OBA). An OBA is a device used to permit the fire fighter to breathe. Unlike the gas mask, which only filters outside air and therefore cannot be used for fire fighting,

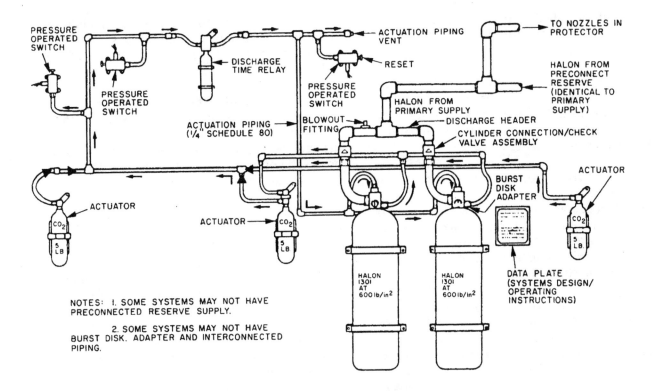

Figure 4-4.—A typical Halon system.

the OBA generates its own oxygen and circulates it through a closed system. Essential components of the OBA are an airtight faceplate with eyepieces and a speaking diaphragm, exhalation and inhalation tubes, an oxygen-producing canister, and a breathing bag. Exhaled air flows through the exhalation tube to the bottom of the canister (where it loses carbon dioxide and moisture and gains oxygen), into the breathing bag, and then into the lungs through the inhalation tube.

STEAM.—Steam can be used as an extinguishing agent for class A and B fires where available. Steam extinguishes fires by smothering them. Boiler air casings have installed steam smothering systems in case of fire. Steam should only be used when other extinguishing agents are not available.

Fire-fighting Systems and Equipment

To fight fires effectively, you must have a thorough knowledge of fire-fighting systems and equipment. This part of the chapter will deal with the systems and equipment that are commonly used aboard naval ships to apply the extinguishing agents to fires. This information is general in nature and should be supplemented with a thorough understanding of the systems and equipment available to you on your ship.

FIREMAIN SYSTEMS.—The firemain system receives water pumped from the sea. It distributes this water to fireplugs, sprinkling systems, flushing systems, machinery cooling-water systems, countermeasure washdown systems, and other systems as required. The firemain system is used primarily to supply the fireplug and the sprinkling systems; the other uses of the systems are secondary.

Naval ships have three basic types of firemain systems: the single-main system, the horizontal-loop system, and the vertical-loop system. The type of firemain system in any particular ship depends upon the characteristics and functions of the ship. Small ships generally have straight-line, single-main systems. Large ships usually have one of the loop systems or a composite system, which is some combination or variation of the following three basic types:

● The single-main firemain system consists of one main that extends fore and aft. The main is generally installed near the centerline of the ship, extending forward and aft as far as necessary.

• The horizontal-loop firemain system consists of two single fore-and-aft, cross-connected mains. The two mains are installed in the same horizontal plane but are separated athwartships as far as practical.

• The vertical-loop firemain system consists of two single fore-and-aft, cross-connected mains. The two mains are separated both horizontally and vertically. As a rule, the lower main is located below the lowest complete watertight deck, and the upper main is located below the highest complete watertight deck.

• A composite firemain system consists of two mains installed on the damage control deck and separated athwartships. A bypass main is installed at the lower level near the centerline. Cross connections are installed alternately between one service main and the bypass main.

Firemain outlets or fireplugs are of two different sizes, 2 1/2 or 1 1/2 inches in diameter. A wye gate adapter is available to connect two 1 1/2-inch hoses to a 2 1/2-inch fireplug. To prevent all-purpose nozzles from becoming clogged, all fireplugs have a quick-cleaning strainer (fig. 4-5) attached to collect foreign material, such as marine growth and encrustation. The strainer handle is connected to a ball valve. When the handle is parallel to the strainer, the ball valve is closed and water flows through the

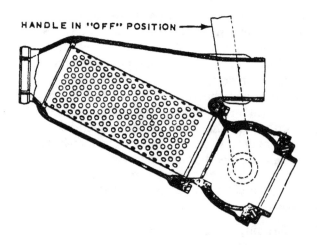

Figure 4-5.—Quick-cleaning strainer.

strainer and into the fire hose. When the handle is in the off position, the ball valve is open, and the water flows into the strainer and out of the ball valve.

Fire Hoses and Fittings.—The standard fire hose used by the Navy in the past was a double-jacketed, rubber-lined cotton hose. Now, it's an orange polypropylene-jacketed, rubber-lined hose. The orange hoses will phase out the older ones. The standard hose comes in both 1 1/2- and 2 1/2-inch diameter size, both in a standard length of 50 feet. The 50-foot hose is referred to as a length; longer hoses are described in terms of how many lengths they contain. Thus, a 100-foot hose is described as two lengths.

A standard length of hose has two fittings—a male fitting at one end and a female fitting at the other end. Double female couplings are available to connect two male fittings. This is necessary when making up a jumper line assembly to bypass a damaged section of firemain. Double male couplings can be used to join two female fittings (for example, a nozzle to a female fitting). Reduction couplings are used to connect 2 1/2-inch hoses to a 1 1/2-inch outlet, or to attach a 1 1/2-inch hose to a 2 1/2-inch outlet.

Figure 4-6 shows properly rigged hoses at a fire station. Fire stations will have one or more racks provided. Each rack will usually have one 100-foot line faked out on it so that it is free-running with ends hanging down so that the couplings are ready for use. Fire stations should have a spanner wrench available for tightening and breaking apart connections, and an applicator stowed for use.

Hoses should be handled and stowed with care. Spare hoses are rolled up and stored in repair lockers. To roll a hose, lay it out straight, then fold it back so the male end is on top and its length 4 feet less than the female end. Roll the hose starting at the folded end so that the male end stays inside and that its threads are protected. To use the rolled hose, set it on the deck (female end down) and give it a shove to unroll it. Then take the male end to the scene of action. Before connecting two lengths of hose make sure the gaskets are in place.

After use, the hoses should be thoroughly drained and properly dried. If this is not done, salt water reacts with the hose lining and produces an acid that can deteriorate the hose. Hoses should be hydrostatically tested annually according to PMS to 250 pounds per square inch (psi) for 5 minutes.

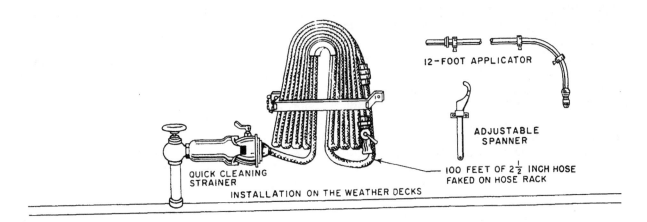

QUICK CLEANING STRAINER

INSTALLATION ON THE WEATHER DECKS

12-FOOT APPLICATOR

ADJUSTABLE SPANNER

100 FEET OF 2½ INCH HOSE FAKED ON HOSE RACK

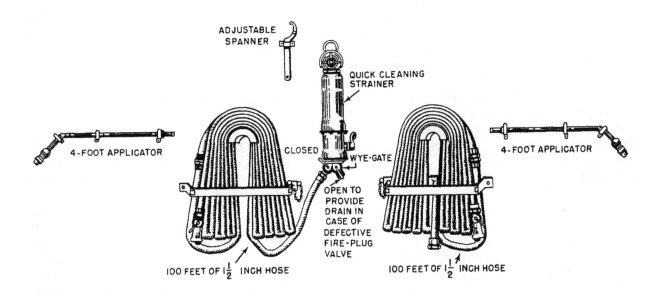

ADJUSTABLE SPANNER

QUICK CLEANING STRAINER

CLOSED

WYE-GATE

OPEN TO PROVIDE DRAIN IN CASE OF DEFECTIVE FIRE-PLUG VALVE

4-FOOT APPLICATOR

4-FOOT APPLICATOR

100 FEET OF 1½ INCH HOSE

100 FEET OF 1½ INCH HOSE

Figure 4-6.—Fire station equipment.

Nozzles and Applicators.—The Navy all-purpose nozzle, shown in figure 4-7, comes in two standard sizes: 2 1/2 inches and 1 1/2 inches. The valve on the nozzle can be set in one of three positions—SHUT, FOG, or OPEN. The nozzle can provide the fire fighter with three different water flow patterns—solid stream, high-velocity fog, or low-velocity fog.

Fog is generated by either a high-velocity nozzle tip or a low-velocity fog head on the end of an insertable applicator. A number of small hole outlets are drilled at converging angles so that streams impinge upon each other and break up into fog particles. The size, design, and placement of the outlets determine velocity with which the

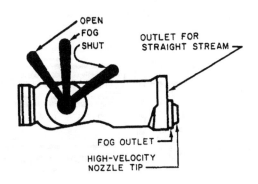

OPEN
FOG
SHUT

OUTLET FOR STRAIGHT STREAM

FOG OUTLET

HIGH-VELOCITY NOZZLE TIP

Figure 4-7.—All-purpose nozzle.

fog particles emerge from the nozzle. If you want high-velocity fog, leave the high-velocity nozzle tip in place in the fog outlet on the nozzle. Figure 4-8 shows the different size applicators available. If you want low-velocity fog, remove the tip and insert a low-velocity applicator in the fog outlet on the nozzle. Both the nozzle tip and applicator are held in place by a bayonet joint. When the high-velocity nozzle tip is removed, it remains attached to the nozzle by a short chain.

The solid stream comes out of the top of two openings on the nozzle. Use of the solid stream should be limited to breaking up burning material at a distance. Solid stream is always available to the fire fighter.

The high-velocity fog has some reach or throw in an umbrella-shaped pattern. It discharges 117 gallons per minute (gpm) with a 2 1/2-inch nozzle and a firemain pressure of 100 psi. The solid stream has the most reach and power. It discharges 250 gpm with a 2 1/2-inch nozzle and a firemain pressure of 100 psi.

In comparison, the low-velocity fog is a wider shaped pattern that is easily thrown about by drafts or wind currents. The low-velocity fog is used by the No. 2 hose of a fire-fighting team and is used to cool and protect the fire-fighting team from the heat. The fog also dilutes some vapors, smoke, and fumes.

Fog is a very effective cooling agent. In compartments fully involved in fires, fog should be used to reduce the heat and flames before entering. Fog can be discharged through a door or air port and directed above the fire. It may be necessary to cut a small hole in the bulkhead to insert an applicator to cool the fire or space. The fog will be heated to steam by the fire, and the steam will assist in smothering the fire. Caution should be used around openings because of the possibility of outrushing gases. Fog is also used in setting fire boundaries to prevent heat transfer by conduction through bulkheads to adjacent compartments.

In combating a class B fire, water could be used. However, a solid stream of water should not be used, as it would only spread the fuel and fire around. Use of fog can possibly put the fire out, but the danger of reflash is present until all the fuel is cooled below the ignition point.

A variable nozzle (fig. 4-9) is used to regulate the AFFF hose. A variable nozzle has an adjustable spray pattern. It is adjusted by twisting or turning the nozzle tip/head. The spray pattern can be varied from a solid stream to a wide pattern with the spray similar to a low-velocity fog. Some variable nozzles have a ring just behind the nozzle head that can adjust the flow rate. Different variable nozzles have different methods of activating. Some have a bail handle on top, similar to that on an all-purpose nozzle. Others have a trigger-operated valve forward of the grip.

MAGAZINE SPRINKLER SYSTEMS.— Sprinkler systems are used for emergency cooling of, and fire fighting in, magazines, ready-service rooms, and ammunition and missile handling areas. A magazine sprinkler system consists of a network of pipes. These pipes are secured to the overhead and connected by a sprinkler system control valve to the ship's firemain system. The pipes are fitted with spray heads or sprinkler-head valves. They are arranged so that the water forced through them showers all parts of the magazines or ammunition and missile handling areas. Magazine sprinkler systems can completely flood their designated spaces within an hour. To prevent unnecessary flooding of adjacent areas, all compartments equipped with sprinkler systems are watertight. Upper deck handling and ready-service rooms are equipped with drains that limit the water level to a few inches.

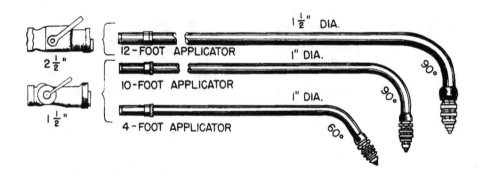

Figure 4-8.—Standard applicators.

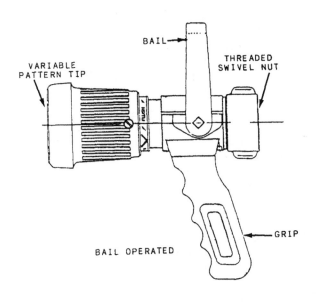

VARIABLE PATTERN TIP

BAIL

THREADED SWIVEL NUT

GRIP

BAIL OPERATED

Figure 4-9.—Variable nozzle.

The two basic types of hydraulically controlled sprinkler systems are the firemain-operated dry type and the firemain-operated wet type. Both types may be found on the same ship. However, the dry type is normally installed in gun ammunition magazines and the wet type in missile magazines. The Gunner's Mates maintain these sprinkler systems.

AFFF SYSTEMS.—AFFF is one of the most familiar and widely used fire-fighting agents. Because of this, you need a basic understanding of the AFFF systems and their components and equipment. Different ships will use different AFFF setups. As stated earlier, AFFF is produced by supplying a mixture of foam concentrate and firemain pressure to a mixing device. The ideal mixing ratio for AFFF is 6 percent concentrate and 94 percent water. Some mixing devices will fluctuate in the mixing ratio, depending on demand rates and water pressure.

The AFFF systems components include the AFFF concentrate tanks, various valves used in the system, and the single agent hose reel.

AFFF is stored in tanks of 50- to 1200-gallon capacity. The tanks are rectangular or cylindrical in shape and are fabricated out of 90/10 copper-nickel or corrosion-resistant steel. Each service tank is located inside the AFFF station and is fitted with a gooseneck vent, drain connection, fill connection, gauge glass and glass guard or liquid level indicator, recirculating line, and an access manhole for tank maintenance. The gooseneck vent prevents an access buildup of pressure within the tank during storage and prevents a vacuum when the system is in operation.

The AFFF systems require a variety of valves with different capabilities and functions. All installed AFFF systems have a hytrol or hycheck valve as the water control valve. This is a diaphragm-type, hydraulically operated, check valve. The valve is held closed by pressure to the diaphragm from the firemain. The system is activated by a draining pressure from this valve. On some systems, the line to the water control valve goes to each AFFF station, where pressure can be drained. On newer systems, each station has an electric switch which opens the solenoid operated pilot valve (SOPV), which drains pressure and allows the water control valve to open. All AFFF systems also have an AFFF concentrate control valve. This is also a diaphragm-type, hydraulically operated valve that is normally closed. This valve is opened by water pressure from downstream of the water control valve. Systems with an injection pump or a centrifugal pump have the pump started by the same electrical signal that opens the SOPV.

The AFFF hose reels provide a swivel connection to the piping, which permits the hoses to be unwound from the reels while still connected to the piping. A geared/hand crank system is incorporated into the reels to facilitate rewinding of the hoses. A manually operated friction brake is attached to the reel pinion shaft. The brake is used to lock the hose reel in place and prevent it from turning. The brake should always be slightly engaged when the reel is in ready condition so the hose will not unreel itself due to vibration.

There is different equipment used to deliver and mix AFFF. Some of the equipment is old and is being replaced by newer equipment as it becomes available. The equipment being phased out includes the Navy pickup unit (NPU) and the FP-180 water motor proportioner (both portable and installed). This equipment is being replaced by the in-line foam eductor.

In-line Foam Eductor.—The in-line foam eductor (fig. 4-10) is becoming standard on all naval ships. The in-line foam eductor is simple in design, operation, and maintenance. It uses a venturi principle to induct AFFF concentrate into a water stream. As water enters the inlet end of the in-line foam eductor, the velocity of the water is increased by the use of a flow restriction. This creates a negative pressure area in the venturi throat. A pickup tube with a ball check valve and a crow's foot is connected to the venturi throat. The ball check valve stops water backflow when the flow is restricted or when the nozzle is shut off. The pickup tube is placed into a container of AFFF concentrate. Atmospheric pressure forces the concentrate up the tube and into the water stream, thus producing the AFFF/water solution. The AFFF nozzle and in-line foam eductor that are used together must have the same flow rating. If you do not match the nozzle and the in-line foam eductor, you will get either poor quality AFFF/water solution or no solution at all. The in-line foam eductor currently used in the Navy is rated at 95 gpm with a 1 1/2-inch hose inlet and discharge connection. The authorized nozzle is a 95-gpm, variable-pattern nozzle with a trigger-operated shutoff.

The in-line foam eductor has several advantages over the FP-180 and the NPU. First, it is lighter in weight and has simpler maintenance requirements than the FP-180. The weight factor is significant because it allows you to respond to a fire quickly. The efficiency of the in-line foam eductor does not decrease as much as the efficiency of the FP-180 does with age and wear. In addition, the in-line foam eductor does not have to be relocated when the nozzleman moves, as is the case with the NPU. This allows containers of AFFF concentrate to be left at one location away from the scene of the fire. This reduces confusion and gives the nozzleman a greater freedom of movement.

The in-line foam eductor has a few disadvantages. A firemain pressure of 100 psig must be maintained to the in-line foam eductor to produce a good quality AFFF/water solution. There may be only one length of hose (50 feet) between the fireplug discharge and the in-line foam eductor inlet. You can also connect the in-line foam eductor directly to the fireplug. This limits the maximum length of hose authorized between the in-line foam eductor discharge and the nozzle inlet to 150 feet. Finally, you should not operate the nozzle any higher than one deck above the location of the in-line foam eductor.

Balanced-pressure Proportioners.—There are several installed AFFF systems in the fleet. The older systems use the FP-180 or the FP-1000 to mix the AFFF concentrate with firemain water. The newer systems use injection pumps or a balanced-pressure proportioner. The FP-180 has two internally attached turbines. A flow of firemain water through the FP-180 turns one turbine, which turns the other turbine, which pumps in the AFFF concentrate. The FP-1000 operates in the same way, but it also has an electrically driven centrifugal pump to provide a positive pressure of concentrate to the FP-1000. The newer balanced proportioners systems also have this same pump.

There are two types of balanced proportioners, the type II and the type III. Both use the venturi effect to mix AFFF concentrate and water together. The type II proportioner has internal moving parts to control the amount of AFFF concentrate entering the proportioner. The body houses a water sleeve, a water float with a guide ring, a water float return spring, a concentrate sleeve, and a concentrate float with a guide rod. The water float is moved forward by the flow of water. Because the concentrate float is connected to the water float, any movement of the water float will cause the concentrate float to move proportionately. The type III proportioner is essentially a hollow tube with a flow restriction and no moving parts. It resembles an eductor

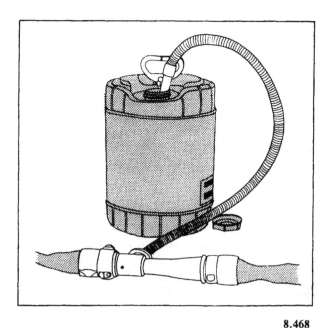

8.468

Figure 4-10.—In-line foam eductor.

(discussed later in this chapter and also in chapter 19). There is an orifice plate to control the amount of AFFF concentrate entering the proportioner.

Both types of proportioners have a back pressure, or balancing, valve (fig. 4-11) installed in the system. This valve is a diaphragm-type, hydraulically operated, globe valve. The diaphragm has connections to sense both AFFF concentrate pump pressure and firemain pressure. The valve regulates the recirculation from the AFFF concentrate pump to maintain AFFF concentrate pressure to the proportioner equal to the water pressure.

AFFF Injection Pumps.—AFFF concentrate injection pumps are used to mix AFFF with water.

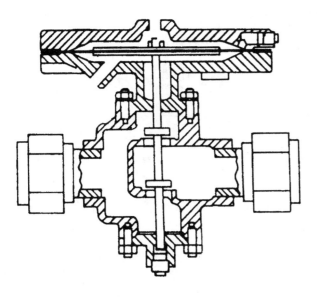

Figure 4-11.—Back pressure, or balancing, valve.

These pumps are permanently mounted, positive-displacement, sliding-vane, electrically driven pumps. The pump is rated at 175 psi and has an internal relief valve set at 210 psi. The pump unit consists of a pump, motor, and reduction gear coupled together on a steel base. Some pumps may be belt driven instead of driven by a reduction gear and flexible coupling. There are both single- and two-speed injection pumps. The installed FP-180 is being replaced by a 12-gpm injection pump. The single-speed pump (fig. 4-12) can also be of 27-, 35-, 60-, or 65-gpm capacity, depending upon the size of the AFFF installed system. For two-speed injection pumps, remote control stations are segmented into high- or low-demand stations. Each station will start the pump at the appropriate speed. The two-speed injection pump injects AFFF into the water stream at 27 gpm at low speed and 65 gpm at high speed. The injection pumps are sized to produce a 6 percent nominal concentration at peak demand. The concentration will exceed 6 percent in most cases. The system has no sensing to adjust to flow demand changes, so the concentration percentages will vary. The system also has a hose connection to allow use of the pump to transfer concentrate to another tank.

TWIN AGENT SYSTEM.—Although twin agent systems are being phased out, a brief discussion is provided to explain the ones out in the fleet. When installed, there are at least two fixed twin agent stations in each shipboard machinery space. Each station consists of twinned AFFF and PKP hoses mounted on reels, and one 125-pound PKP cylinder. The twin agent concept gives the fire party the ability to combat fuel oil

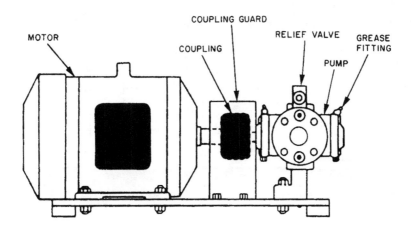

Figure 4-12.—A single-speed AFFF concentrate injection pump.

spray fires and liquid pool fires. The AFFF should be used on liquid pool fires. The PKP (dry chemical) should be used on fuel oil spray fires. When both types of fires exist simultaneously, both AFFF and PKP should be discharged at the same time. On the damage control deck, you will find the AFFF tank and a fixed FP-180 proportioner for mixing the AFFF with water from the firemain (fig. 4-13).

TWIN AGENT UNIT (TAU).—The TAU is a portable flight deck system, mounted on a tractor (fig. 4-14). It is used primarily for the rescue of pilots. One person can operate this unit. There is a sphere containing 200 pounds of PKP

and a cylinder containing 80 gallons of mixed AFFF. The AFFF discharge hose is 1 inch in diameter, and the PKP discharge hose is 3/4 inch in diameter. The discharge hoses are joined in parallel with the nozzles arranged side by side and can be operated independently or concurrently.

Fire Party Organization

The organization of a fire-fighting party depends on the number of personnel available. Figure 4-15 shows the basic organization of a small fire party. Variations are authorized as required by the needs of a particular ship.

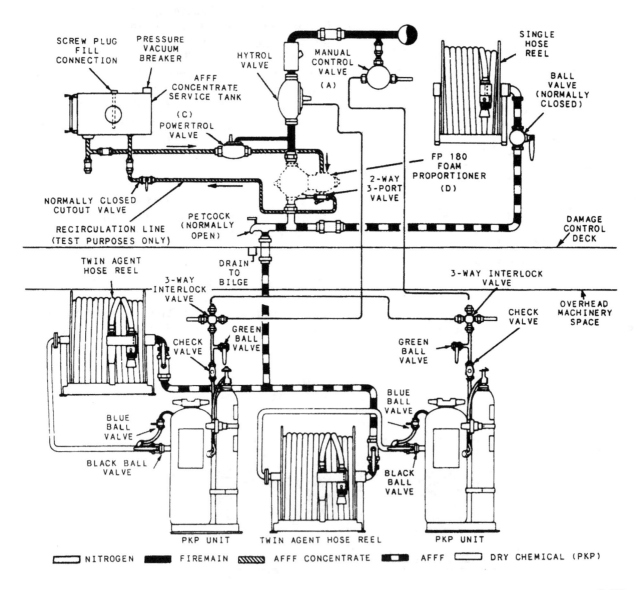

8.492

Figure 4-13.—Typical AFFF/PKP layout.

4-18

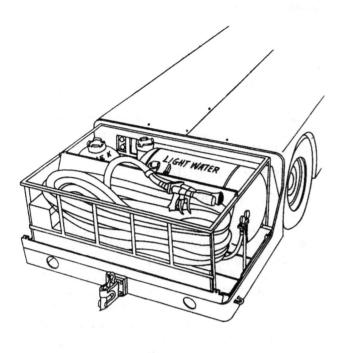

103.147

Figure 4-14.—The aviation TAU.

NUMBER OF MEN	FUNCTION	TEAM
1	Scene Leader	
1	Team Leader—OBA	Attack Team
2	Nozzleman—OBA	Attack Team
2	Hoseman—OBA	Attack Team
2	Plugman	
2	Investigator—OBA	
4	Boundaryman	
2	Messenger/Phone Talker	
1	Electrician	

Figure 4-15.—A fire party organization.

The scene leader is in charge at the scene. When a fire is called away, the scene leader goes directly to the scene of the fire and "sizes up the situation." The scene leader determines the method of extinguishment, the best direction and method of attack, and where to establish initial fire and smoke boundaries. The scene leader stations himself where he may best control the fire team and establish communications with DCC (via repair locker at general quarters) using the best means available. The scene leader receives reports from the team leader and investigators and makes reports to DCC. The scene leader ensures that members of the attack team bring a naval fire-fighter's thermal imager (NFTI) to the scene of the fire, and directs its use. The NFTI is a thermal-imaging device that you can use to see through smoke and detect small temperature differences. Using the NFTI, you can quickly check a compartment for personnel casualties. The NFTI allows fire fighters to spot the source of a fire.

The team leader is in charge of the attack team, which as a minimum consists of four people—two nozzlemen and two hosemen. The team leader will wear an OBA (fitted with a voice amplifier, if available) and will be fully dressed in fire-fighting clothing. The team leader directs the nozzlemen in the use of their hoses and reports to the scene leader.

The nozzlemen are members of the attack team. They will be fully dressed for fire fighting and will wear an OBA. The nozzlemen employ their hoses (or other extinguishers) under the direction of the team leader.

The hosemen are members of the attack team. They will be fully dressed for fire fighting and will wear an OBA. The hosemen run the attack hose from the fireplug to the scene, and keep the hose from getting fouled while fighting the fire. Other duties include relaying spoken messages and orders between the scene leader and the nozzlemen.

The plugmen stand by the fireplug for hose charging, securing, and cleaning the strainer under the direction of the scene leader. For fires using extinguishments other than water, the plugmen provide CO_2 or PKP bottles or they operate the in-line foam eductor.

The investigators constantly rove around the perimeter of the fire. They direct the boundarymen in setting fire boundaries, and they report to the scene leader. The investigators will wear OBAs.

The boundarymen proceed directly to the scene when a fire is called away. They set fire boundaries as directed by the investigators.

The messengers/phone talkers take message blanks and sound-powered phones to the scene and report to the scene leader when a fire is called away. They relay messages between the scene leader and DCC.

The electrician secures all electrical power to and through the affected space. The electrician must be qualified to wear an OBA so that he may make checks of ventilation wiring prior to desmoking.

For a more detailed breakdown of the duties of the fire-fighting party, refer to chapter 9 of NWP 62-1 (Rev C).

Fire-fighting Operations

Each fire is different and the deployment of personnel and equipment will vary. Fire-fighting operations must be directed by the scene leader and the attack team leader. All other personnel must remain quiet and orderly for effective fire fighting. The primary objective must be to extinguish the fire. Fire boundaries must be established and precautions carried out to prevent the spread of fire. Investigators will check on all sides, above and below the fire. In addition, attention should be given to ensure fire, smoke, or fumes are not spreading through the ventilation ducting.

When the fire has been extinguished, steps should be taken to remove the smoke and fumes that remain. Smoke and fumes are hazardous to personnel and are avoided by the use of an OBA. The reduction in visibility by smoke is a hazard as well as a nuisance, but it must be endured and all ventilation systems kept closed until the fire has been completely extinguished to avoid feeding additional oxygen to the fire. Also, since a single ventilation system aboard ship frequently serves a number of compartments, premature use might result in spreading the fire beyond established boundaries. The additional fire hazard resulting from the use of ventilating systems or ducts during a fire is considered of greater importance than the doubtful improvement in visibility resulting from their use.

NSTM, chapter 555, contains detailed procedures for ventilation control during main space (firerooms, engine rooms, or machinery rooms) fires. For other fires, all ventilation system closures (supply, exhaust, and recirculation) must be secured in the area of the fire, including the electrical systems to blowers and similar devices.

When the fire has been completely extinguished and the area is cool, natural and forced ventilation can be used for clearing compartments of smoke and fumes.

NSTM, chapter 555, also contains an extensive discussion of the considerations and equipment used in desmoking operations. Desmoking must be accomplished with great caution due to the possible presence of explosive gases. No one should enter the smoke boundaries without respiratory protection until the area has been declared safe to enter.

CONTROL OF FLOODING

Flooding may occur from a number of different causes. Underwater or waterline damage, ruptured water piping, the use of large quantities of water for fire fighting or counterflooding, and the improper maintenance of boundaries are all possible causes of flooding aboard ship. Ballasting fuel oil tanks with sea water after the oil has been removed is not considered flooding; ballasting merely consists of replacing one liquid with another to maintain the ship's stability.

If a ship suffers such extensive damage that it never stops listing, trimming, and settling in the water, the chances are that it will go down within a few minutes. If, on the other hand, a ship stops listing, trimming, and settling shortly after the damage occurs, it is not likely to sink at all unless progressive flooding is allowed to occur. Thus, there is an excellent chance of saving any ship that does not sink immediately. There is no case on record of a ship sinking suddenly after it has stopped listing, trimming, and settling, except in cases where progressive flooding occurred.

The control of flooding requires that the amount of water entering the hull be restricted or entirely stopped. The removal of flooding water cannot be accomplished until flooding boundaries have been established. Pump capacity should never be wasted on compartments that cannot be quickly and effectively made tight. If a compartment fills rapidly, it is a sign that pumping capacity will be wasted until the openings have been plugged or patched. The futility of merely circulating water should be obvious.

Once flooding boundaries have been established, the removal of water should be undertaken on a systematic basis. Loose water—that is, water with a free surface—and water that is located high in the ship should be removed first. Compartments that are solidly flooded and are low in the ship are generally dewatered last, unless the flooding is sufficiently off center to cause a serious list. Compartments must always be dewatered in a sequence that will contribute to the overall stability of the ship. For example, a ship can capsize if low, solidly flooded compartments were dewatered while water still remained in high, partially flooded compartments.

To know which compartments to dewater first, it is necessary to know the effects of flooding of different compartments on ship's stability. This information is given in the flooding effect diagram in the ship's damage control book. The flooding effect diagram consists of a series of plan views for each deck of the ship, showing all watertight, airtight, fumetight, and fire-retarding subdivisions. Compartments on the flooding effect diagram are color coded in the following way:

- If flooding the compartment results in a decrease in stability because of high weight, free surface effect, or both, the compartment is colored pink.

- If flooding the compartment improves stability, even though free surface exists, the compartment is colored green.

- If flooding the compartment improves stability, when the compartment is solidly flooded but impairs stability when free surface effect exists, the compartment is colored yellow.

- If flooding the compartment has no definite effect on stability, the compartment is uncolored.

The flooding effect diagram also shows the weight of salt water (in tons) required to fill the compartment; this is indicated by a numeral in the upper left-hand corner. In addition, the transverse moment of the weight (in foot-tons) about the centerline of the ship is indicated for all compartments that are not symmetrical about the centerline.

Facilities for dewatering compartments consist of the fixed drainage system of the ship and portable equipment, such as electric submersible pumps, P-250 Mod 1 (and P-250E and P-250) pumps, and eductors. A hole in the hull, with an area of only 1 square foot, 15 feet below the surface, will admit water at the rate of 13,900 gpm. The total pumping capacity of the fixed drainage systems in a large combatant ship, for example, is only 12,200 gpm. The amount of water that comes into a ship through the hole or flows from one compartment to the next varies directly with the area of the hole and the square root of its depth.

The amount and location of water must be considered in order to estimate the number of pumps required to handle a flooding situation. The capacity and availability of the installed drainage system and portable pumps must also be considered. It is necessary to know the rate of flooding in the compartment; that is, how much water is coming in. To know this, you must know the number, size, and location of holes in the compartment; the status of the holes; and to what effect they have been plugged.

Electrical Submersible Pumps

Electrical submersible pumps used aboard naval ships are centrifugal pumps, driven by a water-jacketed, constant-speed alternating current (ac) motor. When a submersible pump is used to dewater a compartment, it is lowered into the water, and the discharge hose is led to the nearest point of discharge. Since the delivery of the pump increases as the discharge head is decreased, dewatering can be done faster if the lowest possible discharge point is used. When a high discharge is necessary, two pumps can be used in tandem, with the first or lower pump discharging to the suction of the second pump. This is shown in figure 4-16.

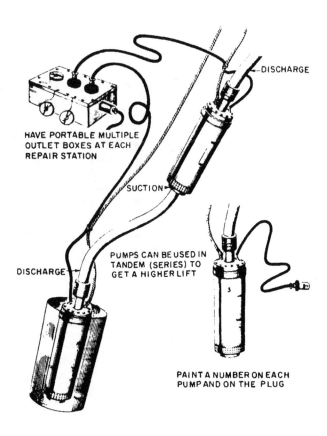

Figure 4-16.—Tandem connection for submersible pumps.

P-250 Mod 1 Pumps

At this time there are three different portable, gasoline-driven pumps used in the Navy. They are the P-250, the PE-250, and the P-250 Mod 1 (fig. 4-17). All three pumps are rated at a capacity of 250 gpm for fire-fighting operations at 100+ psi and are self-priming up to a suction lift of between 16 to 20 feet. The P-250 Mod 1 is the most modern design, and it will replace the other two as more become available. Refer to NSTM, chapter 555, and the specific technical manual for specific pump operating procedures and parameters.

All three gasoline engine-driven pumps produce carbon monoxide and carbon dioxide. When used below decks, their exhaust must be led outside the ship through ports provided for that purpose. The engine must never be run in a space containing explosive vapors. These engines are cooled by the fluid it pumps and should not be used to pump fuel or oil.

Eductors

One of the hazards of fire fighting aboard ship is the possibility of sinking the ship while putting out the fire, if using water to fight the fire. A 2 1/2-inch nozzle operating at 100-psi pressure can discharge almost a ton of water per minute. All water pumped into the ship must be removed before it accumulates to an amount large enough to impair stability.

The P-250 can be used to dewater by a straight pumping action by placing the suction in the flooded space. With a lower discharge pressure requirement for dewatering, the P-250 will pump at 300 gpm. To dewater at a faster rate, an eductor may be used. An eductor is a jet pump with no moving parts. An eductor moves liquid from one place to another by entraining the pumped liquid in a rapidly flowing stream of water. The eductor can perform low-head dewatering operations at a greater rate of discharge than can be obtained

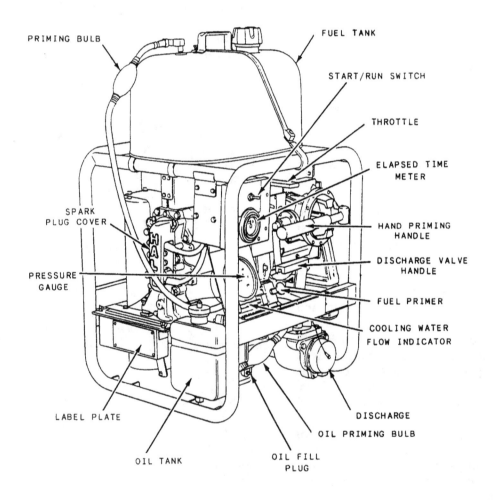

PRIMING BULB

FUEL TANK

START/RUN SWITCH

THROTTLE

ELAPSED TIME METER

HAND PRIMING HANDLE

DISCHARGE VALVE HANDLE

FUEL PRIMER

COOLING WATER FLOW INDICATOR

SPARK PLUG COVER

PRESSURE GAUGE

LABEL PLATE

OIL TANK

OIL FILL PLUG

OIL PRIMING BULB

DISCHARGE

3.164

Figure 4-17.—P-250 Mod 1 pump.

4-22

by straight pumping with available emergency pumps. Eductors are used to pump liquids that cannot be pumped by other portable pumps. Also, liquids that contain fairly small particles of foreign matter can be pumped by using an eductor. It uses firemain pressure as the motive force (using Bernoulli's principle, as discussed in chapter 19). Figure 4-18 shows how to rig a P-250 and an eductor to dewater a flooded space at a rate of 500 gpm. The pump will dewater at 250 gpm by direct action with the foot valve (suction) in the flooded space. With the eductor in the flooded space, it removes an additional 250 gpm to the pump's direct action. In some cases, such as the flooded space having fuel or oil present, the P-250 suction hose must be placed in any source of uncontaminated water. This practice eliminates the chance of damaging the P-250 or of igniting the flammable liquid if the liquid were to run through the pump. Also a space could be dewatered by an eductor run from firemain pressure.

REPAIR OF STRUCTURAL DAMAGE

The kind of damage that may require prompt repairs includes holes above and below the waterline; cracks in steel plating; punctured, weakened, or distorted bulkheads; warped or sprung hatches; weakened or distorted beams, supports, decks, and other strength members; ruptured or cracked piping; severed electrical cables; broken or distorted machinery foundations; damaged machinery; and a wide variety of miscellaneous wreckage that interferes with access or ship's operations.

An important thing to remember is that a series of insignificant looking structural damage can just as easily sink a ship. A natural tendency is to attack the large, obvious damage first and overlook smaller damage. Damage control personnel may waste time trying to patch large holes in a flooded compartment, and disregard smaller holes through which progressive flooding is occurring. Large holes in the underwater hull, as a general rule, cannot be repaired by the damage control personnel but require the drydocking of the ship. Concentrating on smaller interior holes will prevent progressive flooding.

Holes in the hull or just above the waterline should be given immediate attention. These holes may appear relatively harmless; however, they are extremely hazardous. As the ship rolls, lists, or loses buoyancy, these holes become submerged and admit water at a level above the ship's center of gravity.

The methods and materials used to repair holes above the waterline are also used for underwater holes. The greatest difficulty in

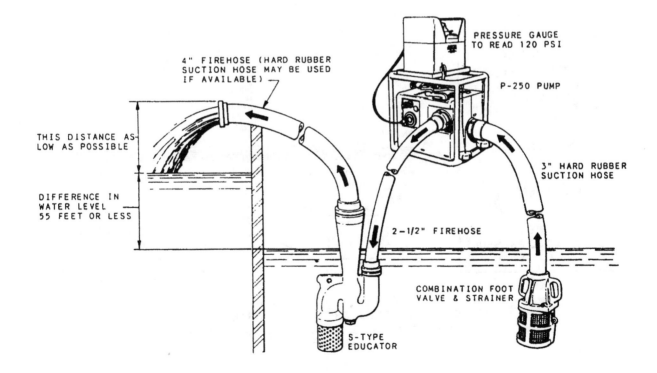

Figure 4-18.—P-250 and eductor rigged to dewater.

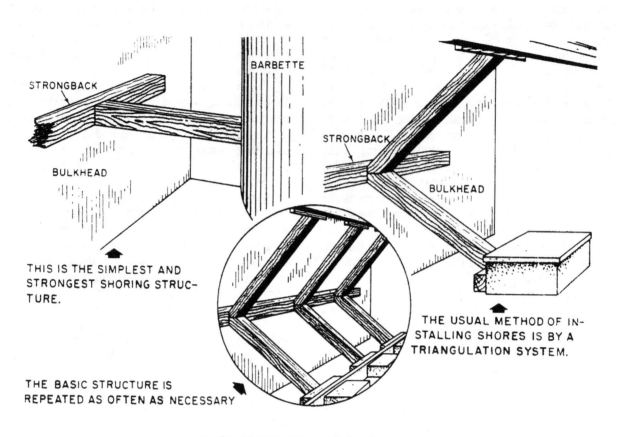

STRONGBACK

BARBETTE

BULKHEAD

THIS IS THE SIMPLEST AND
STRONGEST SHORING STRUC-
TURE.

STRONGBACK

BULKHEAD

THE BASIC STRUCTURE IS
REPEATED AS OFTEN AS NECESSARY

THE USUAL METHOD OF IN-
STALLING SHORES IS BY A
TRIANGULATION SYSTEM.

WHEN OBSTRUCTIONS PREVENT USE
OF THE TRIANGULATION SYSTEM
THIS METHOD MAY BE USED.

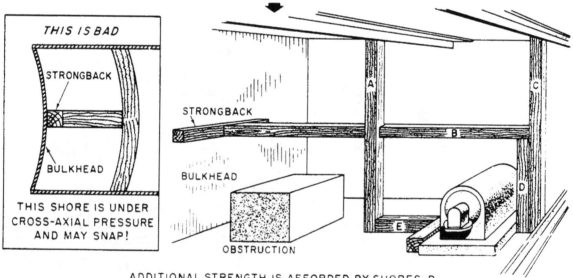

THIS IS BAD

STRONGBACK

BULKHEAD

THIS SHORE IS UNDER
CROSS-AXIAL PRESSURE
AND MAY SNAP!

STRONGBACK

BULKHEAD

OBSTRUCTION

ADDITIONAL STRENGTH IS AFFORDED BY SHORES B
AND C. HORIZONTAL SHORE B IS SUPPORTED BY D
AND A, AND IS BRACED AGAINST A UNIT OF MACHINERY BY
MEANS OF E.

Figure 4-19.—Examples of shoring.

repairing underwater holes is this inaccessibility of the damage. If an inboard compartment is flooded, opening doors and hatches to get to the damage would result in further flooding of other compartments. In such a case, a person wearing a shallow water diving apparatus needs to go down into the compartment. Repair work in flooded compartments is often hampered by wreckage, absence of light to work by, and difficulty keeping buoyant repair material submerged.

Shoring is often used aboard ship to support buckled, weakened, distorted decks and bulkheads, to build up temporary decks and bulkheads against the sea, to support hatches and doors, and to provide support for equipment that has broken loose.

The basic materials required for shoring are shores, wedges, sholes, and strongbacks. A shore is a portable beam. A wedge is triangular on the sides and rectangular on the butt end. A shole is a flat block used under the end of a shore to distribute pressure. A strongback is a piece used to distribute pressure or to serve as an anchor for a patch. Although wooden shoring material is predominantly used, there are metal shores developed for increasing strength.

There is not any one set of rules on when to shore. Sometimes the need for shoring is obvious, as in the case of damaged hatches or doors; but, sometimes dangerously weakened strength members under armament or machinery are not readily noticeable. Although shoring is occasionally done when not really necessary, the best general rule is to shore in case of doubt rather than rely or gamble on the strength of an important strength member, deck, bulkhead, hatch, or door.

There are several methods or shapes of shoring that can be done. Some are shown as examples in figure 4-19. In addition, a steel shore is shown in figure 4-20.

CBR DEFENSE

Because of the possibility of chemical warfare, biological warfare, or a nuclear attack, you should be prepared to protect yourself and your ship and station if such an attack occurs.

CBR defense is both an individual and a group responsibility. What you do before, during, and after an attack will affect your own and your ship's chances of survival.

In studying CBR defense, remember that weapons are always being developed and new

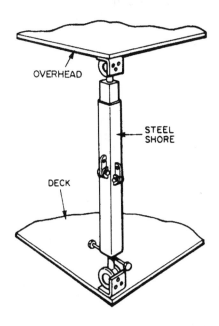

Figure 4-20.—A steel shore.

defensive measures are being established to deal with them. You will need to keep up to date with these changes.

The basic guidelines for defense and protective actions to be taken in the event of CBR attack are set forth in the CBR defense bill contained in NWP 62-1 (Rev. C), *Surface Ship Survivability*. Aboard ship the engineer officer is responsible for maintaining this bill and ensuring that it is current and ready for immediate execution.

CBR defense measures may be divided into two phases: (1) preparatory measures taken in anticipation of an attack, and (2) active measures taken immediately after an attack.

Preparatory measures to be taken before an attack include the following:

- Thorough indoctrination and training of ship's force

- Removal of material that may constitute contamination hazards

- Intermittent activation of the countermeasure washdown system as a pre-attack measure to coat the ship with a film of water to help prevent surface contamination

- Masking of exposed personnel (and other personnel, as ordered)

- Establishment of ship closure, including closing of CIRCLE WILLIAM fittings

- Donning of protective clothing by exposed personnel, as ordered

- Evasive action by the ship

Active measures to be taken immediately following an attack include the following:

- Evasive and self-protective action by personnel

- Evasive maneuvering of the ship

- Activation of the countermeasure washdown system

- Evacuation and remanning of exposed topside stations, as ordered

- Decontamination of personnel

- Detection and prediction of contaminated areas

- Decontamination of material

- Ventilation of contaminated spaces as soon as the ship is in a clean atmosphere.

CBR defense is an enormously complex and wide ranging subject, one in which policies and procedures are subject to constant change. The present discussion is limited to a few aspects of CBR defense that are of primary importance aboard ship. More detailed information on CBR defense may be obtained from NSTM, chapters 070 and 470, and the *CBR Defense Handbook for Training*, NAVSEA S-5080-AA-HBK-010.

PROTECTIVE EQUIPMENT

All naval personnel should be familiar with the protective equipment designed for CBR defense. The following paragraphs cover protective clothing and masks.

Protective Clothing

Types of special clothing used aboard ship for protection against CBR contamination include impregnated clothing, the CBR protective suit, wet-weather clothing, and ordinary work clothing.

IMPREGNATED CLOTHING.—Impregnated clothing, sometimes called permeable

clothing, is supplied to ships in quantities sufficient to outfit 25 percent or more of the ship's complement. Impregnated clothing is olive green. A complete outfit includes impregnated socks, gloves, trousers with attached suspenders, and a jumper (parka) with attached hood. Impregnated clothing is treated with a chemical agent that neutralizes chemicals; a chlorinated paraffin is used as a binder. The presence of these chemicals gives the impregnated clothing a slight odor of chlorine and a slightly greasy or clammy feel. The impregnated treatment should remain effective for 5 to 10 years or possibly longer, if the clothing is stowed in unopened containers in a dry place with cool or warm temperatures and if protected from sunlight or daylight.

Impregnated clothing should not be worn longer than necessary, especially in hot weather. Prolonged wearing may cause a rash to develop where the skin comes in contact with the impregnated material. Impregnated clothing is being phased out and will be replaced by the CBR protective suit.

CBR PROTECTIVE SUIT.—The CBR protective suit (fig. 4-21) consists of two pieces—trousers and a smock (also called a parka).

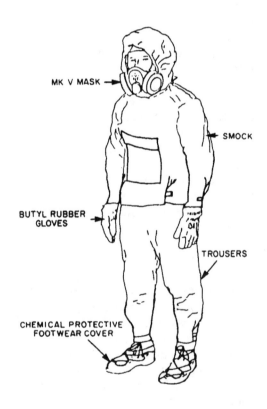

MK V MASK

SMOCK

BUTYL RUBBER GLOVES

TROUSERS

CHEMICAL PROTECTIVE FOOTWEAR COVER

Figure 4-21.—CBR protective suit.

Overboots and gloves are worn in addition to the suit. The CBR suit is issued in a plastic envelope that is pressure packeted, air evacuated, and heat sealed. It is then placed in a polyethylene bag and heat sealed. The CBR suit has a shelf life of 5 years when unopened. Once removed from its plastic envelope, it protects against all CBR agents for 14 days, if kept dry. Once exposed to chemical contamination, the suit provides 6 hours of continuous protection from agent penetration, after which it should be discarded. The gloves are issued in a clear polyethylene bag with an instruction sheet. The overboots are issued in a clear polyethylene bag with two pairs of laces and an instruction sheet. The gloves and overboots provide protection for 6 days, starting with the first day of wear. Once exposed to chemical contamination, they also provide at least 6 hours of protection from agent penetration. Gloves and overboots, if in good condition, may be decontaminated and used again.

WET-WEATHER CLOTHING.—When worn over other types of clothing, wet-weather clothing protects the other clothing from penetration by liquid chemical agents and radioactive particles. It also reduces the amount of vapors that penetrates the skin. It also protects under wet/high wind conditions. Wet-weather clothing includes a parka and overalls. Wet-weather clothing will protect against chemical agents for 6 hours in a contaminated environment and may be decontaminated and reissued.

ORDINARY WORK CLOTHING.—Ordinary work clothing (including long underwear, field socks, coveralls, field boots, and watch cap) provides only limited protection in preventing droplets of chemical agents and vapors from reaching the skin. Ordinary work clothing is not as effective as other types of clothing in preventing contamination. If other protective clothing is not available, two layers of ordinary work clothing (with improvised neck and sleeve closures) will provide some protection.

Protective Masks

The protective mask is a very important item of protective equipment, since it protects the vulnerable areas such as the eyes, the face, and the respiratory tract. The reason it is so important is because inhaling CBR agents is much more dangerous than getting them on the outside of your body. Without filtration of some kind, a large amount of contamination could be inhaled in a short time. The protective mask provides protection against CBR contamination by filtering the air before it is inhaled.

In general, all protective masks operate on the same principles. As the wearer inhales, air is drawn into a filtering system. This system consists of a mechanical filter that clears the air of solid or liquid particles and a chemical filling (usually activated charcoal) that absorbs or neutralizes toxic and irritating vapors. The purified air then passes to the region of the mask where it can be inhaled. Exhaled air is expelled from the mask through an outlet valve which is constructed so that it only opens to permit exhaled air to escape.

Protective masks do not afford protection against ammonia or carbon monoxide, nor are they effective in confined spaces where the oxygen content is too low (less than 16 percent) to sustain life. When it is necessary to enter spaces where there is a deficiency of oxygen, you should use an OBA. Information on masks can be found under the subject heading in various chapters of the NSTM.

DETECTION OF CBR CONTAMINATION

The very nature of CBR contamination makes detection and identification difficult. Chemical agents are usually colorless and odorless. Biological agents are microscopic in size and have no characteristic color or odor to help in identification. Radiation cannot be seen, heard, felt, or otherwise perceived through senses.

It is obvious, then, that with contamination that you cannot see, smell, feel, taste, or hear, specialized methods of detection are required. Mechanical, chemical, and electrical devices are available or under development for the detection of CBR contamination.

Detection of Chemical Agents

Various detection devices have been identified for the detection and identification of chemical agents. Most of these devices indicate the presence of chemical agents by color changes that are chemically produced. To date, no single detector has been developed which is effective under all conditions for all chemical agents. A number of devices, including air sampling kits, papers, crayons, silica gel tubes, and indicator solutions, are in use in the Navy. Some of these devices are also useful in establishing the completeness of decontamination and in estimating the hazards of operating in contaminated areas.

Detection of Biological Agents

There are two possible approaches to the problem of detecting biological agents. Physical detection is based on the measurement of particles within a specific size range (and possibly the simultaneous measurement of other physical properties of the particles). Biological detection involves growing the organisms, examining them under a microscope, and subjecting them to a variety of biochemical and biological tests. Although positive identification can frequently be made by biological detection methods, the process is difficult, exacting, and relatively slow. By the time a biological agent has been detected and identified in this fashion, personnel may well be showing symptoms of their illness.

Biological detection may be divided into two phases: the sampling phase and the laboratory phase. The sampling phase may be a joint responsibility of damage control personnel and of the medical department. During this phase, samples of the materials are taken from over a wide area. These materials include air, surfaces of bulkheads and decks, clothing, equipment, water, food, or anything else suspected of being contaminated. The samples are then shipped to a medical laboratory for identification of the agent, if present. The laboratory phase is obviously a medical department responsibility.

Detection of Radiation

Nuclear radiation cannot be detected by any of the five senses. Therefore special instruments and devices have been developed to do this job. From the military standpoint, we not only need to detect radioactivity, but also need to know where the radiation is and what the intensity of the radiation is. Radiation detection, indication, and computation (radiac) instruments serve both of these needs. There are several types of radiacs: (1) intensity meters, (2) laboratory counters, and (3) dosimeters. There are three types of particles emitted during radioactive decay: (1) alpha particles, (2) beta particles, (3) and gamma rays. The basic types of radiacs are discussed here. More information on radiac instruments can be found in NSTM, chapter 070; *Damage Controlman 3 & 2,* NAVEDTRA 10572; or the manufacturer's technical manuals for each radiac.

The radiac instruments discussed in the following paragraphs are designed for CBR contamination detection. There are other radiacs designed for nuclear safety with nuclear power plants that will not be discussed in this manual.

INTENSITY METERS.—Intensity meters are used to measure gamma radiation field and/or surface radiation attributed to beta particles. The intensity is measured in roentgens per hour (R/hr) or rads per hour (rad/hr). The roentgen is a unit of exposure dose of radiation. A milliroentgen is 1/1000 of a roentgen. The rad is a unit of absorbed dose. It represents the absorption of radiation of a defined amount of energy per unit mass of the absorbing material. A rad relates radiation dose to biological effects. A millirad is 1/1000 of a rad.

Intensity meters have a sensitive element (detector) that ionizes when radiation enters it. The rate of ionization is proportional to the current through the meter which corresponds to the reading on the dial in unit time (intensity).

Intensity meters are defined as being either high range or low range based on their scale of measurement. High-range radiacs will measure radiation in either roentgens or rads; low-range radiacs will measure radiation in either milliroentgens or millirads.

Current Navy inventory has three types of intensity meters for shipboard CBR use: the AN/PDR-27, the AN/PDR-43, and the AN/PDR-65. The AN/PDR-27 is a portable low-range survey meter. The AN/PDR-43 is a portable high-range survey meter. A removable beta shield on the sensor allows both portable radiacs to detect beta radiation. The AN/PDR-65 is a fixed system to detect external gamma radiation. This system has sensors mounted on the mast on the outside the ship. The meter readout is installed on the bridge and/or DCC. This meter also has an accumulated dose counter.

LABORATORY METER.—A laboratory meter is a device that counts individual radiation particles entering a sensitive element. These types of instruments measure radiation in counts per minute. The AN/PDR-56 is the only laboratory counter radiac currently in inventory for shipboard CBR use. This counter measures alpha particle radiation. Due to an alpha particle's short range, the meter must be held 1/8 to 1/4 inch from the surface being surveyed.

DOSIMETERS.—Dosimeters will only provide information on accumulated dose. There are two types of dosimeters: self-reading and nonself-reading. The person wearing a self-reading dosimeter may read the radiation dose. The nonself-reading dosimeter requires a special instrument to read the accumulated dose. The three CBR dosimeters are the IM-9/PD, the IM-143/PD, and the DT-60/PD. The IM-9/PD is a low-range self-reading pocket dosimeter. The IM-143/PD is identical except it is high-range. The DT-60/PD is a high-range, nonself-reading dosimeter in the form of a locket designed to be worn around the neck. The CP-95/PD is the radiac computer indicator that is used to read the amount of radiation the DT-60/PD has been exposed to.

MONITORING AND SURVEY

The monitoring and surveying of any area contaminated is a vital part of CBR defense. In general, monitoring and surveying are done for the purpose of locating the hazards, isolating the contaminated areas, recording the results of the survey, and reporting the findings through the appropriate chain of command.

Specifically, the purpose of a radiological monitoring survey is to determine the location, type, and intensity of radiological contamination. This type of monitoring survey made at any given time depends on the radiological situation and the tactical situation. Gross or rapid surveys are made as soon as possible after a nuclear weapon has been exploded to get a general idea of the extent of contamination. Detailed surveys are made later to obtain a more complete picture of the radiological situation.

Aboard ship, two main types of survey would be required after a nuclear attack. Ship surveys (first gross, then detailed) include surveys of all weather decks, interior spaces, machinery, circulating systems, equipment, and so forth. Personnel safety surveys (usually detailed) are concerned with protecting personnel from skin contamination and internal contamination. Personnel safety surveys include the monitoring of skin, clothing, food, and water, and the measurement of concentrations of radioactive material in the air (aerosols). Both ship surveys and personnel safety surveys are made aboard ship by trained damage control personnel.

Detailed instructions for making monitoring surveys cannot be specified for all situations, since a great many factors (type of ship, distance from blast, extent of damage, tactical situation, and so on) must be considered before monitoring procedures can be decided upon. However, certain basic guidelines that apply to monitoring situations may be stated as follows:

1. Monitors must be thoroughly trained before the need for monitoring arises. Learning to operate radiacs takes time. Simulated practice such as walking through a drill may teach general movement made by the monitoring team, but it does not train the operator in the use of the radiac. All personnel who perform monitoring operations with a radiac should be thoroughly trained in the use of available radiacs.

2. Standard measuring techniques must be used. A measurement of radiation is useless unless the distance between the source of radiation and the point of measurement is known. For example, a radiac held 2 feet away from a source of known radiation will indicate only 1/4 as much radiation as the same instrument would indicate if held 1 foot away from the source. A radiac held 3 feet away from a source of known radiation will indicate only 1/9 as much radiation as the same instrument would indicate if held 1 foot away from the source.

3. All necessary information must be recorded and reported. The information obtained by monitoring parties is forwarded to DCC, where the measurements are plotted according to time and location. To develop an accurate overall picture of the radiological condition of the ship, DCC must have precise and complete information from all monitoring parties. Each monitoring party must record and report the object or area monitored, the location of the object or area in relation to some fixed point, the intensity and type of radiation, the distance between the radiac and the source of radiation, the time and date of the measurement, the name of the petty officer in charge of the monitoring team, and the type and serial number of the radiac used.

CONTAMINATION MARKERS

Areas or objects that are contaminated by CBR attack must be clearly marked to warn personnel approaching the area. The markers should outline dangerous areas and thus establish boundaries within which you may exercise safety control. Radiation hot spots should be identified. These are areas where radiation intensities are significantly greater than the radiation level of surrounding areas.

The standard North Atlantic Treaty Organization (NATO) system for marking areas that are contaminated by CBR attack is used by the U.S. Navy. The standard survey markers are shown in figure 4-22. Each marker is in the shape of a right triangle; one side is 11 1/2 inches long, the other two sides are 8 inches long. Each type of contamination is readily identified by the color of the marker. The markers are made of a rigid material, such as wood, metal, or plastic. The front of the marker faces away from the contaminated area, and the back of the marker faces the contaminated area.

The fronts and backs are identified as follows:

● The chemical contamination marker has a yellow front with the word *GAS* painted or written in red. The back of the marker is yellow. The date and time of the detection of the chemical agent are written on the back, together with the name of the agent (if known).

● The biological contamination marker has a blue front with the word *BIO* painted in red. The back of the marker is blue. The date and time of the detection of the biological agent are written on the back, together with the name of the agent (if known).

● The radiological contamination marker has a white front with the word *ATOM* written or painted in black. The back of the marker is white. The dose rate, the time of the reading, and the time of the burst (if known) are written on the back.

CBR DECONTAMINATION

The basic purpose of decontamination is to remove or neutralize CBR contamination so that the mission of the ship or activity can be carried out without endangering the life or health of assigned personnel.

Decontamination operations may be both difficult and dangerous. Personnel engaged in these operations must be thoroughly trained in the proper techniques. Certain operations, such as the decontamination of food and water, should be done only by experts qualified in such work. However, all ship's personnel should receive adequate training in the elementary principles of decontamination so they can perform emergency decontamination operations.

After an attack, data from CBR surveys is used to determine the extent and degree of contamination. Contaminated personnel must be decontaminated as soon as possible. Before decontamination is undertaken, DCC must appraise the tactical situation and determine the priority of decontamination.

Chemical Decontamination

The purpose of chemical decontamination is to remove or neutralize the chemical agents so they no longer present a hazard to personnel. The methods generally used are natural weathering, chemical action, heat, and physical removal.

NATURAL WEATHERING.—Natural weathering relies on the effect of sun, rain, and wind to dissipate, evaporate, or decompose chemical agents. Weathering is by far the simplest and most common method of chemical decontamination. In some cases, it offers the only practical means of neutralizing the effects of chemical agents, particularly where large areas are contaminated.

CHEMICAL ACTION.—Decontamination by chemical action involves a chemical reaction between the chemical agent and a chemical decontaminant. The reaction usually results in the formation of a harmless new compound that can be removed more easily than the original agent. Neutralization of chemical agents can result from chemical reactions of oxidation, chlorination, reduction, or hydrolysis.

HEAT.—Expendable objects or objects of little value may be burned if they become contaminated. This procedure should not be used except as an emergency measure, or as a means of disposing of material that has been highly contaminated. If this method is used, a very hot fire must be used. Intense heat is necessary for destruction of chemical agents; moderate or low heat may only volatilize the agent and spread it by secondary aerosols. When a large amount of highly contaminated material is being burned, downwind areas may contain a dangerous concentration of toxic vapors. Therefore, you must keep personnel away from such areas.

Hot air may be blown over a contaminated surface to decontaminate it. Steam, especially high-pressure steam, may be used to decontaminate chemical agents in interior compartments of the ship. The steam hydrolyzes and evaporates chemical agents and flushes them away from the surface. The effectiveness of steam increases with temperature—the hotter the steam, the faster the neutralizing action. Chemical decontamination may also be accomplished by sealing off porous surfaces to prevent the absorption of chemical

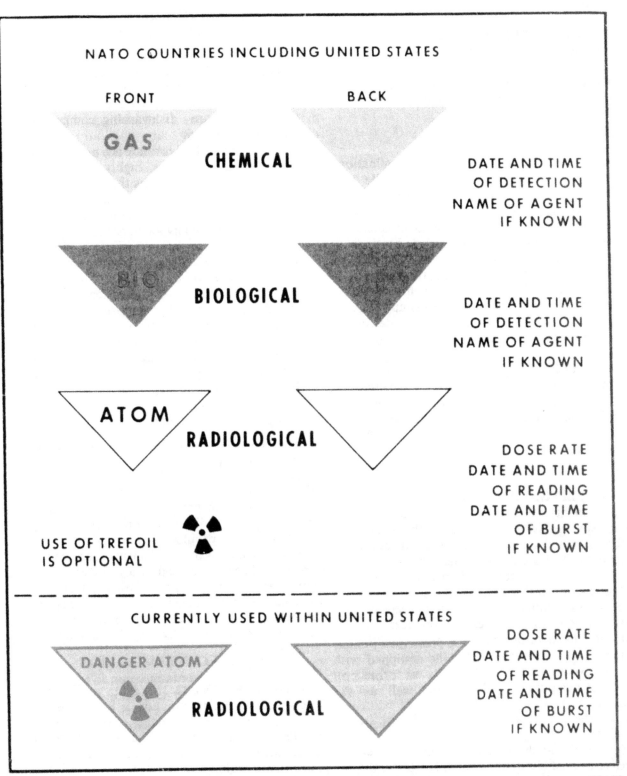

Figure 4-22.—CBR contamination markers.

agents or to prevent volatilization of agents already on the surface.

PHYSICAL REMOVAL.—Chemical decontamination can also be carried out by physically removing toxic agents from the contaminated surface. This is done by washing or flushing the surface with water, steam, or various solvents.

Biological Decontamination

The purpose of biological decontamination is to destroy the biological agents. The most common methods used in biological decontamination are flushing, scrubbing, and heating. Sterilizing gases and disinfectant vapors are used in industrial decontamination, but they are not recommended for shipboard use. The method to be used in any particular case depends upon the nature of the area or equipment to be decontaminated, if this is known.

COUNTERMEASURE WASHDOWN.—For the ship's exterior surfaces the countermeasure washdown system (together with hosing down) will be the most effective countermeasure. After washdown and hosing, thorough scrubbing or swabbing with approved compounds will decompose any remaining biological agents.

SCRUBBING.—Use liquid disinfectants or decontaminants to decontaminate the ship's interior spaces. However, it is difficult to reach all contaminated surfaces to be sure that liquid contacts the biological organisms. To overcome this problem, try to apply liquid disinfectants or decontaminants to all surfaces with a course spray device or swabs until the surface is completely wet. Then rub or scrub all accessible surfaces with swabs or brushes to bring the organisms and the liquid into close contact. After this treatment, close the compartment for a prescribed period. Operating personnel should be equipped with masks and protective clothing as protection against the decontaminant as well as the organisms.

Several liquid decontaminants have proven effective against biological agents. However, the only one generally available for shipboard use is calcium hypochlorinate. It should be used in a 1 percent solution for general decontamination and a 9 percent solution for heavily contaminated areas.

WARNING

Calcium hypochlorinate and oil form an explosive hazard. Be careful when it is used.

This solution is more effective if a half-percent detergent called decontaminating compound is added. If decontaminating compound is not available, you can substitute laundry detergents in the same percentages. If calcium hypochlorinate is not available, you can substitute laundry detergent, soaps, dishwashing compounds, or boiler compound.

Calcium hypochlorinate is a powerful oxidizing agent, and it is highly corrosive. The corrosiveness increases as the temperature goes up. Therefore, calcium hypochlorinate should not be used to decontaminate sensitive electrical or electronic equipment on aircraft, weapons material, navigation equipment, or similar equipment. This is especially true when steam is used.

In using calcium hypochlorite or other cleaning agents for biological decontamination, you should start at the highest point and work downward. Change the wiping cloths, swabs, and scrub brushes at frequent intervals. Also, dispose of the waste at the same intervals.

Radiological Decontamination

Radiological decontamination neither neutralizes nor destroys the contamination; instead, it merely removes the contamination from one particular area and transfers it to an area in which it presents less of an hazard. At sea, radioactive waste is disposed of directly over the side. At shore installations, the problem is more difficult.

Several methods of radiological decontamination have been developed; they differ in effectiveness in removing contamination, in applicability to given surfaces, and in the speed with which they may be applied. Some methods are particularly suited for rapid gross decontamination, others are better suited for detailed decontamination.

GROSS DECONTAMINATION.—The purpose of gross decontamination is to reduce the radiological intensity as quickly as possible to a safe level—or at least a level that is safe for a limited period of time. In gross decontamination, speed is the major consideration.

Flushing with water, preferably with water under high pressure, is the most practical way of accomplishing gross decontamination. Aboard ship the countermeasure washdown system is used to wash down all the ship's surfaces, from high to low and from bow to stern. The counter-measure washdown system consists of piping and a series of nozzles that are specially designed to throw a large spray pattern on weather decks and other surfaces. If the countermeasure washdown system is activated before the ship is exposed to contamination, a film of water prevents heavy contamination of the ship by coating the surface to prevent contaminating material from sticking to the surfaces. Figure 4-23 shows a countermeasure washdown system in operation aboard an aircraft carrier. On flight decks, the nozzles are recessed into the deck to prevent interfering with flight operation. On aircraft carriers, this system is also capable of discharging

AFFF instead or water and is an extremely effective fire-fighting system.

Manual methods may be used to accomplish gross decontamination, but they are slower and less effective than the ship's countermeasure washdown system. Manual methods of decontamination include (1) firehosing the surfaces with salt water, and (2) scrubbing the surfaces with detergent, then firehosing the surfaces and flushing the contaminant over the side.

Steam is also a useful agent for gross decontamination, particularly where it is necessary to remove greasy or oily films. Steam decontamination is usually followed with hosing with hot water and detergent.

5.50.1

Figure 4-23.—Countermeasure washdown system in operation.

DETAILED DECONTAMINATION.—As time and facilities permit, detailed decontamination is carried out. The main purpose of detailed decontamination is to reduce the contamination to a level of minimal radiological hazard to personnel.

Each square foot of contaminated surface receives careful treatment. These detailed methods include caustic paint stripping, wet sandblasting on the hull, refinishing of decking, and acid descaling of saltwater piping. The choice of method depends on the type of surface or object to be decontaminated.

DAMAGE CONTROL PRECAUTIONS

The urgent nature of damage control operations can lead to a dangerous neglect of certain safety precautions. Driven by the need to act rapidly, personnel sometimes take chances they would not even consider in less hazardous situations. This is unfortunate, since there are few areas in which safety precautions are as important as in damage control. Failure to observe safety precautions can lead and, in fact has led, to loss of ships.

Because damage control includes so many different operations and involves the use of so many different items of equipment, it is not feasible to list all the detailed precautions that must be observed. Some of the basic precautions that apply to practically all damage control work are noted briefly in the following paragraphs.

No one should be allowed to take any action to control fires, flooding, or other damage until the situation can be investigated and analyzed. Although speed is essential for effective damage control, correct action is even more important.

The extent of damage must not be underestimated. It is always necessary to remember that hidden damage may be even more severe than visible damage. Very real dangers may exist from damage that is not giving immediate trouble. For example, small holes at or just above the waterline may appear to be relatively minor, but they have been known to sink a ship.

It is extremely dangerous to assume that damage has been prematurely controlled merely because fires are extinguished, leaks plugged, and compartments dewatered. Fires may flare up again, plugs may work out of holes, and compartments may spring new leaks. Damage requires constant checking even after it appears controlled.

Doors, hatches, and other accesses should be kept closed, unless absolutely necessary while making repairs. Records show many cases of progressive flooding through open doors or hatches that should have been closed.

No person should attempt to be a one-person damage control organization. All damage must be reported to DCC or a repair party immediately. The damage control organization is the key to successful damage control. Separate, uncoordinated actions by individuals may do more harm than good.

Many actions taken to control damage can have a definite effect on a ship's characteristics, such as watertight integrity, stability, and weight and moment. The dangers involved in pumping large amounts of water into the ship to combat fires should be obvious. Less obvious, perhaps, is the fact that the repair of structural damage may also affect a ship's characteristics. For example, the addition of high- or off-center weight produces the same effect as high- or off-center flooding.

While most repairs will not individually amount to much in terms of weight shifts or additions, it is possible that a number of relatively small changes could add up sufficiently to endanger an already damaged unstable ship. The only way to control this kind of hazard is by fully and accurately reporting all damage and actions to DCC. Ship stability problems are worked out in DCC, but DCC must receive information required from the repair personnel.

In all aspects of damage control, it is imperative to make full use of all available devices for detection of hazards. Several types of instruments are available for detecting explosive concentrations, flammable or toxic gases, and the amount of oxygen present. Personnel should thoroughly be trained in using these devices before entering potentially hazardous compartments or voids.

REFERENCES

Damage Controlman 3 & 2, NAVEDTRA 10572, Naval Education and Training Program Management Support Activity, Pensacola, Florida, 1986.

Naval Ships' Technical Manual (NSTM), NAVSEA S9086-S3-STM-010, Chapter 555, "Shipboard Firefighting," Naval Sea Systems Command, Washington, D.C., 1988.

NWP 62-1 (Rev. C), "Surface Ship Survivability," Chief of Naval Operations, Washington, D.C., 1989.

CHAPTER 5

FUNDAMENTALS OF SHIP PROPULSION AND STEERING

LEARNING OBJECTIVES

Upon completion of this chapter, you should be able to do the following:

1. Identify causes of resistance encountered by a moving ship.

2. Identify definitions of terms and nomenclature used in describing propulsion and steering equipment.

3. Identify bearing classifications and the characteristics of bearings.

4. Recognize the construction features of reduction gear bearings, line shaft bearings,

stern tubes, stern bearings, and propulsion bearings.

5. Recognize the operating principles of the main thrust bearing.

6. Recognize maintenance and operating procedures of reduction gears, shafting, and bearings.

7. Identify sources of trouble in reduction gears, shafting, and bearings.

INTRODUCTION

The ability to move through the water and to control the direction of movement are among the most fundamental of all ship requirements. Ship propulsion is achieved through the conversion, transmission, and use of energy in a sequence of events. These events include the development of power in a prime mover, the transmission of power to the propellers, the development of thrust on the working surfaces of the propeller blades, and the transmission of thrust to the ship's structure in such a way as to move the ship through the water. Control of the direction of movement is achieved partially by steering devices that receive their power from the steering engines and partially by the arrangement, speed, and direction of rotation of the ship's propellers.

This chapter is concerned with the basic principles of ship propulsion and steering and with the propellers, bearings, shafting, reduction gears, rudders, and other devices required to move the ship and to control its direction of movement. The

prime movers, which are the source of propulsive power, are discussed in detail in other chapters of this text and are therefore mentioned only briefly in this chapter.

RESISTANCE

The movement of a ship through the water requires the expenditure of sufficient energy to overcome the resistance of the water and, to a lesser extent, the resistance of the air. The components of resistance may be considered as (1) skin or frictional resistance, (2) wave-making resistance, (3) eddy resistance, and (4) air resistance.

SKIN/FRICTIONAL RESISTANCE

Skin or frictional resistance occurs because liquid particles in contact with the ship are carried along with the ship, while liquid particles a short distance away are moving at much lower velocities. Frictional resistance is therefore the result of fluid shear between adjacent layers of water. Under most conditions, frictional resistance constitutes a large part of the total resistance.

WAVE-MAKING RESISTANCE

Wave-making resistance results from the generation and propagation of wave trains by the ship in motion. Figure 5-1 illustrates bow, stern, and transverse waves generated by a ship in motion. When the crests of the waves make an oblique angle with the line of the ship's direction, the waves are known as diverging waves. These waves, once generated, travel clear of the ship and give no further trouble. The transverse waves, which have crest lines at a 90-degree angle to the ship's direction, do not have visible, breaking crests. The transverse waves are actually the invisible part of the continuous wave train, which includes the visible divergent waves at the bow and the stern. The wave-making resistance of the ship is a resistance that must be allowed for in the design of ships, since the generation and propagation of wave trains require the expenditure of a definite amount of energy.

EDDY RESISTANCE

Eddy resistance occurs when the flow lines do not close in behind a moving hull, thus creating a low-pressure area in the water behind the stern of the ship. Because of this low-pressure area, energy is dissipated as the water eddies. Most ships are designed to minimize the separation of the flow lines from the ship, thus minimizing eddy resistance. Eddy resistance is relatively minor in naval ships.

AIR RESISTANCE

Air resistance, although small, also requires the expenditure of some energy. Air resistance may be considered as frictional resistance and eddy resistance, with most of it being eddy resistance.

THE DEVELOPMENT AND TRANSMISSION OF PROPULSIVE POWER

Figure 5-2 shows the general principles of ship propulsion and shows the functional relationships of the units required for the development and transmission of propulsive power. The geared-turbine installation is chosen for this example because it is the propulsion plant most commonly used in naval service today. The same basic principles apply to all types of propulsion plants.

The units directly involved in the development and transmission of propulsive power are the prime mover, the shaft, the propelling device, and the thrust bearing. The various bearings used to support the shaft and the reduction gears may be regarded as necessary accessories.

The prime mover provides the mechanical energy required to turn the shaft and drive the propelling device. The prime mover shown in figure 5-2 may be a steam turbine, a diesel engine, a gas turbine engine, or an electric motor.

The propulsion shaft provides a means of transmitting mechanical energy from the prime mover to the propelling device and transmitting thrust from the propelling device to the thrust bearing.

The propelling device imparts velocity to a column of water and moves it in the direction opposite to the direction in which it is desired to move the ship. A reactive force (thrust) is thereby developed against the velocity-imparting device. This thrust, when transmitted to the ship's structure, causes the ship to move through the water. In essence, then, we may think of propelling devices as pumps that are designed to move a column of water to build up a reactive

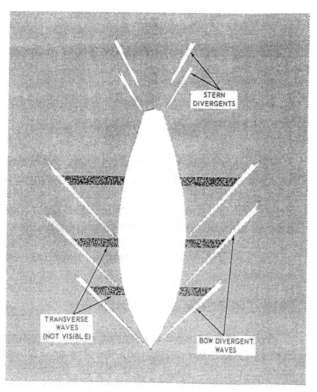

Figure 5-1.—Bow, stern, and transverse waves.

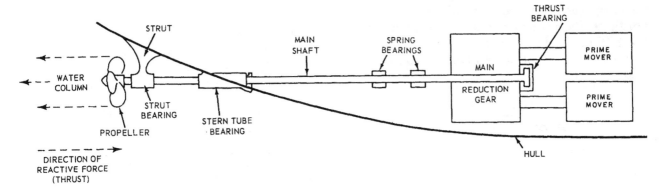

Figure 5-2.—Principles of ship propulsion.

force sufficient to move the ship. The screw propeller is the propelling device used on practically all naval ships.

The thrust bearing absorbs the axial thrust that is developed on the propeller and transmitted through the shaft. Since the thrust bearing is firmly fixed in relation to the ship's structure, any thrust developed on the propeller must be transmitted to the ship in such a way as to move the ship through the water.

The purpose of the bearings that support the shaft is to absorb radial thrust and to maintain the correct alignment of the shaft and the propeller.

The reduction gear, shown in figure 5-2, allows the turbines to operate at high rotational speeds while the propellers operate at lower speeds, thus providing for the most efficient operation of both turbines and propellers.

The propellers, bearings, shafting, and reduction gears that are directly or indirectly involved in the development and transmission of propulsive power are discussed in more detail following a general discussion of power requirements for naval ships.

POWER REQUIREMENTS

The power output of a marine engine is expressed in terms of horsepower. One horsepower is equal to 550 foot-pounds (ft-lb) of work per second or 33,000 ft-lb of work per minute. Different types of engines are rated in different kinds of horsepower. Steam reciprocating engines are rated in indicated horsepower (ihp); internal combustion engines are usually rated in brake horsepower (bhp); and steam turbines are rated in shaft horsepower (shp).

Indicated horsepower is the power measured in the cylinders of the engine. Brake horsepower is the power measured at the crankshaft coupling by a mechanical, hydraulic, or electric brake. Shaft horsepower is the power transmitted through the shaft to the propeller. Shaft horsepower can be measured with a torsion meter. It can also be determined by computation. Shaft horsepower may vary from time to time within the same plant because of variations in the condition of the bottom, the draft of the ship, the state of the sea, and other factors. Shaft horsepower may be determined by the following formula:

$$shp = \frac{2\pi\,NT}{33,000}$$

Where

shp = shaft horsepower
N = rpm
T = torque (in ft-lb) measured with a torsion meter

The amount of power that the propelling machinery must develop to drive a ship at a desired speed may be determined by direct calculation or by calculations based on the measured resistance of a model having a definite size relationship to the ship.

When the latter method of calculating power requirements is used, ship models are towed at various speeds in long tanks or basins. The most elaborate facility used for testing models in this way is the Navy's David W. Taylor model basin at Carderock, Maryland. The main basin is 2775 feet long, 51 feet wide, and 22 feet deep. A powered carriage spanning this tank and riding on machine rails is equipped to tow an attached model directly below it. The carriage carries instruments to measure and record the speed of travel and the resistance of the model. From the resistance, the effective horsepower (ehp) (among other things) may be calculated. Effective horsepower is the horsepower required to tow the ship. Therefore,

$$ehp = \frac{\dfrac{6080V}{R_T\,60}}{33,000}$$

Where

ehp = effective horsepower

R_T = tow rope resistance, in pounds

V = speed, in knots

The relationship between effective horsepower and shaft horsepower is called the propulsive efficiency or the propulsive coefficient of the ship. It is equal to the product of the propeller efficiency and the hull efficiency.

The variation of hull resistance at moderate speeds of any well-designed ship is approximately proportional to the square of the speed. The power required to propel a ship is proportional to the product of the hull resistance and speed. Therefore, it follows that under steady running conditions, the power required to drive a ship is approximately proportional to the cube of the propeller speed. While this relationship is not exact enough for actual design, it serves as a useful guide for operating the propelling plant.

Since the power required to drive a ship is approximately proportional to the cube of the propeller speed, 50 percent of full power will drive a ship at about 79.4 percent of maximum speed attainable when full power is used for propulsion, and only 12.5 percent of full power is needed for about 50 percent of maximum speed attainable. This relationship is shown graphically in figure 5-3.

The relation of speed, torque, and horsepower to ship's resistance and propeller speed under steady running conditions can be expressed in the following equations:

$$S = k_1 \times (rpm)$$

$$T = k_2 \times (rpm)^2$$

$$shp = \frac{2\pi\,k_2}{33,000} \times (rpm)^3$$

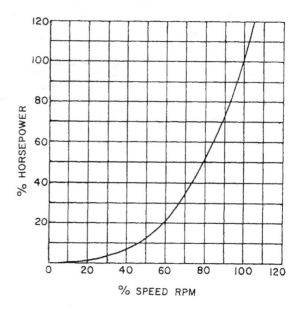

Figure 5-3.—Horsepower-speed cube relationship.

Where

S = ship's speed, in knots

T = torque required to turn propeller, in ft-lb

shp = shaft horsepower

rpm = propeller revolutions per minute

k_1, k_2 = proportionality factors

The proportionality factors depend on many conditions, such as displacement, trim, condition of hull and propeller with respect to fouling, depth of water, sea and wind conditions, and the position of the ship. Conditions that increase the resistance of the ship to motion cause k_1 to be smaller and k_2 to be larger.

In a smooth sea, the proportionality factors can be considered reasonably constant. In rough seas, however, a ship is subjected to varying degrees of immersion and wave impact, which cause these factors to fluctuate over a considerable range. It is to be expected, therefore, that peak loads in excess of the loads required in smooth seas will be imposed on the propulsion plant to maintain the ship's rated speed. Thus, propulsion plants are designed with sufficient reserve power to handle the fluctuating loads that must be expected.

There is no simple relationship for determining the power required to reverse the propeller when the ship is moving ahead or the power required to turn the propeller ahead when the ship is moving astern. To meet Navy requirements, a ship must be able to reverse from full speed ahead to full speed astern within a prescribed period of time. The propulsion plant of any ship must be designed to furnish sufficient power for meeting the reversing specifications.

PROPELLERS

The propelling device most commonly used for naval ships is the screw propeller, so called because it advances through the water in somewhat the same way that a screw advances through wood or a bolt advances when it is screwed into a nut. With the screw propeller, as with a screw, the axial distance advanced with each complete revolution is known as the pitch. The path of advance of each propeller blade section is helicoidal (having the approximate shape of a flattened spiral).

There is, however, a difference between the way a screw propeller advances and the way a bolt advances in a nut. Since water is not as solid a medium, the propeller slips or skids; hence, the actual distance advanced in one complete revolution is less than the theoretical advance for one complete revolution. The difference between the theoretical and the actual advance per revolution is called the slip. Slip is usually expressed as a ratio of the theoretical advance per revolution (the pitch) and the actual advance per revolution. Thus,

$$\text{Slip ratio} = \frac{E - A}{E}$$

Where

E = shaft rpm × pitch = engine distance per minute

A = actual distance per minute

Screw propellers may be broadly classified as fixed-pitch propellers or controllable-pitch propellers. The pitch of a fixed-pitch propeller cannot be altered during operation; the pitch of a controllable-pitch propeller can be changed continuously, subject to bridge or engine-room control.

A screw propeller consists of a hub and several (usually three or four) blades spaced at equal angles about the axis. Where the blades are integral with the hub, the propeller is known as a solid propeller. Where the blades are separately cast and secured to the hub by studs and nuts, the propeller is referred to as a built-up propeller.

Solid propellers may be further classified as having constant pitch or variable pitch. In a constant-pitch propeller, the pitch of each radius is the same. On a variable-pitch propeller, the pitch at each radius may vary.

Propellers are classified as being right-hand or left-hand propellers, depending upon the direction of rotation. When viewed from astern, with the ship moving ahead, a right-hand propeller rotates in a clockwise direction and a left-hand propeller rotates in a counterclockwise direction. The great majority of single-screw ships have right-hand propellers. Multiple-screw ships have right-hand propellers to port. Reversing the direction of rotation of a propeller reverses the direction of thrust and, consequently, reverses the direction of the ship's movement.

Some of the terms used in connection with screw propellers are identified in figure 5-4. The

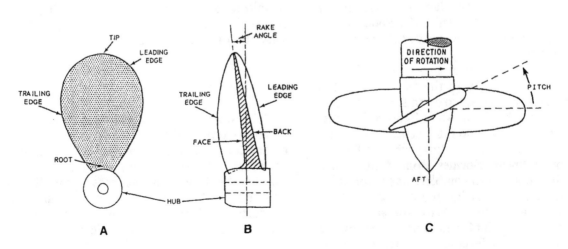

Figure 5-4.—Propeller blade.

term *face* (or *pressure face*) identifies the after side of the blade, when the ship is moving ahead. The term *back* (or *suction back*) identifies the surface opposite the face. As the propeller rotates, the face of the blade increases the pressure on the water near it and gives the water a positive astern movement. The back of the blade creates a low pressure or suction area just ahead of the blade. The overall thrust is derived from the increased water velocity that results from the total pressure differential thus created.

The tip of the blade is the point most distant from the hub. The root of the blade is the area where the blade arm joins the hub. The leading edge is the edge that first cuts the water when the ship is going ahead. The trailing edge (also called the following edge) is opposite the leading edge. A rake angle exists when there is a rake either forward or aft—that is, when the blade is not precisely perpendicular to the long axis of the shaft.

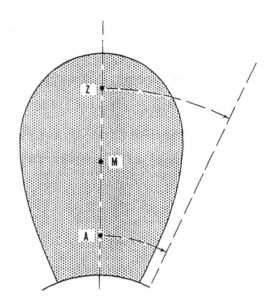

Figure 5-5.—Linear velocity and reactive thrust.

Blade Angle

The blade angle (or pitch angle) of a propeller may be defined as the angle included between the blade and a line perpendicular to the shaft center line. If the blade angle were 0°, no pressure would be developed on the blade face. If the blade angle were 90°, the entire pressure would be exerted sideways and none of it aft. Within certain limits, the amount of reactive thrust developed by a blade is a function of the blade angle.

Blade Velocity

The sternward velocity imparted to the water by the rotation of the propeller blades is partially a function of the speed at which the blades rotate. In general, the higher the speed, the greater the reactive thrust.

However, every part of a rotating blade does not give equal velocity to the water unless the blade is specially designed to do this. For example, consider the flat blade shown in figure 5-5. Points A and Z move about the shaft center with equal angular velocity (rpm) but with different instantaneous linear velocities. Point Z must move farther than point A to complete one revolution; hence, the linear velocity at point Z must be greater than at point A. With the same pitch angle, therefore, point Z will exert more pressure on the water and so develop more reactive thrust

than point A. The higher the linear velocity of any part of a blade, the greater will be the reactive thrust.

Real propeller blades are not flat but are designed with complex surfaces (approximately helicoidal) to permit every infinitesimal area to produce equal thrust. Since point Z has a higher linear velocity than point A, the thrust at point Z must be decreased by decreasing the pitch angle at point Z. Point M, lying between points Z and A, would have (on a flat blade) a linear velocity less than Z but greater than A. In a real propeller, then, point M must be set at a pitch angle that is greater than the pitch angle at point Z but less than the pitch angle at point A. Since the linear velocity of the parts of a blade varies from root to tip, and since it is desired to have every infinitesimal area of the blade produce equal thrust, it is apparent that a real propeller must vary the pitch angle from root to tip.

Propeller Size

The size of a propeller—that is, the size of the area swept by the blades—has a definite effect on the total thrust that can be developed on the propeller. Within certain limits, the thrust that can be developed increases as the diameter and the total blade area increase. Since it is impractical to increase propeller diameter beyond a certain point, propeller blade area is usually made as great as possible by using as many blades as are feasible under the circumstances. Three-bladed and four-bladed marine propellers are commonly used.

Thrust Deduction

Because of the friction between the hull and the water, water is carried forward with the hull and is given a forward velocity. This movement of adjacent water is called the wake. Since the propeller revolves in this body of forward-moving water, the sternward velocity given to the propeller is less than if there were no wake. Since the wake is traveling with the ship, the speed of advance over the ground is greater than the speed through the wake.

At the same time, a propeller draws water from under the stern of the ship, thus creating a suction that tends to keep the ship from going ahead. The increase in resistance that occurs because of this suction is known as thrust deduction.

Number and Location of Propellers

A single propeller is located on the ship's centerline as far aft as possible to minimize the thrust deduction factor. Vertically, the propeller must be located deep enough so that in still water the blades do not draw in air but high enough so that it can benefit from the wake. The propeller must not be located so high that it will be likely to break the surface in rough weather, since this would lead to racing and perhaps a broken shaft.

A twin-screw ship has the propellers located one on each side, well aft, with sufficient tip clearance to limit thrust deduction.

A quadruple-screw ship has the outboard propellers located forward of and above the inboard propellers to avoid propeller stream interference.

Controllable Reversible Pitch Propellers

Controllable-pitch propellers (fig. 5-6) are used in some naval ships. Controllable-pitch propellers give a ship excellent maneuverability and allow the propellers to develop maximum thrust at any given shaft rpm. Ships with controllable-pitch

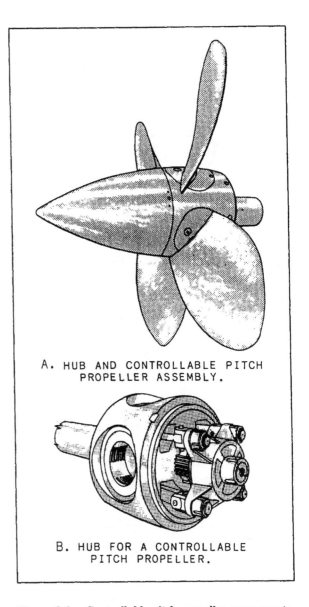

A. HUB AND CONTROLLABLE PITCH PROPELLER ASSEMBLY.

B. HUB FOR A CONTROLLABLE PITCH PROPELLER.

Figure 5-6.—Controllable-pitch propeller components.

propellers require no reversing gear since the direction of the propeller thrust can be changed without changing the direction of shaft rotation. Controllable-pitch systems are widely used on diesel-driven ships, while gas turbine-powered ships use controllable pitch as the only means available for providing reverse thrust.

Controllable-pitch propeller systems may be controlled from the bridge or from the engine room through piping inside a hollow propulsion shaft to the propeller hub. Hydraulic or mechanical controls are used to apply the actuating force required to change the position, or angle of the pitch, of the propeller blades.

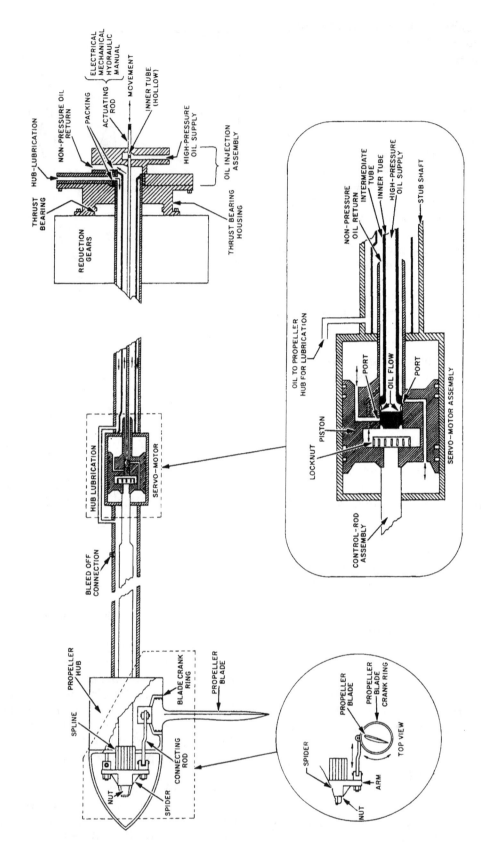

Figure 5-7.—Hydraulic system of a controllable-pitch propeller.

The hydraulic system, shown in figure 5-7, is the most widely used means of providing the force required to change the pitch of a controllable-pitch propeller. In this type of system, a valve-positioning mechanism actuates an oil control valve. The oil control valve permits hydraulic oil, under pressure, to be introduced to either side of a piston (which is connected to the propeller blade) and at the same time allows for the controlled discharge of hydraulic oil from the other side of the piston. This action repositions the piston and thus changes the pitch of the propeller blades.

Some controllable-pitch propellers have mechanical means for providing the blade actuating force necessary to change the pitch of the blades. In these designs, a worm screw and crosshead nut are used instead of the hydraulic devices for transmitting the actuating force. The torque required for rotating the worm screw is supplied either by an electric motor or by the main propulsion plant through pneumatic brakes.

In most installations, propeller pitch and engine power are controlled through a single lever. Movement of the lever causes both engine speed and pitch to change to suit the powering condition ordered. In emergencies, and in ships without single lever control, the propeller pitch may be changed independently of the engine power setting. Under this condition, overspeeding of the engine can result if the pitch is set too low, or overtorquing of the engine can result if the pitch is set too high. Controllable pitch propellers are also discussed in chapter 16.

Propeller Problems

One of the major problems encountered with propellers is known as cavitation. Cavitation is the formation of a vacuum around a propeller that is revolving at a speed above a certain critical value (which varies, depending upon the size, number, and shape of the propeller blades). The speed at which cavitation begins to occur is different in different types of ships; the turbulence increases in proportion to the propeller rpm. Specifically, a propeller rotating at a high speed will develop a stream velocity that creates a low pressure. This low pressure is less than the vaporization point of the water, and from each blade tip there appears to develop a spiral of bubbles (fig. 5-8). The water boils at the low-pressure points. As the vapor bubbles of cavitation move into regions where the pressure is higher, the bubbles collapse rapidly and produce a high-pitched noise.

The net result of cavitation is to produce (1) a high level of underwater noise; (2) erosion of propeller blades; (3) vibration, with subsequent blade failure from metallic fatigue; and (4) overall loss in propeller efficiency, requiring a proportionate increase in power for a given speed.

In naval warfare, the movements of surface ships and submarines can be plotted by sonar bearings on propeller noise. Because of the high static water pressure at submarine operational depths, cavitation sets in when a submarine is operating at a much higher rpm than when near the surface. For obvious reasons, a submarine that

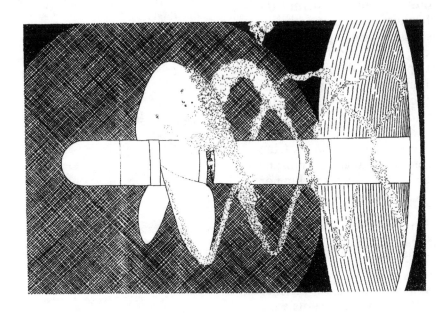

71.23

Figure 5-8.—Cavitating propeller.

5-9

is under attack will immediately dive deep so that it can use high propeller rpm with the least amount of noise.

A certain amount of vibration is always present aboard ship. Propeller vibration, however, may also be caused by a fouled blade or by seaweed. If a propeller strikes a submerged object, the blades may be nicked.

Another propeller phenomenon is the "singing" propeller. The usual cause of this noise is that the trailing edges of the blades have not been properly prepared before installation. The flutter caused by the flow around the edges may induce a resonant vibration. A singing propeller can be heard for a great distance.

BEARINGS

From the standpoint of mechanics, the term *bearing* may be applied to anything that supports a moving element of a machine. However, this section is concerned only with those bearings that support or confine the motion of sliding, rotating, and oscillating parts on revolving shafts or movable surfaces of naval machinery.

Since naval machinery is constantly exposed to varying operating conditions, bearing material must meet rigid standards. A number of non-ferrous alloys are used as bearing metals. In general, these alloys are tin-base, copper-base, or aluminum-base alloys. The term *babbit metal* is often used for lead-base and tin-base alloys.

Bearings must be made of materials that will withstand varying pressures and yet permit the surfaces to move with minimum wear and friction. In addition, bearings must be held in position with very close tolerance, permitting freedom of movement and quiet operation. Because of these requirements, good bearing materials must possess a combination of the following characteristics for a given application:

—The compressive strength of the bearing alloy at maximum operating temperature must be able to withstand high loads without cracking or deforming.

—Bearing alloys must have high fatigue resistance to prevent cracking and flaking under varying operating conditions.

—Bearing alloys must have high thermal conductivity to prevent localized hot spots with resultant fatigue and seizure.

—Bearing materials must be capable of retaining an effective oil film.

—Bearing materials must be highly resistant to corrosion.

Classifications of Bearings

The reciprocating and rotating elements or members, supported by bearings, may be subject to external loads, which can be resolved into components having normal, radial, or axial directions, or a combination of the two. Bearings are generally classified as sliding surface (friction) or rolling contact (antifriction) bearings.

Sliding surface bearings may be defined broadly as those bearings that have sliding contact between their surfaces. In these bearings, one body slides or moves on the surface of another and sliding friction is developed if the rubbing surfaces are not lubricated. Examples of friction bearings are thrust bearings and journal bearings (fig. 5-9), such as the spring or line shaft bearings installed aboard ship.

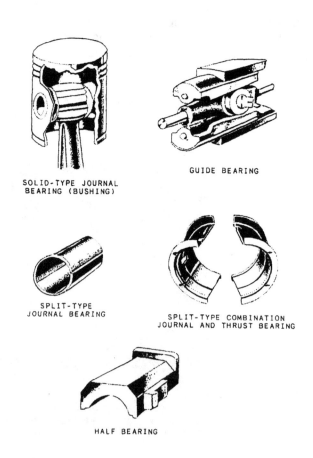

SOLID-TYPE JOURNAL BEARING (BUSHING)

GUIDE BEARING

SPLIT-TYPE JOURNAL BEARING

SPLIT-TYPE COMBINATION JOURNAL AND THRUST BEARING

HALF BEARING

Figure 5-9.—Various types of friction bearings.

Journal bearings are extensively used aboard ship. Journal bearings may be subdivided into different styles or types, the most common of which are solid bearings, half bearings, and two-part or split bearings. A typical solid-type journal bearing application is the piston bearing, more commonly called a bushing. An example of a solid bearing is a piston rod wristpin bushing, such as found in compressors. Perhaps the most common application of the half bearing in marine equipment is the propeller shaft bearing. Since the load is exerted only in one direction, a half bearing obviously is less costly than a full bearing of any type. Split bearings are used more frequently than any other friction-type bearing. A good example is the turbine bearing. Split bearings can be made adjustable to compensate for wear.

Guide bearings, as the name implies, are used for guiding the longitudinal motion of a shaft or other part. Perhaps the best illustrations of guide bearings are the valve guides in an internal-combustion engine.

Thrust bearings are used to limit the motion of or to support a shaft or other rotating part longitudinally. Thrust bearings sometimes are combined functionally with journal bearings.

Antifriction, or rolling contact, bearings are so called because their design takes advantage of the fact that less energy is required to overcome rolling friction than is required to overcome sliding friction. These bearings may be defined broadly as bearings that have rolling contact between their surfaces. These bearings may be classified as roller bearings or ball bearings according to the shape of the rolling elements. Both roller and ball bearings are made in different types, some being arranged to carry both radial and thrust loads. In these bearings, the balls or rollers generally are assembled between two rings or races, the contacting faces of which are shaped to fit the balls or rollers.

The basic difference between ball and roller bearings is that a ball at any given instant carries the load on any two tiny spots diametrically opposite, while a roller carries the load on two narrow lines (fig. 5-10). Theoretically, the area of the spot or line of contact is infinitesimal. Practically, the area of contact depends on how much the bearing material will distort under the applied load. Obviously, rolling contact bearings must be made of hard materials because if the distortion under load is appreciable, the resulting friction will defeat the purpose of the bearings. Bearings with small, highly loaded contact areas must be lubricated carefully if they are to have

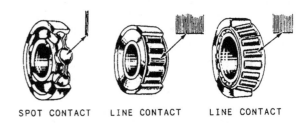

SPOT CONTACT   LINE CONTACT   LINE CONTACT

Figure 5-10.—Load-carrying areas of ball and roller bearings.

the antifriction properties they are designed to provide. If improperly lubricated, the highly polished surfaces of the balls and rollers soon will crack, check, or pit, and the failure of the complete bearing follows.

Both sliding surface and rolling contact bearings may be further classified by their function: radial, thrust, and angular-contact (actually a combination of radial and thrust) bearings. Radial bearings, designed primarily to carry a load in a direction perpendicular to the axis of rotation, are used to limit motion in a radial direction. Thrust bearings can carry only axial loads; that is, a force parallel to the axis of rotation, tending to cause endwise motion of the shaft. Angular-contact bearings can support both radial and thrust loads.

The simplest forms of radial bearings are the integral and the insert types. The integral type is formed by surfacing a part of the machine frame with the bearing material, while the insert bearing is a plain bushing inserted into and held in place in the machine frame. The insert bearing may be either a solid or a split bushing, and may consist of the bearing material alone or be enclosed in a case or shell. In the integral bearing, there is no means of compensating for wear, and when the maximum allowable clearance is reached the bearing must be resurfaced. The insert bearing, like the integral type, has no means for adjustment due to wear, and must be replaced when maximum clearance is reached.

The pivoted shoe is a more complicated type of radial bearing. This bearing consists of a shell containing a series of pivoted pads or shoes, faced with bearing material.

The plain pivot or single disk thrust bearing consists of the end of a journal extending into a cup-shaped housing, the bottom of which holds the single disk of the bearing material.

The multidisk thrust bearing is similar to the plain pivot bearing except that several disks are placed between the end of the journal and the housing. Alternate disks of bronze and steel are generally used. The lower disk is fastened in the bearing housing and the upper one to the journal, while the intermediate disks are free.

The multicollar thrust bearing consists of a journal with thrust collars integral with or fastened to the shaft; these collars fit into recesses in the bearing housing, which are faced with bearing metal. This type of bearing is generally used on horizontal shafts carrying light thrust loads.

The pivoted shoe thrust bearing is similar to the pivoted shoe radial bearing except that it has a thrust collar fixed to the shaft, which runs against the pivoted shoes. This type of bearing is generally suitable for both directions of rotation.

Angular loading is generally taken by using radial bearing to restrain the radial load and some form of thrust bearing to handle the axial load. This may be accomplished by using two separate bearings or a combination of radial and thrust (radial thrust). A typical example is the multicollar bearing, which has its recesses entirely surfaced with bearing material. The faces of the collars carry the thrust load and the cylindrical edge surfaces handle the radial load.

Main Reduction Gear and Propulsion Turbine Bearings

Reduction gear bearings support the weight of the gears and their shafts. Reduction gear

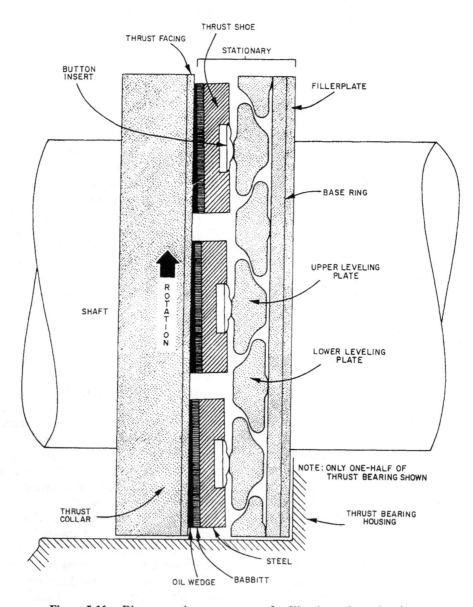

Figure 5-11.—Diagrammatic arrangement of a Kingsbury thrust bearing.

bearings of the babbitt-lined split type are rigidly mounted into the bearing housings by dowels. These bearings are split in halves, but the split is not always in a horizontal plane. On many pinion and bull gear bearings, the pressure is against the cap and not always in a vertical direction. The bearing shells are so secured in the housing that the point of pressure on both ahead and astern operation is as nearly midway between the joint faces as practicable.

Turbine bearings are pressure-lubricated by the same forced-feed system that lubricates the reduction gear bearings.

Main Thrust Bearings

The main thrust bearing is usually located in the reduction gear casing. It absorbs the axial thrust transmitted through the shaft from the propeller.

Kingsbury bearings, also called segmental pivoted-shoe thrust bearings (fig. 5-11), are commonly used for main thrust bearings. This bearing consists of pivoted segments or shoes (usually six or eight) against which the thrust collar revolves. The action of the thrust shoes against the thrust collar restrains the ahead or astern axial motion of the shaft to which the thrust collar is secured. These bearings operate on the principle that a wedge-shaped film of oil is more readily formed and maintained than a flat film,

and that it can, therefore, carry a heavier load for any given size.

The upper leveling plates, upon which the shoes rest, and the lower leveling plates equalize the thrust load among the shoes. The base ring, which supports the lower leveling plates, holds the plates in place. This ring transmits the thrust on the plates to the ship's structure via housing members that are bolted to the foundation. Shoe supports (hardened steel buttons or pivots) located in the shoes separate the shoes and the upper leveling plates. This separation enables the shoe segments to assume the angle required to pivot the shoes against the upper leveling plates. Pins and dowels hold the upper and lower leveling plates in position. This allows for ample play between the base ring and the plates and ensures freedom of movement (oscillation only) of the leveling plates. The base ring is prevented from turning by its keyed construction, which secures the ring to its housing.

Line Shaft Bearings or Spring Bearings

Most of the line shaft bearings or spring bearings are of the ring- or disk-oiled, babbitt-faced, spherical seat, or shell type. This type of bearing (fig. 5-12) is designed primarily to align itself to support the weight of the shafting.

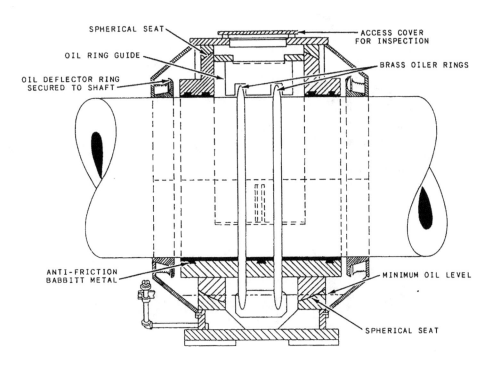

Figure 5-12.—Line shaft bearing (ring-oiled).

The brass oiler rings, shown in figure 5-12, are a loose fit. The rings are retained in an axial position by guides or grooves in the outer bearing shell. As the shaft rotates, friction between the rings and the shaft is enough to cause the rings to rotate with the shaft. The rings dip into the oil in the sump. Oil is retained on the inside diameter of the rings and is carried to the upper bearings by the rings. The action of the oil ring guides and the contact of the rings on the upper shaft cause the oil to be removed from the rings and to lubricate the bearings.

The disk-oiled spring bearing (fig. 5-13) is basically the same as the ring-oil type except it uses an oil disk to lubricate the bearing. The oil disk is attached to, and rotates with, the shaft. Oil is removed from the disk by a scraper, located at the top of the bearing. The oil then runs into a pocket at the top of the upper bearing shell. (Fig. 5-13 also shows a detailed picture of the scraper

arrangement.) From here, the oil enters the bearing through drilled holes.

Tests have shown that the disk delivers more oil at all speeds than the ring discussed earlier, especially at turning gear speeds. The disk is also more reliable than the ring. Some ships have line shaft bearings that are force-lubricated by a pump.

You should check line shaft bearing temperatures and oil levels at least once an hour during normal operation. Inspection and maintenance should be done according to PMS requirements.

Stern Tubes and Stern Tube Bearings

The stern tube is located where the shaft passes through the hull of the ship. The shaft is supported in the stern tube by two stern tube bearings; one on the inner side, and one on the outer side of the hull. The construction of the

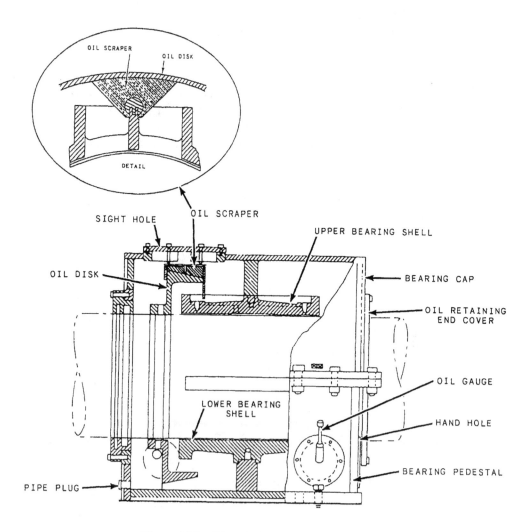

Figure 5-13.—Line shaft bearing (disk-oiled).

stern tube bearing is basically the same as that of the strut bearing, which is described later in this chapter.

The point where the shaft passes through the hull must be sealed to prevent seawater from entering the ship. This is accomplished primarily by using either packing or mechanical seals. Stern tube packing (fig. 5-14) is used only on older ships as a primary sealer.

This packing method uses a stuffing box that is flanged and bolted to the stern tube. Its casting is divided into two annular (ring-shaped) compartments. The forward space is the stuffing box. The after space contains a flushing connection to provide a constant flow of water

through the stern tube (from inside the ship to outside the ship) for lubricating, cooling, and flushing the bearings. This flushing connection is supplied by the firemain. A drain connection is provided both to test the presence of cooling water in the bearing and to allow seawater to flow through the stern tube to lubricate and cool the packing when the ship is underway. This is done where natural seawater circulation is used.

The gland for the stuffing box is divided longitudinally into two parts. The packing material used, Teflon-impregnated asbestos (PTE), is according to MIL-P-24377 and has replaced all previously used packing material.

Gland leakage is required to prevent packing from heating up, crystallizing, and scoring the shaft sealing surface within the gland. Usually, the gland is tightened and the flushing connection is closed to eliminate leakage when the ship is in port. It is loosened just enough to permit a slight trickle of water for cooling purposes when the ship is underway. Whenever packing is added to a stern tube, the gland is drawn up evenly by using a rule to measure the distance between the gland and the stuffing box.

The inflatable sealing ring is used to repair or replace the prime sealing elements when the ship is waterborne. The split inflatable rubber seal ring, shown in figure 5-15, is installed aft of the prime seal assembly rig. When the seal is needed, it is

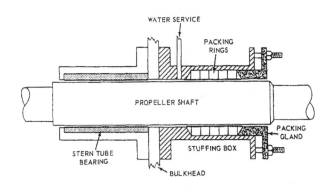

Figure 5-14.—Stern tube stuffing box and gland.

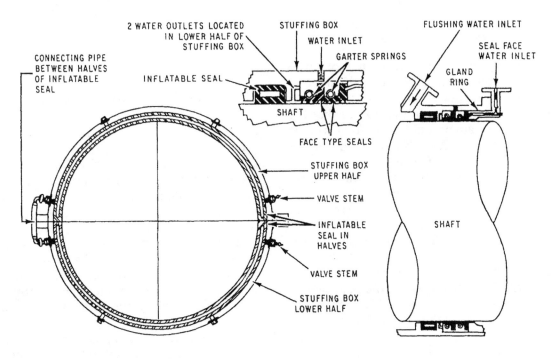

Figure 5-15.—Inflatable sealing ring.

blown up with nitrogen and expands against the shaft, making a seal. Nitrogen is used because it is dry and will not deteriorate the seal as rapidly as air.

In some installations, mechanical seals are used to seal the stern tube. Two of the major advances of these seals are (1) they will operate maintenance free for extended periods, and (2) they accommodate gross misalignment and allow for excessive bearing wear, high vibration, and large radial and axial movement of the shaft.

Strut Bearings

Strut bearings, as well as the stern tube bearings, are equipped with composition bushings that are split longitudinally into two halves (fig. 5-16). The outer surface of the bushing is machined with steps to bear on matching landings in the bore of the strut.

Since it is usually not practical to use oil or grease as a lubricant for underwater bearings, some other frictionless material must be used. There are certain materials that become slippery when wet. They include synthetic rubber, lignum vitae (a hard tropical wood with excellent wearing qualities), and laminated phenolic material consisting of layers of cotton fabric impregnated and bonded with phenolic resin. Strips made from any of these materials are fitted inside the bearing. Most Navy installations use rubber composition strips as shown in figure 5-16.

MAIN PROPULSION SHAFTING

The main propulsion shafting may be up to 30 inches in diameter. It is divided into two environmental sections: the inboard or line shafting and the outboard or waterborne shafting (fig. 5-17). The main propulsion shafting is made of forged steel and is usually hollow in sizes of more than 6 inches in diameter.

The line shafting consists of several sections of shaft. These include the thrust shaft when the main thrust bearing is not located in the reduction gear casing. These sections are usually joined together by bolts through flange couplings, which are forged integral with the shaft sections.

The outboard shafting consists of the propeller shaft, or tailshaft, upon which the propeller is mounted; the stern tube shaft, which penetrates the hull of the ship and thus makes the transition between inboard and outboard shafting; and, in some cases, an intermediate, or dropout, shaft located between the propeller shaft and the stern

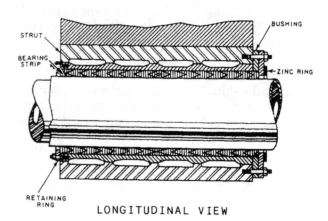

LONGITUDINAL VIEW

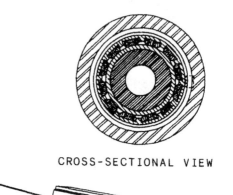

CROSS-SECTIONAL VIEW

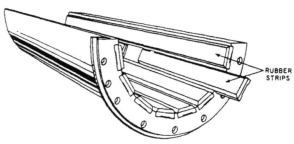

RUBBER STRIPPING

Figure 5-16.—Details of a strut bearing.

tube shaft. In some single-shaft surface ships and submarines, the propeller shaft also functions as the stern tube shaft. The outboard shafting is protected from seawater corrosion by a covering of either plastic or rubber. Outboard shafting sections are usually joined to each other by integral flanges, although some older ships may have a removable outboard shaft coupling known as a muff-type outboard coupling. The stern tube shaft is connected to the line shafting by a removable coupling known as the inboard stern tube coupling.

Circular steel or composition shields, known as fairwaters, are secured to the after end of the stern tube and to the struts. They are "faired in"

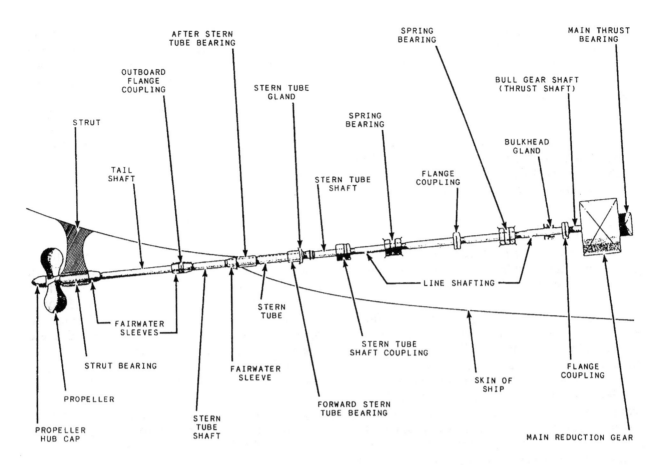

Figure 5-17.—Propulsion shafting.

to the shafting to reduce turbulence. This "fairing in" is accomplished by the gradual reduction in diameter from that of the stern tube or strut to that of the shaft. In some cases, a flange is adjacent to a strut or the aft end of the stern tube. In these cases, the fairwater may either extend beyond the flange (either forward or aft as applicable) and fair into the shaft. Or, it may fair into a rotating coupling cover, which is attached to the flange and shaft. Some ships have watertight rotating coupling covers filled with tallow, which seals the flanged coupling from seawater.

REDUCTION GEARS

Reduction gears are coupled to the turbine shaft though various arrangements of gearing. These gears reduce the speed of the turbine to the low speed required by the propulsion shaft and propeller. Reduction gears may be driven by one or more turbines. A combination of gears is

known as a *train*, and that term will be used in this chapter.

Reduction gears are classified according to the number of steps used to reduce speed and the arrangement of the gearing. When two gears are meshed and the driving gear (called a pinion) is larger than the mating gear, the driving gear is called a speed increaser. A driving gear that is smaller than the mating gear is called a speed decreaser. The ratio of the speeds is proportional to the diameters of the pinion and the gear. When there are just two gears, the train is known as a single-reduction gear (single-speed increaser or decreaser). Double-reduction gears have more than two gears working together. They keep the size of the bull gear (the large gear attached to the propeller shaft) from becoming too large.

Propulsion Reduction Gears

Turbine drives normally use double-reduction gears. The articulated locked train, double-reduction gear (fig. 5-18) is the most common. All propulsion reduction gears in combatant ships use double-helical gears, which are shown in figure 5-18. These gears produce smoother action of the reduction gearing and avoid tooth shock. A double-helical gear has two sets of teeth at complementary angles to each other; therefore, axial thrust is eliminated. Each member of a double-helical gear set should be capable of axial float to prevent excessive tooth loads caused by mismatch of meshing elements. (Axial float means "capable of free motion, neither supporting nor supported by other gears axially.") There is a groove around the center of the gear where the teeth sets come together. This groove provides a path for oil flow, so that a hydraulic pressure is not created between the gears where they mesh.

When the first-reduction gear and the second-reduction pinion each have two bearings and are connected by a quill shaft and flexible coupling(s), the design is called articulated. A quill shaft is essentially a gear coupling. It has two shaft rings with internal teeth and a shaft with external teeth around each end. The shaft rings are bolted on the far ends of each of the two gears to be connected. The floating member, now called the quill shaft (fig. 5-19), passes through the hollow centers of both gears. It is supported only at the ends, where its teeth mesh with the shaft ends.

An articulated locked train gear has two high-speed pinions. Each pinion drives two first-reduction gears connected by quill shafts to two separate second-reduction pinions. Each gear and pinion is mounted in its own two bearings.

The term *locked train* means that the two first-reduction pinions are locked between the two first-reduction gears and transmit the power from the turbines equally to the two gears. This method causes the load on each gear tooth to be reduced by 50 percent.

Construction of Main Reduction Gears

The main rotating elements in a propulsion gear unit (main gears and pinions) operate at high rotational speeds. They transmit tremendous power loads. Very slight unevenness of tooth contour and tooth spacing will cause the gears to operate noisily or even to fail. Therefore, these gears are manufactured to very close tolerances. They are cut in rooms in which the temperatures and humidity are closely controlled. Expansion and contraction of the gear-blank, during machining, are negligible. Oxidation, due to moisture in the air, is nearly eliminated. In addition, all gears are carefully checked for errors.

CASINGS.—Except for some small units, gear casings are of welded construction. Most

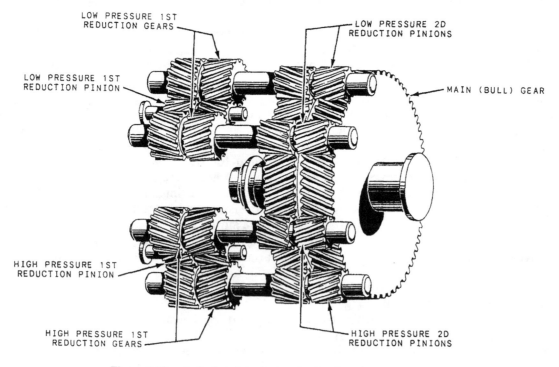

Figure 5-18.—Articulated, locked train double-reduction gear.

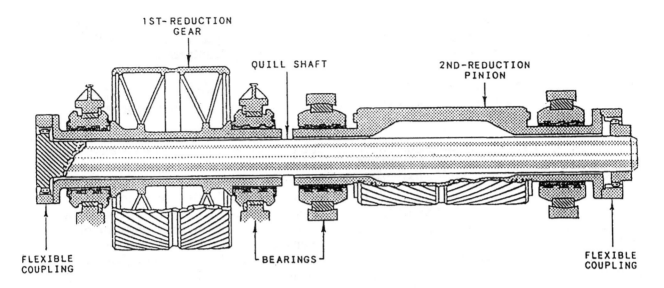

Figure 5-19.—Quill shaft assembly.

gears have an upper and lower casing and gear case covers. These casings are of box-girder construction with integral-bearing blocks. The low-speed bull-gear bearing housing is an integral part of the lower casing. Gear case covers are bolted and securely locked to the upper casing. They are arranged so that they can be removed for access to gearing and bearing caps. Inspection plates are usually provided in covers and gear cases so that the rotating parts may be inspected.

Casings are arranged for access to oil spray fittings so that the fittings may be cleaned. Turning gears, tachometer drives, lube-oil pump drives, sight flow and thermometer fittings, thermocouples or resistance temperature element (RTE) junction boxes, and electrostatic precipitator vent connections are attached to gear casings.

GEARS.—In general, gears are of one-piece construction and are made of steel forgings. The gear wheel construction and materials depend on the size of the gear. For small gears, the entire gear wheel may be made from a single steel forging. Large gears are generally built up by welding. These gear sections usually are the shaft, the hub or center (which may be omitted when the webs are welded to the shaft), the webs, and the rim in which teeth are cut. The shaft is always of forged steel. When propeller-thrust bearings are located within the propulsion gear casing, a collar is located at the forward end of the low-speed gear shaft. Or, an integral flange for attachment of thrust bearing facing collars is

located on the other end of the shaft, forward of the line shaft flange. Other types of construction consist of a close-grained, cast-steel body. The body is welded to a shaft, with teeth cut directly in the casting, or a cast-steel body is pressed on a shaft and secured by fore and aft shrink rings and a locknut.

In a double-reduction, articulated, divided-power path gear set, the high-speed and second-reduction pinions are usually machined from forgings. The first- and second-reduction gears are usually fabricated. The gear shaft and rim are made of steel forgings. The rim and shaft are assembled with steel webs, welded to the shaft and rim. In wide-faced gears, the position of the steel webs, with respect to the gear teeth, is important in that it may affect gear tooth wear patterns. This assembly is stress-relieved and heat-treated for desired hardness.

Some gears are rough cut before heat treatment, and the final finish operation brings them to the proper size. The journals are then cut slightly oversize to permit a final finishing operation. The teeth are cut in a temperature-controlled room. The cutting operation is continuous. This prevents heat that is generated in cutting from affecting the roundness of the finished gear. The air temperature control prevents changes in the ambient temperature from affecting the roundness of the gear. When the tooth cutting and the finishing operations are completed, the journals are finished so that they are concentric with the shaft axis. The assembly is then balanced. The bull gear is made in a similar fashion. Some first-reduction gears and bull gears in naval use are keyed and locked to the shaft with a locking device. This is done before the teeth are cut and before balancing. When the gears are all completed, the contact between pinion and gear is usually checked in a gear rolling machine before they are assembled.

FLEXIBLE COUPLINGS.—Flexible couplings provide longitudinal and angular flexibility between the turbine shafts and the pinion shafts. This permits each shaft to be adjusted axially to its proper position to obtain total axial float.

Most installations have gear-tooth type of flexible couplings. Power is transmitted through a floating intermediate member with external teeth. These teeth mesh with the internal teeth of the shaft rings (sleeves) mounted on the driving and driven shafts.

Figure 5-20 shows a design of the gear-type flexible couplings that connect the main turbines to the high-speed pinions of the main reduction gear. The couplings also allow for expansion of the turbine shafts. This takes care of any slight misalignment between the main turbines and the reduction gears, such as thermal expansion and hull movements. Another type of flexible coupling is shown in figure 5-21. In this coupling, the floating member is the two sleeves that are bolted together. The internal teeth of the sleeve mesh with the external teeth of the hubs mounted on

the shaft. This type of coupling is used most often in diesel engine gears and where self-contained lubrication is advantageous.

CARE OF REDUCTION GEARS, SHAFTING, AND BEARINGS

The main reduction gear is one of the largest and most expensive units of machinery found in the engineering department. Main reduction gears that are installed properly and operated properly will give years of satisfactory service. However, a serious casualty to main reduction gears will either put the ship out of commission or force it to operate at reduced speed. Extensive repairs to the main reduction gear can be very expensive because they usually have to be made at a shipyard.

Some things are essential for the proper operation of reduction gears. Proper lubrication includes supplying the required amount of oil to the gears and bearings, plus keeping the oil clean and at a proper temperature. Locking and unlocking the shaft must be done according to the manufacturer's instructions. Abnormal noises and vibrations must be investigated and corrective action taken. Gears must be inspected according to instructions issued by NAVSEA, the type commander, or other proper authority. Preventive and corrective maintenance must be conducted according to the 3-M Systems.

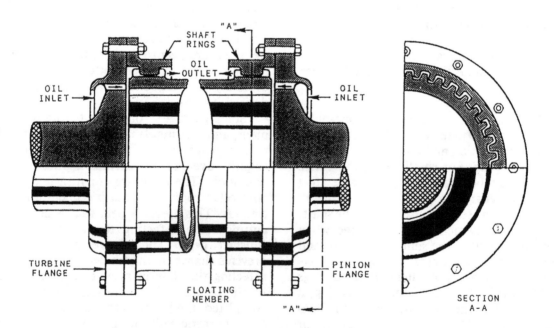

Figure 5-20.—Gear-type flexible coupling.

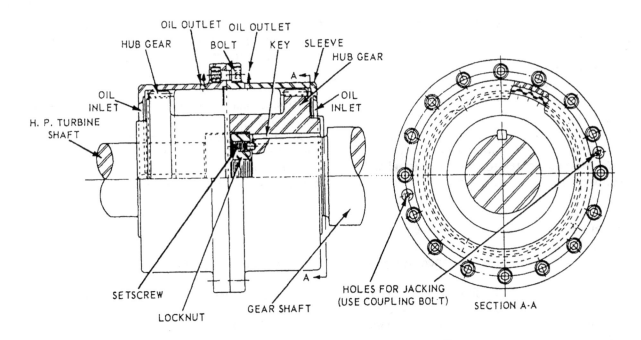

Figure 5-21.—Another gear-type flexible coupling.

PROPER LUBRICATION

Proper lubrication of reduction gears and bearings is extremely important. The correct quantity and quality of lubricating oil must, at all times, be available in the main sump. The oil must be clean and must be supplied to the gears and bearings at the proper pressure and temperature.

To accomplish proper lubrication of gears and bearings, several conditions must be met. The lube-oil service pump must deliver the proper discharge pressure. All relief valves in the lube-oil system must be set to function at their designed pressure. On most older ships, each bearing has a needle valve to control the amount of oil delivered to the bearing. On newer ships, the quantity of oil to each bearing is controlled by an orifice in the supply line. The needle valve setting or the orifice opening must meet manufacturers' specifications or the supply of oil will be affected. Too small a quantity of oil will cause the bearing to run hot. If too much oil is delivered to the bearing, the excessive pressure may cause the oil to leak at the oil seal rings. Too much oil may also cause a bearing to overheat.

Lube oil must reach the bearings at the proper temperature. If the oil is too cold, one of the effects is insufficient oil flow for cooling purposes. If the oil supply is too hot, some lubricating capacity is lost.

For most reduction gears, the normal temperature of oil leaving the lube-oil cooler should be between 120°F and 130°F. For full power operation, the temperature of the oil leaving the bearings should be between 140°F and 160°F. The maximum temperature rise of oil passing through any gear or bearing should not exceed 50°F and the final temperature of oil leaving the gear or bearing must not exceed 180°F. The temperature rise and limitation may be determined by installed thermometers or RTEs.

Cleanliness of lubricating oil cannot be overstressed. Oil must be free from impurities, such as water, grit, metal, and dirt. Clean out metal flakes and dirt when new gears are wearing in or when gears have been opened for inspection. Lint or dirt left in the system may clog the oil spray nozzles. The spray nozzles must be kept open at all times, and they must never be altered without NAVSEA approval.

The lube-oil strainers perform satisfactorily under normal operating conditions, but they cannot trap particles of metal and dirt that are fine enough to pass through the mesh. These fine particles may become embedded in the bearing metal and cause wear on the bearings and journals. In addition, these fine abrasive particles passing through the gear teeth act like a lapping compound and may remove metal from the teeth.

LOCKING AND UNLOCKING THE MAIN SHAFT

In an emergency, or in the event of a casualty to the main propulsion machinery, you may have to stop and lock a propeller shaft to prevent damage to the machinery. Casualty control actions for main propulsion machinery are outlined in the *Engineering Operational Casualty Control (EOCC) Manual* written for each individual ship. Some procedures discussed in the following paragraphs are general in nature and may not apply to your ship.

When a shaft is stopped, the most expeditious way to lock a propeller shaft while underway is to engage the turning gear and then apply the brake.

Engine-room personnel should be trained on how to safely lock and unlock the main shaft by carrying out actual drills. Each steaming watch should have a sufficient number of trained personnel available to stop and lock the main shaft. During drills, the shaft should not be locked for more than 5 minutes. The ahead throttle should NEVER be opened when the turning gear is engaged. The torque produced by the ahead engines is in the same direction as the torque of the locked shaft; to open the ahead throttle would result in damage to the turning gear.

The maximum safe operating speed of a ship with a locked shaft can be found in the manufacturer's technical manual. Additional information in the safe maximum speed that your ship can steam with a locked shaft can be found in *NSTM*, chapters 231 and 9420. If the shaft has been locked for 5 minutes or more, the turbine rotors may become bowed, and special precautions are recommended. Before the shaft is allowed to turn, personnel must be stationed at the turbines to check for unusual noises and vibration. When the turning gear is disengaged, the astern throttle should be slowly closed. The torque produced by the propeller passing through the water will start the shaft rotating. If, when the propeller starts to turn, vibration indicates a bowed rotor, the ship's speed should be reduced to the point where little or no vibration of the turbine is noticeable; this speed should be maintained until the rotor is straightened. If operation at such low speed is not practicable, the turbines should be slowed by use of the astern throttle to the point of least vibration but with the turbines still operating in the ahead direction. When the turbines are slowed to the point of little or no vibration, the shaft should be operated at

that speed and the ahead throttle should be opened slightly to permit some steam flow through the affected turbine. The heat from the steam will warm the shaft and aid in straightening it. Lowering the main condenser vacuum will add additional heat to the turbines; this will increase the exhaust pressure and temperature.

As the vibration decreases, the astern throttle can be closed gradually, allowing the speed of the shaft to increase. The shaft speed should be increased slowly and a check for vibration should be maintained. The turbine is not ready for normal operation until the vibration has disappeared at all possible speeds.

NOISES AND VIBRATION

On steam turbine-driven ships, noises may occur at low speeds or when maneuvering or when passing through shallow water. Generally, these noises do not result from any defect in the propulsion machinery and will not occur during normal operation. A rumbling sound which occurs at low shaft rpm is generally due to the low-pressure turbine gearing floating through its backlash. The rumbling and thumping noises which may occur during maneuvering or during operation in shallow water are caused by vibrations initiated by the propeller. These noises are characteristic only of some ships and should be regarded as normal sound for these units. These sounds will disappear with change of propeller rpm or when the other causes mentioned are no longer present. These noises can usually be noticed when the ship is backing, especially in choppy seas or ground swells.

A properly operating reduction gear has a definite sound which an experienced watch stander can easily recognize. The operator should also be familiar with the normal operating sound of the reduction gear at different speeds and under various operating conditions.

If any abnormal sounds occur, an investigation should be made immediately. In making an investigation, much will depend on how the operator interprets the sound or noise.

The lube-oil temperature and pressure may or may not help an operator determine the reasons for the abnormal sounds. A badly wiped bearing may be indicated by a rapid rise in oil temperature for the individual bearing. A certain sound or noise may indicate misalignment or improper meshing of the gears. If unusual sounds are caused by misalignment of gears or foreign matter passing through the gear teeth, the shaft should be stopped and a thorough investigation should be made before the gears are operated again.

For a wiped bearing, or any other bearing casualty that has caused a very high temperature, the following procedure should be followed: If the temperature of the lube oil leaving any bearing has exceeded the permissible limits, slow or stop the unit and inspect the bearing for wear. The bearing may be wiped only a small amount and the shaft may be operated at reduced speed until the tactical situation allows sufficient time to inspect the bearing.

The most common causes of vibration in a main reduction gear installation are faulty alignment, bent shafting, damaged propellers, and improper balance.

A gradual increase in the vibration in a main reduction gear that has been operating satisfactorily for a long period of time can usually be traced to a cause outside the reduction gears. The turbine rotors, rather than the gears, are more likely to be out of balance.

When reduction gears are built, the gears are carefully balanced (both statically and dynamically). A small amount of unbalance in the gears will cause unusual noise, vibration, and abnormal wear of bearings.

When the ship has been damaged, vibration of the main reduction gear installation may result from misalignment of the turbine, the main shafting, the main shaft bearings, or the main reduction gear foundation. When the vibration occurs within the main reduction gears, damage to the propeller should be one of the first things to be considered. The vulnerable position of the propellers makes them more liable to damage than other parts of the plant. Bent or broken propeller blades will transmit vibration to the main reduction gears. Propellers can also become fouled with line or cable, which will cause gears to vibrate. No reduction gear vibration is too trivial to overlook. Always make a complete investigation.

MAINTENANCE AND INSPECTION

Under normal conditions, major repairs and major items of maintenance on main reduction gears should be accomplished by a shipyard. When a ship is deployed overseas and at other times when shipyard facilities are not available, emergency repairs should be accomplished, if possible, by a repair ship or an advanced base. Inspections, checks, and minor repairs should be accomplished by ship's force.

Under normal conditions, the main reduction gear bearings and gears will operate for an indefinite period. If abnormal conditions occur, the shipyard will normally perform the repairs. Spares are carried aboard sufficient to replace 50 percent of the number of bearings installed in the main reduction gear. Usually each bearing is interchangeable for the starboard or port installation. The manufacturer's technical manual must be checked to determine interchangeability of gear bearings.

Special tools and equipment needed to lift main reduction gear covers, to handle the quill shaft when removing bearings from it, and to take required readings and measurements are normally carried aboard. Special tools and equipment should always be aboard in case emergency repairs have to be made by repair ships or bases not required to carry these items.

The manufacturer's technical manual is the best source of information concerning repairs and maintenance of any specific reduction gear installation. Inspection requirements for reduction gears, shafting, bearings, and propellers can be found in *NSTM*, chapters 245, 9420, and 9430.

The inspections mentioned in this section are the minimum requirements only. Where defects are suspected, or conditions so indicate, inspections should be made at more frequent intervals.

To open any inspection plates or other fittings of the main reduction gears, you should first obtain permission from the engineer officer. Before replacing an inspection plate, connection, or cover which permits access to the gear casing, an officer of the engineering department must make a careful inspection to ensure that no foreign matter has entered or remains in the casing or oil lines. If a repair activity does the work, an officer from the repair activity must also inspect the gear casing. An entry of the inspections and name of the officer or officers must be made in the Engineering Log.

The importance of proper gear tooth contact cannot be overemphasized. Any abnormal condition that may be revealed by operational sounds or by inspections should be corrected as soon as possible. Any abnormal condition that is not corrected will cause excessive wear, which may result in general disintegration of the tooth surfaces.

If proper tooth contact is obtained when the gears are installed, little wear of teeth will occur. Excessive wear cannot take place without metallic contact. Proper clearances and adequate lubrication will prevent most gear tooth trouble.

If proper contact is obtained when the gears are installed, the initial wearing, which takes place under conditions of normal load and adequate lubrication, will smooth out rough and uneven places on the gear teeth. This initial wearing-in is referred to as normal wear or running-in. As long as operating conditions remain normal, no further wear will occur.

Small shallow pits starting near the pitch line will frequently form during the initial stage of operation; this process is called initial pitting. Often the pits (about the size of a pinhead or even smaller) can be seen only under a magnifying glass. These pits are not detrimental and usually disappear in the course of normal wear.

Pitting that is progressive and continues at an increasing rate is known as destructive pitting. The pits are fairly large and are relatively deep. Destructive pitting is not likely to occur under proper operating conditions, but could be caused by excessive loading, too soft material, or improper lubrication. This type of pitting is usually due to misalignment or to improper lubrication.

The condition in which groups of scratches appear on the teeth (from the bottom to the top of the tooth) is termed *abrasion*, or *scratching*. This condition may be caused by inadequate lubrication, or by the presence of foreign matter in the lubricating oil. When abrasion or scratching is noted, the lubricating system and the gear spray fixtures should be examined immediately. If dirty oil is found to be the cause, the system must be thoroughly cleaned and the whole charge of oil centrifuged.

The term *scoring* denotes a general roughening of the whole tooth surface. Scoring marks are deeper and more pronounced than scratching and they cover an area of the tooth, instead of occurring haphazardly, as in scratching or abrasion. Small areas of scoring may occur in the same position on all teeth. Scoring, with proper alignment and operation, usually results from inadequate lubrication and is intensified by the use of dirty oil. If these conditions are not corrected, continued operation will result in a general disintegration of the tooth surface.

Under normal conditions, all alignment inspections and checks, plus the necessary repairs, are accomplished by naval shipyards. Incorrect alignment will be indicated by abnormal vibration, unusual noise, and wear of the flexible couplings or main reduction gears. When misalignment is indicated, shipyard personnel should make a detailed inspection.

Two sets of readings are required to get an accurate check of the propulsion shafting. One set of readings is taken with the ship in drydock and another set of readings is taken with the ship waterborne—under normal loading conditions. The main shaft is disconnected, marked, and turned so that a set of readings can be taken in four different positions. Four readings are taken (top, bottom, and both sides). The alignment of the shaft can be determined by studying the different readings taken. The naval shipyard will decide whether or not corrections in alignment are necessary.

During shipyard overhauls, the ship's force should make the following inspections:

—Inspect the condition and clearance of thrust shoes to ensure proper position of gears. Blow out thrusts with dry air after the inspection. Record the readings. Inspect the thrust collar, nut, and locking device.

—If turbine coupling inspection has indicated undue wear, check alignment between pinions and turbines.

—Inspect and clean oil sump.

When conditions warrant or if trouble is suspected, a work request may be submitted to a naval shipyard to perform a "10-year" inspection of the main reduction gears. This inspection includes clearances and condition of bearings and journals; alignment checks and readings; and any other tests, inspections, or maintenance work that may be considered necessary.

Naval Sea Systems Command authorization is not necessary for lifting reduction gear covers. Covers should be lifted when trouble is suspected. An open gear case is a serious hazard to the main plant; therefore, careful consideration of the dangers of uncovering a gear case must be balanced against the reasons for suspecting internal trouble before deciding to lift the gear case. The 10-year interval may be extended by the type commander if conditions indicate that a longer period between inspections is desirable.

The correction of any defects disclosed by regular tests and inspections, and the observance of manufacturer's instructions, should ensure that the gears are ready for full power at all times.

In addition to inspections which may be directed by proper authority, you should open the inspection plates to check the operation of the spray nozzles and to examine the tooth contact and the condition of teeth to note changes that may have occurred during full power trials. Running for a few hours at high power will show any possible condition of improper contact or abnormal wear that would not have shown up in months of operation at lower power. You should also check the clearance of the main thrust bearing. Do not open gear cases, bearings, and thrusts immediately before full power trials.

STEERING

The direction of movement of a ship is controlled partly by steering devices that receive their power from steering engines and partially by the arrangement, speed, and direction of rotation of the ship's propellers.

The steering device is called a rudder. The rudder is more or less a rectangular metal blade (usually hollow on large ships) that is supported by a rudder stock. The rudder stock enters the ship through a rudder post and a watertight fitting, as shown in figure 5-22. A yoke, or quadrant, secured to the head of the rudder stock, transmits the motion imparted by the steering mechanism.

Basically, a ship's rudder is used to attain and maintain a desired heading. The force necessary to accomplish this is developed by the dynamic pressure against the flat surface of the rudder. The magnitude of this force and the direction and degree to which it is applied produces the rudder effect, which controls stern movement and thus controls the ship's heading.

To function most effectively, a rudder should be located aft of and quite close to the propeller. Many modern ships have twin rudders, each set directly behind a propeller to receive the full thrust of water. This arrangement tends to make a ship highly manueverable.

Three types of rudders are in general use—the unbalanced rudder, the semibalanced rudder, and the balanced rudder. These three types are shown in figure 5-23. The Navy also uses other types of rudders.

The type of rudder is based on the location of the center of the area of the rudder in respect to the rudder stock. The center of the area is also the face of the force that turns the ship. As this force is developed, opposite force must be maintained by the steering engine to maintain the rudder angle. The type of rudder used has a direct relation to the power requirement for the steering engine.

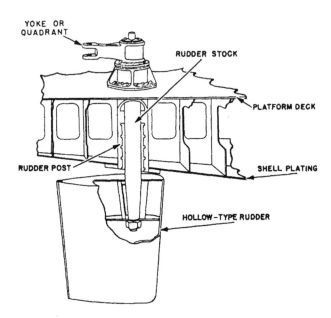

Figure 5-22.—Rudder assembly.

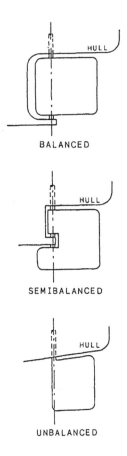

Figure 5-23.—Types of rudders.

REFERENCES

Engineman 3, NAVEDTRA 10539, Naval Education and Training Program Management Support Activity, Pensacola, Florida, 1989.

Machinist's Mate 3 & 2, NAVEDTRA 10524-F, Naval Education and Training Program Management Support Activity, Pensacola, Florida, 1987.

Naval Ships' Technical Manual, Chapter 244, "Bearings," NAVSEA S9086-HN-STM-000, Change 8, Naval Sea Systems Command, Washington, D.C., 1989.

Naval Ships' Technical Manual, Chapter 245, "Propellers," NAVSEA S9086-HP-STM-000, Naval Sea Systems Command, Washington, D.C., 1988.

Naval Ships' Technical Manual, Chapter 9420, "Propulsion Reduction Gears, Couplings, and Associated Components," NAVSEA 0901-LP-420-0002, Change 15, Naval Sea Systems Command, Washington, D.C., 1985.

Naval Ships' Technical Manual, Chapter 9430, "Shafting, Bearings, and Seals," NAVSEA 0901-LP-430-0012, Change 13, Naval Sea Systems Command, Washington, D.C., 1989.

CHAPTER 6

THEORY OF LUBRICATION

LEARNING OBJECTIVES

Upon completion of this chapter, you should be able to do the following:

1. Explain the fundamentals of lubrication.

2. Identify the types of friction.

3. Identify symbols used to classify and describe the properties of lubricating oils and greases.

4. Explain the fundamentals of lube oil purification.

5. Identify the types of lube-oil purifiers used in lube oil purification.

INTRODUCTION

Lubrication reduces friction between moving parts by substituting fluid friction for sliding or rolling friction. Without lubrication, it would be difficult to move a 100-pound weight across a rough surface; however, with lubrication and properly designed bearing surfaces, a very small motor could move a 1-million-pound load. By reducing friction, lubrication lowers the amount of energy required to perform mechanical actions and causes less heat to be produced.

Lubrication is a matter of vital importance throughout the shipboard engineering plant. Moving surfaces must be steadily supplied with the proper kind of lubricants. Lubricants must be maintained at specified standards of purity and at designed pressures and temperatures. Without proper lubrication, many units of shipboard machinery would grind to a screeching halt.

The lubrication requirements of shipboard machinery are met in various ways, depending on the machinery. In this chapter, we will discuss the basic theories of lubrication, the lubricants used aboard ships, and the lubrication systems installed for many shipboard units. We will also discuss the devices used to maintain lubricating oils in the required condition of purity.

THEORY OF LUBRICATION

Friction is the natural resistance to motion caused by surface contact, and the purpose of lubrication is to reduce this friction. The friction that exists between a body at rest and the surface upon which it rests is called static friction. The friction that exists between moving bodies (or between one moving body and a stationary surface) is called kinetic friction. Static friction is greater than kinetic friction. Static friction and inertia must be overcome to put a body in motion. To keep a body in motion, kinetic friction must be overcome.

There are three types of kinetic friction: sliding friction, rolling friction, and fluid friction. Sliding friction occurs when one solid body slides across another solid body. Rolling friction occurs when a curved body, such as a cylinder or sphere, rolls across a surface. Fluid friction is the resistance to motion exhibited by a fluid.

Fluid friction occurs because of two properties of a lubricant: cohesion and adhesion. COHESION is the molecular attraction between particles that tends to hold a substance together. ADHESION is the molecular attraction between particles that tends to cause unlike surfaces to stick together. If a paddle is used to stir a fluid, for example, cohesion between the particles of the fluid tends to hold the molecules together. This retards the motion of the fluid. But adhesion of fluid particles causes the fluid to stick to the paddle. This further causes friction between the paddle and the fluid. In the theory of lubrication, cohesion and adhesion have major roles. Adhesion is the property of a lubricant that causes it to stick (or adhere) to the parts being lubricated; cohesion is the property that holds the lubricant together and enables it to resist breakdown under pressure. Other important properties of a lubricant will be discussed later in this chapter.

Different materials have varying degrees of cohesion and adhesion. In general, solid bodies are highly cohesive but slightly adhesive. Most fluids are highly adhesive but only slightly cohesive.

FLUID LUBRICATION

One of the qualities of a liquid is that it cannot be forced into a smaller space than it already occupies. A liquid is incompressible. This fact allows moving metal surfaces to be separated from each other. Because of this, liquid is used for most lubrication needs. As long as the lubricant film remains unbroken, fluid friction replaces sliding friction and rolling friction.

In any process involving friction, some power is consumed and some heat is produced. Overcoming sliding friction consumes the greatest amount of power and produces the greatest amount of heat. Overcoming rolling friction consumes less power and produces less heat. Overcoming fluid friction consumes the least amount of power and produces the least amount of heat.

LANGMUIR THEORY

A present accepted theory of lubrication is based on the Langmuir theory of the action of fluid films of oil between two surfaces, one or both of which are in motion. Theoretically, three or more layers of oil film exist between two lubricated bearing surfaces. Two of the films are boundary films (indicated as I and V in fig. 6-1, view A); one clings to the surface of the rotating journal and the other clings to the stationary lining of the bearing. Between these two boundary films are one or more fluid films (indicated as II, III, and IV in fig. 6-1, view A).

When the rotating journal is set in motion, a wedge of oil is formed (fig. 6-1, view B). Contact between the two metal surfaces is prevented when

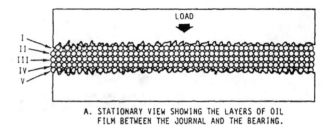
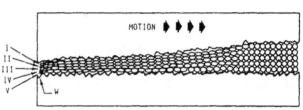

A. STATIONARY VIEW SHOWING THE LAYERS OF OIL FILM BETWEEN THE JOURNAL AND THE BEARING.

B. MOTION VIEW SHOWING THE FORMATION OF THE WEDGE OF OIL (W) AS THE FLUID FILMS (II, III, AND IV) SLIDE BETWEEN THE JOURNAL AND THE BEARING.

Figure 6-1.—Theory of oil film lubrication showing boundary and fluid oil films.

oil films II, III, and IV (fig. 6-1, view A) slide between the two boundary films. The theory is again illustrated in figure 6-2. The position of the oil wedge, W, is shown with respect to the position of the journal as it starts and continues in motion.

The views shown in figure 6-2 represent a journal or shaft rotating in a solid bearing. The clearances are enlarged in the drawing to show the formation of the oil film. The shaded portion represents the clearance filled with oil. The stationary view, A, shows the film in the process of being squeezed out while the journal is at rest. As the journal begins to turn and to increase speed, oil adhering to the surfaces of the journal is carried into the film. The film increases in thickness and tends to lift the journal, as shown in the starting view, B. As the speed increases, the journal takes the position shown in the running view, C. Varying temperatures cause changes in oil viscosity. These changes modify the film thickness and position of the journal. Viscosity will be discussed later in this chapter.

If conditions are correct, the two surfaces are properly separated. A momentary contact may occur at the time the motion is started.

FACTORS AFFECTING LUBRICATION

A number of factors determine the effectiveness of oil film lubrication. They include pressure, temperature, viscosity, speed, alignment,

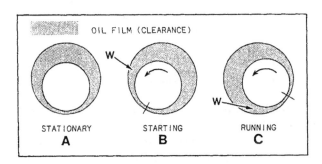

Figure 6-2.—Journal rotation in a solid bearing showing the distribution of the oil film.

condition of the bearing surfaces, running clearances, and the purity of the lubricant. Many of these factors are interrelated and inter-dependent. For example, the viscosity of any given oil is affected by temperature, and the temperature is affected by running speed. Therefore, the viscosity is partially dependent on the running speed.

A lubricant must stick to the bearing surfaces and support the load at operating speeds. More adhesiveness is required to make a lubricant adhere to bearing surfaces at high speeds than at low speeds. At low speeds, greater cohesiveness is required to keep the lubricant from being squeezed out from between the bearing surfaces.

Large clearances between bearing surfaces require high viscosity and cohesiveness in the lubricant to ensure maintenance of the lubricating oil film. The larger the clearance, the greater must be the lubricant's resistance to being pounded out, with consequent destruction of the lubricating oil film.

High unit load on a bearing requires high viscosity of the lubricant. A lubricant subjected to high loading must be sufficiently cohesive to hold together and maintain the oil film.

LUBRICANTS

Although synthetic lubricants are used today, the Navy uses petroleum as its main source of oils and greases. By various refining processes, lubricating oils are extracted from crude petroleum and blended into a number of products. Sometimes additives (chemical compounds) are included in the process. Lubricating oils have to meet a wide range of lubrication requirements.

Lubricating Oils

Lubricating oils approved for shipboard use are limited to those grades and types that are necessary to provide proper lubrication under all anticipated operating conditions.

Diesel engines use a detergent-dispersant type of additive oil to keep the engines clean. These lubricating oils must be fortified with oxidation and corrosion inhibitors. This allows long periods between oil changes and prevents corrosion of bearing materials.

Steam turbines use an oil of high initial film strength. This oil is fortified with antifoaming additives and additives that control oxidation and corrosion. Also, extreme pressure (EP) additives are used. These additives help carry the very high loading found in the reduction gear.

For general lubrications and in hydraulic systems using petroleum lubricants, the Navy must use certain oils. These special viscosity series of oils are strengthened with oxidation and corrosion inhibitors and antifoam additives. Deck machinery uses compounded oils, which are mineral oils with additives.

Special lubricating oils are available for a wide variety of services. The *Federal Supply Catalog* has a list of these oils. Among the important specialty oils are those used for lubricating refrigerant compressors. These oils must have a very low pour point and must be maintained with a high degree of freedom from moisture.

The main synthetic lubricants in naval use are (1) a phosphate-ester type of fire-resistant hydraulic fluid, used chiefly in the aircraft elevators of carriers; (2) a water-base glycol hydraulic fluid, used chiefly in the catapult retracting gears; and (3) the lubricating fluid used in gas turbine engines.

CLASSIFICATION OF LUBRICATING OILS.—The Navy identifies lubricating oils by number symbols. Each identification symbol consists of four digits and, in some cases, appended letters. The first digit shows the class of oil according to type and use; the last three digits show the viscosity of the oil. The viscosity digits are actually the number of seconds required for 60 milliliters (ml) of oil to flow through a standard orifice at a certain temperature. Symbol 3080, for example, shows that the oil is in the 3000 series. It also shows that a 60-ml sample flows through a standard orifice in 80 seconds when the oil is at a certain temperature (210°F, in this instance). Another example is symbol 2135 TH.

This symbol shows that a 60-ml sample flows through a standard orifice in 135 seconds when the oil is at a certain temperature (130°F, in this case). The letters H, T, TH, or TEP added to a basic number show that the oil contains additives for special purposes.

PROPERTIES OF LUBRICATING OILS.— Lubricating oils used by the Navy are tested for a number of properties. These include (1) viscosity, (2) pour point, (3) flash point, (4) fire point, (5) autogenous ignition point, (6) demulsibility, (7) neutralization number, and (8) precipitation number. Standard test methods are used for making all tests. The properties of lube oil are briefly explained in the following paragraphs.

Viscosity.—The viscosity of an oil is its tendency to resist flow or change of shape. A liquid of high viscosity flows very slowly. In variable climates, automobile owners, for example, change oils according to prevailing seasons. Oil changes are necessary because heavy oil becomes too sluggish in cold weather, and light oil becomes too thin in hot weather. The higher the temperature of an oil, the lower its viscosity becomes; lowering the temperature increases the viscosity. The high viscosity or stiffness of the lube oil on a cold morning makes an automobile engine difficult to start. The viscosity must always be high enough to keep a good oil film between the moving parts; otherwise, friction will increase, resulting in power loss and rapid wear on the parts.

Oils are graded by their viscosities at a certain temperature. Grading is set up by noting the number of seconds required for a given quantity (60 ml) of oil at the given temperature to flow through a standard orifice. The right grade of oil, therefore, means oil of the proper viscosity.

Every oil has a viscosity index based on the slope of the temperature-viscosity curve. The viscosity index depends on the rate of change in viscosity of a given oil with a change in temperature. A low index figure means a steep slope of the curve, or a great variation of viscosity with a change in temperature; a high index figure means a flatter slope, or lesser variation of viscosity with the same changes in temperatures. When using an oil with a high viscosity index, its viscosity or body will change less when the temperature of the engine increases.

Pour Point.—The pour point of an oil is the lowest temperature at which the oil will barely flow from the container. At a temperature below the pour point, oil congeals or solidifies. Lube oils used in cold weather operations must have a low pour point. (**NOTE:** The pour point is closely related to the viscosity of the oil. In general, an oil of high viscosity will have a higher pour point than an oil of low viscosity.)

Flash Point.—The flash point of an oil is the temperature at which enough vapor is given off to flash when a flame or spark is present. The required flash point minimums vary from 300°F for the lightest to 510°F for the heaviest (forced-feed) oils. The temperature of oils are always far below that under normal operating conditions.

Fire Point.—The fire point of an oil is the temperature at which the oil will continue to burn when ignited.

Autogenous Ignition Point.—The autogenous ignition point of an oil is the temperature at which the flammable vapors given off from the oil will burn. This kind of burning will occur without the application of a spark or flame. For most lubricating oils, this temperature is in the range of 460°F to 815°F.

Demulsibility.—The demulsibility, or emulsion characteristic, of an oil is its ability to separate cleanly from any water present— an important factor in forced-feed systems. Water (fresh or salt) should be always kept out of oils.

Neutralization Number.—The neutralization number of an oil is the measure of the acid content. The number of milligrams of potassium hydroxide (KOH) required to neutralize 1 gram of oil defines the neutralization number. All petroleum products oxidize in the presence of air and heat. The products of this oxidation include organic acids. High amounts of organic acids have harmful results on galvanized surfaces and on alloy bearings at high temperatures. The demulsibility of the oil with respect to fresh water and seawater also relies on the amount of organic acids. High organic acid levels may cause decreased demulsibility. The formation of sludge and emulsions too stable to be broken by available means may result. This last problem may occur in turbine installations. An increase in acidity is a sign that lubricating oil is breaking down.

Precipitation Number.—The precipitation number of an oil is a measure of the amount of solids classified as asphalts or carbon residue contained in the oil. The number is reached by diluting a known amount of oil with naphtha and separating the precipitate by centrifuging—the volume of separated solids equals the precipitation number. This test helps you find out quickly the presence of foreign materials in used oils. An oil with a high precipitation number may cause trouble in an engine. It could leave deposits or plug up valves and pumps.

Lubricating Greases

Lubricating greases are gels formed with lubricating oil and a thickener. The lubricating oil is thickened by adding a soap or a clay, such as bentonite. The soaps are chemical compounds formed by combining fatty acids with various alkali metals, such as calcium, sodium, aluminum, zinc, barium, lithium, lead, or potassium. The characteristics of the finished grease are determined by the type of alkali metal, the type of fatty acid, the quality of the base lubricating oil, the manufacturing process, and the additives.

Lubricating grease hardness (consistency) is determined by measuring, in tenths of a millimeter, the depth to which a standard cone penetrates a sample of grease. The sample is brought to a temperature of 77°F, subjected to 60 double strokes in a standard grease worker, then tested for penetration. This test simulates the work performed on a grease during operation and provides a common starting point for measurement.

Navy specifications have been drawn to cover the several grades of lubricating greases. The grades most common in engine room use are ball and roller bearing grease and EP grease.

HIGH-PERFORMANCE BALL AND ROLLER BEARING GREASE.—Ball and roller bearing grease is for general use in equipment designated to operate at temperatures up to 300°F. For temperature applications above 300°F, high-temperature, electric-motor, ball and roller bearing grease must be used.

HIGH-TEMPERATURE GREASE.—High-temperature grease is used in equipment that operates at temperatures above 300°F.

WATER-RESISTANT GREASE.—There are two types of water-resistant greases: general-purpose and wire rope. General-purpose grease is used for lubricating bearings that operate in areas exposed to water contamination, such as periscope bearings and stern tube bearings. Wire rope grease protects wire rope surfaces from seawater contamination, lubricates the wires in the rope, and protects exposed gears.

EXTREME PRESSURE (EP) GREASE.—Certain bearing surfaces are so heavily loaded that ordinary grease lubricants cannot maintain a film to prevent contact of rubbing surfaces. For such applications, solid lubricants are incorporated into conventional greases as additives to provide additional load-carrying capability. Levels and types of EP lubricants required will vary from specification to specification depending upon intended use.

LUBE OIL PURIFICATION

The forced-feed lubrication systems in modern naval ships rely on pure oil. Oil that stays pure can be used for a long time. Lube oil does not wear out, it is merely robbed of its lubricating properties by foreign substances.

Contaminants interfere with the ability of the oil to maintain a good lubricating film between metal surfaces. These contaminants must be removed or the oil will not meet lubrication requirements. Dirt, sludge, and other contaminants will act as abrasives to score and scratch the rubbing metal surfaces within engines, generators, pumps, and blowers. Water is the greatest source of contamination. Strainers, filters, settling tanks, and centrifugal purifiers are used in lubrication systems to keep the oil pure.

STRAINERS/FILTERS

Strainers or filters are used in many lubricating systems to prevent the passage of grit, scale, dirt, and other foreign material. Duplex strainers are used in lubricating systems in which an uninterrupted flow of lubricating oil must be maintained; the flow may be diverted from one strainer basket to the other while one is being cleaned. Filters may be installed directly in pressure lubricating systems or they may be installed as bypass filters.

The use of strainers and filters does not solve the problem of water contamination of lubricating oil. Even a very small amount of water in lubricating oil can be extremely damaging to machinery, piping, valves, and other equipment. Water in lubricating oil can cause widespread pitting and corrosion. Water also increases frictional resistance and can cause the oil film to break prematurely. Every effort must be made to prevent the entry of water into any lubricating system.

Lubricating oil piping is generally arranged to permit two methods of purification: batch purification and continuous purification. The batch process uses settling tanks, while the continuous process uses centrifugal purifiers.

SETTLING TANKS

In the batch process, the lube oil is transferred from the sump to a settling tank by a purifier or a transfer pump. Settling tanks permit oil to stand while water and other impurities settle out. A number of layers of contaminants may form in the bottom of the tank. The number of layers depends on the specific gravity of the various contaminating substances. For example, a layer of

metal may form on the bottom, followed by a layer of sludge, a layer of water, and then the clean oil on top. After the oil is heated and allowed to settle for several hours, water and other impurities that have accumulated in the settling tanks are removed. The oil that is left in the tanks is then centrifuged and returned to the sump or storage tank.

CENTRIFUGAL PURIFIERS

When a ship is at sea or when time does not permit batch purification in the settling tanks, the continuous purification process is used. Centrifugal purifiers are used in this process. The purifier takes the oil from the sump in a continuous cycle. Before entering the purifier, the oil is heated to help remove the impurities.

A purifier may be used to remove water and/or sediment from oil. When water must be removed, the purifier is called a separator. When the main source of contamination is sediment, the purifier is used as a clarifier.

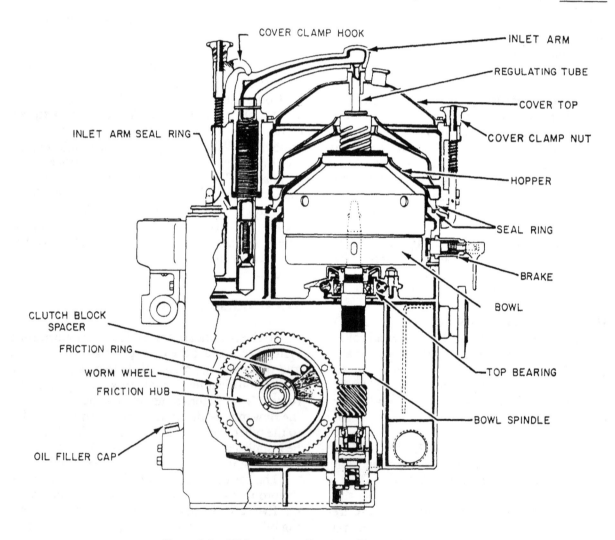

Figure 6-3.—Disk-type centrifugal purifier (DeLaval).

When used to purify lubricating oil, a purifier may be used as either a separator or a clarifier. Aboard ship, a purifier is almost always operated as a separator.

Two types of purifiers are used in Navy installations. Both operate on the same principle. The principal difference is in the design of the rotating units. In one type, the rotating element is a bowl-like container that encases a stack of disks. This is the disk-type DeLaval purifier. In the other type, the rotating element is a hollow, tubular rotor and is the tubular-type Sharples purifier.

Disk-Type Purifier

A sectional view of a disk-type centrifugal purifier is shown in figure 6-3. The bowl is mounted on the upper end of the vertical bowl spindle, which is driven by a worm wheel and friction clutch assembly. A radial thrust bearing at the lower end of the bowl spindle carries the weight of the bowl spindle and absorbs any thrust created by the driving action. The parts of a disk-type bowl are shown in figure 6-4.

The flow of oil through the bowl and additional parts are shown in figure 6-5. Contaminated

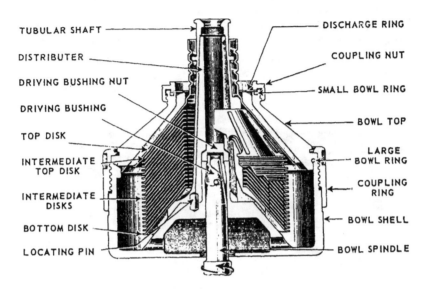

Figure 6-4.—Parts of a disk-type purifier bowl (DeLaval).

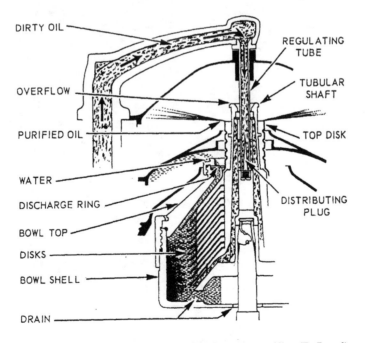

Figure 6-5.—Path of oil through a disk-type purifier (DeLaval).

oil enters the top of the revolving bowl through the regulating tube. The oil then passes down inside the tubular shaft and out at the bottom into the stack of disks. As the dirty oil flows up through the distribution holes of the disks, the high centrifugal force exerted by the revolving bowl causes the dirt, sludge, and water to move outward. The purified oil flows inward and upward, discharging from the neck of the top disk. The water forms a seal between the top disk and the bowl top. (The top disk is the dividing line between the water and the oil.) The disks divide the space within the bowl into many separate narrow passages or spaces. The liquid confined within each passage is restricted so that it can flow only along that passage. This arrangement minimizes agitation of the liquid as it passes through the bowl. It also makes shallow settling distances between the disks.

Most of the dirt and sludge remains in the bowl and collects in a more or less uniform layer on the inside vertical surface of the bowl shell. Any water, along with some dirt and sludge, separated from the oil, is discharged through the discharge ring at the top of the bowl.

Tubular-Type Purifier

A cross section of a tubular-type centrifugal purifier is shown in figure 6-6. This type of purifier consists essentially of a hollow rotor or bowl that rotates at high speeds. The rotor has an opening in the bottom to allow the dirty lube oil to enter. It also has two sets of openings at the top to allow the oil and water (separator) or the oil by itself (clarifier) to discharge (see insert in fig. 6-6). The bowl, or hollow rotor, of the purifier is connected by a coupling unit to a

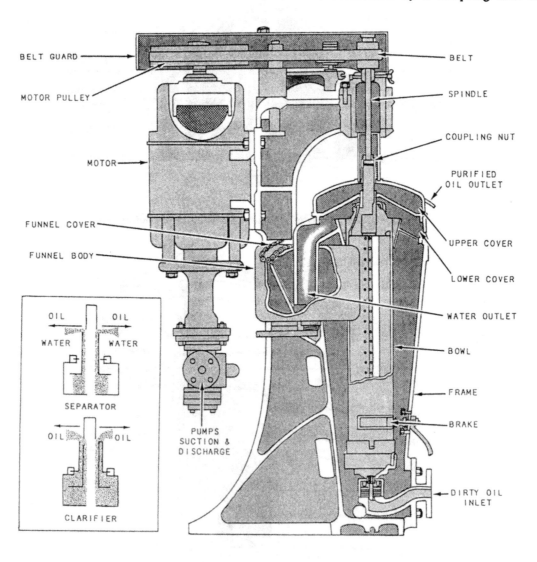

Figure 6-6.—Tubular-type centrifugal purifier (Sharples).

spindle that is suspended from a ball bearing assembly. The bowl is belt-driven by an electric motor mounted on the frame of the purifier.

The lower end of the bowl extends into a flexibly mounted guide bushing. The assembly, of which the bushing is a part, restrains movement of the bottom of the bowl but allows enough movement so that the bowl can center itself about its axis of rotation when the purifier is in operation. Inside the bowl is a device that consists of three flat plates equally spaced radially. This device is commonly referred to as the three-wing device or as the three-wing. The three-wing rotates with the bowl and forces the liquid in the bowl to rotate at the same speed as the bowl. The liquid to be centrifuged is fed into the bottom of the bowl through the feed nozzle, under pressure, so that the liquid jets into the bowl in a stream.

When the purifier is used as a lube-oil clarifier, the three-wing has a cone on the bottom. The feed jet strikes against this cone to bring the liquid smoothly up to bowl speed without making an emulsion. This type of three-wing device is shown in figure 6-7.

The process of separation is basically the same in the tubular-type purifier as in the disk-type purifier. In both types, the separated oil assumes the innermost position and the separated water moves outward. Both liquids are discharged separately from the bowl, and the solids separated from the liquid are retained in the bowl.

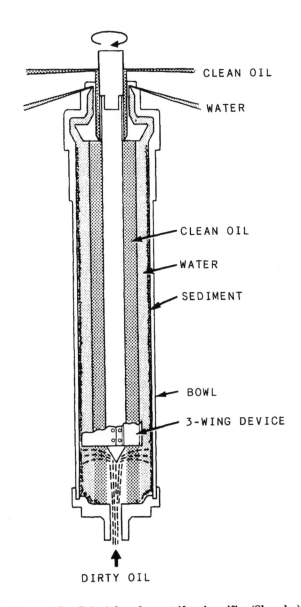

Figure 6-7.—Principles of a centrifugal purifier (Sharples).

REFERENCES

Engineman 3, NAVEDTRA 10539, Naval Education and Training Program Management Support Activity, Pensacola, Florida, 1989.

Machinist's Mate 3 & 2, NAVEDTRA 10524-F1, Naval Education and Training Program Management Support Activity, Pensacola, Florida, 1987.

Naval Ships' Technical Manual, Chapter 262, "Lubricating Oils, Greases, Hydraulic Fluids, and Lubrication Systems," NAVSEA S9086-H7-STM-010, Naval Sea Systems Command, Washington, D.C., 1987.

CHAPTER 7

PRINCIPLES OF MEASUREMENT

LEARNING OBJECTIVES

Upon completion of this chapter, you should be able to do the following:

1. Discuss the basic principles of measurement, including units and standards of measurement.

2. Identify the basic principles of and instruments used in measuring temperature, pressure, fluid flow, liquid level, rotational speed, specific gravity, and viscosity.

INTRODUCTION

Measurement is the language of engineers. The shipboard engineering plant contains numerous gauges and instruments that tell operating personnel the plant's operating condition. The gauges and instruments also provide information for the recorded instrument readings and for other engineering records and reports.

This chapter describes some of the basic types of gauges and instruments used in shipboard engineering plants. These gauges and instruments measure important variables such as temperature, pressure, fluid flow, liquid level, and rotational speed. Because of the wide variety of gauges and instruments used in connection with shipboard engineering equipment, only the basic principles of measurement and commonly used types of gauges are discussed in this chapter. Unusual or highly specialized instruments or ones that apply to a single or unique piece of equipment are discussed in this chapter only as a means of bringing out some interesting or important aspects of measurement.

THE CONCEPT OF MEASUREMENT

One of the primary ways we extend our knowledge and understanding of the universe and of the world around us is by the measurement of various quantities. Because we live in a world in which practically everything seems to be in some way measured or counted, we tend to assume that measurement is basically simple. In reality, however, it may be quite difficult to develop an appropriate mode of measurement even after we have recognized the need. Without an appropriate mode of measurement, we may fail to recognize the significance of the phenomena we observe. Thus, the development of scientific and engineering principles has been and undoubtedly will continue to be inextricably tied to the concept of measurement.

Many of our views on the nature of things are profoundly influenced by the procedures we devise for measurement. In the history of science, the application of a new instrument or the refinement of a measuring technique has led to new ideas about the universe or about the nature of the thing being measured. Until the seventeenth century, it was commonly believed that water rose in a suction pump because "nature abhors a vacuum." This belief persisted in spite of Galileo's observation that water would not rise more that 32 feet in a suction pump. The concept of a "sea of air" surrounding the earth and exerting pressure upon it was very closely related to Torricelli's experiments with a column of mercury in a glass tube (a barometer). Pascal's experiments proved this hyphotesis in 1648. Pascal suggested using Torricelli's barometer at the base of a mountain and again at the top of the mountain. He reasoned that if the air exerts a pressure, the mercury should stand higher in the column at the base of the mountain than it should be at the top. Further experimentation and measurement by Robert Boyle and others led to the development of many important concepts concerning the nature of air and other gases, and led eventually to an understanding of the relationship between the volume and the pressure of a gas (Boyle's law).

Another example of the effects of measurement upon our basic concepts of the nature of things is found in the study of heat. Quantitative studies of heat were not possible before the invention of the thermometer. Before the mid-nineteenth century, heat was thought of as an invisible, weightless fluid called caloric. The caloric theory persisted partly due to faulty interpretations of experimental results which were in part due to difficulties of measurement. The downfall of the caloric theory was necessary before we could conceive of heat as energy and understand the relationship between heat and work. The relationship between heat and work is, of course, basic to the entire field of engineering and thermodynamics (see chapter 8). The discovery that heat tends to flow from hotter to colder bodies until a state of equilibrium is reached could not have been accomplished without the thermometer.

SYSTEMS, UNITS, AND STANDARDS OF MEASUREMENT

Practically all units of measurement are derived from a few basic quantities or fundamental dimensions. In all commonly used systems of measurement, length and time are taken as two of the fundamental dimensions. The third fundamental dimension is mass or force (weight) depending on the system. In all systems, temperature is the fourth fundamental dimension.

The first three fundamental dimensions, length, time, and either mass or force, are sometimes called mechanical quantities or dimensions. All other important mechanical quantities can be defined in terms of these three fundamentals. Temperature, the fourth fundamental dimension is in a different category because it is not a mechanical quantity. By using the three mechanical fundamental quantities and the quantity of temperature, practically all quantities of any importance is derived.

It is often said that there are two systems of measurement –a metric system and a British system. Actually, however, there are several metric systems and several British systems. A more meaningful classification of systems of measurement can be made by saying that some systems are gravitational and others are absolute. In gravitational systems, the units of force are defined in terms of the effects of the force of gravity upon a standard sample of matter at a specified location on the surface of the earth. In absolute systems, units of force are defined in terms that are completely independent of the effects of the force of gravity. Thus, a metric system could be either gravitational or absolute, and a British system could be gravitational or absolute,

depending upon the terms by which force is defined in the particular system.

MASS AND WEIGHT

To understand what is meant by gravitational and absolute systems of measurement, you must have a clear understanding of the difference between mass and weight. Mass, a measure of the total quantity of matter in an object or body, is completely independent of the force of gravity, so the mass of any given object is always the same, no matter where it is located on the surface of the earth or practically anywhere. Weight, on the other hand, is a measure of the force of attraction between the mass of the earth and the mass of another body or object. Since the force of attraction is not identical in all places, the weight of a body depends upon the location of the body with respect to the earth.

The relationship between mass and weight can be understood from the equation

$$w = mg$$

where

w = weight

m = mass

g = acceleration due to gravity

The value for acceleration due to gravity (normally represented by the letter g) is almost constant for bodies at or near the surface of the earth. This value is approximately 32 feet per second per second in British systems of measurement, 9.8 meters per second per second in one metric system, and 980 centimeters per second per second in another metric system. More precise values of g, including variations that occur with changes in latitude and elevation, may be obtained from physics and engineering textbooks and handbooks.

BASIC MECHANICAL UNITS

Table 7-1 shows the basic mechanical quantities of length, mass or force, and time, together with a number of derived units, used in several systems of measurement. By examining some of the units, we may see how force is defined and, thus, see why each system is called absolute or gravitational.

In the metric absolute meter-kilogram-second (MKS) system of measurement, the unit of mass is the kilogram, the unit of length is the meter, the unit of time is the second, and the unit of acceleration is meters per second per second. (This is sometimes written as

Table 7-1.--Units of Measurement in Several Common Systems

	MKS Metric Absolute System	CGS Metric Absolute System	British FPS Absolute System	British Engineering Gravitational System	British FPS Gravitational System
Length	meter (m)	centimeter (cm)	foot (ft)	foot (ft)	foot (ft)
Area	square meter (m^2)	square centimeter (cm^2)	square foot (ft^2)	square foot (ft^2)	square foot (ft^2)
Volume	cubic meter (m^3) or liter (l)	cubic centimeter (cm^3 or cc) or milliliter (ml)	cubic foot (ft^3)	cubic foot (ft^3)	cubic foot (ft^3)
Mass	kilogram (kg)	gram (g)	pound (lb)	slug	pound (lb)
Force (Weight)	newton (n, new, or nt)	dyne	poundal (pdl)	pound (lb) or pound-force (lbf)	pound (lb) or pound-force (lbf)
Time	second (sec)	second (sec)	second (sec)	second (sec)	second (sec)
Velocity	m/sec	cm/sec	ft/sec	ft/sec	ft/sec
Acceleration	m/sec^2	cm/sec^2	ft/sec^2	ft/sec^2	ft/sec^2
Pressure	nt/m^2	$dynes/cm^2$	pdl/ft^2	lb/ft^2 or lbf/ft^2	lb/ft^2 or lbf/ft^2
Energy	joule	erg	foot-poundal (ft-pdl)	foot-pound (ft-lb) or foot-pound-force (ft-lbf)	foot-pound (ft-lb) or foot-pound-force (ft-lbf)
Power	watt	ergs per second (erg/sec)	ft-pdl/sec	ft-lb/sec or ft-lbf/sec	ft-lb/sec or ft-lbf/sec

m/sec^2.) The unit of force is called a Newton. By definition, 1 Newton is the force required to accelerate a mass of 1 kilogram at the rate of 1 m/sec^2. In other words, the unit of force is defined in such a way that unit force gives unit acceleration to unit mass.

The same thing holds true in the metric absolute centimeter-gram-second (CGS) system of measurement, where the gram is the unit of mass, the centimeter is the unit of length, the second is the unit of time, and centimeters per second per second (cm/sec^2) is the unit of acceleration. In this system, the unit of force is called a dyne. By definition, 1 dyne is the force required to accelerate a mass of 1 gram at the rate of 1 c/sec^2. Again, force is defined in such a way that unit force gives unit acceleration to unit mass.

The same applies to the British absolute foot-pound-second (FPS) system of measurement, where the pound is the unit of mass, the foot is the unit of length, the second is the unit of time, and feet per second per second (ft/sec^2) is the unit of acceleration. In

this system, the unit of force is called a poundal. By definition, 1 poundal is the amount of force required to give a mass of 1 pound an acceleration of 1 ft/sec^2. Again, force is defined in such a way that unit force gives unit acceleration to unit mass.

Now let us look at a British gravitational system–the foot-pound-second (FPS) gravitational system that we use in the United States for most everyday measurements. The foot is the unit of length, the pound is the unit of mass, the second is the unit of time, and feet per second per second is the unit of acceleration. In this system, the unit of force is called the pound. (Actually, it should be called pound-force, but this usage is rarely followed.) In this system, a force of 1 pound acting upon a mass of 1 pound produces an acceleration of 32 ft/sec^2. Note that unit force does not produce unit acceleration when acting on unit mass; rather unit force produces unit acceleration when acting on unit weight. Since force is defined in gravitational terms, rather than in absolute terms, we say that this is a gravitational system of measurement.

Table 7-2.–Decimal System Prefixes

THESE PREFIXES MAY BE APPLIED
TO ALL SI UNITS

MULTIPLES AND SUBMULTIPLES

	Prefixes	Symbols
$1\ 000\ 000\ 000\ 000 = 10^{12}$	tera (tĕr′á)	T
$1\ 000\ 000\ 000 = 10^{9}$	giga (jǐ′gá)	G
$1\ 000\ 000 = 10^{6}$	mega (mĕg′á)	M*
$1\ 000 = 10^{3}$	kilo (kǐl′ŏ)	k*
$100 = 10^{2}$	hecto (hĕk′tŏ)	h
$10 = 10$	deka (dĕk′á)	da
$0.1 = 10^{-1}$	deci (dĕs′ǐ)	d
$0.01 = 10^{-2}$	centi (sĕn′tǐ)	c*
$0.001 = 10^{-3}$	milli (mǐl′ǐ)	m*
$0.000\ 001 = 10^{-6}$	micro (mī′krŏ)	μ*
$0.000\ 000\ 001 = 10^{-9}$	nano (năn′ŏ)	n
$0.000\ 000\ 000\ 001 = 10^{-12}$	pico (pē′kŏ)	p
$0.000\ 000\ 000\ 000\ 001 = 10^{-15}$	femto (fĕm′tŏ)	f
$0.000\ 000\ 000\ 000\ 000\ 001 = 10^{-18}$	atto (ăt′tŏ)	a

* MOST COMMONLY USED

The gravitational system that is usually called the British Engineering System also uses the pound (or more precisely, the pound-force) as the unit of force. But this system has its own unit of mass–the slug. By definition, 1 slug is the quantity of mass that is accelerated at the rate of 1 ft/sec² when acted on by a force of 1 pound. In other words, 1 slug equals 32 pounds, 2 slugs equal 64 pounds, and so forth. By using the slug as the unit of mass, the British Engineering System sets up consistent units of measurement in which unit force acting upon unit mass produces unit acceleration. Note however, that this is still a gravitational system rather than an absolute system.

By this time it should be obvious that the relationships expressed in one system of measurement do not necessarily hold when a different system is used. When using any particular system, you must understand the precise meaning of all terms used in that system. This is not always a simple matter, since there is an enormous number of possible combinations of units, and, in many cases, the same word is used to express quite different ideas. Another source of confusion is the way in which the various systems of measurement are used. In everyday life, we use the British gravitational FPS system. In scientific work, we use one of the metric systems. In engineering and other technical fields, we use a British system or a metric system, depending upon the field involved. The only way to avoid total confusion in the use of measurement terms is to make sure that you understand the precise meaning of each term, as it relates to the particular system being used.

The units shown in table 7-1 are only a few of the units that may be derived in each system. For example, the unit of pressure shown for both the British gravitational systems is pounds per square foot (lb/ft^2 or psf). However, the unit pounds per square inch (lb/in^2 or psi) is equally acceptable and is very commonly used. Similar conversions can be made for any of the other units, as long as the basic relationships of the system are accurately maintained.

When converting values from a metric system to a British system (or vice versa), you must understand the units used in each system. Most of us know quite a bit about the units commonly used in British systems, but less about the units used in metric systems. The following paragraphs describe the basic structure of the metric system.

All metric systems of measurement are decimal systems–that is, the size of the units vary by multiples of 10. This makes computation very simple. Another handy thing about the metric systems is that the prefixes for the names of the units tell you the relative size of the units. Take the prefix kilo-, for example. Kilo indicates 1000; so a kilogram is 1000 grams, a kilometer is 1000 meters, and so forth. Or take the prefix milli-, for another example; it indicates thousandth. So 1 millimeter is a thousandth of a meter, 1 milligram is a thousandth of a gram, and so forth. The best way to become familiar with the units in the metric systems is to associate the more commonly used prefixes with the positive and negative powers of 10, as shown in table 7-2.

Table 7-3 gives selected values for mechanical units in British systems of measurement. Table 7-4 gives selected values for mechanical units in metric systems of measurement. Table 7-5 gives some British-metric and metric-British equivalents. The examples given in these tables are chosen primarily to help you develop an

Table 7-3.–Selected Values of Mechanical Units in British Systems of Measurement

TYPE OF MECHANICAL UNIT	SELECTED VALUES
LENGTH	12 inches (in.) = 1 foot (ft) 3 ft = 1 yard (yd) 5280 ft = 1 mile (mi) 1760 yd = 1 mi
AREA	144 square inches = 1 square foot (sq in. or in.^{2}) (sq ft or ft^2) 9 ft^2 = 1 yd^2
VOLUME	1728 cubic inches = 1 cubic foot (cu in. or in.^{3}) (cu ft or ft^3) 27 ft^3 = 1 yd^3
FORCE (WEIGHT)	16 ounces (oz) = 1 pound (lb) 2000 lb = 1 ton
VELOCITY	60 miles per hour = 88 feet per second (ft/sec) (mi/hr or mph)

Table 7-4.–Selected Values of Mechanical Units in Metric Systems of Measurement

TYPE OF MECHANICAL UNIT	SELECTED VALUES
LENGTH	10 millimeters (mm) = 1 centimeter (cm) 10 cm = 1 decimeter (dm) 10 dm = 1 meter (m) 100 cm = 1 meter 1000 mm = 1 meter 10 m = 1 dekameter (dkm) 100 m = 1 hectometer (hm) 1000 m = 1 kilometer (km)
AREA	100 square millimeters (mm^2) = $1\ cm^2$ $100\ cm^2$ = $1\ dm^2$ $100\ dm^2$ = $1\ m^2$
VOLUME	1000 cubic millimeters (mm^3) = $1\ cm^3$ $1000\ cm^3$ = $1\ dm^3$ $1000\ dm^3$ = $1\ m^3$
	1 milliliter (ml) = $1\ cm^3$ 1000 ml = 1 liter (l) 100 centiliters (cl) = 1 liter 10 deciliters (dl) = 1 liter
MASS	1000 milligrams (mg) = 1 gram (g) 100 centigrams (cg) = 1 gram 1000 grams = 1 kilogram (kg)

understanding of the relative sizes of the mechanical units. More complete tables are available in many physics and engineering textbooks and handbooks.

STANDARDS OF MEASUREMENT

The importance of having precise and uniform standards of measurement is recognized by all the major countries of the world, and international conferences on weights and measures are held from time to time. The International Bureau of Weights and Measures is in France. Each major country has its own bureau or office charged with the responsibility of maintaining the required measurement standards, including the basic standards of length, mass, and time. In the United States, the National Bureau of Standards (NBS) is responsible for maintaining basic standards and for prescribing precise measuring techniques.

Length

Previously, the international standard of length was a platinum-iridium alloy bar kept at the International Bureau of Weights and Measures in France. By definition, the standard meter was the distance between two parallel lines marked on this bar, measured at 0°C. Copies of this international standard were maintained by other countries; the United States standard meter bar was maintained at the NBS in Washington, D.C.

Note that the standard of length was the meter even for countries that were not on the metric system. The yard was defined in terms of the meter, 1 yard being equal to 0.9144 meter.

Since 1960, the standard of length by international agreement is an atomic constant: the wavelength of the orange-red light emitted by individual atoms of krypton-86 in a tube filled with krypton gas in which an electrical discharge is maintained. By definition, 1 meter is equal to 1,650,763.73 wavelengths of the orange-red

Table 7-5.–Selected British-Metric and Metric-British Conversions

TYPE OF MECHANICAL UNIT	BRITISH-METRIC CONVERSIONS	METRIC-BRITISH CONVERSIONS
LENGTH	1 inch = 2.540 centimeters 1 foot = 0.3048 meter 1 yard = 0.9144 meter 1 mile = 1.6093 kilometers 1 mile = 1609.3 meters	1 centimeter = 0.3937 inch 1 meter = 39.37 inches 1 kilometer = 0.62137 mile
AREA	$1\ in.^2$ = $6.452\ cm^2$ $1\ ft^2$ = $929\ cm^2$ $1\ yd^2$ = $0.8361\ m^2$ $1\ mi^2$ = $2.59\ km^2$	$100\ mm^2$ = $0.15499\ in.^2$ $100\ cm^2$ = $15.499\ in.^2$ $100\ m^2$ = $119.6\ yd^2$ $1\ km^2$ = $0.386\ mi^2$
VOLUME	$1\ in.^3$ = $16.387\ cm^3$ $1\ ft^3$ = $0.0283\ m^3$ $1\ yd^3$ = $0.7646\ m^3$ $231\ in.^3$ = 3.7853 liters	$1000\ mm^3$ = $0.06102\ in.^3$ $1000\ cm^3$ = $61.02\ in.^3$ $1\ m^3$ = $35.314\ ft^3$ 1 liter = 1.0567 liquid quarts
WEIGHT	1 grain = 0.0648 gram 1 ounce = 23.3495 grams 1 pound = 453.592 grams 1 pound = 0.4536 kilograms	1 gram = 15.4324 grains 1 gram = 0.03527 ounce 1 gram = 0.002205 pound 1 kilogram = 2.2046 pounds

light of krypton-86, and 1 inch is equal to 41,929.399 wavelengths. A device called an optical interferometer is used to determine the number of the wavelengths of the orange-red light of krypton-86 in an unknown length.

Mass

The standard of mass is the mass of a cylinder of platinum-iridium alloy defined as having a mass of 1 kilogram. The standard kilogram mass is kept at the International Bureau of Weights and Measures in France. The United States standard kilogram mass is kept at the NBS.

The standard of mass is kept in a vault. No more than once a year, the standard is removed from the vault and used for checking the values of smaller standards. The United States standard kilogram mass is taken to France every few years for comparison with the international standard. Every precaution is taken to keep the kilogram mass in perfect condition, free of nicks,

scratches, and corrosion. The standard is always handled with forceps; it is never touched by human hands.

When the national standard is compared on a precision balance with high precision copies, the copies are found to be accurate to within 1 part in 100 million.

Time

Originally, the standard of time was the mean solar second–that is 1/86,400 of a mean solar day, as determined by successive appearances of the sun overhead, averaged over a year. In 1960, the standard of time was changed to the tropical year 1900, which is the time it took the sun to move from a designated point back to the same point in the year 1900. By definition, 1 second was equal to 1/31,556,925.9747 of the tropical year 1900. The subdivision of the tropical year 1900 into smaller time intervals was accomplished by means of laboratory-type pendulum clocks, together with observation of natural phenomena such as the nightly movement of the star and the moon.

ce 1967, the basic standard of time is the time
l for the transition between two energy states of
um-133 atom. According to this standard, 1
is defined as 9,192,631,770 cycles of this
particular transition in the cesium-133 atom.

The United States standard of time is maintained by
a cesium clock that is kept at the NBS laboratories in
Boulder, Colorado. The time signals operated by the
NBS are based on this cesium clock.

MEASUREMENT OF TEMPERATURE

Temperature is measured by bringing a measuring
system (such as a thermometer) into contact with the
system in which we need to measure the temperature.
We then measure some property of the measuring
system–the expansion of a liquid, the pressure of a gas,
electromotive force, electrical resistance, or some other
mechanical, electrical, or optical property that has a
definite and known relationship with temperature. Thus,
we infer the temperature of the measured system by the
measurement of some property of the measuring system.

But the measurement of a property other than
temperature will take us only so far in using the
measurement of temperature. For convenience in
comparing temperatures and in noting changes in
temperature, we must be able to assign a numerical value
to any given temperature. For this we need temperature
scales.

Originally, temperature scales were constructed
around the boiling and the freezing points of pure water
at atmospheric pressure. These two fixed and
reproducible points were used to define a fairly large
temperature interval that was then subdivided into the
uniform smaller intervals called degrees. The two most
familiar temperature scales constructed in this manner
are the Celsius scale and the Fahrenheit scale.

The Celsius scale is often called the centigrade scale
in the United States and Great Britain. By international
agreement, however, the name was changed from
centigrade to Celsius in honor of the eighteenth-century
Swedish astronomer, Anders Celsius. The symbol for a
degree on this scale (no matter whether it is called
Celsius or centigrade) is °C. The Celsius scale takes 0°C
as the freezing point and the 100°C as the boiling point
of pure water at atmospheric pressure. The Fahrenheit
scale takes 32°F as the freezing point and 212°F as the
boiling point of pure water at atmospheric pressure. The
interval between the freezing and boiling points is
divided into 100 degrees on the Celsius and divided into
180 degrees on the Fahrenheit scale.

Since the actual value of the interval between the
freezing and boiling points is identical, the numerical
readings on the Celsius and Fahrenheit thermometers
have no absolute significance and the size of the degree
is arbitrarily chosen for each scale. The relationship
between degrees Celsius and degrees Fahrenheit is
given by the following formulas:

$$°F = (9/5°C) + 32$$

$$°C = 5/9 (°F - 32)$$

Many people have trouble remembering these
formulas, with the result that they may either get them
mixed up or have to look them up in a book every time
a conversion is necessary. If you concentrate on trying
to remember the basic relationships given by these
formulas, you may find it easier to make conversions.
These essential points to remember are as follows:

1. Celsius degrees are larger than Fahrenheit
degrees. One Celsius degree is equal to 1.8 Fahrenheit
degrees, and each Fahrenheit degree is only 5/9 of a
Celsius degree.

2. The zero point on the Celsius scale represents
exactly the same temperature as the 32-degree point on
the Fahrenheit scale.

3. The temperatures 100°C and 212°F are
identical.

In some scientific and engineering work,
particularly where heat calculations are involved, an
absolute temperature scale is used. The zero point on an
absolute temperature scale is the point called absolute
zero. Absolute zero is determined theoretically, rather
than by actual measurement. Since the pressure of a gas
at constant volume is directly proportional to the
temperature, we assume that the pressure of a gas is a
valid measure of its temperature. On this assumption,
the lowest possible temperature (absolute zero) is
defined as the temperature at which the pressure of a gas
would be zero.

Two absolute temperature scales have been in use
for many years. The Rankine absolute scale is an
extension of the Fahrenheit scale; it is sometimes called
the Fahrenheit absolute scale. Degrees on the Rankine
scale are the same size as degrees on the Fahrenheit
scale, but the zero is at -459.67°F. In other words,
absolute zero is zero on the Rankine scale and -459.67°
on the Fahrenheit scale.

A second absolute scale, the Kelvin, is more widely
used than the Rankine. The Kelvin scale was originally
conceived as an extension of the Celsius scale, with

degrees of the same size but with the zero point shifted to absolute zero. Absolute zero on the Celsius scale is -273.15°.

In 1954, a new international absolute scale was developed. The new scale was based upon one fixed point, rather than two. The one fixed point was the triple point of water–that is, the point at which all three phases of water (solid, liquid, and vapor) can exist together in equilibrium. The triple point of water, which is 0.01°C above the freezing point of water, was chosen because it can be reproduced with much greater accuracy than either the freezing point or the boiling point. On this new scale, the triple point was given the value 273.16K. Note that neither the word *degrees* nor the symbol ° is used; instead, the units are referred to as Kelvin and the symbol K is used rather than the symbol °K.

In 1960 the triple point of water was finally adopted as the fundamental reference for this temperature scale.

The scale now in use is the International Practical Temperature Scale of 1968 (IPTS-68). However, you often see this scale referred to as the Kelvin scale.

Although the triple point of water is considered the basic or the fundamental reference for IPTS-68, five other fixed points are used to help define the scale. These are the freezing point of gold, the freezing point of silver, the boiling point of sulfur, the boiling point of water, and the boiling point of oxygen.

Figure 7-1 is a comparison of the Kelvin, Celsius, Fahrenheit, and Rankine temperatures. All of the temperature points listed above absolute zero are considered as fixed points on the Kelvin scale except for the freezing point of water. The other scales are based on the freezing and boiling points of water.

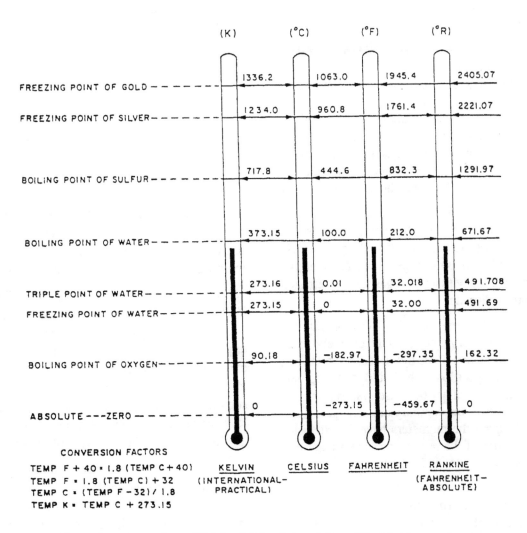

Figure 7-1.–Comparison of Kelvin, Celsius, Fahrenheit, and Rankine temperatures.

TEMPERATURE MEASURING DEVICES

Since temperature is one of the basic engineering variables, temperature measurement is essential to the proper operation of a shipboard engineering plant. The temperature of steam, water, lubricating oil, and other vital fluids must be monitored constantly and the results of this measurement entered in engineering records and logs at frequent intervals.

Devices used for measuring temperature may be classified in various ways. In this chapter, we will discuss two major categories–expansion thermometers and pyrometers.

Expansion Thermometers

Expansion thermometers operate on the principle that the expansion of solids, liquids, and gases has a known relationship to temperature changes. The types of expansion thermometers discussed here are liquid-in-glass thermometers, bimetallic expansion thermometers, and filled-system expansion thermometers.

LIQUID-IN-GLASS-THERMOMETERS.– Liquid-in-glass thermometers are probably the oldest, simplest, and most widely used devices for measuring temperature. A liquid-in-glass thermometer (fig. 7-2) has a bulb and a very fine bore capillary tube containing alcohol, or some other liquid that expands uniformly as the temperature rises or contracts uniformly as the temperature falls. The selection of liquid is based on the temperature range in which the thermometer is to be used.

Almost all liquid-in-glass thermometers are sealed so atmospheric pressure does not affect the reading. The space above the liquid in this type of thermometer may be a vacuum or filled with an inert gas such as nitrogen, argon, or carbon dioxide.

The capillary bore may be either round or elliptical. In any case, it is very small, so a relatively small expansion or contraction of the liquid causes a relatively large change in the position of the liquid in the capillary tube. Although the capillary bore itself is very small in diameter, the walls of the capillary tube are quite thick. Most liquid-in-glass thermometers have an expansion chamber at the top of the bore to provide a margin of safety for the instrument if it should accidentally be overheated.

Liquid-in-glass thermometers may have graduations etched directly on the glass stem or placed on a separate strip of material located behind the stem.

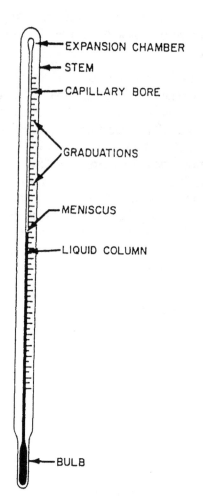

Figure 7-2.–Liquid-in-glass thermometer.

Many thermometers used in shipboard engineering plants have the graduations marked on a separate strip; this type is generally easier to read than the type that has the graduations marked directly on the stem.

BIMETALLIC EXPANSION THERMOMETER.–Bimetallic expansion thermometers make use of different metals having different coefficients of linear expansion. The essential element in a bimetallic expansion thermometer is a bimetallic strip consisting of two layers of different metals fused together. When such a strip is subjected to temperature changes, one layer expands or contracts more than the other, thus tending to change the curvature of the strip.

Figure 7-3 shows the basic principle of a bimetallic expansion thermometer. When one end of a straight bimetallic strip is fixed in place, the other end tends to curve away from the side that has the greater coefficient of linear expansion when the strip is heated.

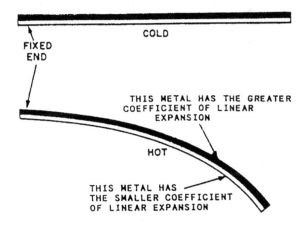

Figure 7-3.–Effect of unequal expansion of a bimetallic strip.

For use in thermometers, the bimetallic strip is normally wound into a spiral (fig. 7-4), a single helix, or a multiple helix. The end of the strip that is not fixed in position is fastened to the end of a pointer that moves over a circular scale. Bimetallic thermometers are easily adapted for use as recording thermometers; a pen is attached to the pointer and positioned in such a way that it marks on a revolving chart.

FILLED-SYSTEM THERMOMETERS.–In general, filled-system thermometers are designed for use in locations where the indicating part of the instrument must be placed some distance away from the point where the temperature is to be measured. For this reason, they are often called distant-reading thermometers. However, this is not true of all filled-system thermometers. In a few designs, the capillary tubing is extremely short, and in a few, it is nonexistent. In general, however, filled-system thermometers are designed to be distant-reading

thermometers. Some distant-reading thermometers may have capillaries as long as 125 feet.

There are two basic types of filled-system thermometers. One has a Bourdon tube that responds primarily to changes in the volume of the filling fluid; the other is one which the Bourdon tube responds primarily to changes in the pressure of the filling fluid. Clearly, some pressure effect will exist in volumetric thermometers, and some volumetric effect will exist in pressure thermometers.

A distant-reading thermometer (fig. 7-5) consists of a hollow metal sensing bulb at one end of a small-bore capillary tube, which is connected at the other end to a Bourdon tube or other device that responds to volume changes or to pressure changes. The system is partially or completely filled with fluid that expands when heated and contracts when cooled. The fluid may be a gas, an organic liquid, or a combination of liquid and vapor.

The device usually used to indicate temperature changes by its response to volume changes or to pressure changes is called a Bourdon tube. A Bourdon tube is a curved or twisted tube that is open at one end and sealed

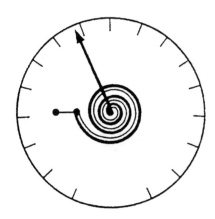

Figure 7-4.–Bimetallic thermometer (flat, spiral strip).

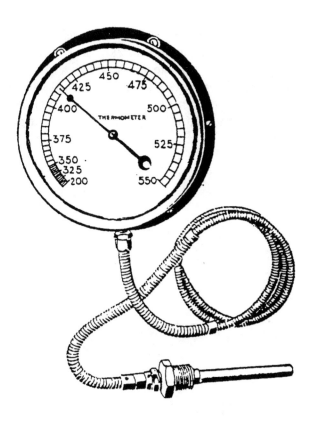

Figure 7-5.–A distant-reading Bourdon-tube thermometer.

at the other end (fig. 7-6). The open end of the tube is fixed in position, and the sealed end is free to move. The tube is more or less elliptical in cross section; it does not form a true circle. It becomes more circular when there is an increase in volume or in internal pressure of the contained fluid; this tends to straighten the tube. Opposing this action, the spring action of the tube metal tends to coil the tube. Since the open end of the Bourdon tube is rigidly fastened, the sealed end moves as the volume or pressure of the contained fluid changes. A pointer is attached to the sealed end of the tube through appropriate linkages; the assembly is placed over an appropriately calibrated dial. The result is a Bourdon-tube gauge that may be used for measuring temperature or pressure, depending upon the design of the gauge and the calibration of the scale.

Bourdon tubes are made in several shapes for various applications. The C-shaped Bourdon tube shown in figure 7-6 is the most commonly used type; spiral and helical Bourdon tubes are used where design requirements include the need for a longer length Bourdon tube.

Pyrometers

The term *pyrometer* is used to include a number of measuring devices that, in general, are suitable for use at relatively high temperatures; some pyrometers, however, are also suitable for use at low temperatures. The types of pyrometers we are concerned with here include thermocouple pyrometers, resistance thermometers, radiation pyrometers, and optical pyrometers.

THERMOCOUPLE PYROMETER.–The operation of a thermocouple pyrometer (sometimes called a thermoelectric pyrometer) is based on the observed fact that an electromotive force is generated when two junctions of two dissimilar metals are at different temperatures. A simple thermocouple is illustrated in figure 7-7. Since the electromotive force generated is proportional to the temperature difference between the measuring junction (hot junction) and the reference junctions (cold junctions), the indicating instrument can be marked off to indicate degrees of temperature even though it is actually measuring emfs. The indicating instrument is a millivoltmeter or some other electrical device capable of measuring and indicating small direct-current emfs. The strips or wires of dissimilar metals are welded, twisted, fused, or otherwise firmly joined together. The extension leads are usually of the same metals as the thermocouple itself.

RESISTANCE THERMOMETERS.–Resistance thermometers are based on the principle that the electrical resistance of a metal changes with changes in temperature. A resistance thermometer is, thus, actually an instrument that measures electrical resistance but is calibrated in degrees of temperature rather than in units of electrical resistance.

The sensitive element in a resistance thermometer is a winding of small diameter nickel, platinum, or other metallic wire. The resistance winding is located in the lower end of a bulb (sometimes called a stem); it is electrically but not thermally insulated from the stem. The resistance winding is connected by two, three, or four leads to the circuit of the indicating instrument. The circuit is a Wheatstone bridge or some other simple circuit that contains known resistances with which the resistance of the thermometer winding is compared.

RADIATION AND OPTICAL PYROM-ETERS.–Radiation and optical pyrometers are used to measure very high temperatures. Both types of pyrometers measure temperature by measuring the amount of energy radiated by the hot object. The main

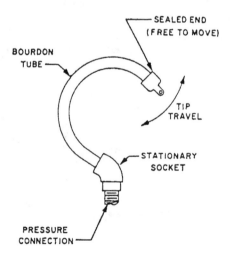

Figure 7-6.–C-shaped Bourdon tube.

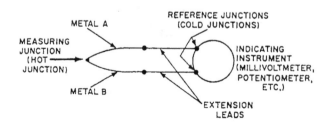

Figure 7-7.–A simple thermocouple.

difference between the two types is in their range of sensitivity; radiation pyrometers are (theoretically) sensitive to the entire spectrum of radiant energy, while optical pyrometers are sensitive to only one wavelength or to a very narrow band of wavelengths.

Figure 7-8 shows schematically the general operating principle of a simple radiation pyrometer. Radiant energy from the hot object is concentrated on the detecting device by a lens or, in some cases, a conical mirror or a combination of mirror and lens. The detecting device may be a thermocouple, a thermopile (a group of thermocouples in series), a photocell, or other element in which some electrical quantity (emf, resistance, and so on) varies as the temperature of the hot object varies. The meter or indicated part of the instrument may be a millivoltmeter or some similar device.

An optical pyrometer measures temperature by comparing visible light emitted by the hot object with the light from a standard source. This instrument consists of an eyepiece, a telescope that contains a filament similar to the filament of an electric light bulb, and a potentiometer.

The person operating the optical pyrometer looks through the eyepiece and focuses the telescope on the hot object, meanwhile also observing the tin glowing filament across the field of the telescope. While watching the hot object and the filament, the operator adjusts the filament current (and consequently the brightness of the filament) by turning a knob on the potentiometer until the filament seems to disappear and to merge with the hot object. When the filament current

has been adjusted so that the filament just matches the hot object in brightness, the operator turns another knob slightly to balance the potentiometer. The potentiometer measures filament current but the dial is calibrated in degrees of temperature. As may be noted from this description, this type of optical pyrometer requires a certain amount of skill and judgment on the part of the operator. In some other types of optical pyrometers, automatic operation is achieved by use of photoelectric cells arranged in a bridge network.

MEASUREMENT OF PRESSURE

Pressure, like temperature, is one of the basic engineering variables and one that must be frequently measured aboard ship. Before discussing the devices that measure pressure, let us consider certain definitions that are important in any discussion of pressure measurement.

PRESSURE DEFINITIONS

Pressure is defined as force per unit area. The simplest pressure units are ones that indicate how much force is applied to an area of a certain size. These units include pounds per square inch (psi), pounds per square feet (psf), ounces per square inch, Newtons per square centimeter, and dynes per square centimeter, depending upon the system you use.

You will also use another kind of pressure unit that involves length. These units include inches of water (in.H_2O), inches of mercury (in.Hg), and inches of some other liquid of a known density. Actually, these units do not involve length as a fundamental dimension. Rather, length is taken as a measure of force or weight. For example, a reading of 1 in.H_2O means that the exerted pressure is able to support a column of water 1 inch high, or that a column of water in a U-tube would be displaced 1 inch by the pressure being measured. Similarly, a reading of 12 in.Hg means that the measured pressure is sufficient to support a column of mercury 12 inches high. What is really being expressed (even though it is not mentioned in the pressure unit) is the fact that a certain quantity of material (water, mercury, and so on) of known density exerts a certain definite force upon a specified area. Pressure is still force per unit area, even if the pressure unit refers to inches of some liquid.

It is often necessary to convert from one type of pressure to another. Complete conversion tables may be found in many texts and handbooks. Conversion factors for pounds per square inch, inches of mercury, and inches of water are as follows:

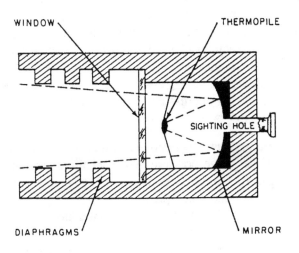

Figure 7-8.–A simple radiation pyrometer.

$$1 \text{ in.Hg} = 0.49 \text{ psi}$$
$$1 \text{ psi} = 2.036 \text{ in.Hg}$$
$$1 \text{ in.H}_2\text{O} = 0.036 \text{ psi}$$
$$1 \text{ psi} = 27.68 \text{ in.H}_2\text{O}$$
$$1 \text{ in.H}_2\text{O} = 0.074 \text{ in.Hg}$$
$$1 \text{ in.Hg} = 13.6 \text{ in.H}_2\text{O}$$

In interpreting pressure measurements, a great deal of confusion arises because the zero point on most pressure gauges represents atmospheric pressure rather that zero absolute pressure. Thus, it is often necessary to specify the kind of pressure being measured under any given conditions. To clarify the numerous meanings of the word *pressure*, the relationships among gauge pressure, atmospheric pressure, vacuum, and absolute pressure are shown in figure 7-9.

GAUGE PRESSURE is the pressure actually shown on the dial of a gauge that registers pressure at or above atmospheric pressure. An ordinary pressure gauge reading of zero does not mean that there is no pressure in the absolute sense; rather, it means that there is no pressure in excess of atmospheric pressure.

ATMOSPHERIC PRESSURE is the pressure exerted by the weight of the atmosphere. At sea level, the average pressure of the atmosphere is sufficient to hold a column of mercury at the height of 76.0 millimeters or 29.92 in.Hg. Since a column of mercury

1 inch high exerts a pressure of 0.49 psi, a column of mercury 29.92 inches high exerts a pressure that is equal to 29.92 x 0.49, or approximately 14.7 psi. Since we are dealing now in absolute pressure, we say that the average atmospheric pressure at sea level is 14.7 pounds per square inch absolute (psia). It is zero on the ordinary pressure gauge.

Notice, however, that the figure of 14.7 psia represents the average atmospheric pressure at sea level; it does not always represent the actual pressure being exerted by the atmosphere at the moment that a gauge is being read.

BAROMETRIC PRESSURE is the actual atmospheric pressures that exists at any given moment. Barometric pressure may be measured by a simple mercury column or by a specially designed instrument called an aneroid barometer.

A space in which the pressure is less than atmospheric pressure is said to be under partial vacuum. The amount of vacuum is expressed in terms of the difference between the absolute pressure in the space and the pressure of the atmosphere. Most commonly, vacuum is expressed in in.Hg with the vacuum gauge scale marked from 0 to 30 in.Hg. When a vacuum gauge reads zero, the pressure in the space is the same as atmospheric pressure or, in other words, there is no vacuum. A vacuum gauge reading of 29.92 in.Hg would indicate a nearly perfect vacuum. In actual practice, it is impossible to obtain a perfect vacuum even under laboratory conditions. A reading between 0 and 29.92 is a partial vacuum.

Absolute pressure is atmospheric pressure plus gauge pressure or minus vacuum. For example, a gauge pressure of 300 psig equals an absolute pressure of 314.7 psia (300 + 14.7). Or, for example, consider a space in which the measured vacuum is 10 in.Hg; the absolute pressure in this space is figured by subtracting the measured vacuum (10 in.Hg) from the nearly perfect vacuum. The absolute pressure then will be 19.92 or approximately 20 in.Hg absolute. It is important to note that the amount of pressure in a space under vacuum can only be expressed in terms of absolute pressure.

You may have noticed that sometimes we use psig to indicate gauge pressure and other times we merely use psi. By common convention, gauge pressure is always assumed when pressure is given in psi, psf, or similar units. The letter *g* (for gauge) is added only when there is some possibility of confusion. Absolute pressure, on the other hand, is always expressed as psia, pounds per square foot absolute (psfa), and so forth. It

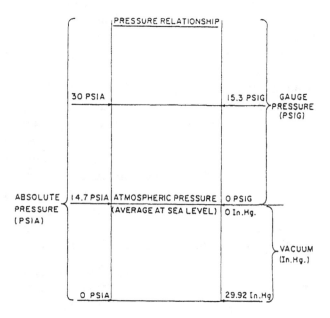

Figure 7-9.–Relationships among gauge pressure, atmospheric pressure, vacuum, and absolute pressure.

is always necessary to establish clearly just what kind of pressure we are talking about, unless this is very clear from the nature of the discussion.

To this point, we have considered only the most basic and most common units of measurement. It is important to remember that hundreds of other units can be derived from these units; remember also that specialized fields require specialized units of measurement. Additional units of measurement are introduced in appropriate places throughout the remainder of this training manual. When you have more complicated units of measurement, you may find it helpful to first review the basic information given here.

PRESSURE-MEASURING DEVICES

Operating pressures of shipboard machinery are constantly monitored. As with temperature readings, pressure readings provide you with an indication of the operating condition of the equipment. The pressure-measuring instruments discussed in this chapter are of two types–mechanical and electrical.

Mechanical Pressure-Measuring Instruments

Two of the primary mechanical pressure-measuring instruments found aboard ship are the Bourdon tube and the bellows elements. Bourdon-tube elements are suitable for the measurement of very high pressures up to 100,000 psig. The upper limits for bellows elements is about 800 psig.

BOURDON-TUBE ELEMENTS.–Bourdon-tube elements used in pressure gauges are essentially the same as those described for use in filled-system thermometers. Bourdon tubes for pressure gauges are made of brass, phosphor bronze, stainless steel, beryllium-copper, or other metals, depending upon the requirements of service.

Bourdon-tube pressure gauges are often classified as simplex or duplex, depending upon whether they measure one pressure or two. A simplex gauge has only one Bourdon tube and measures only one pressure. Figure 7-10 shows a simplex Bourdon-tube pressure gauge with views of the dial and gear operating mechanisms. The pointer marked RED HAND in view A is a manually positioned hand that is set to the normal working pressure of the machinery or equipment on which the gauge is installed; the hand marked POINTER is the only hand that moves in response to pressure changes.

When two Bourdon tubes are mounted in a single case, with each mechanism acting independently but with two pointers mounted on a common dial, the

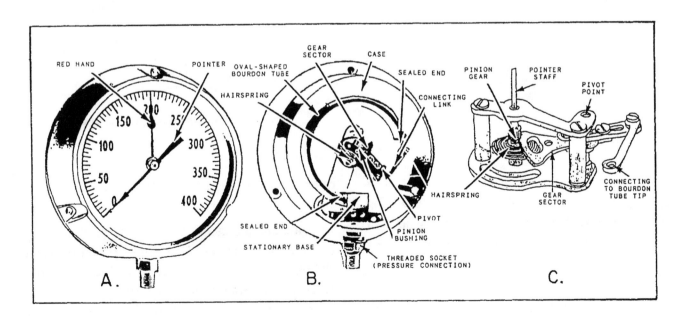

Figure 7-10.–Simplex Bourdon-tube pressure gauge.

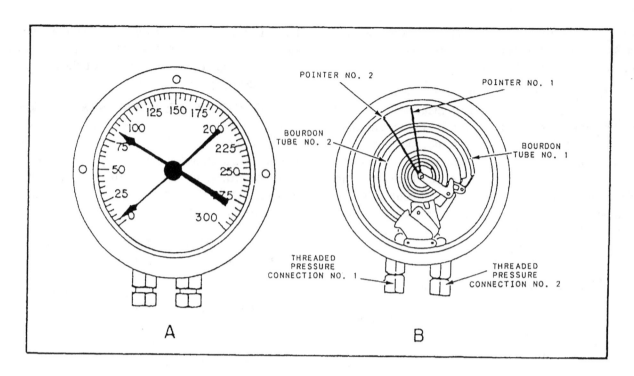

Figure 7-11.–Duplex Bourdon-tube pressure gauge.

Figure 7-12.–Bourdon-tube vacuum gauge.

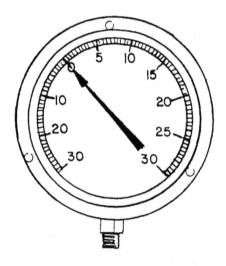

Figure 7-13.–Compound Bourdon-tube gauge.

assembly is called a duplex gauge. Figure 7-11 shows a duplex gauge with views of the dial and operating mechanism. Note that each Bourdon tube has its own pressure connection and its own pointer. Duplex gauges are used to give simultaneous indication of the pressure at two different locations.

Bourdon-tube vacuum gauges are marked off in in.Hg, as shown in figure 7-12. When a gauge is designed to measure both vacuum and pressure, it is

called a compound gauge; it is marked off both in in.Hg and psig, as shown in figure 7-13.

Differential pressure may also be measured with Bourdon-tube gauges. One kind of Bourdon-tube differential pressure gauge is shown in figure 7-14. This gauge has two Bourdon tubes but only one pointer. The Bourdon tubes are connected in such a way that they are the pressure difference, rather than either of the two actual pressures indicated by the pointer.

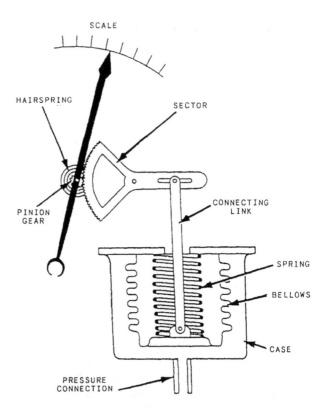

Figure 7-15.–Simple bellows gauge.

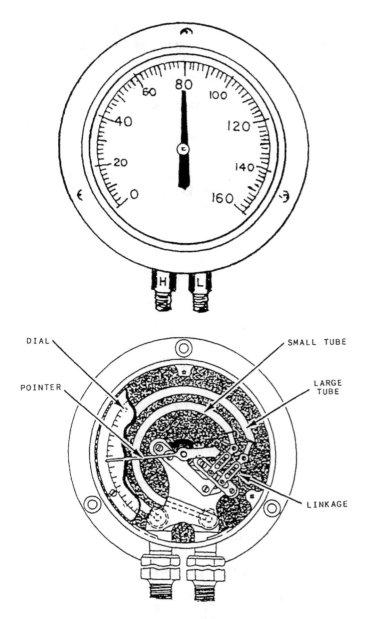

Figure 7-14.–Bourdon-tube differential pressure gauge.

BELLOWS ELEMENT.–A bellows elastic element is a convoluted unit that expands and contracts axially with changes in pressure. The pressure to be measured can be applied to the outside or to the inside of the bellows. In practice, most bellows measuring devices have the pressure applied to the outside of the bellows, as shown in figure 7-15. Like Bourdon-tube elements, bellows elastic elements are made of brass, phosphor bronze, stainless steel, beryllium-copper, or other metal suitable for the intended service of the gauge.

Most bellows gauges are spring-loaded–that is, a spring opposes the bellows and thus prevents full expansion of the bellows. Limiting the expansion of the bellows in this way protects the bellows and prolongs its life. In a spring-loaded bellows element, the deflection is the result of the force acting on the bellows and the opposing force of the spring.

Although some bellows instruments can be designed for measuring pressures up to 800 psig, their primary application aboard ship is in the measurement of quite low pressures or small pressure differentials.

Many differential pressure gauges are of the bellows type. In some designs, one pressure is applied to the inside of the bellows and the other pressure is applied to the outside. In other designs, a differential pressure reading is obtained by opposing two bellows in a single case.

Bellows elements are used in various applications where the pressure-sensitive device must be powerful enough to operate not only the indicating pointer but also some type of recording device.

Electrical Pressure-Measuring Instruments (Transducers)

Transducers are devices that receive energy from one system and retransmit it to another system. The energy retransmitted is often in a different form than that received. A pressure transducer receives energy in the form of pressure and retransmits energy in the form of electrical current.

Transducers allow monitoring at remote locations in some propulsion plants. Mechanical gauges provide pressure readings at the machinery locations or on gauge panels in the immediate area. At a central location, remote readings are provided by using transducers in conjunction with signal conditioners. Transducers provide the capability of sensing variable pressures and transmitting them in proportional electrical signals. Pressure transducers are widely used in ship propulsion and auxiliary spaces. They can also be used to monitor alarms and machinery operation.

Pressure transducers are generally designed to sense absolute, gauge, or differential pressure. The typical unit (fig. 7-16) receives pressure through the pressure ports. It transmits an electrical signal, proportional to the pressure input, through the electrical connector. Pressure transducers are available in pressure ranges from 0 to 6 inches water differential to 0 to 10,000 psig. Regardless of the pressure range of a specified unit, the electrical output is always the same. The electrical signal is conditioned by the signal conditioners before being displayed on an analog meter or a digital readout located on a centrally located console.

PRESSURE GAUGE INSTALLATION

Bourdon-tube pressure gauges used for steam service are always installed in such a way that steam cannot actually enter the gauge. This type of installation is necessary to protect the Bourdon tube from very high

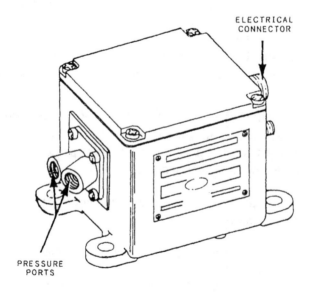

Figure 7-16.–Pressure transducer assembly.

temperatures. An exposed uninsulated coil is provided in the line leading to the gauge, and the steam condenses into water in this exposed coil. Thus, there is always a condensate seal between the gauge and the steam line.

Pressure gauge connections are normally made to the top of the pressure line or the highest point on the machinery in which the pressure is to be measured. Pressure gauges are usually mounted on flat-surfaced gauge boards in such a way as to minimize vibration; this is a matter of considerable importance since ships experience great structural vibrations from screws and machinery. Some gauges are designed to withstand vibrations that may be normally expected from machinery. Pressure gauges designed to withstand shock and vibration frequently use small size capillary tubing between the connection and the elastic elements to protect the gauge mechanism and the pointer; small size tubing is used between the test connection or gauge valve and the gauge so that piping deflections will not cause errors in the gauge readings.

MEASUREMENT OF FLUID FLOW

A great many devices, many of them quite ingenious, are used in the measurement of fluid flow. The discussion here is concerned primarily with the types of fluid flow measuring devices that find relatively wide application in shipboard engineering. These devices may be classified as positive-displacement meters, head meters, and area meters.

POSITIVE-DISPLACEMENT METERS

Positive-displacement meters are used for measuring fluid flow. In a meter of this type, each cycle or complete revolution of a measuring element displaces a definite, fixed volume of liquid. Measuring elements used in positive-displacement meters include disks, pistons, lobes, vanes, and impellers. The motion of these devices may be classified as reciprocating, rotating, oscillating, or nutating, depending upon the type of measuring element used and the general design of the meter readout or register.

A positive-displacement meter of the nutating-piston type is shown in figure 7-17. The flow of oil through the meter causes the piston (also called a disk) to move with a nutating motion. Understanding the nature of this motion is the key to understanding the operation of the meter. A nutating motion might be described as a "rocking around" motion; it is similar to the motion of a spun coin just before the coin settles on its side. Since the piston is nutating on a lower spherical

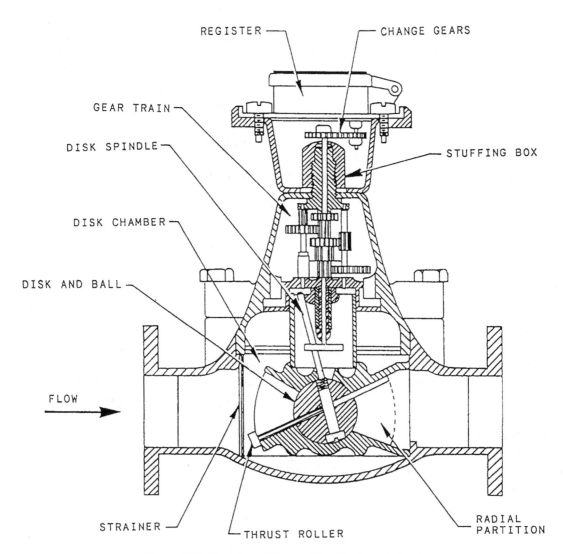

REGISTER —

— CHANGE GEARS

GEAR TRAIN —

DISK SPINDLE —

— STUFFING BOX

DISK CHAMBER —

DISK AND BALL —

FLOW

STRAINER —

— THRUST ROLLER

RADIAL
PARTITION

Figure 7-17.–Nutating-piston positive-displacement meter.

bearing surface, the piston in the meter cannot settle on its side like a spun coin.

The piston cannot rotate because it is held in place by a fixed vane or guide that runs vertically through a slot in the piston. However, the nutating motion of the piston imparts a rotary motion to the pin that projects from the upper spherical surface. The rotary movement of the pin rotates the gears, and the movement of the gears actuates a counting device or register at the top of the meter.

Although the action of the nutating piston is smooth and continuous, there is nevertheless a definite cycle involved in the measurement of liquid flow through this meter. The nutating action of the piston seals the measuring chamber off into separate compartments, and these compartments are alternately filled and emptied. Since each compartment holds a definite volume of the

liquid, the meter is properly classed as a positive-displacement meter.

Totalizing meters have a readout in gallons or pounds of liquid; however, they may also indicate rate of flow in gallons per minute (gpm) or in other flow units.

HEAD METERS

Head meters measure fluid flow by measuring the pressure differential across a specially designed restriction in the flow line. The restriction may be an orifice plate, a flow nozzle, a venturi tube, an elbow, a pitot tube, or some similar device. As the fluid flows toward the restriction, the velocity decreases and the pressure (or head) increases; as the fluid flows through the restriction, the velocity increases and the pressure decreases. Figure 7-18 shows the pressure changes that occur as a fluid flows through a line that contains an orifice plate or similar restriction. Note that there is a slight increase in pressure just ahead of the restriction and then a sudden drop in pressure at the restriction. The point of minimum pressure and maximum velocity is slightly downstream from the restriction; this point is called the vena contracta. Beyond the vena contracta, the velocity decreases and the pressure increases until eventual normal flow is reestablished.

The flow nozzle, orifice plate, or other restriction in the line is called the primary element of the head meter. A high-pressure tap upstream from the restriction and a low-pressure tap downstream from it are connected to a differential bellows, a diaphragm, or some other device for measuring differential pressure.

The pressure drop occurring in a fluid flowing through a restriction varies as the square of the fluid velocity; or, to put it another way, the square root of the pressure differential is proportional to the rate of fluid flow. Because of the square-root relationship between the pressure differential and the rate of fluid flow, a square-root extracting device is usually included so that the scale can be graduated in even steps or increments. Without a device for extracting the square root of the pressure differential, the scale would have to be unevenly divided, with wider divisions at the top of the scale than at the bottom.

AREA METERS

An area meter indicates the rate of fluid flow with an orifice that is varied in area by variations in the fluid flow. The variations in the area of the orifice are produced by some type of movable device which is positioned by the pressure of the flowing fluid. Since the fluid itself positions the movable device and, thus, varies the area of the orifice, there is no significant pressure drop between the upstream side and the downstream side of the variable orifice. Since there is an essentially linear relationship between the area of the orifice and the rate of flow, there is no need for a square-root extracting device in an area meter.

Most area meters are classified as rotameters or piston-type meters, depending upon the type of device used to vary the area of the orifice.

Figure 7-19 shows a simple rotameter-type area meter used in some shipboard distilling plants. A tapered glass tube is installed vertically, with the smaller end at the bottom. Water flows in at the bottom, upward through the tube, and out at the top. A rod, supported by the end fittings of the tube, is centered in the tube. A movable rotor rides freely on the rod and is positioned by the fluid. Variations in the pressure of the fluid lead to variations in the position of the rotor; and, since the glass tube is tapered, the size of the annular orifice between the rotor and the tube is different at each position of the rotor. An increase in flow is thus indicated by the rotor rising on the rod. The glass tube is so calibrated that the flow may be read directly, with the reading being taken at the top of the rotor.

In a piston-type area meter, a piston is lifted by fluid pressure. As the piston is lifted, it uncovers a port through which the fluid flows. The port area uncovered by the lifting of the piston is directly proportional to the rate of fluid flow. Therefore, the position of the piston

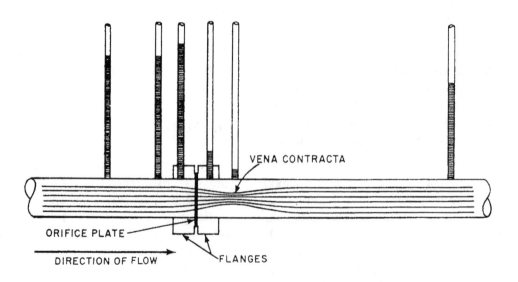

Figure 7-18.–Pressure changes in the fluid flowing through a restriction in the line.

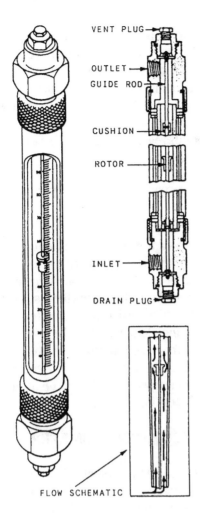

Figure 7-19.–Area meter (rotameter type) for measuring fluid flow.

provides a direct indication of the rate of flow. The means by which the position of the piston is transmitted to an indicating dial varies according to the design of the particular meter and the service for which it is intended.

MEASUREMENT OF LIQUID LEVEL

In the engineering plant aboard ship, it is frequently necessary for operating personnel to know the level of various liquids in various locations. The level of the water in a ship's boiler is a prime example of a liquid level that must be known at all times. However, there are other water levels that are also important–the level of fuel oil in service and storage tanks, the level of water in the DFT, the level of lubricating oil in the oil sumps of main and auxiliary machinery, and drains in various drain tanks, to name a few.

A wide variety of devices, some of them simple and some complex, are available for measuring liquid level. Some measure liquid level quite directly by measuring the height of a column of liquid. Others measure

pressure, volume, or some property of the liquid from which we may then infer liquid level.

The gauge glass is one of the simplest kinds of liquid level measuring devices and one that is commonly used. Gauge glasses are used on boilers, DFTs, in lube-oil storage tanks, and on other shipboard machinery. Basically, a gauge glass is just one leg of a U-tube, with the other leg being the tank, drum, or other vessel in which the liquid level is to be measured. The liquid level in gauge glass is thus the same as the liquid level in the tank or drum, and the reading can be made directly by direct visual observation. Gauge glasses vary in details of construction, depending upon the pressure, temperature, and other service conditions they must withstand.

The measurement of liquid level in tanks aboard ship may be accomplished by simple devices such as direct-reading sounding rules and gauging tapes or by some form of permanently installed, remote-reading gauging system. Remote-reading gauging systems are often referred to as tank level indicators (TLIs). TLIs will tell you the exact amount of liquid in a tank. The scale of the TLIs may be calibrated to indicate level, volume, or weight.

MEASUREMENT OF ROTATIONAL SPEED

The rotational speed of propeller shafts, turbines, generators, blowers, pumps, and other kinds of shipboard machinery is measured with tachometers. For most shipboard machinery, rotational speed is expressed in revolutions per minute (rpm). The tachometers most commonly used aboard ship are the centrifugal type, the chronometric type, and the resonance type. Stroboscopic tachometers are also used occasionally.

Some types of machinery are equipped with permanently mounted tachometers, but portable tachometers are used for checking the rpm of many units. A portable tachometer of the centrifugal type or of the chronometric type is applied manually to a depression or a projection at the center of a moving shaft. Each portable tachometer is supplied with several hard rubber tips; to use the instrument, the operator selects a tip of the proper shape, fits it over the end of the tachometer drive shaft, and holds the tip against the center of the moving shaft. Some tachometers are also supplied with a small wheel that can be fitted to the end of the drive shaft and used to measure the linear velocity (in feet per second) of a wheel or a journal; with this type of instrument, the wheel is held against the outer surface of the moving object. Portable tachometers are used only for intermittent reading, not for continuous operation.

CENTRIFUGAL TACHOMETERS

As the name implies, a centrifugal tachometer uses centrifugal force for its operation. The main parts of a centrifugal tachometer are shown in figure 7-20, and the dial of the instrument is shown in figure 7-21. Centrifugal force acts upon weights or flyballs that are connected by linkage to an upper and a lower collar. The upper collar is fixed to the drive shaft, but the lower collar is free to move up and down the shaft. A spring that fits over the drive shaft connects the upper collar and the lower collar. As the drive shaft begins to rotate, the flyballs spin around with it. Centrifugal force tends to pull the flyballs away from the center, thus raising the lower collar and compressing the spring. The lower collar is connected to the pointer; the upward movement of the collar causes the pointer to move to a higher rpm reading on the dial. The centrifugal tachometer registers rpm of a rotating shaft as long as it is in contact with the shaft. For this reason it is also called a constant-reading tachometer.

CHRONOMETRIC TACHOMETER

A chronometric tachometer (fig. 7-22) is a combination of a watch and a revolution counter that measures the average number of rpm of a rotating shaft. The device is not a constant-reading instrument; the outer drive shaft runs free when the instrument is applied to a rotating shaft until a starting button is depressed to start the timing element. After the drive shaft has been disengaged from the rotating shaft, the pointer remains in position on the dial until it is returned to the zero position by the operation of a reset button (could be the same as the starting button).

STROBOSCOPIC TACHOMETER

A stroboscopic tachometer is a device that allows rotating, reciprocating, or vibrating machinery to be viewed intermittently, under flashing light, in such a way that the movement of the machinery appears to be

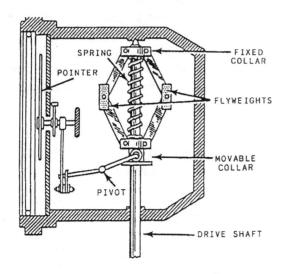

Figure 7-20.–Main parts of a centrifugal tachometer.

Figure 7-21.–Dial of a centrifugal tachometer.

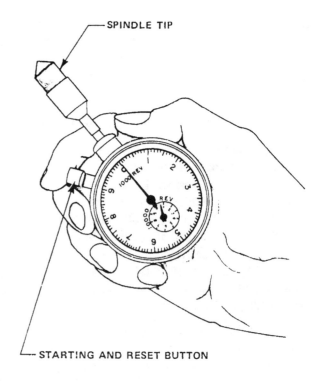

Figure 7-22.–A chronometric tachometer.

slowed, stopped, or reversed. Because the illumination is intermittent, rather than steady, the eye receives a series of views rather than one continuous view.

When the speed of the flashing light coincides with the speed of the moving machinery, the machinery appears to be motionless. This effect occurs because the moving object is seen each time at the same point in its cycle of movement. If the flashing rate is decreased slightly, the machinery appears to be moving slowly in the true direction of movement; if the flashing rate is increased slightly, the machinery appears to be moving in the reverse direction. To measure the speed of a machine, therefore, it is only necessary to find the rate of intermittent illumination to which the machinery appears to be motionless. To observe the operating machinery in slow motion, it is necessary to adjust the stroboscope until the machinery appears to be moving at the desired direction.

The stroboscopic tachometer furnished for shipboard use is a small, portable instrument. It is calibrated so that the speed can be read directly from the control dial. The flashing rate is determined by a self-contained electronic pulse generator that can be adjusted, with the direct-reading dial, to any value between 600 and 14,400 rpm. The relationship between rotational speed and flashing rate may be illustrated by an example. If an electric fan is operating at a rate of 1800 rpm, it will appear to be motionless when it is viewed through a stroboscopic tachometer that is flashing at the rate of 1800 times per minute.

Because the stroboscopic tachometer is never used in direct contact with moving machinery, it is particularly useful for measuring the speed or observing the operation of machinery that is run by a relatively small power input. It is also very useful for measuring the speed of machinery that is installed in inaccessible places.

MEASUREMENT OF SPECIFIC GRAVITY

The specific gravity of a substance is defined as the ratio of the density of the substance to the density of a standard substance. The standard of density for liquids and solids is pure water; for gases, the standard is air. Each standard (water or air) is considered to have a specific gravity of 1.00 under standard conditions of pressure and temperature. (For engineering purposes, the standard pressure and temperature conditions for water as a standard of specific gravity are atmospheric

pressure and 60°F.) For a solid or liquid substance then, we may say that

$$Sp. \ gr. \ of \ soild \ or \ liquid = \frac{density \ of \ substance}{density \ of \ water}$$

Density is sometimes defined as the mass per unit volume of a substance and sometimes as the weight per unit volume. In engineering, fortunately, this difference in definitions rarely causes confusion because we are usually interested in relative densities–or, in other words, specific gravity. Since specific gravity is the ratio of two densities, it really does not matter whether we use mass densities or weight densities; the units cancel out and give us a pure number that is independent of the system of units used.

Aboard ship, it is sometimes necessary to measure the specific gravity of various liquids. This is usually done by using a device called a hydrometer. A hydrometer measures specific gravity by comparing the buoyancy (or loss of weight) of an object in water with the buoyancy of the same object in the liquid being measured. Since the buoyancy of an object is directly related to the density of the liquid, then

$$Sp. \ gr. \ of \ liquid = \frac{buoyant \ force \ of \ liquid}{buoyant \ force \ of \ water}$$

A hydrometer is merely a calibrated rod that is weighted at one end so that it floats in a vertical position in the liquid being measured. Hydrometers are calibrated in such a way that the specific gravity of the liquid may be read directly from the scale; in other words, the comparison between the density of the liquid being measured and the density of water is built in by the calibration of the hydrometer.

For fuel oil and other petroleum products, it is customary to measure degrees API, rather than specific gravity, according to a scale developed by the American Petroleum Institute (API). The relationship between specific gravity and API gravity is given by the following formula:

$$Sp. \ gr. = \frac{141.5}{131.5 + degrees \ API}$$

MEASUREMENT OF VISCOSITY

The viscosity of a liquid is a measure of its resistance to flow. A liquid is said to have high viscosity if it flows sluggishly, like cold molasses. It is said to have low viscosity if it flows freely, like water. The viscosity of most liquids is greatly affected by temperature; in general, liquids are less viscous at higher temperatures. The viscosity of an oil is usually expressed as the number of seconds required for a given amount of oil to flow through an orifice of a specified size when the oil is at a specified temperature. Devices used to measure the rate of flow (and hence the viscosity) are called viscosimeters.

The viscosimeter more commonly used is a Saybolt viscosimeter with two orifices. The larger orifice is called the Saybolt Furol orifice; the smaller one is called the Saybolt universal orifice. The Furol orifice is used for measuring the viscosity of relatively heavy oils; the Universal orifice is used for measuring the viscosity of relatively light oils.

A Saybolt viscosimeter consists of an oil tube, a constant-temperature oil bath that maintains the correct temperature of the sample in the tube, a 60 cc (cubic centimeter) graduated receiving flask, thermometers for measuring the temperature of the oil sample and of the oil bath, and a timing device. A Saybolt viscosimeter is shown in figure 7-23; figure 7-24 shows details of the viscosimeter oil tube.

The oil to be tested is strained and poured into the oil tube. The tube is surrounded by the constant-temperature oil bath. When the oil sample is at the correct temperature, the cork is pulled from the lower end of the tube and the sample flows through the orifice and into the graduated receiving flask. The time (in

Figure 7-23.–A Saybolt viscosimeter.

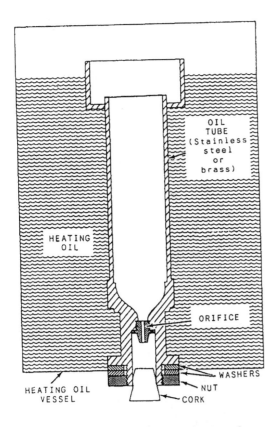

Figure 7-24.–Details of a viscosimeter tube.

seconds) required for the oil to fill the receiving flask to the 60 cc mark is noted.

The viscosity of the oil is expressed by indicating three things; first, the number of seconds required for 60 cc of oil to flow into the receiving flask; second, the type of orifice used; and third, the temperature of the oil sample at the time the viscosity determination is made. For example, suppose that a sample of a lubricating oil is heated to 122°F and that 132 seconds are required for 60 cc of the sample to flow through a Saybolt universal orifice and into the receiving flask. The viscosity of this oil is said to be 132 seconds Saybolt universal at 122°F. This is usually expressed in shorter form as 132 SSU at 122°F.

Saybolt Furol viscosities are obtained at 122°F. The same temperature is used for obtaining Saybolt universal viscosities of fuel, but various other temperatures are used for obtaining Saybolt universal viscosities of oils other than fuel oil. Thus, it is important that the temperature be included in the statement of viscosity.

OTHER TYPES OF MEASUREMENT

Thus far in this chapter we have been largely concerned with basic principles of measurement and with widely used kinds of measuring devices. We have taken up many of the devices used to measure the fundamental variables of temperature, pressure, fluid flow, liquid level, and rotational speed. We have considered the measurement of specific gravity and viscosity. For the most part, we have dealt with measuring devices that might be considered as basically mechanical in nature.

Many other kinds of measurement are required in shipboard engineering plants. While it is true that many of the principles of measurement discussed in this chapter apply to measuring devices other than those described here, it is also true that a specific application may require a measuring device that is not precisely the same as any device that we have considered. Where appropriate, other types of measuring devices are discussed in other chapters of this text, as they relate to some particular type of machinery or equipment. In some instances, you may find it helpful to come back to this chapter to renew your understanding of measurement as we have discussed here.

REFERENCES

Gas Turbine Systems Technician (Electrical) 3/Gas Turbine Systems Technician (Mechanical) 3, Volume 1, NAVEDTRA 10563, Naval Education and Training Program Management Support Activity, Pensacola, Florida, 1989.

CHAPTER 8

INTRODUCTION TO THERMODYNAMICS

LEARNING OBJECTIVES

Upon completion of this chapter, you should be able to do the following:

1. Recognize the basic principles and laws concerning the conservation of energy and the exchange and transformation of energy.

2. Identify the different types of state changes.

3. Discuss the cycles of thermodynamics and the second law of thermodynamics.

4. Identify the purposes and components of energy balances for shipboard engineering plants.

INTRODUCTION

The shipboard engineering plant may be thought of as a series of devices and arrangements for the exchange and transformation of energy. The greatest, most important transformation of energy is done in the shipboard plant that produces mechanical work from thermal energy. Ships depend upon this transformation to make them move through the water. On steam-driven ships, steam serves the vital purpose of carrying energy to the engines. The source of this energy may be the combustion of a conventional fuel oil or nuclear fission. In either case, the steam that is generated is the medium by which thermal energy is carried to the ship's engines, where it is converted into mechanical energy that propels the ship. In addition, energy transformations related directly or indirectly to the basic propulsion plant energy conversion provide steam to ship's service turbine-driven generators (SSTGs), which provide electrical power for many vital services, such as steering, ventilation, heating, air conditioning, and refrigeration; for the operation of various electrical and electronic devices; and for the loading, aiming, and firing of the ship's weapons.

To acquire a basic understanding of the design of shipboard engineering plants, you must have some understanding of certain concepts in the field of thermodynamics. In the broadest sense of the term, thermodynamics is the physical science that deals with energy and energy transformations. The branch of thermodynamics that is of primary interest to engineers is usually referred to as applied thermodynamics or engineering thermodynamics; it deals with fundamental design and operational considerations of boilers,

turbines, internal combustion engines, air compressors, refrigeration and air-conditioning equipment, and other machinery in which energy is exchanged or transferred to produce some desired effect.

This chapter deals with certain thermodynamic concepts that are particularly necessary as a basis for understanding the shipboard engineering plant. The information given here is introductory in nature; obviously, it is not in any sense a complete or thorough exploration of the subject. We will as much as possible depend on verbal description rather than mathematical analysis to develop our understanding of the laws and principles of energy exchanges and transformations. Many of the terms used in this chapter, including such basic terms as *energy* and *heat,* have more specialized and more precise meanings in the study of thermodynamics than they do in everyday life or even in the study of general physics. Thermodynamics is the science concerned with the interrelationship between thermal energy and mechanical energy. Heat transfer is the science that deals with the methods or modes by which thermal energy is able to translate (move from one location to another).

ENERGY

Although energy has a general meaning to almost everyone, it is commonly defined as the ability to do work. However, this is not entirely adequate as a definition, since work is not the only effect that is produced by energy. Sunlight striking a glass of water and causing its temperature to rise is an example of energy though no work is involved (this is called conversion). Heat can flow from one body to another without doing any work at all, but the heat must still be considered as energy and the process of heat transfer must be recognized as a process that has produced an effect. Electromagnetic energy in the form of light produces the noticeable effect of raising the temperature of the water in the glass as the light is converted into thermal energy.

Energy exists in many forms. MECHANICAL ENERGY is the energy associated with large bodies or objects–usually things that are big enough to see. THERMAL ENERGY is energy associated with molecules. CHEMICAL ENERGY is energy that arises from the forces that bind the atoms together in a molecule. Chemical energy is demonstrated whenever combustion or any other chemical reaction takes place. ELECTRICAL ENERGY, light, X rays, and radio waves are examples of energy associated with particles that are even smaller than atoms.

Each type of energy can be broken down into the subclassifications of stored energy or transitional energy. Stored energy can be thought of as energy that is "contained in" or "stored in" a substance or system. We may further subclassify stored energy into two kinds: (1) potential energy and (2) kinetic energy. When energy is stored in a system because of the relative positions of two or more objects or particles, we call it potential energy. When energy is stored in a system because of the relative velocities of two or more objects or particles, we call it kinetic energy. Remember, all stored energy is either potential energy or kinetic energy.

Energy in transition is, as the name implies, energy that is in the process of being transferred from one object or system to another. ALL energy in transition begins and ends as stored energy.

To understand any form of energy, you need to know the relative size of the bodies or particles in the energy system and whether the energy is stored or in transition. Bearing in mind the two modes of classification, let us now examine mechanical energy and thermal energy–the two forms of energy that are of particular interest in practically all aspects of shipboard engineering.

MECHANICAL ENERGY

Energy associated with a system composed of relatively large bodies is called mechanical energy. The two forms of stored mechanical energy are mechanical potential energy and mechanical kinetic energy.[1] Mechanical energy in transition is called work; thermal energy in transition is called heat.

Mechanical potential energy is stored in a system by virtue of the relative positions of the bodies that make up the system. The mechanical potential energy associated with the gravitational attraction between the earth and another body provides us with many everyday examples. A rock resting on the edge of a cliff in such a position that it will fall freely if pushed has mechanical potential energy. Water at the top of a dam has mechanical potential energy. A sled that is being held at the top of an icy hill has mechanical potential energy. Note that in each of these examples the energy resides neither in the earth alone nor in the other object alone but rather in an energy system of which the earth is merely one component.

Mechanical kinetic energy is stored in a system by virtue of the relative velocities of the component parts of the system. Push that rock over the edge of the cliff, open the gate of the dam, or let go of the sled, and something will move. The rock will fall, the water will flow, the sled will slide down the hill. In each case the mechanical potential energy will be changed to mechanical kinetic energy. Since it is customary to ascribe zero velocity to an object that is at rest with respect to the earth, it is also customary to think of kinetic energy as though it pertained only to the object that is in motion with respect to the earth. You should remember, however, that kinetic energy, like potential energy, is properly assigned to the system rather than to any one component of the system.

In these examples of mechanical potential energy and mechanical kinetic energy, we have used an external source of energy to get things started. Energy from some outside source is required to push the rock, open the gate of the dam, or let go of the sled. All real machines and processes require this kind of boost from an energy source outside of the system; similarly, the energy from any one system is bound to affect other energy systems, since no one system can be completely isolated as far

---

1 Although all forms of energy may be stored as potential energy or as kinetic energy, these refer, in common usage, to mechanical potential energy and mechanical kinetic energy, unless some other form of energy, such as thermal or chemical, is specified.

as energy is concerned. However, it is easier to understand the basic energy concepts if you disregard all the other energy systems that might be involved in or affected by each energy process. Hence, we will generally consider one system at a time, disregarding energy boosts that may be received from an outside source and the energy transfers that may take place between the system we are considering and any other system.

It should be emphasized that mechanical potential energy and mechanical kinetic energy are both stored forms of energy. Some confusion arises because mechanical kinetic energy is often referred to as the "energy of motion," thus leading to the false conclusion that "energy in transition" is somehow involved. This is not the case, however. Mechanical work (work) is the only form of mechanical energy that can properly be considered as energy in transition.

Mechanical potential energy and mechanical kinetic energy are mutually convertible. To take the example of the rock resting on the edge of the cliff, let us suppose that some external force pushes the rock over the edge so that it falls. As the rock falls, the system loses potential energy but gains kinetic energy. By the time the rock reaches the ground at the base of the cliff, all the potential energy of the system has been converted into kinetic energy. The sum of the potential energy and the kinetic energy is identical at each point along the line of fall, but the proportions of potential energy and kinetic energy are constantly changing as the rock falls.

To take another example, consider a baseball that is thrown straight up into the air. The ball has kinetic energy while it is in upward motion, but the amount of kinetic energy is decreasing and the amount of potential energy is increasing as the ball travels upward. When the ball has reached it's uppermost position, before it starts to fall back toward the earth, it has only potential energy. Then, as the ball falls back toward the earth, the potential energy is converted into kinetic energy again.

The magnitude of the mechanical potential energy stored in a system by virtue of the relative positions of the bodies that make up the system is proportional to (1) the force of attraction between the bodies and (2) the distance between the bodies. In the case of the rock that is ready to fall from the edge of the cliff, we are concerned with (1) the force of attraction between the earth and the rock–that is, the force of gravity acting upon the rock, or the weight of the rock, and

(2) the linear separation between the two objects. If we measure the weight in pounds and the distance in feet, the amount of mechanical potential energy stored in the system by virtue of the elevation of the rock is measured in the unit called the foot-pound (ft-lb). Specifically,

$$E_p = W \times D$$

where

E_p = mechanical potential energy, in ft-lb

W = weight of body, in pounds

D = distance between earth and body, in feet

The magnitude of mechanical kinetic energy is proportional to the mass and to the square of the velocity of an object that has velocity with respect to another object, or

$$E_k = \frac{MV^2}{2}$$

where

E_k = mechanical kinetic energy, in ft-lb

M = mass of body, in pounds

V = velocity of body relative to the earth, in feet per second

Where it is more convenient to use the weight of the body, rather than the mass, the equation becomes

$$E_k = \frac{WV^2}{2g}$$

where

W = the weight of the body, in pounds

g = the acceleration due to gravity, generally taken as 32.2 feet per second

Work, as we have seen, is mechanical energy in transition–that is, it is a transitory form of mechanical energy that occurs only between two or more other forms of energy. Work is done when a tangible body or substance is moved through a tangible distance by the action of a tangible force. Thus, work may be defined as the energy that is transferred by the action of a force through a distance, or

$$E_{wk} = F \times D$$

where

E_{wk} = work, in ft-lb

F = force, in pounds

D = distance (or displacement), in feet

In the case of work being done against gravity, the force is numerically the same as the weight of the object or body that is being displaced. A man pushing a lawnmower, for example, is exerting some force that acts in the direction in which the lawnmower is moving; but he is also exerting some force that acts downward, at right angles to the direction of displacement. In this case, only the forward component of the exerted force results in work in the forward motion of the lawnmower.

You should note that no work is done unless something is displaced from its previous position. The amount of work done is independent of the time it takes to do it. When a person lifts a mass of 1 pound through a distance of 1 foot, the person has done 1 ft-lb of work, whether the person does it in half a second or half an hour. The rate at which work is done is called power. In the field of mechanical engineering, the common unit of measurement for power is horsepower (hp) in the English system or joules per second (j/s) in the SI system. One horsepower is equal to 33,000 ft-lb of work per minute or 550 foot-pounds of work per second. Thus, a machine that is capable of doing 550 ft-lb of work per second is said to be a 1-hp machine.

Suppose we move an object in such a way that it returns to its original position. Have we done work or haven't we? Let us consider the example of lifting a 5-pound weight to the top of a 3-foot table. By this action, we have performed 15 ft-lb of work. Now suppose we let the weight fall back to the floor, so that it ends up in the same position it had originally. Displacement is zero, so work must be zero. But what has happened to the 15 ft-lb of work we put into the system when we lifted the weight to the top of the table? By doing this work, we gave the system 15 ft-lb of mechanical potential energy. When the weight fell back to the floor, the mechanical potential energy was converted into mechanical kinetic energy. In one sense, therefore, we say that our work was "undone" and that no net work has been done.

On the other hand, we may choose to regard the two actions separately. In such a case, we say that we have done 15 ft-lb of work by lifting the weight and that the force of gravity acting upon the weight has done 15 ft-lb of work to return the weight to its original position on the floor. However, we must regard one work as positive and the other work as negative. The two cancel each other out, so there is again no net work. But in this case, we have recognized that 15 ft-lb of work was performed twice, in two separate operations, by two different agencies. If one has trouble with the idea that mechanical kinetic energy is stored rather than in

transition, we can think of it in another way. This example has been elaborated at some length because we may draw several important inferences from it. First, it may help to clarify the concept of work as a form of energy that must be accounted for. Also, it may help to convey the real meaning of the statement that work is mechanical energy in transition. Work is energy in transition because it occurs only temporarily, between other forms of energy, and because it must always begin and end as stored energy. And finally, the example suggests the need for arbitrary reference planes in connection with the measurement of potential energy, kinetic energy, and work. The quantitative consideration of any form of energy requires a frame of reference that defines the starting point and the stopping point of any particular operation; the reference planes are practically always relative rather than absolute.

THERMAL ENERGY

Energy associated primarily with systems of molecules is called thermal energy. Like other kinds of energy, thermal energy may exist in stored form (in which case it is called internal energy) or as energy in transition (in which case it is called heat).

In common usage, the term heat is often used to include all forms of thermal energy. However, this lack of distinction between heat and the stored forms of thermal energy can lead to serious confusion. In this text, therefore, the term internal energy is used to describe the stored forms of thermal energy, and the term heat is used only to describe thermal energy in transition.

Internal Energy

Internal energy, like all stored forms of energy, exists either as potential energy or as kinetic energy. Internal potential energy is the energy associated with the force of attraction that exists between molecules. The magnitude of internal potential energy is dependent upon the mass of the molecules and the average distance by which they are separated in much the same way that mechanical potential energy depends upon the mass of the bodies in the system and the distance by which they are separated. The force of attraction between molecules is greatest in solids, less in liquids and yielding substances, and least of all in gases and vapors. Whenever something happens to change the average distance between the molecules of a substance, there is a corresponding change in the internal potential energy of the substance.

Internal kinetic energy is the energy associated primarily with the activity of molecules, just as mechanical kinetic energy is the energy associated with the velocities of relatively large bodies. It is important to note that the temperature of a substance arises from and is proportional to the molecular activity with which internal kinetic energy is associated.

For most purposes, we will not need to distinguish between the two stored forms of internal energy. Therefore, instead of referring to internal potential energy and internal kinetic energy, we may often simply use the term internal energy. When used in this way, without qualification, the term internal energy should be understood to mean the sum total of all internal energy stored in the substance or system by virtue of the motion of molecules or by virtue of the forces of attraction between molecules.

Heat

Although the term heat is more familiar than the term internal energy, it may be more difficult to arrive at an accurate definition of heat. Heat is thermal energy in transition. Like work, heat is a transitory energy form existing between two or more other forms of energy. Since the flow of thermal energy can occur only when there is a temperature difference between two objects or regions, it is apparent that heat is not a property or attribute of any one object or substance. If a person accidentally touches a hot stove, he or she may understandably feel that heat is a property of the stove. More accurately, however, he or she might reflect that his or her hand and the stove constitute an energy system and that thermal energy flows from the stove to his or her hand because the stove has a higher temperature than the hand.

As another example of the difference between heat and internal energy, consider two equal lengths of piping, made of identical materials and containing steam at the same pressure and temperature. One pipe is well insulated, one is not. From everyday experience, we expect more heat to flow from the uninsulated section of pipe than from the insulated section. When the two pipes are first filled with steam, the steam in one pipe contains exactly as much internal energy as the steam in the other pipe. We know this is true because the two pipes contain equal volumes of steam at equal pressures and temperatures. After a few minutes, the steam in the uninsulated pipe will contain much less internal energy than the steam in the insulated pipe, as we can tell by reading the pressure and temperature gauges on each pipe. What has happened? Stored thermal energy (internal energy) has moved from one place to another; first from the steam to the pipe, then from the uninsulated pipe to the air. It is this movement, or this flow, of energy that should properly be called heat. (Heat flow, or the transfer of thermal energy from one body, substance, or region to another, always takes place from a region of higher temperature to a region of lower temperature.) Temperature is a reflection of the amount of internal kinetic energy possessed by an object or a substance, and it is therefore an attribute or property of the substance. The movement or flow of thermal energy (heat) is an attribute of the energy system rather than of any one component of it.[2]

Units of Measurement

In engineering, heat is commonly measured in the unit called the British thermal unit (Btu). Originally, 1 Btu was defined as the quantity of heat required to raise the temperature of 1 pound of water through 1 degree on the Fahrenheit scale. A similar unit called the calorie was originally defined as the quantity of heat required to raise the temperature of 1 gram of water through 1 degree on the Celsius scale. These units are still used, but the original definitions have been abandoned by international agreement. Several reasons contributed to the abandonment of the original definitions of the Btu and the calorie. For one thing, precise measurements indicated that the quantity of heat required to raise a specified amount of water through 1 degree on the appropriate scale was not constant at all temperatures.

---

2 The correct definition of heat is emphasized here to avoid subsequent misunderstanding in the study of thermodynamic processes. It is obvious that "heat" and related words are sometimes used in a general way to indicate temperature. For example, we have no simple way of referring to an object with a large amount of internal kinetic energy except to say that it is "hot." Similarly, a reference to the heat of the sun may mean either the temperature of the sun or the amount of heat being radiated by the sun. Even heat flow or heat transfer–the terms quite properly used to describe the flow of thermal energy–are sometimes used in such a way as to imply that heat is a property of one object or substance rather than an attribute of an energy system. To a certain extent, such inaccurate use of heat and related words is really unavoidable; we must continue to "add heat" and " remove heat" and perform other impossible operations, verbally, unless we wish to adopt a very stuffy and long-winded form of speech. It is essential, however, that we maintain a clear understanding of the true nature of heat and of the distinction between heat and the stored forms of thermal energy.

Second, and perhaps even more important, the recognition of heat as a form of energy makes the Btu and the calorie unnecessary. It has been suggested that the calorie and the Btu could be given up entirely and that heat could be expressed directly in joules, ergs, ft-lb, or other established energy units. Some progress has been made in this direction, but not much; the Btu and the calorie are still the units of heat most widely used in engineering and in the physical sciences generally. The Btu and the calorie are now defined in terms of the unit of energy called the joule. The following relationships have thus been established by definition or derived from established definitions. The values given here are considerably more precise than those normally required in engineering calculations.

$$1 \text{ calorie} = \frac{1}{860}$$

$$= 4.18605 \text{ joules}$$

$$= 3.0883 \text{ ft-lb}$$

$$1 \text{ Btu} = 251.88 \text{ calories}$$

$$= 778.26 \text{ ft-lb}$$

$$= 1054.996 \text{ joules}$$

When large amounts of thermal energy are involved, it is often more convenient to use multiples of the Btu or the calorie. For example, we may wish to refer to thousands or millions of Btu, in which case we would use the unit kB (1 kB = 1000 Btu) or the unit mB (1 mB = 1,000,000 Btu). Similarly, the kilocalorie may be used when we wish to express calories in thousands (1 kilocalorie = 1000 calories). The kilocalorie, also called the "large calorie," is the unit normally used for indicating the thermal energies of various foods. Thus, a portion of food that contains 100 calories actually contains 100 kilocalories or 100,000 ordinary calories.

HEAT TRANSFER.—Heat flow, or the transfer of thermal energy from one body, substance, or region to another, takes place always from a region of higher temperature to a region of lower temperature. (Energy exchanges between molecules may be thought of as being random, in the statistical sense; therefore, some exchanges of thermal energy may indeed "go" in the wrong direction—that is, from a colder region to a warmer region. On the average, the flow of heat is always from the higher to the lower temperature.) In thermodynamics, the high-temperature region may be called the source or the emitting region; the low-temperature region may be called the sink, the receiver, or the receiving region.

There are three modes of heat transfer: conduction, radiation, and convection. It will be easier for you to understand heat transfer if we make a distinction between conduction and radiation, on the one hand, and convection, on the other. Conduction and radiation may be regarded as the primary modes of heat flow. Convection may best be thought of as a related but basically different and special kind of process that involves the movement of a mass of fluid from one place to another.

Conduction.—Conduction is the mode by which heat flows from a hotter to a colder region when there is physical contact between the two regions. For example, consider a metal bar that is held so that one end of it is in boiling water. In a very short time the end of the bar that is not in the boiling water will become too hot to hold. We say that heat has been conducted from molecule to molecule along the entire length of the bar. The molecules in the layer nearest the source of heat become increasingly active as they receive thermal energy. Since each layer of molecules is bound to the adjacent layers by cohesive forces, the motion is passed on to the next layer which, in turn, sets up increased activity in the next layer. The process of conduction continues as long as there is a temperature difference between the two ends of the bar. The total quantity of heat conducted depends upon a number of factors. Let us consider a bar of homogeneous material that is uniform in cross-sectional area throughout its length. One end of the bar is kept at a uniformly high temperature; the other end is kept at a uniformly low temperature. After a steady and uniform flow of heat has been established, the total quantity of heat that will be conducted through this bar depends upon the following relationships:

- The total quantity of heat passing through the conductor in a given length of time is directly proportional to the cross-sectional area of the conductor. The cross-sectional area is measured normal to (that is, at right angles to) the direction of heat flow.

- The total quantity of heat passing through the conductor in a given length of time is proportional to the thermal gradient—that is, to the difference in temperature between the two ends of the bar, divided by the length of the bar.

- The quantity of heat is directly proportional to the time of heat flow.

- The quantity of heat depends upon the thermal conductivity of the material of which the bar is made. Thermal conductivity (k) is different for each material.

These relationships may be expressed by the following equation:

$$Q = kTA\frac{t_1 - t_2}{L}$$

where

Q = quantity of heat, in Btu or calories

k = coefficient of thermal conductivity (charac teristic of each material)

T = time during which heat flows

A = cross-sectional area, normal to the path of heat

t_1 = temperature at the hot end of the bar

t_2 = temperature at the cold end of the bar

L = distance between the two ends of the bar

This equation, which is sometimes called the general conduction equation, applies whether we are using a metric system or a British system. Consistency in the use of units is, of course, vital.

The quantity is called the thermal gradient or the $\frac{t_1 - t_2}{L}$ temperature gradient. In the metric CGS system, the temperature gradient is expressed in degrees Celsius per centimeter of length; the cross-sectional area is expressed in square centimeters; and the time is expressed in seconds. In British units, the temperature gradient is expressed in Btu per inch (or sometimes per foot) of length; the cross-sectional area is expressed in square feet; and the time is expressed in seconds or in hours. (Caution is required in using the British units; you must know whether the temperature gradient indicates per inch or per foot, and whether the time is expressed in seconds or in hours.)

From the general conduction equation, we may infer that the coefficient of thermal conductivity (k) represents the quantity of heat that will flow through unit cross section and unit length of a material in unit time when there is unit temperature difference between the hotter and the colder faces of the material.

Thermal conductivity is determined experimentally for various materials. We may perhaps visualize the process of conduction more clearly and understand its quantitative aspects more fully by examining an apparatus for the determination of thermal conductivity and by setting up a problem.

Figure 8-1 shows a device that could be used for determining thermal conductivity. Assume that we have a bar of uniform diameter, made of an unknown metal. (If we knew the kind of metal, we could look up the thermal conductivity in a table; since we do not know the metal, we will find k experimentally.) One end of the bar is inserted into a steam chest in which a constant temperature is maintained; the other end of the bar is inserted into a water chest. The quantity of water flowing through the water chest and the entrance and exit temperature of the water are measured. Also, the temperature of the bar itself is measured at two points by thermometers inserted into holes in the bar; we may choose any two points along the bar, provided they are reasonably far

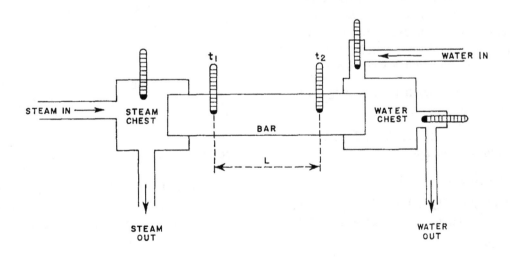

Figure 8-1.–Device for measuring thermal conductivity.

apart and provided they are some distance away from the steam chest and the water chest.

We will assume that the entire apparatus is perfectly insulated so that the temperature difference between t_1 and t_2 is an accurate reflection of the heat conducted along the bar and so that the amount of heat absorbed by the circulating water in the water chest is a true indication of the heat conducted from the hotter end of the bar to the colder end. We will assume that the following data are known at the outset or learned by measurement or determined in the course of the experiment:

Specific heat of water = 1.00

Temperature of water entering water chest = 20°C

Temperature of water leaving water chest = 30°C

Mass of water passing through water chest = 1300 grams

t_1 (temperature at hotter end of bar) = 80°C

t_2 (temperature at cooler end of bar) = 60°C

A (cross-sectional area of bar) = 20 square centimeters

L (distance between points of temperature measurement on bar) = 10 centimeters

T (time of heat flow) = 6 minutes = 360 seconds

To determine the thermal conductivity of our unknown metal, we will use two equations. One is the general conduction equation

$$Q = kTA\frac{t_1 - t_2}{L}$$

where, as we have seen, Q may be expressed in calories or Btu. In this example, we are using the metric CGS system and must therefore express Q in calories. The second equation we will use gives us a second way of calculating Q–that is, by determining the amount of heat absorbed by the circulating water. Thus,

Q = mass of water x temperature change of water x specific heat of water

Substituting some of our known values in this second equation, we find that

Q = (1300) (10) (1) = 13,000 calories

Using this value of Q and substituting other known values in the general conduction equation, we find that

13,00 = k (360) (20)

= k (360) (20) (2)

= 14,400 k

09 = k

The general conduction equation applies only when there is a steady state thermal gradient–that is, after a uniform flow of heat has been established. It should be noted also that k varies slightly as a function of temperature, although for many purposes the rise in k that goes with a rise in temperature is so slight that it can safely be disregarded.

In considering the experimental determination of thermal conductivity, why do we include "Specific heat of water = 1.00" as one of the known data? What is specific heat, and what is its utility? Specific heat (also called heat capacity or specific heat capacity) is, like thermal conductivity, a thermal property of matter that must be determined experimentally for each substance. In general, we may say that specific heat is the property of matter that explains why the addition of equal quantities of heat to two different substances will not necessarily produce the same temperature rise in the two substances. We may define the specific heat of any substance as the quantity of heat required to raise the temperature of unit mass of that substance 1 degree. (Specific heat as defined here should not be confused with the relatively useless concept of specific heat ratio, by which the heat capacity of each substance is compared to the heat capacity of water [taken as 1.00]. The specific heat ratio is, obviously, a pure number without units.) In the metric CGS system, specific heat is expressed in calories per gram per degree Celsius; in the metric MKS system, it is expressed in kilocalories per kilogram per degree Celsius; and in British systems, it is expressed in Btu per pound per degree Fahrenheit. The specific heat of water is 1.00 in any system, and the numerical value of specific heat for any given substance is the same in all systems (although the units are, of course, different).

Specific heat is determined experimentally by laboratory procedures that are extremely complex and difficult in practice, although basically simple in theory. One of the most common methods of determining specific heat is known as the method of mixtures. In this procedure, a known mass of finely divided metal is heated and then mixed with a known mass of water. The temperatures of the metal before mixing, of the water before mixing, and of the mixture just as it reaches thermal equilibrium are measured. Then, on the simple premise that the heat lost by one substance must be gained by the other substance, the specific heat of the metal can be found by using the following equation:

$$m_1c_1 (t_1 - t_3) = m_2c_2 (t_3 - t_2)$$

where

m_1 = mass of metal

m_2 = mass of water

c_1 = specific heat of metal

c_2 = specific heat of water (known to be 1.00)

t_1 = temperature of metal before mixing

t_2 = temperature of water before mixing

t_3 = temperature at which water and metal reach thermal equilibrium

In words, then, we may say that the mass times the specific heat times the temperature change of the first substance must equal the mass times the specific heat times the temperature change of the second substance. In this equation and in this verbal statement, we are ignoring the thermal energy absorbed by the apparatus, by the stirring rods, and by the thermometers. In actually determining specific heats, it is often necessary to account for all thermal energy, even that relatively minute quantity that is absorbed by the equipment. In such a case, the heat absorbed by the equipment is merely added to the right-hand side of the equation. Specific heat is primarily useful in that it allows us to determine the quantity of heat added to a substance merely by observing the temperature rise, when we know the mass and the specific heat of the substance. And this, in fact, is precisely what we did in the thermal conductivity problem, where we calculated the amount of heat that had been absorbed by the water in the water chest by using the equation

Q = mass x temperature change x specific heat

Specific heat varies, in greater or lesser degree, according to pressure, volume, and temperature. Specific heat values quoted for solids and liquids are obtained through experimental procedures in which the substance is kept at constant pressure. The specific heat of any gas may vary tremendously, having in fact an almost infinite variety of values because of the almost infinite variety of processes and states during which energy is transferred to or by a gas. For convenience, specific heats of gases are given as specific heat at constant volume (c_v) and specific heat at constant pressure (c_p).

Radiation.–Thermal radiation is a mode of heat transfer that does not involve any physical contact between the emitting region and the receiving region. A person sitting near a hot stove is warmed by thermal radiation from the stove, even though the air in between remains relatively cold. Thermal radiation from the sun warms the earth without warming the space through which it passes. Thermal radiation passes through any transparent substance–air, glass, ice–without warming it to any extent because transparent materials are very poor absorbers of radiant energy. All substances–solids, liquids, and gases–emit radiant energy at all times. We tend to think of radiant energy as something that is emitted only by extremely hot objects, such as the sun, a stove, or a furnace, but this is a very limited view of the nature of radiant energy. The earth absorbs radiant energy emitted by the sun, but the earth in turn radiates energy to the stars. A stove radiates energy to everything surrounding it, but at the same time all the surrounding objects are radiating energy to the stove. A child standing near a snowman may well believe that the snowman is "radiating cold" rather than emitting radiant energy; actually, however, both the child and the snowman are emitting radiant energy. The child, of course, is radiating far more energy than the snowman, so the net effect of this energy exchange is that the snowman grows warmer and the child grows colder. We are literally surrounded by, and a part of, such energy exchanges at all times. As we consider these energy exchanges, we may arrive at a new view of thermal equilibrium: when objects are radiating precisely as much thermal energy as they are receiving, in any given period of time, they are in thermal equilibrium.

Thermal radiation is an electromagnetic wave phenomenon, differing from light, radio waves, and other electromagnetic phenomenon merely in the wavelengths involved. When the wavelengths are in the infrared part of the electromagnetic spectrum (when they are just below the range of visible light waves), we refer to the radiated energy as thermal radiation. You should note, however, that all electromagnetic waves transport energy that can be absorbed by matter and that can, in many cases, result in observable thermal effects. For example, one energy unit of light absorbed by a substance produces the same temperature rise in that substance as is produced by the absorption of an equal amount of thermal (infrared) energy.

When radiant energy falls upon a body that can absorb it, some of the energy is absorbed and some is reflected. The amount absorbed and the amount reflected depend in large part upon the surface of the receiving body. Dark, opaque bodies absorb more thermal radiation than shiny, bright, white, or polished bodies. Shiny, bright, white, or polished bodies reflect more thermal radiation than dark, opaque bodies. Good radiators are also good absorbers, and poor radiators are poor absorbers. In general, good reflectors are poor radiators and poor absorbers.

In considering thermal radiation, the concept of black body radiation is frequently a useful construct. A black body is conceived of as an ideal or theoretical body that, being perfectly black, is a perfect radiator, a perfect absorber, and a perfect nonreflector of radiant energy. The thermal radiation emitted by such a perfect black body is proportional to T^4–that is, to the absolute temperature raised to the fourth power. Because of the fourth power relationship, doubling the absolute temperature increases the radiation 16 times, tripling the absolute temperature increases the radiation 64 times, and so forth. The thermal radiation emitted by real bodies is also proportional to the fourth power of the absolute temperature, although the total radiation emitted by a real body depends also upon the surface of the body. Consideration of the relationship between the thermal radiation of a body and the fourth power of the absolute temperature of that body explains why the problem of thermal insulation against radiation losses increases so enormously as the temperature increases.

Convection.–Although convection is often loosely classified as a mode of heat transfer, it is more accurately regarded as the mechanical transportation of a mass of fluid (liquid or gas) from one place to another. In the process of this transportation, all the thermal energy stored within the fluid remains in stored form unless it is transferred by radiation or by conduction. Since convection does not involve thermal energy in transition, we cannot in the most fundamental sense regard it as a mode of heat transfer.

Convection is the transportation or the movement of some portions within a mass of fluid. As this movement occurs, the moving portions of the fluid transport their contained thermal energy to other parts of the fluid. The effect of convection is thus to mix the various portions of the fluid. The part that was at the bottom of the container may move to the top, or the part that was at one side may move to the other side. As this mixing takes place, heat transfer occurs by conduction and radiation from one part of the fluid to another and between the fluid and its surroundings. In other words, convection transports portions of the fluid from one

place to another, mixes the fluid, and thus provides an opportunity for heat transfer to occur. But convection does not, in and of itself, "transfer" thermal energy.

Convection serves a vital purpose in bringing the different parts of a fluid into close contact with each other so that heat transfer can occur. Without convection, there would be little heat transfer from, to, or within fluids, since most fluids are very poor at transferring heat except when they are in motion.

There are two kinds of convection: natural convection and forced convection. Natural convection occurs when there are differences in the density of different parts of the fluid. The differences in density are usually caused by unequal temperatures within the mass of fluid. As the air over a hot radiator is heated, for example, it becomes less dense and therefore begins to rise. Cooler, heavier air is drawn in to replace the heated air that has moved upward, and convection currents are thus set up. Another example of natural convection, and one that may be quite readily observed, may be found in a pan of water that is being heated on a stove. As the water near the bottom of the pan is heated first, it becomes less dense and moves upward. This displaces the cooler, heavier water and forces it downward; as the cooler water is heated in turn, it rises and displaces the water near the top. By the time the water has almost reached the boiling point, a considerable amount of motion can be observed in the water. Forced convection occurs when some mechanical device, such as a pump or a fan, produces movement of a fluid. Many examples of forced convection may be observed in the shipboard engineering plant: feed pumps transporting water to the boilers, fuel-oil pumps moving fuel through the fuel-oil system, lubricating-oil pumps forcing lubricating oil through coolers, and forced draft blowers pushing air into the boiler air casing, to name just a few.

The mathematical treatment of convection is extremely complex, largely because the amount of heat gained or lost through the convection process depends upon so many different factors. Empirically determined convection coefficients that take account of these many factors are available for most kinds of engineering equipment.

SENSIBLE HEAT AND LATENT HEAT.–The terms sensible heat and latent heat are often used to indicate the effect that the transfer of heat has upon a substance. The flow of heat from one substance to another is normally reflected in a temperature change in each substance–that is, the hotter substance becomes cooler and the cooler substance becomes hotter. However, the flow of heat is not reflected in a temperature change in a substance that is in process of changing from one physical state to another.[3] When the flow of heat is reflected in a temperature change, we say that sensible heat has been added to or removed from a substance. When the flow of heat is not reflected in a temperature change but is reflected in the changing physical state of a substance, we say that latent heat has been added or removed. Since heat is defined as thermal energy in transition, we must not infer that sensible heat and latent heat are really two different kinds of heat. Instead, they serve to distinguish between two different kinds of effects produced by the transfer of heat; and at a more fundamental level, they indicate something about the manner in which the thermal energy was or will be stored. Sensible heat involves internal kinetic energy and latent heat involves internal potential energy.

The three fundamental physical states of all matter are solid, liquid, and gas (or vapor). The physical state of a substance is closely related to the distance between molecules. The molecules are closest together in solids, farther apart in liquids, and farthest apart in gases. When the flow of heat to a substance is not reflected in a temperature change, we know that the energy is being used to increase the distance between the molecules of the substance and thus change it from a solid to a liquid or from a liquid to a gas. In other words, the addition of heat to a substance that is in process of changing from solid to liquid or from liquid to gas results in an increase in the amount of internal potential energy stored in the substance, but it does not result in an increase in the amount of internal kinetic energy. Only after the change of state has been fully accomplished does the addition of heat result in a change in the amount of internal kinetic energy stored in the substance; hence, there is no temperature change until after the change of state is complete.

In a sense, we may think of latent heat as the energy price that must be paid for a change of state from solid to liquid or from liquid to gas. But the energy is not lost; rather, it is stored in the substance as internal potential energy. The energy price is "repaid" so to speak, when

the substance changes back from gas to liquid or from liquid to solid; during these changes of state, the substance gives off heat without any change in temperature.

The amount of latent heat required to cause a change of state–or, on the other hand, the amount of latent heat given off during a change of state–varies according to the pressure under which the process takes place. For example, it takes about 970 Btu to change 1 pound of water to steam at atmospheric pressure (14.7 psia), but it takes only 62 Btu to change 1 pound of water to steam at 3200 psia.

Figure 8-2 shows the relationship between sensible heat and latent heat for one substance, water, at atmospheric pressure. (The same kind of chart could be drawn up for other substances, but different amounts of thermal energy would of course be required for each change of temperature or of physical state.) If we start with 1 pound of ice at 0°F, we must add 16 Btu to raise the temperature of the ice to 32°F. We call this adding sensible heat. To change the pound of ice at 32°F, we must add 144 Btu (the latent heat of fusion). The will be no change in temperature while the ice is melting. After all the ice has melted, however, the temperature of the water will be raised as additional heat is supplied. Again, we are adding sensible heat. If we add 180 Btu–that is, 1 Btu for each degree of temperature between 32°F and 212°F–the temperature of the water will be raised to the boiling point. To change the pound of water at 212°F to a pound of steam at 212°F, we must add 970 Btu (the latent heat of vaporization). After all the water has been converted to steam, the addition of more heat will cause an increase in the temperature of the steam. If we add 42 Btu to the pound of steam that is at 212°F, we can superheat[4] it to 300°F.

The same relationships apply when heat is being removed. The removal of 42 Btu from the pound of steam that is at 300°F will cause the temperature to drop to 212°F. As the pound of steam at 212°F changes to a pound of water at 212°F, 970 Btu are given off. When a

---

3 In thermodynamics, the physical state of a subtance (solid, liquid, or gas) is usually described by the term phase, while the term *phase,* while the term state is used to describe the substance with respect to all of its properties–phase, pressure, temperature, specific volume, and so forth. Thus, the phase of a substance may be considered as merely one of the serveral properties that fix the state of the substance. While the precision of this usage as some obvious advantages, it is not instandard use among engineers. In this text, therefore, the term *physical state* (or sometimes *state*) is used to denote the molecular condition of a substance that determines whether the substance is a solid, a liquid, or a gas.

4 A vapor or gas is said to be superheated when its temperature has been raised above the temperature of the liquid from which the vapor or gas is being generated. As may be inferred from the discussion, it is impossible to superheat a vapor or gas as long as it is in contact with the liquid from which it is being generated.

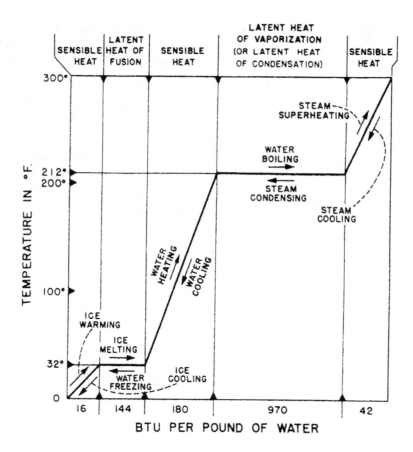

Figure 8-2.–Relationship between sensible heat and latent heat for water at atmospheric pressure.

gas or vapor is changing to a liquid, we usually use the term latent heat of condensation; numerically, of course, the latent heat of condensation is exactly the same as the latent heat of vaporization. (A vapor or gas is said to be superheated when its temperature has been raised above the temperature of the liquid from which the vapor or gas is being generated. As may be inferred from the discussion, it is impossible to superheat a vapor or gas as long as it is in contact with the liquid from which it is being generated.) The removal of another 180 Btu will lower the temperature of the pound of water from 212°F to 32°F. As the pound of water at 32°F changes to a pound of ice at 32°F, 144 Btu are given off without any accompanying change in temperature. Further removal of heat causes the temperature of the ice to decrease.

HEAT TRANSFER APPARATUS.–Any device or apparatus designed to allow the flow of thermal energy from one fluid to another is called a heat exchanger. The shipboard engineering plant contains an enormous number and variety of heat exchangers, ranging from large items such as boilers and main condensers to relatively small items such as lubricating-oil coolers.

As a basis for understanding something about heat transfer in real heat exchangers, you need to visualize the general configuration of the most commonly used type of heat exchanger. With few exceptions, like the deaerating feed tanks that are basically described as direct-contact heat exchangers, rather than surface heat exchangers, because heat transfer is accomplished by the actual mixing of the hotter and the colder fluids, heat exchangers used aboard ship are of the indirect or surface type, which means that heat flows from one fluid to another through some kind of tube, plate, or other "surface" that separates the two fluids and keeps them from mixing. Most surface heat exchangers are of the shell and tube types, consisting of a bundle of metal tubes that fit inside a shell. One fluid flows through the inside of the tubes and the other flows through the shell, around the outside of the tubes. The exchanges of thermal energy that take place in even a simple heat exchanger are really quite complex. The processes of conduction, radiation, and convection are involved in practically all heat exchangers.

Processes involving latent heat–that is, the processes of evaporation, condensation, melting, and solidification–may contribute to the heat transfer problem. In all cases, heat transfer is affected by physical properties of the fluids that are exchanging thermal energy and by physical properties of the metal through which the change is being effected. The temperature differences involved, the extent and nature of the fluid films, the thickness and nature of the metals through which heat transfer takes place, the length and area of the path of heat flow, the types of surfaces involved, the velocity of flow, and other factors also define the amount of heat transferred in any heat exchanger.

Because heat transfer is such a complex phenomenon, heat transfer calculations are necessarily complex. For some purposes, heat transfer problems are simplified by the use of an overall coefficient of heat transfer (U), which may be defined experimentally for any specific set of conditions. Tabulated values of U are available for various kinds of heat exchanger metal tubes, for building materials, and for other materials; in most cases, the values of U are approximate, since various conditions such as temperature, velocity of flow, condition of the heat transfer surfaces, and the physical properties of the fluids have a profound effect upon the amount of heat transferred.

The transfer of heat in a heat exchanger involves the flow of heat from the hot fluid to the tube metal and from the tube metal to the cold fluid. In addition, heat must also be transferred through two layers of fluid (one on the inside and one on the outside of the tube) that are not flowing with the remainder of the fluid but are almost motionless. These relatively stagnant layers, known as BOUNDARY LAYERS or FLUID FILMS, are extremely small in size but have an extremely important effect on heat transfer. As previously noted, most fluids are very poor transferrers of heat. As a fluid is flowing, however, convection and mechanical mixing of the fluid brings the molecules into such intimate contact that heat transfer can and does occur. Other things being equal, increasing the velocity of fluid flow increases heat transfer.

Since the fluid is almost motionless, heat transfer through the film is very poor. The effect of fluid films on heat transfer is shown in figure 8-3. The temperature line indicates the changes in temperature that occur as heat is transferred from the hot fluid to the fluid film, from this fluid film to the tube metal, from the tube metal to the other fluid film, and from this fluid film to the cold fluid.

It is important here to maintain the distinction, previously established, between heat and temperature. Increasing the velocity of flow increases the amount of heat that is transferred, but decreasing the velocity

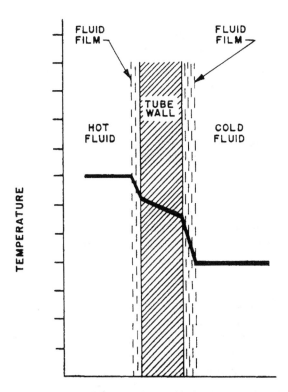

Figure 8-3.–Effect of fluid film on heat transfer.

increases the temperature of the fluid. This fact is of considerable practical importance in the design and operation of heat exchangers. In a heat exchanger designed for high velocity flow, stagnation of the flow is likely to cause severe overheating of the heat exchanger metal. As may be seen, the major part of the temperature drop occurs in the fluid films rather than in the tube metal. Note, also, that the thicker fluid film is more resistant to heat transfer than the thinner fluid film.

The velocity of flow and the amount of turbulence in the flow affect heat transfer by altering the thickness of the fluid film. Increasing the velocity of flow diminishes the thickness of the fluid film and thus increases heat transfer. Turbulent flow breaks up the fluid film and thus increases heat transfer. Although there are some obvious disadvantages to excessive turbulence, many heat exchangers are designed to operate with a certain amount of turbulence so that the fluid films will be kept to a minimum.

In real heat exchangers, the accumulation of deposits of scale, soot, or dirt on the inside or the outside of the tubes has a profound and detrimental effect upon heat transfer. Such deposits not only reduce the efficiency of the heat exchanger but also tend to cause overheating of the tube metal.

In surface heat exchangers, the components may be arranged so as to provide parallel flow, counter flow, or cross flow of the two fluids. In parallel flow (fig. 8-4), both fluids flow in the same direction. Parallel-flow heat exchangers are rarely used for naval service, largely because they would require an impossibly long heat transfer surface to achieve the required amount of heat transfer. In counter flow (fig. 8-5), the two fluids flow in opposite directions. Many heat exchangers used aboard ship are of the counter-flow type. In cross flow (fig. 8-6), one fluid flows at right angles to the other. Cross flow is used particularly where the purpose of the heat exchanger is to remove latent heat and thus change the physical state of a substance. Main and auxiliary condensers are typically of the cross-flow type, as are several other smaller shipboard condensers.

Surface heat exchangers are referred to as single-pass units, if each fluid passes the other only once, or as multipass units, if one fluid passes the other more than once. Multipass flow may be obtained by the arrangement of the tubes and of the fluid inlets and outlets, or it may be obtained by using baffles to guide a fluid so that it passes the other fluid more than once before it leaves the heat exchanger.

THE FIRST LAW OF THERMODYNAMICS

In the previous discussion of energy, we occasionally assumed a general principle that must now

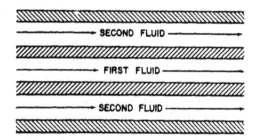

Figure 8-4.–Parallel flow in a heat exchanger.

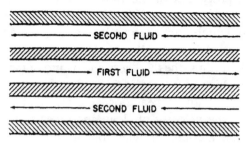

Figure 8-5.–Counter flow in a heat exchanger.

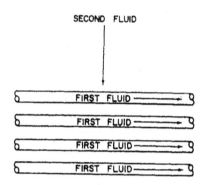

Figure 8-6.–Cross flow in a heat exchanger.

be stated. This principle is called the principle of the conservation of energy. The principle may be stated in several ways. Most commonly, perhaps, it is stated as energy can be neither destroyed nor created, but only transformed. Another statement is that energy may be transformed from one form to another, but the total energy of any body or system of bodies is a quantity that can neither be increased nor diminished by the action of the body or bodies. Still another way of stating this principle is by saying that the total quantity of energy in the universe is always the same. Regardless of the mode of expression, the principle of the conservation of energy applies to all kinds of energy.[5]

In spite of the mutual convertibility of energy and mass, the principle of the conservation of energy may still be regarded separately as the cornerstone of the science of thermodynamics. Machinery designed under

---

5 The principle of the conservation of energy and the principle of the conservation of mass have been basic to the development of modern science. Until the establishment of the theory of relativity, with its implication of the mutual convertability of energy and mass, the two principles were considered quite separate. According to the theory of relativity, however, they must be considered merely as two phases of a single principle, which states that mass and energy are interchangeable and the total amount of matter and energy in the universe is constant. Nuclear fission, a process in which atomic nuclei split into fragments with the release of enormous quantities of energy, is a dramatic example of the actual conversion of matter into energy. Even in the familiar process of combustion, modern techniques of measurement have led to the discovery that a very minute quantity of matter is converted into energy; for example, about 0.00007 ounce of matter is converted into energy when 6 tons of carbon are burned with 16 tons of oxygen.

this principle alone still functions in an orderly and predictable fashion.]

Energy equations for many thermodynamic processes are based directly upon the principle of the conservation of energy. When the principle of the conservation of energy is written in equation form, it is known as the general energy equation and is expressed as

energy in = energy out

or, in more detail, it may be stated that the energy entering a system equals the energy leaving the system plus any accumulation and minus any dimunition in the amount of energy stored within the system.

The first law of thermodynamics, a special statement of the principle of the conservation of energy, deals with the transformation of mechanical energy to thermal energy and of thermal energy to mechanical energy. The first law is commonly stated as Thermal energy and mechanical energy are mutually convertible, in the ratio of 778 ft-lb to 1 Btu.

The ratio of conversion between mechanical energy and thermal energy is known as the mechanical equivalent of heat, or joule's equivalent. It is symbolized by the letter J and, according to the first law of thermodynamics, it is expressed as

$$J = 778 \text{ ft-lb per Btu}$$

or

$$J = \frac{778 \text{ ft-lb}}{1 \text{ Btu}}$$

The mechanical equivalent of heat provides us, directly or by extension, with a number of useful numerical values relating to heat, work, and power. Some of the most widely used values are given here; others may be obtained from engineering handbooks and similar publications.

$1 \text{ Btu} = 778 \text{ ft-lb}$

$1 \text{ hp} = 33,000 \text{ ft-lb per min} = 550 \text{ ft-lb per sec}$

$1 \text{ kw} = 1.341 \text{ hp}$

$1 \text{ hp} = 2,545 \text{ Btu per hr} = 42.42 \text{ Btu per min}$

$1 \text{ kw} = 3,413 \text{ Btu per hr}$

$1 \text{ kw} = 44,256 \text{ ft-lb per min}$

$1 \text{ hp-hr} = 2,545 \text{ Btu}$

$1 \text{ kw-hr} = 3,413 \text{ Btu}$

The first law of thermodynamics is often written in equation form as follows:

$$U_2 - U_1 = Q - W$$

where

U_1 = internal energy of the system at the beginning of the process

U_2 = internal energy of the system at the end of the process

Q = net heat flowing into the system during the process

W = net work done by the system during the process

Another common statement of the first law of thermodynamics is that a perpetual motion machine of the first class is impossible. To understand the significance of this statement, you need to understand the classification of perpetual motion machines. Although no perpetual motion machine exists–or, indeed, has ever been constructed–it is possible to conceive of three different categories. A perpetual motion machine of the first class is one that would put out more energy in the form of work than it absorbed in the form of heat. Since such a machine would actually create energy, it would violate the first law of thermodynamics and the principle of the conservation of energy. A perpetual motion machine of the second class would permit the reversal of irreversible processes and would, thus, violate the second law of thermodynamics, as discussed presently. A machine of the third class would be one in which absolutely no friction existed. Interestingly enough, there are no theoretical grounds for declaring that a machine of the third class is completely impossible; however, such a machine would be entirely contrary to our experience and would violate some of our profoundest convictions about the nature of energy and matter.

THERMODYNAMIC SYSTEMS

A thermodynamic system may be defined as a bounded region that contains matter. The boundaries may be fixed or they may vary in shape, form, and location. The matter within a system may be matter in any form–solid, liquid, or gas–or in some combination of forms. For some purposes, devices such as engines, pumps, boilers, and so forth may be regarded as being matter included within a thermodynamic system; for other purposes, each such device may be considered as a system in itself. A thermodynamic system may be

..irely real, entirely imaginary, or a mixture of real and imaginary. A thermodynamic system may be capable of exchanging energy, in the form of heat and/or work, with its environs; or it may be an isolated system, in which case no heat can flow to or from the system and no work can be done on or by the system.

If a thermodynamic system appears to be a flexible thing, consider the further statement that "...a system may be said to be whatever one is talking about, and its environs are everything else." (Kiefer, Kinney, and Stuart, *Principles of Engineering Thermodynamics*, 2nd ed., John Wiley & Sons, New York, 1954 [p.32].) Such flexibility of definition is entirely reasonable for most purposes. When we must account for energy, however, we will find it necessary to rigidly define and limit the system or systems under consideration. It is in terms of energy accounting, then, that the concept of a thermodynamic system is most useful.

A thermodynamic system requires a working substance to receive, store, transport, and deliver energy. In most systems, the working substance is fluid–liquid, vapor, or gas. (Some writers use the term gas to indicate a gaseous substance that can be liquefied only by very large changes in pressure or temperature, reserving the term *vapor* for a gaseous substance that can be liquefied more easily, by slight changes of pressure or temperature. Other writers define a vapor as a gas that is in equilibrium with its liquid. For a great many purposes, the properties of a vapor are essentially the same as the properties of real gases; hence, the distinction is not always important.)

The state of a thermodynamic system is specified by giving the values of two or more properties. These properties, which are called state variables or thermodynamic coordinates, include such common properties as pressure, temperature, volume, and mass, as well as more complex properties such as enthalpy and entropy. Although some systems are adequately described by giving the value of only two variables, many systems require the specification of three or more variables.

THERMODYNAMIC PROCESSES

A thermodynamic process may be defined as any physical occurrence during which an effect is produced by the transformation or redistribution of energy. The occurrence of a thermodynamic process is evidenced by changes in some or all of the state variables of the system. The processes of most interest in engineering are those involving heat and work.

In connection with any process, you should consider the physical character of the process; the manner in which energy is transformed or redistributed as the process takes place; the kind and amount of energy that is stored in the system before energy; and the changes that are brought about in the system as the result of the process. You should also consider the energy exchanges that occur between the system and its surroundings during the process, since such energy exchanges will have an effect on the final state of the system.

The lifting of an object–as for example, the lifting of a rock from the base of a cliff to the top of the cliff–is a simple example of a process involving work against gravity. Before the process begins, the energy that will be required to lift the rock is stored in some form in some other energy system. While the process is occurring, energy in the form of work flows from the external system to the earth-rock system. At the end of the process, the energy is stored in the earth-rock system in the form of mechanical potential energy. The change that has been brought about by this process is manifested by the separation of the rock and the earth.

Now suppose we push the rock off the top of the cliff and allow it to fall freely toward the base of the cliff. Disregarding the push (which is actually an input of energy from some external system), the process that now takes place is an example of work done by gravity. The work done by gravity converts the mechanical potential energy of the system into mechanical kinetic energy. Thus, it is clear that energy in transition–work, in this case–begins and ends as stored energy.

When the rock hits the earth, other processes occur. Some work will be expended in compressing the earth upon which the rock falls, and some energy will then be stored as internal kinetic energy in the rock and in the earth. The increase in internal kinetic energy will be manifested by a rise in the temperature of the rock and of the earth, and still another process will then take place as heat flows from the rock and from the earth. Some energy may also be stored as internal potential energy because of molecular displacements in the rock and earth.

The compression of a spring provides an example of a process involving elastic deformation. As force is applied to compress the spring, work is done. The major effect of the energy thus supplied as work is to decrease the distance between molecules in the spring, thus increasing the amount of internal potential energy stored in the spring. If we suddenly release the spring, the stored internal potential energy is suddenly released and the spring shoots away.

The turning of a shaft is another example of a process involving elastic deformation. Suppose that a strong twisting force is applied to a shaft at rest. The first part of this process will cause an elastic deformation of the shaft. The distance between molecules in the shaft is changed, and there is a storage of internal potential energy before the shaft begins to turn. When the applied force becomes great enough to turn the shaft, there will also be a storage of mechanical kinetic energy. As long as the applied force remains constant and the shaft continues to turn, these stored forms of energy will remain stored in unchanging amount. Meanwhile, a great deal of mechanical energy in transition (work) will continuously flow through the shaft to some other system.

When a solid body is dragged across a rough horizontal surface, the process is one of work against friction. The work done in moving the object will be equal to the force required to overcome the friction multiplied by the distance through which the object is moved. In this process, the energy supplied as work is transformed very largely into internal kinetic energy, as evidenced by an increase in temperature. Some of the energy may be transformed into internal potential energy because of molecular displacements in the object and in the surface over which it is being moved.

A propeller rotating in water is an example of a process in which work causes fluid turbulence. The first effect of the movement of the propeller is to impart various motions to the water, thus causing turbulence. For a short time this movement of the water represents mechanical kinetic energy, but the energy is rapidly transformed into internal kinetic energy, as evidenced by a rise in the temperature of the water.

The addition of thermal energy to a piece of metal is a simple example of a process involving heat. As the metal is heated, the temperature rises, indicating a storage within the metal of internal kinetic energy. Also, the metal expands; thus, we know that some part of the energy delivered as heat is transformed into work as the metal expands against the resistance of its surroundings. If we continue heating the metal to its melting point, we will note a process in which the flow of heat results in a change in the physical state of the substance but does not, at this point, result in a further rise in temperature. Because of the enormous number and variety of processes that may occur, some basic classification of processes involving heat and work is desirable. The following paragraphs will discuss the classification of processes according to the type of flow and types of state

change. Discussion of processes as "reversible" or "irreversible" is reserved for a later section.

TYPE OF FLOW

When classified according to type of flow of the working fluid, thermodynamic processes may be considered under the general headings of (1) nonflow processes and (2) steady flow processes. A nonflow process is one in which the working fluid does not flow into or out of its container in the course of the process. The same molecules of the working fluid that were present at the beginning of the process are therefore present at the end of the process. Nonflow processes occur in reciprocating steam engines, air compressors, internal combustion engines, and other kinds of machinery. Since a piston and cylinder arrangement is typical of most nonflow processes, let us examine a nonflow process such as might occur in the cylinder shown in figure 8-7.

Suppose that we move the cylinder from position 1 to position 2, thereby compressing the fluid contained in the cylinder above the piston. Suppose, further, that we imagine this to be a completely ideal process, and one that is thus entirely without friction. The aspects of this process that we might want to know about are (1) the heat added or removed in the course of the process, (2) the work done on the working fluid or by the working fluid, and (3) the net change in the internal energy of the working substance.

From the general energy equation, we know that energy in must equal energy out. For the nonflow process, the general energy equation may be written as

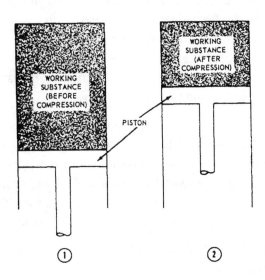

Figure 8-7.–Piston and cylinder arrangement for a nonflow process.

$$Q_{12} = (U_2 - U_1) + \frac{wk_{12}}{J} \; Btu$$

where

Q_{12} = total heat transferred, in Btu (positive if heat is added during the process; negative if heat is removed during the process)

U_1 = total internal energy, in Btu, at state 1

U_2 = total internal energy, in Btu, at state 2

$U_2 - U_1$ = net change in internal energy from state 1 to state 2

wk_{12} = work done between state 1 and state 2, in ft-lb (positive if work is done by the working substance; negative if work is done on the working substance)

J = the mechanical equivalent of heat, 778 ft-lb per Btu

total work done by or on the working substance, in Btu (positive if work is done by the substance; negative if work is done on the substance)

This equation deals with total heat, total work, and total internal energy. If it is more convenient to make calculations in terms of 1 pound of the working substance, we would write the equation as follows:

$$Q_{12} = (U_2 - U_1) + \frac{wk_{12}}{J} \; Btu \; per \; lb$$

where the value of J remains the same and where Q, U, and wk have the general meanings noted previously but refer to the values for 1 pound of the working substance rather than to the values for the total quantity of the working substance. In both equations, note that the subscripts 1 and 2 refer to a separation in time rather than to a separation in space.

EXAMPLE: Four pounds of working substance are compressed in the cylinder shown in figure 8-7. The process is accomplished without the addition or removal of any heat but with a net increase in total internal energy of 120 Btu. Find the work done on or by the working substance, in Btu per pound and in ft-lb per pound.

SOLUTION: First arrange the equation to fit the problem, as follows:

$$\frac{wk_{12}}{J} + (U_2 - U_1) + Q_{12}$$

Since no heat is added or removed, $Q_{12} = 0$. Since $U_2 - U_1$, or the net increase in total internal energy, is equal to 120 Btu, and since we are dealing with 4 pounds of the working substance,

$$U_2 - U_1 = \frac{120}{4} = 30 \; Btu \; per \; pound.$$

The work done on or by the working substance, in Btu per pound, is given by the expression

$\frac{wk_{12}}{J}$. Thus $\frac{wk_{12}}{J} = (-30) + 0$ Btu per lb

$$= -30 \; Btu \; per \; lb$$

The answer is negative, indicating that the work is done on the working substance rather than by the working substance. To find the work done on the working substance in ft-lb per pound, we merely solve the equation for wk_{12} rather than for

$$\frac{wk_{12}}{J}$$

and substitute. Thus,

$$wk_{12} = (-30)(778) = -23,340 \; ft\text{-}lb \; per \; lb$$

Again, the negative answer indicates that work is done on the working substance rather than by the working substance. A steady-flow process is one in which a working substance flows steadily and uniformly through some device. Boilers, turbines, condensers, centrifugal pumps, forced draft blowers, and many other actual machines are designed for steady-flow processes. In an ideal steady-flow process, the following conditions exist:

1. The properties–pressure, temperature, specific volume, and so on–of the working fluid remain constant at any particular cross section in the flow system, although the properties obviously must change as the fluid proceeds from section to section.

2. The average velocity of the working fluid remains constant at any selected cross section in the flow system, although it may change as the fluid proceeds from section to section.

3. The system is always completely filled with the working fluid, and the total weight of the fluid in the system remains constant. Thus, for each pound of working fluid that enters the system during a given period of time, there is a discharge of 1 pound of fluid during the same period of time.

4. The net rate of heat transfer and the work performed on or by the working fluid remains constant.

In actual machinery designed for steady-flow processes, some of these conditions are not entirely satisfied at certain times. For example, a steady-flow machine, such as a boiler or a turbine, is not actually going through a steady-flow process until the warming up period is over and the machine has settled down to steady operation. For most practical purposes, minor fluctuations of properties and velocities caused by load variations do not invalidate the use of steady-flow concepts. In fact, even such piston and cylinder devices as air compressors and reciprocating steam engines may be considered as steady-flow machines if there are enough cylinders or if some other arrangement is used to smooth out the flow so that it is essentially uniform at the inlet and the outlet.

The equations for steady-flow processes are based on the general energy equation—that is, energy in must equal energy out. Steady-flow equations are written in various ways, depending upon the forms of energy that are involved in the process under consideration. The forms of energy that, to greater or lesser degree, enter into any general equation for steady-flow processes are as follows:

- Internal energy
- Heat
- Mechanical potential energy
- Mechanical kinetic energy
- Work
- Flow work

The first five of these energy terms are familiar, but the last one may be new. Flow work, sometimes called displacement energy, is the mechanical energy necessary to maintain the steady flow of a stream of fluid. The numerical value of flow work may be calculated by finding the product of the absolute pressure (in pounds per square feet) and the volume of the fluid (in cubic feet). Thus,

$$\text{flow work} = pV \text{ ft-lb}$$

or, more conveniently, using specific volume rather than total volume,

$$\text{flow work} = pV \text{ ft-lb per lb}$$

The product pV will, of course, have a numerical value even when there is no flow of fluid. However, this value represents flow work only when there is a steady, continuous flow of fluid. Flow work may also be expressed in terms of Btu per pound, as

$$\text{flow work} = \frac{pV}{J} \text{ Btu per lb}$$

As mentioned before, the steady-flow equations take various forms, depending upon the nature of the process under consideration. However, the terms for internal energy and flow work almost invariably appear in any steady-flow process. For convenience, this combination of internal energy and flow work has been given a name, a symbol, and units of measurement. The name is enthalpy (accent on second syllable). The symbol is H for total enthalpy or h for specific enthalpy–that is, enthalpy per pound. Total enthalpy, H, may be measured in Btu or in ft-lb. Specific enthalpy (enthalpy per pound) may be measured in Btu per pound or in ft-lb per pound. The enthalpy equation may be written as follows:

$$H = \frac{pV}{J} = U$$

where

H = total enthalpy, in Btu

U = total internal energy, in Btu

p = absolute pressure, in pounds per square foot

V = total volume, in cubic feet

J = the mechanical equivalent of heat, 778 ft-lb per Btu

Since it is frequently more convenient in thermodynamics to make calculations in terms of 1 pound of the working substance, we should note also the equation for specific enthalpy:

$$h = u + \frac{pV}{J} \text{ Btu per lb}$$

Here h, u, and V are specific enthalpy, specific internal energy, and specific volume, respectively. When you want to calculate enthalpy in ft-lb, rather than in Btu, drop the J from the equations and convert u to ft-lbs. The terms *heat content* and *total heat* are sometimes used to describe this property that we have designated as enthalpy. However, these terms tend to be misleading because the change in enthalpy of a working fluid does not always measure the amount of energy transferred as heat, nor is it necessarily caused by the transfer of energy in the form of heat. Also, the transferred energy that causes a change in enthalpy is not entirely "contained in" the working fluid, as these terms tend to imply; although the internal energy, u, is stored in the working fluid, the pV cannot in any way be considered as "contained in" the fluid.

TYPES OF STATE CHANGE

Thus far we have considered processes classified as nonflow or steady flow. The nature of the state changes undergone by a working fluid provides us with another useful way of classifying processes. The terms used to identify certain common types of state changes are defined briefly in the following paragraphs.

Isobaric State Changes

An isobaric state change is one in which the pressure of and on the working fluid is constant throughout the change. In other words, an isobaric change is a constant pressure change. Isobaric changes occur in some piston and cylinder devices in which the piston operates in such a fashion as to maintain a constant pressure. Isobaric state changes are not typical of most steady-flow processes, but they are approximated in some steady-flow processes in which friction and shaft work are of insignificant magnitude.

An isobaric state change involves changes of enthalpy. One equation that has frequent application to isobaric state changes is written as follows:

$$(q_{12})_p = h_2 - h_1$$

where

$(q_{12})_p$ = heat transferred between state 1 and state 2, with subscript p indicating constant pressure

h_1 = enthalpy of working fluid at state 1

h_2 = enthalpy of working fluid at state 2

Isometric State Changes

A state change is said to be isometric when the volume (and specific volume) of the working fluid is maintained constant. In other words, an isometric change is a constant-volume change. Isometic changes involve changes in internal energy, according to the following equation:

$$q_V = u_2 - u_1$$

where

q_V = heat transferred, with subscript V indicating constant volume

u_1 = specific internal energy of working substance at state

u_2 = specific internal energy of working substance at state

Isothermal State Changes

An isothermal change is one in which the temperature of the working fluid remains constant throughout the change.

Isenthalpic State Changes

When the enthalpy of the working fluid does not change during the process, the change is said to be isenthalpic. Throttling processes are basically isenthalpic–that is, $h_1 = h_2$.

Isentropic State Changes

An isentropic state change is one in which there is no change in the property known as entropy. The significance of entropy and of isentropic state changes is discussed later in this chapter.

Adiabatic State Changes

An adiabatic state change is one that occurs in such a way that there is no transfer of heat to or from the system while the process is occurring. In many real processes, adiabatic changes are produced by performimg the process rapidly. Since heat transfer is relatively slow, any rapidly performed process can approach being adiabatic. Compression and expansion of working fluids are frequently achieved adiabatically. For an adiabatic process, the energy equation may be written as follows:

$$U_2 - U_1 = Wk$$

where

U_1 = internal energy of working fluid at state 1

U_2 = internal energy of working fluid at state 2

Wk = work performed on or by the working fluid

In words, we may say that the net change of internal energy is equal to the work performed in an adiabatic process. The work may be either positive or negative, depending upon whether work is done on the working substance, as in compression, or by the working substance, as in expansion.

THERMODYNAMIC CYCLES

A thermodynamic cycle is a recurring series of thermodynamic processes through which an effect is produced by the transformation or redistribution of energy. In other words, a cycle is a series of processes repeated over and over again in the same order.

All thermodynamic cycles may be classified as being open cycles or closed cycles. An open cycle is one in which the working fluid is taken in, used, and then discarded. A closed cycle is one in which the working fluid never leaves the cycle, except through accidental leakage; instead, the working fluid undergoes a series of processes that are of such a nature that the fluid is returned periodically to its initial state and is then used again.

The open cycle is exemplified by the internal combustion engine, in which atmospheric air supplies the oxygen for combustion and in which the exhaust products are to the atmosphere. In fact, another way to describe an open cycle is to say that it is one that includes the atmosphere at some point.

The closed cycle is exemplified by the condensing steam power plant used for ship propulsion on many naval ships. In such a cycle, the working substance (water) is changed to steam in the boilers. The steam performs work as it expands through the turbines, and is then condensed to water again in the main condenser (main engine) and auxiliary condenser (ship's service turbine generator). The water is returned to the boilers as boiler feed, and is thus used over and over again. (This is known as the basic steam cycle.)

Thermodynamic cycles are also classified as heated-engine cycles or as unheated-engine cycles, depending upon the point in the cycle at which heat is added to the working substance. (The terms *heated engine* and *unheated engine* should not be confused with the term *heat engine*. Any machine that is designed to convert thermal energy to mechanical energy in the form of work is known as a heat engine. Thus, both internal combustion engines and steam turbines are heat engines; but the first has a heated-engine cycle and the second has an unheated-engine cycle). In a heated-engine cycle, heat is added to the working substance in the engine itself. In an unheated-engine cycle, the working substance receives heat in some device that is separate from the engine. The condensing steam power plant has an unheated engine cycle, since the working substance is heated separately in the boilers and then piped to the engines (steam turbines). There are five basic elements in any thermodynamic cycle:

- The working substance

- The engine

- A heat source, or high-temperature region

- A heat receiver, or low-temperature region

- A pump

The working substance is the medium by which energy is carried through the cycle. The engine is the device that converts the thermal energy of the working substance into useful mechanical energy in the form of work. The heat source supplies heat to the working substance. The heat receiver absorbs heat from the working substance. The pump moves the working substance from the low-pressure side of the cycle to the high-pressure side.

The essential elements of a closed, unheated-engine cycle are shown in figure 8-8. This is the basic plan of the typical condensing steam power plant.

In an open, heated-engine cycle, such as that of an internal combustion engine, the essential elements are all present but are arranged in a somewhat different order. In this type of cycle, atmospheric air and fuel are both drawn into the cylinder of the engine. Combustion takes place in the cylinder, either by compression or by spark, and the resulting internal energy of the working substance is transformed into work by which the piston is moved. Since the space above the piston is a high-pressure area when the piston is near the top of its stroke and a low-pressure area when the piston is near the bottom, the piston may be thought of as a pump in the sense that it "pumps" the working fluid from the low-pressure to the high-pressure side of the system. Thus, in terms of function, the piston-and-cylinder

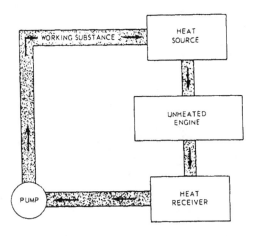

Figure 8-8.–Essential elements of closed unheated-engine cycle.

arrangement may be thought of as including the heat source, the engine, and the pump. An open, heated-engine cycle might therefore be represented as shown in figure 8-9.

THE CONCEPT OF REVERSIBILITY

When we put a pan of water on the stove and turn on the heat, we expect the water to boil rather than to freeze. After we have mixed hot and cold water, we do not expect the resulting mixture to resolve itself into two separate batches of water at two different temperatures. When we open the valve on a cylinder of compressed air, we expect compressed air to rush out; we would be quite surprised if atmospheric air rushed into the cylinder and compressed itself. When a shaft is rotating, we expect a temperature rise in the bearings; when the shaft has been stopped, we would be truly amazed to observe internal energy from the bearings flowing to the shaft and causing it to start rotating again. When we drag a block of wood across a rough surface, we expect some of the mechanical energy expended in this act to be converted into thermal energy–that is, we expect a storage of internal energy in the wooden block and the rough surface, as evidenced by temperature rises in these materials. But if this stored internal energy should suddenly turn to and move the wooden block back to its original position, our incredulity would know no bounds.

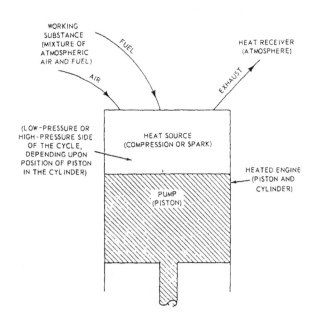

Figure 8-9.–Essential elements of open heated-engine cycle.

All of which merely goes to show that we have certain expectations, based on experience, as to the direction in which processes will move. The reasonableness of our expectations is attested by the fact that in all recorded history there is no report of water freezing instead of boiling when heat is applied; there is no report of a lukewarm fluid unmixing itself and separating into hot and cold fluids; there is no report of a gas compressing itself without the agency of some external force; there is no report of the heat of friction being spontaneously used to perform mechanical work.

Are these actions really impossible? The first law of thermodynamics says that mechanical energy and thermal energy are mutually convertible, but it says nothing about the direction of such conversions. If we consider only the first law, all the improbable actions just mentioned are perfectly possible and all processes could be thought of as being reversible. In an absolute sense, perhaps, we cannot guarantee that water will never freeze instead of boil when it is placed on a hot stove; but we are certainly safe in saying that this or any other completely reversible thermodynamic process is at the outer limits of probability. For all practical purposes, then, we will say that there is no such thing as a completely reversible process.

Nevertheless, the concept of reversibility is extremely useful in evaluating real thermodynamic processes. At this point, therefore, let us define a reversible thermodynamic process as one that would have the following characteristics:

— The process could be made to occur in precisely reverse order, so that the energy system and all associated systems would be returned from their final condition to the conditions that existed before the process started.

— All energy that was transformed or redistributed during the process would be returned from its final to its original form, amount, and location.

THE SECOND LAW OF THERMODYNAMICS

Since the first law of thermodynamics does not deal with the direction of thermodynamic processes, and since experience indicates that actual processes are not reversible, it is apparent that the first law must be supplemented by some statement of principle that will limit the direction of thermodynamic processes. The second law of thermodynamics is such a statement. Although the second law is perhaps more empirical than the first law, and perhaps something less of a "law" in an absolute sense, it is of enormous practical value in the study of thermodynamics.

The second law of thermodynamics may be stated in various ways. One statement, known as the Clausius statement, is that no process is possible where the sole result is the removal of heat from a low-temperature reservoir and the absorption of an equal amount of heat by a high-temperature reservoir. Among other things, this statement indicates that water will not freeze when heat is applied. Note that the Clausius statement includes and goes somewhat beyond the common observation that heat flows only from a hotter to a colder substance. The statement that no process is possible where the sole result is the removal of heat from a single reservoir and the performance of an equivalent amount of work is known as the Kelvin-Planck statement of the second law. Among other things, this statement says that we cannot expect the heat of friction to reverse itself and perform mechanical work. More broadly, this statement indicates a certain one-sidedness that is inherent in thermodynamic processes. Energy in the form of work can be converted entirely to energy in the form of heat; but energy in the form of heat can never be entirely converted to energy in the form of work. A very important inference to be drawn from the second law is that no engine, actual or ideal, can convert all the heat supplied to it into work, since some heat must always be rejected to a receiver that is at a lower temperature than the source. In other words, there can be no heat flow without a temperature difference and there can be no conversion to work without a flow of heat. A further inference from this inference is sometimes given as a statement of the second law: No thermodynamic cycle can have a thermal efficiency of 100 percent.

We must say, then, that the first law of thermodynamics deals with the conservation of energy and with the mutual convertibility of heat and work, while the second law limits the direction of thermodynamic processes and the extent of heat-to-work energy conversions.

THE CONCEPT OF ENTROPY

Entropy is the theoretical measure of energy, such as steam, that cannot be transformed into mechanical work in a thermodynamic system. The concept of reversibility and the second law of thermodynamics are closely related to the concept of entropy. In fact, the second law may be stated as No process can occur in which the entropy of an isolated system decreases; the total entropy of an isolated system can theoretically remain constant in some reversible (ideal) processes, but in all irreversible (real) processes the total entropy of an isolated system must increase.

From other statements of the second law, we know that the transformation of heat to work is always dependent upon a flow of heat from a high-temperature region to a low-temperature region. The concept of the unavailability of a certain portion of the energy supplied as heat to any thermodynamic system is clearly implied in the second law, since it is apparent that some heat must always be rejected to a receiver that is at a lower temperature than the source, if there is to be any conversion of heat to work. The heat that must be so rejected is therefore unavailable for conversion into mechanical work.

Entropy is an index of the unavailability of energy. Since heat can never be completely converted into work, we may think of entropy as a measure or an indication of how much heat must be rejected to a low-temperature receiver if we are to use the rest of the heat for the production of useful work. We may also think of entropy as an index or measure of the reversibility of a process. All real processes are irreversible to some degree, and all real processes involve a "growth" or increase of entropy. Irreversibility and entropy are closely related; any process in which entropy has increased is an irreversible process.

The entropy of an isolated system is at its maximum value when the system is in a state of equilibrium. The concept of an absolute minimum–that is, an absolute zero value of entropy–is sometimes referred to as the third law of thermodynamics or Nernst's law. This principle states that the absolute zero of entropy would occur at the absolute zero of temperature for any pure material in the crystalline state. By extension, therefore, it should be possible to assign absolute values to the entropy of pure materials, if such absolute values were needed. For most purposes, however, we are interested in knowing the values of the changes in entropy rather than the absolute values of entropy. Hence an arbitrary zero point for entropy has been established at 32$F.

Entropy changes depend upon the amount of heat transferred to or from the working fluid, upon the absolute temperature of the heat source, and upon the absolute temperature of the heat receiver. Although actual entropy calculations are complex beyond the scope of this text, one equation is given here to indicate the units in which entropy is measured and to give the relationship between entropy and heat and temperature.

Note that this equation applies only to a reversible isothermal process in which $T_1 = T_2$.

$$S_2 - S_1 = \frac{Q}{T}$$

where

S_1 = total entropy of working fluid at state 1, in Btu per °R

S_2 = total entropy of working fluid at state 2, in Btu per °R

Q = heat supplied, in Btu

T = absolute temperature at which the process takes place, in °R

The fact that the total entropy of an isolated system must always increase does not mean that the entropy of all parts of the system must always increase. In many real processes, we find increases in entropy in some parts of a system and, at the same time, decreases in entropy in other parts of the system.

But the important thing to note is that the increases in entropy are always greater than the decreases; therefore, the total entropy of an isolated system must always increase.

Each increase in entropy is permanent. In a universal sense, entropy can be created but it can never be destroyed or gotten rid of, although it may be transferred from one system to another. Every natural process that occurs in the universe increases the total entropy of the universe, and this increase in entropy is irreversible. The concept of the universe eventually "running down" might be expressed in terms of entropy by saying that the entropy of the universe is constantly "building up." The so-called "heat death of the universe" is envisioned as the ultimate result of all possible natural processes having taken place and the universe being in total equilibrium, with entropy at the absolute maximum. Such a statement need not imply a total lack of energy remaining in the universe; but any energy that might remain would be completely unavailable and therefore completely useless.

THE CARNOT PRINCIPLE

According to the second law of thermodynamics, no thermodynamic cycle can have a thermal efficiency of 100 percent–that is, no heat engine can convert into work all of the energy that is supplied as heat. The question now arises as to how much heat must be rejected to a receiver that is at a lower temperature than the source? Looking at it another way, what is the maximum thermal efficiency that could theoretically be achieved by a heat engine operating without friction and without any other of the irreversible processes that must occur in all real machines?

To answer this question, Carnot, a French engineer, developed an imaginary and completely reversible cycle. In the Carnot cycle, all heat is supplied at a single high temperature and all heat that must be rejected is rejected at a single low temperature. The cycle is fully reversible. When proceeding in one direction, the Carnot cycle takes in a certain amount of heat, rejects a certain amount of heat, and puts out a certain amount of work. When the cycle is reversed, the quantity of work that was originally the output of the cycle is now put into the cycle; the amount of heat that was originally taken in is now the amount rejected; and the amount of heat that was originally rejected is now the amount taken in. When thus reversed, the cycle is called a Carnot refrigeration cycle.

Obviously, no real machine is capable of such complete reversibility, but the concept of the Carnot cycle is nonetheless an extremely useful one. By analysis of the Carnot cycle, it can be proved that no engine, actual or ideal, can be more efficient than an ideal, reversible Carnot cycle. The thermal efficiency of the Carnot cycle is given by the following equation:

$$\text{thermal efficiency} = \frac{\text{work output}}{\text{heat input}} \quad \frac{T_s - T_r}{T_s}$$

where T_s equals the absolute temperature at which heat flows from the source to the working fluid and T_r equals the absolute temperature at which heat is rejected to the receiver.

The implications of this statement are of profound importance, since it establishes the fact that thermal efficiency depends only upon the temperature difference between the heat source and the heat receiver. Thermal efficiency does not depend upon the properties of the working fluid, the type of engine used in the cycle, or the nature of the process–combustion, nuclear fission, and so on–that produces the heat at the heat source. The basic principle thus established by analysis of the Carnot cycle is called the Carnot principle, and may be stated as follows: The motive power of heat is independent of the agents employed to realize it, its quantity being fixed solely by the temperatures of the bodies between which the transfer of heat occurs.

WORKING SUBSTANCES

As previously noted, a thermodynamic system requires a working substance to receive, store, transport, and deliver energy. The working substance is almost always a fluid and is therefore frequently referred to as the working fluid. Water (together with its vapor and steam) is one of the most commonly used working fluids, although air, ammonia, carbon dioxide, and a wide variety of other fluids are used in certain kinds of systems. A working substance may change its physical state during the course of a thermodynamic cycle or it may remain in one state, depending upon the nature of the cycle and the processes involved.

To understand the behavior of working fluids, you should have some understanding of the laws of perfect gases, of the relationships between liquids and their vapors, and of the ways in which the properties of working fluids may be represented and tabulated. These topics are discussed in the following sections.

LAWS OF PERFECT GASES

The relationships of the volume, the absolute pressure, and the absolute temperature in the hypothetical substances known as "perfect gases" were stated by the physicists Boyle and Charles in the form of various gas laws. The laws thus established may be combined and summarized in the general statement: For a given weight of any gas, the product of the absolute pressure and the volume, divided by the absolute temperature, is a constant. Or, in equation form,

$$\frac{pV}{T} = \frac{p_1 V_1}{T_1} = \frac{p_2 V_2}{T_2} = R$$

where

p = absolute pressure

V = total volume

T = absolute temperature

R = the gas constant

Although the laws of perfect gases were developed on the basis of experiments made with air and other real gases, later experiments showed that these relationships do not hold precisely for real gases over the entire range of pressures and temperatures. However, air and other gases used as working fluids may be treated as perfect gases over quite a wide range of pressures and temperatures without any appreciable error being introduced. Values of the gas constant for some common gases are as follows:

Air 53.3

Oxygen 48.3

Nitrogen 55.0

Hydrogen 766.0

Helium 386.0

LIQUIDS AND THEIR VAPORS

When heat is transferred to a liquid, the average velocity of the molecules is increased and the amount of internal kinetic energy stored in the liquid is increased. As the average velocity of the molecules increases, some molecules that are at or near the surface of the liquid momentarily achieve unusually high velocities; and some of these escape from the liquid and enter the space above, where they exist in the vapor state. As more and more of the molecules escape and come into the vapor state, the probability increases that some of the vapor molecules will momentarily have unusually low velocities; these molecules will be captured by the liquid. As a result of this exchange of molecules between the liquid and the vapor, a condition of equilibrium is reached and an equilibrium pressure is established. The equilibrium pressure depends upon the molecular structure of the fluid and upon its temperature. For any given fluid, therefore, there is a definite relationship between the temperature and the pressure at which a liquid and its vapor may exist in equilibrium contact with each other.

As long as the vapor is in contact with the liquid from which it is being generated, the liquid and the vapor will remain at the same temperature. If the liquid and the vapor are in a closed container (such as a boiler with all steam stop valves closed) both the temperature and the pressure of the liquid and its vapor will increase as heat is added. If the vapor is permitted to leave the steam space at a rate equal to the evaporation rate, an equilibrium will be established at the equilibrium pressure for the particular temperature.

The pressure and the temperature that are related in the manner just described are known as the saturation pressure and the saturation temperature. For any specified pressure there is a corresponding temperature of vaporization known as the saturation temperature; and for any specified temperature there is a corresponding saturation pressure.

A liquid that is under any specified pressure and at the saturation temperature for that particular pressure is called a saturated liquid. A liquid that is at any temperature below its saturation temperature is said to be a subcooled liquid. For example, the saturation temperature that corresponds to atmospheric pressure (14.7 psia) is 212°F for water. Therefore, water at 212°F and under atmospheric pressure is said to be a saturated liquid. Water flowing in a river or standing in a pond is also under atmospheric pressure, but it is at a much lower temperature; hence, this water is said to be subcooled.

A vapor that is under any specified pressure and at the saturation temperature corresponding to that pressure is said to be a saturated vapor. Water at 14.7 psia and 212°F produces a vapor known as saturated steam. As previously noted, it is impossible to raise the temperature of a vapor above the temperature of its liquid as long as the two are in contact. If the vapor is drawn off into a separate container, however, and additional heat is supplied to the vapor, the temperature of the vapor is raised. A vapor that has been raised to a temperature that is above its saturation temperature is called a superheated vapor, and the vessel or container in which the saturated steam is superheated is called a superheater. The elementary boiler and superheater illustrated in figure 8-10 show the general principle of generating and superheating steam. Practically all naval propulsion boilers have superheaters for superheating the saturated steam generated in the generating sections of the boiler; the steam is then called superheated steam. The amount by which the temperature of a superheated vapor exceeds the temperature of a saturated vapor at the same pressure is known as the degree of superheat. For example, if saturated steam at a pressure of 600 psia and a corresponding saturation temperature of 486°F is superheated to 786°F, the degree of superheat is 300°F.

For any substance there is a critical point at which the properties of the saturated liquid are exactly the same as the properties of the saturated vapor. For water, the critical point is reached at 3206.2 psia (critical pressure) and 704.40°F (critical temperature). At the critical point, the vapor and the liquid are indistinguishable. No change of physical state occurs when the pressure is increased or when additional heat is supplied; the vapor cannot be made to liquefy and the liquid cannot be made to vaporize as long as the substance is at or above its critical pressure and critical temperature. At this point, we could no longer refer to water and steam since we cannot tell the water and the steam apart; instead, the substance is now merely called a fluid or a working fluid. Boilers designed to operate above the critical point are called supercritical boilers. Supercritical boilers are not used at present in the propulsion plants of naval ships; however, some boilers of this type are used in stationary steam power plants.

REPRESENTION OF PROPERTIES

The condition of a working fluid at any point within a thermodynamic cycle or system is established by the properties of the substance at that point. The properties that are of special interest in engineering thermodynamics include pressure, temperature, volume, enthalpy, entropy, and internal energy. These properties have been discussed at some length in this chapter. At this point we are concerned less with the properties themselves than with the way in which they are tabulated and the way in which they are represented graphically.

Steam Tables

In the region near a change of physical state, the behavior of a gaseous substance becomes too complex for the relatively simple energy calculations that apply to perfect gases and to many real gases over a wide range of pressures and temperatures. Because of the complicated equations needed to describe the properties of vapors, engineers customarily depend upon tables of vapor properties of liquids and their vapors. The vapor tables that are perhaps most commonly used are those that give the thermodynamic properties of steam. The most authoritative tables of thermodynamic properties

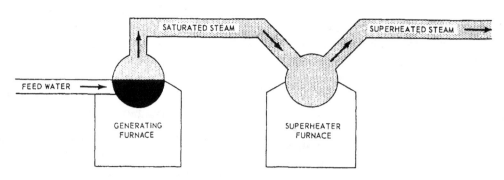

Figure 8-10.–Elementary boiler and superheater.

of steam are those prepared by Keenan and Keyes under the title of *Thermodynamic Properties of Steam.* Figure 8-11 is excerpted from Table III of the Keenan and Keyes steam tables; it is included here chiefly to show the general arrangement of information in these tables, rather than to provide any significant amount of data concerning the thermodynamic properties of saturated steam. The information given in each column of the steam table for the properties of saturated steam is described briefly in the sections that follow. Note that the subscript f is commonly used to denote properties of the saturated liquid, g to denote properties of the saturated vapor, and fg to denote property change between the two states.

Column 1 gives the saturation pressure of the saturated water and the saturated steam at the temperature given in column 2. Note that the pressure is absolute pressure, not gauge pressure.

Column 2 gives the saturation temperature for the pressure shown in column 1. This temperature is what is commonly referred to as "the boiling point" of the liquid at the pressure shown in column 1. In the steam tables, the temperature is usually given in degrees Fahrenheit rather than in degrees of an absolute temperature scale. However, the absolute temperature can always be obtained by simple computation if it should be needed.

Column 3 gives the specific volume of the saturated liquid (water) at the pressure shown in column 1 and the temperature shown in column 2. Specific volume is expressed in cubic feet per pound.

Column 4 gives the specific volume of the saturated vapor (steam) at the pressure and temperature shown.

Note that some portions of the steam tables have another column for the increase in specific volume that occurs during evaporation. This column is labelled Specific Volume, Evapo symbolized by $V_{fg}°$.

Column 5 gives the enthalpy per pound of the saturated liquid at the pressure and temperature shown. The enthalpy of saturated water at 32°F and the corresponding saturation pressure of 0.08854 psia is taken as zero; hence, all enthalpy figures indicate enthalpy with respect to this arbitrarily assigned zero point. For example, the enthalpy of 1 pound of saturated water at 190 psia and 377.51°F is 350.79 Btu more than the enthalpy of 1 pound of saturated water at 0.08854 psia and 32°F.

Column 6 gives the enthalpy of evaporation per pound of working fluid–that is, the change in enthalpy that occurs during evaporation. This column is of particular significance since it indicates the Btu per pound that must be supplied to change the saturated liquid (water) to the saturated vapor (steam) at the pressure and temperature shown. In other words, the enthalpy of evaporation is what we formerly described as the latent heat of vaporization.

PROPERTIES OF SATURATED STEAM											
ABS. PRESS. (PSIA)	TEMP.°F	SPECIFIC VOLUME		ENTHALPY			ENTROPY			INTERNAL ENERGY	
		SAT. LIQUID	SAT. VAPOR	SAT. LIQUID	EVAP.	SAT. VAPOR	SAT. LIQUID	EVAP.	SAT. VAPOR	SAT. LIQUID	SAT. VAPOR
p	t	v_f	v_g	b_f	b_{fg}	b_g	s_f	s_{fg}	s_g	u_f	u_g
190	377.51	0.01833	2.404	350.79	846.8	1197.6	0.5381	1.0116	1.5497	350.15	1113.1
200	381.79	0.01839	2.288	355.36	843.0	1198.4	0.5435	1.0018	1.5453	354.68	1113.7
250	400.95	0.01865	1.8438	376.00	825.1	1201.1	0.5675	0.9588	1.5263	375.14	1115.8
(1)	(2)	(3)	(4)	(5)	(6)	(7)	(8)	(9)	(10)	(11)	(12)

Courtesy of Joseph H. Keenan and Fredrick G. Keyes, Thermodynamic Properties of Steam, New York, John Wiley & Sons, Inc., 1937

147.X

Figure 8-11.–Excerpts from Keenan and Keyes steam tables.

Column 7 gives the enthalpy per pound of the saturated vapor at the pressure and temperature shown. Note that this is the sum of the enthalpy of the saturated liquid and the enthalpy of evaporation.

Column 8 gives the entropy per pound of the saturated liquid at the pressure and temperature shown. The zero point for entropy, like the zero point for enthalpy, is arbitrarily established at 32°F and the corresponding saturation pressure of 0.08854 psia.

Column 9 gives the entropy of evaporation per pound of working fluid at the indicated pressure and temperature. In other words, this column shows the change of entropy that occurs during evaporation.

Column 10 gives the entropy per pound of the saturated vapor at the pressure and temperature shown. Note that this is the sum of the entropy of the saturated liquid and the entropy of evaporation.

Columns 11 and 12 give the internal energy of the saturated liquid and the internal energy of the saturated vapor, respectively. In addition to giving properties of the saturated liquid and the saturated vapor, the steam tables include data on the superheated vapor and other pertinent information.

GRAPHICAL REPRESENTATION OF PROPERTIES

Thermodynamic graphs or diagrams are frequently useful for showing the relationship between two or more properties of a working fluid.

The relationships among pressure, volume, and temperature of a perfect gas are sometimes represented by a three-dimensional diagram of the type shown in figure 8-12. The p-V-T surface of a real substance may also be represented in this way, but the diagrams become much more complex because the relationships among the properties are more complex. A simplified p-V-T surface for water is shown in figure 8-13. The significance of some of the lines on this diagram will become clearer as we consider some of the related two-dimensional diagrams.

Three-dimensional diagrams are extremely useful in giving an overall picture of the p-V-T relationships, but they are difficult to construct and are somewhat difficult to use for detailed analysis. Two-dimensional graphs are frequently projected from the three-dimensional p-V-T surfaces. Even on two-dimensional diagram, a great many relationships of properties can be indicated by contour lines or superimposed curves.

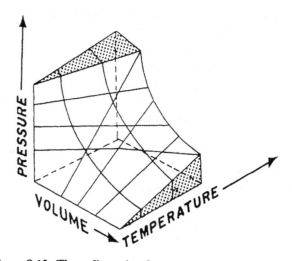

Figure 8-12.–Three-dimensional representation of the p-V-T surface for perfect gas.

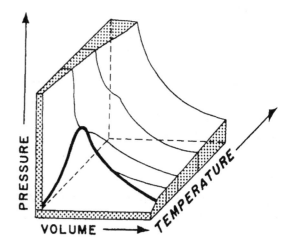

Figure 8-13.–Three-dimensional representation of the p-V-T surface for water.

The p-V diagram is made by plotting known values of pressure (p) along the ordinate and values of specific volume (V) along the abscissa. The p-V diagram, as it applies to internal combustion engines, is discussed in chapter 17 of this text. To illustrate the construction of a p-V diagram, let us consider the isothermal compression of 1 pound of air from an initial pressure of 1000 pounds per square foot absolute (psfa) to a final pressure of 6000 psfa. Let us assume that the air is at a temperature of 90°F, or 550°R. Since we may treat air as a perfect gas under these conditions of pressure and temperature, we may use the laws of perfect gases and the following equation:

$$pV = RT$$

where

p = absolute pressure, psfa

V = specific volume, cu ft per lb

R = gas constant (53.3 for air)

T = absolute temperature, °R

Since the compression is isothermal, T is constant and the expression RT is equal to 53.3 x 550, or 29,315. From the equation, we can see that p and V must vary inversely; as p goes up, V goes down. Hence, for any given value of p we may find a value of V merely by dividing 29,315 by p. Choosing six values of p and computing the values of V, we obtain the following values:

STATE A: p = 1000, V = 29.3

STATE B: p = 2000, V = 14.7

STATE C: p = 3000, V = 9.8

STATE D: p = 4000, V = 7.3

STATE E: p = 5000, V = 5.9

STATE F: p = 6000, V = 4.9

By plotting these values on graph paper, we obtain the p-V diagram shown in figure 8-14.

The curve applies only to the indicated temperature—that is, it is an isothermal curve. The values of p and V may be calculated for the same process at other temperatures, and plotted as before; in this case

we obtain a series of isothermal curves (or isotherms) such as those shown in figure 8-15.

A p-V diagram for water and steam is shown in figure 8-16. This diagram and, in fact, most diagrams

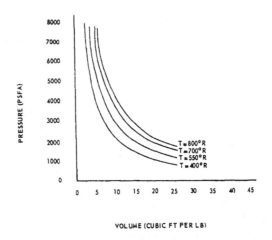

VOLUME (CUBIC FT PER LB)

Figure 8-15.–Group of isothermal curves a on p-V diagram.

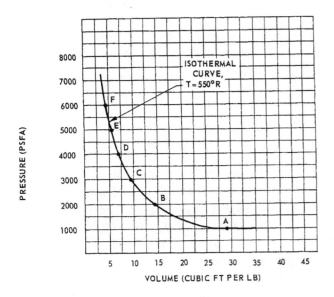

Figure 8-14.–Constant temperature (isothermal) line on a p-V diagram.

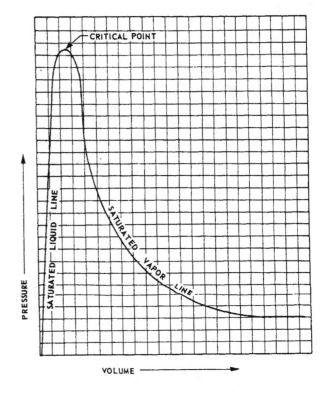

VOLUME

Figure 8-16.–A p-V diagram for water.

for real substances in the region of a state change is not drawn to scale because of the very great difference in the specific volume of the liquid and the specific volume of the vapor. Even though it is not drawn to scale, the p-V diagram serves a useful purpose in indicating the general configuration of the saturated liquid line and the saturated vapor line. These lines, which are called process lines, blend smoothly at the critical point. The shape formed by the process lines is characteristic of water and will be observed on all p-V diagrams of this substance.

A two-dimensional pressure-temperature (p-T) diagram of the type shown in figure 8-17 is useful because it indicates the way in which the phase of a substance depends upon pressure and temperature. The solid-liquid curve, for example, indicates the effects of pressure on the melting (or freezing) point; the liquid-vapor curve indicates the effects of pressure on the boiling point; and the solid-vapor indicates the effects of pressure on the sublimation point. The intersection of these three equilibrium curves shows the triple point–that is, the single pressure and temperature at which all three phases can coexist. The termination of the liquid-vapor equilibrium curve indicates the critical point, the point at which the liquid and the vapor are no longer distinguishable because their properties are identical.

Other two-dimensional diagrams that find application in engineering include the temperature-entropy (T-s) diagram; the enthalpy-entropy (h-s) diagram, also called the Mollier diagram; the pressure-enthalpy (p-h) diagram; and the enthalpy-volume (h-v) diagram. Of these, the Mollier diagram is probably of major importance in the study of

steam engineering. Mollier diagrams are included in many steam tables and are also available in engineering handbooks and some thermodynamics texts.

ENERGY RELATIONSHIPS IN THE SHIPBOARD PROPULSION CYCLE

At the beginning of this chapter, we stated that the shipboard engineering plant may be thought of as a series of devices and arrangements for the exchange and transformation of energy. Many of these transformation and energy exchanges have been discussed in this chapter, but they have not been taken up in sequence. Figure 8-18 illustrates the basic propulsion cycle of a conventional steam-driven ship with geared turbine drive, and it shows some of the major energy transformations that take place.

The first energy transformation occurs when fuel oil is burned in the boiler furnace. By the process of combustion, the chemical energy stored in the fuel oil is transformed into thermal energy. Thermal energy flows from the hot combustion gases to the water in the boiler. While the boiler stop valves are still closed, steam begins to form in the boiler; the volume of the steam remains constant but the pressure and temperature increase, indicating a storage of internal energy. When operating pressure is reached and the steam stop valves are opened, the high pressure of the steam causes it to flow to the turbines. The pressure of the steam thus provides the potential for doing work; the actual conversion of heat to work takes place in the turbines. The changes in internal energy between the boiler and the condenser (as evidenced by changes in pressure temperature) indicate that heat has been converted to work in the turbines. The work output of the turbines turns the shaft and so drives the ship.

Two main energy transformations are involved in converting thermal energy to work in the turbines. First, the thermal energy of the steam is transformed into mechanical kinetic energy as the steam flows through one or more nozzles. And second, the mechanical kinetic energy of the steam is transformed into work as the steam impinges upon the projecting blades of the turbine and thus causes the turbine to turn. The turning of the turbine rotor causes the propeller shaft to turn also, although at a slower speed, since the turbine is connected to the propeller shaft through the reduction gears. The steam exhausts from the turbine to the condenser, where it gives up its latent heat of condensation to the circulating sea water.

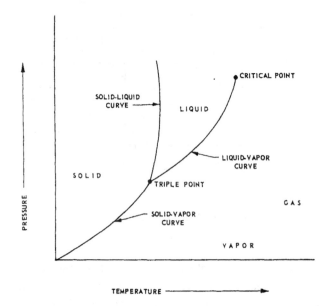

Figure 8-17.–A p-T diagram.

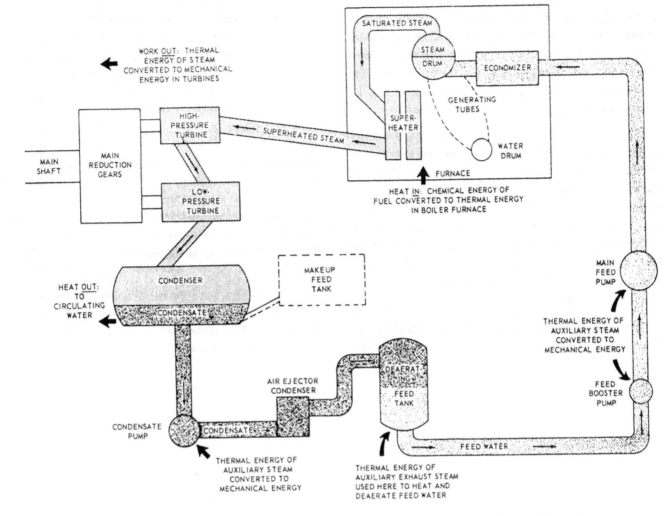

Figure 8-18.–Energy relationships in the basic propulsion cycle of a conventional steam-driven ship.

For the remainder of this cycle, energy is required to get the water (condensate and feed water) back to the boiler where it will again be heated and changed into steam. The energy used for this is generally the thermal energy of the auxiliary steam. In the case of turbine-driven feed pumps, the conversion of thermal energy to mechanical energy occurs in the same way as it does in the case of the propulsion turbines. In the case of motor-driven pumps, the energy conversion is from thermal energy to electrical energy (in a turbogenerator) and then from electrical energy to mechanical energy (work) in the pumps.

ENERGY BALANCES

From previous discussion, we know that putting 1 Btu in at the boiler furnace does not mean that 778 ft-lb of work will be available for propelling the ship through the water. Some of the energy put in at the boiler furnace is used by auxiliary machinery, such as pumps and

forced draft blowers, to supply the boiler with feedwater, fuel oil, and combustion air. Distilling plants, turbogenerators, steering gears, steam catapults, heating systems, galley and laundry equipment, and many other units throughout the ship use energy derived directly or indirectly from the energy put in at the boiler furnace.

In addition, there are many "energy losses" throughout the engineering plant. As we have seen, energy cannot actually be lost. But when it is transformed into a form of energy that we cannot use, we say there has been an energy loss. Since no insulation is perfect, some thermal energy is always lost as steam travels through piping. Friction losses occur in all machinery and piping. Some heat must be wasted as the combustion gases go up the stack. Some heat must be lost at the condenser as the steam exhausted from the turbines gives up heat to the circulating seawater. We cannot expect all of the heat supplied to be converted into work; even in the most efficient possible cycle, we know that some heat must always be rejected to a receiver that is at a lower temperature than the source. Thus, each Btu that is theoretically put in at the boiler furnace must be divided up a good many ways before the energy can be completely accounted for. But the energy account will always balance. Energy in must always equal energy out.

Designers of engineering equipment use energy balances to analyze energy exchanges and to compute the energy requirements for proposed equipment or plants. Operating engineers use energy balances to evaluate plant performance. The engineering officer of a naval ship may find it necessary to make energy balances to find out whether the plant is operating at designed efficiency or whether defects are causing unnecessary waste of steam, fuel, and energy.

An energy balance for an entire engineering plant is usually made up in the form of a flow diagram similar to (but more detailed than) the one shown in figure 8-18. A number of numerical values are entered on the flow diagram, the most important of which are the quantities of the working fluid flowing per hour at various points and the thermodynamic states of the working fluid at various points. The quantity of fluid flowing per hour may be obtained by direct measurement of flow through flow meters or nozzles or by calculation; in some instances, we may have to estimate steam consumption of pumps and other units on the basis of available test data. Data on the state of the working fluid is obtained from pressure and temperature readings. Enthalpy calculations are made and noted at various points on the diagram. The complete energy balance includes tabular data as well as the data shown on the flow diagram.

NOTE: We have covered steam in this chapter of thermodynamics, but keep in mind that the principles of thermodynamics also works the same in all engineering plants, including nuclear or gas turbine.

CHAPTER 9

MACHINERY ARRANGEMENT AND PLANT LAYOUT

LEARNING OBJECTIVES

Upon completion of this chapter, you should be able to do the following:

1. Identify some arrangements of propulsion machinery in naval ships.

2. Recognize the engineering symbols and markings used with engineering piping systems.

3. Describe the modes of operation of a propulsion plant.

4. Recognize the piping systems aboard ship and the functional relationships of the systems to the overall operation of the propulsion plant.

INTRODUCTION

To understand a shipboard propulsion plant, it is necessary to visualize the general components of the plant as a whole and to understand the physical relationship among the various units. This chapter provides general information on the distribution and arrangement of propulsion machinery in a conventional steam turbine propulsion plant. This chapter also provides information on the arrangement of the major engineering piping systems that connect and serve the various auxiliary machinery.

As you study this chapter, remember that the information is general, rather than specific, in nature. No two ships—not even sister ships—are exactly alike in their arrangement of machinery and piping. The examples in this chapter give some idea of the variety of arrangements you may find on steam-driven surface ships, and they give the basic functions of the machinery and piping. These examples cannot, however, give an exact picture of the machinery and piping on any one ship.

ARRANGEMENT OF PROPULSION MACHINERY

The propulsion machinery on conventional steam-driven surface ships includes (1) the propulsion boilers, (2) the propulsion turbines, (3) the condensers, (4) the reduction gears, and (5) the pumps, forced draft blowers, deaerating feed tanks, and other auxiliary machinery units that directly serve the major propulsion units. On most steam-driven ships, other than oilers, tankers, and certain auxiliaries, the propulsion machinery is located amidships. Turbogenerators and their auxiliary condensers are usually located in the propulsion machinery spaces. Engineering equipment that is not directly associated with the operation of the major propulsion units may be located either in or near the propulsion machinery spaces, or in other parts of the ship, as space permits.

Before going further in our discussion, let us define some of the terminology we will use.

The terms *propulsion boilers* and, occasionally, *main boilers* identify the boilers in a propulsion plant. The terms are used to distinguish between them and the auxiliary boilers that are installed on some ships. The term *propulsion turbines* identifies the turbines in a propulsion plant. These turbines must be distinguished from the many auxiliary turbines that are used on all steam-driven ships to drive pumps, forced draft blowers, and auxiliary units. The propulsion turbines are also sometimes referred to as main engines. The term *propulsion unit* identifies the combination of propulsion turbines, main reduction gears, and main condenser in any one propulsion plant; however, propulsion unit may also be used in a more general sense to indicate any major unit in the propulsion plant.

9-1

Each propulsion shaft has an identifying number that is based on the location of the shaft. Working from starboard to port, the shaft nearest the starboard side is the No. 1 shaft, the one next inboard is the No. 2 shaft, and so forth. On newer construction ships, the propulsion machinery that serves each shaft is given the same number as that of the shaft. For example, the No. 2 shaft is served by the No. 2 propulsion unit and the No. 2 boiler. When two similar units serve one shaft, the identifying number is followed by a letter. If, for example, two boilers serve the No. 2 propulsion unit and the No. 2 shaft, the boilers are identified as boiler No. 2A and boiler No. 2B. When letters

are used, they follow in sequence, working from starboard to port and then from forward to aft.

Figure 9-1 gives a basic overview of the steam propulsion plant. This drawing does not indicate the physical layout of the plant; however, it provides enough information for a thermodynamic summary of this system. It is one that is worth your effort to be able to reproduce from memory. It presents the functional relationships for use in considering the plant layout. The drawing only shows the major piping systems. Of note is the auxiliary steam system. This drawing shows that auxiliary steam is provided to run the forced draft blower, main feed pump, condensate pump, main

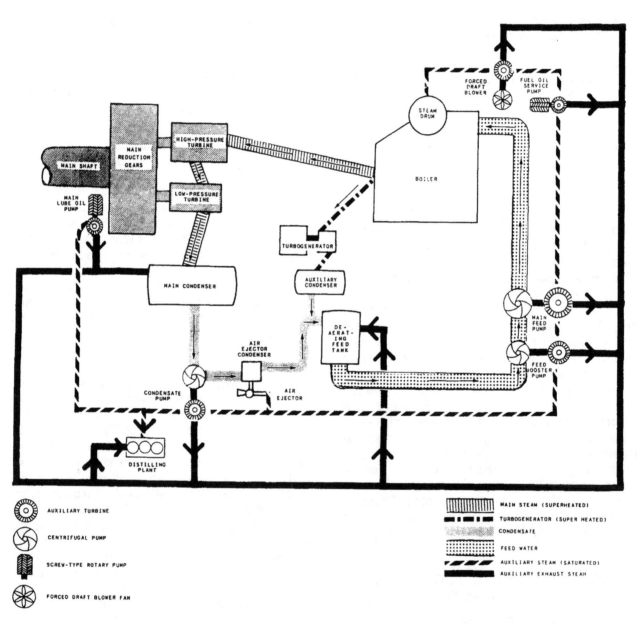

147.71

Figure 9-1.—Basic overview of the steam propulsion plant.

lube-oil pump, and the fuel-oil service pump. On newer ships, the main feed booster pump, condensate pump, main lube-oil pump, and fuel-oil service pump may be run by an electric motor. On some ships, there may be a split—one run by a steam turbine, the other by an electric motor. On ships having 1200-psi main steam, the main feed pump is run by main steam.

The propulsion machinery spaces may be physically arranged in several ways. Some ships have firerooms and engine rooms. Firerooms contain boilers and the stations for operating them, and engine rooms contain propulsion turbines and the stations for operating them. On some ships, one fireroom serves one engine room; on others, two firerooms serve one engine room. Instead of firerooms and engine rooms, many ships of recent design have spaces that are called machinery rooms (or main machinery rooms). Each machinery room contains both the boilers and the propulsion turbines that serve a particular shaft. On ships with automatic controls, the propulsion machinery is mostly operated from separate, enclosed, operating stations located within the machinery room, fireroom, or engine room.

No matter which arrangement is used in the machinery spaces, the propulsion machinery is usually on two levels. The condensers and the main reduction gears are on the lower level. The propulsion turbines and the high-speed pinion gears to which they are connected are on the upper level. The low-pressure turbine exhaust is located directly above the condenser. Boilers occupy spaces on both the lower level and the upper level. The stations for firing the boilers (sometimes referred to as the firing aisle) are on the lower level, while the stations for operating the valves that admit feedwater to the boiler are on the upper level. The boilers are usually located on the centerline of the ship or distributed symmetrically about the centerline. The long axis of the boiler drums runs fore and aft rather than athwartships. Other machinery, including the propulsion auxiliaries, is arranged in various ways as space and weight considerations permit.

Figure 9-2 shows the general arrangement of propulsion machinery on destroyers and cruisers. The machinery on these ships is arranged so that the forward fireroom and forward engine room can be operated together as one completely independent plant, while the after fireroom and the after engine room can be operated together as another completely independent plant. Each propulsion plant includes the propulsion machinery and the auxiliaries required for independent operation. A nonmachinery space may exist between the plants.

Figures 9-3, 9-4, 9-5, and 9-6 show the arrangement of machinery in the No. 1 fireroom and the No. 1 engine room of the destroyers of class DDG-45 and class DDG-46. The arrangement shown in these illustrations is also typical of that in destroyers DDG-39 through -46. The forward (No. 1) fireroom and engine room may be operated together as a separate plant, as may the after (No. 2) fireroom and engine room. This is defined as split-plant operation.

METHODS OF PROPULSION PLANT OPERATION

The major engineering systems on most naval ships are provided with cross-connections that allow the engineering plants to be operated either independently (split-plant) or together (cross-connected). In cross-connected operation, boilers may supply steam to propulsion turbines that they

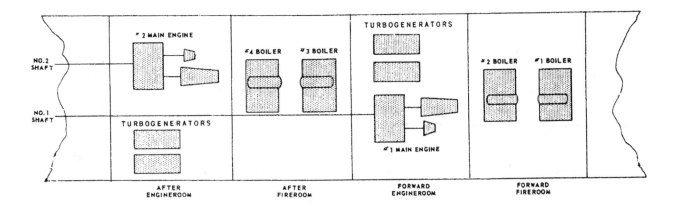

Figure 9-2.—Propulsion machinery arrangement on destroyers and cruisers.

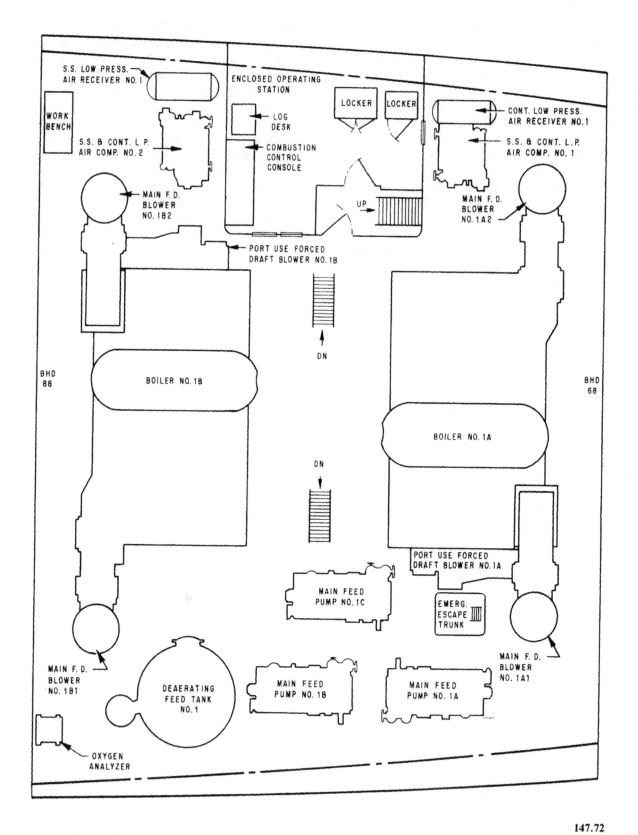

147.72

Figure 9-3.—Arrangement of machinery on upper level of No. 1 fireroom, DDG-45 and DDG-46.

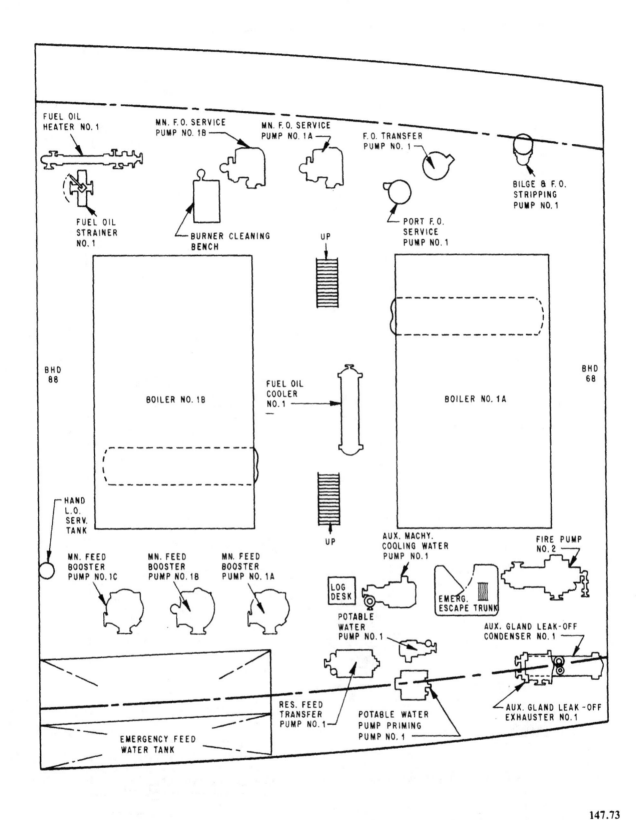

147.73

Figure 9-4.—Arrangement of machinery on lower level of No. 1 fireroom, DDG-45 and DDG-46.

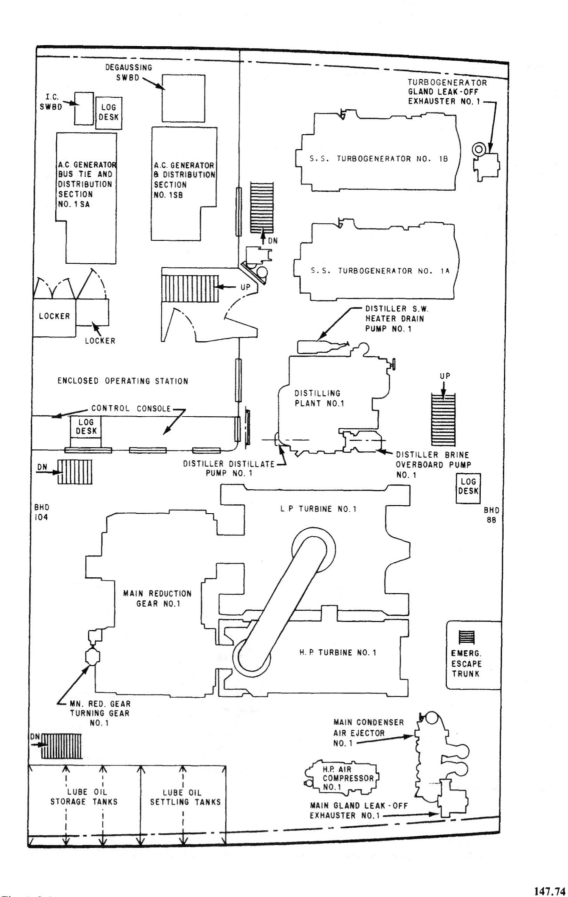

147.74

Figure 9-5.—Arrangement of machinery on upper level of No. 1 engine room, DDG-45 and DDG-46.

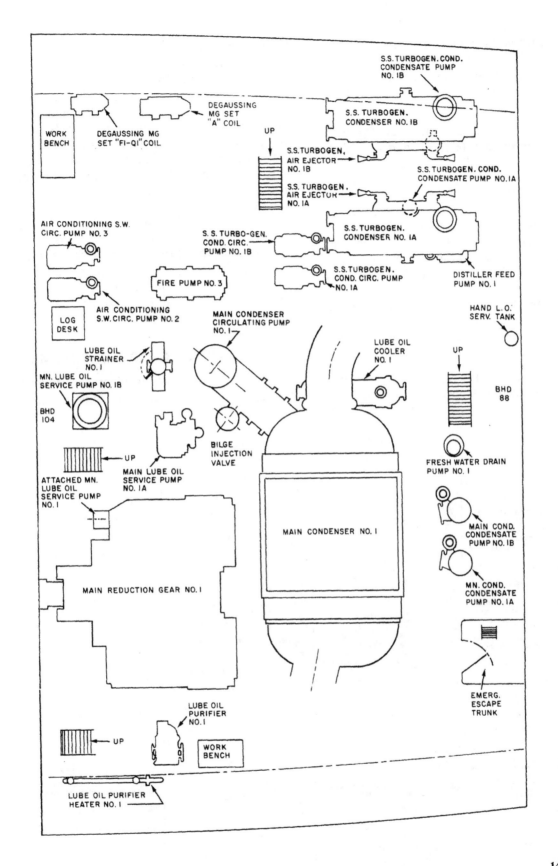

147.75

Figure 9-6.—Arrangement of machinery on lower level of No. 1 engine room, DDG-45 and DDG-46.

do not serve when the plant is split. In split-plant operation, the boilers, turbines, pumps, blowers, and other machinery are so divided that there are two or more separate and complete engineering plants.

Cross-connected operation was formerly standard for peacetime steaming, and split-plant operation was used only when maximum reliability was required; for example, when a ship was operating in enemy waters in time of war, operating in heavy seas, maneuvering in restricted waters, or engaged in underway fueling. However, the greater reliability of split-plant operation has led to its increasing use. At the present time, split-plant operation is the standard method of underway operation for most naval ships. Cross-connected operation is used for in-port steaming and at other times, such as when economy or maintenance considerations assume greater importance.

Aboard some ships the engineering plants can be operated by a method known as group operation.

For example, the USS *Forrestal* (CV-59) has four separate propulsion plants. The two forward plants (No. 1 and No. 4) constitute the forward group, and the two after plants (No. 2 and No. 3) constitute the after group. Although each of these four plants is normally used for the independent (split-plant) operation of one shaft, the boilers in any one plant can be cross-connected to supply steam to the turbines in the other plant in the same group. While underway, therefore, the boilers in the No. 1 plant can be cross-connected to supply steam to the No. 4 plant; however, they cannot be cross-connected to supply steam to the two plants in the other group (No. 2 and No. 3). For in-port operation, any boiler can be cross-connected to supply steam to any turbogenerator and to all other steam-driven auxiliaries.

ENGINEERING PIPING SYSTEMS

The various units of machinery and equipment aboard ship are connected by miles of piping.

Table 9-1.—Engineering Symbols

Symbol	Description	Symbol	Description	Symbol	Description	Symbol	Description
	GLOBE, GATE & ANGLE VALVE		ANGLE V NORMALLY CLOSED		UNLOADING VALVE WITH MANUAL OPERATION		STRAINER, REGULAR & INTERNAL
	GLOBE, GATE & ANGLE VALVE LOCKED OPEN		COMBINED EXHAUST & RELIEF VALVE		PRESS. RGLT VALVE – AIR PILOT OPERATED INCREASE PRESSURE CLOSES VALVE		STRAINER, DUPLEX
	GLOBE, GATE & ANGLE VALVE LOCKED CLOSED		BOILER SAFETY VALVE		AIR PILOT VALVE – AIR PRESSURE OPENS VALVE		STRAINER, DUPLEX MAGNETIC TYPE
	GLOBE, GATE & ANGLE HOSE VALVE		BOILER SAFETY VALVE, PILOT ACTUATED W/EASING GEAR		CONTROL PILOT		STRAINER, SIMPLEX — OFFSET
	GATE, HOSE VALVE, LOCKED OPEN		SUPERHEATER SAFETY V, PILOT OPERATED WITH EASING GEAR		AIR OIL OPERATED TRIP VALVE		STRAINER, Y TYPE
	GATE HOSE VALVE, LOCKED CLOSED		RELIEF OR SENTINEL VALVE		PRESS. RGLT DIAPHRAGM V, AIR PILOT ACTUATED, CONN ABOVE DIAPHRAGM		STRAINER, SELF CLEANING
	GATE VALVE OPERATED AT VALVE & ADJACENT COMPT		GLOBE & ANGLE STOP CHECK VALVE		PRESS RGLT DIAPHRAGM V, AIR PILOT ACTUACTED, CONN BELOW DIAPHRAGM		STRAINER, STEAM
	F.O. CONTROL V, INCREASED AIR PRESSURE OPENS V		GLOBE STOP CHECK VALVE, LOCKED IN CHECK POSITION		PRESSURE REDUCING VALVE		FILTER
	GLOBE, GATE & ANGLE V PNEU OPR – CLOSING ONLY		GLOBE STOP CHECK VALVE, LOCKED IN CHECK POSITION		UNLOADING VALVE		AIR PURIFIER & FILTER
	GLOBE, GATE & ANGLE VALVE - REMOTELY OPERATED		GLOBE STOP CHECK VALVE -- REMOTELY OPERATED		PRESSURE REDUCING VALVE EXTERNALLY SENSED PILOT		BIMETALLIC TRAP
	BUTTERLY VALVE		SWING CHECK VALVE		PRESSURE REDUCING VALVE INTERNALLY SENSED PILOT		OIL REMOVAL FILTER
	BUTTERFLY V LOCKED OPEN		LIFT CHECK VALVE		BOILER FEED REGULATOR V		SIGHT FLOW & GAGE GLASS
	LOCK SHIELD VALVE		BALL CHECK VALVE		AUTO. RECIRCULATING VALVE		SIGHT GLASS
	GLOBE, FLOAT OPERATED V		ANGLE LIFT CHECK VALVE		BACK PRESSURE VALVE		THERMOMETER
	GLOBE, VALVE – AIR OPERATED – SPRING CLOSING		COMBINATION REGULATOR & MOISTURE SEPARATOR (RELIEF TYPE) WITH HOSE & NOZZLE		SOLENOID VALVE – 3 WAY 2 PORT		THERMOMETER, REMOTE
	GLOBE VALVE – AIR OPERATED – SPRING OPENING		PRESSURE REGULATOR VALVE, AIR PILOT OPERATED WITH MANUAL OPERATION		SOLENOID VALVE		THERMOMETER, REMOTE LOCAL
	QUICK-ACTING VALVE				THREE WAY THERMOSTATIC V		DIFFERENTIAL PRESSURE GAGE
	MOTOR-OPERATED VALVE		3-WAY, 2-PORT, VALVE		3 WAY 2 PORT VALVE		PRESSURE GAGE
	BOILER BOTTOM BLOW VALVE		AUTO. PRESSURE REGULATOR		MANUAL CONTROL 3 WAY 2 PORT SPRING RETURN, SELF CLOSING		VACUUM GAGE
	GLOBE & ANGLE NEEDLE V GLOBE NEEDLE VALVE LOCKED OPEN		MICROMETER VALVE		SUCTION DISCHARGE MANIFOLD		VACUUM PRESSURE GAGE
	GLOBE NEEDLE VALVE – REMOTELY OPERATED		DIAPHRAGM OPERATED V		IMPULSE TRAP		DEHYDRATOR (REFRIG TYPE)
	GLOBE & ANGLE VALVE – BOILER SURFACE BLOW VALVE		THERMOSTATICALLY CONTROLLED VALVE		IMPULSE TRAP & STRAINER		THERMO ELEMENT

Each piping system consists of sections of pipe or tubing, fittings to join the sections, and valves to control the flow of fluid. Most piping systems also include a number of other fittings and accessories, such as vents, drains, traps, strainers, relief valves, and gauges and instruments. Piping system components are discussed later in chapter 18. In this chapter, we are concerned with the general arrangement and layout of the major engineering piping system aboard ship. You can obtain more detailed information on the piping systems of your ship from the plant and system diagrams in the Engineering Operational Sequencing System (EOSS), from the ship's Propulsion Operating Guide (POG), from the Ship's Information Book, or from the ship's file of blueprints or plans. You can learn the location of all machinery, piping, valves, and other units by tracing out the systems and drawing your own sketches. If you use the symbols shown in table 9-1 when making your skeches, you can learn the symbols that are generally used in engineering

blueprints and drawings at the same time that you are learning the arrangement of machinery and piping. As you get used to reading engineering drawings and blueprints, you will probably notice some variations in the symbols used. However, most of the symbols are easy enough to recognize after you have had some practice in reading and drawing engineering sketches. If you are in doubt about the meaning of a particular engineering symbol, look at the drawing to see whether there is a legend or list of symbols used in that drawing.

For the purpose of identification, each shipboard piping system is marked at suitable intervals along the length of the system. The markings may be applied with paint and stencils or with prepainted vinyl cloth markers. The markings are black letters on a white background for all systems except oxygen. Oxygen systems are marked with white letters on a dark background.

The piping identification markings include the functional name of the system and, where necessary, the specific service of the system. Markings

Table 9-1.—Engineering Symbols—Continued

SALINITY CELL	SIAMESE HOSE CONNECTION	EXPANSION JOINT	CALORIMETER
AIR GUN	AIR COCK	AIR PILOT (DIRECT ACTING)	LOUVER
DESUPERHEATER	CONSTANT HEAD CHAMBER	AIR PILOT (REVERSE ACTING)	AIR FILTER GAGE
FOOT VALVE	SPRAY	AIR PILOT WITH THERMOSTAT (DIRECT ACTING)	DAMPER
FLANGED SPOOL PIECE	LUBRICATOR	ORIFICE	COMPRESSED AIR PIPING
BULKHEAD JOINT, FIXED	HOSE RACK FOAM OUTLET	SPECTACLE FLANGE	ROOM THERMOSTAT VENT CONTROLLER
DRIP PAN	SLIP JOINT W/CUP & DRAIN	PRESSURE BREAKDOWN ORIFICE	CONSOLE AIR SUPPLY
SPANNER WRENCH	TEST CONNECTION	SPECTACLE FLG W/ORIFICE OPERATED FOR CLOSING ONLY	FIELD AIR SUPPLY
LOW PRESSURE FOG APPLICATOR	NATURAL SUPPLY OR EXHAUST THRU BULKHEAD	SPECTACLE FLG WITH TURBINE	AIR LOCK HEADER
SOUNDING TUBE STRIKING PLATE	DOOR TO REMAIN OPEN DURING SHUT DOWN CONDITION	AXIAL FLOW FAN (VERT)	PNEU SIGNAL PRESSURE LINE
SPLASH PLATE	AIR CONN AT CONSOLE	DUCT HEATER STEAM	GAGE IMPULSE LINE
CO2 HOSE REELS & CYLINDERS	AIR CONN AT PANEL	FIRE PLUG HOSE STATION	LUBRICATING OIL PIPING
FLUSH CONNECTION	CENTRIFUGAL PUMP	PRESSURE SWITCH	FUEL OIL PIPING
SOUNDING TUBE CAP	PUMP MOTOR	DIFFERENTIAL INDICATOR	RAW WATER
SOUND ISOLATION FITTING	HAND PUMP	15 LB CO2 CYLINDER	FRESH WATER PIPING
HOSE ADAPTER	FUEL OIL TRANSFER & STRIPPING PUMP	18 LB DRY CHEMICAL CYLINDER	EXHAUST PIPING
FIELD MAKE-UP CONNECTION		AIR FAILURE (PRESSURE SWITCH)	FUEL OIL VENT LINE
CAPPED TEST CONNECTION	FUNNEL, OPEN	LOW AIR SUPPLY (PRESSURE SWITCH)	EXHAUST LINE
FLOW NOZZLE	FUNNEL, STEAM—TELL TALE	HIGH AIR SUPPLY (PRESSURE SWITCH)	VALVE NORMALLY CLOSED
FLOW MEASURING NOZZLE	FUNNEL, CLOSED	AXIAL FLOW FAN (HORIZ)	VALVE NORMALLY OPEN
RESERVOIR		COOLING COIL DUCT TYPE	FOAM PIPING
WATER METER	FUNNEL WITH SWING COVER	UNIT COOLER	
SOOT BLOWER ELEMENT	FLEXIBLE CONNECTION	UNIT HEATER	
DESUPERHEATER NOZZLE	FLEXIBLE HOSE CONNECTION	FIRE HOSE RACK	
RETRACTABLE SOOT BLOWER	SLIP JOINT WITH CAP, DRAIN & FLEXIBLE ELEMENT	LINE I.D. NUMBER	
		CONVECTOR HEATER	

must also include arrows to show the direction of flow.

Piping identification markings are not required on piping in tanks, voids, cofferdams, bilges, and other unmanned spaces. All other piping must be marked at least once in each manned space and at least twice in each machinery space. Systems that serve propulsion plants and systems that convey flammable or toxic fluids must be marked at least twice in each space. Where feasible, piping identification markings are placed near the entry and near the exit to any space and at the junction of interconnecting systems. Short runs of piping that serve an immediately obvious purpose, such as short vents or drains, need not be marked. As a rule, piping on the weather decks does not require marking; if it does require marking, label plates are used, rather than stenciled paint or prepainted vinyl labels.

Most shipboard piping is painted to match and blend with its surrounding bulkheads, overheads, or other structures. In a very few systems, color is used in a specified manner to aid in the rapid identification of the systems. For example, JP-5 piping in interior spaces is painted purple, saltwater piping is painted green, and fireplugs are painted red.

Each valve is marked on the rim of the handwheel, on a circular label plate secured by the handwheel nut, or on a label attached either to the ship's structure or to adjacent piping. The valve label gives the name and purpose of the valve, when this information is not immediately apparent from the piping system marking, and the location of the valve. The location is indicated by three numbers that give, in order, the vertical (deck) level, the longitudinal (frame) position, and the transverse (port or starboard) position. Consider, for example, a drain bulkhead stop valve that bears the label 2-85-1.

The location of this valve is indicated by three groups of numbers. The first group of numbers indicates the vertical position—the second deck. The second group of numbers indicates the longitudinal position—frame 85. The third number indicates the transverse position— starboard side because it is an odd number. (Port side is indicated by an even number.) The numbers indicating transverse position begin at the centerline of the ship and progress outward toward the sides. For example, a second drain

bulkhead stop installed on the same level and at the same frame, but farther to starboard, would be identified as 2-85-3.

If the piping system identification does not clearly indicate the system the valve belongs to, the valve must be labeled to indicate the system (drain bulkhead stop, in these examples).

In vital engineering piping systems, a somewhat different system of marking is used to identify main line valves, cross-connection or split-plant valves, and remote-operated valves. Instead of being identified by location, these valves are assigned casualty control identification numbers, by system, such as the following:

Main steam	MS1, MS2, MS3, and so on
Auxiliary steam	AS1, AS2, AS3, AS4, and so on
Main condensate	MCN1, MCN2, and so on
Auxiliary exhaust	AE1, AE2, AE3, and so on
Fuel-oil service	FOS1, FOS2, and so on

In the vital engineering systems aboard newer ships, the system for marking valves is slightly different. It consists of a three-part designation in the following sequence: first, a number designating the shaft or plant number; second, letters designating the system; and third, a number, or a combination of a number and a letter, indicating the individual valve. Individual valve numbers are assigned in sequence, beginning at the origin of a system and following in order to the end of the system, excluding branch lines. In other words, the first valve in the main line is No. 1, the second is No. 2, and so forth. Since parallel flow paths frequently exist, it is often necessary to assign a shaft number and a system designation to the parallel flow paths as well as to the basic main line of the system. The valves in the parallel flow paths are then numbered in sequence; identical numbers are used for valves that perform like functions in each of the parallel flow paths, but a letter suffix is added to distinguish between the similar valves.

All engineering personnel must be familiar with the valve markings used in the vital engineering systems. Use of the identification numbers tends to prevent confusion and error when the plant is being split or cross-connected or when damaged sections are being isolated. The identification numbers can be used to indicate which valve or valves must be opened or closed; thus, the actual physical location of the valve or valves need not be described. However, the identification markings cannot serve their intended purpose unless all engineering personnel are thoroughly familiar with the physical location and the identification number of each valve they may be required to operate.

MAIN STEAM SYSTEMS

The main steam system is the shortest of all the major engineering piping systems aboard ship. This statement is true regardless of the steam pressures involved.

Although the main steam systems are approximately the same for a 1200-psi system as for a 600-psi pressure system, we must not jump to the conclusion that the plants as a whole are identical. Important differences exist between 1200-psi plants and lower-pressure plants.

Any piping that carries superheated steam is considered part of the main steam system. Aboard many ships, the main steam system includes only the piping that carries superheated steam from the boilers to the propulsion turbines, the turbo-generators, and the main feed pumps.

Aboard some ships (both 600-psi and 1200-psi), the main steam system supplies superheated steam to several other units as well. For example, aircraft carriers use superheated steam to run the steam catapult systems. Ships that have 1200-psi main steam systems use superheated steam to operate the main feed pump. Also, some ships use superheated steam to operate forced draft blowers, main circulating pumps, and other auxiliaries. On most ships that have 1200-psi main steam systems, the steam for soot blowers is NOT supplied from the main steam system; instead, steam for the soot blowers is taken from a 1200-psi auxiliary steam system.

Figure 9-7 illustrates the main steam system of DDG-45 and DDG-46. Each boiler is provided

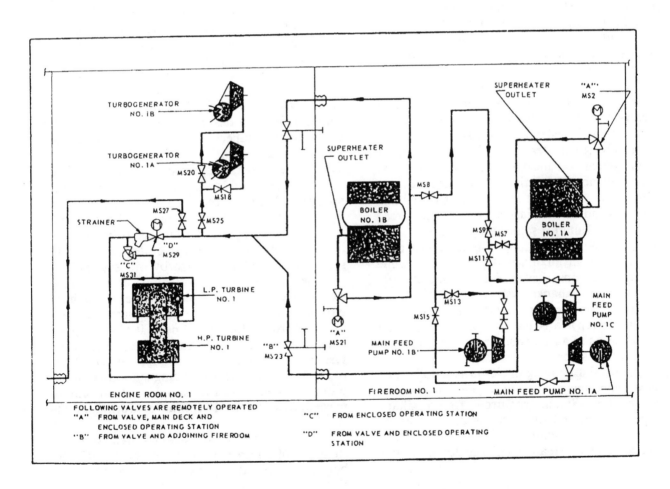

Figure 9-7.—Main steam system, DDG-45 and DDG-46.

with a boiler stop valve that can be operated either locally from the fireroom or remotely from the damage control deck. A second line stop valve in each fireroom or engine room provides two-valve protection for the boiler when it is not in use and permits effective isolation in case of damage. This type of two-valve protection is standard for all boilers installed aboard U.S. Navy ships.

For ahead operation, the superheated steam passes through a main steam strainer, a guarding valve, and a throttle valve before entering the high-pressure turbine. From the high-pressure turbine, the steam passes through a crossover pipe to the low-pressure turbine; then, it exhausts to the condenser. For astern operation, the super-heated steam passes through the steam strainer and through a stop valve (the astern throttle); then, it goes to the steam chest of the astern element. The astern elements are located at each end of the low-pressure turbine.

AUXILIARY STEAM SYSTEMS

Auxiliary steam systems supply steam at the pressures and temperatures required for the operation of many systems and units of machinery, both inside and outside the engineering spaces. Auxiliary steam is often called saturated steam or desuperheated steam.

Constant and intermittent service steam systems, steam smothering systems, whistles, air ejectors, forced draft blowers, and a wide variety of pumps are typical of the systems and machinery that receive their steam supply from auxiliary steam systems on most steam-driven ships. As previously mentioned, the units are not the same on all ships. Some of the newer ships use main steam instead of auxiliary steam for the forced draft blowers and for some pumps. Aboard some ships, turbine gland sealing systems receive their supply from an auxiliary steam system, while aboard other ships, they receive their supply from the auxiliary exhaust system. In general, an increasing use of electrically driven (rather than turbine-driven) auxiliaries has led to the simplification of auxiliary steam systems on newer construction ships.

Aboard ships having single-furnace boilers, all steam generated in the boiler goes through the superheater. Steam required for auxiliary purposes is passed through the desuperheater. Aboard ships having 600-psi single-furnace boilers, auxiliary steam leaves the desuperheater approximately 100°F above saturation temperature. Aboard ships having 1200-psi single-furnace boilers, the desuperheated auxiliary steam may still contain quite a bit of superheat.

Most ships that have 600-psi main steam systems have a 600-psi auxiliary steam system, a 150-psi auxiliary steam system, plus some lower-pressure service systems. The 600-psi auxiliary steam system serves some machinery directly and supplies the 150-psi system through reducing valves or reducing stations. The 150-psi auxiliary steam system serves some units directly and provides auxiliary steam for units or systems that require auxiliary steam at even lower pressures.

Ships that have a 1200-psi main steam system have a 1200-psi auxiliary steam system, usually a 600-psi auxiliary steam system, a 150-psi auxiliary steam system, and several constant and intermittent steam service systems. We will discuss the auxiliary steam systems of the DDG-45 and DDG-46 in some detail as examples of auxiliary steam systems aboard ships having 1200-psi main steam systems.

The 1200-psi and the 600-psi auxiliary steam systems for the forward plant of the DDG-45 and DDG-46 are shown in figure 9-8. A similar arrangement exists in the after plant. The 1200-psi auxiliary steam system for each plant is completely separate and independent. The 600-psi systems can be cross-connected but are not normally operated that way. Each plant has two boilers, both of which supply steam to the 1200-psi auxiliary steam system of that plant. The steam comes from the desuperheater outlet of each boiler; it is desuperheated from approximately 905°F (the operating temperature at the superheater outlet) to approximately 700°F. Note that the steam in this auxiliary steam system still has somewhat more than 100°F of superheat, so it is not strictly saturated steam. The 1200-psi auxiliary steam lines from each boiler are interconnected so that either boiler can provide steam for everything served by this system.

The 1200-psi auxiliary steam system supplies steam directly to the soot blowers, the forced draft blowers, and the reducing stations that reduce the pressure from 1200 to 600 psig. It also supplies augmenting steam to the auxiliary exhaust steam system. This is done by a 1200-psi to 12-psi reducing station. This reducing station is an air pilot operated valve that supplies steam only if needed. (See chapter 18 for more information on this.)

The 600-psi auxiliary steam system supplies steam at 600 psig and approximately 650°F to both fireroom and engine room equipment. In the fireroom, the 600-psi system supplies steam to the

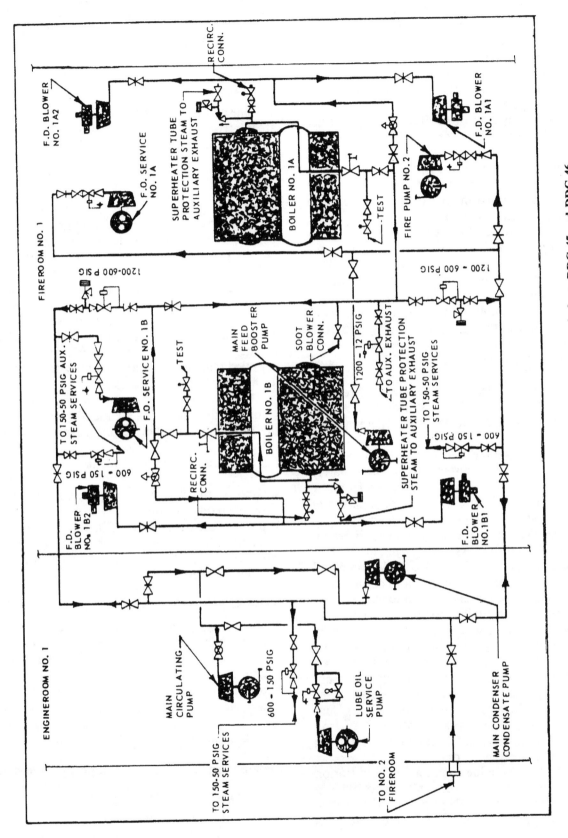

Figure 9-8.—1200-psi and 600-psi auxiliary steam system, forward plant, DDG-45 and DDG-46.

fuel-oil service pumps, the main feed booster pump, the fire pump, and the reducing stations that reduce the pressure from 600 to 150 psig. In the engine room, the 600-psi system supplies steam to the standby lube-oil service pump, the main condensate pump, the main circulating pump, and a reducing station that reduces the pressure from 600 to 150 psig.

Aboard ships whose auxiliaries are primarily electrically driven, there are no 600-psig auxiliary steam systems. They are not needed.

The 150-psi and the 50-psi auxiliary steam systems for the after plant of the DDG-45 and DDG-46 are shown in figure 9-9. The forward plant has similar systems.

The 150-psi auxiliary steam system in each plant supplies all machinery, equipment, and connections that require 150-psi steam. This system also supplies steam to other reduced pressure systems, via reducing stations, and may deliver steam to other ships or receive steam from outside sources through special piping and deck connections. Another function of the 150-psi system is to augment the auxiliary exhaust system.

Steam for the 150-psi system in the fireroom is supplied from the reducing stations that reduce the pressure from 600 to 150 psig. There are two such stations in each fireroom. A spray-type in-line desuperheater reduces the temperature of the fireroom's 150-psi system to approximately 400°F. Services and auxiliaries operated from the 150-psi system in the fireroom include superheater protection steam systems, service steam systems, boiler casing steam smothering systems, fireroom bilge steam smother systems, bilge and fuel-oil tank stripping pumps (in the No. 1 fireroom and No. 2 engine room only), and hose connections for boiling out boilers. In emergencies, the fireroom's 150-psi auxiliary steam system can also supply steam for some units that are normally supplied by the engine room's 150-psi system.

The reducing station that reduces steam from 600 to 150 psig in the engine room supplies steam at 150 psig and 610°F to the main and auxiliary air ejectors, the distilling plant air ejectors, and the turbine gland seal systems. In-line desuperheaters are NOT installed in the 150-psi system in the engine room.

In the No. 2 engine room, a reducing station reduces steam pressure from 150 psig to 100 psig and supplies steam at 100 psig and 385°F to the ship's laundry and tailor shop equipment. This 100-psi auxiliary steam system is called the 100-psi constant service system.

The 150-psi system also supplies two 50-psi systems—a constant service system and an intermittent service system. Both of these systems are shown in figure 9-9.

AUXILIARY EXHAUST STEAM SYSTEMS

The auxiliary exhaust system receives exhaust steam from pumps, forced draft blowers, and other auxiliaries that do not exhaust directly to a condenser. Auxiliary exhaust steam is used in various units, including deaerating feed tanks, distilling plants, and (aboard many ships) turbine gland seal systems. The pressure in the auxiliary exhaust system is maintained at approximately 15 psig, the exact pressure depending on the ship. If the pressure becomes too high, automatic unloading valves (dumping valves) allow the excess steam to go to the main or auxiliary condensers. In the event the unloading valves fail, relief valves allow the steam to escape to the atmosphere. If the pressure in the auxiliary exhaust system drops too low, makeup steam is supplied from an auxiliary steam system through an augmenting valve.

There are usually two auxiliary exhaust augmenting valves. The first augmenting valve is from the 150-psi system. The second augmenting valve is from the 1200-psi auxiliary steam system (unless the ship has 600-psi boilers, in which case the augmenting valve is from the 600-psi auxiliary steam system). The preferred augmenting steam valve is from the higher-pressure system. The 150-psi augmenting valve is installed to provide auxiliary exhaust steam to operate the evaporators when receiving shore steam or when operating auxiliary boilers (if installed).

STEAM ESCAPE PIPING SYSTEM

Steam escape piping is installed to provide an unobstructed passage for the escape of steam from boiler safety valves. The auxiliary exhaust steam system is also connected to the escape piping to allow steam in the system to "unload" to the atmosphere if the pressure becomes excessively high. Since the escape piping is usually shown on the plans or drawings that show the auxiliary exhaust piping, this system has come to be called the auxiliary exhaust and escape steam system.

GLAND SEAL AND GLAND EXHAUST SYSTEMS

Gland sealing steam is supplied to the shaft glands of propulsion turbines and generator

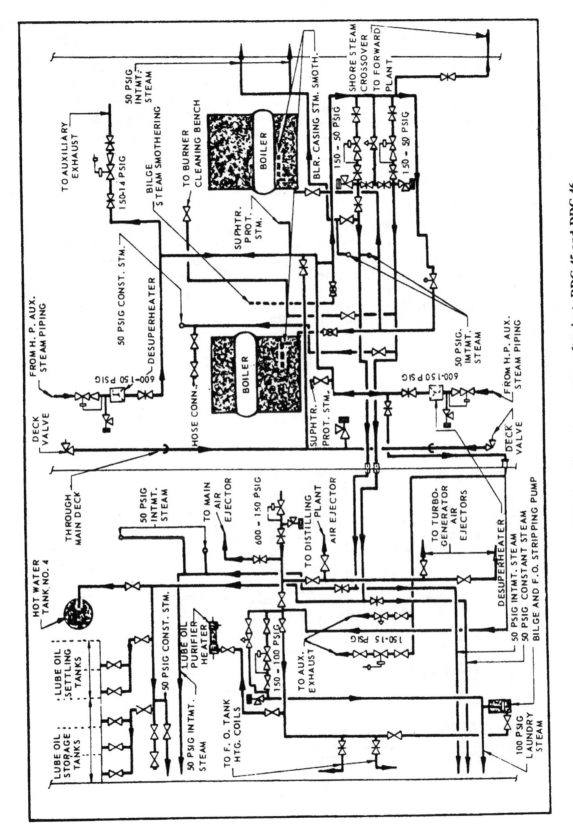

Figure 9-9.—150-psi and 50-psi auxiliary steam system, afterplant, DDG-45 and DDG-46.

turbines to seal the shaft glands against two kinds of leakage: (1) air leaking into the turbine casings, and (2) steam leaking out of the turbine casings. These two kinds of leakage may seem rather contradictory; however, each kind of leakage could occur under some operating conditions if the shaft glands were not sealed.

Pressures in the gland seal system are low, ranging from approximately 1/2 psig to 2 psig, depending on the conditions of operation. Gland exhaust piping carries the steam and air from the turbine shaft glands to the gland exhaust condenser, where the steam is condensed and returned to the condensate system.

Aboard most ships, gland sealing steam is supplied from the auxiliary exhaust system. Aboard some ships, however, it is supplied from the 150-psi auxiliary steam system. In either case, the steam is supplied through reducing valves or reducing stations. Figure 9-10 shows a typical gland seal and gland exhaust system for propulsion turbines.

CONDENSATE AND FEED SYSTEMS

Condensate and feed systems include all the piping that carries water from the condensers to the boilers and from the feed tanks to the boilers. The condensate system includes the main and auxiliary condensers, the condensate pumps, and the piping from the condensers to the deaerating feed tank (DFT). The boiler feed system includes the feed booster pump, the main feed pump, and the piping required to carry water from the DFT to the boilers. Together, the condensate and feed systems begin at the condensers and end at the boiler steam drum.

It is a little difficult to say whether the DFT is part of the condensate system or part of the boiler feed system, since the water level in the tank is generally taken as the dividing line between the two systems. The water is called condensate between the condenser and the DFT. It is called feedwater or boiler feed between the DFT and the boiler steam drum.

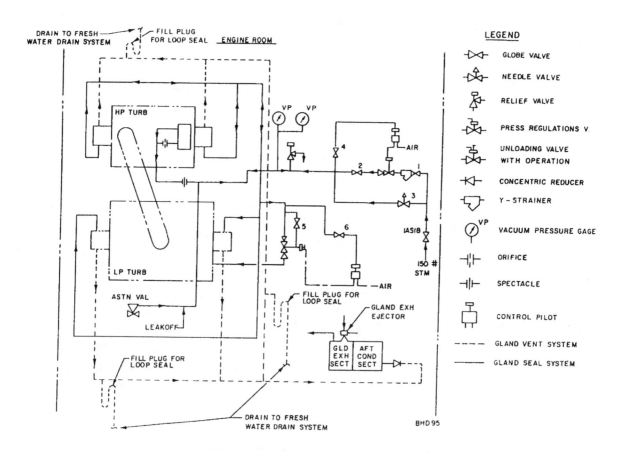

Figure 9-10.—Main engine gland seal and gland exhaust system.

Four main types of feed systems have been used aboard naval ships: (1) the open feed system, (2) the semiclosed feed system, (3) the vacuum-closed feed system, and (4) the pressure-closed feed system. The development of these systems, in the sequence listed, has gone along with the development of boilers. As boilers have been designed for higher and higher operating pressures and temperatures, the removal of dissolved oxygen from the feedwater has become increasingly important because the higher pressures and temperatures speed up the corrosive effects of dissolved oxygen. Each new type of feed system represents an improvement over the one before it in reducing the amount of oxygen dissolved or suspended in the feedwater.

Since all modern naval ships have pressure-closed feed systems, we will discuss only this type.

Pressure-closed feed systems are used aboard all naval ships that have boilers operating at and above 600 psi. They are also used aboard most ships that have lower boiler operating pressures.

In a pressure-closed feed system, all condensate and feed lines throughout the system (except for the very short line between the condenser and the suction side of the condensate pump) are under positive pressure. The system is closed to prevent the entrance of air.

A pressure-closed system is shown in figure 9-11. Follow this illustration as we discuss the system.

The main condenser is the beginning of the condensate system. The main condenser is a heat exchanger in which exhaust steam from the low-pressure turbines is condensed as it comes in contact with tubes through which cool seawater

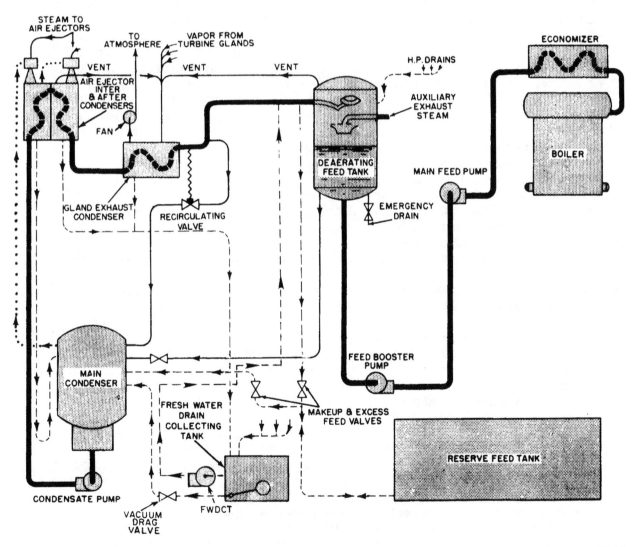

38.16

Figure 9-11.—Pressure-closed feed system.

is flowing. The main condenser also receives and condenses auxiliary exhaust steam that is not required elsewhere in the plant.

The main condenser is the means by which feedwater is recovered and returned to the feed system. The exhaust steam enters at the top of the condenser and the water (now called condensate) collects at the bottom of the condenser in the hotwell below the tubes. If we imagine a shipboard propulsion plant in which there is no condenser and the turbines simply exhaust to atmosphere, and if we consider the vast quantities of feedwater that would be required to support even one boiler that generates 150,000 pounds of steam per hour, we see immediately the vital function the main condenser serves in recovering feedwater.

The main condenser operates under a vacuum of approximately 25 to 28.5 inches of mercury. The basic cause of vacuum is the condensation of the steam. This is because the specific volume of steam is enormously greater than the specific volume of water. Since the condenser is not totally airtight, air ejectors are installed to remove air and other noncondensable gases from the condenser.

The designed vacuum varies according to the design of the turbine installation and according to such operational factors as the load on the condenser, the temperature of the outside seawater, and the tightness of the condenser. As condensate travels to the DFT, it passes through the air-ejector condensers and vent condensers. The condensate cools and condenses the steam from the steam-air mixtures. The condensate does not mix with the steam-air mixtures here, nor is it deaerated. Air ejectors, air-ejector condensers, vent condensers, and the DFT will be further discussed in chapter 13.

In the DFT, water is heated by direct contact with the auxiliary exhaust steam and is deaerated. Next, it is collected in the lower portion of the tank. At this point, it is now feedwater. Feedwater is pumped to the boiler by the main feed pumps, with the main feed booster pumps providing a positive suction for the main feed pump.

Makeup feed is added to the system as necessary. This will be required as the system loses steam and water by leaks and nonrecovered drains. Demand for makeup feed is determined by the water level in the DFT. Makeup feed is vacuum dragged from feedwater tanks or the feedwater drain collecting tank into the main condenser. Drains collected in the freshwater drain collecting tank are pumped by an electric

motor-driven pump to the condensate line. The freshwater drain tank has a float that is the input to controls for the pump motors and the vacuum-drag valve. The excess and makeup feed valves are operated by levelmatic controls on the DFT.

STEAM AND FRESHWATER DRAIN SYSTEMS

Since a shipboard steam plant is a "closed" system, most of the feedwater is recovered and is used over and over again to generate steam. So far, you have learned that steam is condensed in the main and auxiliary condensers and that the condensate is returned to the feed system for reuse. You also learned that the auxiliary exhaust steam is used in the DFT, which returns it to the feed system for reuse.

Steam is used throughout the ship to operate a variety of machinery, equipment, and piping, which does not exhaust either to a condenser or to the auxiliary exhaust system. Therefore, steam and freshwater drain systems are provided so that feedwater can be recovered and put back into the feed system after it has been used (as steam) in distilling plants, steam catapult systems, water heaters, whistles and sirens, and many other units and systems throughout the ship. The piping systems, as well as the water that they carry to the feed systems, are known as drains.

There are three steam and freshwater drain systems that recover feedwater from machinery and piping. They are as follows:

- High-pressure steam drainage system

- Service steam drainage system

- Freshwater drain collecting system

In addition, a fourth system collects contaminated drains that cannot be returned to the feed system. Briefly, let us describe these drainage systems.

High-Pressure Steam Drainage System

The high-pressure steam drainage system generally includes drains from superheater headers, throttle valves, main and auxiliary steam lines, steam catapults (on carriers), and other steam equipment or systems that operate at pressures of 150 psi or more. The high-pressure drains aboard some ships are led directly into the DFT. Aboard some newer ships, the high-pressure drains empty into the auxiliary exhaust line just before the auxiliary exhaust steam enters the DFT. In either case, of course, the high-pressure drains end up in the same place—the DFT.

Service Steam Drainage System

The service steam drainage system collects the uncontaminated drains from low-pressure (below 150 psi) steam piping systems and steam equipment outside the machinery spaces. Space heaters as well as equipment used in the laundry, the tailor shop, and the galley are typical sources of drains for the service steam drainage system. Aboard some ships, these drains discharge into the most conveniently located freshwater drain collecting tank. Aboard other ships, particularly on large combatant ships, such as carriers, the service steam drains discharge into special service steam drain collecting tanks located in the machinery spaces. The contents of the service steam drain collecting tanks are discharged to the condensate system. In addition, each tank has gravity drain connections to the freshwater drain collecting tank and to the bilge sump tank located in the same space.

Note that the service steam drainage system collects only clean drains that are suitable for use as boiler feed. Contaminated service steam drains, such as those from laundry presses, are discharged overboard.

Freshwater Drain Collecting System

The freshwater drain collecting system, often called the low-pressure drain system, collects drains from various piping systems, machinery, and equipment that also operates within the engineering space at steam pressures less than 150 psi. Both the service steam drainage system and the oil heating drainage system can discharge to the freshwater drain collecting tank, although they normally discharge directly to the feed system. In general, the freshwater drain collecting system collects gravity drains (open-funnel or sight-flow drains), turbine gland seal drains, auxiliary exhaust drains, air-ejector aftercondenser drains, and a variety of other low-pressure drains, which result from the condensation of steam during warming up or operation of steam machinery and piping, including the main steam system.

Freshwater drains are collected in freshwater drain collecting tanks located in the machinery spaces. The contents of these tanks may enter the feed system in one of two ways—they may be

drawn into the condenser by vacuum drag, or they may be pumped to enter the condensate system just ahead of the DFT.

Contaminated Drainage System

A contaminated drainage system is installed in each main and auxiliary machinery space where dry bilges must be maintained. The contaminated drainage system collects oil and water from machinery and piping that normally has some leakage, and also collects drainage from any other services, which may at times be contaminated. The contaminated drains are collected in a bilge sump tank located in the machinery space from which the drains are being collected. The contents of the bilge sump tank are removed by the bilge drainage system; they do not go to the feed system.

Aboard newer ships, the contaminated drains are further separated into a waste water drain system, which collects saltwater drains, and an oily water drain system, which collects steam drains that are contaminated with oil.

FUEL-OIL SYSTEMS

Boiler fuel-oil systems aboard ship include fuel-oil tanks, fuel-oil piping, fuel-oil pumps, and the equipment that is used to strain, measure, and burn the fuel oil. We will limit our discussion chiefly to the general arrangement of fuel-oil tanks and fuel-oil piping.

Fuel-Oil Tanks

Four kinds of tanks are used for holding boiler fuel oil: (1) storage tanks, (2) overflow tanks, (3) service tanks, and (4) contaminated oil settling tanks.

FUEL-OIL STORAGE TANKS.—The main fuel-oil storage tanks are an integral part of the ship's structure. They may be located forward and aft of the machinery spaces, abreast of these spaces, and in double-bottom compartments. However, fuel-oil storage tanks are NEVER located in double-bottom compartments directly under boilers. Some fuel-oil storage tanks, or ballast tanks, have connections that allow them to be filled with either fuel oil or seawater from the ballasting system.

FUEL-OIL OVERFLOW TANKS.—Other fuel-oil tanks are designated as fuel-oil overflow tanks. These tanks receive the overflow from fuel-oil storage tanks that are not fitted with independent overboard overflows. Overflow tanks can also be filled with seawater from the ballasting system and are also called ballast tanks.

FUEL-OIL SERVICE TANKS.—Fuel oil is taken aboard at the deck connection and is piped to the storage tanks through the fuel-oil filling and transfer. From the storage tanks, oil is pumped to the fuel-oil service tanks. All fuel for immediate use is drawn from the service tanks. The fuel-oil service tanks are considered part of the fuel-oil service system.

CONTAMINATED OIL SETTLING TANKS.—The contaminated oil settling tanks are used to hold oil that is contaminated with water or other impurities. After the oil has settled, the unburnable material, such as water and sludge, is pumped out through low-suction connections. The burnable oil remaining in the tanks is transferred to a storage tank or a service tank.

The contaminated oil settling tanks also receive and store oil, or oily water, until it can be discharged overboard without violation of the Environmental Quality Improvement Act. This act prohibits the overboard discharge of oil, or water that contains oil, in port and in prohibited zones in oceans and seas throughout the world. Information on when to empty the contaminated oil settling tanks either overboard or to barges can be found in OPNAV Instruction 5090.1, *Environmental and Natural Resources Protection Manual.*

Fuel-oil tanks are vented to the atmosphere through pipes leading from the top of the tank to a location above decks. The vent pipes allow vapor to escape when the tank is being emptied. Most fuel-oil tanks are equipped with manholes, overflow lines, sounding tubes, liquid-level indicators, and lines for filling, emptying, and cross-connecting.

Fuel-Oil Piping System

The fuel-oil piping system includes (1) the fuel-oil filling and transfer system, (2) the fuel-oil tank stripping system, and (3) the fuel-oil service system. The fuel-oil systems are arranged in such a way that different fuel-oil pumps take suction from the tanks at different levels. Stripping system pumps have low level suction connections. Fuel-oil booster and transfer pumps take suction above the stripping system pumps. Fuel-oil service pumps have high suction connections from the fuel-oil service tanks.

FUEL-OIL FILLING AND TRANSFER SYSTEM.—The fuel oil filling and transfer system receives fuel oil aboard and (1) fills the fuel-oil storage tanks, (2) fills the fuel-oil service tanks, (3) changes the list of the ship by transferring oil between port tanks and starboard tanks, (4) changes the trim of the ship by transferring oil between forward and after tanks, (5) discharges oil for fueling other ships, and, (6) in emergencies, transfers fuel oil directly to the suction side of the fuel-oil service pumps.

Ships have pressure filling systems that are connected to the transfer mains in such a way that the filling lines and deck connections can be used for both receiving and discharging fuel oil.

In general, the filling and transfer system consists of large mains running fore and aft; transfer mains; cross-connections; fuel-oil booster and transfer pumps; risers for taking on or discharging fuel oil; and lines and manifolds arranged so that the fuel-oil booster and transfer pumps can transfer oil from one tank to another and, when necessary, deliver fuel oil to the suction side of the fuel-oil service pumps.

FUEL-OIL TANK STRIPPING SYSTEM.—The fuel-oil tank stripping system clears fuel-oil storage tanks and fuel-oil service tanks of sludge and water before oil is pumped from these tanks by the fuel-oil booster and transfer pumps or by the fuel-oil service pumps. The stripping system is connected through manifolds to the bilge pump or, in some installations, to special stripping system pumps. The stripping system discharges the contaminated oil, sludge, and water overboard or to the contaminated oil settling tanks.

FUEL-OIL SERVICE SYSTEM.—The fuel-oil service system used aboard any ship depends partly on the type of fuel-oil burners installed on the boilers. The fuel-oil service system includes the fuel-oil service tanks, a service main, manifolds, piping, and fuel-oil service pumps.

Fuel-oil service pumps take suction from the service tanks through independent tailpipes, cutout valves or manifolds, suction mains, and pump connections. The suction arrangements for fuel-oil service pumps allow rapid changes in pump suction from one service tank to another. The pump suction piping is arranged to keep to a minimum any contamination that might result if one service pump takes suction from a service tank that is contaminated with water. The service suction main, which is common to all pumps in one particular space, has connections to the fuel-oil transfer main through stop-check valves that are normally locked in the closed position.

Fuel-oil service pumps are usually screw-type rotary pumps. Some ships have electric motor-driven fuel-oil service pumps, while other ships have turbine-driven fuel-oil service pumps. Ships that have turbine-driven fuel-oil service pumps also have a small capacity electric-driven port and cruising fuel-oil service pump that is used for light-off and in-port steaming. You will find additional information on fuel-oil pumps in chapter 19 of this manual.

The fuel-oil service system also needs fuel-oil strainers, burner lines, and other such items to deliver fuel-oil to the boiler fronts at the required pressures. Figure 9-12 shows the fuel-oil service system for the forward plant of destroyer DDG-5.

Aboard some ships, JP-5 can be used as boiler fuel for emergencies. The JP-5 systems are arranged so that they can discharge to the fuel-oil service system.

JP-5 SYSTEMS

Most ships have a separate JP-5 system installed to provide fuel for helicopters. The JP-5 system will be quite similar to the boiler fuel-oil system. The system has more stringent requirements on preventing contaminants, sludge, or water in fuel provided to helicopters. JP-5 storage systems will not have ballast systems and should not be filled with seawater or salt water. There is normally an arrangement to provide JP-5 to the boiler fuel-oil system in an emergency. All ships with steam propulsion will have emergency generators, usually diesel, in case of an emergency or a casualty. The diesels will normally burn the same fuel as the boilers, having it provided from the same filling and transfer system and storage tanks. The diesels will have their own service tanks and fuel-oil service system.

Fuels

The standard fuel used for shipboard propulsion plants is distillate fuel, Marine (DFM). This fuel is standardized with the North Atlantic Treaty Organization (NATO) countries, and identified by a code number, F-76. Prior to the 1970's, the U.S. Navy used a fuel known as Navy special fuel oil (NSFO) to run boilers. This oil had a dark, black appearance, thick and high viscosity, and a high sulfur content. This fuel required heaters, usually steam, in the service tanks and the service

system piping to decrease viscosity and improve flow. The high sulfur content was responsible for lots of fireside deposits and soot. This fuel is still in use in some foreign countries. Upon close examination of your fireroom, you may be able to locate these old heaters or be able to deduct where they were located. The advent of DFM removed the need for a separate diesel fuel system for the emergency diesel generators. DFM is used in all propulsion plants, including boilers, diesels, and gas turbines, and it simplifies and reduces the cost of the logistics.

BALLASTING SYSTEM

The ballasting system allows controlled flooding of certain designated tanks when such flooding is required for stability control. All tanks that are designated as fuel-oil and ballast tanks (and also certain voids) may be flooded by the ballasting system. Seawater is used for ballasting, and it may be taken from the firemain or directly from sea chests.

Combined ballasting and drainage systems are arranged so that all designated compartments and tanks can be ballasted either separately or together. Drainage pumps or eductors are used to remove ballast water.

MAIN LUBRICATING OIL SYSTEMS

Main lubricating oil systems aboard steam-driven ships provide lubrication for the turbine bearings and the reduction gears. The main lube-oil system usually includes a filling and transfer system, a purifying system, and separate service systems for each propulsion plant. Aboard most ships, each lube-oil service system includes three positive-displacement lube-oil service pumps: (1) a shaft-driven pump, (2) a turbine-driven pump, and (3) a motor-driven pump. The shaft-driven pump, attached to and driven by either the propulsion shaft or the quill shaft of the reduction gear, is used as the regular lube-oil service pump when the shaft is turning fast enough so that the pump can supply the required lube-oil pressure. The turbine-driven pump is used while the ship is getting underway and then as standby at normal speeds. The motor-driven pump serves as standby for the other two lube-oil service pumps.

The FF-1052/1078 class ships have an electric motor-driven standby pump and emergency pump. An electric control arrangement actuates the standby pump when lube-oil pressure falls to a specific value below normal. If the pressure

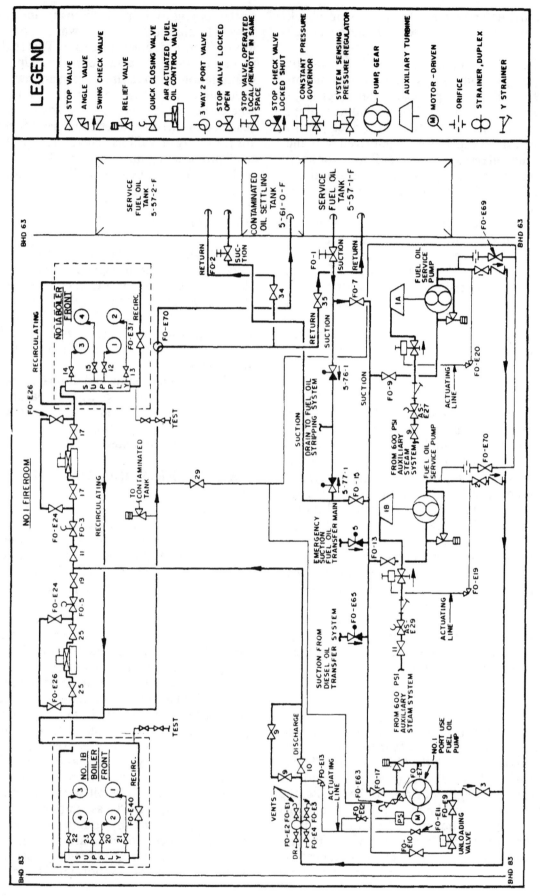

Figure 9-12.—Fuel-oil service system.

continues to drop, the emergency pump is activated. Then, if the pressure continues to drop, the low lube-oil pressure alarm sounds. Figure 9-13 shows a typical lube-oil service system for the main engine on a Knox class frigate.

COMPRESSED AIR SYSTEMS

Completely independent compressed air systems with individual compressors include the high-pressure air system, the ship's service air system, the aircraft starting and cooling air system, the combustion control system, the air deballasting system, and the oxygen-nitrogen generator system. For other services, air is taken from the high-pressure system or from the ship's service air system, as required.

The high-pressure air system is designed to provide air above 600 psi and up to 5000 psi for charging air banks and, at required pressures, for services such as missiles, diesel engine starting and control, torpedo charging, and torpedo workshops. When air is required for these services at less than the system pressure, a reducing valve is used. This valve is located at the outlet from the high-pressure air system.

Air for diesel engine starting and control is provided on some ships, either by a medium-range compressor at a pressure of 600 psi or from the high-pressure system through appropriate reducing valves.

The ship's service compressed air system is a low-pressure system that is installed aboard practically all surface ships. This system provides compressed air at the required pressure to operate pneumatic tools as well as oil-burning forges and furnaces, to charge pump air chambers, to clean equipment, and for a variety of other uses. The normal working pressure of the ship's service air system is 100 psi. However, for tenders and repair ships that have a greater demand for air, the system produces a higher working pressure (usually about 125 psi). The ship's service air system is normally supplied from a low-pressure air compressor. For some ships, however, the system may be supplied from a high-pressure system through reducing valves.

An aircraft starting and cooling air system, installed aboard aircraft carriers, provides air at varying temperatures (50° to 500°F) and pressures (48 to 62 psia) to meet the conditions required for starting and cooling aircraft being served. The air compressor is driven by gas turbine.

Combustion control air systems, installed aboard most ships, provide air to the pneumatic units in the boiler control systems. A boiler

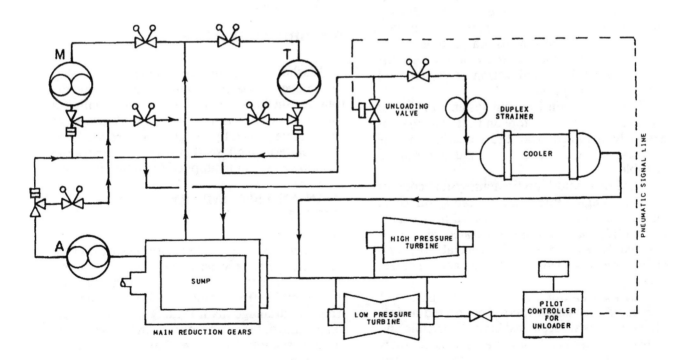

147.193

Figure 9-13.—Main lube-oil system.

control air system usually consists of an air compressor, an air receiver, the piping, and the filters required to supply clean air to all units of the boiler control system.

An air deballasting system is provided for some amphibious transport docks (LPDs) for deballasting by air. This system provides large quantities of air (7500 cubic feet per minute) at low pressure (20 psi). All compressors discharge to a common air loop distribution, which feeds all ballast tanks.

Oxygen-nitrogen generator systems are installed aboard aircraft carriers and submarine tenders. The air is supplied by high-pressure air compressors, through oil filters and moisture separators, directly to the oxygen-nitrogen generator. Compressed air plants are discussed in chapter 23.

FIREMAIN SYSTEMS

The firemain system receives water pumped from the sea and distributes it to fireplugs, sprinkling systems, and other related systems as required.

Naval ships use three basic types of firemain systems—the single-main system, the horizontal-loop system, and the vertical-loop system. The type of firemain system installed aboard any particular ship depends on the characteristics and functions of the ship. Small ships generally have single-main firemain systems. Large ships usually have one of the loop systems or a composite system, which is a combination or variation of the three basic types.

The single-main firemain system consists of one main that extends fore and aft. The main is generally installed near the centerline of the ship, extending as far forward and as far aft as possible.

The horizontal-loop firemain system consists of two single fore-and-aft, cross-connected mains. The two mains are installed in the same horizontal plant but are separated as far as practicable athwartship. In general, the two mains are installed on the damage control deck.

The vertical-loop firemain system consists of two single fore-and-aft, cross-connected mains. The two mains are separated both athwartship and vertically. As a rule, the low main is located below the lowest complete watertight deck, and the upper main is located below the highest complete watertight deck.

SPRINKLING SYSTEMS

Sprinkling systems are installed aboard ships in magazines, handling rooms, hangar decks, missile spaces, and other spaces where flammable materials are stowed. Water for these systems is supplied from the firemain through branch lines. Most sprinkling systems aboard ship are of the dry type; that is, they are not charged with water beyond the sprinkling control valves except when they are in use.

Sprinkling systems in magazines that contain missiles are of the wet type. The sprinkling control valves in magazine sprinkling systems are operated automatically by heat-actuated devices.

Other sprinkling control valves are operated manually or hydraulically, either locally or from remote stations. In those areas of the ship in which major flammable liquid fires could occur, such as in aircraft hangars, foam sprinkling systems are provided.

STEAM SMOTHERING SYSTEMS

Permanently fitted steam smothering lines are provided aboard most naval ships for smothering fires in the spaces between the inner and outer casings of each boiler. Steam for the double-casing steam smothering system is provided by the 150-psi auxiliary steam system.

WASHDOWN SYSTEMS

Washdown systems are installed aboard ship to remove radioactive contamination from the topside surfaces of the ship. Essentially, a washdown system is a dry-pipe sprinkler system with nozzles specially designed to throw a large spray pattern on all weather decks. Water for the washdown system is supplied from the firemain.

MAIN AND SECONDARY DRAIN SYSTEMS

Every ship is provided with fixed drainage systems to remove water from within the hull. A fixed drainage system consists of (1) a length of suction piping (the drainage main), (2) branches from the drainage main leading into the spaces to be drained, (3) one or more pumps or eductors to take suction from the drainage main, (4) discharge piping leading overboard from the pumps, and (5) the necessary valves, strainers, and manifolds.

Main Drainage Systems

Drainage of the machinery spaces aboard most ships is handled by the main drainage system, also known as the bilge drainage system. This system runs throughout the main machinery spaces. Aboard some ships, the main drainage system may also extend forward and aft of the machinery compartments.

For smaller ships, the main drainage system consists of a single drainage main running fore and aft, usually along the centerline. For larger ships, the main drainage system is a loop system (ends joined together), which extends along both sides of the engineering spaces. Aboard many newer construction ships, the main drainage system may be used to drain voids that are used for counterflooding and to empty fuel-oil tanks that have been ballasted with sea-water.

The types of pumps used in the main drainage system include steam-driven reciprocating pumps (on older ships), motor-driven centrifugal pumps, and eductors. These and other types of pumps are discussed in chapter 19 of this manual.

In emergencies, an engine room can be drained by the circulating pump that circulates seawater through the main condenser. Each circulating pump has a secondary bilge suction connection to the bilges in the engine room where the circulating pump is located. The use of the main circulating pump for drainage is strictly an emergency procedure, since all drains must be discharged through the saltwater side of the main condenser.

NOTE: The engine room in which the circulating pump is located is the only space that can be drained by the pump.

Secondary Drainage Systems

For many ships, secondary drainage systems, also called compartment drainage systems, are used to drain spaces forward and aft of the main machinery spaces. The piping of secondary drainage systems is smaller than that used in main drainage systems. Secondary drainage systems are independent systems, each with its own eductor or pump. The types of pumps used in secondary drainage systems may be motor-driven centrifugal pumps, eductors, or portable electric-submersible pumps.

POTABLE WATER SYSTEMS

Potable water systems are designed to provide a constant supply of potable (drinkable) water for all ship's service requirements. Potable water is stored in various tanks throughout the ship. The system is pressurized either by a pump and pressure tank or by a continuously operating circulating pump. The potable water system supplies scuttlebutts, sinks, showers, the scullery, and the galley. It also provides makeup water for various freshwater cooling systems.

HYDRAULIC SYSTEMS

Hydraulic systems aboard ship operate steering gear, anchor windlasses, hydraulic presses, remote control valves, and other such units. Hydraulic systems operate on the principle that since liquids are noncompressible, a force exerted at any point on an enclosed liquid is transmitted equally in all directions. Hence, a hydraulic system permits a great amount of work to be done with relatively little effort on the part of shipboard personnel.

Some hydraulic systems, for example steering systems, still use a petroleum-based product. Recently, other fluids have been developed for hydraulic systems in an effort to improve safety and reduce the likelihood of fire or explosion. For example, catapult retracting engines, jet blast deflectors, and weapons elevators use a fluid called Houghto-safe (MIL-H-22072). This fluid is water-based and contains a fluid that is a potential carcinogen in animals. Aircraft elevators use a fluid called Frugoel (MIL-H-19457). This fluid is a phosphate ester and contains a controlled amount of neurotoxic material. When working with either of these fluids, you should follow the safety precautions as prescribed in *NSTM*, chapter 262, and the material safety data sheets.

REFERENCES

Boiler Technician 3 & 2, NAVEDTRA 10535-H, Chapter 3, Naval Education and Training Program Management Support Activity, Pensacola, Florida, 1983.

Naval Ships' Technical Manual (NSTM), NAVSEA S9086-H7-STM-010, Chapter 262, "Lubricating Oils, Greases, Hydraulic Fluids, and Lubricating Systems," Naval Sea Systems Command, Washington D.C., 1987.

CHAPTER 10

PROPULSION BOILERS

LEARNING OBJECTIVES

Upon completion of this chapter, you should be able to do the following:

1. Recognize the meanings of standard boiler terms.

2. Identify the principles of boiler classification.

3. Identify boiler components and their functions.

4. Recognize the principles of identifying boiler tubes.

5. Describe the boiler water and steam circulation path in the boiler.

6. Describe the path of boiler air and combustion gases in the boiler.

7. Define terms used to describe the water in a steam water cycle.

8. Identify the effects of waterside deposits, waterside corrosion, and carryover on a steam propulsion plant.

9. Recognize general principles of combustion to fuel oil in a boiler furnace, and identify factors that contribute to heat loss.

INTRODUCTION

In the conventional steam turbine propulsion plant, the boiler is the source, or high temperature region, of the thermodynamic cycle. The steam that is generated in the boiler is led to the propulsion turbines, where its thermal energy is converted into mechanical energy. This mechanical energy drives the ship and provides power for vital services.

In essence, a boiler is merely a container in which water can be boiled and steam generated. A tea kettle on a stove is basically a boiler, although a rather inefficient one. In designing a boiler to produce a large amount of steam, a large heat transfer surface is necessary. In a modern boiler, there are more than a thousand tubes to provide a large heat transfer area in a small space. Boiler tubes are provided with water flow from the bottom from headers that distribute the boiler water. A water steam mixture will exit the tubes at the top and enter the steam drum. Insulating refractories surround the boiler inner casing when

combustion occurs. In addition, the boiler has downcomers installed between the inner and outer casings of the boiler to distribute boiler water to the water drum and headers. This relatively cool water maintains boiler circulation. The steam drum has internal fittings to extract near moisture-free steam and distribute incoming feedwater. The internal fittings are discussed in the next chapter. The boiler is a rather complicated piece of machinery when all its associated fittings, piping, and accessories are considered.

All propulsion boilers produce superheated steam for use by the turbines. As discussed in the chapter on thermodynamics, steam must be separated and drawn off before it can be superheated. The component of the boiler where this occurs is called the superheater. On most modern boilers, the superheater has an inlet header and an outlet header connected by a number of superheater tubes. By having a large number of tubes, again the heat transfer area is increased. In the superheater, steam temperature is raised without any significant change in pressure.

Modern boilers have the superheater tubes in the same part of the same furnace as the steam-generating tubes. Some older boilers had the superheater tubes in a different furnace, with its own fuel-oil burners where the amount of super-heat could be controlled.

The use of superheated steam improves thermodynamic efficiency. Superheater steam has more energy per unit mass. Maximum possible efficiency depends upon absolute temperature of source (superheated steam) and receiver (condenser). Moisture content of steam has different variables in operation for machinery. Steam turbines can suffer erosion and corrosion due to the moisture content of steam. Although only used in a few applications in modern ships, reciprocating steam machinery requires some moisture content in the steam for lubricating moving parts of the machinery.

This chapter deals primarily with the types of propulsion boilers commonly in use in the Navy. NSTM, chapter 221, is the basic doctrine reference on boilers. For detailed information on the boilers in any particular ship, consult the manufacturer's technical manuals furnished with the boilers.

BOILER DEFINITIONS

Some of the terms used in connection with boilers are defined here. It is important that you understand these definitions and that you use the terms correctly.

BOILER FULL-POWER CAPACITY—Boiler full-power capacity, as specified in the contract specifications of the ship, is expressed as the number of pounds of steam generated per hour at the pressure and temperature required for all purposes to develop contract shaft horsepower of the ship divided by the number of boilers installed. Boiler full-power capacity is listed in the design data section of the manufacturer's technical manual for the boilers in each ship. It may be listed either as the capacity at full power or as the designed rate of actual evaporation per boiler at full power.

BOILER OVERLOAD CAPACITY—Boiler overload capacity is specified in the design of the boiler. It is given in terms of either steaming rate or firing rate, depending on the individual installation. Boiler overload capacity is usually 120 percent of boiler full-power capacity.

SUPERHEATER OUTLET PRESSURE—Superheater outlet pressure is the actual pressure at the superheater outlet at any given time.

STEAM DRUM PRESSURE—Steam drum pressure is the actual pressure in the boiler steam drum at any given time.

OPERATING PRESSURE—Operating pressure is the constant pressure at which the boiler is being operated. The constant pressure may be carried at either the steam drum or the superheater outlet, depending on the design features of the boiler. Therefore, operating pressure is the same as either the steam drum or the superheater outlet pressure (depending on which is used as the controlling pressure). Operating pressure is specified in the design of a boiler and is indicated in the manufacturer's technical manual.

DESIGN PRESSURE—Design pressure is a pressure specified by the boiler manufacturer as a criterion for boiler design. Design pressure is NOT the same as operating pressure. It is somewhat higher than operating pressure. Design pressure is given in the manufacturer's technical manual for the boiler.

DESIGN TEMPERATURE—Design temperature is the intended maximum operating temperature at the superheater outlet at some specified rate of operation. For combatant ships the specified rate of operation is normally full-power capacity.

OPERATING TEMPERATURE—Operating temperature is the actual temperature at the superheater outlet. Operating temperature is the same as design temperature ONLY when the boiler is operating at the rate specified in the definition of design temperature.

FIREROOM—A fireroom is a compartment that contains boilers, associated auxiliary equipment, and the operating station for operating them. In some ships, the boilers and propulsion turbines are in the same compartment, in which case the compartment is called a machinery room.

BOILER OPERATING STATION—The station from which a boiler or boilers are operated is referred to as a boiler operating station.

STEAMING HOURS—The term *steaming hours* indicates the time during which the boiler has fires lighted for raising steam and the time during which it is generating steam. Time during which fires are not lighted is NOT included in steaming hours.

TOTAL HEATING SURFACE—The total heating surface of any steam-generating unit consists of that portion of the heat transfer apparatus (tubes) that is exposed on one side to the gases of combustion and on the other side to the water or steam being heated. Thus, the total heating surface equals the sum of the generating (tubes) surface, superheater (tubes) surface, and economizer (tubes) surface, as measured on the combustion gas side.

SUPERHEATER SURFACE—The superheater surface is that portion of the total heating surface where the steam is heated after leaving the boiler steam drum.

ECONOMIZER SURFACE—The economizer surface is that portion of the total heating surface where the feedwater is heated before it enters the boiler steam drum.

MAXIMUM STEAM DRUM PRESSURE—Maximum steam drum pressure is that point at which the first boiler safety valve lifts.

MINIMUM STEAM DRUM PRESSURE—Minimum steam drum pressure is usually 85 percent of the boiler operating pressure.

BOILER CLASSIFICATIONS

Although boilers may vary considerably in details of design, most of them may be classified and described in terms of a few basic features or characteristics. Some knowledge of the methods of classification provides a useful basis for understanding the design and construction of the various types of naval boilers.

INTENDED SERVICE

A good place to begin in classifying boilers is to consider their intended service. By this method of classification, naval boilers are divided into two classes: propulsion boilers and auxiliary boilers. Propulsion boilers are used to provide steam for ship propulsion and for vital auxiliary services.

Auxiliary boilers, when installed, are designed to provide steam for hotel services (run evaporators, heating, laundry, galley and scullery machinery). Gas turbine-powered ships have water-tube boilers, referred to as waste-heat boilers, that use the hot exhaust gases of the gas turbines as the heat source. Many diesel-driven ships have auxiliary boilers, and some ships with propulsion boilers have auxiliary boilers installed to allow the propulsion boilers to be secured while in port.

LOCATION OF FIRE AND WATER SPACES

One of the basic classifications of boilers is according to the relative location of the fire and water spaces. By this method of classification, boilers are divided into two classes: fire-tube boilers and water-tube boilers. In the fire-tube boilers, the gases of combustion flow through the tubes and thereby heat the water that surrounds the tubes. In water-tube boilers, the water flows through the tubes and is heated by the gases of combustion that fill the furnace and heat the outside metal surfaces of the tubes.

All propulsion boilers used in modern naval ships are of the water-tube type. Auxiliary boilers may be either fire-tube or water-tube boilers.

TYPES OF CIRCULATION

Water-tube boilers are further classified according to the method of water circulation. Water-tube boilers may be classified as natural circulation boilers or as forced circulation boilers.

Natural Circulation

In natural circulation boilers, the circulation of water depends on the difference between the density of an ascending mixture of hot water and steam and a descending body of relatively cool and steam-free water. The difference in density occurs because the water expands as it is heated, and thus becomes less dense. Another way to describe natural circulation is to say that it is caused by convection currents, which result from the uneven heating of the water contained in the boiler.

Natural circulation may be either free or accelerated. In a boiler with free natural circulation, the generating tubes are installed almost horizontally, with only a slight incline toward the vertical. When the generating tubes are installed at a much greater angle of inclination, the rate of water circulation is definitely increased. Therefore, boilers in which the tubes slope quite steeply from steam drum to water drum are said to have accelerated natural circulation.

Most modern naval boilers are designed for accelerated natural circulation. In such boilers, large tubes (3 inches or more in diameter) are installed between the steam drum and the water drum. These large tubes, or downcomers, are located outside the furnace and away from the heat of combustion. They serve as pathways for the downward flow of relatively cool water. When a sufficient number of downcomers are installed, all small tubes can be generating tubes, carrying steam and water upward, and all downward flow can be carried by the downcomers. The size and number of downcomers installed varies from one type of boiler to another, but downcomers are installed in all modern naval boilers.

Forced Circulation

Forced circulation boilers are, as their name implies, quite different in design from the boilers that use natural circulation. Forced circulation boilers depend upon pumps, rather than upon natural differences in density, for the circulation of water within the boiler. Because forced circulation boilers are not limited by the requirement that hot water and steam must be allowed to flow upward while cooler water flows downward, a great variety of arrangements may be found in forced circulation boilers.

Forced circulation boilers are not used in propulsion boilers. The auxiliary waste-heat boilers installed on the DD-963, DDG-993, and CG-72 classes of ships are of the forced circulation design.

ARRANGEMENT OF STEAM AND WATER SPACES

Natural circulation water-tube boilers are classified as drum-type boilers or header-type boilers, depending upon the arrangement of the steam and water spaces. Drum-type boilers have one or more water drums (and usually one or more water headers as well). Header-type boilers have no water drum; instead, the tubes enter a great many headers which serve the same purpose as water drums.

What is a header, and what is the difference between a header and a drum? The term *header* is commonly used in engineering to describe any tube, chamber, drum, or similar piece to which tubes or pipes are connected in such a way as to permit the flow of fluid from one tube (or group of tubes) to another. Essentially, a header is a type of manifold or collection point. As far as boilers are concerned, the only distinction between a drum and a header is size. Drums may be entered by a person while headers cannot. But both serve basically the same purpose.

Boilers are often classified by the shape of the furnace formed by the steam and water spaces— that is, by the tubes. For example, double-furnace boilers are often called M-type boilers because the arrangement of tubes is roughly M-shaped. Single-furnace boilers are often called D-type boilers because the tubes form a shape that looks like the letter D.

NUMBER OF FURNACES

All boilers commonly used in the propulsion plants of naval ships may be classified as either single-furnace boilers or double-furnace boilers. The D-type boiler is a single-furnace boiler; the M-type boiler is a double-furnace (divided-furnace) boiler.

M-type boilers were installed in all combatant ships designed during World War II. No new ships have been built with M-type boilers since 1955. At this time, the only ships in service with M-type boilers are the four battleships and the aircraft carriers—USS *Coral Sea* and *Midway*.

BURNER LOCATION

Another recent development in naval boilers makes it convenient to classify boilers on the basis of where the burners are located. Most burners in naval propulsion plants are located at the front of the boiler. However, some ships (AO-177 and LKA-113 classes) have their burners located on the top of the boilers. These are called topfired boilers.

FURNACE PRESSURE

Almost all propulsion boilers operate with only a slight air pressure in the furnace. This pressure at most might be 5 psig, and is more frequently measured in inches of water pressure. In the past, the U.S. Navy operated 16 frigates of the *Garcia* and *Brooke* class that had pressurized furnaces. Pressure in the furnace varied with the boiler's firing rate and would reach 50 psig at full power. The furnace in these boilers was pressurized by a supercharger or air compressor (quite similar to that found in a gas turbine engine), rather than a forced draft blower. The boiler construction of pressurized fired boilers was significantly different, having a vertical cylindrically shaped furnace. These boilers had a higher heat transfer rate and a higher rate of changeover of boiler water and combustion gases. In addition, these boilers were very lightweight and compact, being only half the size and weight of a nonpressurized fired boiler of the same capacity.

TYPES OF SUPERHEATERS

On most propulsion boilers in use, the superheater tubes are protected from radiant heat by water screen tubes. The water screen tubes absorb the intense radiant heat of the furnace, and the superheater tubes are heated by convection currents rather than by direct radiation. Hence, these superheaters are referred to as convection-type superheaters.

In some older designed boilers, the superheater tubes are exposed directly to the radiant heat of the furnace without the protection of water screen tubes. Superheaters of this kind are called radiant-type superheaters. At present, radiant-type superheaters are rarely used.

CONTROL OF SUPERHEAT

A boiler that provides some means of controlling the degree of superheat independently of the rate of steam generation is said to have controlled superheat. A boiler in which such separate control is not possible is said to have uncontrolled superheat.

Normally, the term *superheat control boiler* is used to identify a double-furnace boiler where the degree of superheat is controlled in the second furnace. The term *uncontrolled superheat boiler* or *noncontrolled superheat boiler* is used to identify a single-furnace boiler where all steam generated is passed through the superheater, which does not have a separate means to control the degree of superheat.

OPERATING PRESSURE

For some purposes, it is convenient to classify boilers according to operating pressure. Classifications of this type are approximate rather than exact. The term *600-psi boiler* applies to boilers with operating pressures ranging from 600 to 700 psi. Some older tenders have propulsion boilers with operating pressures ranging from 400 to 500 psi.

The term *1200-psi boilers* applies to boilers with operating pressures ranging from 1200 to 1250 psi. Almost all combatants (aircraft carriers, cruisers, destroyers, and frigates) have 1200-psi boilers. Some personnel refer to 1200-psi boilers as high-pressure boilers.

Although the U.S. Navy does not operate any boilers with operating pressures ranging from 700 to 1200 psi, future designs may have boilers that operate in this range. However, several foreign navies and shore-based electrical power generation stations operate boilers in this range of operating pressure.

BOILER COMPONENTS

Most propulsion boilers now used by the Navy have essentially the same components: steam and water drums, generating and circulating tubes, superheaters, economizers, and a number of accessories and fittings required for boiler operation and control.

DRUMS AND HEADERS

Boilers are installed in the ship in such a way that the long axis of the boiler drum runs fore and aft, rather than athwartships, so that the water will not surge from one end of the drum to the other as the ship rolls.

Steam Drum

The steam drum is located at the top of the boiler. It is cylindrical in shape, except on some boilers the steam drum may be slightly flattened along its lower curved surface. The steam drum receives feedwater and serves as a place for the accumulation of the saturated steam which is generated in the tubes. The tubes enter the steam drum below the normal water level of the drum. The steam and water mixture from these tubes goes through separators where the water is separated from the steam.

Figure 10-1 shows the construction of a steam drum. Two sheets of steel are rolled or bent to the required semicircular shape and are then welded together. The upper sheet is called the wrapper sheet; the lower sheet is called the tube sheet. Notice that the tube sheet is thicker than the wrapper sheet. The extra thickness is required in the tube sheet to ensure adequate strength of the tube sheet after the holes for the generating tubes have been drilled.

The ends of the drums are closed with drumheads that are welded to the shell (as shown in fig. 10-2). At least one of the drumheads contains a manhole that permits access to the drum for inspection, cleaning, and repair; recently built boilers have a manhole in each drumhead.

The steam drum either contains or is connected to many of the important fittings and instruments required for boiler operation. Boiler fittings and instruments are discussed in chapter 11.

Water Drums and Headers

Water drums and headers equalize the distribution of water to the generating tubes and provide a place for the accumulation of loose scale and other solid matter that may be present in the boiler water. In drum-type boilers with natural circulation, the water drums and water headers are at the bottom of the boiler. Water drums are usually round. Headers may be round, oval, or square. Headers are provided with access openings, called handholes, of the type shown in figure 10-3. Water drums are usually provided

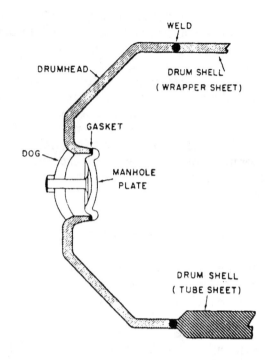

Figure 10-2.—Drumhead secured to steam drum.

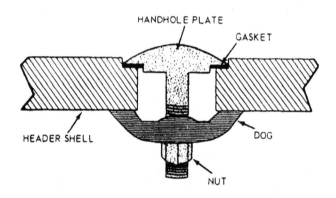

Figure 10-3.—Header handhole and handhole plate.

with manholes similar to the manholes in steam drums.

GENERATING AND CIRCULATING TUBES

Most of the tubes in a boiler are generating and circulating tubes. There are four main kinds of generating and circulating tubes: (1) generating tubes in the main generating tube bank, (2) water wall tubes, (3) water screen tubes, and (4) downcomers. These tubes are made of steel similar to the steel used for the drums and headers. Most

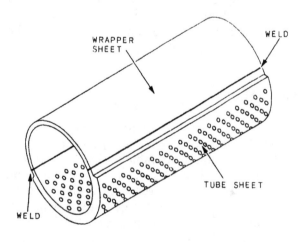

Figure 10-1.—Boiler steam drum.

tubes in the main generating bank are about 1 inch or 1 1/4 inches in outside diameter. Water wall tubes, water screen tubes, and the two or three rows of generating tubes next to the furnace are generally a little larger. Downcomers are larger still, being approximately 3 to 11 inches in outside diameter.

Since the steam drum is at the top of the boiler and the water drums and headers are at the bottom, the generating and circulating tubes must be installed more or less vertically. In most cases, each tube enters the steam drum and the water drum, or header, normal to (that is, at right angles to) the drum surfaces. This means that all the tubes in any one row are curved in exactly the same way, but the curvature of different rows is not the same. Tubes are usually installed normal to the drum surfaces to allow the maximum number of tube holes to be drilled in the tube sheets with a minimum weakening of the drums. However, non-normal installation is permitted if certain advantages in design can be achieved.

What purpose do all these generating and circulating tubes serve? Generating tubes provide the area in which most of the saturated steam is generated. The function of water wall tubes is to protect the furnace refractories, thus allowing higher heat release than would be possible without this protection. The water wall tubes are also generating tubes at high firing rates. Water screen tubes protect the superheater from direct radiant heat. Water screen tubes, like water wall tubes, are generating tubes at high firing rates. Downcomers are installed between the inner and outer casings of the boiler to carry the downward flow of relatively cool water and thus maintain the boiler circulation. Downcomers are not designed to be generating tubes.

In addition to the four main types of generating and circulating tubes just mentioned, there are a few large superheater support tubes which, in addition to providing partial support for the steam drum and for the superheater, serve as downcomers at low firing rates and as generating tubes at high firing rates.

Since a modern boiler is likely to contain between 1000 and 2000 tubes, some system of tube identification is essential. Generating and circulating tubes are identified by lettering on the rows of tubes and numbering on the individual tubes in each row. A tube row runs from the front to the rear of the boiler. The row of tubes next to the furnace is row A, the next row is row B, the next is row C, and so forth. If there are more than 26 rows in a tube bank, the rows after Z are lettered AA, BB, CC, DD, EE, and so forth. Each tube in each row is then designated by a number, beginning with 1 at the front of the boiler and numbering back toward the rear.

The letter that identifies a tube row is often preceded by an R or an L, particularly in the case of water screen tubes, superheater support tubes, and furnace division wall tubes. These letters may indicate that the tube is bent for either a right-hand or left-hand boiler. You should check your boiler manufacturer's technical manual to determine which boilers on your ship are right- or left-handed boilers.

Figure 10-4 shows part of a boiler tube renewal sheet for a single-furnace boiler and illustrates the method used to identify tubes.

The boiler tube renewal sheet, which is normally carried aboard ship, is one source of information on tube identification. Another source of information is the manufacturer's technical manual furnished with the boilers; some older technical manuals and most of the newer ones include a tube identification diagram. If precise identification of the boiler tubes cannot be made from either of these sources, you will have to consult the boiler plans to obtain the correct identification. You must also consult the boiler plans before ordering replacement tubes.

In all reports and correspondence concerning boiler tubes, proper identification and exact locations of the tubes is essential.

SUPERHEATERS

Propulsion boilers now in naval service have convection-type superheaters, with water screen tubes installed between the superheater tubes and the furnace to absorb the intense radiant heat and protect the superheater.

Most convection-type superheaters have U-shaped tubes, which are installed horizontally in the boiler, and two headers, which are installed more or less vertically at the rear of the boiler. One end of each U-shaped tube enters one superheater header, and the other end enters the other header. The superheater headers are divided internally by one or more division plates, which act as baffles to direct the flow of steam. Some superheater headers are divided externally as well as internally.

The superheaters of some boilers have a walk-in or cavity-type feature—an access space or cavity in the middle of the superheater tube bank. This cavity, which runs the full length and height of

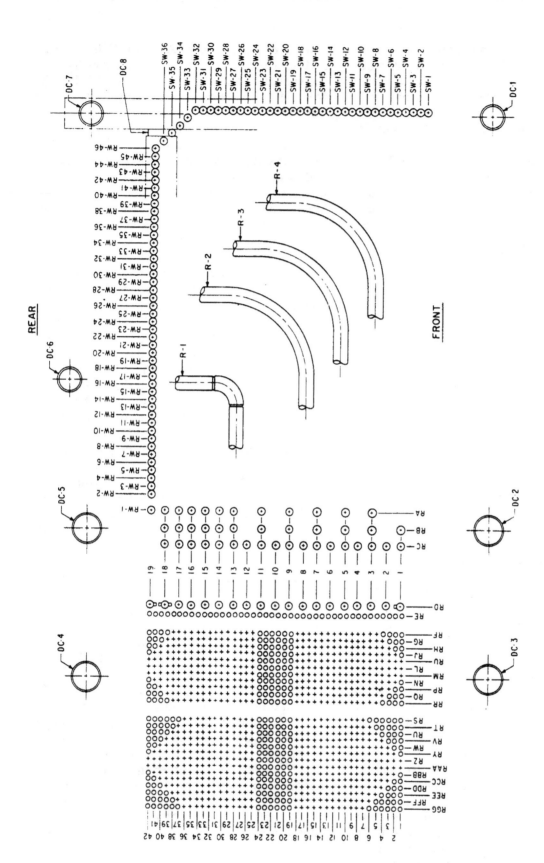

Figure 10-4.—Boiler tube renewal sheet, showing tube identification.

10-8

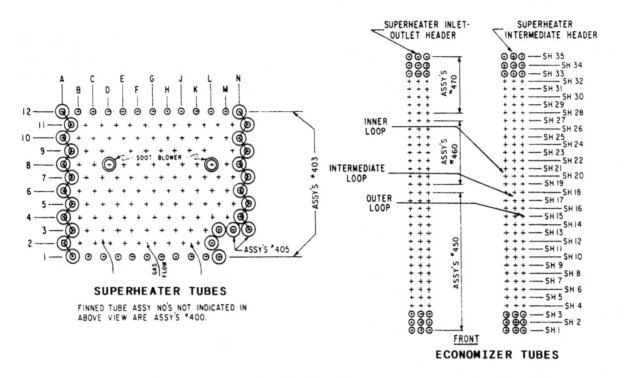

SUPERHEATER TUBES

FINNED TUBE ASSY NO'S NOT INDICATED IN
ABOVE VIEW ARE ASSY'S #400.

ECONOMIZER TUBES

Figure 10-4.—Boiler tube renewal sheet, showing tube identification—Continued.

the superheater, provides accessibility for cleaning, maintenance, and repair. Some of the walk-in superheaters have U-shaped tubes; others have W-shaped tubes.

Some boilers have vertical, rather than horizontal, convection-type superheaters. In these boilers, the U-bend superheater tubes are installed almost vertically, with the U-bends near the top of the boiler; the tubes are approximately parallel

to the main bank of generating tubes and the water screen tubes. Two superheater headers are near the bottom of the boiler, running horizontally from the front of the boiler to the rear.

Figures 10-5 and 10-6 illustrate the U- and the W-shaped tubes. U-shaped tubes are used with a

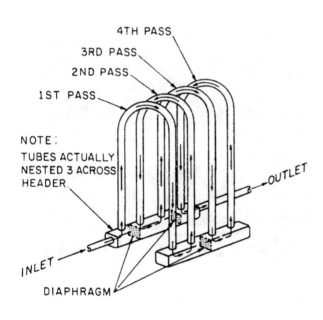

Figure 10-5.—Vertical U-type four-pass superheater.

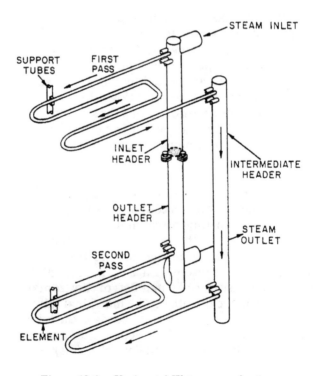

Figure 10-6.—Horizontal W-type superheater.

10-9

vertical superheater, while the W-shaped tubes are used with a horizontal superheater. The arrows in the illustrations trace the steam flow through the superheater and show the number of passes the steam makes through the super-heater.

DESUPERHEATERS

On most boilers, all steam is superheated, but a small amount of steam is redirected through a desuperheater line. The desuperheater can be located in either the water drum or the steam drum, but most generally the desuperheater will be found in the steam drum below the normal water level. The purpose of the desuperheater is to lower the superheated steam temperature back to or close to saturated steam temperature for the proper steam lubrication of the auxiliary machinery. The desuperheater tube bundle is flanged to the superheater outlet on the inlet side and the auxiliary steam stop on the outlet side.

The desuperheater tube bundle has different shapes in different boilers, examples are the S, U, and W.

ECONOMIZERS

An economizer is found on practically every propulsion boiler now in use by the Navy. The economizer is an arrangement of tubes installed in the uptake space from the furnace. The economizer tubes are heated by the rising gases of combustion. All feedwater flows through the economizer tubes before entering the steam drum and is warmed by heat, which would otherwise be wasted, as the combustion gases pass up the stack.

One type of economizer commonly used on naval propulsion boilers consists of two headers and a number of tube assemblies (or elements). The headers are at the rear of the economizer, with the inlet header at the top and the outlet header at the bottom. Each tube assembly (element) consists of a number of horizontal tubes, one above the other. The individual tubes are connected in sequence by U-bends at the tube ends, both at the rear of the economizer and at the front. The tube assemblies (elements) are

arranged side by side in a casing. The top tube of each element is welded into the top, or inlet, header; the bottom tube of each element is welded into the bottom, or outlet, header. Thus, each element forms one continuous loop between the inlet header and the outlet header.

Almost all economizer tubes have some sort of metal projections from the outer tube surfaces. These projections are called by various names, including fins, studs, rings, and disks. They are made of aluminum, steel, or other metals, in a variety of shapes. Figure 10-7 illustrates aluminum gill rings on economizer tubes, a common type of projection. These projections serve to extend the heat transfer surface of the tubes on which they are installed.

FUEL-OIL BURNERS

Burners are essential to effective fuel-oil combustion. They provide combustible mixtures of fuel oil and air by reducing the liquid to finely divided particles with an atomizer, and mixing air with these droplets with a register.

The two main parts of a fuel-oil burner are the atomizer assembly and the air register assembly. The atomizer divides the fuel oil into very fine particles; the air register permits combustion air to enter the furnace in such a way that air mixes thoroughly with the finely divided fuel. In addition to the atomizer assembly and the

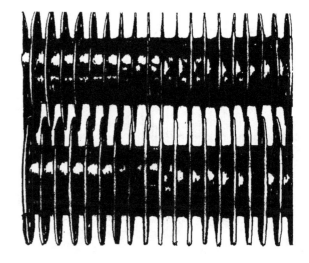

Figure 10-7.—Aluminum gill rings on economizer tubes.

air register assembly, a fuel-oil burner includes various valves, fittings, connections, and burner safety devices.

Atomizers

Mechanical, as well as steam, atomizers are in use on naval ships; with one exception, pressure atomizers. The exception, the use of which is confined to auxiliary boilers, is the rotary atomizer.

In the type of pressure atomizers most commonly found in the Navy, oil is forced under high pressure through passages in the atomizer. The passages are arranged to give the oil a high rotation velocity, thus breaking up the oil by centrifugal force. Rotation is accomplished by either spiral or tangential grooves through which oil is discharged at high pressure into a small cylindrical chamber. The tip end of this chamber is coned out. The orifice through which oil is discharged is at the apex of the cone. As oil leaves the orifice, it breaks up into fog or mist-like particles and forms a hollow cone of finely atomized oil. A blast of air, which has been given a whirling motion in passing through the register, catches the oil mist, mixes with it, and enters the furnace where combustion takes place. There are three most common types of atomizers: (1) the vented plunger, (2) the straight-through-flow, and (3) the steam-assist.

VENTED PLUNGER.—This atomizer is designed to permit wide range operation using the straight mechanical pressure atomization principle but requires a maximum supply oil pressure of approximately 350 psig. Refer to figure 10-8 as we describe the operation of the vented plunger atomizer. Oil flows down the atomizer barrel, around the atomizer cartridge, and enters the whirl chamber through tangential holes drilled around its circumference. The piston in the whirl chamber is spring-loaded and is moved to cover or uncover the tangential holes in the whirl chamber by varying the oil supply pressure. Increasing oil pressure moves the piston to uncover more holes and increase oil rate; decreasing oil pressure does the opposite. Good atomization is maintained at all firing rates due to a high pressure drop maintained through the tangential holes by varying the tangential hole flow area. A hole is drilled through the length of the piston to vent the spring chamber to the furnace, allowing the spring to be compressed. This is possible due to the air core formed in the whirl chamber by the whirling oil. Any oil leakage along the piston into the spring chamber is vented into the furnace in the same manner.

STRAIGHT-THROUGH-FLOW.—In the straight-through-flow atomizer, all oil pumped to the atomizer is sprayed into the furnace. The firing rate of this type of burner is controlled by varying the supply fuel-oil pressure and/or changing sprayer plates. These atomizers have been standardized to obtain maximum performance

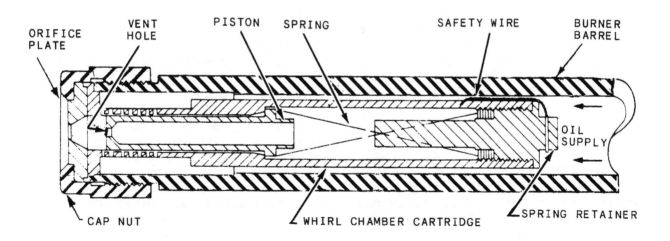

Figure 10-8.—Vented plunger atomizer.

without undue complexity. Standard atomizer parts consist of the following: (see fig. 10-9.)

1. Tip. The tip centers the sprayer plate whirling chamber on the atomizer nozzle box and holds the sprayer plate face tightly against the front face of the atomizer nozzles, preventing passage of oil into the whirling chamber anywhere except through the slots of the sprayer plate.

2. Nozzle. The nozzle leads the oil from the atomizer barrel into the slots of the sprayer plate. The standard nozzle has a 0.5-inch diameter boss and has four oil lead holes.

3. Sprayer plate. The sprayer plate transforms solid streams of oil under high pressure into a cone of fine, foglike particles.

STEAM-ASSIST.—In the steam-assist atomizer, steam breaks up the oil mass into minute particles and projects the oil particles, in the shape of a cone, into the furnace. Two basic types of steam-assisted atomizers are in use in the fleet today: the Y-jet steam atomizer (fig. 10-10) and the LVS steam atomizer (fig. 10-11).

In the Y-jet steam atomizer, steam is supplied through the inner tube of the atomizer to the steam ports of the sprayer plate; oil is delivered to the sprayer plate oil ports through the annulus formed between the inner and outer atomizer tubes. Each steam port meets with an oil port, and the mixture of oil and steam exits from a common hole. Six such exit ports are inclined at an angle of 40° around the axis of the sprayer plate.

In the LVS steam atomizer, a sprayer plate is used to spray oil from the inner tube of the atomizer into a mixing chamber. Steam from the atomizer outer tube also passes through the sprayer plate and into the mixing chamber. The steam and oil blend in the mixing nozzle. The mixture leaves the nozzle as a fine spray through concentrically drilled holes in the mixing nozzle.

When steam is not available, the light-off burner (the one furthest from the superheater) has the capability to substitute low-pressure compressed air for steam. After steam is cut into the atomizing steam system, precaution must be taken to ensure that the steam is thoroughly drained of condensate at the atomizing steam header before the atomizing steam is cut into an individual burner atomizer.

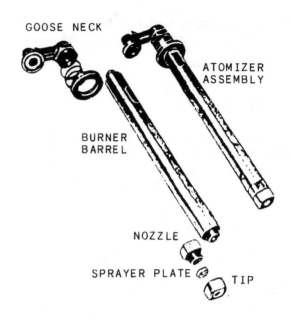

38.17

Figure 10-9.—Straight-through-flow atomizer.

ADVANTAGES OF STEAM ATOMIZATION.—Steam atomization has the following advantages over other types of atomization:

- Clean burning of residual fuel

- Less fireside deposits

- Satisfactory combustion in a cold furnace

- Wide range with lower pressures

- High turndown (see NOTE)

NOTE: Turndown, or turndown ratio, is a measure of the ability of a particular burner to operate satisfactorily over a wide range of firing rates. To the boiler designer, it is the ratio of the maximum satisfactory firing rate to the minimum firing rate of a burner.

CHARACTER OF ATOMIZATION.—Factors affecting the character of atomization, including capacity, spray angle, and particle fineness, are primarily determined by size and relative passage proportions in atomizer parts. They are also determined, to an important extent, by fuel-oil viscosity, steam pressure, steam dryness, and oil pressure. The angle of atomization required differs in various types of registers and boiler furnaces. For this reason, in standardized atomizer parts (all other conditions

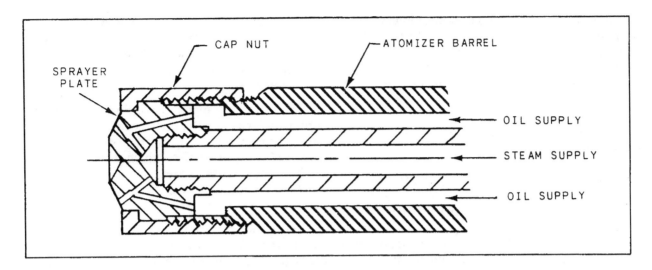

Figure 10-10.—Y-jet steam atomizer.

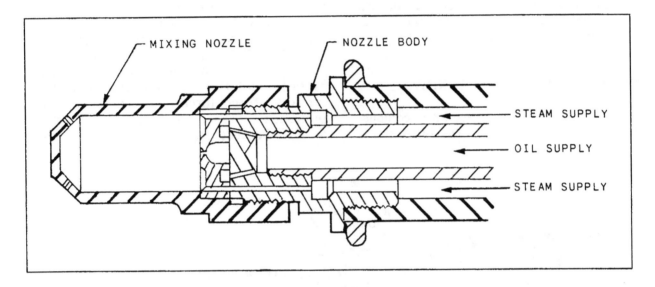

Figure 10-11.—LVS steam atomizer.

remaining unchanged), the relationship between the diameter of the orifice area and the combined cross-sectional area of the tangential slots is highly important because it controls the character of the spray. The size of the orifice of a sprayer plate and the ratio of the combined cross-sectional area of the slots to the area of the orifice are indicated by four numerals stenciled on the face of the plate. The first two digits indicate the bore of the orifice in U.S. standard drill size; the second two numbers are the quotient (omitting the decimal point) of the combined cross-sectional area of the tangential slots divided by the area of the orifice.

Registers

Air registers, when open, direct airflow from the forced draft blowers through the double casing into and around the stream of foglike oil particles produced by the atomizer. When closed, they prevent air from entering the furnace. Air registers employing standard atomizers consist of two principal parts: the diffuser and the air foils. Use of two separate air streams, the primary through the diffuser and the secondary through the air foils, has enabled capacity and flexibility to be increased greatly over that permitted by conical registers. The diffuser causes primary mixing of fuel-oil droplets with air and prevents blowing of the flame from the atomizer; the air foils guide the major quantity of air to mix with the oil particles after they leave the diffuser. Figure 10-12 shows one of the most common air registers in use in the Navy today.

Burner and Atomizer Safety Shutoff Device

The burner and atomizer safety shutoff device (fig. 10-13) is provided on most fuel-oil burner fronts to accept and position the atomizer. It functions solely to prevent personnel from accidentally cutting-in fuel to a burner that does not have an atomizer in place and to prevent uncoupling a fuel-pressurized atomizer, with consequent spraying of fuel around the burner front, which could precipitate a fire. Safety shutoff devices should not be used to admit fuel to the atomizer or to shut off fuel; the burner lead root valves should be used for that purpose.

FURNACES AND REFRACTORIES

A boiler furnace is the space where combustion of air and fuel occurs. The furnace is a rectangular steel casing, which is lined on the floor, front wall, side wall, and rear wall with refractory material. The roof of the furnace is protected from overheating by generating tubes that lie against it.

The refractory material confines the heat of combustion gases to the furnace and protects the casing from overheating. Refractory material is also used to form baffles, which direct the flow of hot combustion gases to protect drums, headers, and tubes from excessive heat and direct contact with the flames.

Refractory materials are used in layers that each have different insulating and heat resistance characteristics. Insulating block is used in the first layer in direct contact with the casing. Insulating block is an excellent thermal insulator that can withstand temperatures up to 1500 °F and is used in 1 or 1 1/2 inch thickness. The next layer is insulating brick, which is not as good a thermal insulator, but can withstand a higher temperature, up to 2500 °F, if there is no direct contact with flames. The sidewalls and roof have water tubes as their third layer of protection. On other surfaces of the furnace, the third layer of protection is firebrick. Firebrick is even less of an insulating material, only 1/10 that of insulating

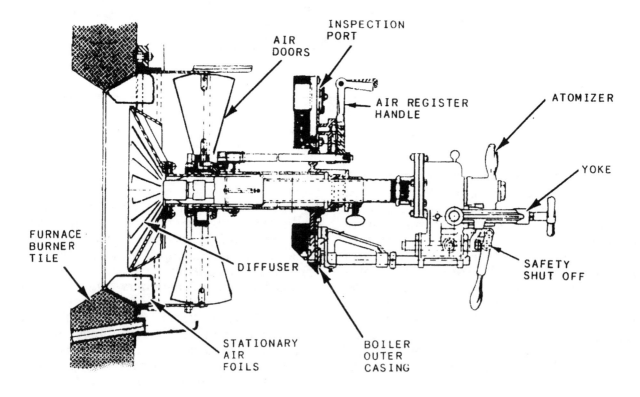

Figure 10-12.—Fuel-oil burner assembly.

10-14

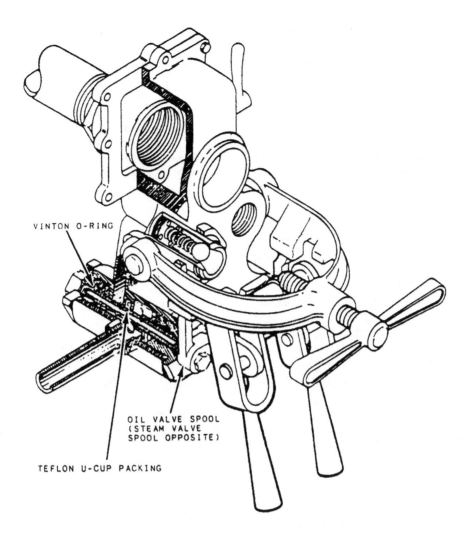

VINTON O-RING

OIL VALVE SPOOL
(STEAM VALVE
SPOOL OPPOSITE)

TEFLON U-CUP PACKING

38.59

Figure 10-13.—Fuel-oil safety shutoff device.

brick, but has excellent fire-resistant qualities. To prevent gas leakage between bricks, an air-setting mortar of finely ground fire clay is used. The mortar also provides some protection against vibrational stresses. Intentional gaps are spaced through the brickwork to serve as expansion joints that allow the brickwork to expand as it is heated without generating stresses. The brickwork is secured to the casing with anchoring devices, which differ depending on the location.

Burner opening cones are built of throat tile, made of the same material as firebrick. Odd-shaped spots and damaged areas are filled with castable or plastic fire-clay refractory. Superheater baffle tiles direct combustion gases over the water screen tubes and the superheater, and are often made of silicon carbide.

CASINGS, UPTAKES, AND SMOKEPIPES

In modern boiler installations, each boiler is enclosed in two steel casings. The inner casing is lined with refractory materials, and the enclosed space forms the boiler furnace. The outer casing extends around most of the inner casing, with an air space in between. At the boiler front, the outer casing usually comes up to a line just above the level of the top burner opening. Air from the forced draft blowers is forced into the space between the inner and the outer casings; from there the air flows through the registers into the furnace.

The inner casing encloses most of the boiler up to the uptakes. The uptakes join the boiler to the smokepipe. As a rule, the uptakes from two or more boilers connect with one smokepipe.

Both the inner and the outer casings are made of steel panels. The panels may be either flanged

10-15

and bolted together (with gaskets used at the joints to make an airtight seal) or they may be of welded construction (except for the access doors).

SADDLES AND SUPPORTS

Each water drum and water header rests upon two saddles, one at the front of the drum or header and one at the rear. The upper flanges of the saddles are curved to fit the curvature of the drum or header and are welded to the drum or header. The bottom flanges are flat and rest on huge beams built up from the ship's structure. The bottom flange of one of the saddles is bolted rigidly to its support. The bottom flange of the other saddle is also bolted to its support, but the bolt holes are elongated in a fore-and-aft direction. As the drum expands or contracts, the saddle that is not rigidly fastened to the support accommodates to the changing length of the drum by sliding forward or backward over a phosphor bronze chock facing or liner. This surface is also greased to minimize friction. The flanges that are not rigidly fastened are known as boiler sliding feet. Figure 10-14 shows the general arrangement of the saddle, support, and sliding feet.

FITTINGS, INSTRUMENTS AND CONTROLS

Internal fittings installed in the steam drum include equipment for (1) distributing the incoming feedwater, (2) separating and drying the steam, (3) giving surface blows to remove solid

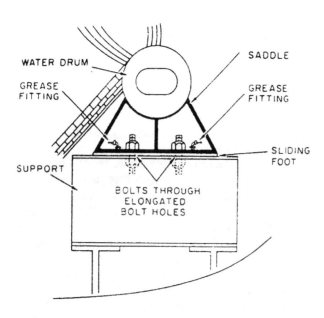

Figure 10-14.—Boiler saddle, support, and sliding feet.

matter from the water, (4) directing the flow of steam and water within the steam drum, and (5) injecting chemicals for boiler water treatment. In addition, uncontrolled superheat boilers have desuperheaters installed to provide steam needed for auxiliary purposes. The desuperheater may be installed in either the steam drum or the water drum and is referred to as an internal fitting.

External fittings and instruments include drains and vents, sampling connections, feed stop and check valves, steam stop valves, safety valves, soot blowers, water gauge glasses and remote water-level indicators, pressure and temperature gauges, superheater temperature alarms, superheater steam flow indicators, smoke indicators, and oil drip detector periscopes. These external fittings, instruments, and controls are discussed in chapter 11 of this manual.

SINGLE-FURNACE PROPULSION BOILERS

Now that we have examined the basic components used in most naval boilers, let's put them together and see how they are arranged to form the boilers now in use in the propulsion plants of naval ships.

Since the majority of propulsion boilers used in the Navy are of the single-furnace type, we will discuss only this type. If your ship has another type, consult the appropriate manufacturer's manual or NSTM, chapter 221.

COMBUSTION AIR AND GAS FLOW

Figure 10-15 shows the combustion air and gas flow in the boiler. Combustion air is discharged from forced air draft blowers through duct work to the rear wall casing of each boiler. There are two blowers per boiler. In the duct work from each blower, there are automatic shutters that permit the flow of air to the boiler but prevent reverse flow, air loss to an idle blower, and windmilling of the idle blower. The combustion air flows between the inner casing and the outer casing of the rear wall. The greater portion of the air travels down under the boiler between the boiler floor and the ship's hull to the front of the boiler. A small portion of the air flows between the two casings, around the side or back of the boiler to the front wall. The air circulating between the casings picks up some of the heat lost by the boiler and reduces the temperature of the outer casing. The heat recovered increases the temperature combustion and helps increase overall boiler efficiency.

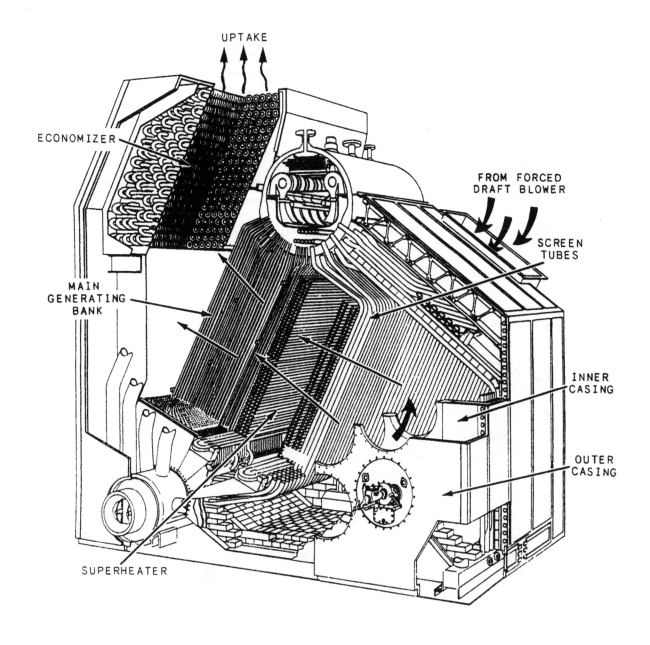

UPTAKE

ECONOMIZER

FROM FORCED
DRAFT BLOWER

SCREEN
TUBES

MAIN
GENERATING
BANK

INNER
CASING

OUTER
CASING

SUPERHEATER

38.41.1

Figure 10-15.—Cutaway view of a boiler, showing combustion air and gas flow.

The primary combustion air passes through the registers and the burner diffuser and is mixed with the atomized oil. The boiler throat ring is equipped with vanes or blades, which give a swirling motion to the air by bypassing the diffuser (secondary air) to provide better mixing with the fuel. When the burner is in operation, the register doors should always be wide open to permit unrestricted airflow. When the burner is secured, the doors should be closed tightly.

It is at this point that ignition takes place. Complete combustion takes place within the furnace. The gases of combustion heat the side and rear wall tubes. Water circulating within the tubes prevents overheating and tube burnup.

The gases of combustion exit the furnace through the screen tubes. The screen tubes greatly reduce direct flame impingement on the superheater tubes when the boiler is operating. From the screen tubes the gas flows through the superheater and its cavity to the main generating bank. From the main bank the gas flows upward through the economizer and into the uptakes.

BOILER WATER AND STEAM CIRCULATION

Figure 10-16 shows the boiler water and steam flow in the boiler. Feedwater enters the economizer inlet header located at the top of the economizer. The inlet header connects to horizontal aluminum finned tubes that run perpendicular to the gas flow. Feedwater travels through these tubes and back through another set of finned tubes by U-bends. Each plane of 13

tubes is referred to as a row of economizer tubes. The feedwater passes through these rows of tubes before exiting the economizer through the outlet header. During full-power conditions, the feedwater temperature is increased approximately 190°F (from 250°F to 440°F).

Feedwater enters the steam drum through a nozzle located in the rear drum head. Inside the steam drum, the water is directed below the normal water level by an internal pipe. This internal pipe, called a feed pipe, has holes that allow the water to disperse evenly along the length of the drum.

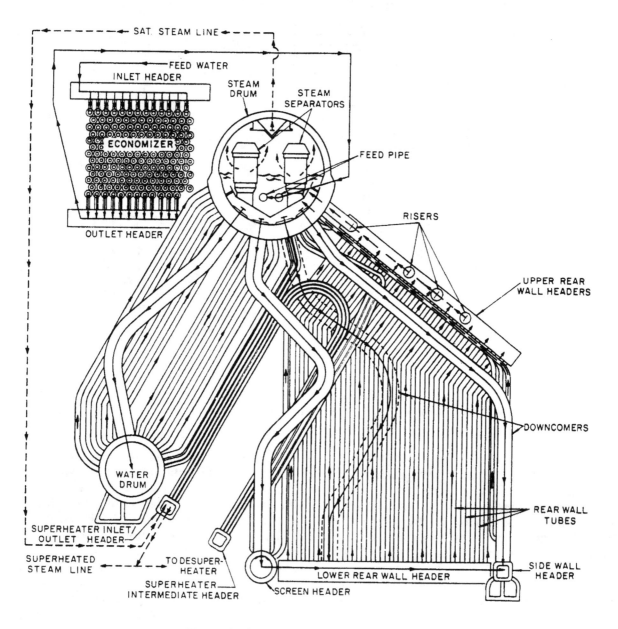

Figure 10-16.—Boiler water and steam flow.

The boiler in figure 10-16 is a natural circulation type. The cooler, denser water flows downward to the water drum, while the heated, less dense mixture of steam and water in the tubes rises to the steam drum.

Downcomers are large distribution pipes. They ensure a positive flow of water to the lower parts of the boiler. Downcomers extend from the steam drum to the water drum, screen header, lower rear wall header, and to the sidewall header. They are located between the inner and outer casings and are kept relatively cool by the flow of combustion air to the burners.

The hot gases of combustion heat the water in the generating, sidewall, rear wall, and screen tubes. As the water is heated in these areas, the density is decreased and the water starts to rise. Continuing up the tubes, a portion of the water changes to steam. The water and steam mixture from the sidewall screen tubes and generating bank enters the steam drum under the deflection plate assembly. The water and steam mixture from the rear wall travels upward to the upper rear wall header to the lower portion of the steam drum through riser pipes.

Steam separators connected to the deflection plate assembly permit steam to flow from under the deflection plate assembly to the top half of the steam drum. The steam separators separate the steam and water mixture by centrifugal action. The water falls back into the lower half of the steam drum while the steam rises to the dry box. The dry box extends longitudinally along the top of the drum. Saturated steam leaves the dry box through the saturated steam outlet nozzle located on top of the steam drum.

The saturated steam is piped from the steam drum nozzle to the superheater inlet/outlet header. The steam is directed through the superheater tubes by diaphragms located in the inlet/outlet and intermediate headers. The steam makes four passes, during which it is heated to superheat temperature.

Once the steam leaves the superheater via the inlet/outlet header, it is split in two directions. The majority of the steam is directed to the main engines. A small portion of the steam is directed to the desuperheater. The desuperheater, located in the water drum, reduces the temperature of the superheated steam to that needed for use with auxiliary equipment.

SUPERHEATER STEAM FLOW AND BOILER LIGHT-OFF

The flow of steam as it occurs after the boiler has been cut-in on the steam line has already been discussed. What happens when a cold boiler is lit-off? How are the superheater tubes protected from the heat of the furnace after fires are lighted but before sufficient steam has been generated to ensure a safe flow through the superheater?

Various methods are used to protect the superheater during this critical period immediately after lighting-off. Very low firing rates are used, and the boiler is warmed up slowly until adequate flow of steam has been established.

The steam flow is established through the superheater by venting the drains to the bilges while warming up the boiler very slowly. Once there is adequate pressure in the superheater, the superheater drains are closed. Next the superheater bleeder valve is opened, if installed. The superheater bleeder valve is located on the outlet of the superheater, in front of the auxiliary steam stop (see fig. 10-17). Steam will flow through the superheater, the desuperheater, the bleeder valve, and into the auxiliary exhaust system. This steam in the auxiliary exhaust system will warm-up the DFT and feedwater it contains if this is the first boiler to be lit-off. Otherwise, this steam is used in the auxiliary exhaust system or condensed to recover the condensate.

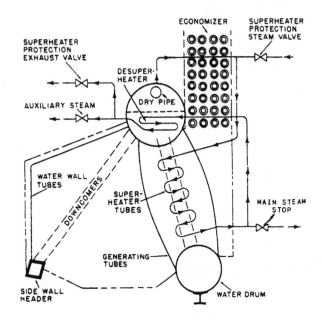

Figure 10-17.—Superheater protection steam arrangement.

The degree of superheat obtained in a single-furnace boiler of the type being considered is primarily dependent upon the firing rate. However, a number of design features and operational considerations also affect the temperature of the steam at the superheater outlet.

Design features that affect the degree of superheat include (1) the type of superheater installed—that is, whether heated by convection, by radiation, or by both; (2) the location of the superheater with respect to the burners; (3) the extent to which the superheater is protected by water screen tubes; (4) the area of superheater heat-transfer surface; (5) the number of passes made by the steam in going through the superheater; (6) the location of gas baffles; and (7) the volume and shape of the furnace.

Operational factors that affect the degree of superheat include (1) the rate of combustion, (2) the temperature of the feedwater, (3) the amount of excess air passing through the furnace, (4) the amount of moisture contained in the steam entering the superheater, (5) the condition of the superheater tube surfaces, and (6) the condition of the water screen tube surfaces. Since these factors may affect the degree of superheat in ways that are not immediately apparent, let us examine them in more detail.

How does the rate of combustion affect the degree of superheat? To begin with, we might imagine a simple relationship in which the degree of superheat goes up directly as the rate of combustion is increased. Such a simple relationship does, in fact, exist—but only up to a certain point. Throughout most of the operating range of this boiler, the degree of superheat goes up quite steadily and regularly as the rate of combustion goes up. Near full power, however, the degree of superheat drops slightly even though the rate of combustion is still going up. Why does this happen? Primarily because the increased firing rate results in an increased generating rate, which in turn results in an increased steam flow through the superheater. The rate of heat absorption increases more rapidly than the rate of steam flow until the boiler is operating at very nearly full power; at this point the rate of steam flow increases more rapidly than the rate of heat absorption. Therefore, the superheater outlet temperature drops slightly.

Suppose the boiler is being fired at a constant rate and the steam is being used at a constant rate. If the temperature of the incoming feedwater increases, what happens to the superheat? Does it increase, decrease, or remain the same?

Surprisingly, the degree of superheat decreases if the feed temperature is increased; more saturated steam is generated from the burning of the same amount of fuel. The increased quantity of saturated steam causes an increase in the rate of flow through the superheater. Since there is no increase in the amount of heat available for transfer to the superheater, the degree of superheat drops slightly.

If the amount of excess air is increased under normal conditions of constant load and a constant rate of combustion, an increase of superheater outlet temperature results. To see why an increase in excess air results in an increase in temperature at the superheater outlet, we must take it step by step:

1. An increase in excess air decreases the average temperature in the furnace.
2. With the furnace temperature lowered, there is less temperature difference between the gases of combustion and the water in the boiler tubes.
3. Because of the smaller temperature difference, the rate of heat transfer is reduced.
4. Because of the decreased rate of heat transfer, the evaporation rate is reduced.
5. The lower evaporation rate causes a reduction in the rate of steam flow through the superheater, with a consequent rise in the superheater outlet temperature.

In addition to this series of events, another factor also tends to increase the superheater outlet temperature when the amount of excess air is increased. Large amounts of excess air tend to cause combustion to occur in the tube bank rather than in the furnace itself. As a result, the temperature in the area around the superheater tubes is higher than usual and the superheater outlet temperature is higher.

Any appreciable amount of moisture in the steam entering the superheater causes a very noticeable drop in superheat. This occurs because steam cannot be superheated as long as it is in contact with the water from which it is being generated. If moisture enters the superheater, therefore, a good deal of heat must be used to dry the steam before the temperature of the steam can rise.

The condition of the superheater tube surfaces has an important effect on superheater outlet temperature. If the tubes have soot on the outside or scale on the inside, heat transfer to the water in these tubes will be retarded. Therefore, there will be more heat available for transfer to the superheater as the gases of combustion flow through the tube bank. Consequently, the superheater outlet temperature will rise.

BOILER WATER REQUIREMENTS

Modern naval boilers cannot be operated safely and efficiently without careful control of boiler water quality. If boiler water conditions are not just precisely right, the high operating pressures and temperatures of modern boilers will lead to rapid deterioration of the boiler metal, with the possibility of serious casualties to boiler pressure parts.

Although our ultimate concern is with the water actually in the boiler, we cannot consider boiler water alone. We must also consider the water in the rest of the system, since we are dealing

with a closed cycle in which water is heated, steam is generated, steam is condensed, and water is returned to the boiler. Because the cycle is continuous and closed, the same water remains in the system except for the water that escapes, either as steam or as water, and is replaced by makeup feed.

WATER IN THE SHIPBOARD WATER CYCLE

In addition to recognizing the continuous, or cyclical, nature of water in a shipboard steam plant, you must be able to distinguish between the kinds of water at different points in the system. This distinction is necessary because different standards are prescribed for the treatment of water at different points in the system. To identify the kinds of water at various points in the shipboard water cycle, look at figure 10-18. Then study and become familiar with the following terms.

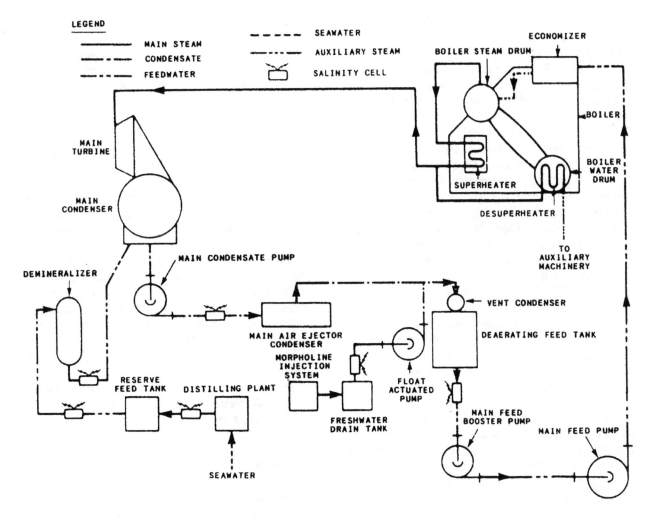

Figure 10-18.—Shipboard water cycle.

DISTILLATE—Distillate is the evaporated water that is discharged from the ship's distilling plant. Water in the shipboard water cycle normally begins as distillate. This distillate is stored in the reserve feedwater tanks until needed in the main steam cycle.

RESERVE FEEDWATER AND MAKEUP FEEDWATER—Distillate, while stored in the feedwater tanks, is called reserve feedwater. In ships without demineralizers, the water in all tanks (including the tank supplying makeup feedwater) is also called reserve feedwater; however, when the water is directed to the condensate system it is called makeup feedwater. In ships equipped with demineralizers, the water entering the demineralizers from any tank is called reserve feedwater, while the water flowing out of the demineralizer to the condensate system is called makeup feedwater.

CONDENSATE—After the steam has done its work, it is returned to the liquid state by cooling in a condenser. The condenser is a heat exchanger in which steam, at a pressure below atmospheric, flows over and is condensed on tubes through which seawater flows. Any water condensed from steam is called condensate. Water from other sources (such as makeup feedwater and low-pressure drains) which is mixed with the condensate becomes part of the condensate.

The principal sources of condensate are the main and auxiliary condensers. Other sources of condensate include the gland exhaust system, heating drains, distilling unit drains, steam coils, and other miscellaneous or service steam system drains. To maintain the proper quantity of water in the system, condensate from various sources is combined with makeup feedwater and is then sent to a DFT where dissolved oxygen and other gases are removed from the water. The discharge from the DFT is called deaerated feedwater. When properly operated, the DFT will maintain the dissolved oxygen content of the deaerated feedwater at or below 15 parts per billion (ppb) (15 micrograms per liter (μg/L)).

FEEDWATER (DEAERATED FEEDWATER) SYSTEM—In a broad sense, the term *feedwater system* includes the makeup feed and transfer systems, the main and auxiliary condensate systems, and the deaerated feedwater system. In a narrow sense, it includes only the deaerated feedwater system. The two usages of the term *feedwater system* should not be confused.

BOILER WATER—Boiler water is the name of the deaerated feedwater as it enters the boiler steam drum. The term *boiler water* describes the water in the steam drum, water drum, headers, and the generating tubes of the boiler.

FRESHWATER—Aboard ship the term *freshwater* generally refers to potable water. However, certain steam drains that are returned to the condensate system are called freshwater drains.

CHEMICAL NATURE OF WATER

To understand water chemistry, we must first understand the concept of molecules. A molecule is the smallest particle into which a compound can be dissolved without changing its chemical nature. A water molecule (H_2O) consists of an oxygen atom and two hydrogen atoms.

Atoms have a nucleus containing positively charged protons and uncharged neutrons, and negatively charged electrons in motion and orbit around the nucleus. Normally, atoms are electrically neutral, having the same number of electrons with a negative charge as protons with a positive charge. When atoms are not electrically neutral, that is, when they have a different number of electrons and protons, they are called ions.

Normally, the number of electrons equals the number of protons contained in the nucleus and their positive and negative charges cancel each other. The electrons are bound in the atom by electrostatic action, so the atom remains neutral unless some external force causes a change in the number of electrons.

Electrons orbit the nucleus in shells. These shells each have an associated energy level and maximum number of electrons to fill it. To chemically form molecules, atoms bond together. In forming the chemical bonds, atoms desire to fill their outer electron shell. In forming ionic bonds, electron shells of two different atoms touch and share electrons.

In some cases, the electrons are not shared but are transferred. When this happens ions are formed. Ions of opposite charges attract each other and form molecules.

pH Value

One important aspect of water is its capability for partial ionization. Pure water may break apart (or ionize) slightly to yield hydrogen

(H$^+$) ions and hydroxyl (OH$^-$) ions. (A hydroxyl ion consists of one atom of hydrogen and one atom of oxygen. The hydroxyl group of atoms may be replaced by one single atom.) The concentration of hydrogen ions and hydroxyl ions is determined by measuring pH.

pH is a measure of the concentration of the hydrogen ion. It is a logarithmic function. The pH scale ranges from 0 to 14. A pH of 7 signifies that the solution is neutral; that is, it has the same amount of hydrogen ions and hydroxyl ions. The pH number indicates the degree of alkalinity or acidity of water. If the pH is below 7, the water has more hydrogen ions and is acid; on the other hand, if the pH is above 7, the water has more hydroxyl ions and is alkaline.

As a logarithmic function, an increase of pH by 1 indicates an increase in the concentration of hydrogen ion concentration by a multiple of 10.

BOILER WATER CONTAMINATION PROBLEMS

Although there are many sources of contamination in feedwater and boiler water, there are only three basic problems, or detrimental effects, of contaminants. These problems are waterside deposits, waterside corrosion, and carryover. The boiler water and feedwater treatment programs are aimed at preventing the occurrence of these three problems.

Waterside Deposits

Waterside deposits will attach themselves to the surface of the boiler tube and act as insulators, reducing the heat transfer where they are located. This reduction of heat transfer will result in an increase in temperature of the boiler tube, which will cause the tube to soften, then blister, and possibly result in tube failure or rupture. Figure 10-19 shows the general manner in which waterside deposits do this. In a boiler operating at 1200 psi, the saturation temperature and the temperature inside the boiler tube is 576°F. The fireside of the tube will typically be 100°F higher. As a deposit restricts heat transfer, the tube temperature will increase and approach 800°F. As the deposits get thicker, the temperature increases to 950°F. At this temperature, the tube metal softens and the tube blisters.

Three contaminants that can be the result of waterside deposits are scale, sludge, and oil. Sludge represents any possible suspended solids in the boiler water. Some sludge is formed as

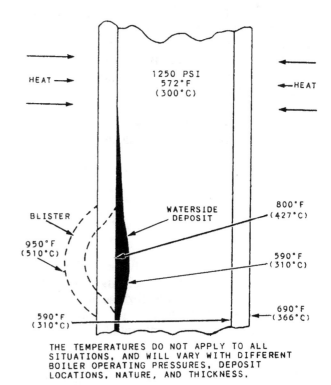

THE TEMPERATURES DO NOT APPLY TO ALL SITUATIONS, AND WILL VARY WITH DIFFERENT BOILER OPERATING PRESSURES, DEPOSIT LOCATIONS, NATURE, AND THICKNESS.

Figure 10-19.—Effect of boiler deposits.

chemical treatment of the boiler water reacts with other contaminants to produce material that will precipitate rather than remain in solution. Sludge is controlled by the requirement to secure the boiler and conduct a bottom blowdown every 168 steaming hours (1 week). The bottom blowdown will remove most sludge, unless it has already deposited itself to the tube surface.

Calcium and magnesium are among the many salts found in the seawater. If present in boiler water, these salts precipitate as a result of chemical reaction with the phosphate ions added to the boiler water as control measures. The precipitated sludge is removed by bottom blowdown. The presence of these salts can be detected by testing the conductivity of boiler water.

Silica is another form of contaminant that may cause scale formation on boiler tubes. Silica sources are from shore water or the desiccant used in dry layups. Silica can be found in seawater, but only minimal concentrations. Silica is present in shore water and shore steam in measurable concentrations. Silica, in comparison to calcium and magnesium, does not contribute to the conductivity of the boiler water. Therefore, detection of silica in the boiler water involves another test.

Fuel oil and 2190 TEP lubricating oil accumulate in the steam drum to cause boiler water and oil to carryover with the steam. They also cause baked-on carbonaceous (carbon containing) deposits on the tubes and oily deposits (sludge balls) in the drum headers. Detergent lube oils like 9250 will not rise into the steam drum, but will emulsify throughout the boiler water and then will cause baked-on carbonaceous deposits, blistering, and eventually tube failure. Since preservatives behave in a manner similar to detergent lube oils, conductivity equipment and pH electrodes will not work when coated with oil.

Waterside Corrosion

Waterside corrosion is usually electrochemical in nature. It may be localized pitting or general corrosion. The boiler tube surface has slight chemical and physical variations. These variations can cause different electrical potentials of the tube surface at different areas. In this case, one area will become an anode (positive terminal) and another becomes the cathode (negative terminal). At the anode, iron tends to dissolve into solution in the boiler water. This dissolved iron may eventually be deposited at the cathode.

In general corrosion, there are many anodes and cathodes formed that tend to constantly change location as corrosion continues. There is a general loss of iron over the whole tube surface as some of the dissolved iron is lost.

This electrolytic action can be minimized by keeping the conductivity of the boiler water to a minimum. Conductivity is the ability to conduct an electrical current and is expressed in michromhos per centimeter (μmho/cm). Ions and salts in the water contribute to its conductivity.

Another method to minimize the electrolytic action of water is to minimize the amount of dissolved oxygen in the water. Oxygen will dissolve readily into water from any surface open to the atmosphere. Dissolved oxygen is removed in the DFT. The DFT removes oxygen, air, and dissolved gases from the feedwater by heating it to saturation temperature and atomizing it. The solubility (maximum possible concentration) of gases in water (or any liquid) decreases to near zero as the temperature approaches the saturation temperature. Also, the solubility of gases in a liquid increases with pressure. Dissolved oxygen will concentrate in an irregular spot in the tube surface. The presence of oxygen will create a strong anode that will not move. Eventually the anode becomes a pit when a significant amount

of iron has gone into solution. Figure 10-20 shows a pit in the boiler tube that indicates an anodic area where iron has gone into solution.

Carryover

Carryover is the circumstance where water is carried over with the steam as it enters the superheater. Normally, this occurs like a fine mist with very small particles of moisture. If it occurs with a carryover of large gulps of slugs, it is referred to as priming. Carryover is dangerous as it can result in severe damage to the superheater, steam lines, turbines, and valves. The biggest danger with priming or high water casualty in the boiler is that the large slug is carried along by the steam pressure. The water has an extremely high density in comparison to the steam and travels at a high velocity. It has such tremendous kinetic energy it can actually rip turbine blades from the rotor. In carryover, the water brings solid material with it. The solid material will be deposited on some surface. This solid material may become a deposit to eventually damage a superheater tube, or it may be deposited on a turbine blade and affect its balance and performance. Figure 10-21 shows a superheater tube in which solid material has been deposited as a result of carryover.

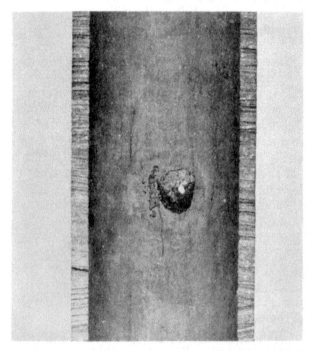

38.138

Figure 10-20.—Localized pit in a boiler tube, caused by dissolved oxygen.

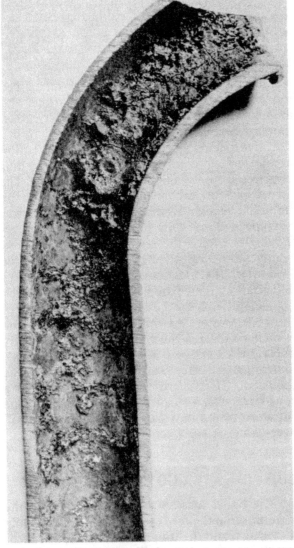

38.140

Figure 10-21.—Evidence of carryover in a superheater tube.

Boiling water produces bubbles that contain steam encircled by the water. The bubbles break at the water surface to release the steam, and the water falls back. In pure water, the bubbles break easily, and an almost complete separation of the water from the steam occurs. Dissolved or suspended solids in the water make the steam bubbles more difficult to break. The effectiveness of the separation of contaminated water from steam then deteriorates. Dissolved solids and suspended solids in boiler water increase in concentration with continued steaming. (They are controlled by blowdown.) As the steam drum area above the surface of the boiler water fills up with these stabilized bubbles, some boiler water is carried along with the saturated steam into the dry pipe. This type of boiler water carryover is termed *foaming*. Since the stabilized bubbles break at high superheat temperatures, the boiler water entrained with the saturated steam evaporates in the superheater. Particles of material that were dissolved and suspended in the boiler water are carried by the velocity of the steam to different parts of the superheater to be deposited on the superheater surfaces, primarily on those in the first pass. Restriction of heat transfer, blistering, and eventual superheater failure can result whenever a deposit builds up. Additionally, one of the materials dissolved in the boiler water, sodium chloride, can cause stress corrosion cracking if it is deposited on stainless steel superheaters.

Another type of boiler water carryover is called priming. Priming is caused by a high water level in the steam drum, extreme rolling and pitching of the ship, or mechanical failure of the steam separators or other steam drum internal parts. Priming consists of slugs of boiler water carried into the dry pipe. A slug of water may reach as far as the turbine, with resultant damage to the unit.

CONTROL OF FEEDWATER CONTAMINANTS

Feedwater conditions must be controlled if corrosion and scale in the feed system are to be minimized. More importantly, corrosion, scale, various deposits, and carryover in high-pressure boiler systems will advance if care of feedwater purity is neglected. Boiler water treatment protects the boiler against feed system upsets from contamination. It does not replace proper preboiler plant operation. Feedwater having low concentrations of dissolved and suspended solids, and no oxygen, is considered good quality feedwater. This type of feedwater is obtained when all units in the preboiler system are functioning properly.

Contaminant Testing

The quality of feedwater is determined by testing samples for chloride, conductivity, salinity, hardness, dissolved oxygen, and pH from various points in the system. The tests that are required are governed by the equipment installed. Chloride tests indicate seawater contamination. Conductivity and salinity both indicate sea and shore water contamination. Hardness tests detect shore water more rapidly than the chloride tests. A dissolved oxygen sample from the DFT indicates the condition of the DFT by measuring the oxygen level of the DFT discharge. The corrosiveness of the water is related to pH. All of these tests are necessary for rapid location and correction of contaminant sources.

Morpholine Condensate/Feedwater Treatment

Morpholine is a water treatment chemical that raises pH and reduces corrosion in the condensate and deaerated feedwater lines of the preboiler system. Preboiler system corrosion is caused principally by the reaction of carbon dioxide with the water in which it is dissolved. There are two sources of carbon dioxide: air containing carbon dioxide, which can enter at almost all points in the preboiler system, and the bicarbonate ion in seawater, which is released when the water is heated in distilling plants, thereby acidifying the reserve feedwater.

Seawater and shore water are the major contributors of bicarbonate. Carbon dioxide reacts with water and forms carbonic acid, which decreases the pH and makes the water more acidic. By raising the pH of the condensate and deaerated feedwater, corrosion in these parts of the preboiler system is reduced.

Morpholine is a neutralizing amine that is added to the freshwater drain-collecting tank. It combines with hydrogen ions produced from the reaction of carbon dioxide in water. It also combines with some hydrogen ions present from water's ionization. As the concentration of hydrogen ions decreases, pH increases and corrosion is reduced. This reduces the amount of sludge that will form in the boiler. The morpholine added to the condensate enters the boiler with the deaerated feedwater; it is volatilized along with the steam; it disperses throughout the steam plant; it is condensed; and finally it is recycled. Although the cycle is continuous, additional morpholine must be fed to the condensate to compensate for losses that occur in the air ejectors, DFT vents, and leaks.

Demineralization of Makeup Feed

Makeup feed of very high purity is produced by the use of a demineralizer. When taking on makeup feed, reserve feedwater is passed through a mixed bed ion exchange resin called a demineralizer. The mixed bed ion exchange resin consists of a chemically balanced mixture of both cation exchange resins and anion exchange resins. The cation exchange resin, whose surface is rich in hydrogen ions, exchanges with cations such as magnesium and calcium. The anion exchange resin, whose surface is rich in hydroxyl ions, exchanges with anions such as chloride and sulfate. As the water passes over the mixed bed ion exchange resin, the dissolved cations and anions exchange with the hydrogen and hydroxyl ions and become attached to the exchange resins. The hydrogen ions displaced from the cation exchange resin combine with the hydroxyl ions displaced from the anion resin to form water. The mixed bed ion exchange resin removes all the cations and anions from the reserve feedwater and exchanges them with pure water. Besides removing the ions, the resin acts as a filter and removes the suspended solids from the reserve feedwater. This demineralization (also called deionization) is an extremely useful process which appreciably reduces the rate of sludge buildup in the boiler water. As the resin bed exhausts its exchange capacity, the makeup feed conductivity quickly rises above 1.0 $\mu mho/cm$ (1.0 $\mu S/cm$), indicating that the resin is spent and has to be replaced. If the reserve feedwater has a consistently high suspended solids content, the filtration action may clog the resin bed and will necessitate an early replacement of the resin.

BOILER WATER CONTROL

The boiler acts as a receiver for all of the materials that the feed system pours into it. Only the water and (if the condensate is treated) morpholine leave the boiler. All contaminants remain behind to advance the damaging conditions (corrosion, scale formation, and carryover) already discussed. To minimize damage, boiler water treatment programs have been established. The Navy treatment is based on a regimen known as coordinated phosphate-pH control (more simply, coordinated phosphate control). The method is designed to serve several purposes:

1. Maintain the pH and phosphate levels so that caustic corrosion cannot occur.

2. Maintain the pH sufficiently high to limit boiler metal corrosion and to protect against acid-forming magnesium reactions. (When magnesium hydroxide forms under alkaline conditions, it is in a sludgelike state unlike the scale that precipitates under neutral or acidic conditions.)

3. Maintain a phosphate residual in the water to precipitate calcium and magnesium as

phosphate sludges which are less adherent than scale.

Since water conditions are maintained in a manner conducive to precipitation of sludges, coordinated phosphate control is referred to as a phosphate precipitating program. Though the boiler water can accommodate some contamination, the boiler should not be considered a safe reaction vessel for the generation of sludge. Sludge is a deposit that has most of the objectionable properties of scale. Another boiler water control method, blowdown, helps to remove it.

Coordinated Phosphate-pH Control

Coordinated phosphate-pH control prevents development of water having excessive hydroxyl ions, which leads to caustic corrosion. The treatment chemicals are trisodium phosphate, dodecahydrate (a crystalline form of trisodium phosphate), and anhydrous disodium phosphate (a powder form of disodium phosphate). The trisodium phosphate reacts with water to form sodium hydroxide and disodium phosphate. The sodium hydroxide contributes hydroxyl ions, which raise pH, while the disodium phosphate provides some of the needed phosphate. Although sodium hydroxide and disodium phosphate could be used in the treatment to produce the same results, the weights of sodium hydroxide needed would be so small that weighing errors would be large. In addition, accidental overaddition of sodium hydroxide would lead to caustic corrosion.

When boiler water containing the correct amount of sodium hydroxide and disodium phosphate is concentrated by heating, or evaporated to dryness, only the trisodium phosphate concentrate remains behind and there is no excess of hydroxyl ions (called free caustic). A primary aim in coordinated phosphate control is the elimination of free caustic, which forms in concentrated boiler water when pH is too high (when there are too many hydroxyl ions).

The calculations for trisodium phosphate yield a curve (the coordinated phosphate curve line presented in fig. 10-22). This curve divides the

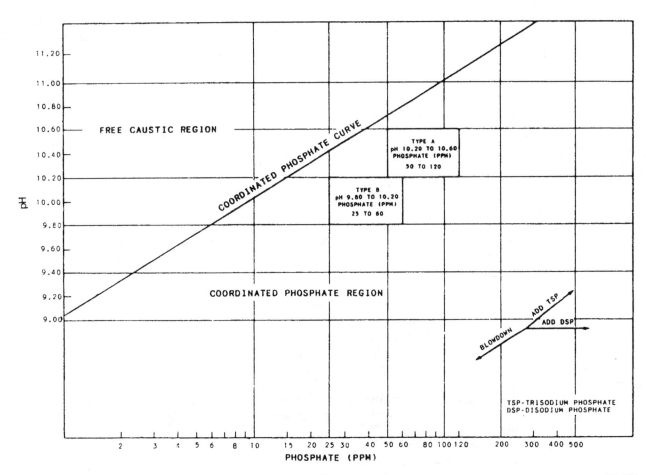

38.170

Figure 10-22.—Coordinated phosphate curve.

10-27

chart into two regions: the free caustic region, on the top, and the coordinated phosphate region, on the bottom. The coordinated phosphate curve line represents what the resulting pH water would be with trisodium phosphate added. As long as boiler water phosphate and pH levels are maintained BELOW the coordinated phosphate curve in figure 10-22, free caustic cannot result. When pH rises above the curve, free caustic is present and caustic corrosion of boiler metal can result.

If trisodium phosphate alone were used in treatment, boiler water control would have to follow exactly along the curve. The addition of extra disodium phosphate does not detectably change the pH of the water and permits assignment of a range for boiler water control. Of significance is the fact that when added to boiler water, neither trisodium phosphate nor disodium phosphate can force boiler water into the free caustic region.

All naval boilers have boiler water controlled under the coordinated phosphate curve. Boilers are classified as either type A or type B for control limits. Control limits are based on the heat transfer rate with the cutoff at 150,000 Btu/hr/ft^2. Type B boilers have the higher heat transfer rate. Type B boilers include all 1200 psi boilers and boilers of the CV-59, AOE-1 through 4, LPD-1 through 15, AGP-3, LSD-28, and FF-1038. All other boilers have lower heat transfer rates and are type A. The high heat transfer boilers have less tolerance to any chemical in the water, whether from treatment or contamination. In addition, the higher pressure boilers cannot tolerate as much conductivity as lower pressure boilers because of carryover. For these reasons, operating limits for type B boilers are lower than for type A boilers.

The Significance of Free Caustic (Corrosion and Brittle Failures)

When more hydroxyl ions are available in boiler water than will completely react with disodium phosphate to form trisodium phosphate on concentrating, the excess hydroxyl ions (free caustic) will concentrate to yield corrosive conditions in crevices, in porous deposits, at surface discontinuities (pits), and at leakage sites. The concentrated solution causes corrosion, cratering, and gouging of tube metal. The concentration of hydroxyl ions results from the temperature difference between the tube metal and the film of boiler water. A 1°C (2°F) drop corresponds to a hydroxyl ion corrosive concentration of about 5 percent regardless of the amount of sodium hydroxide in bulk boiler water. A 3°C (5°F) temperature differential can concentrate hydroxyl ions to about 10 percent. The actual concentration is also determined by the effectiveness of rinsing (washing) boiler tube metal due to natural circulation. Boiler water analyses show that the water is in the free caustic region above the coordinated phosphate curve with a pH of 10.65 and a phosphate of 20 ppm (20 mg/L). The pH at the metal-deposit interface is 13. (See fig. 10-21.) Caustic corrosion takes place at this high pH.

A prime example of caustic corrosion is evidenced by pitting and gouging in the lower bends of screen tubes where deposits tend to collect. A special case of caustic corrosion is created when metal is stressed and is called caustic stress corrosion. This results in a brittle failure similar to and as severe as the hydrogen damage failure caused by excess acid in boiler water.

Reactions in Treated Boiler Water

The behavior of treatment chemicals in general and under various contaminant conditions are described in the following paragraphs.

PRECIPITATION REACTIONS (SLUDGE FORMATION). —The alkaline pH level in boiler water serves to minimize boiler metal corrosion and also provides hydroxyl ions needed to react with the magnesium that would otherwise turn neutral water acidic. The magnesium hydroxide forms a sludge as long as the water remains alkaline. Both the sodium hydroxide and disodium phosphate in the water react with calcium and magnesium to form various phosphate sludges. Sludge in boiler water is an agglomeration of reaction products of boiler water treatment chemicals with calcium and magnesium contaminants as well as the suspended matter entering with the feedwater. The amount of sludge, if allowed to accumulate in the headers and water drum, will grow so large that particles will circulate with the boiler water. As sludge circulates, it begins to adhere to the generating surfaces. At first the adhering sludge is soft and is removable by mechanical cleaning. If allowed to remain on generating surfaces, the soft sludge is converted by heat to hard, baked-on sludge. Although baked-on sludge is physically different from scale (scale forms in place; sludge is carried to high heat transfer areas), it acts just like scale in that it restricts heat transfer and causes blistering and eventual rupture of the tube. Mechanical cleaning of the watersides will not remove baked-on sludge nor scale. Scale is prevented by proper chemical treatment, and sludge is kept low in concentration by maintaining feedwater purity and by effective blowdown, primarily by bottom blowdown of a secured boiler.

EFFECT OF SEAWATER ON BOILER WATER CHEMICALS.—Seawater that enters boiler water contains contaminating chemicals, magnesium, calcium, and chloride, all of which raise the conductivity of the water. Because of the large amount of magnesium, the chemical reaction due to seawater contamination causes pH to drop. Both magnesium and calcium cause phosphate to decrease because the phosphate causes the formation of sludge. Chloride and conductivity increase. The pH and phosphate must always be kept under the coordinated phosphate curve and within specified limits because of the damage that will result if either pH becomes acidic or the concentration of phosphate falls below a safe value.

As a matter of interest, addition of salt alone (sodium chloride) will increase chloride and conductivity but will not affect pH or phosphate. It is neither sodium chloride nor chloride per se that causes scale formation or acid attack.

EFFECT OF SHORE WATER ON BOILER WATER CHEMICALS.—The relatively large amount of calcium in shore water in comparison to magnesium causes phosphate to react first. The pH increases, conductivity increases slowly, and chloride does not change. Since shore water depletes phosphate and increases pH, boiler water control parameters are forced into the free caustic range. To prevent scale formation and caustic corrosion, the pH and phosphate must always be kept under the coordinated phosphate curve and within specified limits.

CHEMICAL HIDEOUT.—The phenomenon of chemical hideout disrupts control of boiler water treatment. Chemical hideout is usually indicated by a diminishing level of phosphate as the boiler steaming rate increases. Phosphate returns when the steaming rate decreases or when the boiler is secured. The reasons for its occurrence are not well defined, but two likely mechanisms have considerable data to support them. Both are probably correct.

One mechanism shows that circumstance must allow concentration of chemicals. The other shows that reaction with the magnetite protective layer occurs.

If tube surfaces are smooth and free of deposits, boiler water circulates freely and is effective in continuously washing tube metal. Porous deposits, crevices, pits, and leakage sites interfere with circulation, creating areas conducive to concentration of boiler water. All of the normally soluble chemicals concentrate. If the interference is severe, localized dry-out of boiler water treatment chemicals and contaminants can occur. When such localized concentration of boiler water occurs, pH, phosphate, and conductivity decrease. A decrease in chloride may or may not be detected. As heat input to the boiler tubes increases, the concentrating effect at the metal increases. As heat transfer rate decreases, the chemicals return to the bulk boiler water.

Phosphates react directly with magnetite to form a solid sodium iron phosphate compound. This is a high-temperature reaction, and the compound decomposes when temperatures are reduced. When a reaction with magnetite is occurring, pH increases when the phosphate decreases; it decreases when the phosphate increases.

Virtually all instances of chemical hideout in naval boilers are reported after each of the following events has occurred:

- Overhaul or boiler repair

- Acid cleaning

- Initiation of morpholine

- Contamination

These events can cause generation of deposits in the boiler and the appearance of chemical hideout. Mechanical cleaning (via water-jet) alleviates this condition when it is caused by soft deposits. However, if contamination causes hard deposits, only acid cleaning will remove them. Hard deposits represent a hazard to boiler operation and in themselves cause hideout.

The reason for development of hideout soon after acid cleaning is not clear, but either or both of the previously described mechanisms may be at work when sensitized metal surfaces and soft deposits are present.

Blowdown

Blowdown provides control of accumulated boiler water solids, both suspended and dissolved.

SURFACE BLOWDOWN.—Surface blowdown serves the purpose of reducing the amounts of materials dissolved in the boiler water. Contamination and treatment chemicals contribute to the dissolved material, and the total amounts of them are measured by conductivity. Within limits, surface blowdown routinely controls conductivity. If contamination or chemical over-addition causes pH or phosphate to exceed upper control limits, both are reduced by surface blowdown as long as the boiler water treatment parameters are below the coordinated phosphate curve. Surface blowdown must never be performed if the pH or phosphate drops below lower limits or if the pH is above the coordinated phosphate curve (except in response to a high water casualty). Chemicals must first be added as necessary to raise pH or phosphate, then the surface blowdown may be accomplished.

Surface blowdown also removes suspended solids and oil. Oil can be removed effectively only by blowing down through the surface blow pipe. Refer to NSTM, chapter 220, volume 2, for detailed surface blowdown procedures.

BOTTOM BLOWDOWN.—The amount of sludge in the boiler is normally controlled by bottom blowdown. For a bottom blowdown to be effective, the boiler should be secured as long as possible prior to initiation of the blowdown to ensure that there is sufficient pressure on the boiler to accomplish the action. If the boiler is being secured for bottom blowdown and is then to be returned to the line, the blowdown is to be initiated when the boiler has been secured for at least 1 hour.

In a boiler with proper chemical treatment, the treatment chemicals will react with and remove contaminants by precipitation reactions (form sludge). Sludge is dealt with by conducting bottom blowdowns. Bottom blowdowns are required to be conducted every 168 steaming hours (1 week). A properly obtained boiler water sample is representable only of dissolved material, not suspended matter or sludge. Refer to NSTM, chapter 220, volume 2, for detailed bottom blowdown procedures.

Idle Boiler Maintenance

The primary consideration of idle boiler maintenance is the prevention of oxygen corrosion of the boiler metal. Wet iron exposed to air (oxygen) will corrode. Corrosion of the iron can be eliminated by removing either the air or the moisture through either one of two forms of idle boiler layup, wet and dry. These are summarized in the following paragraphs. However, for detailed procedures on idle boiler layup, refer to NSTM, chapter 221.

WET LAYUP.—Wet layup of a boiler is accomplished by excluding oxygen from the water. After securing a boiler, one of the following wet layup methods is used to keep the boiler operational:

1. A steam blanket is applied when the boiler pressure reduces to steam blanket pressure. The superheater is drained. (This is the most commonly used boiler layup.)

2. A nitrogen blanket is applied when boiler pressure reduces to nitrogen blanket pressure.

3. Boilers with stainless steel superheaters, except the last boiler to be secured, are backfilled through the superheater with hot deaerated feedwater when the boiler pressure reduces to 15 psi. When the boiler and superheater are completely filled, a pressure not to exceed 100 psi is maintained on the boiler through the backfill connection.

4. After the DFT is secured, and hot deaerated feedwater is not available, the last boiler to be laid up is backfilled with reserve feedwater. A positive pressure not to exceed 100 psi is maintained through the backfill connection.

NOTE: Methods 4 and 5 may be used for all propulsion boilers, including those with stainless steel superheaters.

5. Hot deaerated feedwater is added via the main feed connection until the boiler and superheater are filled and a hydrostatic pressure between 15 and 50 psi is maintained. This method must not be used for boilers having stainless steel superheaters and is not normally recommended for other propulsion boilers.

6. The boiler, the superheater, and the economizer are filled via the feed connection with reserve feedwater containing hydrazine and morpholine. A minimum of 15 psi hydrostatic pressure is maintained on the boiler using the treated water in the reserve feed tank. This method is authorized for use by industrial activities only.

Hydrazine is a liquid oxygen scavenger that does not add to the dissolved solids content of water. Hydrazine and morpholine are added to an empty reserve feed tank which is then filled with cold feedwater. The solution is pumped to the boiler and a minimum 15 psi pressure is maintained. This method is used for complete wet layup of a propulsion boiler which has completed overhaul and has had a successful hydrostatic test. The hydrazine reacts, under alkaline conditions, with oxygen in the feedwater to produce nitrogen and water. The alkaline condition is obtained by the addition of morpholine. If the boiler is steamed with water containing a high concentration of hydrazine, the hydrazine is converted to ammonia which can be harmful to copper alloys in the condensate system. Therefore, prior to light-off, water containing hydrazine is completely drained from the boiler and the reserve feed tank before it is refilled with fresh feedwater.

The positive pressure of the wet layup keeps the air from entering the boiler watersides. An idle boiler under wet layup is not maintained within steaming boiler chemical treatment requirements because a high heat flux is not being applied and contaminated feedwater is not continuously entering the boiler.

DRY LAYUP.—Dry layup of a boiler is accomplished either by using desiccant, in the case of a closed-up boiler, or by blowing warm air through circuits of the open watersides. Both methods prevent water condensation on the watersides and control oxygen corrosion.

In preparing for a dry layup of a propulsion boiler, sodium nitrite is injected into the boiler when the drum pressure drops to less than 100 psi. The sodium nitrite solution reacts with the boiler waterside surfaces to form a passivating film. When dumped, this passivating film remains on the watersides to retard iron oxidation as long as the surfaces are kept perfectly dry.

COMBUSTION REQUIREMENTS

Combustion is a chemical process that releases large amounts of energy, mainly in the form of thermal energy (heat) and some electromagnetic energy, in the form of light. In combustion, the fuel molecule is broken down into its component atoms, which are then oxidized (reformed into molecules with oxygen atoms.)

For combustion to occur, the fuel must be thoroughly mixed with air, which contains the needed oxygen. The fuel must be properly atomized (divided into fine particles) to help facilitate mixing. To atomize, the fuel must be of proper viscosity, which in some cases requires heating the fuel.

Atmospheric air is the source of oxygen for combustion. The composition of air is 23.15 percent oxygen by weight. To obtain 1 pound of oxygen requires 4.32 (1/23.15 percent) pounds of air. The other 3.32 pounds of air is mostly nitrogen.

Chemically, fuel is made up of carbon, hydrogen, and sulfur. When carbon is properly oxidized, it forms carbon dioxide and releases approximately 14,000 Btu of energy per pound of carbon. If there is an inadequate supply of oxygen, carbon will react, but instead forms carbon monoxide. In this case, only 4,000 Btu of energy per pound is released. This incomplete combustion results in a approximate 10,000 Btu of energy loss per pound of carbon. This loss of energy can be avoided by ensuring the fuel and air is mixed in the proper ratio.

Maintaining a proper fuel-air ratio is important to combustion and is also difficult. There are no instruments installed to accurately measure air and fuel flow. The method used to determine if the fuel-air ratio is proper is to observe the boiler exhaust. Dark, or black, smoke indicates an inadequate amount of air. A clear looking stack can be obtained over a wide range of the fuel-air ratio, up to a ratio of 200-300 percent excess air. To obtain the best fuel-air ratio, start with black smoke and adjust combustion until a clear stack is obtained.

Another concern is dealing with the by-product of improper fuel-air ratio. In a situation of excess fuel, sulfur from the fuel is released. The sulfur will undergo a chemical reaction to produce soot, which is an excellent thermal insulator, and will slow down the heat transfer through the tubes to the boiler water.

In the case of excess air, an explosion can result. In this situation, the excess air mixes with some of the fuel to produce an aerosol mixture. The danger is that this aerosol mixture is explosion.

BOILER CAPACITY

The capacity of any boiler is limited by three factors that have to do with the design and operation of the boiler. These limitations, which are known as end points, are (1) the end point for combustion, (2) the end point for moisture carryover, and (3) the end point for water circulation.

Boilers are designed so that these three end points should occur in the following order as boiler load increases: combustion, moisture carryover, then water circulation. Since the end point of combustion occurs first, it is the only end point that is reached in a properly designed and operated boiler. However, if boiler operation is improperly conducted, the other two end points might be reached. Use of sprayer plates larger than those directed by the Naval Sea Systems Command can change the end point of combustion to above the other end points. As the end point of moisture carryover or water circulation is reached, the boiler will suffer great damage, which may also endanger operating personnel.

END POINT FOR COMBUSTION

The end point for combustion for a boiler is reached when the capacity of the sprayer plates, at the designed pressure for the system, is reached or when the maximum amount of air that can be forced into the furnace is insufficient for complete combustion of fuel. If the end point for combustion is actually reached because of insufficient air, the smoke in the uptakes will be black because it will contain particles of unburned fuel. However, this condition should be rare, since the end point for combustion is artificially limited by sprayer plate capacity when the fuel is supplied at the burner manifold at designed operating pressure. As noted before, this artificial limitation upon combustion in the boiler furnace is the factor that would cause the end point for combustion to occur before either of the other two end points.

END POINT FOR MOISTURE CARRYOVER

The rate of steam generation should never be increased to the point at which an excessive amount of moisture is carried over in the steam. In general, naval specifications limit the allowable moisture content of steam leaving the saturated steam outlet to 1/4 of 1 percent.

As you know, excessive carryover can be extremely damaging to piping, valves, and turbines, as well as to the superheater of the boiler. It is not only the moisture itself that is damaging but also the insoluble matter that may be carried in the moisture. This insoluble matter can form scale on superheater tubes, turbine blades, piping, and fittings; in some cases, it may be sufficient to cause unbalance of rotating parts.

As the evaporation rate is increased, the amount of moisture carryover tends to increase also, due to the increased release of steam bubbles. Because modern naval boilers are designed for high evaporation rates, steam separators and various baffle arrangements are used in the steam drum to separate moisture from the steam.

END POINT FOR WATER CIRCULATION

In natural circulation boilers, circulation is dependent upon the difference between the density of the ascending mixture of hot water and steam and the density of the descending body of relatively cool water. As the firing rate is increased, the amount of heat transferred to the tubes is also increased. A greater number of tubes carry the upward flow of water and steam, and fewer tubes are left for the downward flow of water. Without downcomers to ensure a downward flow of water, a point would eventually be reached at which the downward flow would be insufficient to balance the upward flow of water and steam, and some tubes would become overheated and burn out. This condition would determine the end point for water circulation.

The use of downcomers ensures that the end point for water circulation will not be reached merely because the firing rate is increased. Other factors that influence the circulation in a natural circulation boiler are the location of the burners, the arrangement of baffles in the tube banks, and the arrangement of tubes in the tube banks.

Full-power and overload ratings for the boilers in each ship are specified in the manufacturer's technical manual. The total quantity of steam required to develop contract shaft horsepower of the ship, divided by the number of boilers installed, gives boiler full-power capacity. Boiler overload capacity is usually 120 percent of boiler full-power capacity. For some boilers, a specific assigned maximum firing rate is designated.

A boiler should not be forced beyond full-power capacity—that is, it should not be steamed at a rate greater than that required to obtain full-power speed with all the ship's boilers in use. A boiler should NEVER be steamed beyond its overload capacity, or fired beyond the assigned maximum firing rate, except in dire emergency.

BOILER CASUALTY CONTROL

There are many fireroom casualties which require a knowledge of preventive and corrective measures. Some are major, some are minor; but all can be serious. In the event of a casualty, the principal doctrine to be impressed upon operating personnel is the prevention of additional or major casualties. Under normal operating conditions, the safety of personnel and machinery should be given first consideration. Therefore, it is necessary to know instantly and accurately what to do for each casualty. Stopping to find out exactly what must be done for each casualty could mean loss of life, extensive damage to the machinery, and even complete failure of the engineering plant. A fundamental principle of engineering casualty control is split-plant operation. The purpose of the split-plant design is to minimize the damage that might result from any one casualty that affects propulsion power, steering, and electrical power generation.

Although speed in controlling a casualty is essential, action should never be taken without accurate information; otherwise the casualty may be mishandled, and further damage to the machinery may result. Cross-connecting an intact engineering plant with a partly damaged one must be delayed until it is certain that such action will not jeopardize the intact one.

Cross-connecting valves are provided for the main and auxiliary steam systems and other engineering systems so that any boiler or group of boilers, either forward or aft, may supply steam to each engine room. These systems were discussed in chapter 5 of this manual showing the construction of the split-plant design on some types of ships.

The discussion of fireroom casualties in this chapter is intended to give you an overall view of how casualties should be handled. For detailed information on casualty control on your ship, consult the engineering casualty control section of the EOSS. Also, NSTM, chapter 079, volume 3, contains casualty control in depth, and generalized as it applies to all ships.

Most of the casualties discussed in this chapter are usually treated in a step-by-step procedure, but it is beyond the scope of this chapter to give each step in handling each casualty. In the step-by-step procedure, one step is performed, then another, then another, and so forth. In handling actual casualties, however, this step-by-step approach will probably have to be modified. Different circumstances may require a different sequence of steps for control of a casualty. Also, in handling real casualties, several steps will have to be performed at the same time. For example, main control must be notified of any casualty to the boilers or to associated equipment. If "Notify main control" is listed as the third step in controlling a particular casualty, does this mean that the main control is not notified until the first two steps have been completed? Not at all. Notifying main control is a step that can usually be taken at the same time other steps are being taken. It is probably helpful to learn the steps for controlling casualties in the order in which they are given; but do not overlook the fact that the steps may have to be performed simultaneously.

FEEDWATER CASUALTIES

Casualties in the control of water level include low water, high water, feed pump casualties, loss of feed suction, and low feed pressure. These casualties are some of the most serious ones.

Low water is one of the most serious of all fireroom casualties. Low water may be caused by failure of the feed pumps, ruptures in the feed discharge line, defective check valves, low water in the feed tank, or other defects.

However, the most frequent cause of low water is inattention or the diversion of attention. The checkman's sole responsibility is to keep the water in the boiler at a proper level.

Low water is extremely damaging to the boiler and may endanger the lives of fireroom personnel. When the furnace is hot and there is insufficient water to absorb the heat, the heating surfaces are likely to be distorted, the brickwork damaged, and the boiler casing warped by the excessive heat. In addition, serious steam and water leaks may occur as a result of low water.

Disappearance of the water level from the water gauge glasses must be treated as a casualty requiring the immediate securing of the boiler.

Notice that when the water level falls low enough to uncover portions of the tubes, the heat transfer surface is reduced. As a rule, therefore, the steam pressure will drop. Ordinarily a drop in steam pressure is the result of an increased demand for steam, and the natural tendency is to cut-in more burners to fulfill the demand. If the drop in steam pressure is caused by low water, however, increasing the firing rate will result in serious damage to the boiler and possibly an injury to fireroom personnel.

NOTE: The possibility that a drop in steam pressure indicates low water must always be kept in mind. Always check the level in the water gauge glasses before cutting in additional burners, when steam pressure has dropped for no apparent reason.

High water is another serious casualty that is most frequently caused by the inattention of the checkman and the BTOW. If the water level in the gauge glasses goes above the highest visible part, the boiler must be secured immediately By careful observation, it is sometimes possible to distinguish between an empty gauge glass and a full one by the presence or absence of condensate trickling down the inside of the glass. The presence of condensate indicates, of course, an empty glass—that is, a low water casualty. However, the boiler must be secured whether the water is high or low. After the boiler has been secured, the location of the water level can be determined by using the gauge glass cutout valves and drain valves.

The failure of a feed system pump can have drastic consequences. Unless the pump casualty is corrected immediately, the pump failure will lead to low water in the boiler. In addition to the obvious dangers associated with low water, there are some that are equally serious but not so obvious. For example, low water causes complete or partial loss of steam pressure. When steam pressure is lost or greatly reduced, you will lose the services of vital auxiliary machinery—pumps, blowers, and so forth. It is essential, therefore, that feed pump casualties be handled rapidly and correctly.

If the main feed pump discharge pressure is too low, the first three things to be checked are (1) the feed booster pump discharge pressure, (2) the level and pressure in the DFT, and (3) the feed stop and check valves on idle boilers. A failure of the feed booster pump will, of course, cause loss of suction and, therefore, loss of discharge pressure of the main feed pump. If the feed stop check valves on idle boilers have accidentally been left open, the main feed pump discharge pressure may be low merely because water has been pumped to an idle boiler, as well as to the steaming boiler.

Some of the most likely causes of failure of the main feed pump are (1) malfunction of the constant-pressure pump governor, (2) an air-bound or vapor-bound condition of the main feed pump, (3) faulty pump clearances, and (4) malfunction or improper setting of the speed-limiting governor.

In most installations the feed booster pump and the main feed booster pump are in the fireroom. Some of the newer ships have an alarm system to indicate inadequate feed booster pump pressure and a pressure switch to start a standby feed booster pump. In case of problems with the DFT, piping is installed to take a cold suction from the reserve feed tank. Due to the danger involved with thermal shock to the boiler, cold suction should only be done as a last resort.

Some older ships are arranged with the DFT and the main feed booster pump in the engine room and the main feed pump in the fireroom. Casualties to the feed system require close communication between the engine room and the fireroom. In these ships an emergency feed pump is installed in the fireroom. This is a steam-driven reciprocating pump that takes a cold suction from a reserve feed tank.

Another possibility may be to cross-connect and receive feedwater from another fireroom. When doing this, you must be careful that you do not take feedwater from one fireroom that ends up in the other fireroom's DFT where the feedwater casualty exists.

FUEL SYSTEM CASUALTIES

Casualties to any part of the fuel-oil system are serious and must be remedied at once. Common casualties include (1) water in the fuel oil, (2) loss of fuel-oil suction, (3) failure of the fuel-oil service pump, and (4) fuel-oil leaks.

The presence of an appreciable amount of water in the fuel oil is indicated by hissing and sputtering of the fires and atomizers and by racing of the fuel-oil service pump. The situation must be remedied at once; otherwise, choked atomizers, loss of fires, flarebacks, and refractory damage may result.

A loss of fuel-oil suction usually indicates that the oil in the service suction tank has dropped below the level of the fuel-oil service pump suction line. This causes a mixture of air and oil to be pumped to the atomizers. The atomizers begin to hiss and the fuel-oil service pump begins to race. It must be strongly emphasized that the loss of fuel-oil suction can cause serious results. Related casualties may include loss of auxiliary steam and electric power, with the complete loss of all electrically driven and steam-driven machinery.

Failure of the fuel-oil service pump can cause the same progressive series of casualties as those that result from loss of fuel-oil suction.

Fuel-oil leaks are very serious, no matter how small they may be. Fuel-oil vapors are very explosive. Any oil spillage or leakage must be wiped up immediately.

FLAREBACKS

A flareback is likely to occur whenever the pressure in the furnace momentarily exceeds the pressure in the boiler air casing. Flarebacks are caused by an inadequate air supply for the amount of oil being supplied, or by a delay in lighting the mixture of air and oil.

Situations that commonly lead to flarebacks include (1) attempting to light-off or to relight burners from hot brickwork; (2) gunfire or bombing that creates a partial vacuum at the blower intake, thus reducing the air pressure supplied by the blowers; (3) forced draft blower failure; (4) accumulation of unburned fuel oil or combustible gases in furnaces, tubes banks, uptakes, or air casings; and (5) any event that first extinguishes the burners and then allows unburned fuel oil to spray out into the hot furnace. An example of this last situation might be a temporary interruption of the fuel supply, which would cause the burners to go out; when the fuel oil supply returns to normal, the heat of the furnace might not be sufficient to relight the burners immediately. In a few seconds, however, the fuel oil sprayed into the furnace would be vaporized, and a flareback or even an explosion might result.

SUPERHEATER CASUALTIES

Most all modern boilers are uncontrolled superheat. All steam generated from the steam drum then passes through the superheater. Auxiliary steam is then generated by a desuperheater. If there is an inadequate flow of steam through the superheater, combustion gases may damage the superheater tubes by overheating or melting them. One method to ensure a flow of steam through the superheater of the steaming boiler is to lift the superheater safety valve by hand. Another method used during light-off is to open the superheater protection valve, which discharges into this auxiliary exhaust steam.

CASUALTIES TO REFRACTORIES

If brick or plastic falls out of the furnace walls and goes unnoticed, burned casings may result. If brick or plastic falls out of a furnace wall, if

practicable, secure all the burners. If it is not practicable to secure all the burners, secure those burners that are adjacent to the damaged section.

NOTE: It may be necessary to continue operating the boiler until another boiler can be brought in on the line.

CASUALTIES TO BOILER PRESSURE PARTS

When boiler pressure parts, such as tubes, carry away or rupture, escaping steam may cause serious injury to personnel and damage to the boiler. Escaping steam at a high temperature and pressure is invisible and very dangerous. This steam has enough energy to cut and damage steel. The boiler must be secured immediately, relieved of its pressure, and cooled until no more steam is generated. To relieve the boiler of its pressure, lift the safety valves using hand-casing gear. If a boiler pressure part carries away or ruptures, take steps immediately upon discovery of the casualty to minimize and localize the damage as much as circumstances will allow.

Gauge glasses are connected to the water and steam spaces of the steam drum. If a water gauge glass carries away, the mixture of steam and water escaping from the gauge connections may seriously burn personnel in the area. A ball check valve in the high-pressure gauge line functions when the flow is excessive. The hazard of flying glass particles makes this casualty very serious. The particles of glass could lodge in your eyes and blind you, or they could lodge elsewhere in your body and cause serious injury. If a gauge glass casualty occurs, throw a large sheet of asbestos cloth, rubber matting, or similar material over the glass. Then take immediate action to secure the gauge glass.

PRECAUTIONS TO PREVENT FIRES

The following precautions must be taken to prevent fires:

1. Do not allow oil to accumulate in any place. Particular care must be taken to guard against oil accumulation in drip pans under pumps, in bilges, in the furnaces, on the floor plates, and in the bottom of air-casings. Should leakage from the oil system to the fireroom occur at any time, immediate action should be taken to shut off the oil supply, by quick-closing valves, and to stop the oil pump.

2. Absolutely tight joints in all oil lines are essential to safety. Immediate steps must be taken to stop leaks whenever they are discovered. Flange safety shields should be installed on all flanges in fuel-oil service lines to prevent spraying oil on adjacent hot surfaces.

3. No lights should be permitted in the fireroom except electric lights (fitted with steam-tight globes, or lenses, and wire guards), and permanently fitted smoke indicator and water gauge lights. If work is being done in the vicinity of flammable vapors, or if rust-preventive compound is being used, all portable lights should be explosionproof.

BOILER MAINTENANCE

The engineer officer must know the general condition of each boiler and the manner in which each is being operated and maintained. The engineer officer must make periodic inspections to ensure that the exterior and interior surfaces of the boiler are clean; that the refractory linings adequately protect the casing, drums, and headers; that the integrity of the pressure parts are being maintained; and that the operating condition of the burners, safety valves, operating instruments, and other boiler appurtenances are satisfactory.

REFERENCES

Boiler Technician 3 & 2, NAVEDTRA 10535-H, Naval Education and Training Program Management Support Activity, Pensacola, Fla., 1983.

Naval Ships' Technical Manual, NAVSEA 9 59086-GX-STM-020, Chapter 220, Volume 2, ''Boiler Water/Feedwater - Test and Treatment,'' Naval Sea Systems Command, Washington, D.C., December 1987.

CHAPTER 11

BOILER FITTINGS AND CONTROLS

LEARNING OBJECTIVES

Upon completion of this chapter, you should be able to do the following:

1. Identify the internal fittings of boilers and their functions.

2. Identify the external fittings of boilers and their functions.

3. Identify the various automatic boiler control systems, their components, and their operating principles.

INTRODUCTION

The fittings and instruments used on naval boilers are of sufficient number and importance that we will discuss them separately. The term *boiler fittings* describes a number of attachments that are installed in, or closely connected to, the boiler and that are required for the operation of the boiler. Boiler fittings are usually divided into two general classes: (1) internal fittings (or internals), which are those installed inside the steam and water spaces of the boiler; and (2) external fittings, which are those installed outside the steam and water spaces. Boiler instruments and monitoring devices, such as pressure gauges and temperature gauges, are installed to provide personnel information on operating conditions. Most instruments are not external boiler fittings, but the connections (cut-out valves) for these instruments are external boiler fittings.

Automatic boiler control (ABC) systems are installed on all modern boilers to control the fuel and airflow to maintain steam drum pressure and needed steam demand and regulate feedwater inputs to maintain a proper water level in the steam drum. This is done by two separate control systems: an automatic combustion control (ACC) system and a feedwater control (FWC) system.

INTERNAL FITTINGS

The internal fittings installed in the steam drum include equipment for distributing the incoming feedwater, for giving surface blows, for separating steam and water, and for directing the flow of steam and water within the steam drum. Desuperheaters are heat exchangers that lower the temperature of superheated steam so it can be used by auxiliary machinery. They may be located in either the water drum or the steam drum of a boiler.

All boilers require chemical injections for boiler water treatment. Some boilers have a separate chemical feed pipe, others inject the chemicals into the feedwater system to enter the boiler in the feedwater pipe.

The specific design and arrangement of boiler internal fittings will vary in minor detail in each boiler. Different boiler manufacturers have steam separators of different design in their steam drum. Despite the difference in design, they all serve the same function of separating all moisture from the steam generated. This chapter will show the arrangement of boiler internal parts for the three major manufacturers of naval propulsion boilers. These manufacturers are (1) Babcock and Wilcox (B&W), (2) Combustion Engineering (CE), and (3) Foster Wheeler (FW). The next sections will describe the internal parts of each of these manufacturers.

STEAM DRUM INTERNAL FITTINGS—BABCOCK AND WILCOX CO.

Figure 11-1 illustrates the steam drum internals in a B&W boiler. This is typical of the steam drum found onboard the DDG-44 and -45. Follow the figure as we discuss the steam drum internals.

The internal feedwater pipe extends along the entire length of the bottom of the steam drum. Feedwater enters the pipe from the economizer and is evenly distributed throughout the length of the drum.

The chemical feed pipe extends straight through approximately 90 percent of the length of the drum well below the normal water level and connects to the chemical feed nozzle on the end of the drum. A row of holes is drilled along the top center line. The prescribed chemicals,

consistent with the requirements of the periodic feedwater analysis, are injected through the chemical feed pipe into the boiler.

The dry pipe is suspended near the top of the steam drum, along the center line of the drum. Both ends of the dry pipe are closed. Steam enters through perforations in the upper surface of the dry pipe. Thus, the steam must change direction in order to enter the dry pipe. Since moisture is lost whenever steam changes direction, the dry pipe acts as a device to separate steam and moisture. Steam leaves the dry pipe through the main steam outlet and from there goes to the superheater. A few perforations in the bottom of the dry pipe allow water droplets to drain back down to the water in the steam drum.

The surface blow pipe, located just below the normal water level of the drum, also extends the full length of the drum and has a single row of

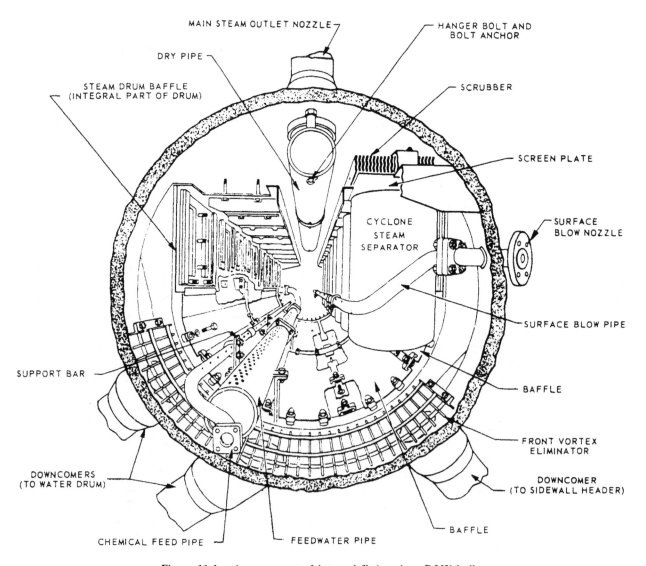

Figure 11-1.—Arrangement of internal fittings in a B&W boiler.

11-2

holes drilled into its top. When the surface blow valve is open, this pipe carries off scum and foam from the surface of the water in the drum.

A continuous baffle is fitted approximately 3 inches from the drum's internal surface. The baffle extends in length from just forward of the generating tubes to just aft of them, but does not extend as far as the downcomers in each end. The forward and after ends of the space left between the baffle and the drum are closed. The baffle is attached at both sides to a flat bar, which is hung from the drum as shown, and extends the full length of the baffle. Ports are cut into each of these flat bars. A cyclone separator is over each of these ports. This arrangement makes the space between the steam drum and the baffle an entirely enclosed space, with the exception of the ports leading to the cyclone separators.

The flow of steam and water through the drum is as follows: water enters the drum through the internal feedwater pipe and flows through the downcomers to the water drum and to the side and rear wall headers. The generating tubes discharge a mixture of steam and water upward into the space behind the baffle in the steam drum. The circulation from the side wall headers is upward through the side wall to the spaces in the drum behind the baffle. The circulation from the rear wall header is up through the tubes to the upper rear wall header where tubes distribute the mixture of steam and water to the space behind the baffle in the steam drum. Since this space is entirely enclosed, the only passage available for the steam and water is through the ports opening to the cyclone separators. The separators on each side of the drum have sufficient capacity to pass all the steam and water that may flow through them. As the mixture passes through the cyclone separators, the steam is separated from the water and passes through the top of the separator. The water discharges from the bottom of the separator. As the steam leaves the cyclone separator, it passes through a scrubber where it is further dried and then, finally, is led to the top of the drum where it enters the dry pipe.

Another interesting feature is illustrated in figure 11-1. It is the vortex eliminator, which reduces the swirling motion of the water as it enters the downcomers. A vortex eliminator consists of a series of gridlike plates arranged in a semicircle to conform to the shape of the lower half of the steam drum. One vortex eliminator is located at the front of the steam drum and another is located at the rear of the drum. In each case, the vortex eliminator is fitted over the ends of the downcomers.

Cyclone Separators (B&W)

The purpose of the cyclone separators is to deliver dry saturated steam to the dry pipe with minimum agitation of the water, thereby reducing the possibility of priming. Priming takes place when boiler water is carried over into the superheater of a boiler.

The details of a cyclone separator with its strainer and scrubber are shown in figure 11-2.

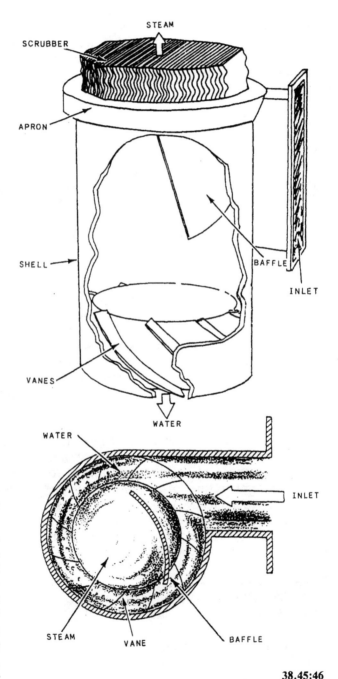

38.45:46

Figure 11-2.—Cyclone steam separator.

11-3

The mixture of steam and water enters the separator through its inlet connection, which is tangent to the separator body. Because of its angle of entry, the mixture of steam and water acquires a rotary motion. As the mixture swirls around, centrifugal force separates or throws outward the heavier water. An internal baffle helps to deflect the steam to the center and the water to the outside. Being lighter, the steam rises through the center through the scrubber element, while the heavy water drops down toward the base of the separator. The scrubber consists of closely spaced corrugated steel plates. As the steam passes through the scrubber, its direction is changed frequently; and with each change of direction, some moisture is lost. The steam passes out of the scrubber element into the top part of the steam drum and then enters the dry pipe.

Curved stationary vanes near the bottom of the separator maintain the rotary motion of the water until it is finally discharged from the bottom of the separator. In the center of the separator, there is a flat plate to which the curved vanes are attached. The plate keeps the steam in the center of the separator and prevents its being carried out of the bottom by the water. The vanes, being located around the periphery of the plate, allow the water to pass from the separator around the outer edges of the body only.

The two end cyclone separators, one on each side, are fitted at their bases with a flat baffle. The baffle directs the flow of water from the separator to the center of the steam drum, where it is mixed thoroughly with the rest of the water in the steam drum and is kept from flowing directly from the separator into the downcomers.

Desuperheater

As mentioned earlier, the desuperheater may be located in either the steam drum or in the water drum. The purpose of the desuperheater is to reduce the temperature of a portion of the superheated steam intended to be used in auxiliary steam systems. For clarity, the desuperheater is not shown in any of the figures of steam drum internal fittings in this chapter.

STEAM DRUM INTERNAL FITTINGS—COMBUSTION ENGINEERING, INC.

Figure 11-3 illustrates the steam drum internals in a CE boiler. By comparing figure 11-1 with figure 11-3, you will find that the principle on

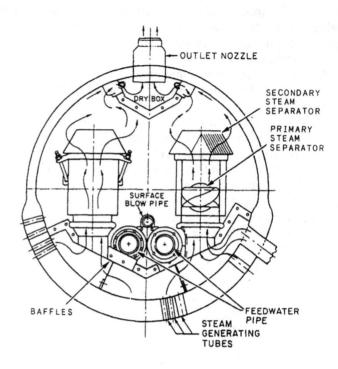

Figure 11-3.—Arrangement of internal fittings in a CE boiler.

which CE internals are built remains the same as that of the B&W, although the locations and construction may differ somewhat. Figure 11-3 is typical of the steam drum installed onboard KNOX class frigates (FF-1078).

To separate the steam and water mixture, a CE boiler uses two primary separators (fig. 11-3). Each primary separator consists of a cylindrical shell with four inside spinner blades secured to dished heads.

The mixture of steam and water is directed into the primary separator through a lower support (not shown). As the steam passes up through the separator, the fixed spinner blades force the steam into a swirling motion. The heavier water is thrown to the sides, where it collects and drops back into the lower half of the drum. The lighter steam rises into the secondary separator.

Secondary separators are made of several corrugated plates in the shape of an inverted basket. They are located on top of, and are bolted to, the primary separators. The corrugated plates further separate the steam and water mixture by creating rapid changes in steam flow direction. The steam can adapt to these rapid changes but the heavier water cannot. The water collects and falls back into the lower portion of the steam drum.

The CE dry box, which serves the same purpose as the B&W dry pipe, is located at the top of the steam drum and is designed to separate any steam and water mixture not separated by the separators. The dry box works on the same principle as the separator (rapid change in direction). At either end of the dry box, there is a drain hole, where collected water may drop back into the lower half of the steam drum.

Since there are other boiler manufacturers, and there are other differences between designs and arrangements which we have not covered, it is important that you be aware of these differences and consult the manufacturers' technical manuals when working on the boilers on your ship.

STEAM DRUM INTERNAL FITTINGS—FOSTER WHEELER

Figure 11-4 illustrates the steam drum internal in an FW boiler. This type of boiler uses horizontal steam separators and chevron dryers to separate steam and water.

A horizontal steam separator (fig. 11-5) is installed in the same manner as cyclone dryers, one row along each side of the length of the steam drum. Each horizontal steam separator has a machined flange bolted to a matching flange on the girth baffle. The mixture of steam and water enter the separator through a tapered inlet section

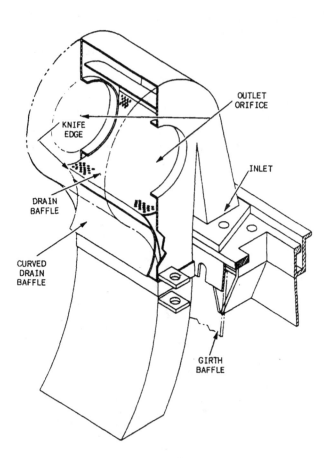

Figure 11-5.—Horizontal steam separator.

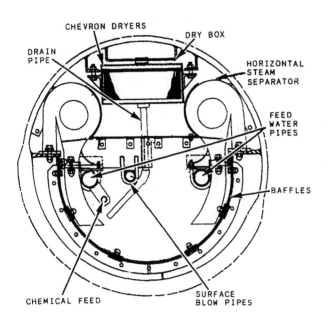

Figure 11-4.—Arrangement of internal fittings in an FW boiler.

on the side near the shell of the steam drum. The steam curves along the shell of the separator. Steam then leaves the separator through outlet orifices at each side. The water that entered with the steam is heavier and continues to follow the separator curve and is discharged through drain holes in the drain baffle. There is a knife edge on the drain baffle to minimize turbulence.

Steam discharged from the horizontal steam separator proceeds to the chevron dryers, located along the top of the steam drum. There are several chevron dryers along the length of the steam drum. Inside the chevron dryers, plates are arranged to resemble chevrons. Water is separated from the steam as it changes directions to follow the chevrons. Steam exits the chevron dryers at the top directly into the dry box.

STEAM DRUM MAINTENANCE

When a boiler is opened to clean watersides, the steam drum internals must be removed. The fittings are bolted into place, and they are removed by unbolting them. When removing the internals, tag and label them so that they will be reinstalled correctly. Access to waterside spaces of a boiler must strictly be controlled to prevent damage or inadvertently leaving foreign material inside when securing. Personnel should remove rings, belts, insignia, or anything that may possibly fall off and into a boiler tube before entering the steam drum. Tools taken inside should be tied off to prevent them from falling into a boiler tube.

When removed, boiler internal fittings should be thoroughly wire brushed and cleaned. Check the dry pipe, internal feed pipe, and surface blow line to ensure all holes are free of obstructions. Inspect the feed pipe and other surfaces for any oil accumulation or residue.

Before reinstalling the fittings, you should wirebrush, clean, and hose down the steam drum. You should also ensure the steam drum is clean and free of any oil or other accumulation.

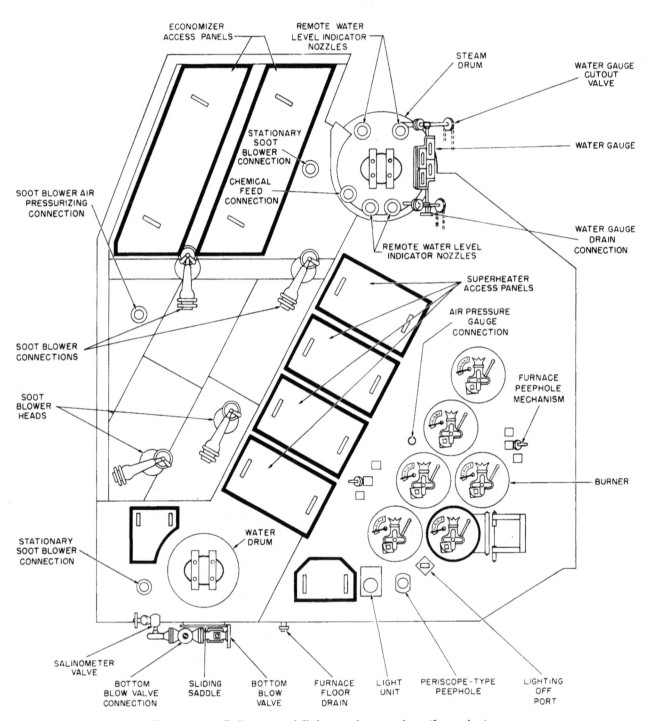

Figure 11-6.—Boiler external fittings and connections (front view).

When replacing the fittings in the steam drum, be sure all bolts are drawn tight. The desuperheater flanges must be thoroughly cleaned before the desuperheater is fitted and bolted into place. New gaskets for these flanges must be installed. The flanges must be evenly and tightly bolted to prevent leakage from these joints.

EXTERNAL FITTINGS AND CONNECTIONS

External fittings and connections commonly used on naval boilers include drains and vents, sampling connections, feed stop and check valves, steam stop valves, safety valves, soot blowers, and blow valves.

External fittings and connections serve purposes that are related to boiler operation. Some of the fittings and connections allow you to control the flow of feedwater and steam. Others serve as safety devices. Still other fittings allow you to perform certain operational procedures that are required for the efficient functioning of the boiler, such as removing soot from the firesides or giving surface blows. The instruments attached to or installed near the boiler give you essential information concerning the conditions existing inside the boiler. To understand the purposes of the external fittings and connections, you must know how each item is related to boiler operation.

Figures 11-6, 11-7, 11-8, and 11-9 show the locations of many external fittings and

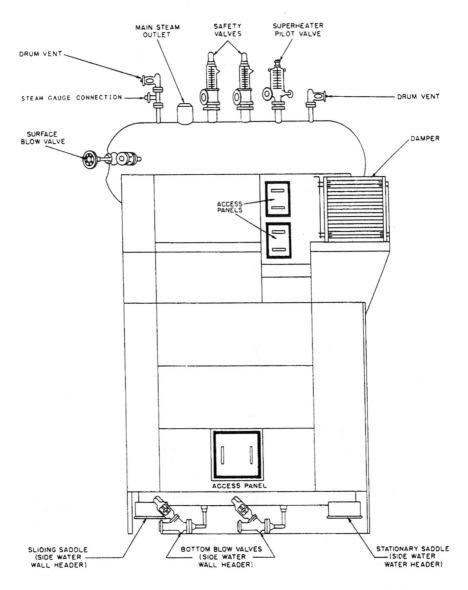

Figure 11-7.—Boiler external fittings and connections (furnance side view).

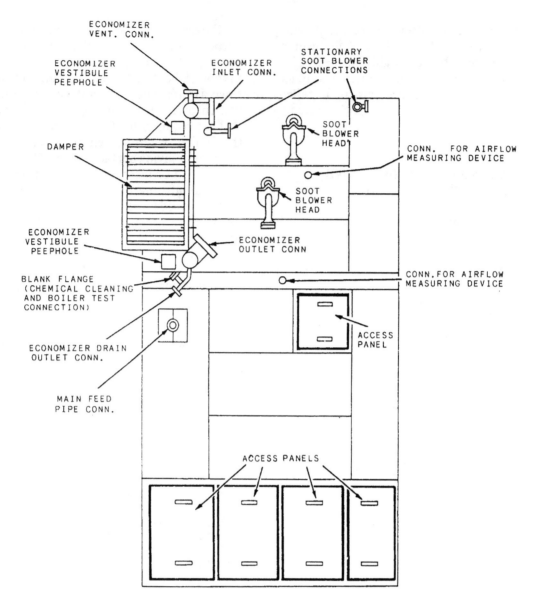

Figure 11-8.—Boiler external fittings and connections (economizer side view).

connections on a single-furnace boiler. As you study the following information on external fittings and connections, you may find it helpful to refer to these figures to see where the various units are installed on, or connected to, the boiler. Keep in mind, however, that the illustrations shown here are for one particular boiler, and differences in boiler design result in differences as to type and location of external fittings and connections. Drawings showing the location of external fittings and connections are usually included in the manufacturer's technical manuals for the boilers aboard each ship.

DRAINS AND VENTS

All the steam and water sections of the boiler, including the main part of the boiler, the economizer, and the superheater, must be provided with drains and vents.

The main part of the boiler may be drained through the bottom blow valve and through water wall header drain valves. Normally, a boiler is vented through the aircock, which is a high-pressure globe valve installed at the highest point of the steam drum. The aircock allows air to escape when the boiler is being filled and when steam is first forming; it also allows air to enter the steam drum when the boiler is being emptied.

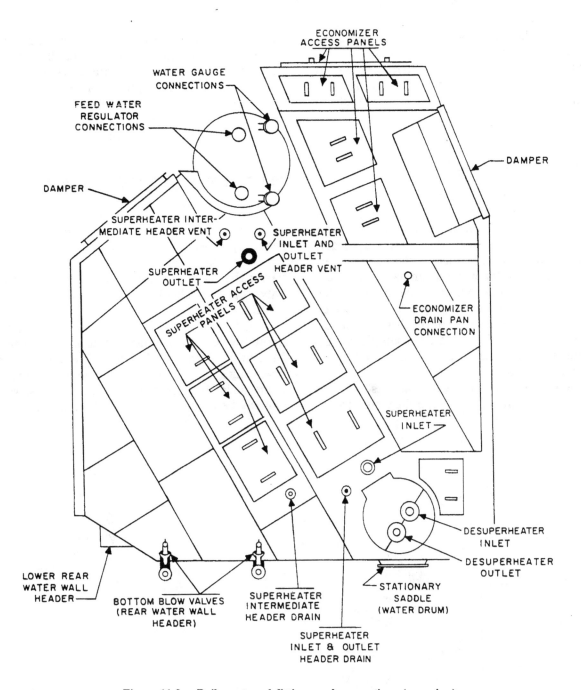

Figure 11-9.—Boiler external fittings and connections (rear view).

The economizer is vented through a vent valve on the economizer inlet piping. It is drained through a drain line from the economizer outlet header. Another drain line, coming from the drain pan installed below the headers, serves as an indicator of possible economizer leakage.

Superheater vents are installed at or near the top of each superheater header or header section; superheater drains are installed at or near the bottom of each header or header section. Thus, each pass of the superheater is vented and drained.

Superheater drains discharge through gravity (open-funnel) drains to the freshwater drain collecting system while steam is being raised in the boiler. After a specified pressure has been reached, the superheater drains are shifted to discharge through steam traps to the high-pressure drain system. The steam traps allow continuous

11-9

drainage of the superheater without excessive loss of steam or pressure.

Figure 11-10 shows the arrangement of superheater vents and drains on a single-furnace boiler, as well as the superheater protection steam connections.

SAMPLING CONNECTIONS

To determine the condition of boiler water, perform tests on a representative sample of the boiler water. One can easily see that a sample can be obtained from the steam drum, the gauge glass, the water drum, or a header. A sample of boiler water taken from the water drum has the greatest probability of being representative of the water in the boiler. However, a sample from the water drum is only representative of the dissolved solids (material) in the boiler water. No sample can be considered representative of the suspended solids (sludge) in the boiler.

Connections for drawing test samples of boiler water are normally connected to the water drum.

A sample cooler is fitted to the outlet side of the sampling connection. The cooler brings the temperature of the sample water down below the boiling point at atmospheric pressure and thus keeps the water from flashing into steam as it is drawn from the higher pressure of the boiler to the lower pressure of the fireroom.

FEED STOP AND CHECK VALVES

Manually operated feed stop and check valves are installed in the feed line to each boiler. The feed stop valve is a regular globe-type stop valve. The so-called check valve is actually a stop-check valve; that is, it functions either as a feed stop valve or as a check valve, depending on the position of the valve stem.

Feed stop and check valves are operated manually. Each valve has a separate handwheel. In addition, the feed stop valve has remote operating gear so that it can be operated from the firing aisle. In normal operation, the feed stop valve is kept fully open, and the check valve is

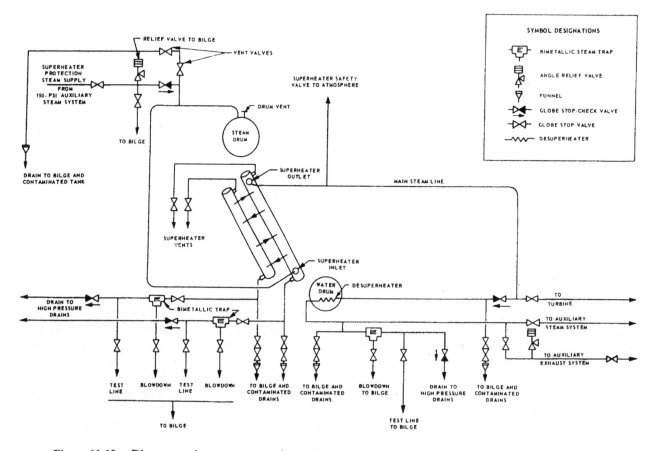

Figure 11-10.—Diagrammatic arrangement of superheater vents, drains, and protection steam connections.

used to regulate the supply of water to the boiler. When automatic feedwater controls are in use, the feed stop and check valves must be fully open so that they will not interfere with the automatic feeding of the boiler. Similarly, the automatic feedwater control valve must be fully open when the boiler is being fed manually through the feed stop valve. The automatic feedwater control valve is designed to fail in the fully open position in case of complete loss of the feedwater control systems.

The feed stop and check valves shown in figure 11-11 are combined in one manifold casting. Note, however, that there are two separate valves—the stop valve and the stop-check or check valve. Note also that each valve has a separate handwheel. In some installations, the two valves are housed in separate flanged castings that are bolted together. No matter which type of installation is used, the feed stop valve is always installed between the feed check valve and the economizer inlet. If the feed check valve fails to function, the feed stop valve can always be closed to prevent backflow from a steaming boiler or from a boiler that has been laid up full of water.

A check valve is installed between the economizer outlet and the steam drum connection to the internal feedwater pipe. This check valve allows water to flow from the economizer to the steam drum but prevents backflow of water and steam from the steam drum to the economizer. Since backflow from the boiler through the economizer would disrupt boiler circulation and cause extremely serious damage to the generating tubes and other pressure parts, the check valve is an important protective device.

STEAM STOP VALVES

Main steam stop valves are used to connect and disconnect boilers to the main steam line. The main steam stop valve, located just after the superheater outlet, is usually called the main steam boiler stop. Figure 11-12 shows a cross-sectional view of a gate-type main steam stop (motor-operated) valve.

The main steam boiler stop is either fully open or fully closed; that is, it is not used as a throttling valve. The valve can be operated manually either at the valve or at a remote operating station or by remote operating gear, which is also provided. Many steam stops are fitted with pneumatic (air) motors that are actuated through toggle operating gear to close the valve from a remote position.

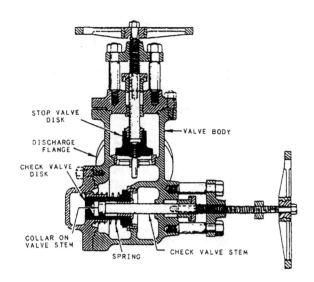

STOP VALVE DISK
VALVE BODY
DISCHARGE FLANGE
CHECK VALVE DISK
COLLAR ON VALVE STEM
SPRING
CHECK VALVE STEM

38.51
Figure 11-11.—Combined feed stop and check valves.

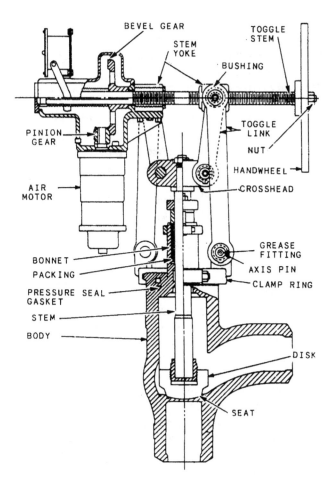

BEVEL GEAR
STEM YOKE
TOGGLE STEM
BUSHING
PINION GEAR
TOGGLE LINK
NUT
HANDWHEEL
CROSSHEAD
AIR MOTOR
GREASE FITTING
BONNET
AXIS PIN
PACKING
CLAMP RING
PRESSURE SEAL GASKET
STEM
BODY
DISK
SEAT

Figure 11-12.—Cross-sectional view of a main steam stop valve.

The toggle operating gear provides the mechanical advantage required to close the valve against boiler pressure. These valves are opened manually.

Main steam boiler stops, like other main steam stop valves, are specifically designed for high-pressure and high-temperature service. The seating surfaces of the disks and seats are usually made of Stellite, a hard, erosion-resistant alloy.

Navy specifications require two-valve protection for each boiler. A second steam stop valve is therefore provided in the main steam line just beyond the main steam boiler stop. This second valve is called the main steam stop guarding valve.

Auxiliary steam stop valves are smaller than main steam stop valves but are otherwise similar.

SAFETY VALVES

Each boiler is fitted with safety valves to allow steam to escape from the boiler (escape steam piping directs steam out through the stack, or mack, to the atmosphere) when the pressure rises above specified limits. The capacity of the safety valves installed on a boiler must be great enough to reduce the steam drum pressure to a specified safe point under all possible operating conditions. The worst possible case, in regards to safety valve capacity, is with the boiler steam stops closed. Safety valves are installed in the steam drum and at the superheater outlet.

Several different kinds of safety valves are used on naval boilers, but all are designed to open completely (pop) when a specified pressure is reached and to remain open until a specified pressure drop (blowdown) has occurred. Safety valves must close tightly, without chattering, and must remain tightly closed after seating.

You must understand the difference between boiler safety valves and ordinary relief valves. The amount of pressure required to lift a relief valve increases as the valve lifts, since the resistance of the spring increases in proportion to the amount of compression. Therefore, a relief valve opens slightly at a specified pressure, discharges a small amount of fluid, and closes at a pressure that is very close to the pressure that causes it to open.

Such an arrangement is not acceptable for boiler safety valves. If the valves were set to lift at anything close to boiler pressure, the valves would be constantly opening and closing, pounding the seats and disks and causing early failure of the valves. Furthermore, relief valves could not discharge the large amount of steam that must be discharged to bring the boiler pressure down to a safe point, since the relief valves would reseat very soon after they opened. To overcome this difficulty, boiler safety valves

are designed to open COMPLETELY at the specified pressure.

Steam Drum Safety Valves

The two types of steam drum safety valves we will discuss are (1) the huddling-chamber type and (2) the nozzle-reaction type.

HUDDLING-CHAMBER SAFETY VALVE.— A steam drum safety valve of the huddling-chamber type is shown in figure 11-13. The initial lift or opening of the valve is caused by the static pressure of the steam in the drum acting upon the bottom of the feather (disk). As soon as the valve begins to open, a projecting lip or ring of larger area is exposed for the steam pressure to act upon. The huddling chamber, which is formed by the position of the adjusting ring, fills with steam as the valve opens. The steam in the huddling chamber builds up a static pressure that acts upon the extra area provided by the projecting lip of the feather. The resulting increase in force overcomes the resistance of the spring, and the

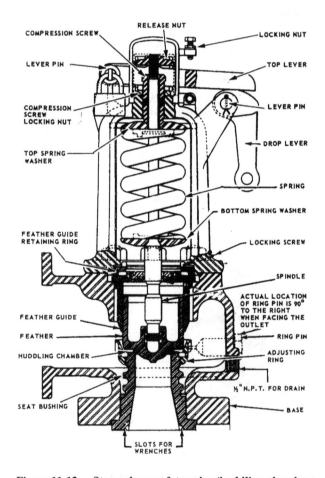

Figure 11-13.—Steam drum safety valve (huddling-chamber type).

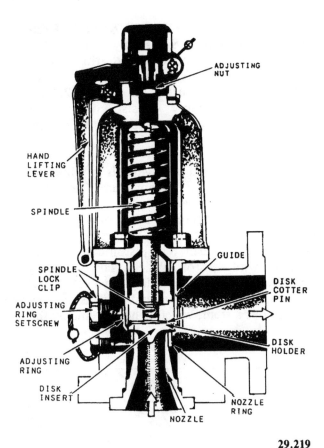

29.219

Figure 11-14.—Steam drum safety valve (nozzle-reaction type).

determines the amount of blowdown that must occur before the valve will reseat.

NOZZLE-REACTION SAFETY VALVE.— A steam drum safety valve of the nozzle-reaction type is shown in figure 11-14. The initial lift of this valve occurs when the static pressure of the steam in the drum acts upon the disk insert with sufficient force to overcome the tension of the spring. As the disk insert lifts, the escaping steam strikes the nozzle ring and changes direction. The resulting force of reaction causes the disk to lift higher, up to approximately 60 percent of rated capacity. Full capacity is reached as the result of a secondary, progressively increasing lift that occurs as an upper adjusting ring is exposed. The ring deflects the steam downward, and the resulting force of reaction causes the disk to lift still higher. Blowdown adjustment in this type of valve is made by raising or lowering the adjusting ring and by raising or lowering the nozzle ring.

Superheater Safety Valves

Safety valves are always installed at the superheater outlet as well as on the steam drum. To ensure an adequate steam flow through the superheater when the steam drum safety valves are lifted, superheater safety valves are set to lift at a lower pressure than that which lifts the steam drum safety valves. We will discuss two kinds of superheater safety valve arrangements: (1) the Crosby two-valve superheater outlet safety valve assembly and (2) the Consolidated three-valve superheater outlet safety valve assembly.

CROSBY TWO-VALVE SUPERHEATER OUTLET SAFETY VALVE ASSEMBLY.— Many single-furnace boilers are equipped with Crosby two-valve superheater outlet safety valve assemblies of the type shown in figure 11-15. In

valve pops; that is, it opens quickly and completely. Because of the larger area now presented for the steam pressure to act upon, the valve reseats at a lower pressure than that which caused it to lift initially. After the specified blowdown has occurred, the valve closes cleanly, with a slight snap.

The amount of compression on the spring determines the pressure at which the valve will pop. The position of the adjusting ring determines the shape of the huddling chamber and thereby

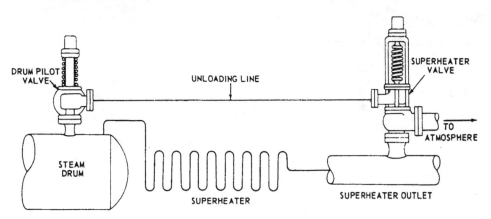

Figure 11-15.—Two-valve superheater outlet safety valve assembly (Crosby).

an assembly of this type, both the valves are spring-loaded. The pilot valve on the steam drum and the superheater valve at the superheater outlet are connected by a pressure transmitting line, which runs from the discharge side of the drum pilot valve to the underside of the piston, which is attached to the spindle of the superheater safety valve.

The superheater valve is set to pop at a pressure approximately 2 percent higher than the pressure that causes the drum pilot valve to pop. When the drum pilot valve pops, the steam pressure is transmitted immediately through the pressure line to the piston; thus, the superheater valve is actuated. If, for any reason, the drum pilot valve should fail to open, the superheater valve will open at a pressure approximately 2 percent higher to protect the superheater.

CONSOLIDATED THREE-VALVE SUPER-HEATER OUTLET SAFETY VALVE AS-SEMBLY.—The Consolidated three-valve superheater outlet safety valve assembly shown in figure 11-16 is used on some 1200-psi boilers. The assembly consists of a pilot valve, a piston actuator, and an unloading valve.

The spring-loaded pilot valve is mounted on the top center line of the steam drum. The piston actuator and the unloading valve are assembled as a unit and mounted on the piping at the superheater outlet; they are connected to each other by a rocker arm. The piston actuator has a cylinder with a piston inside it. The unloading valve contains (1) an actuating valve connected to the rocker arm by a stem and (2) a piston-type disk without a stem, which is held in line by the

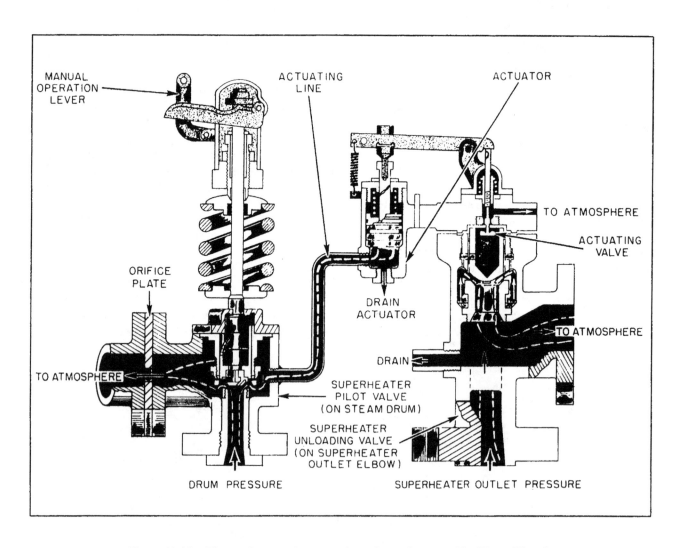

Figure 11-16.—Three-valve superheater outlet safety valve assembly (Consolidated).

cylinder in which it works. The unloading valve is pressure-loaded, not spring-loaded.

Steam from the superheater outlet enters the unloading valve cylinder and gathers around the valve disk above the seat. The steam bleeds through small ports to the space above the disk. When the actuating valve is closed, the steam above the disk of the unloading valve cannot escape, so the pressure above the disk equalizes with the pressure below the disk; that is, the pressure above the disk is equal to superheater outlet pressure.

Excessive pressure in the steam drum—NOT the excessive pressure in the superheater—causes the safety valve in this assembly to lift. When the pilot valve on the steam drum opens, pressure is transmitted from the pilot valve to the cylinder of the piston actuator. Pressure in the piston actuator cylinder is applied under the piston, causing the spring to compress. The rocker arm moves upward at the end over the piston actuator and downward at the end over the unloading valve, thus opening the actuating valve.

When the actuating valve opens, pressure above the piston-type unloading valve disk bleeds off to the atmosphere. The unloading valve therefore opens, allowing steam to flow from the superheater to the atmosphere, and thus protects the superheater tubes from excessive temperatures. When the pilot valve reseats, the actuating valve also reseats. As steam bleeds through the ports to the space above the disk in the unloading valve, pressure builds up and rapidly equals the pressure below the disk. Then the unloading valve closes. In summary, then, the superheater unloading valve always opens immediately after the steam drum pilot valve opens, and closes immediately after the pilot valve closes.

SOOT BLOWERS

Soot blowers are installed on boilers to remove soot from the firesides while the boiler is steaming. The soot blowers must be used regularly and in proper sequence to prevent the accumulation of heavy deposits of soot. Soot is an effective insulator, and any soot deposited on boiler tubes seriously interferes with heat transfer and thus reduces boiler efficiency. If the soot blowers are not used often enough, soot accumulates between the tubes on top of the lower drums and headers.

Soot tends to absorb moisture from the air. The moisture combines with the soot to form sulfuric acid, which corrodes the boiler tubes. Moisture also tends to make the soot pack down into such a solid mass that it cannot be removed by the soot blowers. Another important reason that soot must be removed at regular intervals is that any large accumulation of soot on the boiler firesides constitutes a serious fire hazard.

Soot blowers must be used to blow tubes on all steaming boilers at least once each week while underway, and, when in port or at anchor, just after leaving or prior to entering port. In addition, if practical, tubes should be blown as soon as possible after heavy black smoke has been made.

Soot blowers must be used in the proper sequence so that the soot will be swept progressively toward the uptakes. Normally, the uppermost soot blowers are used at the beginning and then again at the end of the blowing sequence. The exact sequence for blowing tubes should be obtained from the manufacturer's technical manual for the boilers aboard each ship.

Before tubes are blown, due consideration must be given to the effect that the discharge of soot will have upon the upper decks. The engineering officer of the watch (EOOW) must obtain permission from the officer of the deck (OOD) before instructing fireroom personnel to blow tubes.

Some ships are designed with stack dampers that enable the bridge to direct soot from either the port or starboard side as desired.

There are two basic kinds of soot blowers: (1) the rotary (multinozzle) type and (2) the stationary type. Rotary soot blowers are used in most outlet locations of the boiler. The stationary type is used in a few special locations.

The soot blowers commonly used on naval boilers are designed by two different manufacturers (Diamond and Copes Vulcan). Although they are similar in many ways, they differ in certain details of construction. The manufacturer's technical manual for the boiler should be consulted for details of construction, operation, and maintenance of the soot blowers.

A rotary soot blower is shown in figure 11-17. The part of the soot blower that you can see on the outside of the boiler is called the head. The soot blower element, a long pipe with multiple outlets, or nozzles, projects into the tube banks of the boiler. This soot blower is operated by an endless chain, which runs in a sheave wheel. When the chain is pulled, steam is admitted through a steam valve and is discharged at high velocity from the nozzles in the element. The nozzles direct the jets of steam so that they sweep around the tubes, thus preventing direct impingement of steam on the tubes. The soot is loosened so it can be blown out of the boiler.

Each steaming boiler uses its own steam to supply its own soot blowers. However, the steam used for blowing tubes is reduced from boiler pressure to approximately 300 psi. The pressure is reduced either by a pressure control disk above the steam valve or by a pressure control orifice below the steam valve.

Some soot blowers are operated by manual turning of a crank, while others are operated by manual pulling on a chain. Aboard some ships, the soot blowers are operated by push buttons. A press on the button admits air to an air motor that drives the unit.

On some soot blowers, the admission and cutoff of steam is controlled so that the tubes are swept only during a part of each rotation of the element. The part of each rotation during which steam is admitted, and during which the tubes are swept, is called the blowing arcs. Blowing arcs are controlled by cams or stops.

The scavenging air connection shown in figure 11-17 supplies air to the soot blower element and thus keeps combustion gases from backing up into the soot blower heads or the steam piping. A check valve is installed in the scavenging air piping, near the soot blower head. When tubes are being blown, steam enters the short length of air piping between the soot blower head and the check valve, closing the valve. When tubes are not being blown, air pressure in the scavenging air line keeps the check valve open.

The copper tubing connection for the scavenging air is usually at the front of the boiler casing,

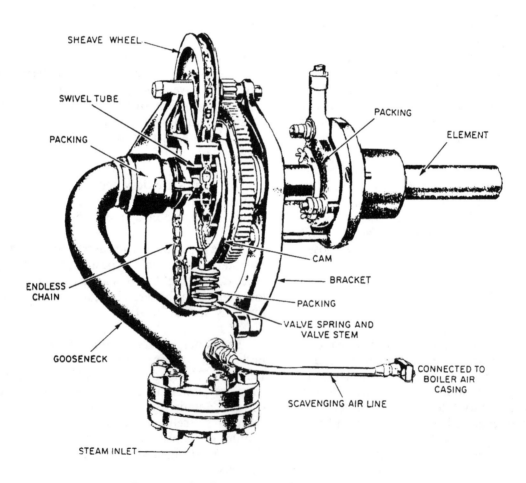

38.54

Figure 11-17.—External view of rotary soot blower.

11-16

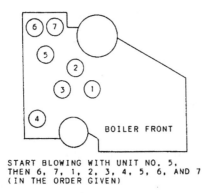

START BLOWING WITH UNIT NO. 5,
THEN 6, 7, 1, 2, 3, 4, 5, 6, AND 7
(IN THE ORDER GIVEN)

Figure 11-18.—Arrangement and sequence for operating a soot blower.

or it may be at some other location so that a shorter length of tubing can be used.

The number of soot blowers installed, the way in which they are arranged, and the blowing arcs for each unit differ from one type of boiler to another. Figure 11-18 indicates the soot blower sequence for a single-furnace boiler.

BLOW VALVES

Some solid matter is always present in boiler water. Since most of the solid matter is heavier than water, it tends to settle in the water drums and headers. Solid matter that is lighter than water rises and forms a scum on the surface of the water in the steam drum. Since most of the solid matter is not carried over with the steam, the concentration

of solids remaining in the boiler water gradually increases as the boiler is steamed. For the sake of efficiency and for the protection of the boiler pressure parts, some of this solid matter must be removed from time to time. Blow valves and blow lines are installed below the boiler for this purpose.

Light solids and scum are removed from the surface of the water in the steam drum by the surface blow line which, as we have already seen, is an internal boiler fitting. Heavy solids and sludge are removed by opening bottom blow valves, which are fitted to each water drum and header. Both surface blow valves and bottom blow valves on modern naval boilers are globe-type stop valves.

Both the surface blow valve and the bottom blow valves discharge to a system of piping called the boiler blow piping. The boiler blow piping system is common to all boilers in any one fireroom. Guarding valves are installed in the line to prevent leakage from a steaming boiler into the blow piping and to prevent leakage from the blow piping into a secured boiler. A guarding valve installed at the outboard bulkhead of the fireroom gives protection against saltwater leakage into the blow piping. After passing through this guarding valve, the water is discharged through an overboard discharge valve (sometimes called a skin valve), which leads overboard below the ship's water line. Figure 11-19 shows the general arrangement of boiler blow piping.

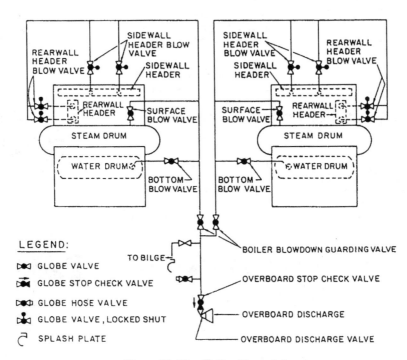

Figure 11-19.—Boiler blow piping.

INSTRUMENTS AND MONITORING DEVICES

Some of the instruments and monitoring devices that provide personnel information on the operating condition of the boiler are water gauge glasses, remote water-level indicators, superheater temperature alarms, smoke indicators, and oil drip detector periscopes.

WATER GAUGE GLASSES

Various combinations of water level indicating devices are used on naval boilers. Every boiler must be equipped with at least two independent devices that show the water level in the steam drum, so that a false water level may be detected through a comparison of the two. At least one of these devices must be a water gauge glass. Some boilers have more than two devices for indicating water level.

Several types of water gauge glasses are used on naval boilers. Gauges are identified by the amount of visibility permitted and usually are installed in pairs: two 10-inch gauges or two 18-inch gauges at the same level, one 10-inch and one 18-inch gauge, or two 10-inch gauges staggered vertically to provide a total visibility of 18 inches.

One type of water gauge glass is shown in figure 11-20. This gauge is assembled with springs, as shown in the illustration. This type of assembly makes it unnecessary to retorque the studs after the gauge has warmed up. (Notice the numbering of the studs in fig. 11-20; the numbers indicate the proper sequence of tightening the studs in assembling the gauge.)

Each water gauge is connected to the steam drum through two cutout valves, one at the top of the drum and one at the bottom. The bottom cutout valve connection contains a ball-check valve. The ball rests on a holder. As long as there is equal pressure on each side of the ball, the ball remains on its holder. But if the water gauge breaks, the sudden rush of water through the bottom connection forces the ball upward onto its seat and thus prevents further escape of hot water. No check valve is installed in the top cutout connection.

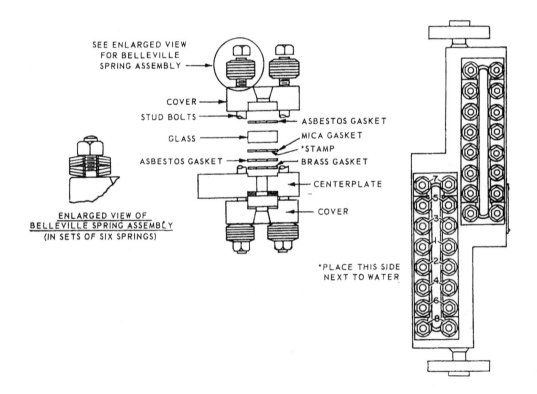

Figure 11-20.—Water gauge glass.

11-18

REMOTE WATER-LEVEL INDICATORS

Remote water-level indicators are used aboard most ships so that the boiler water level can be observed from the lower level of the fireroom or operating station. Two types of remote water-level indicators—the Yarway and the Barton—are the most common in use.

Yarway Remote Water-Level Indicator

The general arrangement of a Yarway remote water-level indicator is shown in figure 11-21. The main parts of the unit are (1) the constant-head chamber, which is mounted on the steam drum at or near the vertical center line of the drum; (2) the graduated indicator, which is usually mounted on an instrument panel; and (3) two reference legs that connect the constant-head chamber to the indicator. The reference legs are marked A and B in figure 11-21.

A constant water level is maintained in leg A (constant pressure leg), since the water level in the constant-head chamber does not vary. The level in leg B (variable pressure leg) is free to fluctuate with changes in the steam drum water level. The upper hemisphere of the constant-head chamber is connected to the steam drum at a point above the highest water level to be indicated; because of this connection, boiler pressure is exerted equally on the water in the two legs. Variable leg B is connected to the steam drum at a point below the lowest water level to be indicated; because of this connection, the water level in leg B is equalized with the water level in the steam drum.

As shown in figure 11-21, each leg is connected by piping to the indicator. In the indicator, the two columns of water terminate at opposite sides of a diaphragm.

Barton Remote Water-Level Indicator

The general arrangement of a Barton remote water-level indicator is shown in figure 11-22. The

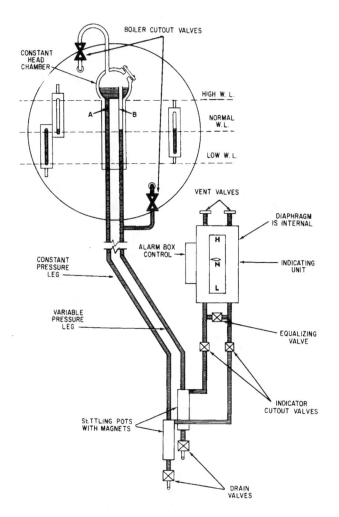

Figure 11-21.—Yarway remote water-level indicator.

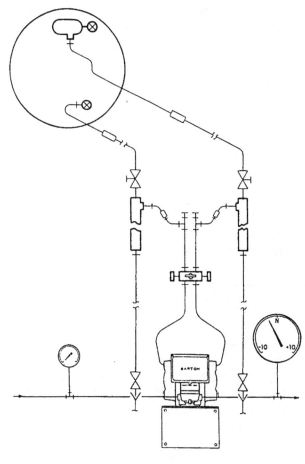

Figure 11-22.—Barton remote water-level indicator.

11-19

operating principles of this device are very similar to the operating principles of the Yarway remote water-level indicator previously discussed.

The main difference is the differential pressure sensing unit (DPU). This unit can be used either as a pneumatic transmitting unit by attaching a pneumatic transmitter to the DPU output shaft, or strictly as an indicator by attaching a gauge to the DPU output shaft. The bellows are liquid-filled and can withstand much more pressure than the diaphragm used in the Yarway remote water-level indicator.

PRESSURE AND TEMPERATURE GAUGES

Boiler operation requires a constant awareness of pressures and temperatures in the boiler and in its associated machinery and systems. Pressure gauges are installed on or near each boiler to indicate steam drum pressure, superheater outlet pressure, auxiliary steam pressure, auxiliary exhaust pressure, feedwater pressure, steam pressure to the forced draft blowers, air casing pressure, and fuel oil pressure. Thermometers are installed to indicate superheated steam temperature, desuperheated steam temperature, feedwater temperature at the economizer inlet and outlet, and, in some ships, uptake temperatures.

In some firerooms, the gauges that indicate steam drum pressure, superheater outlet pressure, superheater outlet temperature, and combustion air pressure are installed on the boiler front. As a rule, however, the indicating units of all pressure gauges are mounted on a boiler gauge board that is easily visible from the firing aisle. Distant-reading thermometers are also installed with the indicating unit mounted on the boiler gauge board. Direct-reading thermometers must, of course, be read at their actual locations. In some installations, a common gauge board is used for all the boilers in one space, instead of having separate gauge boards for each boiler. The ABC console always has a duplicate gauge board for all major boiler operating parameters.

The pressure gauges most commonly used in connection with boilers are of the Bourdon-tube type. However, some manometers and some diaphragm gauges are also used in the fireroom.

The temperature gauges most commonly used in the fireroom are (1) direct-reading liquid-in-glass thermometers and (2) distant-reading Bourdon-tube thermometers. However, bimetallic expansion thermometers are also used. The basic operating principles of pressure and temperature gauges were discussed in the chapter 7 of this manual.

Some thermometers are the bare-bulb type—that is, the bulb is in direct contact with the steam or other fluid being measured. Bare-bulb thermometers must NEVER be removed while the boiler is under pressure. Other thermometers are the well or separable-socket type—that is, the bulb is inserted into a well or socket in the line where the temperature is to be measured. These thermometers are not designed to be used as bare-bulb thermometers; they must always be used in the well or socket.

SUPERHEATER TEMPERATURE ALARMS

Superheater temperature alarms are installed on most boilers to warn operating personnel of dangerously high temperatures in the superheater.

SMOKE INDICATORS

Naval boilers have smoke indicators (sometimes called smoke periscopes) that permit visual observation of the gases of combustion as they pass through the uptakes. Single-furnace boilers have one smoke indicator installed in the uptake. Double-furnace boilers have two smoke indicators to permit visual observation of the combustion gases coming from each boiler.

OIL DRIP DETECTOR PERISCOPES

Some boilers are equipped with oil drip detector periscopes, which permit inspection of the space between the inner and outer casing and the furnace floor under the burners for the accumulation of oil. These periscopes include a reflecting unit with an inner mirror, a vision unit with a mirror that can be seen by operating personnel, and a lamp unit. The reflecting unit can be rotated through 360° by turning the handle. The vision unit can be rotated to any angle to suit the operator. The mirror frame in the vision unit is adjustable to suit the line of sight.

SIMPLE CONTROL SYSTEMS

Any control system, simple or complex, must perform four basic operations: MEASURE, COMPARE, COMPUTE, and CORRECT. It will first measure a value (temperature, pressure, or level) on the output side of a process. Next, the measured value is compared with the desired value. Then, the control system computes the amount and direction of change required to bring the measured output value to the desired output value. Finally, the value on the input side of the process is corrected to bring the output side of the process back to the desired value.

Taken together, these operations constitute a closed loop of action and counteraction by which some quantity or condition is measured and controlled. The closed control loop is often called a feedback loop because it requires a feedback signal from the output side of the process to correct the value on the input side. The closed loop concept is illustrated in figure 11-23.

Any quantity or condition that is measured is a variable. Any quantity or condition that is measured and controlled by a control system is a controlled variable. Any quantity or condition that is varied by a control system to affect the value of a controlled variable is called a manipulated variable.

One must realize that the closed loop of control will not react instantaneously. There are two reasons for this: dead time and lag. Dead time is the time required to transport change from one place to another and can occur in various parts of the process. Lag is the slowing-down effects associated with capacity and resistance of the equipment or substance involved in the process.

AUTOMATIC BOILER CONTROL SYSTEMS

ABC systems are now installed on practically all naval boilers. These systems provide rapid and sensitive response to changes in demand for combustion air, fuel, and feedwater. Boiler control is done by two independent systems: an automatic combustion control system and a feedwater control system. The function of the automatic combustion control (ACC) system is to maintain fuel input and combustion air input to match the steam demand. The function of the feedwater control system is to regulate feedwater input to maintain proper water level in the steam drop. The following paragraphs will discuss some of the ABC systems found aboard ship.

BOILER CONTROL SYSTEMS

Functionally, a boiler control system may be considered as having two parts: a combustion control system and a feedwater control system. The combustion control system maintains the fuel oil input and the combustion air input to the boiler according to the demand for steam. A related function of the combustion control system is to proportion the amount of combustion air to the amount of fuel to provide maximum combustion efficiency. Likewise, the feedwater control system supplies the right amount of feedwater to the boiler under all operating conditions.

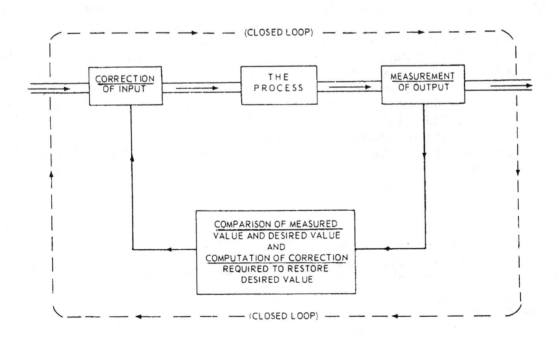

Figure 11-23.—Closed control loop.

Although we may regard the combustion control system and the feedwater control system separately for some purposes, we must not forget that they are closely related and that they usually function together rather than separately.

Control systems used aboard naval ships are Hagan, Bailey, and General Regulator. The most widely used is Hagan. Bailey is being phased out, and in most cases it is removed at overhaul and replaced with either Hagan or General Regulator. Since the trend with recent construction is almost 100 percent Hagan, the rest of this chapter will deal primarily with Hagan controls.

Remember no two systems are identical in all details, even when they are made by the same manufacturer. The only authoritative source of information on the boiler control system of any particular ship is the manufacturer's technical manual for the equipment on that ship.

PNEUMATIC CONTROL SYSTEMS

ABCs installed on naval ships use compressed air as the control medium. The control medium is what is used to send signals between individual components of the control system. There are three reasons why air is the preferred control medium: (1) air is insensitive to shock, (2) the control systems will tolerate small air leaks, and (3) and moisture have little effect on the control system since the air is filtered and dried before use. Pneumatic signals or messages are transmitted in the form of variable air pressure. Other possible control mediums are hydraulic or electrical.

The source of compressed air for ABCs is normally the ship's low-pressure service compressed air system. Some ships will have a separate boiler control air system with an air compressor dedicated to serving only the boiler control system. Air is supplied at 120 psi from the ship's low-pressure service compressed air system. The boiler control system will have an air receiver installed to allow for surges in supply pressure and temporary supply if supply pressure is inadequate.

In considering the action of any pneumatic unit, you must be able to distinguish three kinds of air pressure. Unfortunately, the manufacturers of boiler controls do not always identify these three kinds of pressure signals in quite the same way. For the purposes of this chapter, we will define the three air pressure signals as follows:

SUPPLY AIR PRESSURE is the compressed air pressure that is supplied to each pneumatic unit so that it can develop the appropriate pneumatic "messages" or signals in the form of variable air pressure.

LOADING PRESSURE is the compressed air pressure that goes from one pneumatic unit to another pneumatic unit, EXCEPT when the pressure is imposed upon a motor operator or upon a final control element.

CONTROL PRESSURE is the compressed air pressure that goes from a pneumatic unit to a motor operator of a final control element.

COMBUSTION CONTROL SYSTEM

The combustion control system consists of pneumatic metering devices (transmitters), controllers, relays, and actuators. This equipment is arranged and interconnected to produce an integrated system designed to monitor the steam pressure and control the flow of air and fuel to the steam-generating units to maintain steam pressure at set point. Additionally, a control system is designed to maintain fuel/air (F/A) ratio throughout the operating range at a value that ensures optimum combustion.

The combustion control system serves all boilers located in the same machinery space or fireroom. This system consists of components that (1) sense the steam drum pressure; (2) determine the loading signal for the air and fuel control loop of each boiler, which is necessary to hold steam pressure at set point; and (3) control these variables accordingly.

This system is divided into three groups—the steam pressure control loop, the airflow control loop, and the fuel flow control loop. Component item numbers shown on the control system diagram, appear in parentheses in this discussion (see fig. 11-24). So that you will understand the system, we will discuss each group individually.

Steam Pressure Control Loop

The steam pressure control loop consists of the following components:

- Steam pressure transmitters A and B (1 and 2)

- High signal selector (3)

- Steam pressure controller (4)

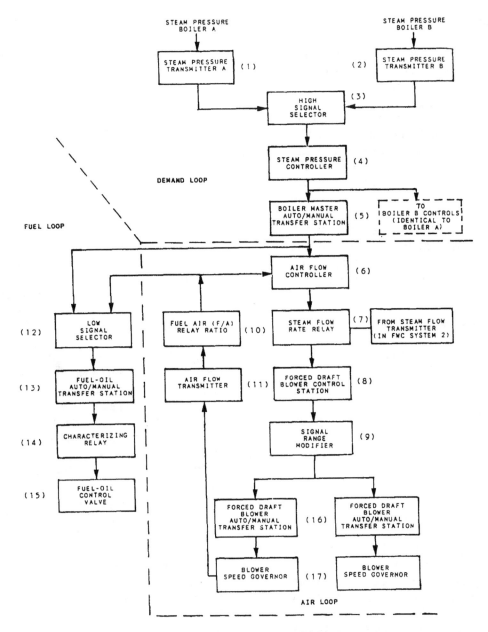

Figure 11-24.—Block diagram of the ACC system.

● Boiler master auto/manual (A/M) station (5)

The steam pressure transmitters (1 and 2) are connected to their respective boiler steam drums. These components sense steam drum pressure and develop an output signal that regulates the rate of combustion required to maintain a constant drum pressure.

The output signals from the direct-acting steam pressure transmitters are transmitted to the high signal selector (3). The high signal selector selects the higher of the pressure signals and sends it on to the steam pressure controller (4).

The steam pressure transmitter is direct-acting. When there is an increase in steam pressure, the steam pressure transmitter sends out an increased signal. Once this signal reaches the steam pressure controller, this increased input signal results in a decreased output.

From the steam pressure controller the signal is fed to the boiler master A/M station (5), where it is passed through and becomes the master demand signal. This completes the steam pressure control loop.

Airflow Control Loop

At the airflow controller (6), two signals are compared—the master demand signal, which is telling the system what it wants, and the airflow feedback signal from the airflow transmitter, which represents what the system has available. If these signals are equal, the output remains constant at the value required. If the signals are different, the controller changes its output in the proper direction to the extent necessary to reduce the difference to zero; and, once again, the system is balanced.

In order for the airflow controller to match the demand signal with the feedback signal, the airflow to the burners must change. To accomplish this, the following components are required:

- Airflow controller (6)

- Steam flow rate relay (7)

- Forced draft blower control station (8)

- Signal range modifier (9)

- Airflow transmitter (11)

- Fuel/air (F/A) ratio relay signal generator (10)

- Forced draft blower A/M stations (16)

The airflow transmitter (11) measures the differential between the wind box and the fire box and sends out a signal representing airflow for the boiler. This airflow signal passes to the F/A ratio relay signal generator (10).

The F/A ratio relay signal generator provides a means of modifying the airflow signal as necessary to provide the F/A ratio needed for optimum combustion throughout the firing range of the boiler. In other words, if some adverse condition results in poor combustion, you have the means to readjust the airflow and restore combustion to an acceptable state until corrective action can be taken or excess air can be added for blowing tubes.

The output of the F/A ratio relay signal generator is called the airflow feedback signal (a loading signal). The airflow feedback signal goes directly to the low signal selector (12) and to the airflow controller (6).

The airflow controller receives the master demand signal and the airflow feedback signal, compares these two signals, and applies proportional plus reset action to correct any difference between them. As long as the difference is zero, the output will remain constant.

The output of the airflow controller is fed to the steam flow rate relay (7). The steam flow rate relay also receives a signal from the steam flow transmitter. The steam flow transmitter is located in the steam flow, or demand, loop of the feedwater control system.

The steam flow is the first measurable signal to change when the boiler demand changes. The steam flow rate relay helps ensure a prompt response of the FDBs to a change in demand. The FDBs have a large amount of inertia and are slow to change speed.

The output of the steam flow rate relay goes to the two-way forced draft blower control station (8). This signal then goes to the signal range modifier (9).

This signal must be modified because the airflow demand signal has a range of 0 to 60 psig while the forced draft blower governor (final control element) has a range of 9 to 60 psig. The signal must be modified so that a given airflow demand signal may produce the same change in the governor.

The signal range modifier sends a signal to the two-way forced draft blower A/M stations (16). From here the signals go to the final control element to control the speed of the forced draft blowers.

Fuel Flow Control Loop

The fuel flow control loop consists of the following components:

- Low signal selector (12)

- Fuel-oil A/M transfer station (13)

- Characterizing relay (14)

- Fuel pressure control valve (15)

The first component in the fuel control loop is the low signal selector (12). The two inputs to this device are the master demand signal and the airflow feedback signal. The low signal selector transmits an output equal to the lower of its two input signals. The low signal selector ensures that the fuel flow never exceeds airflow, thereby preventing black smoke from occurring under all conditions.

If the boiler load is decreased, the master demand signal to the airflow controller and the low signal selector will also decrease. The low signal selector will transmit the lower master demand signal as fuel demand and the air and the oil will decrease together, thus preventing black smoke.

The fuel flow demand signal now passes to the fuel flow A/M selector station (13). Again, you can select the mode of operation; if it is automatic, the signal passes directly to the characterizing relay (14). To maintain efficient combustion, the fuel-oil flow to the burners must be regulated so that it is proportional to the airflow supplied to the burners. The fuel-oil control valve, as in the case of the forced draft blower governor, operates on a different signal range. The signal is modified in the characterizing relay. The output of

the characterizing relay goes directly to the fuel pressure control valve final control element.

This valve receives the output of the characterizing relay and develops an oil pressure that varies in a manner directly proportional to any pneumatic change in its input.

Components Operation

At this point some of the individual components of the automatic combustion control system will be discussed and their operation described.

HAGAN RATIO TOTALIZER.—The Hagan ratio totalizer (fig. 11-25) is used for the

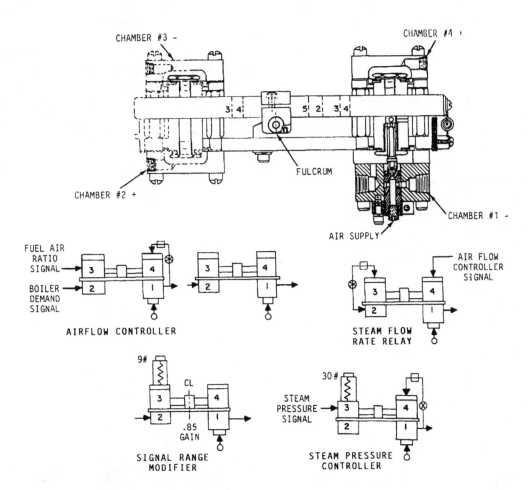

NOTE: A SPRING BARREL AND SPRING ASSEMBLY CAN BE ATTACHED TO CHAMBERS 2, 3, OR 4 TO INDUCE A MECHANICAL INPUT FORCE.

Figure 11-25.—Hagan ratio totalizer.

steam pressure controller, airflow controller, steam flow rate relay, and signal range modifier.

The totalizer consists of four diaphragm chambers, a balance beam, the adjustable fulcrum, and a poppet valve located on chamber #1. Any force (signal) applied to either chamber #2 or #4 will rotate the unit clockwise and open the poppet valve, increasing the output signal. A force applied to either chamber #1 or #3 will rotate the unit counter-clockwise and close the poppet valve, decreasing the output signal. The adjustable fulcrum may be moved to change the ratio of output.

Steam Pressure Controller.—The steam pressure controller is the set point for the system. There is a 30-lb spring attached to chamber #3 which acts as a positive force. The input signal from the high signal selector is piped to chamber #3, so any increase in pressure from the high signal selector results in a decrease in output pressure. The output signal is fed back to chamber #4 through a needle valve and volume tank. This controls the reset or speed at which the controller works.

Airflow Controller.—The airflow controller has two input signals. The output from the boiler A/M station is connected to chamber #2, and the output from the airflow transmitter is connected to chamber #3. The net force is the difference between these two signals. An increase from the boiler A/M station will increase output; an increase from the airflow transmitter will decrease the output. Again, the output is fed back to chamber #4 through a needle valve and volume tank to control reset or controller speed.

Steam Flow Rate Relay.—The steam flow rate relay has two input signals. One signal is from the steam flow transmitter connected to chamber #2 and through a needle valve to chamber #3. When this signal is steady it produces a net force of zero. As this changes, it produces a force, since the needle valve delays the signal to chamber #3. Steam flow is the first measurable variable to change during a change in boiler demand. This input is used to help overcome the lag of the forced draft blower due to its inertia.

Signal Range Modifier.—The signal range modifier is a four-chamber totalizer with a spring mounted on chamber #3. It uses the adjustable fulcrum set off center to the left to give a 0.85 gain in output. The input signal from the FDB

A/M station is applied to chamber #3. Also, a 9-lb spring is attached to chamber #3. The spring will create a 9-lb signal when the input signal is 0 psi. The changed fulcrum position results in a 60-psi output signal for a 60-psi input signal. By having the output signal at 9 psi, the "dead band" associated with the Woodward governor of the forced draft blower is avoided. The governor will not respond to an input signal less than 9 psi.

SIGNAL SELECTORS.—The automatic combustion control system uses two signal selectors (see fig. 11-26). These selectors are identified as Supermite 71H and 71L by the manufacturer, G.W. DAHL. The 71H is the high signal selector which compares the inputs from the two steam drum pressure transmitters. The 71L is the low signal selector with one input from the boiler master station, the other from the F/A ratio relay. Both selectors are physically the same, with O-rings in different positions to change the function—as either a high or low signal selector.

For the high signal selector the neoprene O-rings are placed in the inner grooves on the spool, and for the low signal selector the O-rings are placed on the outer grooves on the spool. Pressure differential between input signals will create a force on the diaphragm and will move the spool. The O-rings will then block one passage and allow the other signal to be passed to the output.

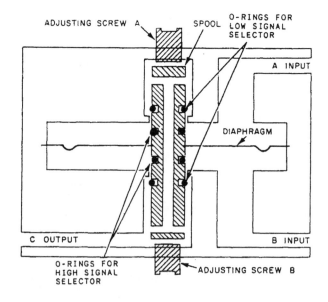

Figure 11-26.—Supermite 71H/71L signal selector.

CHARACTERIZING RELAY.—The characterizing relay compensates for nonlinear relationship between fuel-oil pressure and fuel-oil flow rate. This relationship in actuality is a squared one. In the system, the final control element—the fuel-oil control valve—controls fuel-oil pressure, when the desired effect is to control fuel flow rate in a linear fashion. The characterizing relay modifies the signal to the fuel-oil control valves so that the fuel flow to the burners is proportional to the airflow supplied to the burners.

BOILER MASTER A/M STATION (FOUR-WAY).—This component is used to select the mode of operation, either manual or automatic. When it is set for manual operation, the master demand signal is dead-ended and the manually generated signal is sent out to the rest of the system to control air and oil flow. When it is set for automatic operation, the master demand signal passes through the transfer section to the relay sender (hand generator), there the signal is transmitted through diaphragms and springs to the poppet valve. The output of the poppet valve is then sent back through the transfer station out to the combustion system. In the relay sender, bias can be added when the system is in automatic to compensate for small pressure differences between boilers. A four-way A/M station is shown in figure 11-27.

Bias is a set differential applied to the signal. Its purpose is to parallel machinery to matching speed on forced draft blowers or matching pressure on the boilers.

A two-way A/M station is exactly the same, except it cannot introduce biasing.

FEEDWATER CONTROL SYSTEM

The function of the feedwater control system is to control the amount of feedwater to the boiler. This system is referred to as three element because there are three MEASURED MEANS or VARIABLES in this system: (1) steam flow, (2) feedwater flow, and (3) steam drum water level. Steam flow

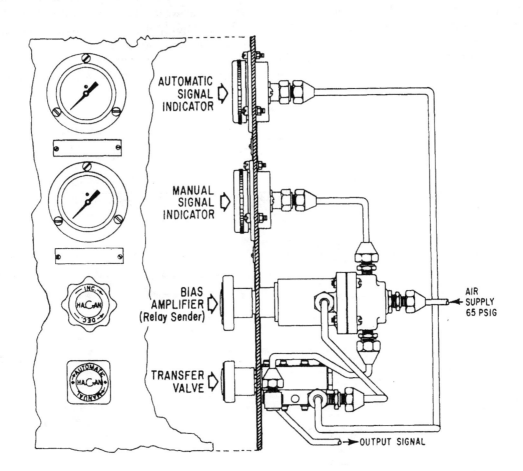

Figure 11-27.—Boiler four-way A/M station.

is the DEMAND ELEMENT, feedwater flow is the RESPONSE or FEEDBACK ELEMENT, and the steam drum water level is a SUPERVISORY ELEMENT. This system will take the difference between the steam flow and the feedwater flow to detect an imbalance. One might think that the difference between these two flow rates is adequate to maintain the proper water level in the steam drum. This is not so. The drum level must also be monitored because of the occurrence of shrink and swell.

Shrink or swell is the temporary descending or rising of the steam drum water level when there is a change in the boiler firing rate. Shrink or swell is associated with the opposite direction of change of boiler demand than expected. To explain this, take for example an increase in boiler demand. With more demand, there is a slight drop in steam drum pressure. With the increased demand there are more steam bubbles and a lower pressure. As the steam bubbles expand, they push the water level up. This is known as swell. All this happens without the feed flow changing. So, if feedwater flow had been immediately increased, as required by the increased steam flow, the possibility of the water level becoming too high may have resulted. By having the steam drum water level as a supervisory element, the feedwater control system will wait until the temporary swell has subsided before

increasing feedwater flow. A decrease in boiler demand will result in shrink, which is exactly the opposite.

Now, take a look at figure 11-28, which is a block diagram of the feedwater control system. The steam and feed flow transmitters develop signals proportional to their representative flows. This is done by measuring differential pressure across a flow nozzle (restriction). The flow transmitter extracts the square root of the differential pressure for a signal directly proportional to the flow rate. The steam flow/feedwater flow relay computes the difference between the two flow rate signals. The output signal from the steam flow/feedwater flow relay is sent to the feedwater flow controller.

The drum level transmitter measures a differential pressure created by a constant and variable water leg from the steam drum. The transmitter develops a pneumatic output that is proportional to the actual water level and transmits the signal to the feedwater flow controller. The needle valve and volume tank create a delay in steam flow/water flow signal that allows the output signal to avoid transient changes and eliminate hunting.

The feedwater controller is a proportional plus reset device that compares its two input signals.

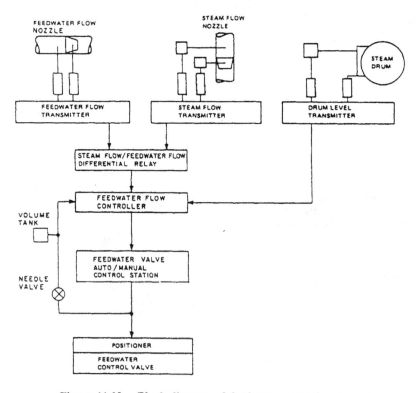

Figure 11-28.—Block diagram of feedwater control system.

The two input signals are from the steam flow/water flow relay and the drum level transmitter. When the two signals differ, the output is adjusted in the opposite direction so that reactions will reduce the difference between the signals and maintain the proper steam drum water level. This component operates exactly the same as the steam flow rate relay in the combustion control system.

Components Operation

The following paragraphs will discuss some of the individual components of the feedwater control system and their operation.

The Hagan ratio totalizer (see fig. 11-25) is also used for the steam flow/water flow relay, the drum level totalizer, and the feedwater flow controller. Refer back to the description of the Hagan ratio totalizer in the section on components operation of the ACC system.

STEAM FLOW/WATER FLOW RELAY.—The steam flow/water flow relay (fig. 11-29, view A) has two input signals. It operates very similar to the ACC airflow controller. There is a 30-lb spring attached to chamber #3. The steam flow signal is piped to chamber #2, and the feedwater

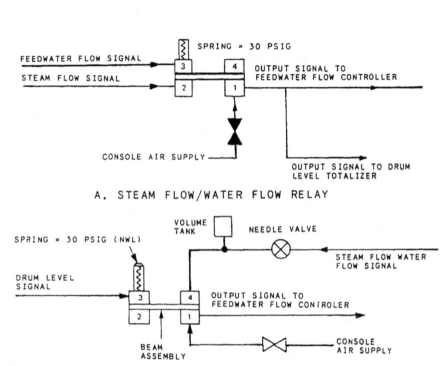

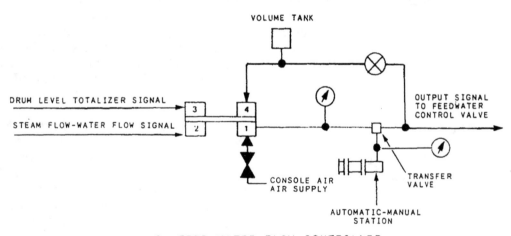

Figure 11-29.—Component parts of feedwater control system.

flow signal is piped to chamber #3. This spring produces a 30-psi output signal when the steam flow and the water flow is matched. If these two flow rates are not matched, the output signal equals 30 psi plus the feedwater flow signal minus the steam flow signal.

DRUM LEVEL TOTALIZER.—The drum level totalizer (fig. 11-29, view B) has the steam flow/water flow relay output piped to chamber #4. There is a needle valve and volume tank in this signal line. The drum level transmitter signal is piped to chamber #3, which has a 30-lb spring attached. A normal drum level is indicated by a 30-psi signal from the drum level transmitter. This signal and the 30-lb signal cancel out and the steam flow/water flow rate relay signal becomes the output signal. The needle valve and volume tank settle down, or smooth out, the steam flow/ water flow relay signal to prevent excessive hunting.

FEEDWATER FLOW CONTROLLER.— The feedwater flow controller (fig. 11-29, view C) operates the same as the airflow controller of the ACC system. The output signal equals the drum level totalizer signal minus the steam flow/water flow relay signal.

AIR LOCKS

An air lock system is provided to lock the settings of the forced draft blower valve operators, the fuel-oil control valve, and the feedwater flow control valve to maintain airflow, fuel flow, and water flow at values that existed at the instant of supply air failure. As a result, these flows are held constant until the control air supply has been restored and the air lock system is reset for normal operation or until the boilers are transferred to local manual operation.

MODES OF OPERATION

There are three modes of operation: automatic, remote manual, and local manual.

Automatic

A system is in complete automatic operation when it operates without human assistance.

Remote Manual

A system is in remote manual when the console operator shifts an A/M station and manually generates the required signal. In this mode, one person may control the boiler from the control console. In addition, the console shows the operator the signals the control system develops, but are not passed to final control elements.

Local Manual

In local manual, each final control element is manned by a watch stander and operated without any assistance from the automatic control system. Local manual control must be used if pneumatic air pressure is not available.

OTHER CONTROL SYSTEMS

There are other control systems of importance to the operation of the boiler. They are the main feed pump control and recirculation systems. These systems ensure the feedwater system operates as required. On some ships, the controls to these systems will be on the same console or board as the ABCs. These systems are similar in that they follow the basic requirements of a closed loop control systems and are pneumatic. Maintenance of these systems is done by the same personnel who maintain the ABCs.

MAINTENANCE

Maintenance of ABC systems will be covered by the PMS system. A pneumatic control system requires a clean and oil-free supply of air. This is obtained only if the air compressor is operating properly and air filters are kept in proper order. The frequency required on checking, draining, and replacing air filters is covered by the PMS system. In any pneumatic system, there are some components that require periodic lubrication, and some components that SHOULD NEVER BE LUBRICATED OR OILED. Unless specifically called for in procedural instructions, DO NOT LUBRICATE OR OIL A COMPONENT OF A PNEUMATIC CONTROL SYSTEM.

In ABC systems, adjustments should NOT be made unless the system is not performing properly, the cause of improper operation is known, and the required adjustment is known. Adjustment should be left to designated ABC technicians.

Every ship should have a designated ABC technician. These technicians are designated by an NEC, which they qualified for by attending the formal training course for that ABC's manufacturer. If your ship does not have an ABC technician qualified in this manner, every effort should be made to qualify a person by sending him or her to the appropriate school. We recommend that the ABC technician oversee all aspects of ABC maintenance.

Boiler Flexibility Test

The boiler flexibility test (flex test) is used on propulsion boilers that have ABCs. The flex test determines if the controls are adjusted properly.

The flex test is conducted by changing boiler load 70 percent in 45 seconds. The steam drum pressure, water level, and combustion must remain within specified limits, as stated by COMNAVSEASYSCOM PMS 301. After completion of 70-percent ramp load changes, all systems must stabilize within 4 minutes from the start of the test. Stability must be observed for 2 minutes. The flex test is conducted within 15 to 95 percent of boiler full power depending on operating requirements.

Procedures and periodicity requirements are covered by the PMS system.

On-line Verification Procedures

The objective in aligning and maintaining the ABC system is to achieve satisfactory overall control system performance. Overall system performance is measured by using the flex test. Failure of the flex test by itself does not provide all information needed to determine why the test was failed. Malfunction or misalignment of almost any component in any of the control loops could cause poor system performance. This, in turn, would cause failure of the flex test. On-line verification (OLV) provides a set of checks. These checks verify proper performance of each of the subsystems or control loops within the ABC system. These checks do two things:

1. They provide a means for quantitatively checking control system performance. Passing all the checks ensures that the controls will give satisfactory overall system performance.

2. They help determine which components need adjustment or repair during poor system performance.

REFERENCES

Boiler Technician 3 & 2, NAVEDTRA 10535-H, Naval Education and Training Program Management Support Activity, Pensacola, Fla., 1983.

Propulsion Plant Manual for FF-1078 to FF-1087, NAVSEA 0941-LP-051-6010, Naval Sea Systems Command, Washington, D.C., December 1974.

CHAPTER 12

STEAM PROPULSION TURBINES

LEARNING OBJECTIVES

Upon completion of this chapter, you should be able to do the following:

1. Recognize the design, operating principles, and functions of steam turbines and their accessories.

2. Identify the different methods of compounding in a turbine.

3. Identify the different methods/modes of steam flow in steam turbines.

4. Identify the different turbine components and accessories.

5. Discuss the two methods of turbine control and the economical methods of operating a propulsion plant.

6. Identify turbine classifications by both class and type.

7. Identify the means used to help ensure the efficient operation of a steam propulsion plant.

8. Discuss proper plant maintenance and safety factors involved in the efficient operation of a propulsion plant.

9. Discuss the mission of casualty control and effective forms of casualty control which prevent plant casualties.

INTRODUCTION

In the study of steam turbines, the first point to be noted is that we have now reached the part of the thermodynamic cycle in which the actual conversion of the thermal energy to mechanical energy takes place.

By simple observation of pressures and temperatures, we can see that the steam leaving a turbine has far less thermal energy than it had when it entered the turbine. It can also be seen that work is performed as the steam passes through the turbine, the work being evidenced by the turning of a shaft and the movement of the ship through the water. Since energy can be transformed but can be neither created nor destroyed, the decrease of thermal energy and the appearance of work cannot be regarded as separate events. Rather, we must infer that thermal energy has been transformed into work—that is, mechanical energy in transition.

Disregarding such irreversible energy losses such as those caused by friction and by heat flow to objects outside the system, it can be shown that two energy transformations are involved. First, there is the thermodynamic process by which thermal energy is transformed into mechanical kinetic energy as the steam flows through one or more nozzles. Second, there is the mechanical process by which mechanical kinetic energy is transformed into work as the steam impinges upon projecting blades of the turbine, thereby turning the turbine rotor.

To understand the process by which thermal energy is converted into mechanical kinetic energy, we must have some understanding of the process that takes place as steam flows through a nozzle. The second energy transformation, from kinetic energy to work, is best understood by considering some basic principles of turbine design.

STEAM FLOW THROUGH NOZZLES

The basic purpose of the nozzle is to convert the thermal energy of the steam into mechanical kinetic energy. This is done by shaping the nozzle in such a way as to cause an increase in steam velocity as it expands from a high-pressure area to a low-pressure area. The nozzle also directs the flow of steam in the direction to impinge upon the turbine blades.

The velocity of steam flow through the nozzle (or any restricted channel) depends on the pressure difference between the nozzle inlet and outlet. If the inlet and outlet pressure are equal, there is no flow. This is a static condition. However, if the outlet pressure is reduced, a flow of steam will begin and velocity will increase as the outlet pressure decreases. There is a point where any further reduction in the outlet pressure will not produce any further increase in steam velocity or flow rate. The ratio of pressure at which this occurs is known as the critical pressure ratio or acoustic pressure ratio. This ratio is about 0.55 for superheated steam. In actuality this limit occurs when the steam velocity is equal to the speed (or velocity) of sound in this steam.

The steam velocity through a nozzle is a function of the pressure differential across the nozzle. However, no further increase in steam flow will occur when the outlet pressure is reduced below 55 percent of the inlet pressure.

When the pressure drop across the nozzle will not exceed the critical pressure ratio, a simple convergent (parallel-wall) nozzle may be used. In this type of nozzle (fig. 12-1), the cross section of the nozzle is the same at the throat and the outlet. This type of nozzle is often referred to as a nonexpanding nozzle because no expansion of steam occurs past the throat of the nozzle.

When design considerations require a higher pressure drop across a nozzle than the critical pressure ratio, a modification of the nozzle is required. In this case, a convergent-divergent nozzle is used. In this type of nozzle (fig. 12-2), the cross-sectional area decreases from the inlet to the throat, then increases from the throat to the outlet. The critical pressure will be reached at the throat of the nozzle, and the gradual increase in area from the throat to the outlet will allow the steam flow to stabilize and minimize turbulence.

The decrease in thermal energy of the steam passing through a nozzle must equal the increase in kinetic energy (disregarding irreversible losses).

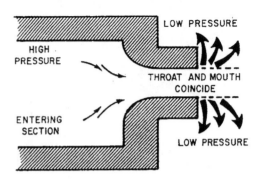

Figure 12-1.—Simple convergent nozzle.

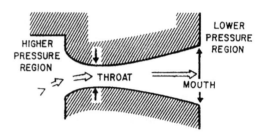

Figure 12-2.—Convergent-divergent nozzle.

The decrease in thermal energy may be expressed as

$$h_1 - h_2$$

where

h_1 = enthalpy of entering steam, in Btu/lb

h_2 = enthalpy of exiting steam, in Btu/lb

You can determine the energy of the steam jet leaving the nozzle using the following equation:

$$\text{Kinetic Energy (ke)} = 1/2 \, mv^2$$

In the equation, m is mass, which is equal to weight (in pounds) divided by the gravity constant (32.2 ft/sec^2). Since enthalpy is in unit per pound, let us assume that there is 1 pound of steam per second flowing from the nozzle. By assuming the flow rate, mass drops out of the equation for kinetic energy, but the gravitational constant must be retained. Also, to match units in both sides of the equation, remember that 1 Btu is equal to 778 foot-pounds. So, by equating the two equations and moving constants around we have the following:

$$v^2 = (h_1 - h_2)(778 \times 64.4)$$

The kinetic energy of the steam leaving the nozzle is directly proportional to the square of the velocity. By causing an increase in velocity, the nozzle causes an increase in the kinetic energy of the steam. Thus, the last equation describes the purpose of a nozzle by equating the decrease in thermal energy with the increase in kinetic energy.

PRINCIPLES OF TURBINE DESIGN

A turbine may be thought of as a bladed wheel or rotor that turns when a jet of steam from the nozzles impinges upon the blades. The basic parts of a turbine are the rotor, which has blades projecting radially from its periphery; a casing, in which the rotor revolves; and nozzles, through which the steam is expanded and directed. The first energy conversion, thermal energy to mechanical kinetic energy, occurs in the nozzles. The second energy conversion of kinetic energy to work occurs in the blades.

The distinction between types of turbines has to do with the manner in which the steam causes the turbine rotor to move. An impulse turbine rotor moves by a direct push or "impulse" from the steam impinging upon the blades. A reaction turbine is moved by the force of reaction.

The distinction between impulse turbines and reaction turbines should not be considered an absolute distinction in real turbines. Both types of turbines use impulse and reactive force to move the rotor, but they are classified by the force which primarily moves the turbine.

IMPULSE TURBINES

When discussing the manner in which kinetic energy is converted to work in the turbine blades, you must consider both the absolute and relative velocity of the steam in relation to the moving blades.

Consider an elementary impulse turbine with blades that are merely flat vanes or plates (fig. 12-3). The rotor moves as the steam jet flows from the nozzle and impinges upon the blades.

The relative velocity of the steam entering the blades must equal the absolute velocity of the steam entering the blades minus the rotational speed of the blades. The relative velocity of the steam exiting the blades must also equal the absolute velocity of the steam entering the blades minus the rotational speed of the blades. Theoretically, the steam velocity does not change as it flows across the blades.

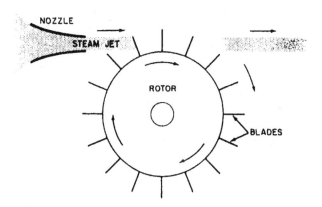

Figure 12-3.—Elementary impulse turbine.

For all the kinetic energy to be converted into work, a curved impulse blade (fig. 12-4) would be required. This blade would be designed so that the steam's exit velocity is zero. It would be curved so that steam enters the blade tangentially rather than at an angle. (In real turbines, the steam enters the blades at an angle rather than tangentially.)

When a curved blade is used, the direction of the steam flow is reversed. Maximum work is obtained from a reversing blade when the blade velocity is exactly one-half the absolute velocity of the steam at the blade entrance. (This statement assumes that the nozzle is tangential to the blades. In actual impulse turbines, maximum amount of work is obtained when the blade speed is one-half the cosine of the nozzle angle times the absolute velocity of the entering steam.) The maximum amount of work obtainable from a reversing blade is twice the amount obtainable from a flat vane.

In actual turbines, it is not feasible to use the complete reversal of steam in the blades, since to do so would require that the nozzle be placed in a position that would also be swept by the

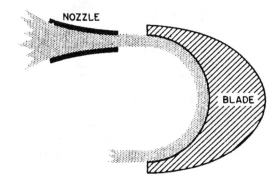

Figure 12-4.—Curved impulse blade.

12-3

blades—an obvious impossibility. Furthermore, if the steam entered the blade tangentially, it would not be carried through the turbine axially (longitudinally). However, it is only the tangential component of the steam velocity that produces work on the turbine blades; hence, the nozzle angle is made as small as possible.

REACTION TURBINES

Reaction turbines are moved by a reactive force rather than by a direct push or impulse. The first reaction turbine—and perhaps the first steam engine of any kind ever made—was developed by the Greek mathematician Hero about 2000 years ago. This turbine, shown in figure 12-5, consisted of a hollow sphere that carried four bent nozzles. The sphere was free to rotate on the tubes that carried steam from the boiler, below, to the sphere. As the steam flowed out through the nozzles, the sphere rotated rapidly in a direction opposite to the direction of steam flow.

Reaction turbines used in modern times use the reactive force of steam in quite a different way. There are no nozzles as such. Instead, the blades that project radially from the periphery of the rotor are formed and mounted in such a way that the spaces between the blades have, in cross section, the shape of nozzles. (The distinction between actual nozzles and the blading which serves the purpose of the nozzles in reaction

Figure 12-5.—Hero's steam turbine.

turbines is mechanical rather than functional. The previous discussion of steam flow through nozzles applies equally well to steam flow through the nozzle-shaped spaces between the blades of reaction turbines.) Since these blades are mounted on the revolving rotor, they are called moving blades.

Fixed or stationary blades of the same shape as the moving blades are fastened to the casing in which the rotor revolves. These fixed blades are installed between successive rows of moving blades. The fixed blades guide the steam into the moving blade system. Since the fixed blades are also shaped and mounted in such a way as to provide nozzle-shaped spaces between the blades, they also act as nozzles.

A reaction turbine is moved by three main forces: (1) the reactive force produced on the moving blades as the steam increases in velocity as it expands through the nozzle-shaped spaces between the blades, (2) the reactive force produced on the moving blades when the steam changes direction, and (3) the push or impulse of the steam impinging upon the blades. A reaction turbine is moved primarily by a reactive force but also to some extent by a direct impulse.

From what we have learned about the function of nozzles, it is apparent that thermal energy is converted into mechanical kinetic energy in the blading of a reaction turbine. The second required energy transformation—that is, from kinetic energy to work—also occurs in the blading.

Since the velocity of the steam is increased in the expansion through the moving blades, the initial velocity of the entering steam must be lower in a reaction turbine than it would be in an impulse turbine with the same speed; or, alternatively, the reaction turbine must run at a higher speed than a comparable impulse turbine in order to operate at approximately the same efficiency.

TURBINE CLASSIFICATION

Thus far, we have classified turbines into two general groups—impulse turbines and reaction turbines—depending on the method used to do useful work. Turbines may be further classified according to the following:

- Type and arrangement of staging

- Direction of steam flow

- Repetition of steam flow

- Division of steam flow

A turbine may also be classified as to whether it is a condensing unit (exhausts to a condenser at a pressure below atmospheric pressure) or a noncondensing unit (exhausts to another system such as the auxiliary exhaust steam system at a pressure above atmospheric pressure).

STAGING ARRANGEMENTS

An impulse turbine stage consists of one set of nozzles and the succeeding row or rows of either fixed or moving blades. Fixed blades in an impulse turbine do nothing more than redirect the steam flow from one row of moving blades to the next. An impulse stage has only one pressure drop, since a pressure drop will occur only in a nozzle. A simple-impulse stage (one set of nozzles and one row of moving blades) is commonly referred to as a RATEAU stage.

A reaction turbine stage consists of one row of fixed blades and a succeeding row of moving blades. Since the fixed blades in a reaction turbine are comparable to the nozzles in an impulse turbine, this definition of a reaction stage seems very similar to the definition of an impulse turbine stage. However, there is an important difference: a reaction stage has two pressure drops, whereas an impulse stage has only one.

To efficiently use the energy of the steam, a turbine must normally have blading and nozzles that will cause more than one pressure drop and/or more than one velocity drop as the steam passes through it. This is called compounding. A turbine that has more than one velocity drop is classified as velocity compounded, and a turbine that has more than one pressure drop is classified as pressure compounded. A combination of pressure and velocity drops is called pressure velocity compounding.

From our definition of stages, we can see that an impulse stage may be velocity compounded but is never pressure compounded; and a reaction stage is always pressure compounded but never velocity compounded. This applies only to stages and NOT to the turbines. Turbines may be designed and constructed to incorporate practically any combination of stages.

Velocity-Compounded Impulse Turbine

A velocity drop in an impulse turbine occurs only in the moving blades; therefore, to obtain more than one velocity drop across an impulse turbine, there must be more than one row of moving blades. (Velocity compounding can also be achieved when only one row of moving blades is used, provided the steam is directed in such a way that it passes through the blades more than once. This point will be explained further in our discussion on the direction of steam flow.) Figure 12-6 shows a velocity-compounded impulse turbine that has two rows of moving blades. (**NOTE:** Two sectional views of the same blading are shown.) This type of arrangement is called a Curtis stage, and, of course, two velocity drops occur across it.

Pressure-Compounded Impulse Turbine

Another method to increase the efficiency of an impulse turbine is to arrange one or more simple-impulse stages in a row. This combination produces a pressure-compounded impulse turbine that has as many pressure drops as it has stages.

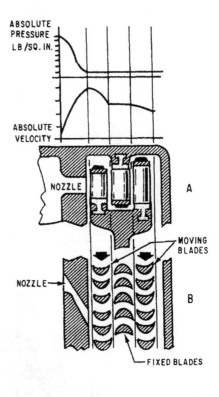

Figure 12-6.—Velocity-compounded impulse turbine, showing pressure-velocity relationship.

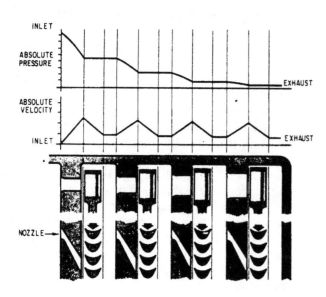

Figure 12-7.—Pressure-compounded impulse turbine, showing pressure-velocity relationship.

Figure 12-7 shows a pressure-compounded impulse turbine.

Pressure-Velocity-Compounded Impulse Turbine

A turbine that has one velocity-compounded stage followed by a series of simple-impulse stages is a pressure-velocity-compounded turbine. This type of turbine (fig. 12-8) is commonly used for ship propulsion. (**NOTE**: Only the upper half of the rotor and blading is shown.)

Pressure-Compounded Reaction Turbine

Because the ideal blade speed in a reaction turbine is so high in relation to the velocity of the entering steam, all reaction turbines are pressure compounded. This means they are arranged so the pressure drop across the turbine from inlet to exhaust is divided into many steps by alternate rows of fixed and moving blades (fig. 12-9). (Note the small curved arrows between the blades and the casing and the stator blades and rotor. The arrows indicate that a small amount of steam leaks around these areas and does no work.) The pressure drop in each set of fixed and moving blades is therefore small, thus causing a lowered steam velocity in all stages and consequently a lowered ideal blade velocity for the turbine as a whole.

Combination Impulse and Reaction Turbine

Another type of turbine is a combination impulse and reaction turbine (fig. 12-10). This

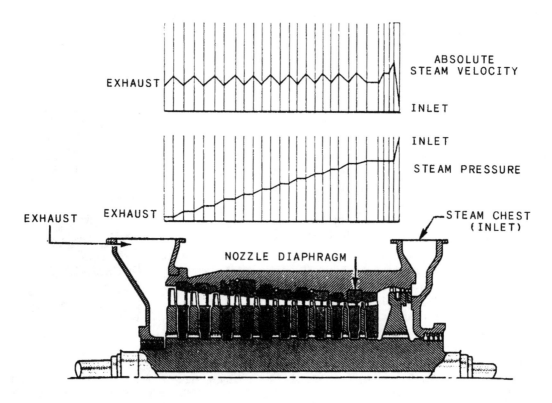

Figure 12-8.—Pressure-velocity-compounded impulse turbine, showing pressure-velocity relationship.

12-6

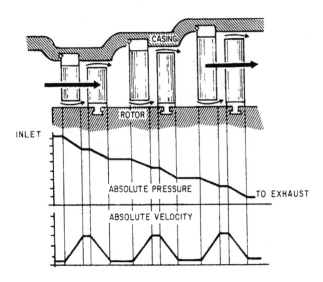

Figure 12-9.—Pressure-compounded reaction blading, showing pressure-velocity relationship.

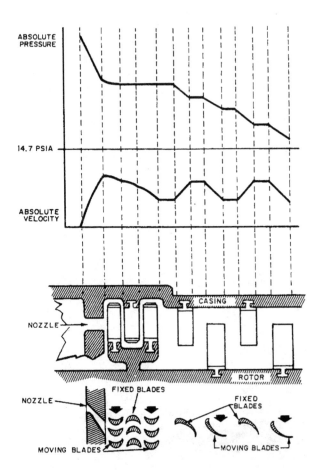

Figure 12-10.—Combination impulse and reaction type turbine, showing pressure-velocity relationship.

turbine design uses a velocity-compounded impulse stage at the high-pressure end of the turbine followed by reaction blading. The impulse blading causes a large pressure and temperature drop at the beginning, which uses a large amount of thermal energy. The reaction part of the turbine is more efficient at the lower steam pressure existing at the low-pressure end of the turbine. Therefore, you can see that for certain applications, this type of turbine would be a highly efficient machine with the advantages of both impulse and reaction blading. (A combination turbine of this type is commonly called a modified Parson's turbine.) Combination impulse and reaction turbines are commonly used as propulsion turbines.

DIRECTION OF STEAM FLOW

The direction of steam flow through a turbine may be axial, radial, or helical. In general, the direction of flow is determined by the relative positions of nozzles, diaphragms, moving blades, and fixed blades.

Axial-Flow Turbines

Most turbines are the axial-flow type. This means that the steam flows in a direction almost parallel to the axis of the turbine shaft. The blades in an axial-flow turbine project outward from the periphery of the rotor.

Radial-Flow Turbines

In radial-flow turbines, the blades are mounted on the side of the rotor near the periphery. The steam enters in such a way that it flows radially toward or away from the long axis of the shaft. Radial-flow turbines are not used for propulsion turbines but are used for some auxiliary turbines.

Helical-Flow Turbines

In a helical-flow turbine, the steam enters at a tangent to the periphery of the rotor and impinges upon the moving blades. The blades are shaped so that the direction of the steam flow is reversed in each blade. Helical-flow turbines are also not used for propulsion turbines but are used for some auxiliary turbines.

NOTE: Radial-flow and helical-flow turbines are discussed further in chapter 14.

REPETITION OF STEAM FLOW

Turbines are classified as single-entry turbines or re-entry turbines, depending on the number of times the steam enters the blades. If the steam passes through the blades only once, the turbine is called a single-entry turbine. If the steam passes through the blades more than once, then it is called a re-entry turbine. Re-entry turbines are used on some auxiliary turbines.

DIVISION OF STEAM FLOW

Turbines are also classified as single-flow or double-flow, depending upon whether the steam flows in one direction or two. In a single-flow turbine (fig. 12-8), the steam enters at the inlet or throttle end, flows through the blading in a more or less axial direction, and emerges at the exhaust end of the turbine. A double-flow turbine consists essentially of two single-flow units mounted on one shaft, in the same casing. The steam enters at the center, between the two units, and flows from the center toward each end of the shaft. The main advantages of the double-flow arrangement are (1) the blades can be shorter than they would have to be in a single-flow turbine of equal capacity, and (2) axial thrust is avoided by having the steam flow in opposite directions. This second point applies primarily to reaction turbines, since impulse turbines develop relatively little thrust in any case. A double-flow reaction turbine, used as a low-pressure (LP) turbine in most propulsion plants, is shown in figure 12-11.

TURBINE COMPONENTS AND ACCESSORIES

Propulsion turbine components and accessories include foundations, casings, nozzles (or equivalent stationary blading), nozzle diaphragms, rotors, blades, turbine materials, shrouding, bearings, shaft packing glands, gland seals, flexible couplings, reduction gears, lubrication systems, and turning gears. The following paragraphs will discuss these components and accessories.

TURBINE FOUNDATIONS

Foundations for propulsion turbines are built up from strength members of the hull to provide a rigid supporting base. The after end of the turbine is secured rigidly to the structural foundation. The forward end of the turbine is secured in such a way as to allow a slight freedom of axial movement that allows the turbine to expand and contract slightly with temperature changes.

The freedom of movement at the forward end is accompanied by one of two methods. Elongated bolt holes or grooved sliding seats may be used

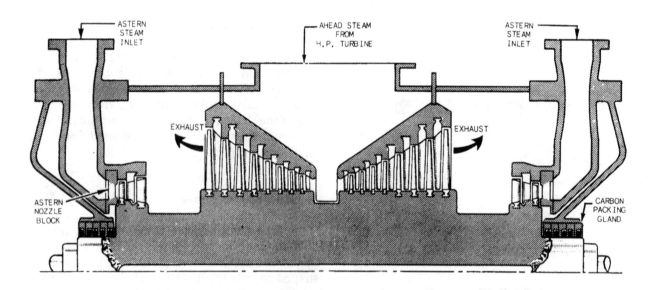

Figure 12-11.—Low-pressure double-axial flow reaction turbine.

12-8

to permit the forward end to slide slightly fore-and-aft, as expansions and contractions occur; or the forward end may be secured to a deep flexible I-beam (fig. 12-12) installed with its longitudinal axis lying athwartship. When the turbine is cold, this I-beam is deflected slightly aft from the vertical position. When the turbine is operating at maximum power, the turbine is deflected forward. This arrangement results in minimum stresses in the I-beam over the complete range of turbine expansion. The fixed end of the turbine is aft so the motion resulting from the expansion cannot be transmitted to the reduction gears, where distortion and serious damage would occur if the after end of the turbine were free to move.

Steam lines connected to turbines are curved or looped in some way (see fig. 12-12) to allow for the expansion of the steam line and to avoid unacceptable strains on the turbine casings that could cause distortion or misalignment.

There are modifications to these two basic methods. In many ships, the main condenser is supported by the LP turbine, which is rigidly mounted on beams that form an integral part of the hull structure. In other ships, the LP turbine may be mounted on the condenser, which, in turn, forms an integral part of the hull. In general, the expansion arrangement of the LP turbine is similar to that of the high-pressure (HP) turbine except that the flexible forward support is seldom used. Instead, some arrangements use keys at the center of the low-pressure casing as the fixed point, with the expansions occurring equally in the forward and aft directions from the keys. Some arrangements fix the after end of the casing

by fitted bolts, while the forward end uses clearance bolts to allow for the expansion motion. Another arrangement uses fitted bolts at each end; these bolts fixed the ends to flexible twin inlet necks on the condenser.

TURBINE CASINGS

Casings for propulsion turbines are divided horizontally to permit access for inspection and repair. Flanged joints on casings are accurately machined to make a steamtight metal-to-metal fit, and the flanges are bolted together. Some HP turbine casings are also split vertically to facilitate manufacture, particularly when different alloys are used for the high-temperature inlet and lower-temperature exhaust end. However, these vertical joints are never unbolted and they are usually seal welded.

Each casing has a steam chest to receive the incoming steam and deliver it to the first-stage nozzles or blades. An exhaust chamber receives the steam from the last row of moving blades and delivers it to the exhaust connection. Openings in the casing include drain connections, steam bypass connections, and openings for the pressure gauges, thermometers, and relief valves.

Turbine casings are fitted with external and internal openings to drain water from low spots into appropriate drain systems or to the main condenser. Proper drainage at light-off is important to dispose of water accumulated during shutdown and water produced when initially introduced steam condenses on the cold metal surfaces. Inadequate drainage could produce local chilling of the rotor and roughness on turbine start-up or damage caused by standing or solid water striking the rotor. Opening the drains during the securing procedure is important to drain or dry out turbines properly and to avoid corrosion of susceptible steel parts during idle periods.

NOZZLES

As previously discussed, the function of the nozzle is to convert the thermal energy of the steam into mechanical kinetic energy. Its secondary function is to direct the steam to the turbine blades. Some turbines have a full arc admission of steam; in this case, the first-stage nozzles extend around the entire circle of the first row of blades. Other turbines have partial arc admission; in this case, only a section of the blade circle is covered by nozzles. In general, the arrangement of nozzles in any turbine depends

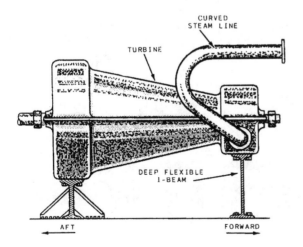

Figure 12-12.—Foundation for a propulsion turbine.

upon the range of power requirements and upon a number of design factors.

A nozzle is essentially an opening or a passageway for the steam. In terms of nozzle construction or arrangement, the actual concern is with the construction or arrangement of the nozzle blocks in which the openings occur. In most turbines, the nozzle blocks are arranged so that the nozzle openings occur in groups, with each group being controlled by a separate nozzle control valve. The quantity of steam delivered to the first stage of the turbine is thus a function of the number of nozzles in use and the pressure differential across the nozzles.

Any throttling of the inlet steam will reduce efficiency. To avoid throttling losses, all nozzle control valves in use are opened fully before any additional valve is opened. Minor variations in speed within any one nozzle control valve combination are taken care of by the throttle.

Most ships employ the nozzle control valve arrangement shown in figure 12-13. The throttle valve is omitted, and steam enters the turbine through nozzle control valves. Speed control is effected by varying the number of nozzle valves that are opened. The variation in the number of nozzle valves is accomplished through the operation of a lifting beam mechanism which consists of a steel beam drilled with holes, which fits over the nozzle valve stems. The valve stems are of varying lengths and are fitted with shoulders at the upper ends. When the beam is lowered, all valves rest upon their seats. When the beam is raised, the valves open in succession, depending upon their stem length—the shorter ones open first, then the longer ones.

NOZZLE DIAPHRAGMS

Nozzle diaphragms (fig. 12-14) are installed in front of the rotating blades for each stage of

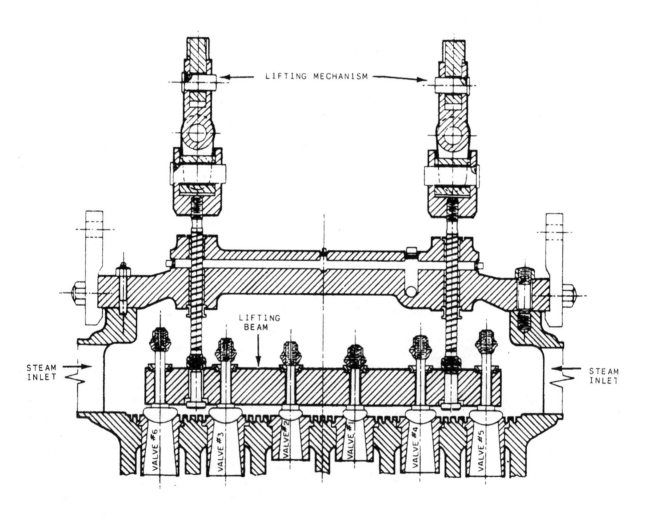

Figure 12-13.—Arrangement of nozzle control valves.

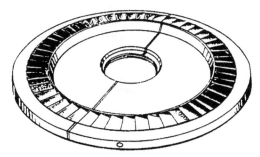

NOZZLE DIAPHRAGM

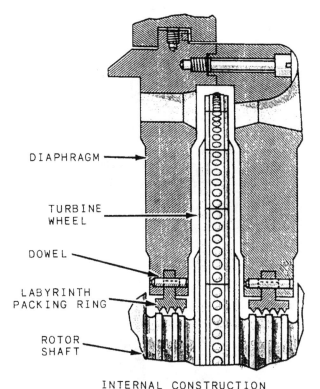

DIAPHRAGM

TURBINE
WHEEL

DOWEL

LABYRINTH
PACKING RING

ROTOR
SHAFT

INTERNAL CONSTRUCTION

Figure 12-14.—Turbine nozzle diaphragms.

a pressure-compounded impulse turbine. The diaphragms contain the nozzles which admit the steam to the rotating blades of each in much the same way as the nozzle groups of the first stage. Partial admission diaphragms admit steam in only a section of the blade circle. Full admission diaphragms have nozzles extending around the entire circle of blades. Because of the pressure drop that exists across each diaphragm, a labyrinth packing ring is placed in a groove in the inner periphery of the diaphragm to minimize the leakage of steam across the diaphragm and along the rotor. Any leakage through the inner periphery of the diaphragm will reduce the amount of the steam thermal energy being converted to mechanical energy, thus reducing the work developed by the stage.

TURBINE ROTORS

The turbine rotor carries the moving blades that receive the steam. In most turbines, particularly those used for ship propulsion, the rotors are forged integrally with the shaft. Figure 12-15 shows an integrally forged turbine rotor to which the blades have not yet been attached.

TURBINE BLADES

The purpose and function of turbine blading has already been discussed. At this point, it is merely necessary to note that the moving blades are fastened securely and rigidly to the turbine

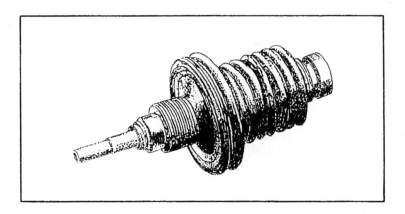

Figure 12-15.—Integrally forged turbine rotor.

rotor. Figure 12-16 shows several ways of fastening blades to turbine rotor wheels.

TURBINE MATERIALS

The materials used to construct turbines will vary somewhat depending on the steam and power conditions for which the turbine is designed.

Turbine casings are generally made of cast carbon steel for nonsuperheated steam applications. Superheated applications use casings made of carbon molybdenum steel. On some submarines, the turbine casings are made of 12 percent chrome stainless steel, which is more resistant to steam erosion than carbon steel.

Turbine rotors (forged wheel and shaft) are made of carbon steel for low-temperature steam (less than 650°F). High-temperature steam designs use carbon molybdenum or some other creep-resistant alloy.

Turbine blades are generally constructed of a corrosion-resistant alloy for a hard, erosion- and corrosion-resistant blade.

TURBINE SHROUDING

Shrouding is a thin metal strip covering the ends of the turbine blades. It is installed to reduce vibration stress, which will cause turbine damage. Shrouding is installed by fitting it over tenons located on the end of each blade. It is then riveted in place.

TURBINE BEARINGS

Turbine rotors are supported and kept in position by bearings. (Bearings are discussed in chapter 5.) The bearings that maintain the correct radial clearance between the rotor and the casing are called radial bearings. Those that limit the axial (longitudinal) movement of the rotor are called thrust bearings.

Propulsion turbines have one radial bearing at each end of the rotor. These bearings are generally known as journal bearings or sleeve bearings. The two metallic surfaces are separated only by a fluid film of oil. The effectiveness of oil-film lubrication depends upon a number of factors, including properties of the lubricant (cohesion, adhesion, viscosity, temperature, and so on) and the clearances, alignment, and surface conditions of the bearing and the journal. Except for the momentary metal-to-metal contact when the turbine is started, the metallic surfaces of the bearing and the journal are constantly separated

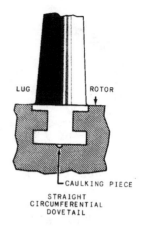

STRAIGHT
CIRCUMFERENTIAL
DOVETAIL

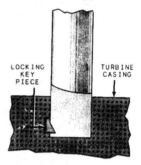

SIDE-LOCKING KEY PIECE

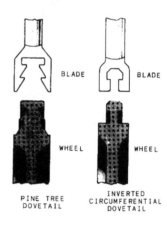

PINE TREE
DOVETAIL

INVERTED
CIRCUMFERENTIAL
DOVETAIL

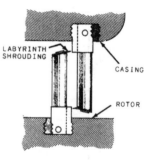

SAWTOOTH SERRATION

Figure 12-16.—Methods of fastening blades to turbine rotor wheels.

by a thin film of oil. (Fluid oil-film lubrication was explained in chapter 6.)

As previously noted, impulse turbines do not, in theory, develop end thrust. In reality, however, a small amount of end thrust is developed which must be absorbed in some way. Kingsbury or pivoted-shoe thrust bearings are usually used on propulsion turbines.

SHAFT PACKING GLANDS

When the pressure inside the turbine casing is greater than atmospheric pressure, shaft packing glands are used to prevent the escape of steam from the casing. The packing glands also prevent air from entering the turbine when pressure within the casing is below atmospheric pressure.

Three types of shaft glands are used on naval turbines: labyrinth packing glands, carbon packing glands, or a combination of both labyrinth and carbon packing glands. Since the labyrinth packing gland is the most widely used method of sealing a turbine shaft in naval turbines, it is the only one that we will discuss here.

Labyrinth packing is used in the glands and interstages of steam turbines. Labyrinth packing consists of machine packing strips for fins mounted on the casing surrounding the shaft to make a very small clearance between the shaft and the strip. Figure 12-17 shows a labyrinth packing gland. The principle of labyrinth packing seals is that as steam leaks through the very narrow spaces between the packing strips and the shaft, the steam pressure drops. As the steam passes from one packing strip to the next, its pressure is gradually reduced and the velocity it might gain through the nozzling effect is lost by the action of the steam as it ricochets back and forth in the gland.

GLAND SEAL

Packing alone will neither stop the flow of steam from the turbine nor prevent the flow of air into the turbine. Gland sealing steam is used to prevent the entrance of outside air into the turbine, which would reduce or destroy the vacuum in the main condenser. Figure 12-17 shows how gland sealing steam of approximately 2 psig (17 psia) is led into a space between two sets of gland packing. (In older ships, steam is supplied from the auxiliary exhaust steam line through a stop valve.) Two weight-loaded valves

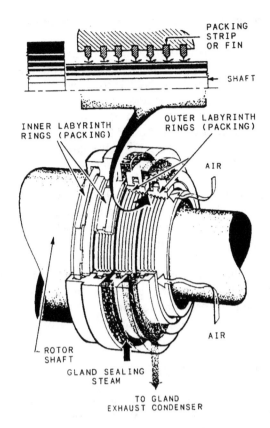

Figure 12-17.—Turbine gland.

operate automatically to maintain a pressure of 1/2 to 2 psig on the glands.

During periods of warming-up, low-speed operation, backdown, and securing, the weight-loaded valves are open. When the turbine is speeded up and the steam flowing from the labyrinth gland of the high-pressure turbine reaches 2 psig, the valve supplying the high-pressure turbine closes automatically, securing the gland sealing steam from the auxiliary exhaust line. The high-pressure turbine gland is then supplying enough steam leakage to seal the low-pressure glands. As the turbine speed is increased further and the pressure in the sealing system rises to approximately 2 1/2 psig, the excess steam is led to the main condenser or to the low-pressure turbine by a manually operated or automatically operated valve.

The leak-off connections are linked to the gland leak-off piping, which collects and condenses the steam that leaks from the gland, thus preventing the escape and loss of steam to the atmosphere.

A fan-type gland seal exhauster puts a slight vacuum on the leak-off piping. The vacuum draws the leak-off steam through a gland exhaust condenser where the steam is condensed and returned to the feed system.

Steam from the 150-psig steam system is admitted to the gland sealing system through an air-operated supply valve. If the pressure in the system becomes excessive, as in high-power operation, excess steam is "dumped" to the condenser through the LP turbine. This is done by another air-operated valve called an excess valve or an excess steam unloading valve. The supply and excess valves both have pilot actuators that sense the pressure in the gland seal steam piping. This system may be used on main propulsion units or steam turbine generators.

Another type of gland seal supply and unloading regulator is the hydraulic type. The hydraulic type senses the gland seal steam pressure in a bellows assembly and uses lube-oil pressure and spring pressure to control the opening and closing of the supply and exhaust (unloading) valves.

FLEXIBLE COUPLINGS

Propulsion turbine shafts are connected to the reduction gears by flexible couplings that are designed to take care of very slight misalignment between the two units. (Flexible couplings were discussed in chapter 5.)

REDUCTION GEARS

Reduction gears for propulsion turbine installations were described and illustrated in chapter 5. At this point, it is important to note that turbines must operate at relatively high speeds for maximum efficiency while propellers must operate at lower speeds for maximum efficiency. Reduction gears are used to allow both turbines and propellers to operate within their most efficient rpm ranges.

LUBRICATION SYSTEMS

Proper lubrication is essential for the operation of any rotating machinery. In particular, the bearings and the reduction gears of turbine installations must be well lubricated at all times. Main lubricating-oil systems were discussed in chapter 9; and the theory of lubrication was discussed in chapter 6.

TURNING GEARS

All geared turbine installations are equipped with a motor-driven jacking or turning gear. The unit is used for turning the turbine during warm-up and securing periods so that the turbine rotor will heat and cool evenly. The rotor of a hot turbine, or one that is in the process of being warmed up, will become bowed and distorted if left stationary for even a few minutes. The turning gear is also used for turning the turbine to bring the reduction gear teeth into view for routine inspection and for making the required daily jacking of the main turbines. The turning gear is mounted on the top of and at the after end of the reduction gear casing as shown in figure 12-18. The brake shown in figure 12-18 is used when it is necessary to lock the shaft after it has been stopped.

TURBINE CONTROL

Turbines operate at different speeds. Therefore, they must have a method of steam inlet control to allow fairly small changes of turbine speed. A system of nozzle control valves performs this function. Although the turbine speed control system aboard your ship may differ from those described here, the basic principles involved are the same as those discussed in the following paragraphs.

NOZZLE CONTROL VALVES

One type of nozzle control valve system is shown in figure 12-13. This type of system controls speed by varying the number of nozzle valves that are opened. A lifting beam mechanism, drilled with holes, fits over the nozzle valve stems. These stems are of varying lengths and are fitted with shoulders or "buttons" at the upper ends. The nozzle valves are commonly referred to as poppets. (A poppet valve has a disc that moves axially, or parallel, with the flow of fluid through the valve.) When the beam is lowered, all valves rest upon their seats. The individual poppets are held tightly shut against their seats by steam pressure. The mushroom shape of the poppets gives the steam pressure adequate surface area to "hold down" the valves. When the beam is raised, the valves open in succession; the shorter ones open first, then the longer ones. Several types of valve linkage arrangements are used to change throttle valve handwheel rotary motion into lifting motion to raise and lower the beam.

Another nozzle control valve arrangement is shown in figure 12-19. In this type, five valves are used to admit steam to the HP turbine. The valves

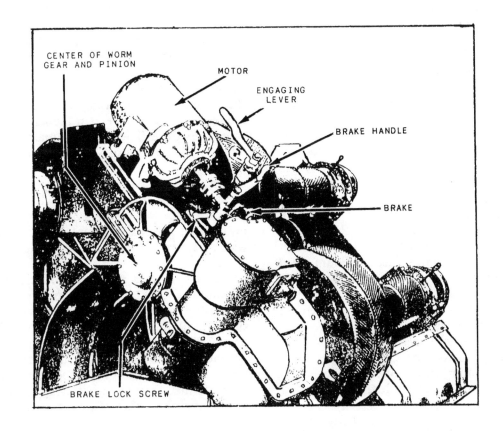

Figure 12-18.—After end of main reduction gear, showing turning gear and propeller locking mechanism.

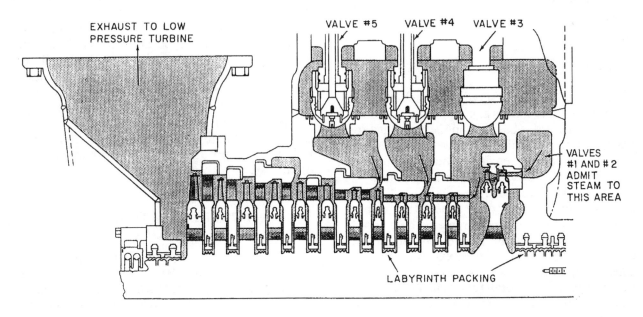

Figure 12-19.—High-pressure turbine showing valve arrangement.

are controlled by a camshaft rotated by a throttle handwheel. The camshaft is arranged to open the valves in succession. Valves No. 1 and No. 2 admit steam to the first-stage chest of the turbine. With valves No. 1 and No. 2 fully open, the pressure in the steam chest is approximately equal to line pressure. The first-stage nozzles pass steam to the first stage at the limit of their capacity. To further increase the power and speed of the turbine, you must increase the amount of steam flowing through the turbine. Valve No. 3 is then opened, admitting steam from the chest through a bypass valve to the second-stage nozzles. With valve No. 3 fully open, it is necessary to open valve No. 4, which admits steam from the chest through a bypass into the fourth-stage nozzles. With valve No. 4 fully open, the turbine should operate at almost its full power rpm. To get maximum speed and power, valve No. 5 must be opened to admit steam to the sixth-stage nozzles.

In some ships, as the handwheel is rotated, an indicator on the throttle valve mounting shows which valves are open. Valve No. 1 begins to open as soon as the indicator moves away from the shut position. The other valves open at the points indicated as the handwheel is turned counterclockwise. When the indicator reaches the open position, all valves are fully open. Turning the handwheel in the clockwise direction closes the valves in the reverse order.

The cam-operated nozzle valve system may also be used when the bypassing of turbine stages does not occur.

CRUISING ARRANGEMENTS

Every ship has a speed at which the fuel consumption per mile is at a minimum. The most economical speed is based upon the combined fuel consumption per shaft horsepower (shp) per hour of the propulsion engines and the auxiliary machinery. However, a considerable quantity of fuel is required to generate steam used by the auxiliary machinery, even when the ship is stopped. As the speed is increased, there is only a very gradual increase in the amount of fuel consumed (per hour) by the auxiliary machinery. Therefore, as the ship's speed is increased, the percentage of the total fuel consumed by the auxiliary machinery actually decreases. For turbine-driven ships, these varying rates of fuel consumption for the auxiliary machinery result in a most economical speed of between 12 and 20 knots, depending on the type of ship.

To conserve fuel and thereby increase the ship's cruising radius, a major part of all steaming is done close to the most economical speed. Therefore, the most economical speed is designated as the cruising speed. The economical speed should be as high as possible. However, there is a practical limit for this speed because of the progressively increasing resistance of water to the ship's hull as the speed of the ship is increased. Increasing the efficiency of the main engines will tend to raise the most economical speed. The propulsion turbines are designated to have their best steam rate at the cruising speed.

Combatant ships should be able to steam at or near full power for long periods of time. This means that propulsion plants must be designed with a relatively high turbine efficiency at high speeds.

A turbine gets maximum efficiency at the optimum ratio of blade speed to steam speed. To obtain the lowest possible fuel consumption per shp per hour at cruising speed AND at full power, propulsion turbines must be designed so the optimum ratio of blade speed to steam speed will be approached at both these speeds. This may be accomplished by using cruising stages: HP, intermediate-pressure (IP), and LP turbine combinations.

ASTERN OPERATION (REVERSING)

Astern elements are provided with steam propulsion turbine units for emergency stopping, backing, and maneuvering. In units using LP turbines (all except type I), there is an astern element in each exhaust end of the double-flow, LP turbine. The single casing turbine units have an astern element in the exhaust end of each turbine. Astern elements are usually velocity-compounded impulse stages, which develop high torque but have low efficiency. An astern turbine is designed to produce from one-fifth to one-half of the ahead turbine full power.

PROPULSION TURBINE CLASSIFICATION

Naval propulsion turbines are classified as class A, class B, and class C turbines according to the type of ship for which they are designed. Class A turbines are designed for use in submarines. Class B turbines are designed for use in amphibious warfare ships, surface combatant ships, mine warfare ships, and patrol ships. Class C turbines are designed for use in auxiliary ships.

Naval propulsion turbines are also classified according to design features. The five major types are as follows:

- Type I (single-casing unit)—The type I propulsion unit consists of one or more ahead elements, each contained in a separate casing and identified as a single casing turbine. Each turbine delivers approximately equal power to a reduction gear. Submarines normally have two type I propulsion turbines operating in parallel with each turbine receiving steam from the main steam line through their own throttle valve and then discharging to a condenser.

- Type II-A (straight-through unit)—The type II-A propulsion unit is a two-element straight-through unit, which consists of two ahead elements, known as an HP element and an LP element. The HP and LP elements are contained in a separate casing and are commonly known as the HP and LP turbines, respectively. The HP and LP turbines deliver power to a single shaft through a gear train and are coupled separately to the reduction gear. Steam is admitted to the HP turbine and flows straight through the turbine axially without bypassing any stage (there is partial bypassing of the first row of blades at high power), and then is exhausted to the LP turbine through a crossover pipe.

- Type II-B (external bypass unit)—The type II-B propulsion unit is similar to the type II-A, except that provision is made for bypassing steam around the first stage or first several stages of the HP turbine at powers above the most economical point of operation. Bypass valves are located in the HP turbine steam chest, with the nozzle control valves.

- Type II-C (internal bypass unit)—The type II-C propulsion unit is also similar to the type II-A, except that provision is made for bypassing steam from the first-stage shell around the next several (one or more) stages of the HP turbine at powers above the most economical point of operation. Bypass valves and steam connections are usually integral with the HP turbine casing; however, some installations have the valves separate, but bolted directly to the casing, with suitable connecting piping between the first-stage shell and valve to the bypass belt.

- Type III (series-parallel unit)—The type III propulsion unit consists of three ahead elements, known as the HP element, IP element, and LP element. The HP and IP elements are combined in a single casing, which is known as the HP-IP turbine. Steam is admitted to the HP-IP turbine and exhausted to the LP turbine through a crossover pipe. For powers up to the most economical point of operation, only the HP element receives inlet steam, with the IP element being supplied in series with the steam from the HP element exhaust. At powers above this point of operation, both elements receive inlet steam in a manner similar to that in a double-flow turbine. During ahead operation, no blading is bypassed. Series-parallel units are used on some of the more recent naval combatant vessels.

All of these types of propulsion turbines have an astern element for backing or reversing. An astern element is located in each end of a double-flow LP turbine casing or can be in either end of each single-casing turbine or single-flow LP turbine.

PROPULSION PLANT OPERATIONS

Operating a ship's propulsion plant requires sound administrative procedures and the cooperation of all engineering department personnel. The reliability and economical operation of the plant is vital to the ship's operational readiness.

A ship must be capable of performing any duty for which it is designed. A ship is considered reliable when it meets all scheduled operations and is in a position to accept unscheduled tasks. To do this, the ship's machinery must be kept in good condition so that the various units will operate as designed.

To obtain economy, the engineering plant, while meeting prescribed requirements, must be operated so as to use a minimum amount of fuel. The fuel performance ratios are good overall indications of the condition of the engineering plant and efficiency of the operating personnel. Fuel performance ratio is the ratio of the amount of fuel used as compared to the amount of fuel allowed for a certain speed or steaming condition. Ship's routinely conduct economy and full power trials to measure and report this performance. In determining the efficiency of a ship's engineering plant, the same consideration is given to the amount of water used on board a ship. Water consumption is computed in (1) gallons of makeup feed per engine mile, (2) gallons of makeup feed per hour at anchor, and (3) gallons of potable water per person per day.

Good engineering practices and safe operation of the plant should never be violated in the interest of economy; furthermore, factors affecting the health and comfort of the crew should meet the standards set by the Navy.

Indoctrination of the ship's crew in methods on conserving water is of the utmost importance and should be given constant consideration.

Aboard naval ships, economy measures cannot be carried to extremes, because several safety factors must be considered. Unless proper safety precautions are taken, reliability may be sacrificed; and in the operation of naval ships, reliability is one of the more important factors. In operating an engineering plant as economically as possible, safety factors and good engineering practice must not be overlooked.

There are several factors that, if given proper consideration, will promote efficient and economical operation of the engineering plant. Some of these factors are (1) maintaining the designed steam pressure, (2) maintaining proper acceleration of the main engines, (3) maintaining high main condenser vacuum, (4) guarding against excessive recirculation of condensate, (5) maintenance of proper insulation and lagging, (6) keeping the consumption of feedwater and potable water within reasonable limits, (7) conserving electrical power, (8) using the correct number of boilers for the best efficiency at the required load levels, and (9) maintaining minimum excess combustion air to the boilers.

MAINTAINING A CONSTANT STEAM PRESSURE is important to the overall efficiency of the engineering plant. Wide or frequent steam pressure fluctuations will affect boiler performance, which could result in considerable loss of economy.

PROPER ACCELERATION AND DE-CELERATION OF THE MAIN ENGINES are important factors in the economical operation of the engineering plant. A fast acceleration will not only interfere with the safe operation of the boilers but will also result in a large waste of fuel oil. The machinist's mate in charge of an engine-room watch, or standing throttle watch, can contribute significantly to the economical and safe operation of the boilers.

A HIGH CONDENSER VACUUM can be obtained only by the proper operation and proper maintenance of the condenser. A low exhaust pressure (high vacuum) is an important factor in obtaining maximum engineering efficiency. Steam exhausting into a low-pressure area has a greater range of expansion and, therefore, is capable of accomplishing more useful work. The total available energy on the steam is much higher per pound of pressure difference in the lower range than in the upper range. This is the most important reason why the condenser vacuum should be maintained as high as possible.

EXCESSIVE RECIRCULATION OF CON-DENSATE should be avoided as it cools the condensate, which then has to be reheated as it enters the deaerating feed tank (DFT). This reheating process causes an excessive amount of steam to be used to maintain the proper temperature in the DFT.

MAINTENANCE OF PROPER INSULA-TION AND LAGGING not only increases the overall economy of the engineering plant but also is a safety measure and increases the comfort of personnel. In every power plant, there is a heat loss as heat flows from heated surfaces, such as piping and machinery, to the surrounding air and cooler objects. This heat loss can be kept to a minimum by proper insulation.

While increasing the economy of the plant, insulation also reduces the quantity of air necessary for ventilating and cooling the space. Proper insulation also reduces the danger of personnel receiving burns from contact with the hot parts of the piping, valves, and machinery. Good insulation, elimination of steam leaks, and a clean ventilation system contribute to good economy and to the comfort and safety of personnel.

CONSERVATION OF FEEDWATER AND POTABLE WATER has a direct bearing on the overall efficiency and economy of the ship. Feedwater and potable water consumption rates are entered in the fuel and water report. Type commanders use these consumption rates as a factor for judging the efficiency of ships operating under their command. Ships having excessive feedwater consumption should take immediate steps to eliminate all steam and water leaks, which contribute to the uneconomical operation of the plant. Improving feedwater consumption rates will also improve fuel-oil performance ratio.

The consumption of potable water by the ship's crew bears a direct relationship to the efficient operation of the engineering plant; the greater the amount of fresh water distilled, the greater the amount of steam used. Conservation of fresh water requires the close cooperation of all personnel aboard ship, since large amounts may be wasted by improper use of laundry, scullery, galley, and showers.

FAILURE TO CONSERVE ELECTRICAL POWER is a very common source of waste aboard ship. Lights are frequently left on when not needed, and bulbs of greater wattage than required are often used. If the ship's ventilation system is improperly operated or improperly maintained, the result is a waste of electrical power. Vent sets are often operated on high speed when low-speed operation would provide adequate ventilation and cooling. Dirty and partially clogged ventilation screens, heaters, cooling units, and ducts will result in inefficient operation and power loss.

In checking the operation of the engineering plant for efficiency, consider the proper operation and maintenance of all units of auxiliary machinery. Economical operation of the distilling plants and of the air compressors will contribute a great deal to the overall efficiency of the plant. Because of the large number and various types of pumps aboard ship, their operation and maintenance are important factors. Units of machinery that operate continuously (or most of the time) must be given careful attention with respect to efficient operation and maintenance.

PROPULSION PLANT MAINTENANCE

The maintenance of maximum operational reliability and efficiency of steam propulsion plants requires a carefully planned and executed program of inspections and preventive maintenance, in addition to strict adherence to prescribed operating procedures and safety precautions. If proper maintenance procedures are followed, abnormal conditions may be prevented.

Preventive inspections and maintenance are vital to successful casualty control, since these activities minimize the occurrence of casualties caused by material failures. Continuous and detailed inspection procedures are necessary not only to discover partly damaged parts which may fail at a critical time, but also to eliminate the underlying conditions which lead to early failure, such as maladjustment, improper lubrication, corrosion, erosion, and other enemies of machinery reliability. Particular and continuous

attention must be paid to the following symptoms of malfunction:

- Unusual noises

- Vibrations

- Abnormal temperatures

- Abnormal pressures

- Abnormal operating speeds

Operating personnel should thoroughly familiarize themselves with the specific temperatures, pressures, and operating speeds of equipment required for normal operation so departures from normal operation will be more readily apparent.

If a gauge, or other instrument for recording operating conditions of machinery, gives an abnormal reading, the cause must be fully investigated. The installation of a spare instrument, or the performance of a calibration test, will quickly indicate whether the abnormal reading is due to instrument error. Any other cause must be traced to its source.

Because of the safety factor commonly incorporated in pumps and similar equipment, considerable loss of capacity can occur before any external evidence is readily apparent. Changes in the operating speeds from normal for the existing load in the case of pressure-governor controlled equipment should be viewed with suspicion. Variations in lubricating oil temperatures and system pressures and from normal pressures indicate either inefficient operation or poor condition of machinery.

In cases where material failure occurs in a unit, a prompt inspection should be made of all similar units to determine if there is danger that a similar failure might occur. Prompt inspection may eliminate a wave of repeated casualties.

Abnormal wear, fatigue, erosion, or corrosion of a particular part may indicate equipment is not being operated within its designed limits or loading, velocity, and lubrication, or it may indicate a design or material deficiency. Unless corrective action can be taken that will ensure that such failures will not occur, special inspections to detect damage should be undertaken as a routine matter.

Strict attention must be paid to the proper lubrication of all equipment. This includes frequent inspection and sampling to determine that the correct quality and quantity of the proper lubricant is in the unit. It is good practice to make a daily check of samples of lubricating oil on all auxiliaries.

CASUALTY CONTROL

The mission of engineering casualty control is to maintain all engineering services in a state of maximum reliability, under all conditions. To carry out this mission, personnel concerned must know the action necessary to prevent, minimize, and correct the effects of operational and battle casualties on the machinery and the electrical and piping installations of their ship. The primary objective of casualty control is to maintain a ship as a whole in such a condition that it will function effectively as a fighting unit. This requires effective maintenance of propulsion machinery, electrical systems, interior and exterior communications, fire control, electronic services, ship control, firemain supply, and miscellaneous services such as heating, air conditioning, and compressed air systems. Failure of any of these services will affect a ship's ability to fulfill its primary objective, either directly by reducing its power, or indirectly by creating conditions which lower personnel morale and efficiency. A secondary objective is the minimization of personnel casualties and of secondary damage to vital machinery.

The details on specific casualties are beyond the scope of this chapter. Detailed information on casualty control can be obtained in *Engineering Operational Casualty Control* (EOCC) manuals (discussed in chapter 25) and *NSTM*, chapter 079, volume 3.

The basic factors influencing the effectiveness of engineering casualty control are much broader than the immediate actions taken at the time of the casualty. Engineering casualty control reaches its peak efficiency by a combination of sound design, careful inspection, thorough plant maintenance (including preventive maintenance), and effective personnel organization and training. CASUALTY PREVENTION IS THE MOST EFFECTIVE FORM OF CASUALTY CONTROL.

REFERENCES

Machinist's Mate 3 & 2, NAVEDTRA 10524-F1, Naval Education and Training Program Management Support Activity, Pensacola, Florida, 1985.

Naval Ships' Technical Manual (NSTM), NAVSEA S9086-G9-STM-000, Chapter 231, "Propulsion Turbines (Steam)," Naval Sea Systems Command, Washington, D. C., January 1982.

CHAPTER 13

CONDENSERS AND OTHER HEAT EXCHANGERS

LEARNING OBJECTIVES

Upon completion of this chapter, you should be able to do the following:

1. Identify the operating principles and construction features of the main condenser.

2. Identify the other heat exchangers in the condensate system, the condensate flow through these components, and the function of these heat exchangers.

3. Identify the three functions, operating principles, and construction features of the deaerating feed tank (DFT).

4. Recognize the possible casualties associated with a condenser and state the casualty control actions taken to prevent further damage.

5. Recognize the safety precautions to observe when a main condenser is opened for cleaning, inspection, or testing.

INTRODUCTION

A heat exchanger is any device or apparatus that allows the transfer of heat from one substance to another. This chapter deals with the major pieces of heat transfer apparatus in the condensate and feed system of the conventional steam turbine propulsion plant. Heat exchangers discussed here include the main condenser, the auxiliary condenser, the air-ejector assemblies, the gland exhaust condenser, and the DFT. The arrangement of piping that connects these units is discussed in chapter 9 of this manual.

MAIN CONDENSER

The main condenser is the heat exchanger in which exhaust steam from the propulsion turbines is condensed as it comes in contact with tubes through which cool seawater is flowing. The main condenser is the heat receiver of the thermodynamic cycle—that is, it is the low-temperature heat sink to which some heat must be rejected. The main condenser is also the means by which feedwater is recovered and returned to the feed system. If we imagine a shipboard propulsion plant in which there is no main condenser and the turbines exhaust to the atmosphere, and if we consider the vast quantities of fresh water that would be required to support even one boiler generating 150,000 pounds of steam per hour, it is immediately apparent that the main condenser serves a vital function in recovering feedwater.

The main condenser is maintained under a vacuum of approximately 25 to 28.5 inches of mercury (in.Hg). The designed vacuum varies according to the design of the turbine installation and to such operational factors as the load of the condenser, the temperature of the outside seawater, and the tightness of the condenser. The designed full-power vacuum for any particular turbine installation may be obtained from the machinery specifications for the plant. Some turbines are designed for a full-power exhaust vacuum of 27.5 in.Hg when the circulating water injection temperature is 75 °F; others are designed to a full-power exhaust vacuum of 25 in.Hg with a circulating water injection temperature of 75 °F.

It is often said that an engine can do a greater amount of useful work if it exhausts to a low-pressure space than if it exhausts against a high-pressure. This statement is undeniably true, but for the condensing steam power plant it may be somewhat misleading because of its emphasis on pressure. The pressure is important because it determines the temperature at which the steam condenses. As noted in chapter 8 of this manual, an increase on the temperature difference between the source (boiler) and the receiver (condenser) increases the thermodynamic efficiency of the cycle. By maintaining the condenser under vacuum, we lower the condensing temperature difference between source and receiver and increase the thermodynamic efficiency of the cycle.

Given a tight condenser and an adequate supply of cooling water, the basic cause of the vacuum in the condenser is the condensation of the steam. This is true because the specific volume of steam is enormously greater than the specific volume of water. Since the condenser is filled with air when the plant is cold, and since some air finds its way into the condenser during the course of operation, the condensation of steam is not sufficient to establish the initial vacuum nor to maintain the required vacuum under all operating conditions. In shipboard steam plants, air ejectors are used to remove air and other noncondensable gases from the condenser. The condensation of steam is thus the major cause of the vacuum, but the air ejectors are required to help establish the initial vacuum and then to assist in maintaining vacuum while the plant is operating.

When the temperature of the outside seawater is relatively high, the condenser tubes are relatively warm and heat transfer is retarded. For this reason, a ship operating on warm tropical waters cannot develop as high a vacuum in the condenser as the same ship could develop when operating in colder waters.

Two basic rules which apply to the operation of single-pass main condensers should be kept in mind. The first is that the overboard temperature should be about 10° higher than the injection temperature. The second rule is that the condensate discharge temperature should be within a few degrees of the temperature corresponding to the vacuum on the condenser. Table 13-1 lists vacuums and corresponding temperatures.

One type of main condenser system is illustrated in figure 13-1. In any main condenser, there are two separate circuits, the vapor condensate circuit and the cooling water circuit.

VAPOR CONDENSATE CIRCUIT

The first circuit in the condenser is the vapor condensate circuit in which the exhaust steam enters the condenser at the top of the shell and is condensed as it comes in contact with the outer surfaces of the condenser tubes. The condensate then falls to the bottom of the condenser, drains

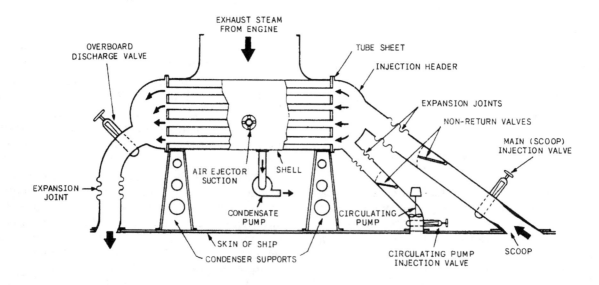

Figure 13-1.—Schematic arrangement of a main condensing system.

Table 13-1.—Properties of Saturated Steam

Absolute Pressure		Vacuum Inches of Mercury	Temperature	
Lb. per Sq. In.	Inches of Hg		°C	°F
0.2	0.4	29.5	11.7	53
0.2	0.5	29.4	15.1	59
0.3	0.6	29.3	18.0	64
0.3	0.7	29.2	20.5	68
0.4	0.8	29.1	22.7	72
0.4	0.9	29.0	24.6	76
0.5	1.0	28.9	26.4	79
0.6	1.2	28.7	29.5	85
0.7	1.4	28.4	32.2	90
0.8	1.6	28.2	34.6	94
0.9	1.8	28.0	36.8	98
1.0	2.0	27.8	38.7	101
1.2	2.4	27.4	42.1	107
1.4	2.8	27.0	45.1	113
1.6	3.2	26.6	47.7	117
1.8	3.6	26.2	50.1	122
2.0	4.0	25.8	52.2	126
2.2	4.4	25.4	54.2	129
2.4	4.8	25.0	56.0	132
2.6	5.2	24.6	57.7	135
2.8	5.7	24.2	59.3	138
3.0	6.1	23.8	60.8	141
3.5	7.1	22.7	64.2	147
4.0	8.1	21.7	67.2	152
4.5	9.1	20.7	69.9	157
5.0	10.1	19.7	72.3	162
5.5	11.2	18.7	74.6	166
6.0	12.2	17.7	76.7	170
6.5	13.2	16.6	78.6	173
7.0	14.2	15.6	80.4	176
7.5	15.2	14.6	80.5	176
8.0	16.2	13.6	83.8	182
8.5	17.3	12.6	85.3	185
9.0	18.3	11.6	86.8	188
9.5	19.3	10.5	88.2	190
10.0	20.3	9.5	89.5	193
11.0	22.4	7.5	92.0	197
12.0	24.4	5.4	94.4	201
13.0	26.4	3.4	96.6	205
14.0	28.5	1.4	98.6	209

into a space called the hotwell, and is removed by the condensate pump. Air and other non-condensable gases that enter with the exhaust steam or that otherwise find their way into the condenser are drawn off by the air ejector through the air-ejector suction opening in the shell of the condenser, above the condensate level.

COOLING WATER CIRCUIT

The second circuit in the condenser is the cooling water circuit. During normal ahead operation, a scoop injection system provides an adequate flow of seawater through the main condenser. The scoop, which is open to the sea, directs the seawater into the injection piping; from there, the water flows into an inlet water chest, flows once through the tubes, and goes overboard through a main overboard sea chest. Scoop injection provides a flow of cooling water at a rate which is proportional to the speed of the ship.

A main circulating pump provides positive circulation of seawater through the condenser at times when the scoop injection system is not effective, such as when the ship is stopped, backing down, or moving ahead at very low speeds.

CONSTRUCTION

All main condensers that have scoop injection are of the straight-tube, single-pass type. They are usually of the construction shown in figure 13-2.

A main condenser may contain from 2,000 to 10,000 copper-nickel alloy tubes. The length and the number of tubes depend upon the size of the condenser. This, in turn, depends upon the capacity requirements. The tube ends are expanded into a tube sheet at the inlet end and expanded or packed into a tube sheet at the outlet end. The tube sheets serve as partitions between the vapor-condensate circuit and the circulating water (seawater) circuit.

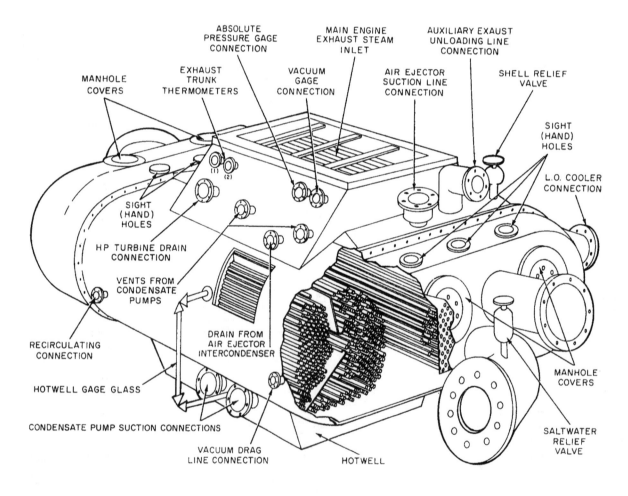

Figure 13-2.—Cutaway view of a main condenser.

Various methods of construction are used to provide for relative expansion and contraction of the shell and the tubes in main condensers. Packing the tubes at the outlet end sometimes makes sufficient provision for expansion and contraction. Where the tubes are expanded into each tube sheet, the shell may have an expansion joint. Expansion joints are also provided in the scoop injection line and in the overboard discharge line.

As shown in figure 13-3, a central steam lane extends from the top of the condenser all the way through the tube bundle, down to the hotwell. The exhaust steam that reaches the hotwell through this steam lane tends to be drawn under the tube bundle towards the sides of the condenser shell, in the general direction of the air-cooling sections, thus sweeping out any air that would otherwise tend to collect in the hotwell. Part of the steam that is drawn through the hotwell under the tube bundle is condensed by the condensate dripping from the condenser tubes. In this process, the condensate (which has been subcooled by its contact with the cold tubes) tends to become reheated to a temperature that approaches the condensing temperature corresponding to the vacuum maintained in the hotwell. The difference between the temperature of the condensate discharge and the condensing temperature corresponding to the vacuum maintained at the exhaust steam inlet to the condenser is called condensate depression. One measure of the efficiency of design and operation of any condenser is its ability to maintain the condensate depression at a reasonably low value under all normal conditions of operation. Excessive condensate depression decreases the operating efficiency of the plant because the subcooled condensate must be reheated in the feed system, with consequent expenditure of steam. Excessive condensate depression also allows an increased absorption of air by the condensate, and this air must be removed to prevent oxygen corrosion of piping and boilers.

Main condensers have various internal baffle arrangements to separate air and steam. This is done so that the air ejectors will not be overloaded by having to pump large quantities of steam along

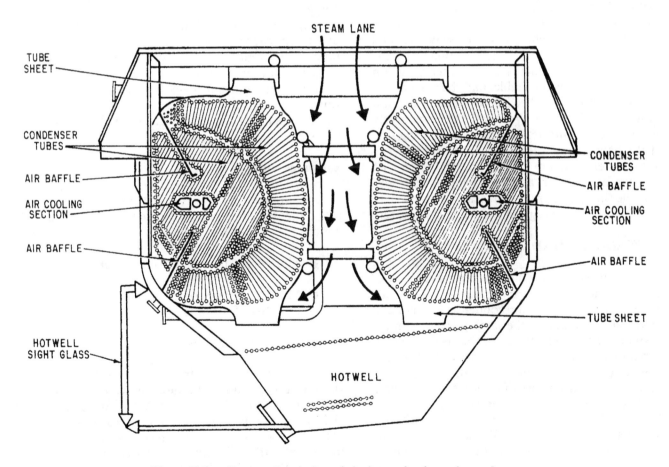

Figure 13-3.—Cross-sectional view of the internals of a main condenser.

with the air. Air cooling sections and air baffles are shown in figure 13-3.

In some installations, the condenser is hung from the LP turbine in such a way that the turbine supports the condenser. Where this type of installation is used, sway braces are used to connect the lower part of the condenser shell with the ship's structure. Spring supports are sometimes used to support part of the weight of the condenser so that it will not have to be entirely supported by the turbine.

Condenser performance may be evaluated by the simple energy balance which takes account of all energy entering and leaving the condenser. In theory, the entering side of the balance should include (1) the mechanical kinetic energy of the entering steam, (2) the thermal energy of the entering steam, (3) the mechanical kinetic energy of the entering seawater, and (4) the thermal energy of the entering seawater. In theory, again, the leaving side of the balance should include (1) the mechanical energy of the leaving condensate, (2) the thermal energy of the leaving condensate, (3) the mechanical kinetic energy of the leaving seawater, and (4) the thermal energy of the leaving seawater. In considering real condensers, however, the entering and leaving mechanical kinetic energies of the seawater tend to be small and tend to cancel each other out, the mechanical kinetic energy of the entering steam is so small as to be negligible, and the mechanical kinetic energy of the leaving condensate is small enough to disregard. With all these relatively insignificant quantities omitted, the entering side of the balance includes only the thermal energy of the entering steam and the thermal energy of the entering seawater, and the leaving side includes only the thermal energy of the leaving condensate and the thermal energy of the leaving seawater.

AUXILIARY CONDENSERS

Condensers into which turbogenerators exhaust are known as auxiliary condensers. Although much smaller than main condensers, auxiliary condensers operate on the same principle. In an auxiliary condenser, however, the cooling water is pumped through the condenser at all times instead of being scoop injected. Also, most auxiliary condensers are of two-pass rather than single-pass construction. The seawater chest is divided into an inlet chamber and a discharge chamber. In other construction features, including the metals used, auxiliary condensers are similar to main condensers.

Auxiliary exhaust steam, in excess of that required for units such as the DFT and the distilling plants, may be directed to the auxiliary condenser or to the main condenser. When the ship is in port, the auxiliary exhaust goes only to the auxiliary condenser. When the ship is preparing to get underway, the auxiliary exhaust goes to the auxiliary condenser until the vacuum in the main condenser is high enough to accept it.

MAINTENANCE OF CONDENSERS

Proper maintenance of condensers is necessary to ensure continuous ship operation and equipment longevity. The Planned Maintenance System (PMS) should be conducted according to the 3-M systems and as described in the sections that follow.

Care of Idle Condensers

Proper lay-up of idle condensers is essential to condenser operation. Improper care can cause degraded performance. The lay-up procedure to be used will depend upon the length of time the condenser will be idle.

Organisms and urchins will grow in an idle unit when seawater is left in it. These growths will block tube, reduce flow, insulate heat transfer surfaces, and lead to corrosion.

If seawater is drained and the condenser is not flushed with fresh water, the growth will dry out and form a hard scale. This scale also reduces heat transfer.

Saltwater Side

Lay-up requirements for the saltwater side of the condensers are divided into the following three conditions:

● Short-term lay-ups up to 1 week

● Midterm lay-ups of more than 1 week and less than 4 months

● Long term lay-ups of 4 months or longer

For short-term lay-ups, the saltwater side should be kept full. Circulate the water once a day for at least 10 minutes by running the circulating pump. If you are unable to circulate the water for 3 or more days in succession, drain the water and refill with fresh water of potable or feedwater quality. For midterm lay-up, drain the saltwater side and immediately fill it with fresh water of potable or feedwater quality. After the first 2 to 3 weeks, drain the condenser and refill with fresh water again. Thereafter, keep the saltwater side filled with fresh water until the ship returns to operation or until the saltwater side is cleaned. For long term lay-ups, open the saltwater side and clean it. After cleaning, keep the saltwater side drained.

Steam Side

The lay-up requirements for the steam side of the condensers are divided into the following conditions:

● Idle periods up to 1 month

● Extended idle conditions in excess of 1 month

For idle periods, empty the hotwells of the condensers and keep them drained.

For extended shutdowns on all condensers, drain and dry out the steam side as soon as possible after the condenser has been secured. This minimizes condenser shell corrosion. The condenser shell can be dried out with an electrically heated air blower discharging into a hotwell opening. After drying, close the condenser openings. Check the condenser weekly and repeat the drying process if moisture is found inside.

Inspecting Condensers

Under normal operating conditions, the saltwater side of a main condenser should be inspected according to the PMS. The steam side should be inspected whenever the inspection covers are removed from the LP turbines.

Conditions may arise that call for more frequent inspections of main condensers. The ship may operate in shallow water or in waters where there are large amounts of seaweeds, schools of fish, or large amounts of oil. When any of these conditions occur, open the condenser for inspection and cleaning.

Carefully maintain the saltwater sides of condensers. This prevents failures caused by deposits on tubes and prevents loss of heat transfer caused by accumulation of these deposits.

Inspect and clean the saltwater sides of condensers at the following times:

● Whenever zinc anodes are checked

● Immediately after grounding of the ship

● Whenever the condensers' performance indicates a possibility of tube fouling

● As soon as practical after operating in shallow or polluted water

Inspect the steam side of the main condenser whenever the manhole plates are removed from the LP turbine. Check the entire steam side for grease and dirt.

Cleaning Condensers

For ordinary cleaning of the saltwater side of a condenser, an air lance should be pushed through each tube. The tube sheets should be washed clean with fresh water and all foreign matters should be removed from the water chest. In cases of more severe fouling, a water lance should be pushed through each tube to remove foreign matter adhering to the tube interior. Extreme fouling, such as that caused by oil or grounding, should be handled in one or two ways. One way is to run a rotating bristle brush through the tubes. Another is to drive soft rubber plugs through the tubes with an air or water gun, followed by a water lance. Be extremely careful that you do not use abrasive tools capable of scratching or marring the tube surface.

The steam side of the condenser is cleaned by boiling out the condenser with a solution of trisodium phosphate. Normally, condensers that serve turbines should not require boiling out more frequently than every shipyard overhaul period.

In cleaning either steam or saltwater side of a condenser, consult the procedures and precautions in chapter 254 of the *NSTM* and appropriate PMS material.

SAFETY PRECAUTIONS FOR CONDENSERS

When opening a main condenser for cleaning, inspection, or testing a main condenser, you must use the following safety precautions. When carried out properly, they will help prevent casualties to personnel and machinery.

1. Before the saltwater side of a condenser is opened, close all sea connections, including the main scoop injection valve, circulating pump suction valve, and main overboard discharge valve tightly. Tag them to prevent accidental opening.

2. On condensers with electrically operated injection and overboard valves, open the electrical circuits serving these motors and tag them to prevent these circuits from being accidentally energized.

3. Before a manhole or handhold plate is removed, drain the saltwater side of the condenser by using the drain valve provided in the inlet water box. This ensures that all sea connections are tightly closed.

4. If practical, replace inspection plates and secure them before you stop work each day.

5. Never subject condensers to a test pressure in excess of 15 psig.

6. When testing for leaks, do not stop because one leak is found. Check the entire surface of both tube sheets, as there may be other leaks. Determine whether each leak is in the tube joint or in the wall so that the proper repairs can be made.

7. There is always the possibility that hydrogen or other gases may be present in the steam side or the saltwater side of a condenser. Do not bring an open flame or tool that might cause a spark close to a newly opened condenser. Do not allow personnel to enter a newly opened condenser until it has been thoroughly ventilated and the space declared safe by a gas-free engineer.

8. Drain the saltwater side of a condenser before flooding the steam side, and keep it drained until the steam side is emptied.

9. The relief valve (set at 15 psig), mounted on the inlet water chest, should be lifted by hand whenever condensers are secured.

10. If a loss of vacuum is accompanied by a hot or flooded condenser, slow or stop the units exhausting into the condenser until the casualty is corrected. Do not allow condensate to collect in condensers and overflow into the turbine or engines.

11. Lift condenser shell relief valves by hand before a condenser is put into service.

12. Do not retain any permanent connection between any condenser and water system that could subject the saltwater side to a pressure in excess of 15 psig.

CONDENSER CASUALTIES

In the event of a casualty to a component part of the propulsion plant, the principal doctrine to be impressed upon operating personnel is the prevention of additional or major casualties. Under normal operating conditions, the safety of personnel and machinery should be given first consideration. Where practicable, the propulsion plant should be kept in operation by standby pumps, auxiliary machinery, and piping systems. The important thing is to prevent minor casualties from becoming major casualties, even if it means suspending operation of the propulsion plant. It is better to stop the main engines for a few minutes than to put them completely out of commission so that major repairs can be accomplished to put them back into operation. In case a casualty occurs, the engineering officer of the watch (EOOW) should be notified as soon as possible. In turn, the EOOW must notify the officer of the deck (OOD) if there will be any effect on the ship's speed or on the ability to answer bells.

The use of the procedures in your ship's *Engineering Operational Casualty Control* (EOCC) manual is mandatory (discussed in chapter 25). It is your principal guide for controlling plant casualties and restoring the plant to normal operations. The procedures discussed in this section may not be applicable to your ship.

Loss of Vacuum

A loss of vacuum on the main condenser that is not handled properly can be very damaging to the turbines and the engineering plant. The major causes of a loss of vacuum are (1) excessive air leakage into the vacuum system, (2) improper operation of the air ejectors, (3) improper operation of the circulating pump (if in operation), (4) a dry make-up feed bottom, (5) improper operation of the freshwater drain collecting pump, (6) improper operation of the main condensate pump, (7) improper line-up of the standby condensate pump, and (8) improper recirculation of condensate.

If a loss of vacuum occurs to the main condenser, you should take the following actions:

1. Inform the bridge of the casualty.
2. Make checks using the EOCC.

If the cause is not found and corrected and vacuum continues to drop, you should take the following actions:

1. At 21 in.Hg, reduce to two-thirds speed.
2. At 18 in.Hg, reduce to one-third speed.
3. At 15 in.Hg, indicate stop on the engine order telegraph (EOT) and isolate the steam side of the main condenser.
4. When the cause of the casualty is found, inform the bridge and tell them the estimated time of repair.
5. When the cause is corrected and vacuum is returning, cut in the exhaust and drains using the EOCC.
6. When the condenser vacuum reaches 16 in.Hg, come to one-third speed.
7. When the condenser vacuum reaches 18 in.Hg, come to two-thirds speed.
8. When the condenser vacuum reaches 22 in.Hg, bring the engine to the ordered speed and report to the bridge that you are ready to answer all bells.

NOTE: All speed changes must be indicated on the EOT.

Hot Condenser

A hot condenser has occurred if the condenser seawater overboard discharge temperature reaches 140°F or higher. The possible causes of this casualty are (1) improper securing of the condenser (normal securing or during a casualty), (2) steam valve(s) leaking through an idle condenser, (3) overboard or injection valve closed during operation, (4) improper operation or failure of the circulating water pump (when in use), (5) plugged sea chest or tubes, (6) a flapper valve stuck open or closed, (7) an air-bound condenser, and (8) loss of way on the ship.

If a hot condenser casualty occurs, you should take the following actions:

1. Report the casualty to the bridge.
2. Ring up STOP on the EOT.
3. Close the throttle valve.
4. Cross-connect the exhausts and drains.
5. Isolate the steam side of the condenser.
6. Stop the circulating water pump (if in operation).
7. Open the air vents on the overboard header.
8. Close the overboard valve.

CAUTION

Do not use the circulating water pump to restore normal or increase seawater flow through the condenser before the condenser overboard temperature is below 140°F. A dangerous situation is generated in this casualty when a steam bubble is formed in the seawater side of the condenser. If a steam bubble exists, the casualty control steps are designed to introduce a slow, limited flow of cooling water through the condenser. If normal seawater circulation is quickly introduced, the cooling effect will condense and collapse the steam bubble, resulting in a vacuum. The vacuum would quickly draw in seawater and generate a shock (water hammer) which could damage condenser tube joints or rupture the rubber expansion joints.

9. Investigate and determine the cause of the casualty.
10. When the cause of the casualty has been corrected and the condenser overboard temperature is less than 140°F, place the condenser back in operation.

Saltwater Leakage Into Condenser

If a condenser salinity indicator shows a rise in the chloride content, the source of contamination must be determined immediately. To locate the source, test the fresh water from different units in the system by checking the proper salinity indicators and by performing chemical tests. Some of the major causes of condenser saltwater contamination are (1) leaky tube(s) in the condenser, (2) make-up feed tank salted up, (3) freshwater drain collecting tank salted up, and (4) leaky feed suction and drain lines which run through the bilges. Each of these possibilities must be investigated to determine the source of the contamination and its elimination.

If it is determined that there is a minor leak in the system, the affected plant will probably be continued in operation. Isolate the condensate system and limit the number of boilers on the engine involved. When operating under these conditions, it will be necessary to blow down the boiler(s) as necessary to keep the boiler salinity within specified limits. However, if the leak is serious, the plant should be secured and the leaks located.

OTHER HEAT EXCHANGERS

The Navy uses several types of heat exchangers. The following paragraphs discuss some of the heat exchangers used aboard ships. These heat exchangers are discussed in relation to the system to which they belong.

AIR-EJECTOR ASSEMBLIES

The air ejector removes air and other noncondensable gases from the condenser. An air ejector is a type of jet pump, that has no moving parts. The flow through the air ejector is maintained by a jet of high-velocity steam passing through a nozzle. The steam is taken from the 140-psi auxiliary steam system on most ships.

The air-ejector assembly (fig. 13-4), used to remove air from the main condenser, usually consists of a first-stage air ejector, an intercondenser, a second-stage air ejector, and an aftercondenser. The two air ejectors operate in series. The intercondenser is under a vacuum of approximately 26 in.Hg. The aftercondenser is at approximately atmospheric pressure.

The first-stage air ejector takes suction on the main condenser and discharges the steam-air mixture to the intercondenser, where the steam content of the mixture is condensed. The resulting condensate drops to the bottom of the intercondenser shell, and from there it drains to the condenser through a U-shaped loop-seal line. The loop-seal condensate line between the intercondenser and the main condenser is necessary to maintain the pressure difference (prevent equalization of vacuum). The air passes to the suction of the second-stage air ejector, where another jet of steam entrains the air and carries it to the aftercondenser. In the aftercondenser, the steam is condensed and returned to the condensate system through the freshwater drain collecting tank, and the air is vented to the atmosphere.

Note that the air ejectors remove air only from the condenser, not from the condensate which passes through the tubes of the intercondenser and the aftercondenser. The condensate merely serves as the cooling medium in these condensers, just as it serves this purpose in the gland exhaust condenser and in the vent condenser.

To provide for continuous operation, two sets of nozzles and diffusers are furnished for each stage of the air ejectors. Only one set is necessary to operate the plant; the other set is maintained ready for use in case of damage or unsatisfactory operation of the set in use. The sets can be used

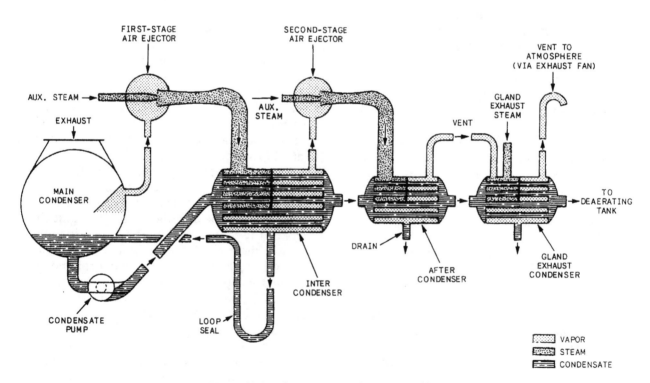

Figure 13-4.—Two-stage air-ejector assembly.

simultaneously when excessive air leakage into the condenser necessitates additional pumping capacity.

Before starting a steam air ejector, the steam line should be drained of all moisture; moisture in the steam will cut the nozzles, and slugs of water will cause unstable operation.

Before cutting steam into the air ejectors, make sure that sufficient cooling water is flowing through the condenser and that the condenser has been properly vented.

The loop seal must be kept airtight, an air leak may cause all water to drain out of the seal.

If you must operate both sets of air ejectors to maintain proper condenser vacuum, air leakage is indicated. Instead of operating two sets of air ejectors, you should eliminate the air leak.

Unstable operation of an air ejector may be caused by any of the following reasons:

- The steam pressure may be lower than the designed amount; the steam temperature and quality may be different than the design condition.

- There may be scale on the nozzle surface.

- The position of the steam nozzle may not be right in relation to the diffuser.

- The condenser drains may be stopped up.

Difficulties due to low pressure are generally caused by improper functioning or improper adjustment of the steam-reducing valve supplying motive steam to the air-ejector assembly. It is essential that DRY steam at FULL operating pressure be supplied to the air-ejector nozzles.

Erosion or fouling of air-ejector nozzles is evidence that wet steam is being admitted to the unit. Faulty nozzles make it impossible to operate the ejector under high vacuum. In some instances, the nozzles may be clogged with grease, boiler compound, or some wet deposit which will decrease the jet efficiency.

GLAND EXHAUST CONDENSER

The gland exhaust condenser receives a steam-air mixture from the propulsion turbine glands and DFT. The steam is condensed and returned to the condensate system through the freshwater drain collecting tank, and the air is discharged to the atmosphere. The atmospheric vent is usually connected to the suction of a small motor-driven fan (gland exhauster), which provides a positive discharge through piping to the atmosphere (gland exhausters normally discharge to a main exhaust vent in the propulsion space). This is necessary to avoid filling the engine room with steam should the air-ejector cooling water supply fail, thereby allowing steam to pass through the intercondenser and aftercondenser without being condensed. The cooling medium on the gland exhaust condenser, as in the air-ejector condensers, is condensate from the main condenser, on its way to the DFT.

In most installations, the gland exhaust condenser appears to be part of the air-ejector assembly, since it is attached to the aftercondenser. However, the gland exhaust condenser is functionally a separate unit even though it is physically attached to the air-ejector aftercondenser.

In serving as the cooling medium in the air ejector condensers and in the gland exhaust condenser, the condensate picks up a certain amount of heat. To some extent this is desirable, since it saves heat which would otherwise be wasted and it reduces the amount of steam required to heat the condensate in the DFT. However, overheating of the condensate could result in inefficient operation of the air ejectors and consequent loss of vacuum in the main condenser. To avoid this difficulty, provision is made for returning some of the condensate to the main condenser when the condensate reaches a certain temperature. As a rule, the recirculating line branches off the condensate just after the gland exhaust condenser. In most installations, the recirculating valve in the recirculating line is thermostatically operated.

DEAERATING FEED TANK

The DFT is a direct-contact heat exchanger which deaerates the condensate and preheats and stores the feedwater.

The water is heated by direct contact with the auxiliary exhaust steam that enters the tank at a pressure just slightly greater than the pressure in the tank. DFTs are usually designed to operate at a pressure of approximately 15 psig and to heat the water to 250°F.

One type of DFT is shown in figure 13-5. Condensate enters the DFT through an inlet pipe and flows into the spray manifold. The spray manifold contains a number of spring-loaded poppet-type spray valves. These valves form the flow stream first into conical sheets, then into jet streams, and finally into fine particles directly in contact with steam. Thus, the condensate is heated to approximately the steam temperature. The efficient heating and the counterflow patterns of the steam and the water in the spray chamber releases all but a small amount of oxygen and other dissolved gases. The water spray surface is partially enclosed by baffles to maintain this strict counterflow relationship and to provide a zone for cooling the released gases before they are ventilated from the tank.

The spray is directed into the water collection cone. The water in the cone drains by gravity through a downtake pipe to the deaerating compartment. From there it is drawn up into the steam scrubbing and mixing passage by the eductor action of the steam leaving the steam passage. This high-velocity steam and water

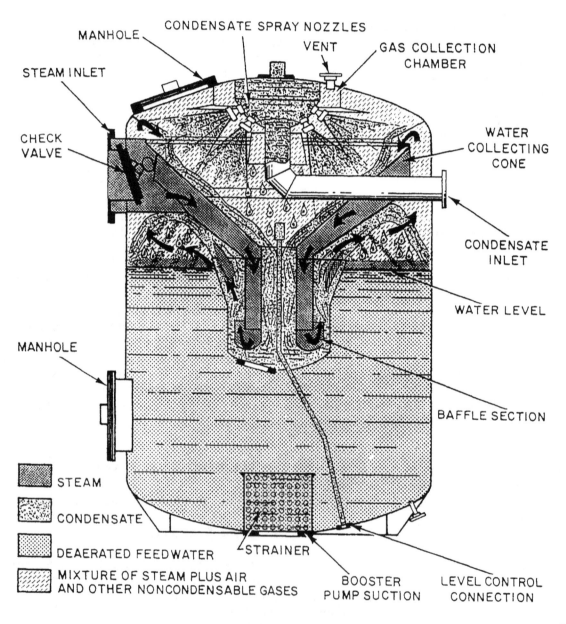

Figure 13-5.—Deaerating feed tank.

38.17

mixture is jetted from the water passage through the baffle space and strikes on the conical impingement baffle. The baffle deflects the flow of water downward into the storage section of the DFT. The water is now deaerated, heated, and ready to be pumped to the boiler.

To enter the initial heating chamber, the steam from the auxiliary exhaust line must break through the baffle section. The steam carries with it any gases released from the water by its jet action; this prevents gas accumulation above the deaerated water in the storage compartment. After starting its upward travel, the steam and released gases break through the water film into the spray chamber. There they come in direct contact with the condensate discharged from the spray nozzles. The spray chamber is where the initial heating and deaeration take place. The portion of steam entering the spray chamber is condensed. The air and noncondensable gases pass to the collection chamber where the released gases and a trace of uncondensed steam is vented out of the DFT.

A strainer is located in the bottom of the tank to prevent any foreign matter from entering the feed booster pump suction.

DFT Operation

During normal operation, the only control necessary is to maintain the water level. If the water level is too high, the tank cannot properly remove the air from the condensate. A low water level may endanger the main feed booster pumps, the main feed pumps, and the boilers.

A common method of DFT level control is the use of automatic makeup feed and excess feed valves. When the tank level drops to a specified point, the level sensing system will transmit an air signal to the makeup feed valve. The valve will open and allow vacuum drag to draw water into the associated condenser from the reserve feedwater tank in service. When the proper DFT level is attained, the makeup feed valve will shut. If the level continues to rise, the level sensing system will cause the excess feed valve to open. This will dump the condensate into the reserve feedwater tank in service.

The DFT removes gases from the feedwater by using the principle that the solubility of gases in the feedwater approaches zero when the water temperature approaches the boiling point. During operation, water is sprayed so that it comes in contact and mixes with the steam from the auxiliary exhaust line. The quantity of steam must always be proportional to the quantity of water. If not, the result will be faulty operation or a casualty.

In most DFTs, a manhole provides access for inspection of spray nozzles; other tanks are designed so that the spray nozzle chamber and the vent condenser (if installed) must be removed for the inspection of the nozzle.

Auxiliary exhaust steam flows directly into the deaerating unit. A check valve is located either in the deaerating unit or in the line leading to the tank. This allows the steam to flow from the auxiliary exhaust line whenever the pressure inside the DFT is less than the pressure in the exhaust line. The check valve also prevents the return flow of water into the auxiliary exhaust line, in the event that the DFT becomes flooded.

DFT Safety

Overfilling the DFT may upset the steam-water balance and cool the water to such an extent that ineffective deaeration will take place. If an excessive amount of cold water enters the DFT, the temperature drop in the tank will cause a corresponding drop in pressure. As the DFT pressure drops, more auxiliary exhaust steam enters the tank. This reduces the auxiliary exhaust line pressure, which causes the augmenting valve (150 psi line to auxiliary exhaust line) to open and bleed live steam into the DFT.

When an excessive amount of cold water suddenly enters the DFT, a serious casualty may result. The large amount of cold water will cool (quench) the upper area of the DFT and condense the steam so fast that the pressure is reduced throughout the DFT. This permits the hot feedwater in the power portion of the DFT and feed booster pump to boil or flash into vapor, causing the booster pump to lose suction. With a loss of feed booster pump pressure, the main feed pump suction is reduced or lost entirely, causing serious damage to the feed pump and loss of feedwater supply to the boiler(s).

Feedwater Cooler

In some installations, the DFT is not located high enough to give sufficient static head for the booster pump suction. Therefore, the booster pump will cavitate. To prevent this cavitation, a feedwater cooler is installed between the feedwater outlet and the booster pump suction. The condensate passes through the feedwater cooler before it enters the DFT. This preheats the condensate and cools the feedwater so that it does not flash to steam as it enters the booster pump suction.

REFERENCES

Machinist's Mate 3 & 2, NAVEDTRA 10524-F1, Naval Education and Training Program Management Support Activity, Pensacola, Florida, 1987.

Naval Ships' Technical Manual (NSTM), NAVSEA S9086-HY-STM-005, Chapter 254, "Condensers, Heat Exchangers, and Air Ejectors," Naval Sea Systems Command, Washington, D.C., October 1989.

CHAPTER 14

AUXILIARY STEAM TURBINES AND FORCED DRAFT BLOWERS

LEARNING OBJECTIVES

Upon completion of this chapter, you should be able to do the following:

1. Identify the types of auxiliary turbines and how they are classified.

2. Identify the two components of a conventional steam propulsion plant that are always powered by an auxiliary steam turbine.

3. Discuss the lubrication systems used on auxiliary steam turbines.

4. Identify the different types of governors used for speed control of auxiliary turbines.

5. Identify the safety devices used on auxiliary turbines.

6. Identify the two types of forced draft blowers (FDBs) used on naval ships.

7. Discuss the types of lubrication systems used on FDBs.

8. Identify the types of governors used to control the speed of FDBs.

9. Identify the types of instruments used to check the speed of an FDB.

INTRODUCTION

Auxiliary steam turbines are used to drive ship's service turbogenerators (SSTGs), main feed pumps (MFPs), and FDBs in all steam propulsion plants. Depending on the particular ship, some additional pumps and auxiliary equipment are also driven by auxiliary steam turbines. Some of the equipment that may be driven by auxiliary turbines include the main condenser circulating pump, main condensate pumps, main feed booster pumps, fuel-oil service pumps, fire pumps, and air compressors.

In many cases, these turbine-driven auxiliaries are duplicated by electrical motor-driven units. These units are more efficient and are present for redundancy and start-up operations until sufficient steam is generated. Motor-driven pumps are usually of a design capacity adequate only for start-up or emergency operations. A turbine-driven unit is more reliable than a motor-driven unit since there is a greater possibility of a loss of electrical power than of steam supply. The presence of noncondensing turbine-driven units also increases overall plant efficiency by providing auxiliary exhaust steam which is used in various ways throughout the plant.

On recent constructions, most auxiliary machinery units outside and also within the engineering spaces are driven by motors. This makes the operation of the plant reliant on continuous electrical power.

The basic principles of steam turbine design, classification, and construction discussed in chapter 12 of this manual apply in general to auxiliary turbines as well as to propulsion turbines, except for specific differences noted in this chapter.

AUXILIARY STEAM TURBINE CLASSIFICATION

Auxiliary steam turbines may be classified according to the following characteristics:

—Speed (constant or variable)

—Exhaust conditions (condensing or non-condensing)

—Shaft position (horizontal or vertical)

—Type (impulse or reaction)

—Steam flow direction (axial, radial, or helical)

—Stages (single or multiple)

—Drive (direct or geared)

—Service (based upon driven auxiliary)

—Power output capacity, limiting speed, and so on

Except for turbine-driven electric generators, auxiliary turbines are usually impulse turbines of either the helical-flow or axial-flow type. They operate against a back pressure of approximately 15 to 17 psig, depending upon the auxiliary exhaust line operating pressure of the ship in which they are installed.

TYPES OF AUXILIARY TURBINES

Many auxiliary turbines are of the impulse type. Reduction gears are used with auxiliary turbines to increase efficiency. Some units that operate at relatively high speeds (FDBs, high-speed centrifugal pumps, and recent SSTGs) are directly coupled to the turbine. Since space requirements frequently demand relatively small units, auxiliary turbines are usually designed with comparatively few stages—often only one. This means a large pressure drop and a high steam velocity in each stage. To obtain maximum efficiency, the blade speed must also be high. Reduction gears are used to reconcile the conflicting speed requirements of the driving and driven units.

DIRECTION OF STEAM FLOW

Most auxiliary turbines are axial-flow units that are similar (except for size and number of stages) to the axial-flow propulsion turbines discussed in chapter 12 of this manual. However, some auxiliary turbines are designed for helical flow and some are designed for radial flow. These types of flow are seldom used in propulsion turbines.

Helical-Flow Turbine

A helical-flow auxiliary turbine is shown in figure 14-1. In this type of turbine, steam enters at a tangent to the periphery of the rotor and impinges upon the moving blades. These blades, which consist of semicircular slots milled obliquely in the wheel periphery, are called buckets. The buckets are shaped so that the direction of steam flow is reversed in each bucket, and steam is directed into a redirecting bucket or reversing chamber mounted on the inner cylindrical surface of the casing. The direction of steam is again reversed in the reversing chamber, and the continuous reversal of the direction of flow keeps the steam moving helically.

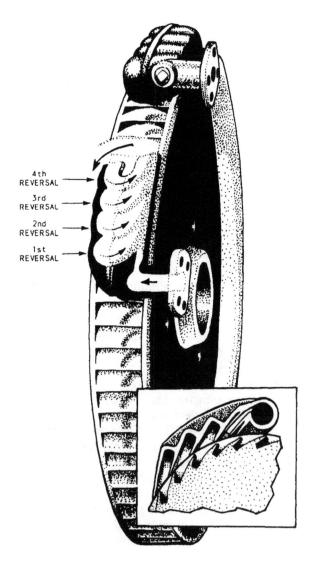

4th REVERSAL
3rd REVERSAL
2nd REVERSAL
1st REVERSAL

33.45

Figure 14-1.—Helical-flow turbine.

Several nozzles are usually installed in this type of turbine. Each nozzle has an accompanying set of redirecting buckets or reversing chambers. Thus, the reversal of steam flow is repeated several times for each nozzle and set of reversing chambers.

A helical-flow turbine is a SINGLE-STAGE turbine because it has only one set of nozzles and therefore only one pressure drop. It is a VELOCITY-COMPOUNDED turbine because the steam passes through the moving blades (buckets) more than once, and the velocity of the steam is therefore used more than once. The helical-flow turbine shown in figure 14-1 corresponds to a turbine in which velocity compounding is achieved by the use of four rows of moving blades.

Helical-flow auxiliary turbines are used for driving some pumps and FDBs.

The helical-flow turbine just discussed is a reentry turbine. Reentry turbines are those in which the steam passes more than once through the blading. Another type of reentry turbine is shown in figure 14-2. This turbine is similar in principle to the helical-flow turbine, except it has one large reversing chamber instead of a number of redirecting chambers. Reentry turbines are sometimes made with two reversing chambers instead of one.

Radial-Flow Turbines

The arrangement of nozzles and blading that provides radial flow in a turbine is shown in

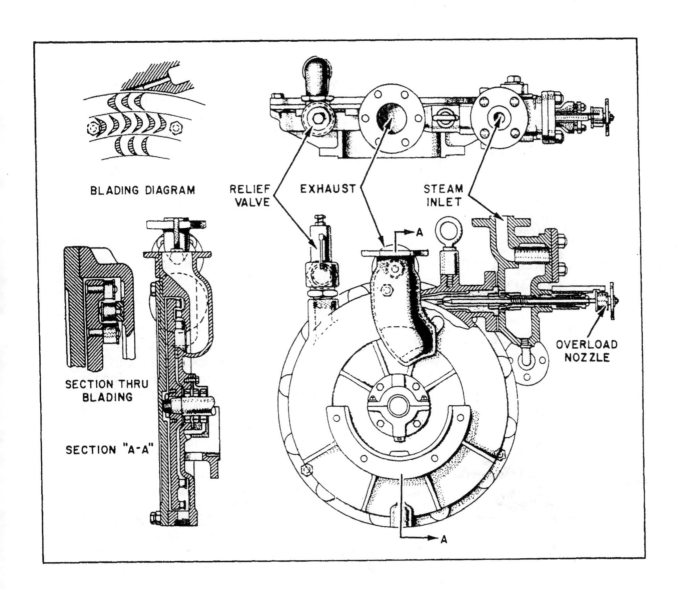

Figure 14-2.—Reentry turbine.

figure 14-3. In a radial-flow turbine, the moving blades are all mounted on a disk, with each row of moving blades at a different radius from the turbine shaft. Radial-flow turbines are also VELOCITY-COMPOUNDED turbines.

The radial-flow turbine shown in figure 14-3 is used to drive main condensate pumps, feed booster pumps, and lubricating-oil service pumps. The basic design of these turbines has changed very little. On newer ships, improved metals are used to withstand the higher pressures and temperatures of steam on a modern warship.

SHIP'S SERVICE TURBOGENERATORS

An SSTG turbine is shown in figure 14-4. Steam is admitted to the turbine through a throttle

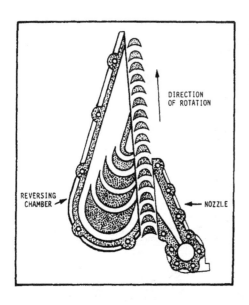

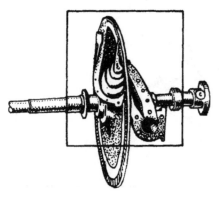

147.194

Figure 14-3.—Radial-flow turbine.

trip valve to the steam chest. The speed is regulated by the number of nozzle control valves under the control of the governor.

Since ship's service generators must supply electricity at a constant voltage and frequency, the turbine must run at a constant speed even though the load varies greatly. Constant speed is maintained through the use of constant-speed governors, which are discussed later in this chapter.

In figure 14-4, the shaft glands of the turbine are supplied with gland sealing steam. The system is the same as that provided for propulsion turbines. Other auxiliary turbines do not require an external source of gland sealing steam since they exhaust to pressures above atmospheric pressure.

Generator turbines vary greatly and should be operated according to the Engineering Operational Sequencing System (EOSS).

MAIN FEED PUMP

The MFP is an important part of the steam propulsion system. The pump is driven by a noncondensing, axial-flow, horizontal-shaft, direct-drive, impulse-type auxiliary steam turbine. The MFP is fitted with constant-pressure pump governors and speed-limiting governors. The pump is a multi-stage centrifugal pump. Centrifugal pumps will be discussed in detail in chapter 18 of this manual.

AUXILIARY TURBINE LUBRICATION

Auxiliary turbines designed to Navy specifications have pressure lubrication systems to lubricate the radial bearings, reduction gears, and governors. Pressure lubrication systems for auxiliary turbines do not provide lubrication for governor linkages or for flexible couplings (except for some turbogenerator sets); these parts of the unit must be lubricated separately.

A pressure lubrication system requires a lube-oil pump. The lube-oil pumps used for auxiliary units are positive-displacement pumps of the simple gear type (discussed in chapter 6 of this manual). The lube-oil pump is generally installed on the turbine end of the FDB unit, but it may be on either the driving or the driven end. The lube-oil pumps for turbogenerators are usually driven by auxiliary gearing connected to the low-speed gear shaft. Some FDBs use a centrifugal pump, supplemented by a viscosity pump for

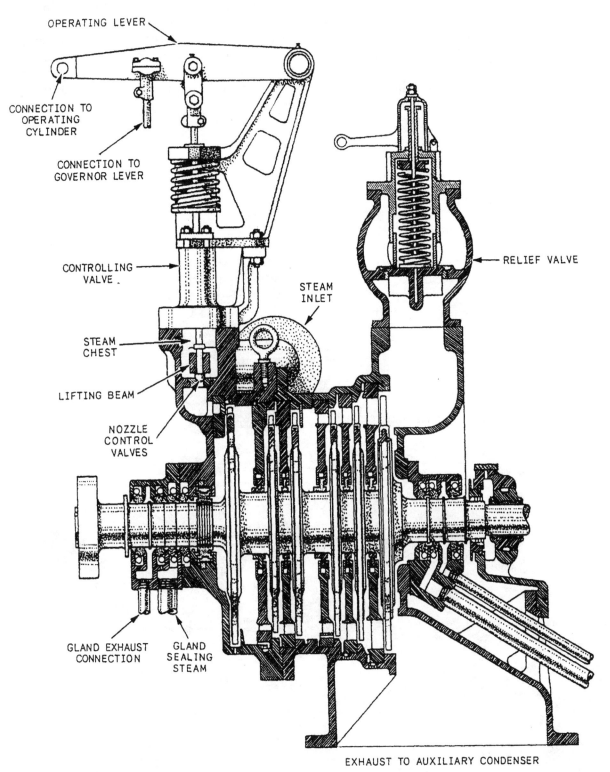

OPERATING LEVER

CONNECTION TO OPERATING CYLINDER

CONNECTION TO GOVERNOR LEVER

CONTROLLING VALVE

STEAM CHEST

LIFTING BEAM

NOZZLE CONTROL VALVES

GLAND EXHAUST CONNECTION

GLAND SEALING STEAM

STEAM INLET

RELIEF VALVE

EXHAUST TO AUXILIARY CONDENSER

47.11

Figure 14-4.—Ship's service turbogenerator steam turbine.

lubrication of the unit. This type of lubrication system is peculiar to FDBs and is discussed later in this chapter.

The pressure lubrication system shown in figure 14-5 is designed for fuel-oil service pumps, fuel-oil booster pumps, and lubricating-oil service pumps. However, it is similar in principle to the lubricating systems of many other units.

In the system illustrated, the bottom section of the gear casing forms the oil reservoir. The reservoir is filled through an oil filler hole in the top of the casing and emptied through a drain outlet at the base of the casing. The shaft, which carries the gear-type oil pump on one end and the governor on the other end, is geared to the pump shaft. The pump shaft is, in turn, geared to the turbine shaft.

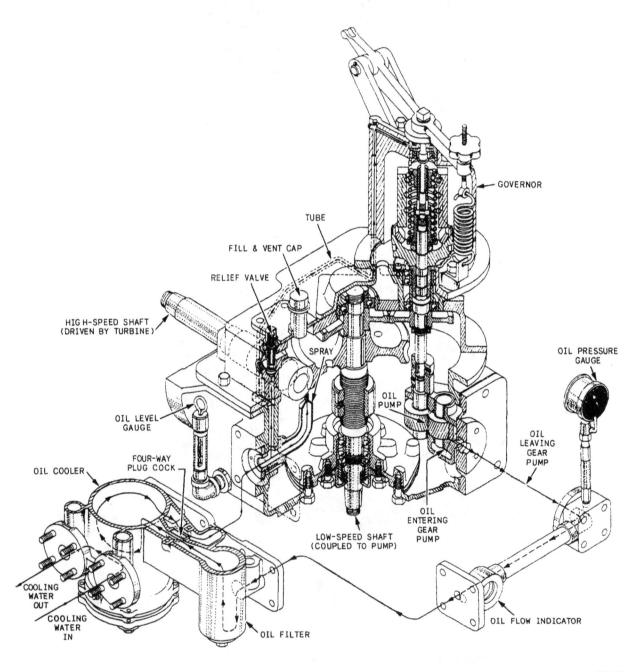

38.97

Figure 14-5.—Isometric diagram of pressure lubrication system.

14-6

The lubricating oil passes through an oil sight flow indicator, a metal edge type of filter, and an oil cooler. Oil is then piped to the bearings on the turbine shaft, to the governor, and to the worm gear on the pump shaft. The bearings and gear on the oil pump and governor shaft are lubricated by oil that drains from the governor and passes back into the oil reservoir. A relief valve is built into the gear casing. This valve protects the system against the development of excessive pressures.

SPEED CONTROL DEVICES

Different types of governors are used for controlling the speed of auxiliary turbines. The discussion here is limited to the constant-speed governors and constant-pressure pump governors, both of which are common in naval use. Additional governing devices that may be encountered on recent ships include hydraulic or electric load sensing governors for turbogenerators and pneumatic, hydraulic, or electric controls for main feed pumps. On ships having automatic combustion and feedwater control systems, the MFP controls may be related in some way to the boiler controls.

CONSTANT-SPEED GOVERNORS

The constant-speed governor, sometimes called the speed-regulating governor is used on constant-speed machines to maintain a constant speed regardless of the load on the turbine. Constant-speed governors are used primarily on generator turbines.

A constant-speed governing system for an SSTG is shown in figure 14-6. The constant-speed governor operates a pilot valve, which controls the flow of oil to an operating cylinder. The operating cylinder, in turn, controls the extent of the opening or closing of the turbine nozzle valves.

With an increased load on the generator, the turbine tends to slow down. Since the governor is driven by the turbine shaft, through reduction gears, the governor also slows. Centrifugal weights (or flyweights) on the governor move inward as the speed decreases, which causes the pilot valve to move upward, permitting oil to enter the operating cylinder. The operating piston rises and through the controlling valve lever, the lifting beam is raised. The nozzle valves open and admit additional steam to the turbines.

The upward motion of the controlling valve lever causes the governor lever to rise, thus raising the bushing. Upward motion of the bushing tends to close the upper port, shutting off the flow of oil to the operating cylinder; this action stops the upward motion of the operating piston. This follow-up motion of the bushing regulates the governing action of the pilot valve. Without this feature, the pilot valve would operate with each slight variation in turbine speed and the nozzle valves would be alternately opened wide and closed completely.

A reverse process occurs when the load on the generator decreases. In this case, the turbine speeds up, the governor speeds up, the centrifugal weights move outward, and the pilot valve moves downward, opening the lower ports and allowing oil to flow out of the operating cylinder. The controlling valve lever lowers the lifting beam, which reduces the amount of steam delivered to the turbine.

Another type of constant-speed governor is the electrohydraulic load-sensing speed governor. This governor produces the same end result as the flyweight type; that is, it maintains constant turbine speed with varying loads on the generator. More detailed information about this type of governor is contained in *Electrician's Mate 1 & C*, NAVEDTRA 10547-E.

CONSTANT-PRESSURE PUMP GOVERNORS

Many turbine-driven pumps are fitted with constant-pressure pump governors. The function of a constant-pressure pump governor is to maintain a constant pump discharge pressure under conditions of varying flow. The governor, which is installed in the steam line to the pump, controls the pump discharge pressure by controlling the amount of steam admitted to the driving turbine.

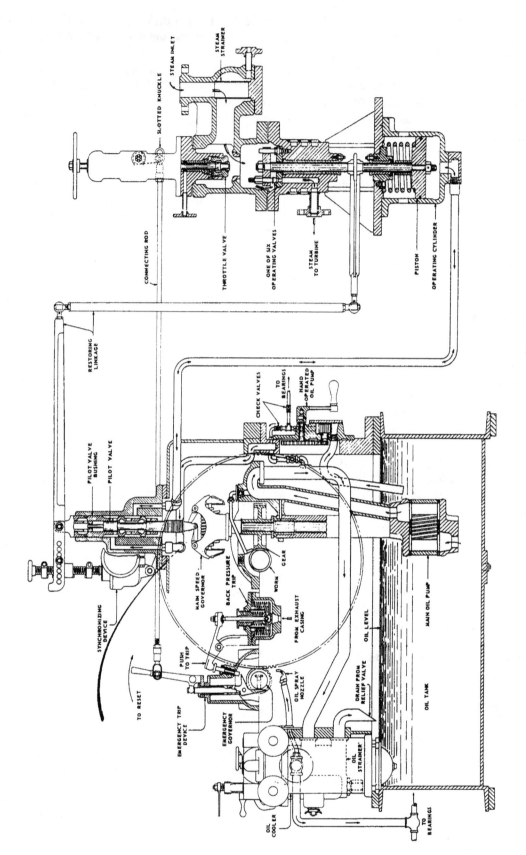

STEAM STRAINER

STEAM INLET

SLOTTED KNUCKLE

CONNECTING ROD

THROTTLE VALVE

ONE OF SIX OPERATING VALVES

STEAM TO TURBINE

PISTON

OPERATING CYLINDER

RESTORING LINKAGE

CHECK VALVES

TO BEARINGS

HAND OPERATED OIL PUMP

PILOT VALVE BUSHING

PILOT VALVE

MAIN SPEED GOVERNOR

BACK PRESSURE TRIP

GEAR

WORM

FROM EXHAUST CASING

OIL LEVEL

MAIN OIL PUMP

SYNCHRONIZING DEVICE

PUSH TO TRIP

TO RESET

EMERGENCY TRIP DEVICE

EMERGENCY GOVERNOR

OIL SPRAY NOZZLE

DRAIN FROM RELIEF VALVE

OIL TANK

OIL COOLER

OIL STRAINER

TO BEARINGS

Figure 14-6.—Constant-speed governing system for ship's service turbogenerator.

A constant-pressure pump governor for a main feed pump is shown in figure 14-7. The governors used on fuel-oil service pumps, lube-oil service pumps, fire and flushing pumps, and various other pumps are almost identical. The chief difference between governors used for different services is in the size of the upper diaphragm. A governor used for a pump that operates with a high discharge pressure has a smaller upper diaphragm than one for a pump that operates with a low discharge pressure.

Two opposing forces are involved in the operation of a constant-pressure pump governor. Fluid from the pump discharge, at discharge pressure is led through an actuating line to the

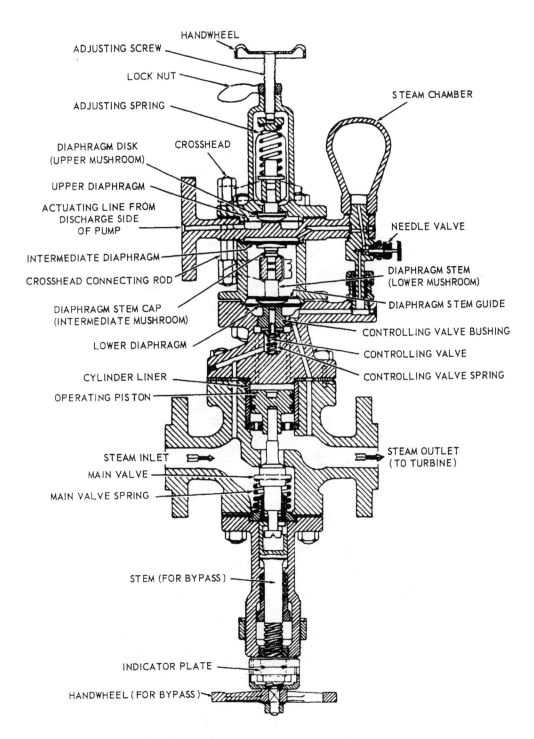

ADJUSTING SCREW
HANDWHEEL
LOCK NUT
ADJUSTING SPRING
STEAM CHAMBER
DIAPHRAGM DISK
(UPPER MUSHROOM)
CROSSHEAD
UPPER DIAPHRAGM
ACTUATING LINE FROM
DISCHARGE SIDE
OF PUMP
NEEDLE VALVE
INTERMEDIATE DIAPHRAGM
CROSSHEAD CONNECTING ROD
DIAPHRAGM STEM
(LOWER MUSHROOM)
DIAPHRAGM STEM GUIDE
DIAPHRAGM STEM CAP
(INTERMEDIATE MUSHROOM)
CONTROLLING VALVE BUSHING
LOWER DIAPHRAGM
CONTROLLING VALVE
CYLINDER LINER
CONTROLLING VALVE SPRING
OPERATING PISTON
STEAM INLET
STEAM OUTLET
(TO TURBINE)
MAIN VALVE
MAIN VALVE SPRING
STEM (FOR BYPASS)
INDICATOR PLATE
HANDWHEEL (FOR BYPASS)

Figure 14-7.—Constant-pressure pump governor for main feed pump.

space below the upper diaphragm. The pump discharge pressure thus exerts an upward force on the upper diaphragm. Opposing this, an adjusting spring exerts a downward force on the upper diaphragm.

When the downward force of the adjusting spring is greater than the upward force of the pump discharge pressure, the spring forces the upper diaphragm and the upper crosshead down. A pair of connecting rods connects the upper crosshead rigidly to the lower crosshead so the entire assembly of upper and lower crossheads moves together. When the crosshead assembly moves down, it pushes the lower mushroom and the lower diaphragm downward. The lower diaphragm is in contact with the controlling valve. When the lower diaphragm is moved down, the controlling valve is forced down and thus opened.

The controlling valve is supplied with a small amount of steam through a port from the inlet side of the governor. When the controlling valve is open, steam passes to the top of the operating piston. The steam pressure acts on the top of the operating piston, forcing the piston down and opening the main valve. The extent to which the main valve is open controls the amount of steam admitted to the driving turbine. Increasing the opening of the main valve therefore increases the

supply of steam to the turbine and so increases the speed of the turbine.

The increased speed of the turbine is reflected in an increased discharge pressure from the pump. This pressure is exerted against the under side of the upper diaphragm. When the pump discharge pressure has increased to the point where the upward force acting on the under side of the upper diaphragm is greater than the downward force exerted by the adjusting spring, the upper diaphragm is moved upward. This action allows the spring to start closing the controlling valve, which allows the main valve spring to start closing the valve against the now reduced pressure on the operating piston. When the main valve starts to close, the steam supply to the turbine is reduced, the speed of the turbine is reduced, and the pump discharge pressure is reduced.

It might seem that the controlling valve and the main valve would be constantly opening and closing and the pump discharge pressure would be varying over a wide range. This does not happen because the governor is designed with an arrangement which prevents excessive opening or closing of the controlling valve. An intermediate diaphragm bears against an intermediate mushroom which bears against the top of the lower crosshead. Steam is led from the

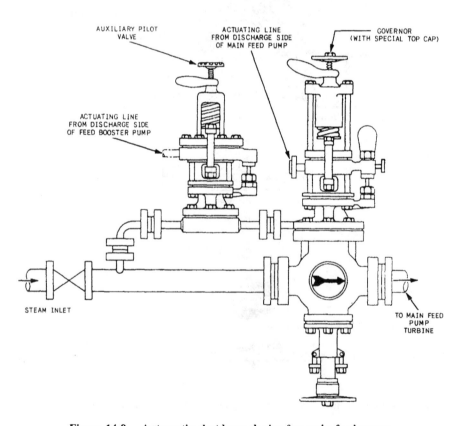

Figure 14-8.—Automatic shutdown device for main feed pump.

governor outlet to the bottom of the lower diaphragm and also through a needle valve to the top of the intermediate diaphragm. A steam chamber is provided to assure a continuous supply of steam at the required pressure to the top of the intermediate diaphragm.

Limiting the movement of the controlling valve in the manner just described reduces the amount of hunting the governor must do to find each new position. Under constant-load conditions, the controlling valve takes a position which causes the main valve to remain open by the required amount. A change in load conditions results in momentary hunting by the governor until it finds the new position required to maintain the pump discharge pressure at the new condition of load.

An automatic shutdown device is used on MFPs. The purpose of the device is to shut down the MFP to protect it from damage in the event of loss of feed booster pump pressure. The shutdown device consists of an auxiliary pilot valve and a constant-pressure pump governor, arranged as shown in figure 14-8. The governor is the same as the constant-pressure pump governor just described except that it has a special top cap. In the regular governor, the steam for the operating piston is supplied to the controlling valve through a port in the governor valve body. In the automatic shutdown device, the steam for the operating piston is supplied to the controlling valve through the auxiliary pilot valve. The auxiliary pilot valve is actuated by the feed booster

pump discharge pressure. When the booster pump discharge pressure is inadequate, the auxiliary pilot valve will not deliver steam to the controlling valve of the governor. Thus, inadequate feed booster pump pressure allows the main valve in the governor to close, shutting off the flow of steam to the MFP turbine.

SAFETY DEVICES

Safety devices used on auxiliary turbines include speed-limiting governors and several kinds of trips. Safety devices differ from speed control devices in that they have no control over the turbine under normal operating conditions. It is only when some abnormal condition occurs that the safety device comes into use to stop the unit or to control its speed.

SPEED-LIMITING GOVERNORS

The speed-limiting governor is essentially a safety device for variable-speed units. It allows the turbine to operate under all conditions from no-load to overload, up to the speed for which the governor is set, but it does not allow operation in excess of 107 percent of rated speed. This type of governor is adjusted to the maximum operating speed of the turbine and has no control over the admission of steam until the upper limit of safe operating speed is reached.

One common type of speed-limiting governor is shown in figure 14-9. This governor is used on

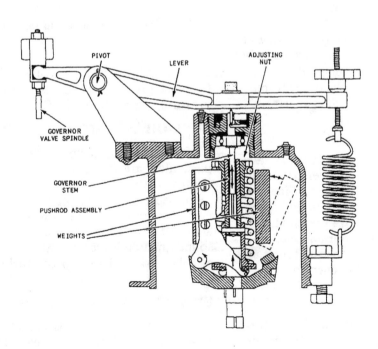

Figure 14-9.—Speed-limiting governor.

main condensate pumps, lube-oil service pumps, and other auxiliaries in the engineering plant. The speed-limiting governor shown in figure 14-9 is designed for use on a pump with a vertical shaft; speed-limiting governors that operate on very much the same principle are used on other auxiliaries that have horizontal shafts.

The governor shaft is driven directly by an auxiliary shaft in the reduction gear and rotates at the same speed as the pump shaft. This speed is proportional to (although lower than) the speed of the driving turbine. Two flyweights are pivoted to a yoke on the governor shaft and carry arms that bear on a pushrod assembly. The pushrod assembly is held down by a strong spring.

Because of centrifugal force, the position of the flyweights is at all times a function of turbine speed. As the turbine speed increases, the flyweights move outward and lift the arms. As the speed of the turbine approaches the speed for which the governor is set, the arms lift against the spring tension. If the turbine speed begins to exceed the speed for which the governor is set, the flyweights move even farther out, thereby causing the governor valve to throttle down on the steam.

When the turbine slows down, as from an increase in load, the centrifugal force on the flyweights is diminished and the governor pushrod spring acts to pull the flyweights inward. This action rotates the lever about its pivot and opens the governor valve, thus admitting more steam to the turbine. The turbine speed increases until normal operating speed is reached.

The speed-limiting governor acts as a constant-speed governor when the turbine is operating at or near rated speed, although it is designed only as a safety device to prevent overspeeding. This governor has no effect on the speed of the turbine at speeds below 95 percent of rated speed.

OVERSPEED TRIPS

Overspeed trips are used on turbines that have constant-speed governors. The overspeed trip shuts off the supply of steam to the turbine and thus stops the unit when a predetermined speed has been reached. Overspeed trips are usually set to trip at about 110 percent of normal operating speed. Figure 14-10 shows the construction of an overspeed trip used on a turbogenerator.

BACK-PRESSURE TRIPS

Back-pressure trips are installed on turbogenerators to protect the turbine by closing the

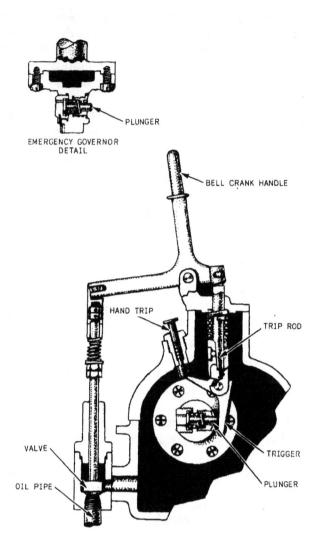

Figure 14-10.—Overspeed trip for a turbogenerator.

throttle automatically when the back pressure (exhaust pressure) becomes too high. A back-pressure trip is shown in figure 14-11.

LUBE-OIL ALARM AND TRIP

Some auxiliary turbines (generator turbines in particular) are fitted with low lube-oil pressure alarms to warn operating personnel when the lube-oil pressure becomes dangerously low. When the lube-oil pressure drops below normal, a pressure-actuated switch completes an electrical circuit to sound an audible alarm. The control system may also be arranged to trip the turbine if lube-oil pressure drops too low.

EMERGENCY HAND TRIPS

Emergency hand trips are installed on turbogenerators to provide a means for closing the

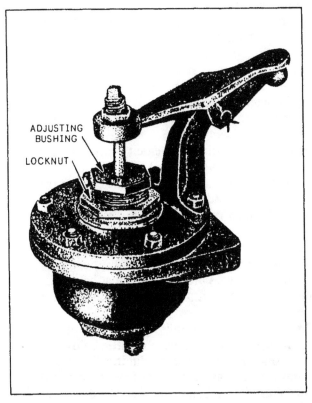

ADJUSTING
BUSHING

LOCKNUT

38.111

Figure 14-11.—Back-pressure trip.

throttle quickly by hand in case of damage to either the turbine or generator. A hand trip is shown in the illustration of an overspeed trip (fig. 14-10).

TESTING SAFETY DEVICES

Speed-limiting governors and safety trips must be tested and maintained according to the requirements set forth by the Planned Maintenance System (PMS) and the EOSS.

FORCED DRAFT BLOWERS

FDBs are gas pumps that provide combustion air to the boiler. These gas pumps are large-volume, low-pressure, high-speed fans. The blowers are either propeller blowers or centrifugal blowers. The main difference between the two types of blowers is the direction of airflow—propeller blowers are axial flow, and centrifugal blowers are radial flow. The vast majority of FDBs are propeller-type blowers. These blowers are usually multi-stage and have either two or three propellers. FDBs can be either horizontal

or vertical shaft, both types are equally common. The turbine is usually a Curtis stage, velocity-compounded impulse turbine. FDBs are usually direct drive and variable speed. Speed control is provided by a speed-limiting governor operated through the boiler automatic combustion control system.

Most propulsion boilers are designed with two main steam-driven FDBs operating in parallel to provide combustion air. If both FDBs are not needed, then one FDB is secured. Balanced automatic shutters are installed in the discharge ducts between each blower and the boiler casing to prevent an idle blower from being rotated in the reverse direction by an operating blower. As a precaution, these shutters are locked in the closed position when the blower is taken out of service. These shutters require periodic lubrication. If not properly lubricated, the automatic shutters may stick open and slam shut with a hard force that can damage the shutters and toggle gear. Some FDBs have a second protection device to prevent damage to the unit. The second device can be a shaft brake or a gear pump that continually pumps oil without a change in the direction of flow when the direction of blower rotation is reversed.

TYPES OF FORCED DRAFT BLOWERS

As discussed earlier, the two main types of FDBs are the centrifugal and propeller blowers, with the propeller type now most commonly used.

Centrifugal blowers may be either vertical or horizontal shaft. In either case, the unit consists of the driving turbine (or other driving unit) at the end of the shaft and the centrifugal fan wheel at the other end of the shaft. Inlet trunks and diffusers are fitted around the blower fan wheel to direct air into the fan wheel and to receive and discharge air from the fan. Centrifugal blowers are fitted with flaps in the suction ducts. In the event of a casualty to one centrifugal blower, air from another blower blows back toward the damaged blower and closes the flaps.

Both horizontal and vertical propeller blowers are used in naval combatant ships. In general, single-stage horizontal blowers are used in older ships and two- or three-stage vertical blowers are used in most combatant ships.

Figure 14-12 shows a single-stage horizontal propeller-type blower. The blower is a complete unit consisting of a driving turbine and a propeller-type fan. The entire unit is mounted on a single-bed plate.

The air intake is screened to prevent the entrance of foreign objects. The blower casing merges into the discharge duct, and the discharge duct is joined to the boiler casing. Diffuser vanes are installed just in front of the blower to prevent rotation of the air stream as it leaves the blower. Additional divisions in the curving sections of the discharge duct also help to control the flow of air.

The shaft that carries both the propeller and the turbine is single forging. The propeller is keyed to the shaft and held to the tapered end of the shaft by a nut and cotter pin. The entire assembly is supported by two main bearings, one on each side of the turbine wheel, outside the turbine casing. The main bearing at the governor end is located in the governor housing, which also contains the thrust bearing. The speed-limiting governor spindle and the lubricating-oil pump shaft are driven by the main shaft of the blower, through a reduction gear.

High-powered two- or three-stage vertical propeller-type blowers are installed in combatant ships. One kind of three-stage vertical propeller-type blower is shown in figure 14-13. There are three propellers at the fan end. Each propeller consists of a solid forged disk to which a number of forged blades are attached. The blades have bulb-shaped roots that are entered in grooves machined across the hub; the blades are kept firmly in place by locking devices. Each propeller disk is keyed to the shaft and secured by locking devices.

The driving turbine is a velocity-compounded impulse turbine (Curtis stage) with two rows of moving blades. The turbine wheel is keyed to the shaft. The lower face of the turbine wheel bears against a shoulder on the shaft; a nut screwed onto the shaft presses against the upper face of the turbine wheel.

The entire rotating assembly is supported by two main bearings. One bearing is just below the propellers and one is just above the thrust bearing in the oil reservoir.

The blower casing is built up of welded plates. From the upper flange down to a little below the lowest propeller, the casing is cylindrical in shape. The shape of the casing changes from cylindrical to cone-shaped and then to square. The discharge opening of the blower casing is rectangular in shape. Guide vanes in the casing control the flow of air and also serve to stiffen the casing. The part of the casing near the propellers is made in sections and is also split vertically to allow removal of the three propellers when necessary. The lower part of the casing, below the air duct, houses the turbine. The lower part of the turbine casing is welded to the oil reservoir structure.

Although all vertical propeller blowers operate on the same principle, and although they may look very much the same from the outside, they are not identical in all details. The major differences to be found among vertical FDBs are in connection with the lubrication systems. Some of these differences are noted in the following section.

BLOWER LUBRICATION SYSTEMS

Since FDBs must operate at very high speeds, correct lubrication of the bearings is essential. A complete pressure lubrication system is an integral part of every FDB. Most FDBs have at least two radial bearings and one thrust bearing. Some blowers have two turbine bearings, two fan bearings, and a thrust bearing.

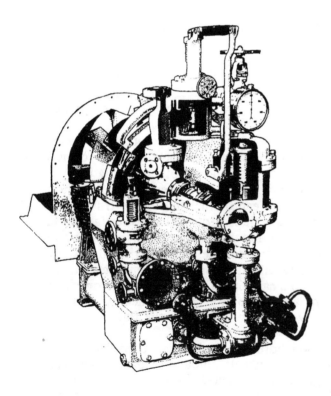

Figure 14-12.—Horizontal propeller-type blower.

14-14

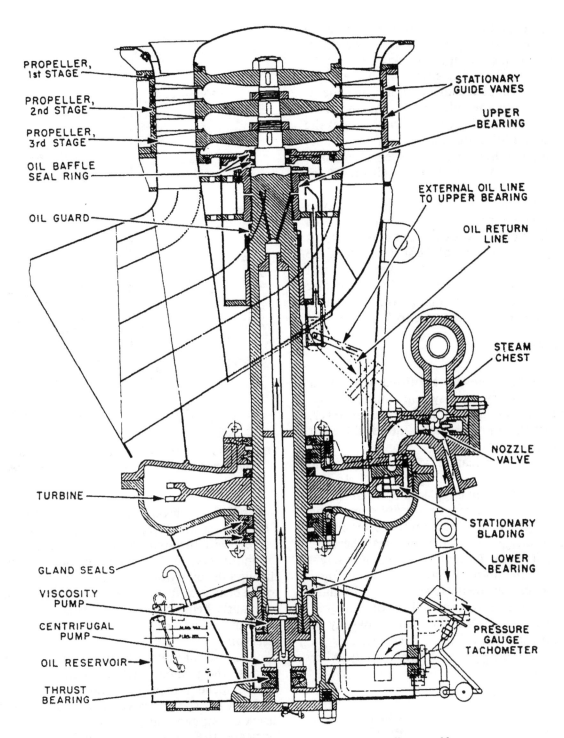

PROPELLER, 1st STAGE

PROPELLER, 2nd STAGE

PROPELLER, 3rd STAGE

OIL BAFFLE SEAL RING

OIL GUARD

TURBINE

GLAND SEALS

VISCOSITY PUMP

CENTRIFUGAL PUMP

OIL RESERVOIR

THRUST BEARING

STATIONARY GUIDE VANES

UPPER BEARING

EXTERNAL OIL LINE TO UPPER BEARING

OIL RETURN LINE

STEAM CHEST

NOZZLE VALVE

STATIONARY BLADING

LOWER BEARING

PRESSURE GAUGE TACHOMETER

Figure 14-13.—Sectional view of a three-stage vertical propeller-type blower.

Horizontal Blowers

The lubrication system for a horizontal FDB includes a pump, an oil filter, an oil cooler, a filling connection, relief valves, oil-level indicators, thermometers, pressure gauges, oil sight flow indicators, and necessary piping.

The pump that is usually turned by the FDB shaft is geared down approximately one-fourth the speed of the turbine. The lube oil is pumped to the bearings from the oil reservoir through the oil filter and the oil cooler. Oil then drains back to the reservoir by gravity.

The gear pump (fig. 14-14) uses an internal gear as the driver, an idler gear as the driver member, and a crescent for sealing between the two members. The idler gear is supported by a rotating part called the idler carrier. The crescent is integral with the idle carrier. As the shaft changes its rotation, the idler carrier rotates the pump cover 180° about the shaft at center line. The rotation of the idler gear is governed by stops, cast in the pump cover, which limit the arc of travel.

In reversing, the crescent always travels through the suction zone of the pump if the boss or the direction arrow cast in the pump cover is pointing in the direction of the suction side of the pump. If this is not the case, the idler carrier will not shift when the shaft reverses. When the idler carrier does not shift, the pump takes suction on the bearings, causing the unit to fail.

Vertical Blowers

Some vertical blowers are fitted with a gear pump and a lubrication system similar to that described for the horizontal blowers. However, most vertical blowers have quite different lubrication systems.

One type of lubrication system used on some vertical FDBs is shown in figure 14-15. In this system, the gear pump is replaced by a centrifugal pump and a helical-groove viscosity pump. The centrifugal pump impeller is on the lower end of the main shaft, just below the lower main bearing. The viscosity pump (also called a friction or screw pump) is on the shaft, just above the centrifugal pump impeller, inside the lower part of the main bearing. As the shaft turns, lubricating oil goes to the lower bearing and from there, by way of the hollow shaft, to the upper bearing. In addition, part of the oil is pumped directly to the

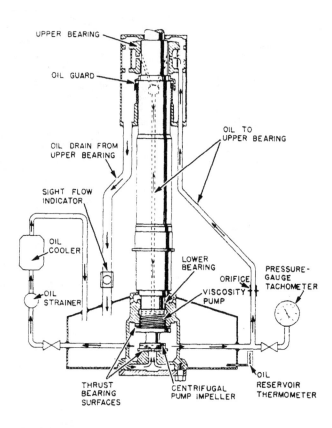

Figure 14-15.—Lubrication system with hollow-shaft oil supply and external oil supply for vertical FDBs.

upper bearing through an external supply line. The oil is returned from the upper bearing to the oil reservoir through an external return line.

In this system, the lubricating oil does not go through the oil strainer or the oil filter on its way to the bearings. Instead, oil from the reservoir is constantly being circulated through an external filter and an external cooler and then back to the reservoir.

The viscosity pump is needed in this system because the pumping action of the centrifugal pump impeller is dependent upon the speed of the shaft. At low speeds, the centrifugal pump cannot develop enough oil pressure to adequately lubricate the bearings. At high speeds, the centrifugal pump alone would develop more oil pressure than is needed for lubrication, and the excessive pressure would tend to cause flooding of the bearings and loss of oil from the lubrication system. The viscosity pump, which is nothing more than a shallow helical thread or groove on the lower part of the shaft, helps to assure sufficient lubrication at low speeds and to prevent the development of excessive oil pressures at high speeds.

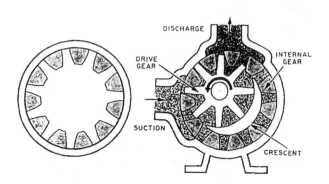

Figure 14-14.—Internal gear pump.

14-16

The hollow-shaft lubrication system is still found in some vertical FDBs. On newer vertical FDBs, oil is pumped to the bearings through an external supply line and passes through an oil filter and an oil cooler on the way to the bearings. Most newer vertical blowers have, in addition to a gear pump, an oscillating-vane hand pump or a small motor-driven gear pump, which is used to establish initial lubrication when the blower is being started. Some of the newer vertical blowers have a centrifugal and a viscosity pump, but these blowers have a completely external oil supply to the bearings instead of a hollow-shaft arrangement.

One of the features of newer vertical blowers is an antirotation device on the shaft of the lubricating pump. This device prevents windmilling of the blower in a reverse direction in the event of leakage through the automatic balance shutters. This antirotation device is continuously lubricated through a series of passageways, which trap some of the leakage from the thrust bearing.

CONTROL OF BLOWER SPEED

FDBs are manually controlled in all naval ships except those that have automatic combustion control systems for their boilers. Speed-limiting governors are fitted to all FDBs to prevent the turbine from exceeding the maximum safe operating speed.

One type of blower speed control device used on ships that have automatic combustion control systems is the Woodward governor. This type of governor regulates the blower speed by determining the position of the steam admission valve in the steam chest. The initial speed settings and subsequent speed adjustments are made by remote manual signal air pressure from the console or automatically, by demands of the ship's combustion and feedwater control systems. Once a setting is made, the governor maintains this setting within close limits.

Blower speed is manually controlled by a valve arrangement that controls the amount of steam admitted to the turbines. In some blowers, a full head of steam is admitted to the steam chest; steam is then admitted to the turbine nozzles through the nozzle valves that are controlled by a manually operated lever or handwheel. The lever or handwheel may be connected by linkages for remote operation. The nozzle valves are arranged so that they open in sequence, rather than all at the same time. The position of the manually operated lever or handwheel determines the number of valves that will open and thus controls the amount of steam that will be admitted to the driving turbine. The steam chest nozzle valve shafts of all blowers serving one boiler are mechanically coupled to provide synchronized operation of the blowers. If only one blower is to be operated, the cutout valve on the steam line to the idle blower must remain closed so that steam will not be admitted to the line.

In other installations, a single throttle valve controls the admission of steam to the steam chest. From the steam chest, the steam enters the turbine casing through all of the nozzles. Thus, the nozzle area is fixed and cannot be changed. Varying the opening of the throttle valve varies the steam pressure to the steam chest and thus varies the speed of the turbine. For emergency control of blowers (and for normal control of blowers in a few older ships) the same throttle valve controls the admission of steam to all blowers serving any one boiler. This throttle valve is installed in the blower steam line. If only one blower is operated, the cutout valve in the branch line to the idle blower must be kept closed.

When the admission of steam is controlled by a multiple-nozzle arrangement, no additional nozzle area is required to bring the blower to maximum speed. In the single-valve arrangement, a special hand-operated nozzle valve is provided for high-speed operations. This nozzle valve, which is sometimes called the overload nozzle valve, is used whenever it is necessary to increase the blower speed beyond that obtainable with the fixed nozzles. The use of the overload nozzle is required only when steam pressure is below normal or when the boiler must be fired above its full power rating.

Most ships have been equipped or refitted with pneumatic diaphragm-actuated steam throttle valves, which provide excellent control of blowers operating in parallel. The most recent FDBs combine the functions of speed-limiting and throttling in a single governor.

CHECKING BLOWER SPEED

Many FDBs are fitted with constant-reading, permanently mounted tachometers for checking on blower speed. Sometimes the tachometer is mounted on top of the governor and is driven by the governor spindle. The governor spindle is driven by the main shaft through a reduction gear and, therefore, does not rotate at the same speed as the main shaft. However, the rpm of the governor spindle is proportional to the rpm of the governor spindle is proportional to the rpm of the main shaft. The tachometer is calibrated to give readings that indicate the speed of the main shaft rather than the speed of the governor spindle.

Some blowers are equipped with a special kind of tachometer called a pressure-gauge tachometer. This instrument is actually a pressure gauge which is calibrated in both psi and rpm. The pressure-gauge tachometer operates on the fact that the oil pressure built up by the centrifugal lube-oil pump has a definite relation to the speed of the pump impeller; and the speed of the impeller is determined by the speed of the main shaft.

Some FDBs are equipped with electric tachometers which have indicating gauges at the blower and at the boiler operating station. The electric tachometer (sometimes called a tachometer generator) consists of a stator and a permanent magnet rotor mounted at the bottom of the turbine shaft. The wire from the generator plugs into a connector inside the pump. Another connector is provided outside the sump for attaching the wire from the generator to the transformer box.

Occasionally, portable tachometers are used to check the speed of FDBs. On some blowers, the tachometer can be applied directly to the end of the main shaft. On others, a portable tachometer must be applied to a separate shaft from which readings may be taken. The small shaft is driven by the main shaft, through gearing. It may rotate at the same speed as the main shaft or its speed may be reduced by the gear arrangement. If the small shaft rotates at a speed different from the speed of the main shaft, the tachometer reading must be converted to obtain the speed of the main shaft.

A stroboscopic tachometer is sometimes used to measure the rpm of an FDB. This instrument may be adjusted so that a mark on the main shaft of the blower appears to be motionless. At this point, the rpm of the shaft may be read directly from the dial.

Vibrating reed tachometers are another type of tachometer used on blowers on which the other types of portable tachometers cannot be used.

OPERATION AND MAINTENANCE OF FDBs

The operation and maintenance of FDBs will not be discussed in this chapter. FDBs must be operated strictly according to the EOSS. Maintenance is dictated by the PMS requirements. EOSS and PMS will be discussed in chapter 25 of this manual.

REFERENCES

Boiler Technician 3 & 2, NAVEDTRA 10535-H, Naval Education and Training Program Management Support Activity, Pensacola, Florida, 1985.

Machinist's Mate 3 & 2, NAVEDTRA 10524-F1, Naval Education and Training Program Management Support Activity, Pensacola, Florida, 1987.

Naval Ships' Technical Manual, NAVSEA S9086-RG-STM-010, Chapter 502, "Auxiliary Steam Turbines," Naval Sea Systems Command, Washington, D.C., January 1990.

Naval Ships' Technical Manual, NAVSEA S9086-S2-STM-000, Chapter 554, "Forced Draft Blowers," Naval Sea Systems Command, Washington, D.C., February 1982.

CHAPTER 15

NUCLEAR POWER PLANTS

LEARNING OBJECTIVES

Upon completion of this chapter, you should be able to do the following:

1. Describe the advantages of nuclear power.

2. Describe an atom, including the composite parts and charge relationships.

3. Identify radioactive atoms, the three frequent radioactive emissions, and their composition.

4. Describe the law of conservation of mass and energy, and identify the variables in Einstein's mass energy equation.

5. Describe the method of calculating energy released in fission.

6. Describe how neutrons are classified by energy levels and how they lose this energy.

7. Identify the basic parts of a reactor core and their characteristics.

8. Describe the physical makeup of the reactor fuels.

9. Describe the function and use of control rods.

10. Describe the purpose of moderators and reactor coolants and identify material used for each.

11. Identify the function of reflectors and shielding.

12. Identify the five types of reactors.

13. Describe the concept of negative temperature coefficient and its relationship to how a reactor reacts to a change in power demands.

14. Describe nuclear poisons and how they are used.

15. Describe the two closed loops used in pressurized water reactors.

INTRODUCTION

Nuclear reactors release nuclear energy by the fission process and transform this energy into thermal energy. The first Navy vessel to use nuclear fission for propulsion was the attack submarine USS *Nautilus* (SSN 571), which became operational in 1955. Since that beginning of the nuclear Navy, the use of nuclear propulsion has expanded to include over 100 attack and ballistic missile submarines, as well as many aircraft carriers and guided missile cruisers.

This chapter is limited to the basic concepts of reactor principles. The discussion of nuclear physics is limited to the fission process, since all nuclear reactors in operation at this time use the fissioning of a heavy element to release nuclear energy.

ADVANTAGES OF NUCLEAR POWER

The fact that a nuclear-powered ship requires no outside source of oxygen from the earth's atmosphere means that the ship can be completely closed off, thereby reducing the hazards of any nuclear attack. This greatly increases the potential of the submarine fleet by giving it the capability of staying submerged for extended periods of time. In fact, prior to the development of the USS *Nautilus*, submarines were merely surface ships that could submerge for only short periods. The USS *Nautilus* was the world's first truly submersible ship. In 1960 the nuclear-powered submarine USS *Triton* completed a submerged circumnavigation of the world, traveling a distance of 35,979 miles in 83 days and 10 hours.

A major advantage of nuclear power for any naval ship is that less logistic support is required. On ships using conventional petroleum fuels as an energy source, the cruising range and strategic value are limited by the amount of fuel that can be stored in their hulls. A ship of this type must either return to port to take on fuel or refuel from a tanker at sea—an operation that is time-consuming and hazardous.

Nuclear-powered ships have virtually unlimited cruising range, since the refueling is done routinely as part of a regular scheduled overhaul. On its first nuclear-fuel load, the USS *Nautilus* steamed 62,562 miles, more than half of this distance fully submerged. To duplicate this feat on a conventionally powered submarine the size of the *Nautilus* would have required over 2 million gallons of fuel oil. The USS *Enterprise* steamed over 200,000 miles before being refueled. In 1963, Operation Sea Orbit, a 30,000-mile cruise around the world in 65 days, completely without logistic support of any kind, proved conclusively the strategic and tactical flexibility of a nuclear-powered task force.

There are other (and perhaps less obvious) advantages of nuclear power for aircraft carriers. For one thing, tanks that would otherwise be used to store boiler fuels can be used on nuclear-powered carriers to store additional aircraft fuels, thus giving the ship a greater striking potential. Another advantage is the lack of stacks; since there are no stack gases to cause turbulence in the flight deck atmosphere, the operation of aircraft is less hazardous than on conventionally powered ships.

FUNDAMENTALS OF NUCLEAR REACTOR PHYSICS

There are more than 100 known elements of which the smallest particle that can be separated

by chemical means is the atom. The Rutherford-Bohr theory of atomic structure (fig. 15-1) describes the atom as being similar to our solar system. At the center of every atom is a nucleus, which is comparable to the sun; moving in orbits around the nucleus are a number of particles called electrons. The electrons have a negative charge and are held in orbit by the attraction of the positively charged nucleus.

Two elementary particles, protons and neutrons, often referred to as nucleons, compose the atomic nucleus. The positive charge of atomic nuclei is attributed to the protons. A proton has an electrical charge equal in magnitude and opposite in sign to that of an electron. A neutron has no charge.

The number of electrons in an atom and their relative orbital positions predict how an element will react chemically; whereas, the number of protons in an atom determines which element it is. An atom that is not ionized contains an equal number of protons and electrons; thus, it is neutral, since the total atomic charge is zero.

As shown in view A of figure 15-1, the hydrogen atom has a single proton in the nucleus and a single orbital electron. Hydrogen, the lightest element, has a mass of approximately 1.

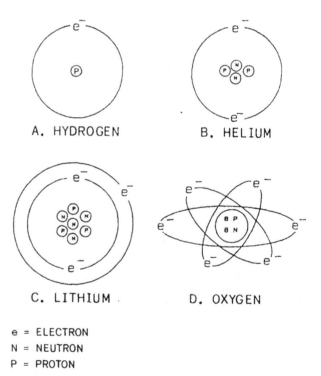

e = ELECTRON
N = NEUTRON
P = PROTON

NOTE: DRAWING NOT TO SCALE

Figure 15-1.—Rutherford-Bohr models of simple atoms.

The next heavier atom is helium. As shown in view B of the figure, each helium atom has two orbital electrons and a nucleus consisting of two protons and two neutrons. The more complex atoms (views C and D) contain more protons and neutrons in the nucleus, with a corresponding increase in the number of planetary electrons. The planetary electrons are arranged in orbits or shells of definite energy levels outside the nucleus.

The characteristics of the elementary atomic particles are compiled in figure 15-2. Note that the mass of a proton is much greater than that of an electron; it takes about 1836 electrons to weigh as much as one proton.

It is possible for atoms of the same element to have different numbers of neutrons, and therefore different masses. Atoms that have the same atomic number (number of protons in the atom) but different masses are called isotopes. Different isotopes of the same element are identified by the atomic mass number, which is the total number of neutrons and protons contained within the nucleus of the atom.

The element hydrogen has three known isotopes, as shown in figure 15-3. The simplest and most common known form of hydrogen (view A) consists of one proton, which is the nucleus, and one orbital electron. Another form of hydrogen, deuterium (view B), consists of one proton and one neutron forming the nucleus and one orbital electron. The third form, tritium (view C), consists of one proton and two neutrons forming the nucleus and one orbital electron.

In scientific notation, the three isotopes of hydrogen are written as follows:

Common hydrogen . . . . $_1^1\text{H}$

Deuterium . . . . $_1^2\text{H}$

Tritium . . . . $_1^3\text{H}$

In this notation, the subscript preceding the symbol of the element indicates the atomic number of the element. The superscript preceding the symbol of the element is the atomic mass number; thus, the superscript indicates which isotope of the element is being referred to.

The general symbol for any atom is thus

$$_Z^A\text{X}$$

where

X = symbol of the element,

Z = atomic number (number of protons),

A = atomic mass number (sum of the number of protons and the number of neutrons).

Of all the known elements, there are more than 2000 isotopes, most of which are radioactive.

Particle	Charge	Mass (amu)*
Proton -------------------	+1	1.00728
Neutron ------------------	0	1.00866
Electron ------------------	−1	0.0005485

*1 amu $= 1.66 \times 10^{-24}$ gram

Figure 15-2.—Characteristics of elementary atomic particles.

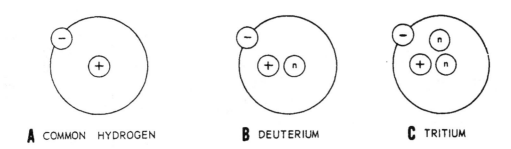

A COMMON HYDROGEN   B DEUTERIUM   C TRITIUM

Figure 15-3.—Isotopes of hydrogen.

Figure 15-4 gives the nuclear composition of various isotopes.

RADIOACTIVITY

All naturally occurring isotopes with atomic number Z greater than 82 are radioactive, and many more isotopes can be made artificially radioactive by bombarding with neutrons that upset the neutron-proton ratio of the normally stable nucleus.

Radioactive isotopes decay, thereby forming lighter and more stable nuclei. Radioactive decay

Element	Symbol	Atomic No. or No. of protons	No. of electrons	No. of neutrons
Hydrogen	^1_1H	1	1	0
Hydrogen (Deuterium)	^2_1H	1	1	1
Hydrogen (Tritium)	^3_1H	1	1	2
Helium	^3_2He	2	2	1
Helium	^4_2He	2	2	2
Helium	^5_2He	2	2	3
Helium	^6_2He	2	2	4
Beryllium	^9_4Be	4	4	5
Sulfur	$^{34}_{16}S$	16	16	18
Chlorine	$^{34}_{17}Cl$	17	17	17
Cobalt	$^{60}_{27}Co$	27	27	33
Copper	$^{63}_{29}Cu$	29	29	34
Yttrium	$^{95}_{39}Y$	39	39	46
Molybdenum	$^{95}_{42}Mo$	42	42	53

Figure 15-4.—Nuclear composition of various isotopes.

Element	Symbol	Atomic No. or No. of protons	No. of electrons	No. of neutrons
Cadmium --	$^{113}_{48}Cd$	48	48	65
Iodine ---	$^{135}_{53}I$	53	53	82
Xenon --	$^{135}_{54}Xe$	54	54	81
Cesuim ---	$^{135}_{55}Cs$	55	55	80
Barium ---	$^{135}_{56}Ba$	56	56	79
Lanthanum ---	$^{139}_{57}La$	57	57	82
Lead --	$^{206}_{82}Pb$	82	82	124
Polonium ---	$^{210}_{84}Po$	84	84	126
Radium ---	$^{226}_{88}Ra$	88	88	138
Thorium --	$^{234}_{90}Th$	90	90	144
Protactinium ---	$^{234}_{91}Pa$	91	91	143
Uranium --	$^{234}_{92}U$	92	92	142
Uranium --	$^{235}_{92}U$	92	92	143
Uranium --	$^{238}_{92}U$	92	92	146
Uranium --	$^{239}_{92}U$	92	92	147
Neptunium ---	$^{239}_{93}Np$	93	93	146
Plutonium --	$^{239}_{94}Pu$	94	94	145

Figure 15-4.—Nuclear composition of various isotopes—Continued.

occurs most frequently through the emission of an alpha particle or a beta particle (positive or negative). One or more gamma rays may also be emitted with the alpha or beta particle.

An alpha particle (symbol α) is composed of 2 protons and 2 neutrons. It is the nucleus of a helium ($_2^4$He) atom, has an electrical charge of +2, and is very stable. In decaying to a more stable element, many unstable nuclei emit an alpha particle. The results of alpha emission can be seen from the following equation:

$$_{92}^{238}U \longrightarrow _2^4\alpha + _{90}^{234}Th$$

In this equation, the parent isotope of uranium ($_{92}^{238}U$) is a naturally occurring, radioactive isotope that decays by alpha emission. Since the A and Z numbers must balance in a nuclear reaction equation, we see that the emission of an alpha particle (2 protons) by a uranium atom (92 protons) produces a daughter atom (thorium, Th) that has 90 protons.

The radioactive isotope of thorium ($_{90}^{234}Th$) produced in the reaction further decays by the emission of a negative beta particle (symbol -β), as indicated in the following equation:

$$_{90}^{234}Th \longrightarrow _{-1}^{0}\beta + _{91}^{234}Pa$$

The negative beta particle is the same as an electron. However, the origin of the beta particle is the nucleus, rather than the orbital shells, of an atom. A negative beta particle is emitted at an extremely high kinetic energy level when a neutron within the nucleus decays to a proton and an electron (the negative beta particle). When this phenomenon occurs, the proton stays within the nucleus, forming an isotope of a different element having the same mass. This negative beta decay occurs in nuclei that have fewer protons than stable nuclei with the same total number of neutrons and protons.

Nuclei that have more protons than stable nuclei with the same total number of neutrons and protons tend to decay by emitting a positively charged beta particle called a positron. The positron is the same as an electron (negative beta particle) except that the positron has a positive charge. Positive beta decay is illustrated by the decay of chlorine-34:

$$_{17}^{34}Cl \longrightarrow _{+1}^{0}\beta + _{16}^{34}S$$

In the illustrated process, a proton changes into a neutron and a positive electron. The latter is then immediately ejected from the nucleus as a positive beta particle.

A radioactive isotope may go through several transformations of these types before reaching a stable state. The radioactive isotope uranium ($_{92}^{238}U$) emits a total of eight alpha particles and six beta particles prior to reaching a stable isotope of lead ($_{82}^{206}Pb$).

The third manner in which a radioactive isotope may reach a more stable configuration is by the emission of gamma rays (symbol τ). The gamma ray is an electromagnetic type of radiation having frequency, high energy, and a short wavelength. Gamma rays are similar to X rays. The distinguishing factor between the two is that gamma rays originate in the nucleus of an atom; whereas, X rays originate from orbital electrons. In general, it can be said that a gamma ray has higher energy, higher frequency, and shorter wavelength than an X ray.

Frequently an isotope that emits an alpha or beta particle in the decay process will emit one or more gamma rays at the same time, as in cobalt ($_{27}^{60}Co$), an isotope that decays by beta emission and at the same time emits two gamma rays of different energy. Some radioactive isotopes reach a stable state by emitting gamma rays only. In the latter case, since gamma rays have neither mass nor electrical charge, the A and Z numbers of the isotope remain unchanged, but the energy of the nucleus is reduced.

An important property of any radioactive isotope is the time involved in radioactive decay. To understand the time element, you need to understand the concept of half-life. Half-life may be defined as the time required for one-half of any given number of radioactive atoms to decay, thus reducing the radiation intensity of that particular isotope by one-half. Half-lives may vary from fractions of microseconds to billions of years. Radioactive isotopes with short half-lives are said to be short-lived, while those with long half-lives are long-lived. Some half-lives of radionuclides of concern in nuclear reactor operation are as follows:

$$^{238}_{92}U \qquad = 4.47 \times 10^9 \text{ years,}$$

$$^{235}_{92}U \qquad = 7.04 \times 10^8 \text{ years,}$$

$$^{135}_{54}Xe \qquad = 9.10 \text{ hours,}$$

$$^{135}_{53}I \qquad = 6.6 \text{ hours.}$$

As stated previously, radioactive isotopes decay most frequently by the emission of alpha particles, beta particles (positive and negative), gamma rays, or a combination of them. However, there are many other phenomena that may occur, including fission and the emission of neutrons, protons, and other forms of energy.

CONSERVATION OF MASS AND ENERGY

The principle of conservation of energy is discussed in chapter 8 of this text. Now, you should consider mass and energy as two phases of the same principle. In so doing, the conservation law becomes the following:

(mass + energy) before =
(mass + energy) after.

The conservation law states the combined total of mass and energy should be the same before and after the reaction. In this equation, we allow mass to be changed to energy or vice versa. Now you need an equation to provide a conversion ratio between mass and energy. This equation is fundamental to the entire subject of nuclear power. The equation used is Einstein's energy equation and is as follows:

$$E = mc^2$$

where

E = energy in ergs,
m = mass in grams,
c = speed of light (3×10^{10} cm/sec).

Mass and energy are not conserved separately, but can be converted into each other.

Several units and conversion factors that have become conventional to the field of nuclear engineering are as follows:

1 eV (electronvolt)	= the energy acquired by an electron as it moves through a potential difference of 1 volt
1 MeV (million electron-volt)	= 10⁶eV = 1.52×10^{-16}Btu
1 amu (atomic mass unit)	= 1/12 the mass of a $^{12}_{6}C$ atom (by definition)
1 amu	= 1.49×10^{-3}erg = 1.66×10^{-24}g = 931 MeV = 1.41×10^{-13} Btu

NUCLEAR ENERGY RELEASED IN FISSION

It was previously stated that the atomic mass number is the total number of nucleons within the nucleus. It can also be said that the atomic mass number is the nearest integer (as found by experiment) to the actual mass of an isotope. In nuclear reaction equations, the entire mass must be accounted for; therefore, the actual mass must be considered.

The atomic mass of any isotope is somewhat less than the sum of the individual masses of the protons, neutrons, and orbital electrons that are the components of that isotope. This difference is termed *mass defect*; it is equivalent to the binding energy of the nucleus. The binding energy of an atomic nucleus is the amount of energy that would be released if that nucleus were formed from its components.

The binding energy of any isotope may be found as in the following example of copper ($^{63}_{29}Cu$), which contains 34 neutrons, 29 protons, and 29 electrons. Using the values given in figure 15-2, we find:

$34 \times 1.00866 = 34.29444$ amu
$29 \times 1.00728 = 29.21112$ amu
$29 \times 0.0005485 = 0.01591$ amu
Total of component masses = 63.52147 amu

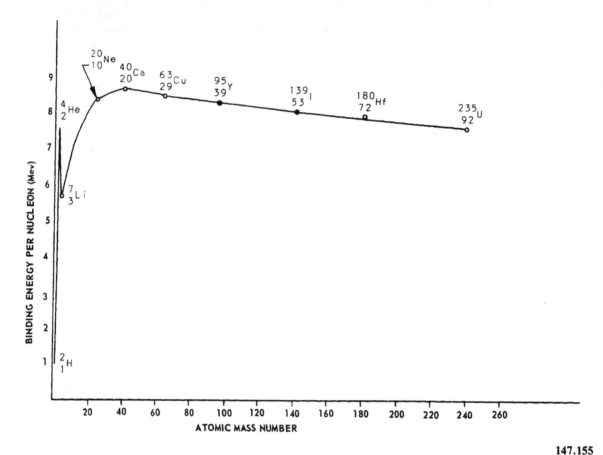

Figure 15-5.—Relationship between atomic mass number and average binding energy per nucleon.

147.155

Less actual mass of atom = 62.92960 amu

Mass defect = 0.59187 amu

Converting to energy, we find:

931 MeV/amu × 0.59187 amu = 551.031 MeV, or

551.031 ÷ 63 = 8.8 MeV/nucleon.

The relationship between mass number and the average binding energy per nucleon is shown in figure 15-5.

In the process of nuclear fission[1], a heavy nucleus splits into two lighter elements called fission fragments plus several neutrons and gamma rays. The energy released in fission is due to the fission fragments having a larger binding energy per nucleon than the original heavy fuel atoms. This additional binding energy is then released in each fission.

There are more than 40 different ways a uranium-235 nucleus may fission, resulting in more than 80 different fission products. For this discussion, let us consider the most probable fission of a uranium-235 nucleus. In slightly more than 6 percent of all fissions, the uranium-235 nucleus will split into fragments having mass numbers of 95 and 139. The following equation is typical:

$$^{235}_{92}U + {}^{1}_{0}n \longrightarrow {}^{95}_{39}Y + {}^{139}_{53}I + 2\ {}^{1}_{0}n + \tau,$$

where the fission fragments, yttrium (Y) and iodine, are both radioactive and decay through negative beta emission to the stable isotopes of molybdenum ($^{95}_{42}$Mo) and lanthanum ($^{139}_{57}$La) respectively.

[1]For a detailed discussion of nuclear fission, refer to Samuel Glasstone and Alexander Sesonske, *Nuclear Reactor Engineering* (New York: Van Nostrand Reinhold, 1981).

One method of determining the energy released from the reaction for the fission of uranium-235 is to use figure 15-5 as follows:

BE/A = 7.6 MeV/nucleon for A = 235,

BE/A = 8.7 MeV/nucleon for A = 95,

BE/A = 8.3 MeV/nucleon for A = 139.

Initial total binding energy is

7.6 MeV/nucleon × 235 nucleons = 1786 MeV.

Final total binding energy is

8.7 MeV/nucleon × 95 nucleons

+ 8.3 MeV/nucleon × 139 nucleons

= 826 MeV + 1154 MeV

= 1980 MeV.

Energy released in fission of one atom of uranium-235 is therefore 1980 MeV − 1786 MeV = 194 MeV.

A second method of determining the energy released from the reaction is to find the difference in atomic mass units of the daughter products and the original nucleus. We also account for the neutron used to bombard the uranium-235 atom and the two neutrons liberated in the fission process:

Mass of uranium-235 atom	= 235.04393	amu
Mass of neutron	= 1.00866	amu
Total original mass	= 236.05259	amu
Mass of molybdenum-95 atom	= 94.90584	amu
Mass of lanthanum-139 atom	= 138.906346	amu
Mass of two neutrons	= 2.01732	amu
Total final mass	= 235.8295	amu

Mass defect = 236.05259 − 235.8295 = 0.22309 amu/fission

Hence,

0.22309 amu/fission × 931 MeV/amu = 207.7 MeV/fission

Thus, we find that fission of one uranium-235 atom releases approximately 200 MeV of energy. Most of this energy (about 80 percent) appears immediately as kinetic energy of the fission fragments. As the fission fragments slow down, they collide with other atoms and molecules; this results in a transfer of kinetic energy to the surrounding particles. The increased molecular motion is manifested as sensible heat. The remaining energy is realized from the decay of fission fragments by beta particle and gamma ray emission, kinetic energy of fission neutrons, and instantaneous gamma ray energy.

In a nuclear reactor, the two neutrons liberated in the reaction are available, under certain conditions, to fission other uranium atoms and assist in maintaining the reactor critical. Each newly born fission neutron is a fast neutron (see next section), which may undergo one of the following processes:

● Leakage from the core

● Absorptions that do not cause fission

● Absorptions that do cause fission

Fast neutrons not undergoing any of these processes are slowed down (thermalized) mainly by elastic scattering collisions with hydrogen atoms in water to become thermal neutrons. The thermal neutrons then undergo one of the three processes and a new generation of neutrons is born.

The ratio of the number of neutrons in any generation to the number of neutrons in the preceding generation is called the effective neutron multiplication factor, K_{eff}:

$$k_{eff} = \frac{\text{number of neutrons in any generation}}{\text{number of neutrons in the preceding generation}}$$

If $k_{eff} > 1$, the number of fission neutrons in the reactor increases with time, and the reactor is said to be supercritical. If $k_{eff} = 1$, the number of neutrons remains constant and the reactor is critical. If $k_{eff} < 1$, the number of neutrons is decreasing and the reactor is subcritical.

NEUTRON REACTIONS

Neutrons may be classified by their kinetic energies. A fast neutron has a kinetic energy of greater than about 0.1 MeV. An epithermal (resonance) neutron in the process of slowing down possesses kinetic energies ranging from about 1 eV to 0.1 MeV. A thermal neutron is in thermal equilibrium with its surroundings and has kinetic energy of less than 1 eV.

Neutrons lose their kinetic energy by interacting (colliding) with atoms in the surrounding area. The probability of a neutron interacting with one atom is dependent upon the effective target area presented by that atom for a particular type of neutron reaction. This effective target area (which is the probability of a neutron reaction occurring) is called cross section. The unit of cross-section measurement is called a "barn." The size of a barn is 10^{-24} square centimeters. The most important neutron reactions occurring in a nuclear reactor are scattering, capture, fission, and absorption, as discussed in the following paragraphs.

Scattering cross section is a measure of the probability of an elastic (billiard ball) collision with a neutron. In this type of collision, part of the kinetic energy of the neutron is imparted to the atom and the neutron rebounds after collision. Neutrons are thermalized (reduced to an energy level below 1 eV) by elastic collisions.

Capture cross section is a measure of the probability of the neutron being captured without causing fission.

Fission cross section is a measure of the probability of fission of the atom after neutron capture.

Absorption cross section is a measure of the probability that an atom will absorb a neutron. The absorption cross section is the sum of the capture cross section and the fission cross section.

The cross section for any given element may vary with the kinetic energy of the approaching neutron. In uranium-235, the fission cross section for a thermal neutron is several hundred times the cross section for a fast neutron. This means that thermal neutrons are much more effective than fast neutrons in causing fission in uranium-235.

REACTOR COMPONENTS

A nuclear reactor must contain at least a critical mass of nuclear fuel. A critical mass contains sufficient fissionable material to enable the reactor to maintain a self-sustaining chain reaction, thereby keeping the reactor critical. A critical mass is dependent upon the species of fissionable material, its concentration and purity, the geometry and size of the reactor, and the matter surrounding the fissionable material[2].

REACTOR FUELS

The components of a reactor that contain fissionable material are called the fuel elements. Present-day reactors use fuel elements that vary considerably in form and disposition. A typical fuel element is comprised of a number of long thin cylinders called fuel rods. These fuel rods are fastened together to form a grid. The main fissionable material in the fuel is $^{235}_{92}U$. To prevent the release of radioactive fission products from the fissionable material into the coolant, the fissionable material is clad with a material such as Zircaloy (an alloy of the element zirconium). This cladding material provides structural strength and corrosion resistance, is a good conductor of heat, and has a small cross section for neutron absorption. To provide operational control of nuclear reactors, collections of fuel elements are combined with control rods. The resulting assembly is called a reactor core.

CONTROL RODS

Control rods are used to control the effective neutron multiplication factor (k_{eff}) of a reactor core. Control rods are fabricated from materials with a high neutron absorption cross section, such as hafnium, cadmium, or boron. Hafnium is particularly suitable because it has a high absorption cross section and good corrosion resistance. In addition, the hafnium isotopes formed by neutron absorption also have high absorption cross sections and are not radioactive. This means that a hafnium control rod does not lose its ability to control the effective neutron multiplication factor (k_{eff}) and does not substantially contribute to the radiation produced

---

[2]For a thorough discussion of the aspects of reactor design, see Samuel Glasstone and Alexander Sesonke, *Nuclear Reactor Engineering* (New York: Van Nostrand Reinhold, 1981).

by core radioactivity. The control rods are moved by specialized motors called control rod drive mechanisms.

When control rods are fully inserted into the core, the core is substantially subcritical and is said to be shut down. To initiate reactor operation, the control rods are partially withdrawn from the core by careful and deliberate operation of the control rod drive mechanisms. When the control rods have been withdrawn to their critical position, the reactor becomes critical. When reactor power is no longer required, the core is shut down by operating the control rod drive mechanisms, inserting the control rods into the core. The control rod drive mechanisms can also provide for an emergency reactor shutdown by inserting the control rods very rapidly. This type of shutdown is called a scram.

MODERATORS

A moderator is the material used to thermalize the neutrons in a reactor. As previously stated, neutrons are thermalized by elastic scattering collisions; therefore, a good moderator must have a high scattering cross section and a low absorption cross section to reduce the speed of a neutron in a small number of collisions. Nuclei whose mass is close to that of a neutron are the most effective in slowing down the fast neutrons. Thus, atoms of low atomic weight generally make the most effective moderators.

Ordinary water makes a good moderator since the cost is low; however it must be free from impurities that may capture the neutrons or add to the radiological hazards.

REACTOR COOLANTS

The primary purpose of a reactor coolant is to absorb heat from the reactor. The coolant may be either a gas or a liquid; it must possess good heat transfer properties, have good thermal properties, be noncorrosive to the system, be nonhazardous if exposed to radiation, and be of low cost. Liquid sodium and carbon dioxide have been used as coolants in experimental reactors. All naval reactors use water as the reactor coolant.

REFLECTORS

The escape of neutrons from the reactor core tends to reduce the reactor's effective neutron multiplication factor and to increase the number of neutrons outside the core. To reduce this leakage of neutrons from the core, a reflector is installed around the outside of the core. The use of a reflector serves to reduce the required core size and the amount of shielding that must be installed to provide for radiation protection. The characteristics required for a reflector are essentially the same as those required for a moderator.

Since ordinary water of high purity is suitable for moderators, coolants, and reflectors, the inference is that it could serve all three functions in the same reactor. This is the case in all naval nuclear reactors.

SHIELDING

The shielding of a nuclear reactor serves the dual purpose of (1) reducing the radiation so that it will not interfere with the necessary instrumentation, and (2) protecting operating personnel from radiation.

The type of shielding material used is dependent upon the purpose of the particular reactor and upon the nature of the radioactive particles being attenuated or absorbed.

The shielding against alpha particles is a relatively simple matter. Since an alpha particle has a positive electrical charge of 2, a few centimeters of air is all that is required for attenuation. Any light material, such as aluminum or plastics, makes a suitable shield for beta particles.

Neutrons and gamma rays have considerable penetrating power; therefore, shielding against them is more difficult. Since neutrons are best attenuated by elastic collisions, any hydrogenous material, such as polyethylene or water, is suitable as a neutron shield. Sometimes polyethylene with boron is used for neutron shields, as boron has a high neutron capture cross section. Gamma rays are best attenuated by a dense material such as lead.

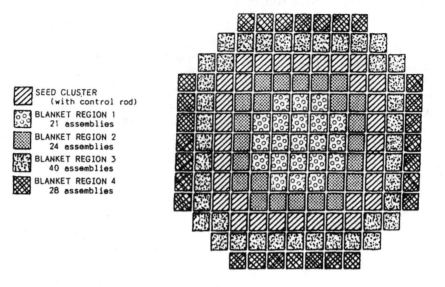

SEED CLUSTER
(with control rod)
BLANKET REGION 1
21 assemblies
BLANKET REGION 2
24 assemblies
BLANKET REGION 3
40 assemblies
BLANKET REGION 4
28 assemblies

147.191

Figure 15-6.—Cross-sectional view of a PWR core.

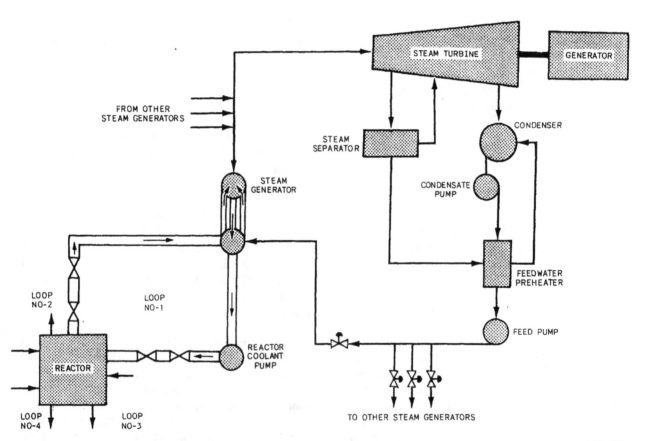

147.158

Figure 15-7.—Schematic diagram of a PWR plant.

NUCLEAR REACTORS

The purpose of any power reactor is to provide thermal energy that can be converted to useful work. Several types of experimental and operational reactors have been designed. They include the pressurized water reactor (PWR), the sodium-cooled reactor, the boiling water reactor (BWR), the light water breeder reactor (LWBR), and the gas-cooled reactor (GCR). The PWR is the only type of reactor in naval use.

The first full-scale nuclear-powered central station in the United States was the pressurized water reactor (PWR) at Shippingport, Pennsylvania.[3] Responsibility for this station, from its initial operation in 1957, through its conversion to the light water breeder reactor (LWBR) in 1977, to its removal from service in 1984 was with the Division of Naval Reactors under the direction of Admiral H. G. Rickover until 1982 and Admiral K. R. McKee thereafter.

The Shippingport PWR was a thermal, heterogeneous reactor fueled with enriched uranium-235 "seed clusters" arranged in a square annulus in the center of the core, surrounded inside and outside by "blanket assemblies" of uranium-238 fuel elements. As shown in figure 15-6, each PWR seed cluster contained a cruciform control rod. The blanket assemblies were divided into four regions according to flow requirements. This type of reactor can be called a converter, since the uranium-238 is converted into the fissionable fuel of plutonium-239.

A schematic diagram of a PWR and its associated steam plant with power output and flow ratings is shown in figure 15-7. The reactor plant consists of a single reactor with four main coolant loops; the plant is capable of maintaining full power on three loops. Each coolant loop contains a steam generator, a pump, associated piping, and valves.

High purity water at a pressure of 2000 psia serves as both moderator and coolant for the plant. At full power the inlet water temperature

to the reactor is 508 °F and the outlet temperature is 542 °F.

The coolant enters the bottom of the reactor vessel (fig. 15-8) where 90 percent of the water flows upward between the fuel plates, with the remainder bypassing the core in order to cool the walls of the reactor vessel and the thermal shield. After having absorbed heat as it goes through the core, the water leaves the top of the reactor vessel

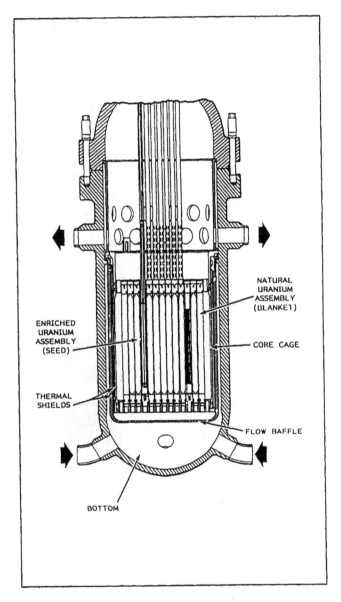

NATURAL URANIUM ASSEMBLY (BLANKET)

ENRICHED URANIUM ASSEMBLY (SEED)

CORE CAGE

THERMAL SHIELDS

FLOW BAFFLE

BOTTOM

147.159

Figure 15-8.—Longitudinal section of a PWR (only two of four inlets and outlets are shown).

---

[3] For a complete description of the Shippingport PWR, consult the text, The *Shippingport Pressurized Water Reactor*, written by personnel from the Division of Naval Reactors, Westinghouse Bettis Atomic Power Laboratory, and Duquesne Light Company (Reading, MA: Addison-Wesley Publication Co., Inc., 1958).

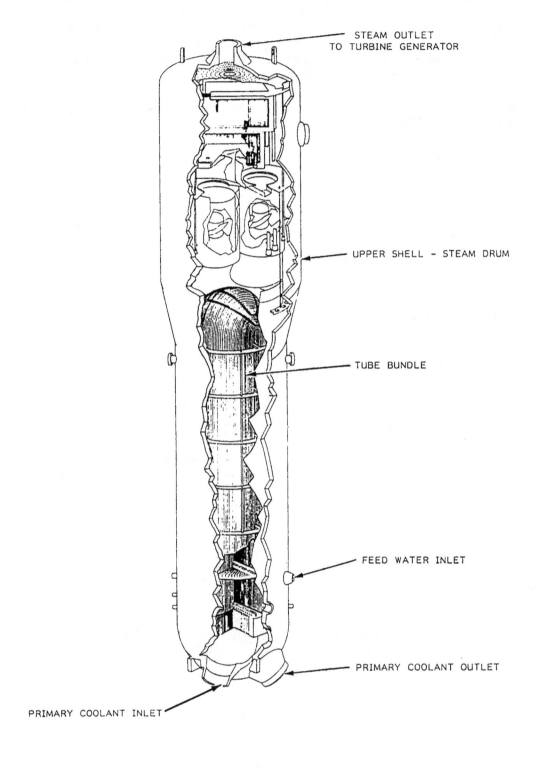

STEAM OUTLET
TO TURBINE GENERATOR

UPPER SHELL - STEAM DRUM

TUBE BUNDLE

FEED WATER INLET

PRIMARY COOLANT OUTLET

PRIMARY COOLANT INLET

147.193

Figure 15-9.—A PWR steam generator.

through the outlet nozzles and flows through connecting piping to the steam generator.

The steam generator (fig. 15-9) is a shell-and-tube type of heat exchanger with the primary coolant (reactor coolant) flowing through the tubes and the secondary water (boiler water) surrounding the tubes. Heat is transferred to the secondary water in the steam generator, producing high quality saturated steam for the use in the turbines.

The primary coolant flows from the steam generator to a hermetically sealed (canned rotor) pump (fig. 15-10) and is pumped through connecting piping to the bottom of the reactor vessel to complete the primary coolant cycle.

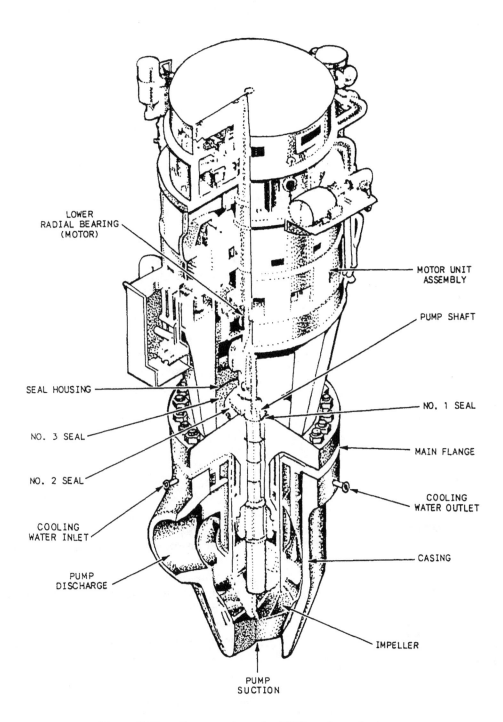

LOWER RADIAL BEARING (MOTOR)

MOTOR UNIT ASSEMBLY

PUMP SHAFT

SEAL HOUSING

NO. 1 SEAL

NO. 3 SEAL

NO. 2 SEAL

MAIN FLANGE

COOLING WATER INLET

COOLING WATER OUTLET

PUMP DISCHARGE

CASING

IMPELLER

PUMP SUCTION

147.160

Figure 15-10.—Cutaway view of a PWR main coolant pump.

15-15

The pressure on the reactor vessel and the main coolant loop is maintained by a pressurizing tank (fig. 15-11) that operates under the saturation conditions of 636°F and 2000 psia. A second function of the pressurizing tank is to act as a surge tank for the primary system. Under no-load conditions the inlet, outlet, and average temperatures of the reactor coolant are nearly equal in value. As the power increases, the average temperature remains constant but the inlet and outlet temperatures diverge. Since the colder leg of the primary coolant is the longer, the net effect in the pressurizer is a decrease in level to make up for the increase in density of the water in the primary loop. The reverse holds true with a decreasing power level. Electrical heaters and a spray valve with a supply of water from the cold leg of the primary coolant assist in maintaining a steam blanket in the upper part of the pressurizer and also assist in maintaining saturation conditions of 636°F and 2000 psia.

PRINCIPLES OF REACTOR CONTROL

Reactor control principles[4] that are of particular interest to this discussion include the negative temperature coefficient, the delayed neutron action, and nuclear poisons.

NEGATIVE TEMPERATURE COEFFICIENT

The term *negative temperature coefficient* is used to express the relationship between temperature and reactivity—as the temperature decreases, the reactivity increases. The negative temperature coefficient is a design requirement and is achieved by the proper ratio of elements in the reactor, the geometry of the reactor, and the physical size of the reactor. The negative temperature coefficient makes it possible to keep a power reactor critical with minimal movement of the control rods.

The concept of the negative temperature coefficient may be most easily understood by use of an example. Assume that, in the PWR plant shown in figure 15-7, the reactor is critical and the machinery is operating at a given power level. Now, if the valve is opened to increase the turbine speed, the rate of steam flow, and the power level of the reactor, the measurable effect with installed instrumentation is a decrease in the temperature of the primary coolant leaving the steam generator. The decrease in temperature is small but significant in that it results in an increase in density of the coolant. As the density of the coolant increases, the probability that a fast neutron will collide with a hydrogen nucleus in a water

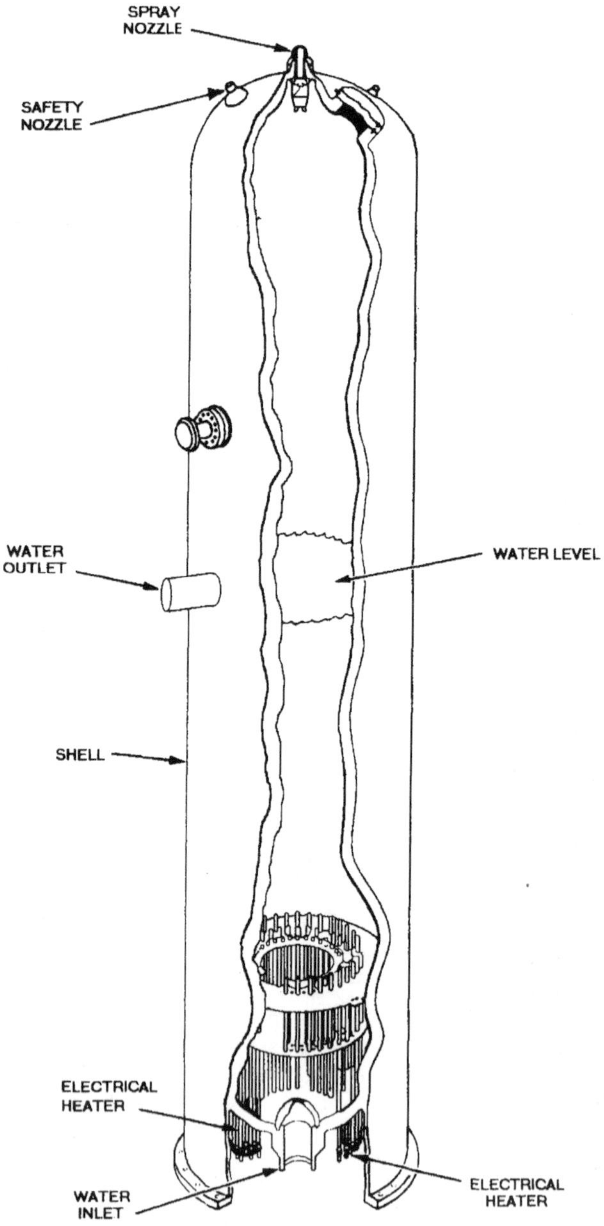

147.161

Figure 15-11.—Cutaway view of a PWR pressurizing tank.

SPRAY NOZZLE

SAFETY NOZZLE

WATER OUTLET

WATER LEVEL

SHELL

ELECTRICAL HEATER

WATER INLET

ELECTRICAL HEATER

[4]For a discussion of reactor kinetics and control, see Samuel Glasstone and Alexander Sesonske, *Nuclear Reactor Engineering* (New York: Van Nostrand Reinhold, 1981).

molecule increases. This higher scattering collision probability allows the coolant, in its capacity as moderator, to thermalize neutrons at a faster rate, supplying more thermal neutrons to be absorbed in the fuel. As more neutrons are absorbed in the fuel, more fissions occur, resulting in a higher power level and more heat being generated by the reactor. The additional heat is removed by the reactor coolant to the secondary water in the steam generator to compensate for the increased steam demand by the turbine. The temperature of the primary coolant leaving the steam generator increases slightly, lowering the scattering cross section of the moderator, and the reactor settles out at a higher power level.

DELAYED NEUTRON ACTION

The delayed neutron action is a phenomenon that simplifies reactor control considerably. Each fission in a nuclear reactor releases on the average between two and three neutrons, which either leak out of the reactor or are absorbed in reactor materials. If the reactor material that absorbs the neutron happens to be the fissionable fuel, and if the neutron is of proper energy level, another fission is likely to result. The majority of the neutrons released in the fission process appear instantaneously and are termed *prompt neutrons*; but other neutrons are born after fission and are termed *delayed neutrons*. The delayed neutrons appear in a time range from less than 1 second to about 2 minutes after the fission takes place. The weighted mean lifetime of the delayed neutrons is approximately 13 seconds. About 0.65 percent of the neutrons produced in the fission process are delayed neutrons.

Should a reactor become prompt critical (critical on prompt neutrons), it would be very difficult to control and any delayed neutrons would tend to make it supercritical. The delayed neutrons have the effect of increasing the reactor period sufficiently to permit reactor control. Reactor period is the time required to change the power level by a factor of e (the base of the system of natural logarithms).

NUCLEAR POISONS

A nuclear poison is material in the reactor that has a high absorption cross section for neutrons. Some poisons are classed as burnable poisons and are placed in the reactor for the purpose of extending the core life; other poisons are generated in the fission process and have a tendency to be a hindrance to reactor operation.

A burnable poison has a relatively high cross section for neutron absorption but is used up in the early part of the core life. By adding a burnable poison to the reactor, more fuel can be loaded into the core, thus extending the life of the core.

Most of the fission products produced in a reactor have a small absorption cross section. The most important one that does have a high absorption cross section for neutrons is xenon-135; this can become a problem near the end of core life. Xenon-135 is a direct fission product a small percentage of the time, but it is mostly produced in the decay of iodine-135, as indicated in the following decay chain:

$$^{135}_{53}\text{I} \quad \xrightarrow{\text{6.585 hrs}} \quad ^{135}_{54}\text{Xe} \quad \xrightarrow{\text{9.10 hrs}}$$

$$^{135}_{55}\text{Cs} \quad \xrightarrow{3 \times 10^6 \text{ yrs}} \quad ^{135}_{56}\text{Ba}$$

Xenon-135 has a high neutron absorption cross section. In normal operation of the reactor, xenon-135 absorbs a neutron and is transformed to the stable isotope of xenon-136, which presents no poison problem to the reactor. Equilibrium xenon is reached after about 40 hours of steady-state operation. At this point, the same amount of xenon-135 is being "burned" by neutron absorption as is being produced by the fission process.

The second, and perhaps the more serious, effect of xenon poisoning occurs near the end of core life. As indicated by the half-lives shown in the xenon decay chain, xenon-135 is produced at a faster rate than it decays. The buildup of xenon-135 in the reactor reaches a maximum about 11 hours after shutdown. Should a scram occur near the end of core life, the xenon buildup may make it impossible to take the reactor critical until the xenon has decayed off. In this unlikely situation, the reactor might have to remain shut down for as much as 2 days before it is capable of overriding the poison buildup.

THE NAVAL NUCLEAR POWER PLANT

Since many aspects of the design and operation of naval nuclear propulsion plants involve classified information, the information presented here is necessarily brief and general in nature.

U.S. naval nuclear propulsion plants use a PWR design that has two basic systems, as illustrated in figure 15-12, the primary system and the secondary system. The primary system circulates ordinary water in a closed loop and consists of the reactor, piping loops, pumps, and steam generators. The heat produced in the reactor is transferred to the water, which is kept under high pressure so it does not boil. This water is pumped through the steam generators, where it gives up its energy. The water is then pumped back into the reactor for reheating.

In the steam generators, the heat from the water in the primary system is transferred through the piping to the water in the secondary system, which creates steam. The secondary system is isolated by piping from the primary system so that the water in the two systems does not intermix.

The secondary system is similar to a conventional steam propulsion plant. Engine-room equipment, which include turbines, turbogenerators, and condensers, are similar in design and fundamental purposes. However, there are some differences—the secondary system has a steam generator, whereas the conventional system has a boiler; and the steam generator produces saturated steam, not superheated steam as in the conventional steam system. Thus, the primary and secondary systems are independent, closed systems in which the water is recirculated and reused. By having the two separate systems, any possible nuclear contamination is limited to the primary system.

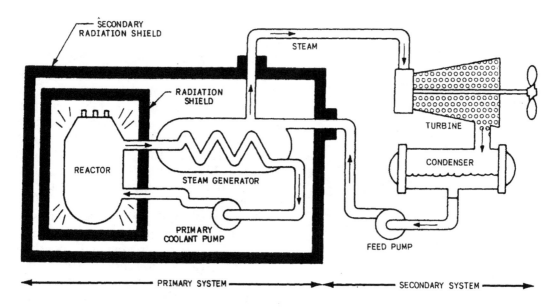

147.192

Figure 15-12.—A naval PWR.

CHAPTER 16

GAS TURBINE ENGINES

LEARNING OBJECTIVES

Upon completion of this chapter, you should be able to do the following:

1. Describe the differences between gas turbine engines, internal combustion engines, and steam turbine engines.

2. Identify the thermodynamic process under which a gas turbine operates.

3. Identify the three major components of a gas turbine.

4. Identify the differences in component parts of the gas turbine in each different shaft-type engine.

5. Identify the two different compressor types in a gas turbine engine.

6. Explain the function of each component part of a gas turbine.

7. Identify the different types and construction of combustion chambers.

8. Identify the function of the turbine and its similarity to a steam turbine.

9. Describe the fuel and speed governing system and the function of its components.

10. Explain the factors that make the lubricating system vitally important to a gas turbine.

11. Describe the air systems associated with a gas turbine and their purposes.

INTRODUCTION

The U.S. Navy entered the marine gas turbine field with the *Asheville* class patrol gunboats. These ships had two diesel engines for cruising and a General Electric LM1500 gas turbine for operating at high speed. As a result of increased reliability and efficiency in gas turbine design, today's Navy uses gas turbines for various ships, both for main propulsion and a number of auxiliary applications. Gas turbines are currently installed as prime movers on cruisers, destroyers, frigates, hovercraft, landing craft, and hydrofoils.

This chapter deals primarily with the general principles of gas turbine engines (GTEs). The LM2500 is the most prevalent GTE in the fleet. It is used as the main engine for the more than 100 ships in the following classes: *Spruance* destroyers (DD-963 through -993), *Kidd* guided

missile destroyers (DDG-994 through -997), *Ticonderoga* guided missile cruisers (CG-47 and up), *Pegasus* hydrofoils (PHM-1 through -6), *Oliver Hazard Perry* guided missile frigates (FFG-7 and up), and *Arliegh Burke* guided missile destroyers (DDG-51 and up). The Allison 501-K17 is used to power the electric generators aboard the *Spruance, Kidd,* and *Ticonderoga* classes of ships. Detailed information of any specific model may be obtained from the manufacturer's technical manual.

GAS TURBINE OPERATION

A gas turbine is composed of three major sections (fig. 16-1):

● Compressor(s)

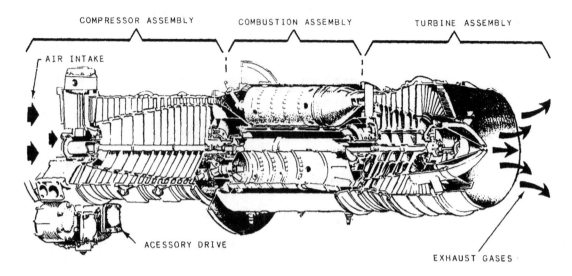

COMPRESSOR ASSEMBLY   COMBUSTION ASSEMBLY   TURBINE ASSEMBLY

AIR INTAKE

ACESSORY DRIVE

EXHAUST GASES

Figure 16-1.—Gas turbine operation.

● Combustion chamber(s)

● Turbine wheel(s)

The operation of a gas turbine involves air being taken in through the air inlet duct by the compressor, which compresses the air and thereby raises pressure and temperature. The air is then discharged into the combustion chamber(s) where fuel is admitted by the fuel nozzle(s). The fuel-air mixture is ignited by igniter(s), and combustion takes place. Combustion is continuous, and the igniters are de-energized after combustion is established. The hot and rapidly expanding gases are directed toward the turbine rotor assembly. Kinetic and thermal energy are extracted by the turbine wheel(s). The action of the gases against the turbine blades causes the turbine assembly to rotate. The turbine rotor is connected to the compressor, which rotates with the turbine. The exhaust gases then are discharged through the exhaust duct.

About 75 percent of the power developed by a gas turbine is used to drive the compressor and accessories, and 25 percent is used to drive a generator or to propel a ship.

There are several pressure, volume, and velocity changes that occur within a gas turbine during operation. The convergent-divergent process is an application of Bernoulli's principle. (If a fluid flowing through a tube reaches a constriction or narrowing of the tube, the velocity of the fluid flowing through the constriction increases and the pressure decreases. The opposite is true when the fluid leaves the constriction; velocity decreases and pressure increases.) Boyle's law and Charles's law also come into play during this process. Boyle's law: The volume of any dry gas varies inversely with the applied pressure, provided the temperature remains constant. Charles's law: If the pressure is constant, the volume of dry gas varies directly with the absolute temperature.

Now, let's apply these laws to the gas turbine. Refer to figure 16-2.

Air is drawn into the front of the compressor. The rotor is so constructed that the area decreases toward the rear. This tapered construction gives a convergent area (area A). Each succeeding stage is smaller, which increases velocity (Bernoulli's law).

Between each rotating stage is a stationary stage or stator. The stator partially converts the high velocity to pressure and directs the air to the next set of rotating blades.

Because of its high rotational speed, the rotor imparts velocity to the air. Each pair of rotor and stator blades constitutes a pressure stage. Also, there is both a pressure increase at each stage and a reduction in volume (Boyle's law).

This process continues at each stage until the air charge enters the diffuser (area B). There is a short area in the diffuser where no further changes take place. As the air charge approaches the end of the diffuser, you will notice that the opening flares (diverges) outward. At this point,

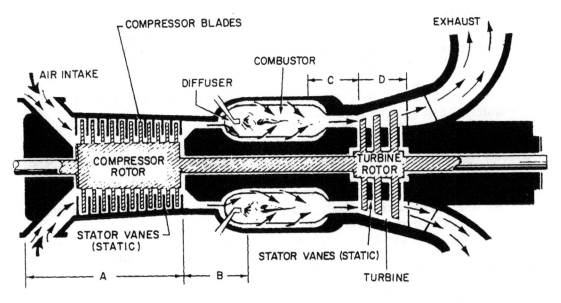

Figure 16-2.—Convergent-divergent process.

the air loses velocity and increases in volume and pressure. Thus, the velocity energy has become pressure energy, while pressure through the diffuser has remained constant. The reverse of Bernoulli's principle and Boyle's law has taken place. The compressor, continuously forcing more air through this section at a constant rate, maintains constant pressure. Once the air is in the combustor, combustion takes place at constant pressure. After combustion there is a large increase in the volume of the air and combustion gases (Charles's law).

The combustion gases go rearward to area C. This occurs partially by velocity imparted by the compressor and partially because area C is a lower-pressure area. The end of area C is the turbine nozzle section. Here you will find a decrease in pressure and an increase in velocity. The high-velocity, high-temperature, low-pressure (LP) gases are directed through the inlet nozzle to the first stage of the turbine rotor (area D). The high-velocity, high-temperature gases cause the rotor to rotate by transferring velocity energy and thermal energy to the turbine blades. Area D is a divergent area. Between each rotating turbine stage is a static stage or nozzle. The nozzles act much the same as the stators in the compressor.

A nozzle is a stator ring with a series of vanes. They act as small nozzles to direct the combustion gases uniformly and at the proper angle to the turbine blades. Due to the design of the nozzles, each succeeding stage imparts velocity to the gases as they pass through the nozzle. Each nozzle converts heat and pressure energy into velocity energy by controlling the expansion of the gas. Each small nozzle has a convergent area.

Each stage of the turbine is larger than the preceding one. The pressure energy drops are quite rapid; consequently, each stage must be larger to use the energy of a lower pressure, lower temperature, and larger volume of gases. If more stages are used, the rate of divergence will be less.

COMPARISON OF GAS TURBINE ENGINES, INTERNAL COMBUSTION ENGINES, AND STEAM TURBINES

The GTE bears some resemblance to an internal combustion engine of the reciprocating type and some resemblance to a steam turbine. However, a brief consideration of the basic principles of a GTE reveals several ways in which the GTE is quite unlike either the reciprocating internal combustion engine or the steam turbine.

WORKING SUBSTANCE

One way in which the three types of engines differ is in the working substance. The working fluid in a steam turbine installation is steam. In both the reciprocating internal combustion engine and the GTE, the working fluid may be considered as being the hot gases of combustion that result from the burning of fuel in air.

OPERATING CYCLE

The GTE operates on the Brayton cycle. The Brayton cycle is one where combustion occurs at constant pressure. In gas turbines, specific components are designed to perform each function separately. These functions are intake, compression, combustion, expansion, and exhaust.

The Brayton cycle can also be graphically explained (fig. 16-3). The compressor draws in air at atmospheric pressure and constant volume (point A). As the air passes through the compressor, it increases in pressure and decreases in volume (line A-B). At point B, combustion occurs at constant pressure while the increased temperature causes a sharp increase in volume (line B-C). The gases at constant pressure and increased volume enter the turbine and expand through it. As the gases pass through the turbine rotor, the rotor turns kinetic and thermal energy into mechanical energy. The expanding size of the passages causes further increase in volume and a sharp decrease in pressure (line C-D). The gases are released through the stack with a large drop in volume and at constant pressure (line D-A). The cycle is continuous in a gas turbine, with each action occurring at all times.

Open and Closed Cycles

Internal combustion engines and GTEs operate on an open cycle. This means the working substance is taken in, passes through the engine once, and is discharged. In comparison, the steam turbine plant operates on a closed cycle. This means that energy is extracted from the working substance in the engine, then the working substance is recovered to have energy added, and it then passes through the engine again in a closed loop.

Heated and Unheated Engines

In addition to being classified as open or closed, operating cycles can be classified as heated or unheated. This classification depends upon the location where energy is added to the working substance. An internal combustion engine is a heated engine in that energy is both added and extracted from the working substance in the engine (in this case the engine cylinder). GTEs and steam plants are unheated engines in that energy is added to the working substance in one location (boiler for the steam plant, combustion chamber for the gas turbine engine) and energy is extracted from the working substance in another location (the turbines).

ADVANTAGES AND DISADVANTAGES OF THE GAS TURBINE PROPULSION SYSTEM

The gas turbine, when compared to other types of engines, offers many advantages. Its greatest asset is its high power-to-weight ratio. The smoothness of the gas turbine and low-frequency vibration in gas turbines makes them preferable to diesel engines because there is less noise for a submarine to pick up at long range. Modern production techniques have made gas turbines economical in terms of horsepower-per-dollar on initial installation, and their increasing reliability makes them a cost-effective alternative to steam turbine or diesel engine installation. In terms of fuel economy, modern marine gas turbines can compete with diesel engines. Compared to a steam propulsion plant, gas turbines are more fuel efficient at high speeds, whereas steam turbines are more efficient at moderate speeds.

However, there are some disadvantages to gas turbines. Since they are high-performance engines, many parts are under high stress. Improper maintenance and lack of attention to details of procedure will impair engine performance and may ultimately lead to engine failure. A pencil mark on a compressor turbine blade or a fingerprint in the wrong place can cause failure of the part. The turbine takes in large quantities of air that may contain substances or objects that can harm the engine.

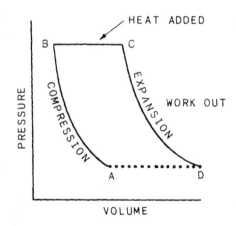

Figure 16-3.—Brayton cycle (open) pressure volume diagram.

Gas turbines produce loud, high-pitched noises that can cause hearing loss. In shipboard installations, special sound-proofing is necessary. This adds to the complexity of the installation and makes access for maintenance more difficult.

From a tactical standpoint, there are two major drawbacks to the GTE. The first is the large amount of exhaust heat produced by the engine. Most current antiship missiles are heat-seekers, and the infrared signature of a GTE makes it an easy target. Countermeasures, such as exhaust gas cooling and infrared decoys, have been developed to reduce this problem.

The second tactical disadvantage is the requirement for depot maintenance and repair of major casualties. The turbines cannot be repaired in place on the ship; they must be removed and replaced with rebuilt engines if anything goes wrong. Ships are designed so the GTE can easily be removed and replaced. This requires the service of a crane. Navy tenders and shore maintenance facilities have cranes and, usually, overhauled engines. With an overhauled gas turbine available, the GTE can be removed and the overhauled engine installed in 1 to 2 days.

TYPES OF GAS TURBINE ENGINES

GTEs are classified by their construction (the type of shafting, compressor, or combustor). The type of shaft used on a GTE may be single shaft, split shaft, or twin spool, with the first two being the two most common used in naval vessels. The compressor may be either centrifugal or axial type. The combustor may be can type, annular type, or can-annular type. The can-type chamber is used primarily on engines with centrifugal compressors. The annular and can-annular chambers are used on engines with axial-flow compressors.

SHAFT TYPES

As stated earlier, the type of shaft used on a GTE may be single shaft, split shaft, or twin spool. The following paragraphs will discuss each type.

Single-Shaft

The GTE shown in figure 16-4 is called a single-shaft type because one shaft from the turbine rotor drives the compressor and an extension of this same shaft drives the load. In most cases, a single-shaft GTE has a speed decreaser or reduction gear between the shaft and the power output shaft; however, there is still a mechanical connection throughout the entire engine.

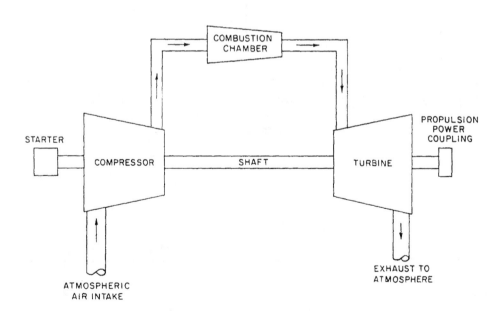

Figure 16-4.—A single-shaft gas turbine engine.

16-5

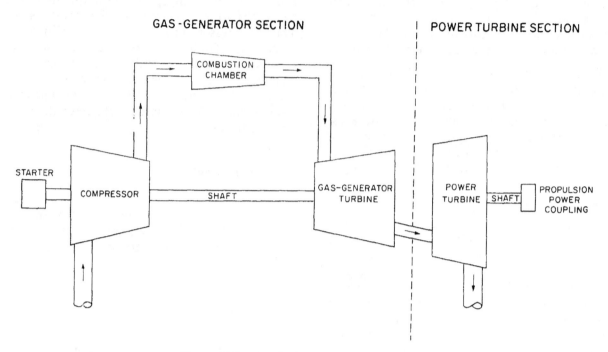

Figure 16-5.—A split-shaft gas turbine engine.

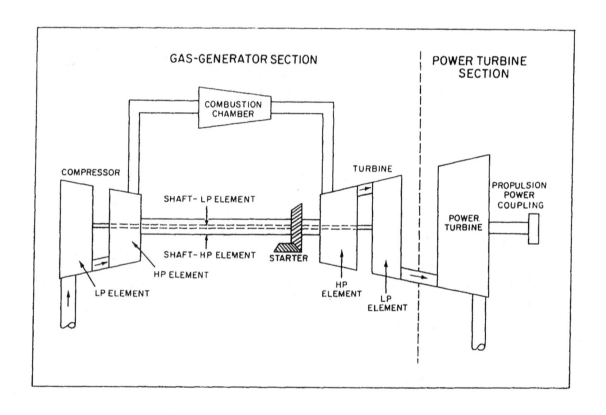

Figure 16-6.—A twin-spool gas turbine engine.

16-6

Split-Shaft

The GTE shown in figure 16-5 is called a split-shaft type. This engine is considered to be split into two sections: the gas-producing section, or gas generator (GG), and the power turbine (PT) section. The GG section, in which a stream of expanding gases is created as a result of continuous combustion, includes the compressor, the combustion chamber (or chambers), and the GG turbine. The PT section consists of a PT and the power output shaft. In this type of GTE, there is no mechanical connection between the GG turbine and the PT. When the engine is operating, the two turbines produce basically the same effect as that produced by a hydraulic torque converter. The split-shaft GTE is well suited for use as a propulsion unit where loads vary, since the GG section can be operated at a steady and continuous speed while the PT section is free to vary with the load. Starting effort required for a split-shaft GTE is far less than that required for a single-shaft GTE connected to the reduction gear, propulsion shaft, and propeller. The LM2500 gas turbine is a split-shaft GTE.

Twin-Spool

In the twin-spool GTE (fig. 16-6), the air compressor is split into two sections or stages, and each stage is driven by a separate turbine element. The LP turbine element drives the LP compressor element and the high-pressure (HP) turbine element drives the HP compressor element. The LP compressor and turbine are connected by a shaft that runs through the hollow shaft that connects the HP turbine to the HP compressor. The starter drives the HP assembly during light off. The PT functions the same as in the split-shaft engine. A larger volume of air can be handled as compared to a single- or split-shaft engine; however, the engine has more moving parts, and the increase in overall dimensions and complexity make the engine less desirable for ship's propulsion than the split-shaft engine. An example of a twin-spool GTE is the Pratt Whitney FT-4 engine used as the main engine on the Coast Guard *Hamilton* class high-endurance cutter.

COMPRESSOR TYPES

Gas turbines may be classified by compressor type, according to the direction of the flow of air through the compressor. The two principal types are centrifugal flow and axial flow. The centrifugal compressor draws in air at the center or eye of the impeller and accelerates it around and outward. In the axial-flow engine, the air is compressed while continuing its original direction of flow. The flow of air is parallel to the axis of the compressor rotor.

Centrifugal Compressor

The centrifugal compressor is usually located between the accessory section and the combustion section. The basic compressor section consists of an impeller, diffuser, and compressor manifold. The diffuser is bolted to the manifold. Often the entire assembly is referred to as the diffuser. For ease of understanding, we will treat each unit separately.

The impeller may be either single entry or dual entry (fig. 16-7). The main differences between

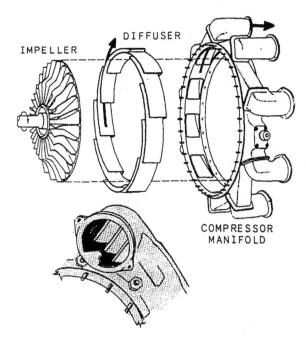

A. SINGLE-ENTRY COMPRESSOR

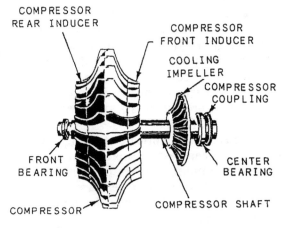

B. DUAL-ENTRY COMPRESSOR

Figure 16-7.—Centrifugal compressors.

16-7

the single entry and dual entry are the size of the impeller and the ducting arrangement. The single-entry impeller permits convenient ducting directly to the inducer vanes. The dual-entry impeller uses a more complicated ducting to reach the rear side. Single-entry impellers are slightly more efficient in receiving air. To provide sufficient air, they must be of greater diameter, which increases the overall diameter of the engine.

Dual-entry impellers are smaller in diameter. They rotate at higher speeds to ensure sufficient airflow. Most gas turbines of present-day design use the dual-entry compressor to reduce engine diameter. The air must enter the engine at almost right angles to the engine axis. Because of this, a plenum chamber is also required for dual-entry compressors. The air must surround the compressor at positive pressure before entering the compressor to give positive flow.

PRINCIPLES OF OPERATION.—The compressor draws in the entering air at the hub of the impeller and accelerates it radially outward by centrifugal force through the impeller. It leaves the impeller at a high velocity with LP and flows through the diffuser (fig. 16-7, view A). The diffuser converts the high-velocity, LP air to low velocity with HP. The compressor manifold diverts the flow of air from the diffuser, which is an integral part of the manifold, into the combustion chambers.

CONSTRUCTION.—In the centrifugal compressor, the manifold has one outlet port for each combustion chamber. The outlet ports are bolted to an outlet elbow on the manifold (fig. 16-7, view A). The outlet ports ensure that the same amount of air is delivered to each combustion chamber.

The outlets are known by a variety of names. Regardless of the names used, the elbows change the airflow from radial flow to axial flow. Then the diffusion process is completed after the turn. Each elbow contains from two to four turning vanes to efficiently perform the turning process. They also reduce air pressure losses by presenting a smooth turning surface.

The impeller is usually fabricated from forged aluminum alloy, heat-treated, machined, and smoothed for minimum flow restriction and turbulence. Some types of impellers are made from a single forging. In other types, the inducer vanes are separate pieces.

Centrifugal compressors may achieve efficiencies of 80 to 84 percent at pressure ratios of 2.5:1

to 4:1 and efficiencies of 76 to 81 percent at pressure ratios of 4:1 to 10:1.

The advantages of centrifugal compressors are that they are rugged, simple in design, relatively light in weight, and they develop an HP ratio per stage.

The disadvantages of centrifugal compressors are a large frontal area, a lower efficiency, and difficulty in using two or more stages due to air loss that will occur between stages and seals.

Axial-Flow Compressors

There are two main types of axial compressors (fig. 16-8). One is the drum type and the other is the disk type.

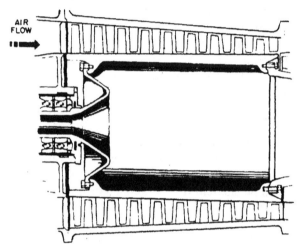

A. DRUM TYPE

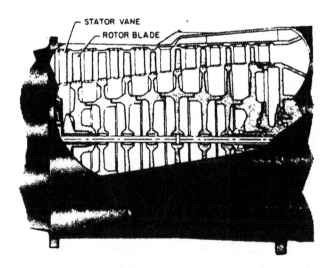

B. DISK TYPE

Figure 16-8.—Axial compressor rotor types.

The purpose of the axial compressor is the same as the centrifugal type. Both take in ambient air and increase the velocity and pressure. They discharge the air through the diffuser into the combustion chamber.

PRINCIPLES OF OPERATION.—The two main elements of an axial-flow compressor are the rotor and the stator (fig. 16-9).

The rotor has fixed blades which force the air rearward much like an aircraft propeller. Behind each rotor stage is a stator. The stator directs the air rearward to the next rotor stage. Each consecutive pair of rotor and stator blades constitutes a pressure stage.

The action of the rotor at each stage increases compression of the air at each stage and accelerates it rearward. By virtue of this increased velocity, energy is transferred from the compressor to the air in the form of velocity energy. The stators at each stage act as diffusers, partially converting high velocity to pressure.

The number of stages required is determined by the amount of air and total pressure rise required. The greater the number of stages, the higher the compression ratio. Most present-day

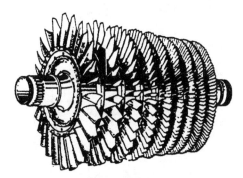

A. ROTOR

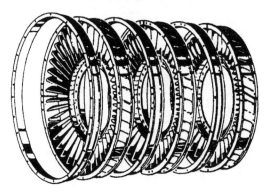

B. STATOR

Figure 16-9.—Axial compressor elements.

engines have 8 to 16 stages, depending on air requirements.

CONSTRUCTION.—The rotor and stators are enclosed in the compressor case. Present-day engines use a case that is horizontally divided into upper and lower halves. The halves are normally bolted together with either dowel pins or fitted bolts. They are located at various points. They ensure proper alignment to each other and in relation to other engine assemblies. The other assemblies bolt to either end of the compressor case.

On some older design engines, the case is a one-piece cylinder open on both ends. The one-piece compressor case is simpler to manufacture; however, any repair or detailed inspection of the compressor rotor is impossible. The engine must be removed and taken to a shop. There it is disassembled for repair or inspection of the rotor or stators. On many engines with the split case, either the upper or lower case can be removed. The engine can remain in place for maintenance and inspection.

The compressor case is usually made of aluminum or steel. The material used will depend on the engine manufacturer and the accessories attached to the case. The compressor case may have external connections made as part of the case. These connections are normally used to bleed air during starting and acceleration or at low-speed operation.

Drum-Type Construction.—The drum-type rotor (fig. 16-8, view A) consists of rings that are flanged to fit one against the other. The entire assembly may then be held together by through bolts. The drum is one diameter over its full length. The blades and stators vary in length from front to rear. The compressor case tapers accordingly. This type of construction is satisfactory for low-speed compressors where centrifugal stresses are low.

Disk-Type Construction.—The disk-type rotor (fig. 16-8, view B) consists of a series of disks constructed of titanium alloys, low-alloy steel, and stainless steel. The blades vary in length from entry to discharge. This is due to a progressive reduction in the annular working space (drum to casing) toward the rear. The working space decreases because the rotor disk diameter increases. The disk-type rotors are used almost exclusively in all present-day, high-speed engines.

TYPES OF COMBUSTION CHAMBERS

The combustion chamber is the location where the fuel-air mixture is burned. The combustion chamber consists of a casing, a perforated inner shell, a fuel nozzle, and a device for initial ignition. The number of combustion chambers in a GTE varies with the type of combustion chamber used.

Combustion chambers are designed to operate with LP losses, high-combustion efficiency, and good flame stability. Additional requirements for the combustion chamber include low rates of carbon formation, light weight, reliability, reasonable length of life, and the ability to mix cold air with the hot combustion gases in such a way as to give uniform temperature distribution to the turbine blades.

Only a small part (perhaps one-fourth) of the air that enters the combustion chamber area is burned with the fuel. The remainder of the air is mixed with combustion gases to keep the temperature low enough to prevent the turbine from overheating.

There are three types of combustion chambers: (1) can, (2) annular, and (3) can-annular. The can-type chamber is used primarily on engines that have a centrifugal compressor. The annular and can-annular types are used on axial-flow compressors.

Can-Type Chamber

The can-type combustion system (fig. 16-10) consists of individual liners and cases mounted around the axis of the engine. Each chamber contains a fuel nozzle. This arrangement makes removing a chamber easy; however, it is a bulky arrangement and makes for a structurally weak engine. The outer casing is welded to a ring that directs the gases into the turbine nozzle. Each of

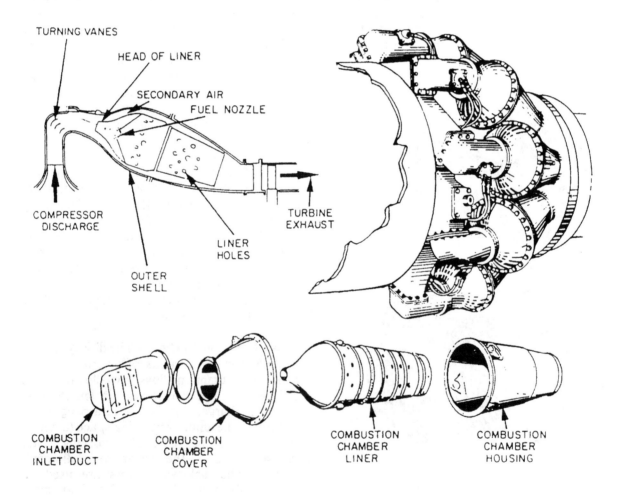

Figure 16-10.—Can-type combustion chamber.

16-10

the casings is linked to the others with a short tube. This arrangement ensures that combustion occurs in all the burners during engine start. Inside each of these tubes is a flame tube that joins an adjacent inner liner.

Annular-Type Chamber

The annular-type combustion chamber (fig. 16-11) is usually found on axial-flow engines. It is probably one of the most popular combustion systems in use. The construction consists of a housing and liner the same as the can type.

The great difference is in the liner. On large engines, the liner consists of an undivided circular shroud extending all the way around the outside of the turbine shaft housing. A large one-piece combustor case covers the liner and is attached at the turbine section and diffuser section.

The dome of the liner has small slots and holes to admit primary air. They also impart a swirling motion for better atomization of fuel. There are also holes in the dome for the fuel nozzles to extend through into the combustion area. In the case of the double-annular chamber, two rows of fuel nozzles are required. The inner and outer liners form the combustion space. The outer liner keeps flame from contacting the combustor case. The inner liner prevents flame from contacting the turbine shaft housing.

Large holes and slots are located along the liners. They (1) admit some cooling air into the combustion space to help cool the hot gases to a safe level, (2) center the flame, and (3) admit the balance of air for combustion. The gases are cooled enough to prevent warpage of the liners.

The annular-type combustion chamber is a very efficient system that minimizes bulk. It can be used most effectively in limited spaces. There are some disadvantages, however. On some engines, the liners are one piece and cannot be removed without engine disassembly. Also, engines that use a one-piece combustor dome must be disassembled to remove the dome.

Can-Annular-Type Chamber

The can-annular-type combustion chamber (fig. 16-12) combines some of the features of both the can and the annular burners.

The can-annular-type chamber design is a result of the split-spool compressor concept. Problems were encountered with a long shaft and with one shaft within the other. Because of these problems, a chamber was designed to perform all the necessary functions.

In the can-annular-type chamber, individual cans are placed inside an annular case. The cans are essentially individual combustion chambers with concentric rings of perforated holes to admit air for cooling. On some models, each can has a round perforated tube that runs down the middle of the can. The tube carries additional air that enters the can through the perforations to

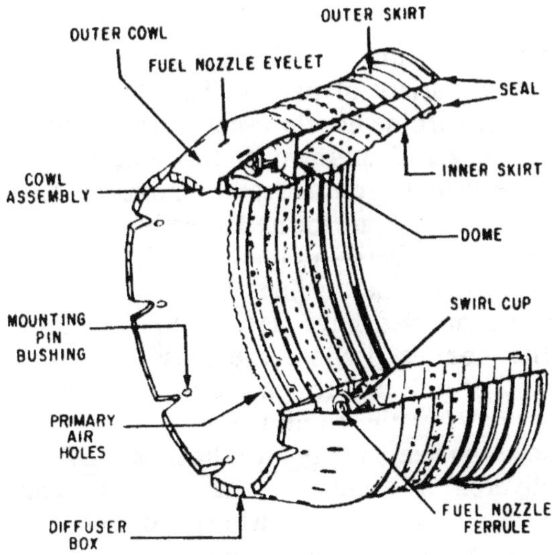

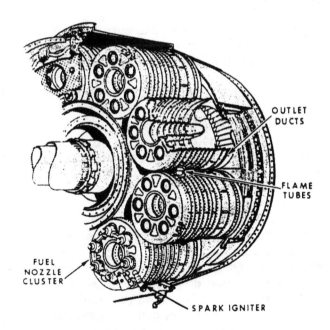

Figure 16-11.—Annular-type combustion chamber.

Figure 16-12.—Can-annual-type combustion chamber.

provide more air for combustion and cooling. The effect is to permit more burning per inch of can length than could otherwise be done.

Fuel nozzle arrangement varies from one nozzle in each can to several nozzles around the perimeter of each can.

The cans have an inherent resistance to buckling because of their small diameter. Each can has two holes that are opposite each other near the forward end of the can. One hole has a collar called a flame tube. When the cans are assembled in the annular case, these holes and their collars form open tubes. The tubes are between adjacent cans so that a flame passes from one can to the next during engine starting.

The short length of the can-annular-type chamber is a structural advantage. It provides minimal pressure drop of the gases between the compressor outlet and the flame area. Another advantage of the can-annular engine is the greater structural strength it gets from its short combustor area. Maintenance is also simple. You can just slide the case back and remove any one burner for inspection or repair. Another good feature is the relatively cool air in the annular outer can. It tends to reduce the high temperatures of the inner cans. At the same time, this air blanket keeps the outer shell of the combustion section cooler.

TURBINE ASSEMBLIES

GTEs are not normally classified by turbine type. However, we will discuss turbines now so you will understand their construction.

In theory, design, and operating characteristics, the turbines used in GTEs are similar to those used in a steam plant. The gas turbine differs from the steam turbine chiefly in (1) the type of blading material used, (2) the means provided for cooling the turbine shaft bearings, and (3) the lower ratio of blade length to wheel diameter.

The terms *gas generator* (GG) *turbine* and *power turbine* (PT) are used to differentiate between the turbines. The GG turbine powers the GG and accessories. The PT powers the ship's propeller through the reduction gear and shafting.

The turbine that drives the compressor of a GTE is located directly behind the combustion chamber outlet. The turbine consists of two basic elements: the stator, or nozzle, and the rotor. Part of a stator element is shown in figure 16-13; a rotor element is shown in figure 16-14.

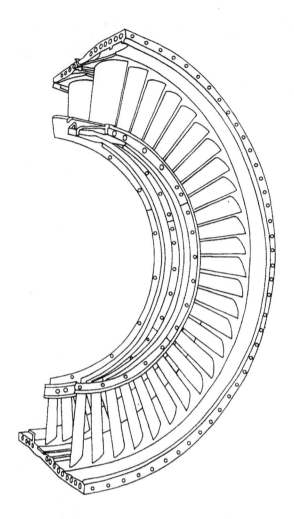

Figure 16-13.—Stator element of a turbine assembly.

Turbine Stators

The stator element of the turbine section is known by a variety of names. The three most common are turbine nozzle vanes, turbine guide vanes, and nozzle diaphragm. In this text, turbine stators are usually referred to as nozzles. The turbine nozzle vanes are located directly aft of the combustion chambers and immediately forward of, and between, the turbine wheels.

Turbine nozzles have a twofold function. First, the nozzles prepare the mass flow for harnessing of power through the turbine rotor. This occurs after the combustion chamber has introduced the heat energy into the mass airflow and delivered it evenly to the nozzles. The stationary vanes of the turbine nozzles are contoured and set at a certain angle. They form a number of small nozzles that discharge the gas as extremely high-speed jets; thus, the nozzle converts a varying portion of the heat and

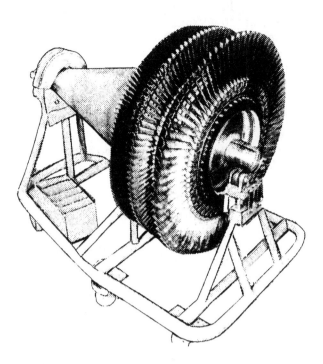

Figure 16-14.—Rotor element of a turbine assembly.

pressure energy to velocity energy. The velocity energy can then be converted to mechanical energy through the rotor blades.

The turbine nozzle functions to deflect the gases to a specific angle in the direction of turbine wheel rotation. The gas flow from the nozzle must enter the turbine blade passageway while it is still rotating. Therefore, it is essential to aim the gas in the general direction of turbine rotation.

Turbine Rotors

The rotor element of the turbine consists of a shaft and bladed wheel(s). The wheel(s) are attached to the main power transmitting shaft of the GTE. The jets of combustion gas leaving the vanes of the stator element act upon the turbine blades. Thus, the turbine wheel can rotate at a high speed. The high rotational speed imposes severe centrifugal loads on the turbine wheel. At the same time the high temperature (1050° to 2300 °F) results in a lowering of the strength of the material.

Consequently, the engine speed and temperature must be controlled to keep turbine operation within safe limits. The operating life of the turbine blading usually determines the life of the GTE.

ENGINE SYSTEMS

The four major systems of a GTE are fuel and speed-governing, lubricating, air, and starter. The major component of these four systems are attached to the accessory drive section.

ACCESSORY DRIVE SECTION

Because the turbine and the compressor are on the same rotating shaft, a popular misconception is that the GTE has only one moving part. This is not the case, however. A GTE requires a starting device (which is usually a moving part), some kind of control mechanism, and power take-offs for the lube-oil and fuel pumps.

The accessory drive section of the GTE takes care of these various accessory functions. The primary purpose of the accessory drive section is to provide space for the mounting of the accessories required for the operation and control of the engine. Secondary purposes include acting as an oil reservoir and/or oil sump and providing for and housing accessory drive gears and reduction gears.

The gear train is driven by the engine rotor through an accessory drive shaft gear coupling. The reduction gearing within the case provides suitable drive speeds for each engine accessory or component. Because the operating rpm of the rotor is so high, the accessory reduction gear ratios are relatively high. The accessory drives are supported by ball bearings assembled in the mounting bores of the accessory case.

Accessories always provided in the accessory drive section include the fuel control, with its governing device; the oil sump; the oil-pressure and scavenging pump or pumps; the auxiliary-fuel pump; and a starter. Additional accessories which may be included in the accessory drive section or which may be provided elsewhere include a starting-fuel pump, a hydraulic-oil pump, a generator, and a tachometer. Most of these accessories are essential for the operation and control of any GTE. However, the particular combination and arrangement of engine-driven accessories depends upon the use for which the GTE is designed.

The three common locations for the accessory drive section are on the air inlet housing, under the compressor front frame, or under the compressor rear frame.

FUEL AND SPEED-GOVERNING SYSTEM

The fuel and speed-governing system regulates and distributes fuel to the combustion section of the GG to control GG speed. The PT speed is not directly controlled, but is established by the gas stream energy level produced by the GG. The fuel and speed-governing system consists of a fuel pump and filter, the main fuel control (MFC), the fuel shutdown valves, the fuel nozzles, and igniters.

To assure an adequate supply of fuel for gas turbine operation, the fuel pump has a higher flow capacity than the gas turbine uses. Within the control the fuel is divided into metered flow and bypass flow. This division maintains a preset pressure drop across the metering valve by the use of a bypass valve. Bypass fuel is ported to the HP element inlet screen of the fuel pump. If an abnormal condition occurs that causes pump outlet pressure to become too high, a relief valve in the pump bypasses fuel back to the HP element inlet screen.

A pressurizing valve, mounted on the fuel control outlet port, maintains back-pressure to ensure adequate fuel pressure for control servo operation. Two electrically operated fuel shutdown valves connected mechanically in series and electrically in parallel provide a positive fuel shutoff. When the fuel shutdown valves are open, metered fuel for gas turbine operation flows from the fuel control, through the pressurizing valve, shutdown valves, fuel manifold, and fuel nozzles. When the fuel shutdown valves are closed, metered fuel is bypassed to the fuel pump inlet; the fuel drain ports in the valves open to allow fuel remaining in the manifold, nozzles, and lines to drain. Thirty fuel nozzles, which project through the compressor rear frame into the

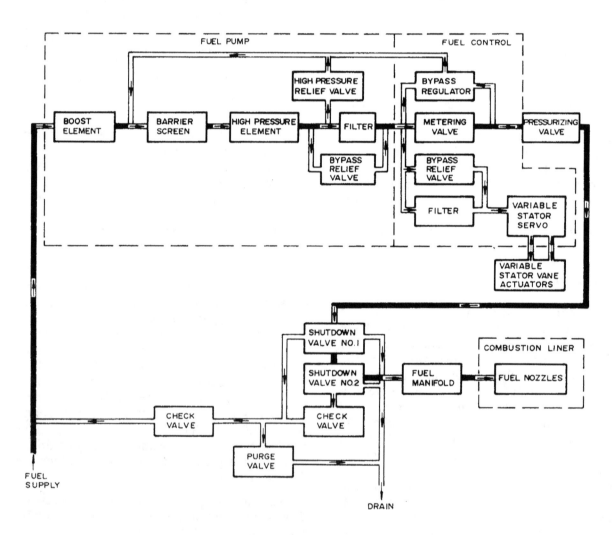

Figure 16-15.—Fuel system block diagram.

combustor, produce an effective spray pattern from start to full power.

The fuel and speed-governing system controls the variable stator vanes (VSVs) to maintain satisfactory compressor performance over a wide range of operating conditions. At high inlet air temperature and low compressor speed, the larger forward stages of the compressor are capable of pumping more air than the smaller aft stages. Because of this characteristic, the aft stages become overloaded. This causes the airflow to stop and possibly reverse. This is known as compressor stall. It is prevented by having the inlet guide vanes (IGVs) and first six stages of stator vanes variable. The fuel and speed-governing system controls the variable vanes, scheduling them toward the closed position when compressor speed drops or inlet temperature rises; thereby, matching the output of the forward stages to that of the rear stages.

NOTE: Figure 16-15 is a block diagram and should be referred to during your study of the fuel system.

Fuel Pump and Filter

Fuel supplied by the fuel-oil service system flows through the base inlet connector of the fuel pump and filter (fig. 16-16).

The fuel pump contains two pumping elements, a centrifugal boost element and an HP gear element. It provides mounting pads and flange

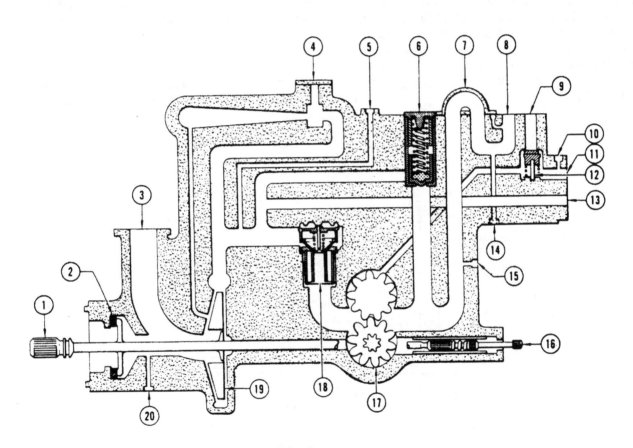

1. Main drive shaft	8. Filter inlet	15. Gear stage discharge pressure tap
2. Shaft seal	9. Filter return	16. Fuel control drive shaft
3. Inlet	10. Pump discharge pressure tap	17. High pressure gear element
4. Vapor in port (capped)	11. Discharge to fuel control	18. Strainer
5. Boost pressure tap	12. Bearing flow regulator	19. Impeller
6. High pressure relief valve	13. Fuel control bypass return	20. Inlet drain
7. Adapter	14. Filter supply pressure tap	

Figure 16-16.—Fuel pump.

ports for the fuel filter and the MFC. This feature reduces the amount of external piping required. The pump also provides a drive shaft for the main fuel control. This eliminates the need for a separate transfer gearbox drive pad.

Fuel from the ship's supply enters the pump through the fuel inlet port and is boosted in pressure by the centrifugal boost element, discharging into a circumferential scroll. The flow passes through a screen which has an integral bypass; it then passes into the HP positive-displacement gear element. The combination of pumping elements is designed to provide improved fuel pump features so that normal operation can be sustained without external boost pumps. The pump incorporates an HP relief valve that lifts at 1350 psia and reseats at 1325 psia. These features protect the pump and downstream components against excessive system pressures.

The fuel filter is an HP filter mounted on the fuel pump and flange-ported to eliminate external piping. The head houses a bypass relief valve; the bowl houses the filter element. The filter element prevents larger contaminants from being carried into the MFC.

HP fuel flows from the fuel pump through the flange port and enters the filter bowl. The fuel then flows from the outside of the filter element to the center, up into the head, out the flange return port, and back into the fuel pump. There, it is routed to the MFC. If the filter becomes clogged, the bypass relief valve opens.

Main Fuel Control

The MFC assembly is the unit that regulates the turbine rpm by adjusting fuel flow from the HP engine-driven pump to the fuel nozzles. Figure 16-17 shows a block diagram of the MFC.

The MFC is basically a speed governor, which senses GG speed and power level position. This control adjusts the fuel flow as necessary to maintain the desired speed set by the power lever. This control is a hydromechanical device that operates by fuel-operated servo valves. It performs the following functions:

1. Controls speed by metering fuel to the fuel nozzles during acceleration, deceleration, and steady-state operation. Excess fuel supplied by the

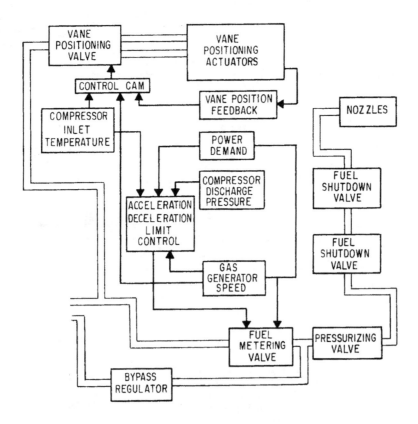

Figure 16-17.—Main fuel control.

fuel pump to the control is returned to the pump downstream of its LP element. The control also uses the fuel from the pump as a hydraulic medium.

2. Alters the fuel schedule automatically to maintain the speed setting and establishes fuel limits for acceleration and deceleration. Three fuel schedules are established by the control: acceleration, deceleration, and minimum fuel schedules. The acceleration schedule limits fuel flow necessary for acceleration to prevent overtemperature and stall. The deceleration schedule limits the rate of fuel flow decrease to prevent combustion flameout during deceleration. The minimum fuel schedule limits fuel flow for starting to prevent overtemperature. The gas turbine parameters vary, so the fuel limits vary to provide optimum acceleration and deceleration schedules. In order for the control to determine the schedules, certain parameters must be sensed. These parameters are compressor discharge pressure (CDP), compressor inlet temperature (CIT), and gas generator speed (N_{GG}). The control, using hydromechanical mechanisms, senses the parameters and computes a limit. The computed limit is compared with actual fuel flow and controls the metering valve should the governor attempt to exceed the limit.

The MFC schedules the VSVs as a function of N_{GG} and CIT. Actual position of the VSVs is sensed by the control via a position feedback cable. One end of the feedback cable is connected to the left master lever arm; the other end is connected to the feedback lever on the MFC.

The VSVs are used to reduce the possibility of compressor stall at low speeds. They modify the amount of airflow and the angle of air attack (pitch) to each set of compressor blades.

The MFC may be divided into a number of working sections, each of which serves a specific function. The operation of these sections is described in the following paragraphs.

FUEL SUPPLY.—Fuel is supplied to the control at fuel pump discharge pressure. The pump supplies more fuel than is required for any engine operating condition. Required fuel flow is determined by a metering valve orifice area. Excess fuel is directed through a bypass valve to the fuel pump. Regulated pressure in the case of the control is used as a reference pressure in the servomechanisms.

One of the three parameters which determine the fuel flow limits is N_{GG}. The fuel control contains a three-dimensional (3-D) cam. It is rotated in proportion to GG speed.

The second of the three parameters is CIT. The 3-D cam is moved axially in relation to CIT.

The third parameter is CDP. Changes in CDP rotate the CDP cam which is mechanically linked to the 3-D cam by levers which combine the output of both cams in adding linkage.

For each N_{GG} and CIT combination, there is a unique position of the 3-D cam. The combination of the 3-D cam position and the CDP cam output in the adding linkage define the allowable acceleration fuel flow limit. A third input represents metered fuel input to the GG. This third input is combined to the adding linkage to determine acceleration and deceleration limits.

Power demand is applied through the power lever angle actuator to a speed governor. The speed governor senses N_{GG}. The governor automatically controls flow of fuel to the GG to maintain a constant N_{GG}. This is done with a flyweight and pilot valve.

To ensure that the GG never exceeds a predetermined safe operating speed, fuel flow is reduced by a force applied to a constant differential pressure valve.

The following external adjustments are provided on the control:

- Fuel Specific Gravity. This affects minor adjustments of minimum and acceleration fuel flow to compensate for specific gravity differences of the various types of fuels.

- Idle Speed. This adjusts minimum steady state (idle) speed.

- Power Trim. This adjusts maximum steady state speed.

- Stator Position Feedback Cable. This is a trim adjustment to open or close the variable stator vanes to the nominal stator schedule.

Minimum speed (idle) adjustment and stator position feedback cable adjustment are normally made only during engine acceptance, testing, or a checkout run. A checkout run follows a component replacement which affects any of these parameters, such as replacement of the MFC.

PRESSURIZING VALVE.—The pressurizing valve pressurizes the fuel system to provide adequate fuel control servo supply pressure and variable stator vane actuation pressure. This is necessary for proper fuel and stator vane scheduling during GG operation at low fuel flow levels. The valve is a fuel pressure-operated, piston valve. The piston is held on its seat (closed) by spring force and fuel pressure (reference pressure) from the MFC. Servo pressure is 110 to 275 pounds per square inch gauge (psig). MFC discharge fuel (metered fuel for combustion) enters the pressurizing valve at the opposite side of the piston. When MFC discharge pressure is 80 to 130 psig greater than reference pressure, the valve opens. Thus, the upstream pressure (including servo supply and stator actuation) is 190 psig or greater before the pressurizing valve opens, which is adequate for proper operation.

Fuel Nozzle

The fuel nozzle (fig. 16-18) is a dual-orifice, swirl atomizer with an internal flow divider.

Thirty fuel nozzles produce the desired spray pattern over the full range of fuel flows. Fuel enters the nozzle through a single tube, flows through a screen, and then the flow divider. When the nozzle is pressurized, primary fuel flows through a drilled passage and tube assembly in the nozzle shank. It then flows through the primary spin chamber and into the combustor. When fuel pressure to the nozzle rises to 330 to 350 psig, the flow divider opens and introduces a secondary fuel flow. The secondary fuel flows through the flow divider, through a passage in the nozzle shank, into the secondary spin chamber, and mixes with the primary flow as it enters the combustor. An air shroud around the nozzle tip scoops a small quantity of air from the main airstream to cool the nozzle tip. This retards the buildup of carbon deposits on its face.

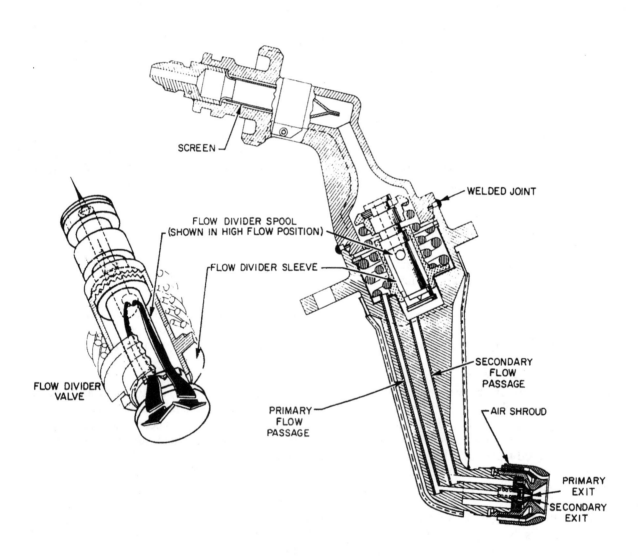

Figure 16-18.—Fuel nozzle.

LUBRICATING SYSTEM

Because of the high operating rpm and the high operating temperatures encountered in gas turbine engines, proper lubrication is of vital importance. The lubricating system is designed to supply bearings and gears with clean lubricating oil at the desired pressures and temperatures. In some installations, the lubricating system also furnishes oil to various hydraulic systems. Heat absorbed by the lubricating oil is transferred to the cooling medium in a lube-oil cooler.

The lubricating system shown in figure 16-19 is the dry-sump type, with a common oil supply from an externally mounted oil tank. The system includes the oil tank, the lubricating-oil pressure pump, the scavenging pumps, the oil cooler, the oil filters, the pressure regulating valve, and the filter and cooler bypass valves. Oil nozzles direct the oil into bearings, gears, and splines. Five separate scavenge elements in the lube and scavenge pump remove oil from A, B, C, and D sumps and the transfer gear box. The scavenged oil is returned to the lube-oil storage and conditioning assembly where it is filtered, cooled, and stored.

Since lubricating oil is supplied to various parts of the system under pressure, there are three

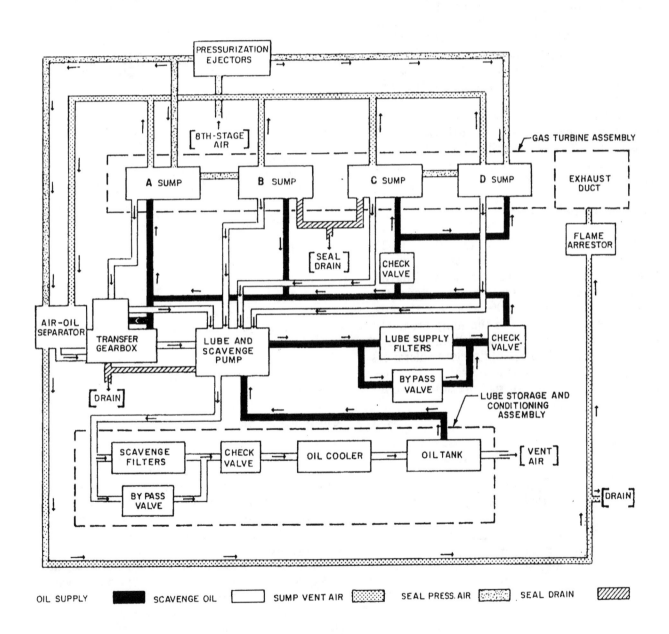

Figure 16-19.—Lubricating system schematic.

16-19

types of rotating oil seals common in gas turbines to prevent oil from leaking to unwanted areas: the lip type, the labyrinth/windback, and the carbon ring. These seals will be discussed in the following paragraphs.

Lip-Type Seal

The lip-type seal (fig. 16-20) is used to prevent leakage in one direction only. A metal frame is covered with a synthetic material, usually neoprene. The neoprene is somewhat smaller than the shaft. The elastic ability of the neoprene will allow the shaft to slide through the seal. The seal is molded with a lip to retain a spring around the center. The spring keeps a snug fit around the shaft. The construction of the lip-type seal allows for some very slight misalignment and for axial movement of the shaft. The lip seals are used where relatively low speeds and temperatures are encountered.

The disadvantages of the lip-type seals are that (1) they will seal against only little or no fluid pressure and (2) they are easily damaged by a burr on the shaft or dirt, either of which will tear the seal and cause leakage.

Labyrinth/Windback Seal

The labyrinth/windback seal (fig. 16-21) combines a rotating seal having oil slingers and a serrated surface with a stationary seal having windback threads and a smooth rub surface. The oil slingers throw oil into the windback threads, which direct the oil back to the sump area. The serrations cut grooves into the smooth surface of the stationary seal to maintain close tolerances throughout a large temperature range. This seal allows a small amount of seal pressurization air

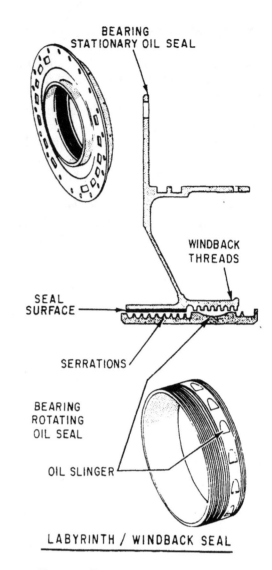

Figure 16-21.—Labyrinth/windback seal.

to leak into the sump, thereby preventing oil leakage.

Carbon Ring Seal

The carbon ring seal (fig. 16-22) consists of a stationary, spring-loaded, carbon sealing ring and a rotating, highly-polished steel mating ring. Carbon seals are used in the transfer gearbox to prevent oil in the gearbox from leaking past the drive shafts of the starter, fuel pump, and auxiliary drive pad.

Another type of carbon seal is also in use. In this type, the carbon rings are not spring-loaded. They move freely around the shaft and seal axially against their housing. When the engine is up to speed, the rings center themselves radially in the housing. Compressor bleed air is forced in

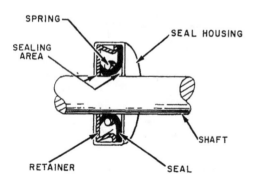

Figure 16-20.—Lip-type seal.

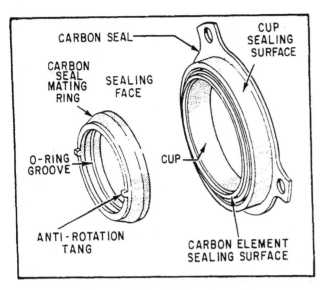

Figure 16-22.—Carbon ring seal.

is sufficient airflow to prevent leakage. However, the leakage is so slight that the engine normally will reach overhaul hours before oil accumulation will have any harmful effects.

Air Seals

The gas turbine air seals (fig. 16-23) are of two types: labyrinth/honeycomb (used in the sump and turbine areas) and fishmouth (used in the combustor and turbine midframe). The labyrinth/honeycomb seal combines a rotating seal having a serrated surface with a stationary seal having a honeycomb surface. The serrations cut into the honeycomb to maintain close tolerances over a large temperature range. The fishmouth seals are circular, stationary, interlocking, and made of sheet metal. The use of these stationary, interlocking seals prevents excessive leakage of hot combustion gas from the primary airflow.

The cavity between the two seals is pressurized from aspirators (air ejectors) that are powered by eighth-stage air. The pressure in the pressurization cavity is always greater than the pressure inside the sump. For this reason, air flowing from the pressurization cavity across the oil seal prevents

between the carbon rings. The air pressure is forced out along the shaft in both directions. The pressure prevents oil from entering the compressor or turbine and combustion gases from reaching the bearings. The main disadvantage of this type of seal is minor oil leakage during start up and run down as the oil pump moves oil before there

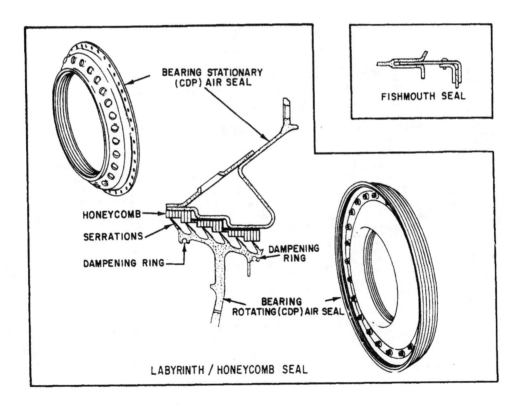

Figure 16-23.—Typical air seals.

16-21

oil from leaking across the seal. This principle is clearly shown in figure 16-24.

AIR SYSTEMS

Air is used for many different functions on the GTE. The primary airflow is used for internal use of the engine. Secondary air is extracted from the engine and is used in various manners as seal air and bleed air.

The GG compressor draws primary air from the ship's inlet, through the enclosure inlet plenum, the inlet screen inlet duct, and the front frame. After being compressed, the primary air enters the combustion section where some of it is mixed with fuel, and the mixture is burned. The remainder of the primary air is used for centering the flame in the combustor and some parts of the GG turbine. The primary air becomes part of the hot combustion gases. Some of the energy in the hot combustion gas is used to turn the GG turbine rotor, which is coupled to and turns the compressor rotor. Upon leaving the GG turbine section, the gas passes into the PT section.

Most of the remaining energy is extracted by the PT rotor, which drives the high-speed, flexible-coupling shaft. The shaft provides the power for the ship's drive system. The gas exits from the PT through the turbine rear frame and passes into the exhaust duct and out through the ship's exhaust.

Secondary air is bled from various locations on the compressor and occasionally from the combustor outer case. The air is fed internally through passages to bearing cavities and seals, and it also cools the GG turbine and nozzles. On some engines, the air is piped externally to seals where shafts extend outside a housing, such as a reduction gear.

The bleed air is taken from various pressure stages due to different pressure requirements at different points in the engine. For example, the front compressor bearing seal may be under a negative atmosphere, and the GG turbine bearing seal may be under 50 psi. Consequently, more air pressure is needed for the turbine bearing than for the compressor bearing.

Bleed air is used for a number of purposes in the following systems:

- Starting

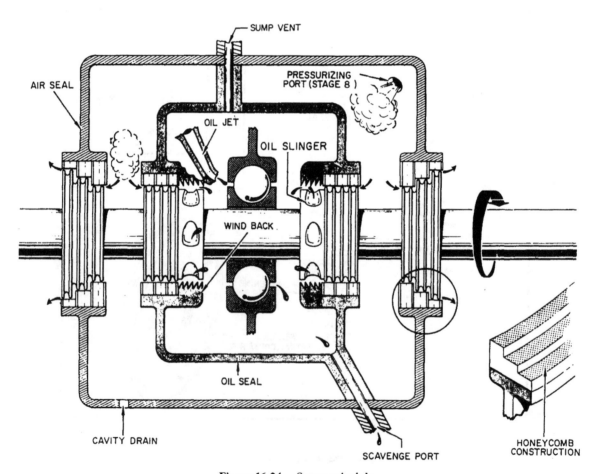

Figure 16-24.—Sump principle.

16-22

- Anti-icing

- Masker air

- Prairie air

These systems differ greatly on different classes of ships; thus, you should consult the specific technical manuals for your ship. A brief explanation of each system is given in the following paragraphs. The starting system is discussed in the next section and therefore is not addressed with the others.

The gas turbine anti-icing systems take hot bleed air from the bleed air header and distributes it to each GTE intake to prevent the formation of ice under an icing condition. An icing condition exists when the inlet air temperature to a GTE is 41°F or less and the humidity of the inlet air is 70 percent or greater.

The masker air system takes hot bleed air from the bleed air system, cools it, and distributes it to the masker emitter rings outside the ship's hull. This reduces or modifies the machinery noise being transmitted through the hull to the water.

The prairie air system modifies the thrashing noise produced by the ship's propellers to disguise the sonar signature of the propellers. The system does this by taking hot bleed air from the bleed air header, cooling it, and distributing it to the leading edges of the propeller blades.

STARTER SYSTEM

The sources of start air differ with the various classes of ships. All ships use the ship's 3000-psi HP air system reduced to 85 psi for the normal source of air for starting. Bleed air from the ship's bleed air main is an alternative method for starting on all ships. This air is provided by ship's gas turbine generators and main engines (LM2500).

On the *Perry* class frigates, there is a third source of start air. This is a start air compressor (SAC). This compressor is hydraulically clutch-coupled to the ship's service diesel generator (SSDG).

Although gas turbines may be started by three basic types of starters (electric, hydraulic, and pneumatic), pneumatic starters are the most common and, therefore, the only type explained in detail within this chapter. Pneumatic starters (fig. 16-25) consist of a small air turbine with reduction gearing and a coupling, driving through the accessory drive gearbox. Pneumatic starters

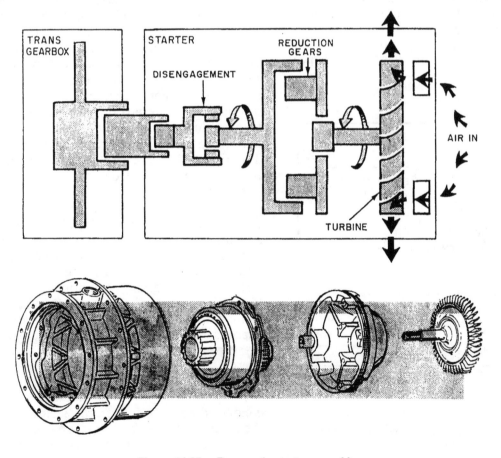

Figure 16-25.—Pneumatic starter assembly.

receive compressed air from an accumulator or the compressor of an engine already running.

With a pneumatic starter, it is important that an adequate volume of air is supplied at sufficient pressure. Otherwise, the starter torque will not produce consistently successful starts. When airbleed is used from another engine that is already operating, the engine being employed as a compressed air supply must be turning over fast enough to supply adequate air to the starter of the engine being started. Intercooling of air bled from an operating engine may be required.

GTEs are started by turning the compressor at sufficient speed to initiate and sustain combustion. Both the compressor and the compressor turbine must be spun. The starter's first requirement is to accelerate the compressor to provide sufficient airflow and pressure to support combustion in the burners.

Once fuel has been introduced and the engine has fired, the starter must continue to accelerate the compressor above the self-sustaining speed of the engine. The starter must provide enough torque to overcome rotor inertia and the friction and air loads of the engine.

Figure 16-26 shows a typical starting sequence for a GTE. When the starter has accelerated the compressor enough to establish airflow through the engine, the ignition is turned on, and then the fuel. The sequence of the starting procedure is important. There must be sufficient airflow through the engine to support combustion at the time the fuel/air mixture is ignited.

After the engine has reached its self-sustaining or self-accelerating speed, the starter can be deactivated. If the starter is cut off below the self-sustaining speed, the engine may decelerate because it doesn't have enough energy to overcome its own friction and operating losses. It may also suffer a "hung start" in which it idles at a speed so low that it is unable to accelerate enough to obtain proper operating parameters. A hung-start engine will overheat because of a lack of cooling air. The starter must therefore continue to boost engine speed well above self-sustaining speed to avoid a hot or hung (false) start, or a combination of both. In a hot start, the engine lights off, but, because of a lack of adequate cooling and combustion air, the exhaust gas temperature exceeds the allowable limit for the engine.

At the proper points in the starting sequence, the starter, and usually the ignition, will cut off. The higher the rpm before the starter cuts, the shorter will be the total time required for the engine to attain idle rpm, because the engine and the starter are working together.

All gas turbine starters must be able to produce sufficient torque to start the engine properly. Gas turbines must get to a certain minimum idle rate for a start to be satisfactory. Hence, the torque characteristics of an acceptable starter exceed by a good margin the amount needed to overcome friction.

Once adequate airflow has been established through the combustion area, fuel can be injected and the spark igniters start the burning process. The spark igniters are high-voltage electrical spark producers powered from the ignition exciter circuits (fig. 16-27).

The ignition exciter derives its input power from the ship's service 60-Hz, 115-volt electrical system. Its function is to produce a high energy spark at the spark igniter in the engine. This must be accomplished with a high degree of reliability under widely varying conditions of internal pressure, humidity, temperature, and vaporization, and in spite of carbonaceous deposits on the spark igniter. To accomplish this, the capacitor discharges a spark of very high energy.

Input voltage is supplied to the exciter, being first led in through a filter that serves to block conducted noise voltage from feeding back into the electrical system. This input voltage is stepped up and applied to a full-wave rectifier. The

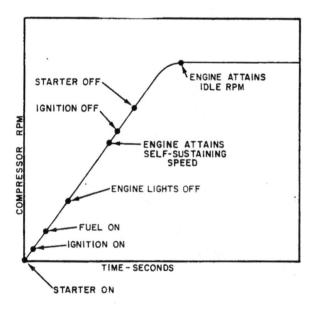

Figure 16-26.—Typical starting sequence for a gas turbine engine.

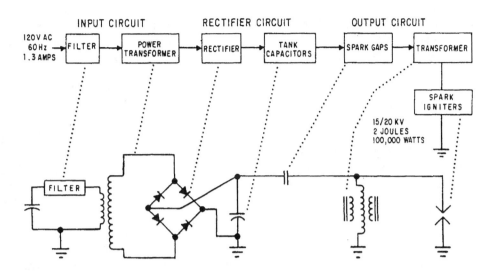

Figure 16-27.—Ignition system diagram.

resulting high-voltage direct current charges a capacitor. The storage capacitor becomes charged up to a maximum of approximately 2 joules. One joule per second equals 1 watt, and the discharge, when a spark is created, takes only a few microseconds.

As the capacitor becomes fully charged, the circuit potential is enough to force current across fixed air gaps between the igniter and the tank capacitors. As the gaps are crossed, the current oscillates at high frequency between a transformer and capacitor in the discharge circuit. These oscillations ionize the air in the gap of the igniter

plug, lowering the resistance in the gap. The current in the tank capacitor is now completely discharged through the oscillator circuits and the spark gap, causing a spark that represents about 100,000 watts of energy. This concentration of maximum energy in minimum time achieves an optimum spark for ignition purposes, capable of blasting carbon deposits and vaporizing globules of fuel.

Spark igniters are of several types, ranging from those resembling common automobile spark plugs to the more common annular gap type shown in figure 16-28. Since they do not operate

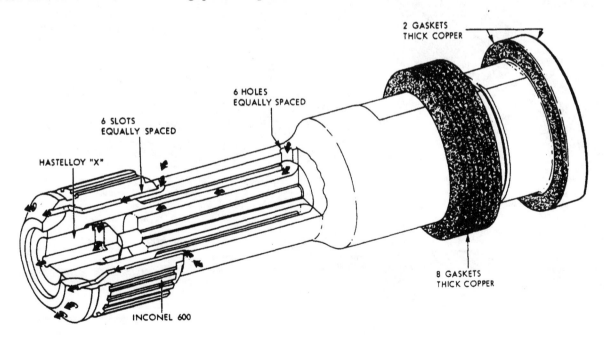

Figure 16-28.—Spark igniter.

continually, they are usually durable and reliable, requiring only occasional cleaning to remove carbon from the tip and the ceramic barrel.

POWER TRANSMISSION SYSTEMS

Like the steam turbine, the GTE operates at a high speed. Likewise, the high speed of the gas turbine must be reduced and the torque must be increased. As with the steam turbine, this is done by reduction gears.

REDUCTION GEARS

The reduction gears installed on board gas turbine ships are very similar to those installed on board steam turbine ships. Figure 16-29 is a diagram of reduction gears on gas turbine ships. They are double-helical, double-reduction, locked-train, reduction gears. On a gas turbine ship, two separate engines are connected to one set of reduction gears. Therefore, power is input into the reduction gears from two separate engines. In order to do this, a clutch is required between each engine and the reduction gear. Almost all gas turbine ships have a synchro self-shifting (SSS) clutch.

SYNCHRO SELF-SHIFTING CLUTCH

The SSS clutch allows one engine to be secured and isolated from the reduction gears. It is a fully automatic freewheel device that transmits power through gear-toothed elements. Figure 16-30 illustrates the operating conditions of the clutch. Clutch engagement is initiated by a pawl and ratchet mechanism. Operation of the clutch is entirely mechanical and depends only on centrifugal and axial force for shifting.

The clutch permits its associated GTE to be brought to rest while the reduction gears and the propeller shaft continue to rotate. The clutch consists of three basic components—a clutch ring

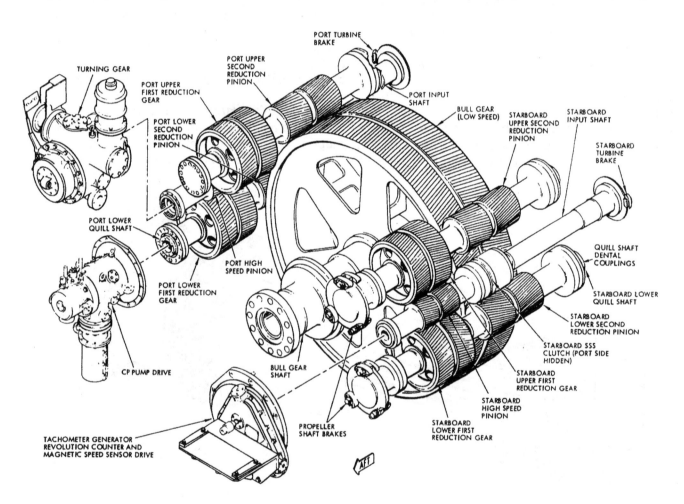

Figure 16-29.—Reduction gear arrangement.

16-26

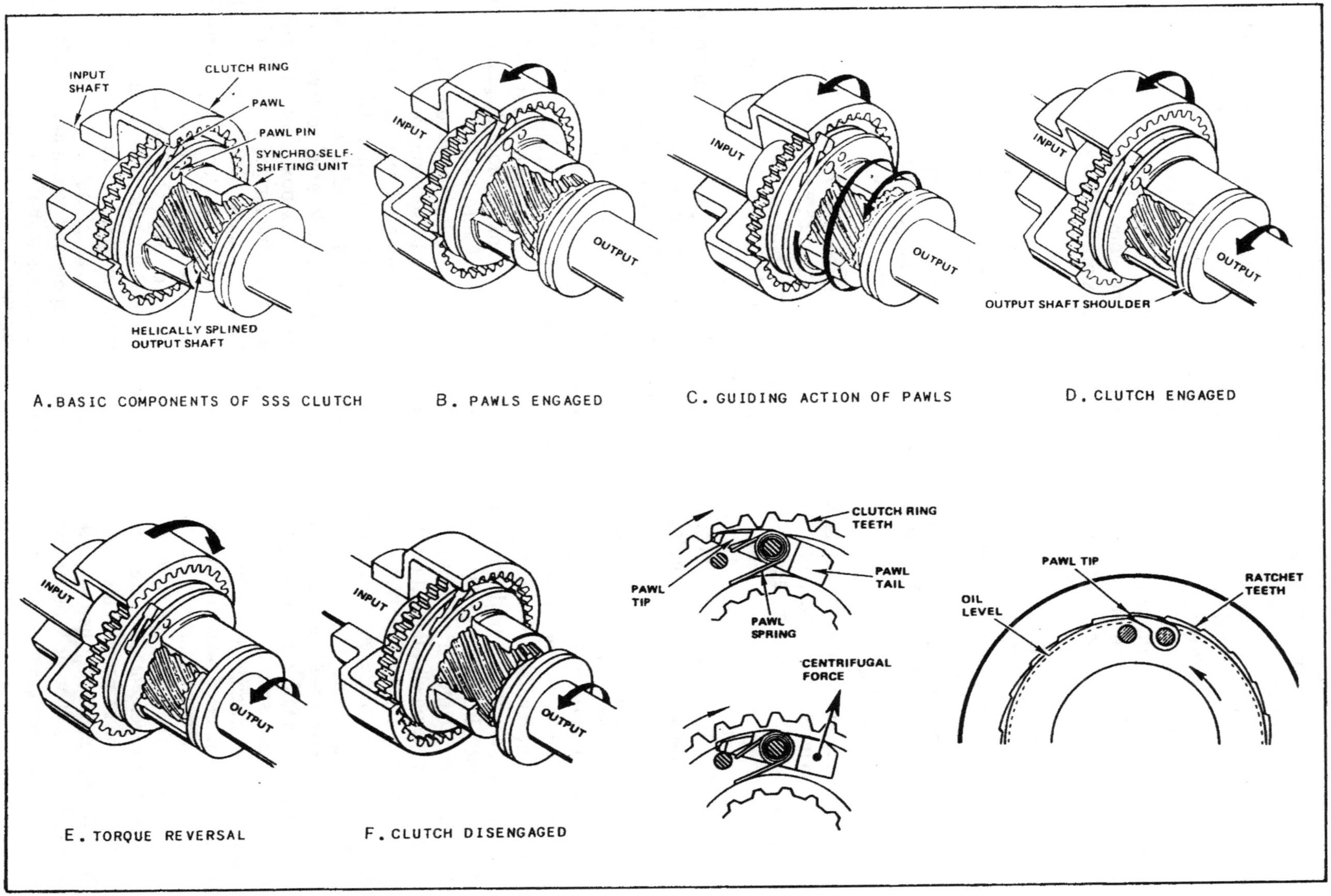

Figure 16-30.—SSS clutch functional diagram.

with internal teeth attached to an input shaft; a helical, externally splined output shaft; and an SSS unit with external teeth and an internal spline to mate with the clutch ring and output shaft, and a set of pawls. These components and their relationship, with the clutch disengaged, are shown schematically in figure 16-30, view A. The SSS unit is in constant mesh with the helical spline of the output shaft and is free to move axially along the spline. The clutch ring is aligned with, but independent of, the output shaft. The pawls are mounted on pins and are spring-loaded to engage the internal teeth of the clutch ring.

When the input shaft and clutch ring start to rotate (fig. 16-30, view B), the internal teeth of the clutch ring and the external teeth of the SSS unit are precisely aligned axially. As the clutch ring continues to rotate with the pawl tips in contact, the pawls and the SSS unit also rotate. The SSS unit then slides axially along the splines (fig. 16-30, view C). The pawls guide the teeth of the clutch ring and the SSS unit into engagement. The SSS unit slides along the splines until it contacts a shoulder on the output shaft. At this point (fig. 16-30, view D), the teeth are in full mesh; the clutch is engaged and transmits the load torque.

The clutch remains engaged as long as torque is transmitted from the input shaft through the SSS unit to the output shaft. When the output shaft overruns the input shaft (fig. 16-30, view E), the torque direction reverses. The SSS unit then slides back along the splines to the disengaged position (fig. 16-30, view F). In this position the output shaft and the clutch ring have no connection except for the pawls that will ratchet against the clutch ring teeth. The pawls are designed to be tail-heavy to prevent continuous ratcheting. The design causes the centrifugal force, at a predetermined speed of the output shaft, to overcome the spring force and swing the pawls out of ratcheting position.

The primary pawls engage the clutch at starting and operate in the speed range from 0 rpm to about 500 rpm. They are symmetrically located about the SSS unit. They are spring-loaded outwardly toward engagement and are mass unbalanced about the pivot point. This causes them to retract from engagement when subjected to the centrifugal forces that develop at high speed. The secondary pawls are mounted on the input clutch ring on axial pivots and engage the clutch at high speed. These pawls are carried by the clutch ring on the input side. They are not spring-loaded and they centrifugally engage.

CONTROLLABLE REVERSIBLE PITCH SYSTEM

The ship's propulsive thrust is provided by a hydraulically actuated propeller or propellers, depending on the class of ship. Each propeller is driven by two GTEs through a reduction gear assembly and line shaft. Since the GTEs cannot be reversed, the controllable reversible pitch (CRP) propellers provide both ahead and astern thrust, thus eliminating the need for a reversing gear.

Since the gas turbine cannot operate at less than approximately 5000 rpm without flaming out or stalling, the CRP propeller is needed to allow the ship to travel at a slow speed. To travel at a slow speed, the propeller pitch is reduced, which reduces the efficiency of the propeller and slows the ship's speed.

The system on the FFG-7 class ship is referred to as a controllable pitch propeller (CPP) and is basically the same. For a more thorough description of your ship's system, check the shipboard technical manuals and ship's information books.

The CRP propeller system for the propulsion shaft consists of the following components: the propeller blades and hub assembly, the propulsion shafting, hydraulic and pneumatic piping and the control rod contained in the shaft, the hydraulic oil power module (HOPM), the oil distribution (OD) box, the control valve manifold block, the electrohydraulic controls, the hydraulic oil sump tank, and the head tank. A block diagram of the system is shown in figure 16-31.

The CRP propeller has five blades. The blade pitch control hydraulic servomotor, mechanical linkage, and hydraulic oil regulating valve are housed in the propeller hub. High-pressure hydraulic control oil is provided for each propeller by an HOPM that is located adjacent to the reduction gear. An OD box, mounted on the forward end of the reduction gear, is mechanically connected to the hydraulic oil regulating valve by a valve control rod. The OD box contains the hydraulic servomechanism that positions the regulating valve rod. The OD box also provides the flow patch connection between the hub servomotor and the HOPM. The regulating valve control rod, prairie air piping, and flow patch for hydraulic oil supply and return are contained in the hollow propulsion shafting.

A control valve's manifold block assembly, mounted on the side of the OD box, contains control valves for both manual and automatic (electronic) pitch control. The manual control

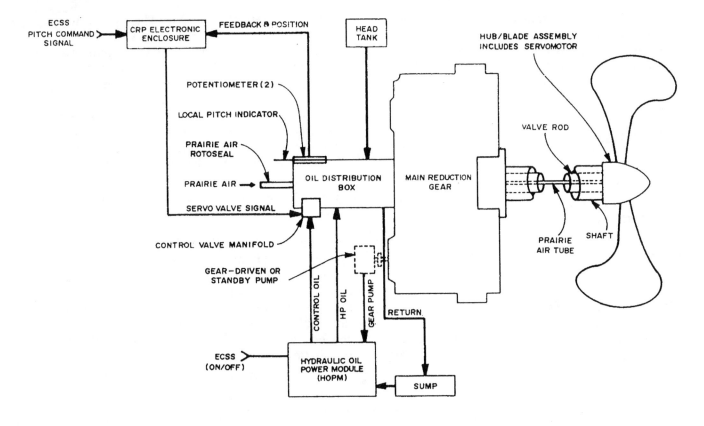

Figure 16-31.—CRP system block diagram.

valves consist of a manual pitch control valve and two manual changeover valves. Automatic pitch control is accomplished through an electro-hydraulic control oil servo valve. This valve responds to an electrical signal generated from the shipboard electronics.

HEAT RECOVERY SYSTEMS

A GTE loses a large amount of energy to the atmosphere that is not used in the turbine. This exhaust heat is a disadvantage of gas turbine ships, since it would make the ship an easy target for infrared-seeking antiship missiles. However, recovery of this energy could increase the efficiency of the GTE.

IN-LINE REGENERATOR

One method to recover this lost energy is through a regenerator (fig. 16-32). A regenerator

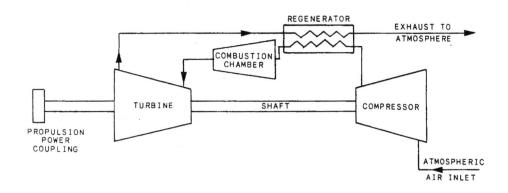

Figure 16-32.—In-line regenerator.

is a heat exchanger that is similar to a boiler's economizer. Compressor discharge air passes through the regenerator and is heated by exhaust gases before it enters the combustion chamber. This preheated air reduces the amount of fuel needed for a given power output.

WASTE-HEAT BOILER

On the DD-963, DD-987 and CG-47 class ships, the gas turbines used to drive the electrical generators have a waste-heat boiler installed. The steam generated by the waste-heat boiler is used to operate the distilling plants, heating, laundry, and scullery.

RANKINE CYCLE ENERGY RECOVERY SYSTEM

Another system developed to recover this energy is the Rankine cycle energy recovery (RACER) system (fig. 16-33). This system was evaluated with favorable results at a shore facility. In this system, the Rankine cycle steam generator and superheater is placed in the exhaust line of the gas turbine. The steam generated is then used to power a turbine that is connected to the same reduction gear as the gas turbine. In comparison to the waste-heat boiler, more steam of a higher pressure is produced. Use of proper engineering design to match power and speed inputs of the gas and steam turbines to the reduction gears maximizes thermal efficiency.

Tests indicate that the addition of the RACER system to a GTE improves fuel performance and could increase the range of a ship by as much as 33 percent. However, at this time, it is too large and heavy for shipboard installation. It is doubtful if it can ever be adapted to meet the space and weight requirements for shipboard installation.

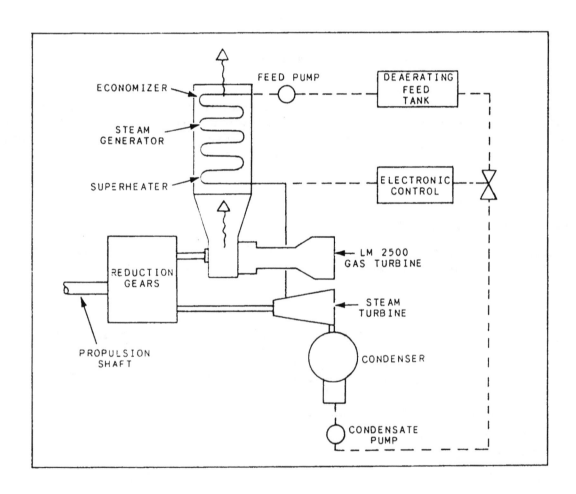

Figure 16-33.—RACER system.

AUTOMATED CENTRAL OPERATING SYSTEM

Naval engineers are constantly striving to design a more reliable engineering plant that provides quick response and requires fewer personnel to operate it. With advances in engineering technology, the use of solid-state devices, and the addition of logic and computer systems, some of these design goals were achieved in the automated central operating system (ACOS).

As shown in figure 16-34, ACOS centralizes the engineering plant with all controls and indicators located at one station, the central control station (CCS), thereby allowing monitoring and operation by fewer watch standers. The use of logic and computer systems reduces the chance of operator error in performing an engineering function. Automatic bell and data loggers reduce the task of hourly readings previously taken by watch standers.

Probably the single, most important function of ACOS is the automatic and continuous monitoring of the engineering plant conditions (parameters) and the subsequent automatic alarm if a condition exceeds a set limit or parameter.

The ACOS provides the means for operating the ship's propulsion plant safely and efficiently. It furnishes the operators with the controls and displays required to start and stop the GTEs. It also furnishes the operators with the controls necessary to change the ship's speed and direction by the gas turbine speed and the pitch of the propeller. These operations are performed at panels or consoles containing the necessary controls and indications for safe operation.

All gas turbine ships are equipped with some type of ACOS. Although each ACOS has different design features, they all contain the basic concepts just discussed. The ACOS also deals with the electric power generation plant, the damage control systems, the auxiliary system, and the fuel transfer system. However, this chapter only describes the portion of ACOS that deals with the GTE, reduction gear, and CRP propeller.

The ACOS provides three different stations that can control the gas turbines power output and ship's speed. The stations are located on the bridge, in the CCS, and in the engine rooms.

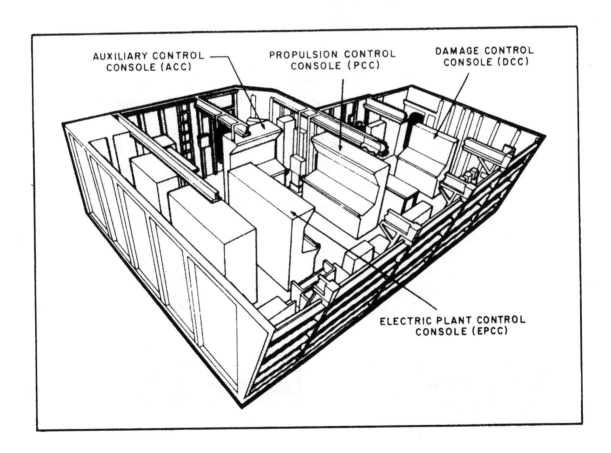

Figure 16-34.—Typical CCS.

Figure 16-35 shows the relationship of the three different stations. However, on the *Perry* class frigate, there is only one engine room and one local control station.

PROPULSION CONTROL CONSOLE

The propulsion control console (PCC) is the primary operating station for the propulsion plant and is located in the CCS. The CCS normally has control of the entire engineering plant, including propulsion, the ship's service generators, and other auxiliary systems. The PCC provides the operator with the necessary controls and displays for starting and stopping the GTEs. Controls on the PCC allow the operator to vary the ship's forward or reverse speed within established design limitations by changing the pitch of the propeller and the speed of the propeller shaft.

The PCC provides two different methods of controlling the ship's progress through the water. The first method requires the operator to individually adjust three levers on the PCC (fig. 16-36). One lever changes the direction and amount of pitch applied at the ship's variable pitch propeller. Each of the remaining two levers controls the speed of one of the GTEs.

The second and primary method of operating the ship's propulsion plant involves the use of a single PCC control lever and a special-purpose digital computer contained in the PCC. This technique for controlling the engines and the propeller pitch with one control and the digital computer is referred to as single-lever programmed control (fig. 16-37).

Single-lever programmed control of the ship's propulsion plant can also be accomplished from

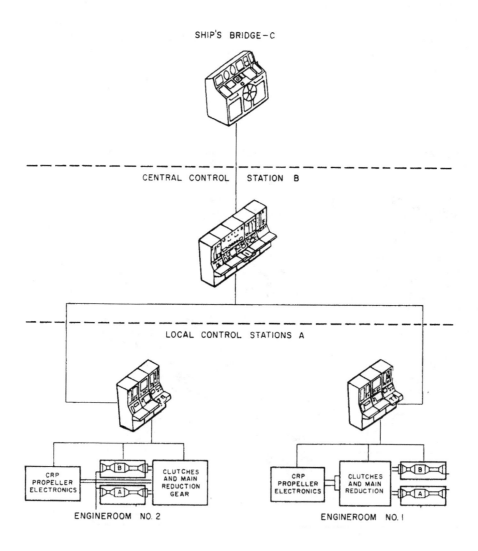

Figure 16-35.—Propulsion control system.

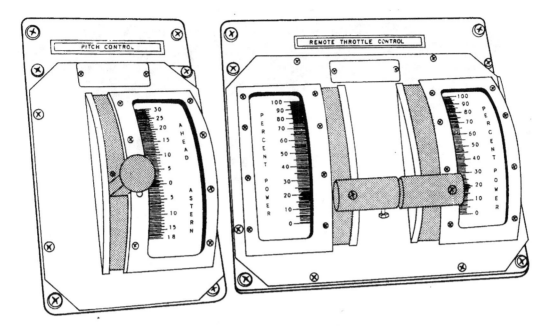

Figure 16-36.—Manual propulsion controls.

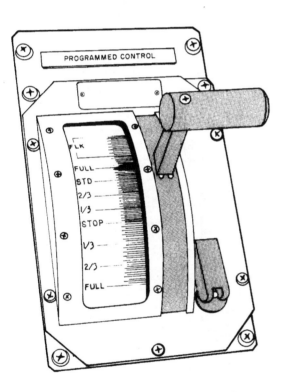

Figure 16-37.—Single-lever programmed control.

the ship control console (SCC) located on the bridge. The lever on the bridge's SCC panel can be operated only after the PCC operator in the CCS relinquishes control. The PCC operator may turn over single-lever programmed control of the

propulsion plant to the ship's bridge after the bridge requests it and after controls on both consoles have been appropriately set.

LOCAL OPERATING PANEL

The local operating panel (LOP) is the secondary operating station and normally is not manned. It is located in the machinery space near the propulsion equipment and contains the necessary controls and indicators to permit direct local manual control of the propulsion equipment, using a throttle control and a pitch control as shown in figure 16-36. The direct local mode of control, although still electronic, permits operation of the equipment independently of the programmed sequencing from the computer and can be used in the event of an emergency or for control during maintenance. The LOP also provides facilities for local control of plant starting and stopping independently of the protective interlocks and logic provided by the automatic start/stop sequencer.

The LOP is the primary control console. This is not to say that the LOP is in control most of the time. What is meant by primary is it may take control from any other remote station. For example, a ship is being operated with the throttle control at the pilot house. However, the engine-room operator places the throttle control to local. Automatically, the LOP assumes control of the throttle operation.

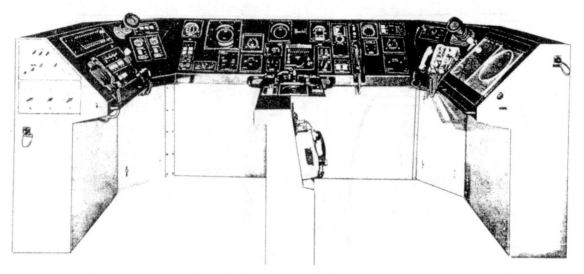

Figure 16-38.—Ship control console.

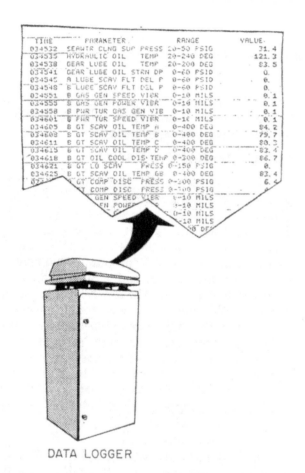

DATA LOGGER

Figure 16-39.—Data logger and sample hard copy printout.

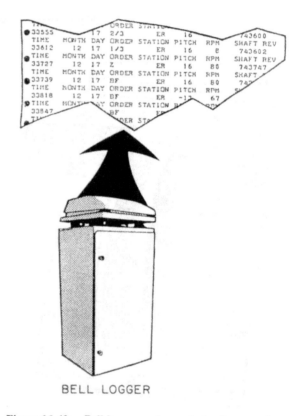

BELL LOGGER

Figure 16-40.—Bell logger and sample hard copy printout.

SHIP CONTROL CONSOLE

The SCC contains the controls and indicators that permit control of the ship's speed from the bridge. This feature provides the OOD with a greater feel for the control of the ship as well as faster response to desired changes. In a twin-screw ship, steering and maneuvering of the ship may be done through this console, using the engines in addition to the rudder. Figure 16-38 is a picture of the SCC installed on FFG-7 class ships.

DATA LOGGER

The data logger (fig. 16-39) provides a hard copy printout of selected monitor points. The printout is initiated automatically once every hour; however, an automatic/demand control permits the operator to demand a printout of data whenever it is needed. In the event of a fault alarm, the data logger also will print out the parameter that caused the alarm. The data logger gives the time in seconds and identifies the monitored sensor (parameter range) as well as the value of the reading.

BELL LOGGER

The bell logger (fig. 16-40) provides an automatic printout each hour or whenever any of the following events occurs:

- Propeller rpm or pitch is changed by more than 5 percent.

- A bell logger printout is demanded by the PCC operator.

- The engine order telegraph is changed.

- The controlling station has been changed (bridge or LOP).

The bell logger prints out the time, month, day, order, station, pitch (angle of CRP pitch), rpm, and shaft revolutions.

REFERENCES

Gas Turbine Systems Technician (Mechanical) 3 & 2, NAVEDTRA 10548-2, Naval Education and Training Program Management Support Activity, Pensacola, FL, 1988.

CHAPTER 17

INTERNAL-COMBUSTION ENGINES

LEARNING OBJECTIVES

Upon completion of this chapter, you should be able to do the following:

1. Explain the basic principles of operation of reciprocating internal-combustion engines.

2. Explain the meaning and significance of such terms as *Otto cycle*, *theoretical diesel cycle*, *actual diesel cycle*, *power stroke*, *compression stroke*, *scavenging*, and *supercharging*.

3. Identify construction differences between single-acting and opposed-piston engines.

4. Describe internal-combustion engine components and systems by function.

5. Identify the various types of drive mechanisms used to transmit engine power to a point where it can be used in performing useful work.

6. Identify the principles of operation of clutches, reverse gears, and reduction gears.

7. Discuss the general rules that apply to the maintenance of internal-combustion engines.

INTRODUCTION

Much of the machinery and equipment discussed in the preceding chapters uses steam as the working fluid in the process of converting thermal energy to mechanical energy. This chapter deals with internal-combustion engines in which air mixed with fuel serves as the working fluid. The internal-combustion engines discussed are those to which thermodynamic cycles of the open and heated engine types apply. In engines which operate on these cycles, the working fluid is taken into the engine, heat is added to the fluid, the energy available in the fluid is used, and then the fluid is discarded. During the process, thermal energy is converted to mechanical energy.

Internal-combustion engines are used extensively in the Navy, serving as propulsion units in a variety of ships, boats, airplanes, and automotive vehicles. Also, internal-combustion engines are used as prime movers for auxiliary machinery. Internal-combustion engines in a majority of the shipboard installations are of the reciprocating type. In naval service today, engines of the gas turbine type serve as power plants.

RECIPROCATING ENGINES

Most of the internal-combustion engines in marine installations of the Navy are of the reciprocating type. This classification is based on the fact that the cylinders in which the energy conversion takes place are fitted with pistons, which employ a reciprocating motion. Internal-combustion engines of the reciprocating type are commonly identified as diesel and gasoline engines. The general practice of the U.S. Navy is to install diesel engines rather than gasoline engines unless special conditions favor the use of gasoline engines.

Most of the information on reciprocating engines in this chapter applies to diesel and gasoline engines. These engines differ in some respects; the principal differences which exist are noted and discussed.

CYCLES OF OPERATION

The operation of an internal-combustion engine involves the admission of fuel and air into a combustion space and the compression and ignition of the charge. The combustion process releases gases and increases the temperature within the space. As temperature increases, pressure increases, and the expansion of gases forces the piston to move. This movement is transmitted through specially designed parts to a shaft. The resulting rotary motion of the shaft is used for work. Thus, the expansion of gases within the cylinder is transformed into rotary mechanical energy. For the process to be continuous, the expanded gases must be removed from the combustion space, a new charge must be admitted, and combustion must be repeated.

Table 17-1.–Sequence of Events in a Cycle of Operation in a Diesel and a Gasoline Engine

DIESEL ENGINE	GASOLINE ENGINE
Intake of air	Intake of fuel and air
Compression of air	Compresion of fuel-air mixture
Injection of fuel	
Ignition and combustion of charge	Ignition and combustion of charge
Expansion of gases	Expansion of gases
Removal of waste	Removal of waste

In the study of engine operation, starting with the admission of air and fuel and following through to the removal of the expanded gases, a series of events or phases takes place. The term *cycle* identifies the sequence of events that takes place in the cylinder of an engine for each power impulse transmitted to the crankshaft. These events always occur in the same order each time the cycle is repeated. The number of events occurring in a cycle of operation depends upon whether the engine is diesel or gasoline. Table 17-1 shows the events and their sequence in one cycle of operation of each of these types of engines.

The principal difference, as shown in table 17-1, in the cycles of operation for diesel and gasoline engines involves the admission of fuel and air to the cylinder. While this takes place as one event in a gasoline engine, it involves two events in a diesel engine. Consequently, there are six main events that take place in the cycle of operation of a diesel engine and five main events that take place in the cycle of a gasoline engine. This is pointed out to emphasize the fact that the number of events that takes place is not identical to the number of piston strokes that occurs during a cycle of operation. Even though the events of a cycle are closely related to piston position and movement, ALL of the events will take place during the cycle regardless of the number of piston strokes involved. We will discuss the relationship of events and piston strokes later in this chapter.

A cycle of operation in either a diesel or gasoline engine involves two basic factors–heat and mechanics. The means by which heat energy is transformed into mechanical energy involves many terms such as *matter, molecules, energy, heat, temperature, the mechanical equivalent of heat, force, pressure, volume, work,* and *power.*

The method by which an engine operates is referred to as the mechanical, or operating, cycle of an engine.

The heat process that produces the forces that move engine parts is referred to as the combustion cycle. Both mechanical and combustion cycles are included in a cycle of operation of an engine.

Mechanical Cycles

We have talked about the events taking place in a cycle of engine operation, but we have said very little about piston strokes except that a complete sequence of events will occur during a cycle regardless of the number of strokes made by the piston. The number of piston strokes occurring during any one cycle of events is limited to either two or four, depending on the design of the engine. Thus, we have a 4-stroke cycle and a 2-stroke cycle. These cycles are known as the mechanical cycles of operation.

Both types of mechanical cycles, 4-stroke and 2-stroke, are used in both diesel and gasoline reciprocating engines. Most gasoline engines in Navy service operate on the 4-stroke cycle. Most diesel engines operate on the 2-stroke cycle. The relationship between the events and piston strokes occurring in a cycle of operation involves some of the differences between the 2-stroke cycle and the 4-stroke cycle.

RELATIONSHIP OF EVENTS AND STROKES IN A CYCLE.–A piston stroke is the distance a piston moves between limits of travel. The cycle of operation in an engine that operates on the 4-stroke cycle involves four piston strokes–intake, compression, power, and exhaust. In the 2-stroke cycle, only two strokes are involved–power and compression.

A check of figure 17-1 will show that the strokes are named to correspond with the events. However, since six events are listed for diesel engines and five events for gasoline engines, more than one event must take place during some of the strokes, especially the 2-stroke cycle. Even so, it is common practice to identify some of the events as strokes of the piston. This is because such events as intake, compression, power, and exhaust in a 4-stroke cycle involve at least a major portion of a stroke and, in some cases, more than one stroke. The same is true of power and compression events and strokes in a 2-stroke cycle. In associating the events with strokes, you should not overlook other events taking place during a cycle of operation. This oversight sometimes leads to confusion when the operating principles of an engine are being considered.

4-STROKE CYCLE DIESEL ENGINE.–To help you understand the relationship between events and strokes, we will discuss the number of events occurring

17-2

during a specific stroke. We will also discuss the duration of an event with respect to a piston stroke and the cases where one event overlaps another. We can demonstrate the relationship of events to strokes by showing the changing situation in a cylinder during a cycle of operation. Figure 17-1 illustrates these changes for a 4-stroke cycle diesel engine.

The relationship of events to strokes is more readily understood if the movements of a piston and its crankshaft are considered first. In figure 17-1, view A, the reciprocating motion and stroke of a piston are indicated and the rotary motion of the crank during two piston strokes is shown. The positions of the piston and crank at the start and end of a stroke are marked "top" and "bottom" respectively. If these positions and movements are marked on a circle (fig. 17-1, view B), the piston position, when at the top of a stroke, is located at the top of the circle. When the piston is at the bottom of a stroke, the piston position is located at the bottom center of the circle. Note in views A and B of figure 17-1 that the top center and bottom center identify points where changes in direction of motion take place. In other words, when the piston is at top center, upward motion has stopped and downward motion is ready to begin. With respect to motion, the piston is "dead." The points which designate changes in direction of motion for a piston and crank are frequently called TOP DEAD CENTER (TDC) and BOTTOM DEAD CENTER (BDC). You should keep TDC and BDC in mind since they identify the start and end of a stroke and since they

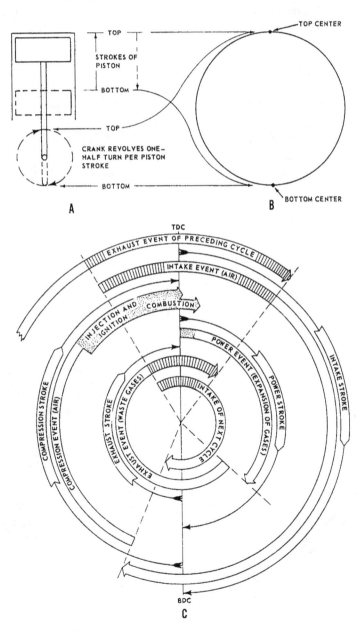

Figure 17-1.–Strokes and events of a 4-stroke cycle diesel engine.

17-3

are the points from which the start and end of events are established.

If the circle illustrated in view B is broken at various points and "spread out" (view C), the events of a cycle and their relationship to the strokes and how some of the events of the cycle overlap can be shown.

By following the strokes and events as illustrated, you can see that the intake event starts before TDC or before the actual downstroke (intake) starts, and it continues on past BDC or beyond the end of the stroke. The compression event starts when the intake event ends, but the upstroke (compression) has been in process since BDC. The injection and ignition events overlap with the latter part of the compression event, which ends at TDC. The burning of the fuel continues a few degrees past TDC. The power event or expansion of gases ends several degrees before the downstroke (power) ends at BDC. The exhaust event starts when the power event ends, and it continues through the complete upstroke (exhaust) and past TDC. Note the overlap of the exhaust event with the intake event of the next cycle. The details on why certain events overlap and why some events are shorter or longer with respect to strokes will be covered later in this chapter.

From the preceding discussion, you can see that the term *stroke* is sometimes used to identify an event that occurs in a cycle of operation. However, you should keep in mind that a stroke involves 180° of crankshaft rotation (or piston movement between dead centers), while the corresponding event may take place during a greater or lesser number of degrees of shaft rotation.

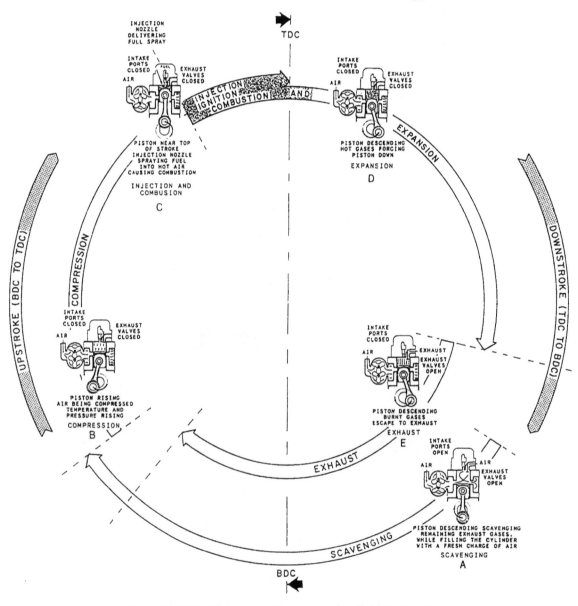

Figure 17-2.–Strokes and events of a 2-stroke cycle diesel engine.

2-STROKE CYCLE DIESEL ENGINE.—The relationship of events to strokes in a 2-stroke cycle diesel engine is shown in figure 17-2. Comparison of figures 17-1 and 17-2 reveals a number of differences between the two types of mechanical or operating cycles. These differences are not too difficult to understand if one keeps in mind that four piston strokes and 720° of crankshaft rotation are involved in the 4-stroke cycle, while only half as many strokes and degrees are involved in a 2-stroke cycle. Reference to the cross-sectional illustrations (fig. 17-2) will aid in associating the event with the relative position of the piston. Even though the two piston strokes are frequently referred to as power and compression, they are identified as the downstroke (TDC to BDC) and upstroke (BDC to TDC) in this discussion to avoid confusion when reference is made to an event.

Starting with the admission of air (fig. 17-2), during the scavenging event (A), we find that the piston is in the lower half of the downstroke and that the exhaust event (E) is in process. The exhaust event started a number of degrees before intake, both starting several degrees before the piston reached BDC. The overlap of these events is necessary so that the incoming air (A) can aid in clearing the cylinder of exhaust gases. Note that the exhaust event stops a few degrees before the

intake event stops, but several degrees after the upstroke of the piston has started. (The exhaust event in some 2-stroke cycle diesel engines ends a few degrees after the intake event ends.) When the scavenging event ends, the cylinder is charged with the air that is to be compressed. The compression event (B) takes place during the major portion of the upstroke. The injection event and ignition and combustion event (C) occur during the latter part of the upstroke. (The point at which the injection ends varies with engines. In some cases, it ends before TDC; in others, a few degrees after TDC.) The intense heat generated during the compression of the air ignites the fuel-air mixture, and the pressure resulting from combustion forces the piston down. The expansion (D) of the gases continues through a major portion of the downstroke. After the force of the gases has been expended, the exhaust valve opens (E) and permits the burned gases to enter the exhaust manifold. As the piston moves downward, the intake ports are uncovered (A) and the incoming air clears the cylinder of the remaining exhaust gases and fills the cylinder with a fresh air charge (A); thus, the cycle of operation has started again.

COMPARISON OF 2-STROKE AND 4-STROKE CYCLE DIESEL ENGINES.—Figure 17-3 shows a comparison of the events that occur during

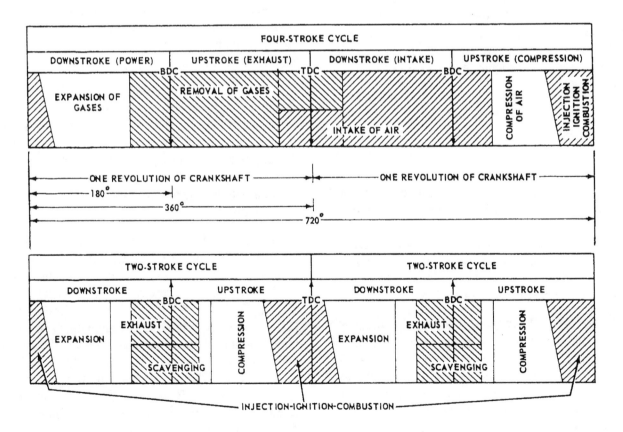

Figure 17-3.—Comparison of the 2-stroke and 4-stroke cycles.

the same length of time for both 2-stroke and 4-stroke cycle engines. The graph shows that, during the same amount of time, a 2-stroke engine will have two power events while a 4-stroke engine will have only one. This might lead you to believe that 2-stroke engines are more efficient than 4-stroke engines; however, that is not the case. To work properly, a 2-stroke engine must have some method of forcing air into and through the cylinders. And, since this air pump (blower) is driven by the engine, it robs some of the horsepower that would otherwise be available to drive the load. Also, the combustion process in a 2-stroke engine is not as complete as it is in a 4-stroke engine. Since each type of engine has certain advantages over the other, the Navy uses both 2-stroke and 4-stroke cycle engines for main propulsion and electrical generating service.

The figures we have used to represent the cycles of operation are for illustrative purposes only. The exact number of degrees before or after TDC or BDC at which an event starts and ends will vary among engines. You can find information on such details in appropriate technical manuals dealing with the specific engine in question.

GASOLINE ENGINES.–Diagrams that show the mechanical cycles of operation in gasoline engines are somewhat similar to those described in this chapter for diesel engines, except that there would be one less event taking place during the gasoline engine cycle. Since air and fuel are admitted to the cylinder of a gasoline engine as a mixture during the intake event, the injection event does not apply.

Combustion Cycles

Up to this point, we have given greater consideration to the strokes of a piston and the related events taking place during a cycle of operation than we have to the heat process involved in the cycle. However, we cannot discuss the mechanics of engine operation without dealing with heat. Such terms as *ignition*, *combustion*, and *expansion of gases* indicate that heat is essential to a cycle of engine operation.

So far, the only difference we have pointed out between diesel and gasoline engines is the number of events that occur during the cycle of operation. We have told you that either the 2-stroke or 4-stroke cycle may apply to both a diesel engine and a gasoline engine. Then, one of the principal differences between these types of engines must involve the heat processes that produce the forces that make the engine operate. The

heat processes are sometimes called combustion or heat cycles.

The two most common combustion cycles associated with reciprocating internal combustion engines are the OTTO cycle (gasoline engines) and the DIESEL cycle (diesel engines). Each of these combustion cycles will be discussed in the following paragraphs.

In talking about combustion cycles, we must bring up another important difference between gasoline and diesel engines–the difference in compression pressure. Compression pressure is directly related to the combustion process in an engine. Diesel engines have a much higher compression pressure than gasoline engines. The higher compression pressure in diesels explains the difference in the methods of ignition used in gasoline and diesel engines.

METHODS OF IGNITION.–When the gases within a cylinder are compressed, the temperature of the confined gases rises. As the compression increases, the temperature rises. In a gasoline engine, the compression temperature is always lower than the point at which the fuel will ignite spontaneously. Thus, the heat required to ignite the fuel must come from an external source spark ignition. On the other hand, the compression temperature in a diesel engine is far above the ignition point of the fuel oil; therefore, ignition takes place as a result of the heat generated by the compression of the air within the cylinder compression ignition.

The difference in the methods of ignition indicates that there is a basic difference in the combustion cycles upon which diesel and gasoline engines operate. This difference involves the behavior of the combustion gases under varying conditions of pressure, temperature, and volume. Since this so, you should be familiar with factors before considering the combustion cycles individually.

RELATIONSHIP OF TEMPERATURE, PRESSURE, AND VOLUME.–The relationship of temperature, pressure, and volume as found in an engine can be illustrated by a description of what takes place in a cylinder that is fitted with a reciprocating piston. Follow views A through D of figure 17-4. In view A, note that the instruments that indicate the pressure within the cylinder and the temperature inside and outside the cylinder show the temperature as approximately 70°F. Assume that it is an airtight container, as it is in our example. Now compare views A and B of figure 17-4. If a force pushes the piston toward the top of the cylinder, the entrapped charge will

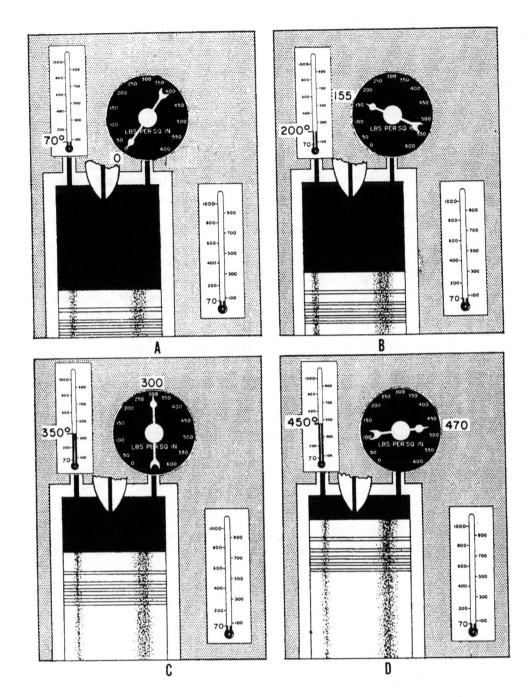

Figure 17-4.—Volume, temperature, and pressure relationship in a cylinder.

be compressed. In views B and C, the compression progresses. The volume of the air decreases, the pressure increases, and the temperature rises. These changing conditions continue as the piston moves. When the piston nears TDC in view D, there has been a marked decrease in volume. Also, both pressure and temperature are much greater than at the beginning of compression. Notice that pressure has gone from 0 psi to 470 psi and the temperature has increased from 70°F to approximately 450°F. These changing conditions indicate that mechanical energy, in the form of force

applied to the piston, has been transformed into heat energy in the compressed air. The temperature of the air has been raised sufficiently to cause ignition of fuel that is injected into the cylinder.

Further changes take place after ignition. Since ignition occurs shortly before TDC, there is little change in volume until the piston passes TDC. However, there is a sharp increase in both pressure and temperature shortly after ignition takes place. The increased pressure forces the piston downward. As the piston moves downward, the gases expand (increase in volume), and pressure and temperature decrease rapidly. These changes in volume, pressure, and temperature are representative of the changing conditions in the cylinder of a diesel engine.

The changes in volume and pressure in an engine cylinder can be illustrated by diagrams similar to those shown in figure 17-5. Such diagrams are made by devices that measure and record the pressures at various piston positions during a cycle of engine operation. Diagrams, such as these, that show the relationship between pressures and corresponding piston positions are called pressure-volume diagrams or indicator cards.

On diagrams that provide a graphic representation of cylinder pressure as related to volume, the vertical line (P) on the diagram represents pressure and the horizontal line (V) represents volume. (Refer to fig. 17-5.) When a diagram is used as an indicator card, the pressure line is marked off in units of pressure and the volume line is marked off in inches. Thus, the volume line can be used to show the length of the piston stroke that is proportional to volume. The distance between adjacent letters on each of the diagrams (views A and B of fig. 17-5) represents an event of a combustion cycle. A combustion cycle includes compression of air, burning of the charge, expansion of gases, and removal of gases.

The diagrams shown in figure 17-5 provide a means by which the Otto and diesel combustion cycles can be compared. Referring to the diagrams as we discuss these combustion cycles will help you to identify the principal differences between these cycles. The diagrams shown are theoretical pressure-volume diagrams. Diagrams representing conditions in operating engines will be given later. Information obtained from actual indicator diagrams may be used in checking engine performance.

OTTO (CONSTANT-VOLUME) CYCLE.–In theory, the Otto combustion cycle is one in which combustion, induced by spark ignition, occurs at constant volume. The Otto cycle and its principles serve as the basis for modern gasoline engine designs.

In the Otto cycle (fig. 17-5, view A), compression of the charge in the cylinder occurs at line AB. Spark ignition occurs at B. Because of the volatility of the mixture, combustion practically amounts to an explosion. Combustion, represented by line BC, occurs (theoretically) just as the piston reaches TDC. During combustion, there is no piston travel. Thus, there is no change in the volume of the gas in the cylinder. This lack of change in volume accounts for the descriptive term *constant volume*. During combustion, there is a rapid rise of temperature followed by a pressure increase. The pressure increase performs the work during the expansion phase, as represented by line CD in the figure. The removal of gases, represented by line DA in the figure, is at constant volume.

THEORETICAL (CONSTANT-PRESSURE) CYCLE.–When discussing diesel engines, we must point out that there is a difference between the

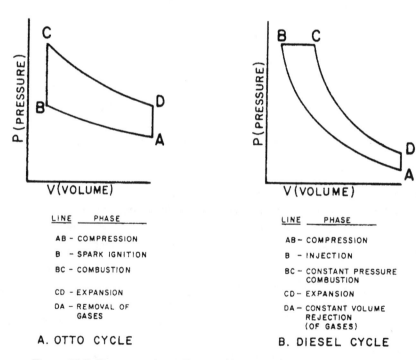

LINE	PHASE
AB –	COMPRESSION
B –	SPARK IGNITION
BC –	COMBUSTION
CD –	EXPANSION
DA –	REMOVAL OF GASES

A. OTTO CYCLE

LINE	PHASE
AB –	COMPRESSION
B –	INJECTION
BC –	CONSTANT PRESSURE COMBUSTION
CD –	EXPANSION
DA –	CONSTANT VOLUME REJECTION (OF GASES)

B. DIESEL CYCLE

Figure 17-5.–Pressure-volume diagrams for theoretical combustion cycles.

theoretical, or "true," diesel cycle and the "actual" diesel cycle that really occurs in an operating diesel engine. The true diesel cycle may be defined as one in which combustion, induced by compression ignition, theoretically occurs at a constant pressure (see fig. 17-5, view B). Adiabatic compression of the air (line AB) increases its temperature to a point that ignition occurs automatically when the fuel is injected. Fuel injection and combustion are controlled to give constant-pressure combustion (line BC). This phase is followed by adiabatic expansion (line CD) and constant volume rejection (line DA).

In the true diesel cycle, the burning of the mixture of fuel and compressed air is a relatively slow process when compared with the quick explosive type of combustion process of the Otto cycle. In the true diesel engine, the injected fuel penetrates the compressed air, some of the fuel ignites, then the rest of the charge burns. The expansion of the gases keeps pace with the change in volume caused by piston travel. Thus, combustion is said to occur at constant pressure (line BC).

ACTUAL COMBUSTION CYCLES.–The preceding discussion covered the theoretical (true) combustion cycles, which serve as the basis for modern engines. In actual operation, modern engines operate on modifications of the theoretical cycles. However, some characteristics of the true cycles are incorporated in the cycles of modern engines as you will see in the following discussion of examples representing the actual cycles of operation in gasoline and diesel engines.

Examples we will use are based on the 4-stroke mechanical cycle so that you may compare the cycles found in both gasoline and diesel engines. We will also point out differences existing in diesel engines operating on the 2-stroke cycle.

The diagrams in figures 17-6 and 17-7 represent the changing conditions in a cylinder during engine operation. Some of the events are exaggerated to show more clearly the changes that take place and, at the same time, to show how the theoretical and actual cycles differ.

The compression ratio situation and a pressure-volume diagram for a 4-stroke Otto cycle are shown in figure 17-6. View A shows the piston on BDC at the start of an upstroke. (In a 4-stroke cycle engine, this stroke could be identified as either the compression stroke or the exhaust stroke.) Study views A and B of figure 17-6. Notice that in moving from BDC to TDC (view B), the piston travels five-sixths of the total distance AB. In other words, the volume has been decreased to one-sixth of the volume when the piston was at BDC. Thus, the compression ratio is 6 to 1.

View C shows the changes in volume and pressure during one complete 4-stroke cycle. Notice that the lines representing the combustion and exhaust phases are not as straight as they were in the theoretical diagram. As in the diagram of the theoretical cycle, the vertical line at the left represents cylinder pressure in psi. Atmospheric pressure is represented by a horizontal line called the atmospheric pressure line. Pressures below this line are less than atmospheric pressure, while pressures above the line are more than atmospheric. The bottom horizontal line represents cylinder volume and piston movement. The volume line is divided into six parts that correspond to the divisions of volume shown in view A. Since piston movement and volume are proportional, the distance between 0 and 6 indicates the volume when the piston is at BDC, and the distance from 0 to 1 indicates the volume with the piston at TDC. Thus, the distance from 1 to 6 corresponds to total piston travel, with the numbers in between identifying changes in volume that result from the reciprocating motion of the piston. The curved lines of view C represent the changes of both pressure and volume that take place during the four piston strokes of the cycle.

To make it easier for you to compare the discussion on the relationship of strokes and events in the diesel 4-stroke cycle (fig. 17-1) with the discussion on the Otto 4-stroke cycle (fig. 17-6), we will begin the cycle of operation at the intake (refer to fig. 17-6). In the Otto cycle, the intake event includes the admission of fuel and air. As indicated earlier, the intake event starts before TDC, or at point A in view C. Note that pressure is decreasing and that after the piston reaches TDC and starts down, a vacuum is created that facilitates the flow of the fuel-air mixture into the cylinder. The intake event continues a few degrees past BDC and ends at point B. Since the piston is now on an upstroke, compression takes place and continues until the piston reaches TDC. Notice the increase in pressure (X to X') and the decrease in volume (F to X). Spark ignition at point C starts combustion, which takes place very rapidly. There is some change in volume since the combustion phase starts before TDC and ends after TDC.

Pressure increases sharply during the combustion phase (curve CD). The increase in pressure provides the force necessary to drive the piston down again. The gases continue to expand as the piston moves toward BDC. The pressure decreases as the volume increases, from D to E. The exhaust event starts at point E, a few degrees before BDC. The pressure drops rapidly until

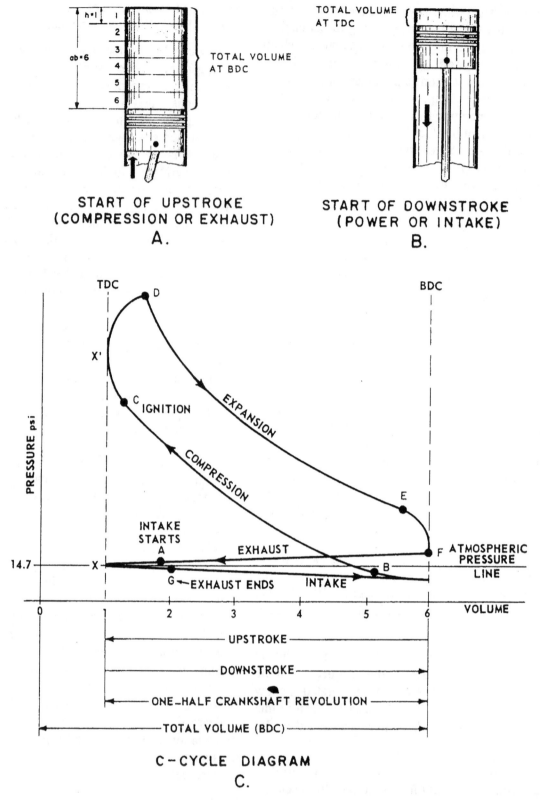

**START OF UPSTROKE
(COMPRESSION OR EXHAUST)
A.**

**START OF DOWNSTROKE
(POWER OR INTAKE)
B.**

TOTAL VOLUME
AT BDC

TOTAL VOLUME
AT TDC

**C - CYCLE DIAGRAM
C.**

Figure 17-6.–Pressure-volume diagram for an Otto (gasoline) 4-stroke cycle.

the piston reaches BDC. As the piston moves toward TDC, there is a slight drop in pressure as the waste gases are discharged. The exhaust event continues a few degrees past TDC to point G so that the incoming charge aids in removing the remaining waste gases.

The actual diesel combustion cycle (fig. 17-7) is one in which the combustion phase, induced by

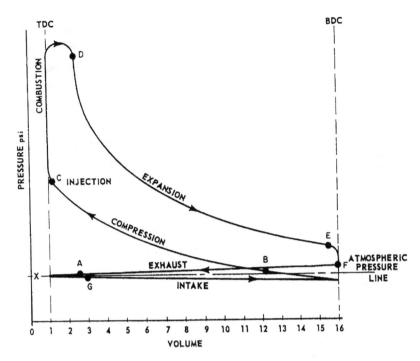

Figure 17-7.—Pressure-volume diagram for a diesel 4-stroke cycle.

compression/ignition, begins on a constant-volume basis and ends on a constant-pressure basis. In other words, the actual cycle is a combination of the Otto and theoretical diesel cycles. The actual cycle is used as the basis for the design of practically all modern diesel engines and is referred to as a modified diesel cycle.

An example of a pressure-volume diagram for a modified 4-stroke cycle diesel engine is shown in figure 17-7. Notice that the volume line (bottom of figure) is divided into 16 units. These units indicate a 16 to 1 compression ratio. The higher compression ratio accounts for the increased temperature necessary to ignite the charge. By comparing this figure with view C of figure 17-6, you can see that the phases of the diesel cycle are relatively the same as those of the Otto cycle, except for the combustion phase. Fuel is injected at point C and combustion is represented by line CD. While combustion in the Otto cycle is at constant-volume practically throughout the phase, combustion in the actual diesel cycle takes place with volume that is practically constant for a short period of time. During this period of time, there is a sharp increase in pressure until the piston reaches a point slightly past TDC. Then, combustion continues at a relatively constant pressure, which drops slightly as combustion ends at point D. For theseto all of them reasons the combustion cycle in modern diesel engines is sometimes referred to as the constant-volume, constant-pressure cycle.

Pressure-volume diagrams for gasoline and diesel engines that operate on the 2-stroke cycle are similar to those just discussed. the only difference is that separate exhaust and intake curves do not exist. They do not exist because intake and exhaust occur during a relatively short interval of time near BDC and do not involve full strokes of the piston as in the 4-stroke cycle. Thus, a pressure-volume diagram for a 2-stroke modified diesel cycle will be similar the diagram formed by the F-B-C-D-E-F cycle illustrated in figure 17-7. The exhaust and intake phases will take place between E and B with some overlap of the events. (Refer again to fig. 17-2.)

The preceding discussion has pointed out some of the main differences between engines that operate on the Otto cycle and those that operate on the diesel cycle. In brief, these differences involve (1) the mixing of fuel and air, (2) the compression ratio, (3) the method of ignition, and (4) the combustion process.

In regard to differences in engines, there is another variation you may find in engines. Sometimes, the manner in which the pressure of combustion gases acts upon the piston is used as a method of classifying engines. This method of classification is discussed in the information that follows.

ACTION OF PRESSURE ON PISTONS

Engines are classified in many ways. You are already familiar with some classifications, such as those based on

- the fuels used (diesel fuel and gasoline),

- the ignition methods (spark and compression),

- the combustion cycles (Otto and diesel), and

- the mechanical cycles (2-stroke and 4-stroke).

Engines may also be classified on the basis of cylinder arrangements (V, in-line, opposed), the cooling media (liquid and air), and the way air enters the cylinder and the exhaust leaves the cylinder (port scavenging and valve scavenging).

Classification of engines according to combustion-gas action is based on whether the pressure created by the combustion gases acts upon one surface of a single piston or against single surfaces of two separate and opposed pistons. The two types of engines under this classification are commonly referred to as single-acting and opposed-piston engines.

Single-Acting Engines

Engines of the single-acting type have one piston per cylinder, with the pressure of combustion gases acting on only one surface of the piston. This is a feature of design rather than of principle, because the basic principles of operation apply whether an engine is single-acting or opposed-piston.

The pistons in most single-acting engines are of the trunk type (length greater than diameter). The barrel or wall of a piston of this type has one end closed (crown) and one end open (skirt). Only the piston crown serves as part of the combustion space surface. Therefore, the pressure of combustion can act only against the crown. Thus, with respect to the surfaces of a piston, pressure is single-acting. All 4-stroke cycle engines and most 2-stroke cycle engines are single-acting.

Opposed-Piston Engines

With respect to the combustion gas action the term *opposed piston* identifies those engines that have two pistons and one combustion space in each cylinder. The pistons are arranged in "opposed" positions; that is, crown to crown, with the combustion space in between. (See fig. 17-8.) When combustion takes place, the gases act against the crowns of both pistons, driving them in opposite directions. Thus, the term *opposed* not only signifies that gases act in opposite directions with respect to pressure and piston surfaces, but also classifies piston arrangement within the cylinder.

In engines that have the opposed-piston arrangement, two crankshafts (upper and lower) are required for transmission of power. Both shafts contribute to the power output of the engine. In opposed-piston engines that are common to Navy service, the crankshafts are connected by a vertical gear drive, which provides the power developed by the upper crankshaft. This power is delivered through the vertical drive shaft to the lower crankshaft. Large roller bearings and thrust bearings support and guide the vertical drive shaft

The cylinders of opposed-piston engines do not have valves. Instead they employ scavenging air ports located near the top of the cylinder. These ports are opened and closed by the upper piston. Exhaust ports, located near the bottom of the cylinder, are closed and opened by the lower piston.

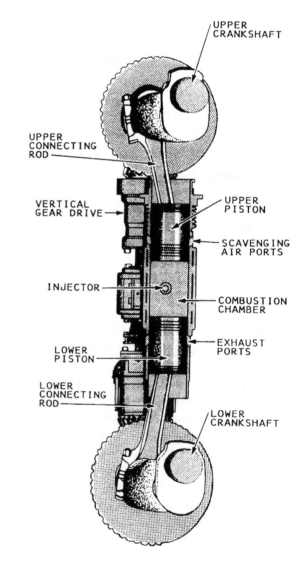

Figure 17-8.--Cylinder and related parts of an opposed-piston engine.

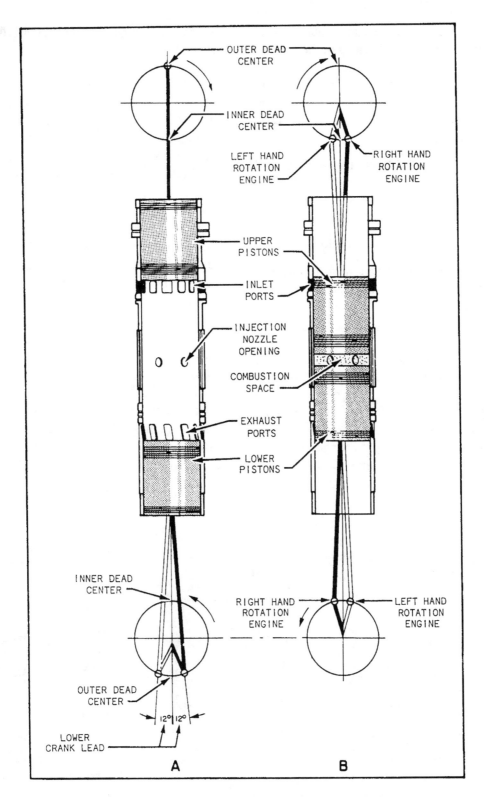

Figure 17-9.—Lower crank lead in an opposed-piston engine.

Movement of the opposed pistons is such that the crowns are closest together near the center of the cylinder. When in this position, the pistons are not at the true piston dead centers. This is because the lower crankshaft operates a few degrees in advance of the upper shaft. The number of degrees that a crank on the lower shaft travels in advance of the corresponding crank of the upper shaft is called lower crank lead. (See fig. 17-9.)

Opposed-piston engines used by the Navy operate on the 2-stroke cycle. In engines of the opposed-piston type, as in 2-stroke cycle single-acting engines, there is an overlap of the various events occurring during a cycle of operation. Injection and the burning of the fuel start during the latter part of the compression event and extend into the power phase. There is also an overlap of the exhaust and scavenging periods. The events in the cycle of operation of an opposed piston, 2-stroke diesel engine are shown in figure 17-10.

Modern engines of the opposed-piston design have several advantages over single-acting engines of comparable rating. Some of these advantages are as follows:

- Less weight per horsepower developed

- Lack of cylinder heads and valve mechanisms (and the cooling and lubricating problems connected with them)

- Fewer moving parts

Single-acting engines have their own advantages, such as not requiring blowers, if they are of the 4-stroke cycle design. These engines are more efficient if they are supercharged with a turbocharger, which is driven by the otherwise wasted energy of exhaust gases. Certain repairs are easier on a single-acting engine since the combustion space can be entered without the removal of an engine crankshaft and piston assembly.

FUNCTIONS OF RECIPROCATING ENGINE COMPONENTS

Most internal-combustion engines of the reciprocating type are constructed in the same general pattern. Although engines are not exactly alike, there are certain features common to all of them, and the main parts of most engines are similarly arranged. Gasoline engines and diesel engines have the same basic structure; therefore, the following discussion of the engine components applies generally to both types of engines. However, differences do exist and we will point these out wherever they occur. The main differences in diesel and gasoline engines exist in the fuel systems and the methods of ignition. The main parts of an internal-combustion engine may be divided into two principal groups: parts and systems. The main parts of an internal-combustion engine may be further divided into stationary parts and moving parts. Stationary parts for the purpose of this discussion include those that do not involve motion, such as the structural frame and its components and related parts. The other group of engine parts are those that involve motion. Many of the principal parts that are within the main structure of an engine are moving parts. These parts convert the thermal energy released by combustion in the cylinder to

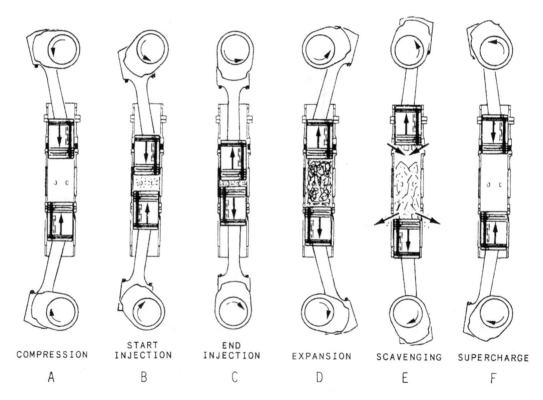

COMPRESSION	START INJECTION	END INJECTION	EXPANSION	SCAVENGING	SUPERCHARGE
A	B	C	D	E	F

Figure 17-10.–Events in the operating cycle of an opposed-piston engine.

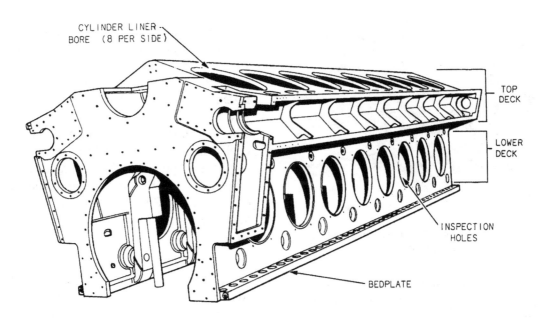

CYLINDER LINER
BORE (8 PER SIDE)

TOP
DECK

LOWER
DECK

INSPECTION
HOLES

BEDPLATE

Figure 17-11.–A cylinder block.

mechanical energy, which is then available for useful work at the crankshaft.

The systems commonly associated with the engine proper are those necessary to make combustion possible and those that minimize and dissipate heat created by combustion and friction. Since combustion requires air, fuel, and heat (ignition), systems providing each may be found on some engines. However, since a diesel engine generates its own heat for combustion within the cylinders, no separate ignition system is required for engines of this type. The problem of heat created as a result of combustion and friction is taken care of by two separate systems: cooling and lubrication. The functions of the parts and systems of engines that operate on the principles already described are discussed briefly in the following paragraphs.

Main Stationary Parts

The stationary parts of an engine maintain the moving parts in their proper relative position so that the gas pressure produced by combustion can "push" the pistons and rotate the crankshaft.

ENGINE FRAME.–The term *frame* is sometimes used to identify a single part of an engine. It is also used to identify several stationary parts that are fastened together. These stationary parts support most of the moving engine parts and engine accessories. When we talk about the frame, we will use the latter meaning in our discussion. As the load-carrying part of the engine,

the frame may include such parts as the cylinder block, base, sump or oil pan, and end plates.

Cylinder Blocks.–A cylinder block is the part of the engine frame that supports the engine's cylinder liners, head (or heads), and crankshaft. The blocks for most large engines are of the welded-steel type of construction. In this type of construction, the block is made of steel forgings and plates that are welded horizontally and vertically for strength and rigidity. These plates are located where loads occur. Deck plates are generally fashioned to house and hold the cylinder liners. The uprights and other members are welded with the deck plates into one rigid unit. Blocks of small high-speed engines are often of cast-iron en bloc (in one piece) construction.

A cylinder block for a large diesel engine that consists of steel plates and forgings is shown in figure 17-11. The steel plates and forgings are welded together to provide the structural support for the stationary and moving components. The upper deck contains the cylinder assemblies and related gear. The lower deck, which forms the crankcase, is mounted with an oil pan to the bedplates.

A cylinder block may contain passages to allow circulation of cooling water around the liners for cooling of the cylinder. However, if the liner is constructed with integral cooling passages, the cylinder block generally will not have cooling passages. Many blocks have drilled lube oil passages. Most 2-stroke cycle engines have air passages in the block.

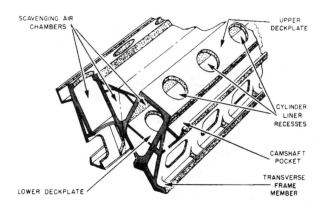

Figure 17-12.–An example of a V-type cylinder block arrangement.

The block shown in figure 17-12 represents blocks constructed for some engines with a V-type cylinder arrangement.

Bases.–In the majority of large engines, a base is used to support the cylinder block. The base shown in figure 17-13 is a welded-steel structure. The base not only supports the cylinder block but also provides a mounting surface for accessories to the engine and serves as a reservoir for the lubricating oil used by the engine. Many of the smaller engines do not have a separate base. Instead, they have an oil pan, which is secured directly to the bottom of the block.

Sumps and Oil Pans.–Since lubrication is essential for proper engine operation, a reservoir that is used for collecting and holding lubricating oil is a necessary part of the structure of an engine. The reservoir may be called a sump or an oil pan, depending on the design of the engine. In most cases, the reservoir is usually attached

directly to the engine. However, in some engines, the oil reservoir may be located at some point relatively remote from the engine. Such engines may be referred to as dry sump engines. As the lubricating oil reaches the oil pan, it immediately drains by gravity flow from the oil pan to a reservoir (that is located apart from the engine) where the lube oil for the engine is collected so it can be cooled and filtered. It is from this reservoir that the engine is supplied with lubricating oil and not from the oil pan. Thus, in dry sump engines, the oil pan remains essentially dry. Regardless of the design of the engine, the oil reservoir serves the same purpose wherever it is located.

End Plates.–Some engines have flat steel plates attached to each end of the cylinder block. End plates add rigidity to the block and provide a mounting for parts such as gears, blowers, pumps, and generators.

Access Openings and Covers.–Many engines, especially the larger ones have access openings in some part of the engine frame (see figs. 17-11 and 17-13). These openings permit access to the cylinder liners, main and connecting rod bearings, injector control shafts, and various other internal engine parts. Access doors (sometimes called covers or plates) for the openings are usually secured with a handwheel or nut-operated clamps and are fitted with gaskets to keep dirt and foreign material out of the interior of the engine.

CYLINDER ASSEMBLIES.–The cylinder assembly completes the structural framework of an engine. As one of the main stationary parts of an engine, the cylinder assembly, along with various related working parts, serves as the area where combustion

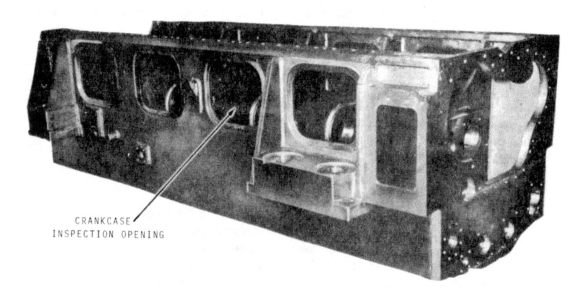

75.333

Figure 17-13.–A cylinder base.

17-16

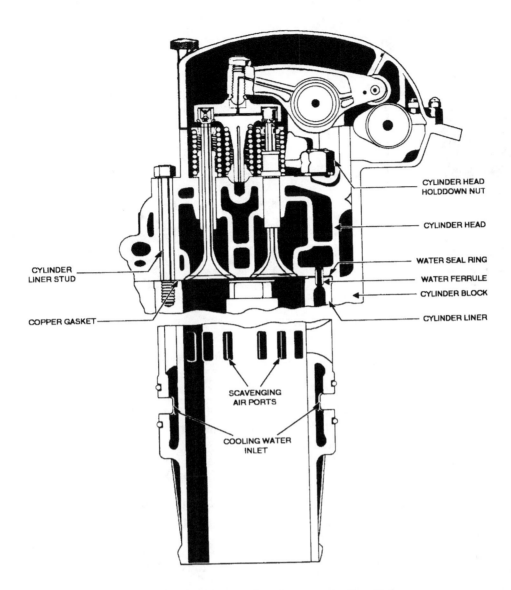

CYLINDER HEAD
HOLDDOWN NUT

CYLINDER HEAD

WATER SEAL RING

WATER FERRULE

CYLINDER BLOCK

CYLINDER LINER

CYLINDER
LINER STUD

COPPER GASKET

SCAVENGING
AIR PORTS

COOLING WATER
INLET

Figure 17-14.–Principal stationary parts of a cylinder.

takes place. For purposes of our discussion, a cylinder assembly consists of the head, the liner, the studs, and the gasket (fig. 17-14).

The design of the parts of the cylinder assembly varies considerably from one type of engine to another. Regardless of differences in design, however, the basic components of all cylinder assemblies function, along with related moving parts, to provide a gastight and a liquidtight space.

Cylinder Liners.–The barrel or bore in which an engine piston moves back and forth may be an integral part of the cylinder block, or it may be a separate sleeve or liner. The first type, common in gasoline engines, has the disadvantage of not being replaceable. Practically all diesel engines are constructed with replaceable cylinder liners.

Six cylinder liners of the replaceable type are shown in figure 17-15. These liners illustrate some of the differences in the design of liners and the relative size of the engines represented

Cylinder Heads.–The liners or bores of an internal-combustion engine must be sealed tightly to form the combustion chambers. In most Navy engines, except for opposed-piston engines, the space at the combustion end of a cylinder is formed and sealed by a cylinder head that is a separate unit from the block.

A number of engine parts that are essential to engine operation may be found in or attached to the cylinder head. The cylinder head may house intake and exhaust valves, valve guides and valve seats, or only exhaust valves and related parts. Rocker arm assemblies are frequently attached to the cylinder head. The fuel injection valve is almost always in the cylinder head or

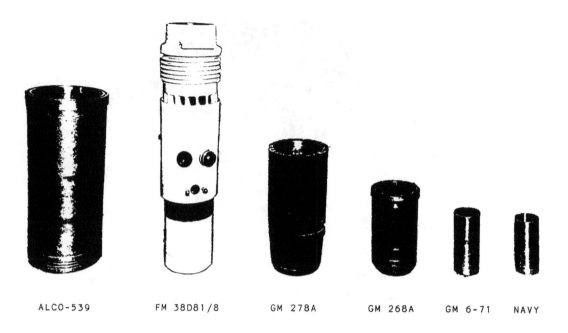

ALCO-539   FM 38D81/8   GM 278A   GM 268A   GM 6-71   NAVY

Figure 17-15.–Cylinder liners of diesel engines.

heads of a diesel engines, while the spark plugs are always in the cylinder head of gasoline engines. Cylinder heads of a diesel engine may also be fitted with air-starting valves, indicator and blowdown valves, and safety valves.

The number of cylinder heads found on diesel engines varies considerably. Small engines of the in-line cylinder arrangement use one head for all cylinders. A single head may cover all the cylinders in each bank of some V-type engines. Large diesel engines generally have one cylinder head for each cylinder, although some engines use one head for each pair of cylinders.

Cylinder Head Studs and Gaskets.–In many engines, the seal between the cylinder head and the block depends principally upon the studs and gaskets. The studs, or stud bolts, secure the cylinder head to the cylinder block. A gasket between the head and the block is compressed to form a seal when the head is properly tightened down. In some engines, gaskets are not used between the cylinder head and block; the mating surfaces of the head and block are accurately machined to form a seal between the two parts.

Principal Moving Parts

Many of the principal parts that are within the main structure of an engine are moving parts. These moving parts convert the thermal energy released by combustion in the cylinder to mechanical energy, which is then available for useful work at the crankshaft. The moving parts included in the conversion process from

combustion to energy output may be divided into the following three major groups:

- The parts that have only reciprocating motion (pistons)

- The parts that have both reciprocating and rotating motion (connecting rods)

- The parts that have only rotating motion (crankshafts and camshafts)

The first two major groups of moving parts may be further grouped under the single heading of piston and rod assemblies. Such an assembly may include a piston, piston rings, piston pin, connecting rod, and related bearings.

PISTONS.–As one of the principal moving parts in the power-transmitting assembly, the piston must be so designed and must be made of such materials that it can withstand the extreme heat and pressure of combustion. Pistons must also be light enough to keep inertia loads on related parts to a minimum. The piston aids in sealing the cylinder to prevent the escape of gases. It also transmits some of the heat through the piston rings to the cylinder wall.

Pistons perform a number of functions. A piston, in addition to serving as the unit that transmits the force of combustion to the connecting rod and conducting the heat of combustion to the cylinder wall, may serve as a valve in opening and closing the ports of a 2-stroke cycle

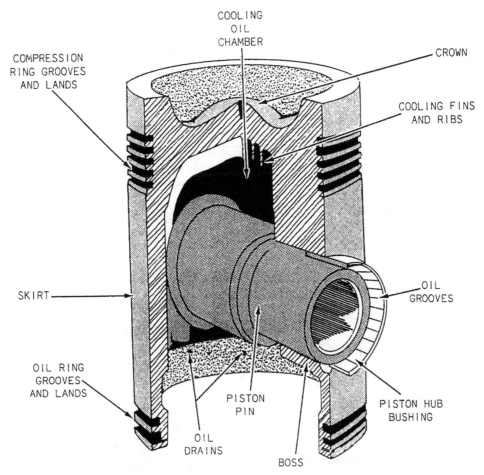

COOLING
OIL
CHAMBER

COMPRESSION
RING GROOVES
AND LANDS

CROWN

COOLING FINS
AND RIBS

SKIRT

OIL
GROOVES

OIL RING
GROOVES
AND LANDS

PISTON HUB
BUSHING

OIL
DRAINS

PISTON
PIN

BOSS

Figure 17-16.–Piston nomenclature for a trunk-type piston.

engine. The nomenclature for the parts of a typical trunk-type piston is given in figure 17-16.

PISTON RINGS.–Piston rings are particularly vital to engine operation in that they must effectively perform three functions: seal the cylinder, distribute and control lubricating oil on the cylinder wall, and transfer heat from the piston to the cylinder wall. All rings on a piston perform the latter function, but two general types of rings–compression and oil–are required to perform the first two functions.

There are numerous types of rings in each of these groups constructed in different ways for particular purposes. Some of the variations in ring design are shown in figure 17-17. The number of rings and their location will also vary considerably with the type and size of pistons.

PISTON PINS AND PISTON BEARINGS.–In trunk-type piston assemblies, the only connection between the piston and the connecting rod is the piston pin (sometimes referred to as the wrist pin) and its bearings. These parts must be of especially strong construction because the power developed in the

cylinder is transmitted from the piston through the pin to the connecting rod. The pin is the pivot point where the straight-line, or reciprocating, motion of the piston changes to the reciprocating and rotating motion of the connecting rod. Thus, the pin is subjected to two principal forces–the forces created by combustion and the side thrust created by the change in direction of motion. (See fig. 17-18.)

CONNECTING ROD.–The connecting link between the piston and the crankshaft or the crankshaft and the crosshead of an engine is the connecting rod. In order that the forces created by combustion can be transmitted to the crankshaft, the rod changes the reciprocating motion of the piston to the rotating motion of the crankshaft.

CAMSHAFT.–The camshaft is a shaft with eccentric projections called cams. The camshaft of an engine is designed to control the operation of valves, usually through various intermediate parts. Originally, cams were made as separate pieces and fastened to the camshaft. However, in most modern engines, the cams are forged or cast as an integral part of the camshaft.

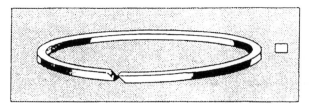

A. DIAGONALLY CUT COMPRESSION RING

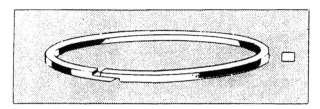

B. LAP-JOINT COMPRESSION RING

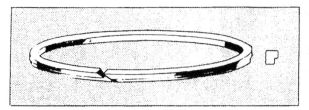

C. OIL RING

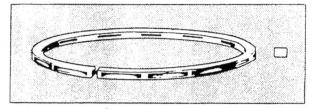

D. SLOTTED OIL RING

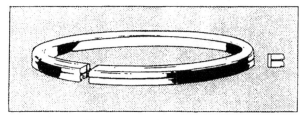

E. THREE-PIECE OIL RING

Figure 17-17.–Types of piston rings.

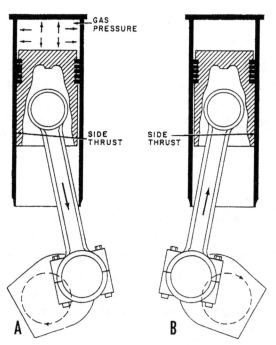

Figure 17-18.–Side thrust of a trunk-type piston in a single-acting engine.

To reduce wear and to withstand repeated shock action, camshafts are made of low-carbon alloy steel with the cam and journal surfaces carburized before the final grinding is done.

The cams are arranged on the shaft so the proper firing order of the cylinders served can take place. If one cylinder is properly timed, the remaining cylinders are automatically in time. All cylinders will be affected if there is a change in timing. The shape of the cam determines the point of opening and closing, the speed of opening and closing, and the amount of the valve lift.

The camshaft is driven by the crankshaft by various means, the most common being by gears or by a chain and sprocket. The camshaft for a 4-stroke cycle engine must turn at one-half of the crankshaft speed; while in the 2-stroke cycle engine, it turns at the same speed as the crankshaft.

The location of the crankshaft differs in various engines. The location of the camshaft depends on the arrangement of the valve mechanism.

CRANKSHAFT.–One of the principal engine parts that has only rotating motion is the crankshaft. As one of the largest and most important moving parts in an engine, the crankshaft changes the movement of the piston and the connecting rod into the rotating motion required to drive such items as reduction gears, propeller shafts, generators, and pumps. As a result of its function, the crankshaft is subjected to all the forces developed in an engine.

While crankshafts of a few larger engines are of the built-up type (forged in separate sections and flanged together), the crankshafts of most modern engines are of one-piece construction. A shaft of this type is shown in figure 17-19. The parts of a crankshaft may be identified by various terms. However, those terms in figure 17-19 are the ones that are most commonly used in the technical manuals for most of the engines used by the Navy.

FLYWHEEL.–The speed of rotation of the crankshaft increases each time the shaft receives a power impulse from one of the pistons. The speed then gradually decreases until another power impulse is received. If permitted to continue unchecked, these fluctuations in speed (their number depending upon the number of cylinders firing on one crankshaft revolution) would result in an undesirable situation with respect to the driven mechanism as well as the engine. Therefore, some means must be provided to stabilize shaft rotation. In some engines, this is accomplished by installing a flywheel on the crankshaft. In other engines, the motion of such engine parts as the connecting rods, webs and lower ends of connecting rods, and such driven units as the clutch and generator serve the purpose. The need for a flywheel decreases as the number of cylinders firing in one revolution of the crankshaft and the mass of the moving parts attached to the crankshaft increases.

A flywheel stores up energy during the power event and releases it during the remaining events of the operating cycle. In other words, when the speed of the shaft tends to increase, the flywheel absorbs energy. When the speed tends to decrease, the flywheel gives up energy to the shaft in an effort to keep shaft rotation uniform. In doing this, a flywheel (1) keeps variations in speed within desired limits at all loads; (2) limits the increase or decrease in speed during sudden changes of load; (3) aids in forcing the piston through the compression event when an engine is running at low or idling speed; and (4) provides leverage or mechanical advantages for a starting motor.

BEARINGS.–An important group of engine parts consists of the bearing. Some bearings remain stationary in performing their function, while others move. One principal group of stationary bearings in an engine is the group that supports the crankshaft. These bearings are generally called main engine bearings. Main bearings in most engines are of the sliding-contact, or plain, type, consisting of two half shells.

Main bearings are subjected to a fluctuating load, as are the connecting rod bearings and the piston pin bearings. However, the manner in which main journal bearings are loaded depends on the type of engine in which they are used.

In a 2-stroke cycle engine, a load is always placed on the lower half of the main bearings and the upper half of the piston pin bearings. In the connecting rod, the load is always placed on the lower half of the main bearings and the upper half of the connecting rod bearings at the crankshaft end of the rod. This is true because the forces of combustion are greater than the inertial forces created by the moving parts.

In a 4-stroke cycle engine, the load is applied first on one bearing shell and then on the other. The reversal of pressure is the result of the large forces of inertia imposed during the intake and exhaust strokes. In other words, inertia tends to lift the crankshaft in its bearings during the intake and exhaust strokes.

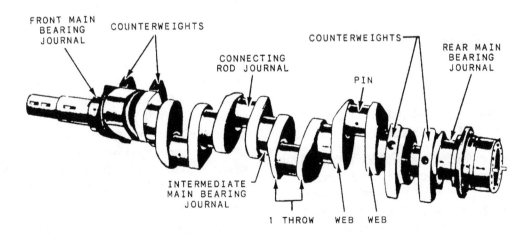

NOTE: 1-THROW=2-WEBS+1 PIN

Figure 17-19.–One-piece, 6-throw crankshaft.

There is a definite reversal of load application on the main bearings of a double-acting engine. In this case, the reversal is caused by combustion taking place first on one end of the piston and then on the other.

The bearings used in connection with most piston pins are of the sleeve bearing or bushing type. These bearings may be further identified according to location–the piston boss, piston pin bearings, and the connecting rod piston bearings.

Bearings or bushings are made of bronze or similar material. Since the bushing material is a relatively hard-bearing metal, surface-hardened piston pins are required. The bore of the bushing is accurately ground in line for the close fit of the piston pin. Most bushings have a number of small grooves cut in their bore for lubrication purposes. Some sleeve bushings have a press fit, while others are "cold shrunk" into the bosses.

Sleeve bushings used in the piston end of connecting rods are similar in design to those used in piston bosses. Generally, bronze makes up the bearing surface. Some bearing surfaces are backed with a case-hardened steel sleeve, and the bushing has a shrink fit in the rod bore. In other bushings, the bushing fit is such that a gradual rotation (creep) takes place in the eye of the connecting rod. In another variation of the sleeve-type bushing, a cast bronze lining is pressed into a steel bushing in the connecting rod.

The types of bearings used for main bearings and in connection with piston-pin assemblies are representative of those used at other points in an engine where bearing surfaces are required.

All of the parts that make a complete engine have by no means been covered in the preceding section of this chapter. Since many engine parts and accessories are commonly associated with the systems of an engine, functions of some of the principal components not covered to this point are considered with the applicable system that they affect.

OPERATING MECHANISMS FOR SYSTEM PARTS AND ACCESSORIES

To this point, consideration has been given only to the main engine parts stationary and moving and two of the systems common to internal-combustion engines. At various points in this chapter, reference has been made to the operation of some of the engine parts. For example, it has been pointed out that the valves open and close at the proper time in the operating cycle and that the impellers or lobes of a blower rotate to compress intake air. However, little consideration has been given to the source of power or to the mechanisms that cause these parts to operate.

In many cases, the mechanism that operates engine valves and blowers may also be mechanisms that operate such items as the governor; fuel, lubricating, and water pumps; and overspeed trips. In some engines, all are operated by the same mechanism. Since mechanisms that transmit power to operate specific parts and accessories may be related to more than one engine system, we will discuss drive mechanisms before getting into the remaining engine systems.

The parts that make up the operating mechanisms of an engine may be divided into two groups: drive mechanisms and actuating mechanisms. The source of power for the operating mechanisms of an engine is the crankshaft.

As used in this chapter, the term *drive mechanism* identifies the group of parts that takes power from the crankshaft and transmits that power to various engine parts and accessories. In engines, the drive mechanism does not change the type of motion, but it may change the direction of motion. For example, the impellers, or lobes, of a blower are driven or operated by a rotary motion from the crankshaft transmitted to the impellers or lobes by the drive mechanism, an arrangement of gears and shafts. While the type of motion (rotary) remains the same, the direction of motion of one impeller is opposite that of the other impeller as a result of the gear arrangement within the drive mechanism.

A drive mechanism may be a gear, chain, or belt. The gear is the most common. Some engines use chain assemblies or a combination of gears and chains as the driving mechanism. Belts are not common on marine engines, but they are used as drive mechanisms on gasoline engines.

Some engines have a single-drive mechanism that transmits power to operate engine parts and accessories. In some engines, there may be two or more separate mechanisms. When separate assemblies are used, the one that transmits power to operate the accessories is called the accessory drive. Some engines have more than one accessory drive. A separate drive mechanism that is used to transmit power to operate engine valves is generally called the camshaft drive or timing mechanism.

The camshaft drive, as the name implies, transmits power to the camshaft of the engine. The shaft, in turn, transmits the power through a combination of parts that causes the engine valves to operate. Since the valves of an engine must open and close at the proper moment (with respect to the position of the piston) and remain in the open and closed positions for definite periods of time, a fixed relationship must be maintained between the rotational speeds of the crankshaft and the camshaft. Camshaft drives are designed to maintain the proper relationship between the speeds of the two shafts. In maintaining this relationship, the drive causes the camshaft to rotate at crankshaft speed in a 2-stroke cycle engine and at one-half crankshaft speed in a 4-stroke cycle engine.

The term *actuating mechanism*, as used in this chapter, identifies that combination of parts that receives power from the drive mechanism and transmits the power to the engine valves. For the intake and exhaust valves, fuel injection, and air starter to operate, there must be a change in the type of motion. The rotary motion of the camshaft must be changed to a reciprocating motion. The group of parts that, by changing the type of motion, causes the valves of an engine to operate is generally referred to as the valve-actuating mechanism. A valve-actuating mechanism may include the cam, cam followers, pushrods, rocker arms, and valve springs. In some

engines, the camshaft is so located that pushrods are not needed. In such engines, the cam follower is a part of the rocker arm. Some actuating mechanisms are designed to transform reciprocating motion into rotary motion, but in internal-combustion engines, most actuating mechanisms change rotary motion into reciprocating motion.

There is considerable variation in the design and arrangement of the parts of operating mechanisms found in different engines. The size of an engine, the cycle of operation, the cylinder arrangement, and other factors govern the design and arrangement of the components as well as the design and arrangement of the mechanisms. Three types of operating mechanisms are shown in figures 17-20, 17-21, and 17-22.

The mechanisms that supply power for the operation of the valves and accessories of gasoline engines are basically the same as those found in diesel engines. Some manufacturers use chain assemblies, while others use gears as the primary means of transmitting power to engine parts. Combination gear-chain drive assemblies are used on some gasoline engines.

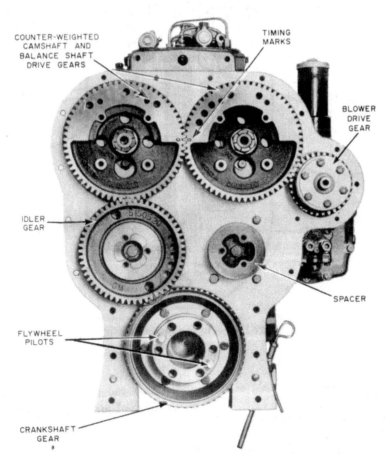

75.98

Figure 17-20.–Camshaft and accessory drive.

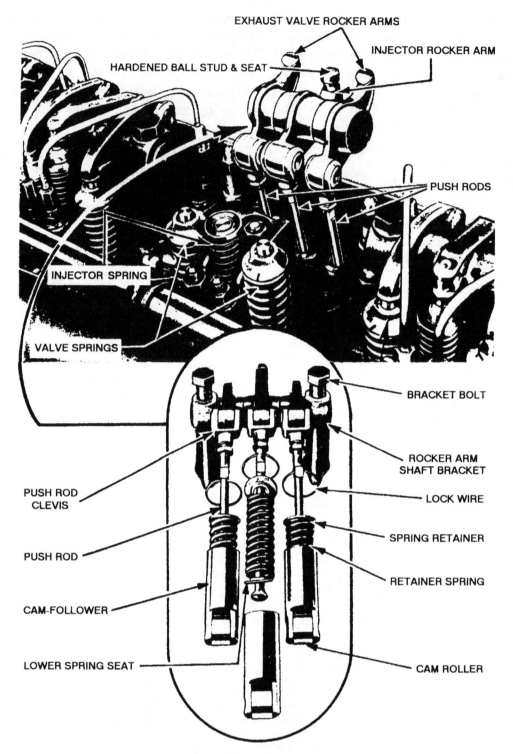

EXHAUST VALVE ROCKER ARMS

INJECTOR ROCKER ARM

HARDENED BALL STUD & SEAT

PUSH RODS

INJECTOR SPRING

VALVE SPRINGS

BRACKET BOLT

ROCKER ARM SHAFT BRACKET

PUSH ROD CLEVIS

LOCK WIRE

PUSH ROD

SPRING RETAINER

CAM-FOLLOWER

RETAINER SPRING

LOWER SPRING SEAT

CAM ROLLER

Figure 17-21.–Valve-actuating mechanism.

ENGINE AIR SYSTEMS

Parts and accessories that supply the cylinders of an engine with air for combustion and remove the waste gases after combustion and the power events are finished are commonly referred to as the intake and exhaust systems. These systems are closely related and, in some cases, are referred to as the air systems of an engine.

TIMING SPROCKET

CRANKSHAFT SPROCKET

CAMSHAFT SPROCKET

TIMING SPROCKET

OVERSPEED GOVERNOR

CAMSHAFT SPROCKET

TIGHTENER SPROCKET

TIMING CHAIN

75.106

Figure 17-22.–Camshaft drive and timing mechanism.

INTAKE SYSTEMS

The following information on air systems deals primarily with the systems of diesel engines; nevertheless, much of the information dealing with the parts of diesel engine air systems is also applicable to most of the parts in similar systems of gasoline engines. However, the intake events in the cycle of operation of a gasoline engine includes the admission of air and fuel as a mixture to the cylinder. For this reason, the intake system of a gasoline engine differs in some respects from a diesel engine.

Although the primary purpose of a diesel engine intake system is to supply the air required for combustion, the system also cleans the air and reduces the noise created by the air as it enters the engine. An intake system may include an air silencer, an air cleaner and screen, an air box or header, intake valves or ports, a blower, an air heater, and an air cooler. Not all of these parts are common to every intake system.

A discussion of the air systems of diesel engines frequently involves the use of two terms that identify processes related to the functions of the intake and exhaust systems. These terms *scavenging* and *supercharging* and the processes they identify are common to many modern diesel engines.

In the intake systems of all modern 2-stroke cycle engines and some 4-stroke cycle engines, a device known as a blower is installed to increase the flow of air into the cylinders. The blower compresses the air and forces it into an air box or manifold (reservoir) that surrounds or is attached to the cylinders of an engine. Thus, more air under constant pressure is available as required during the cycle of operation.

The increased amount of air, a result of blower action, fills the cylinder with a fresh charge of air. During the process, the increased amount of air helps to clear the cylinder of the gases of combustion. This process is called scavenging. Therefore, the intake system of some engines, especially those operating on the 2-stroke cycle, is sometimes called the scavenging system. The air forced into the cylinder is called scavenge air (or scavenging air), and the ports through which it enters are called scavenge ports.

Scavenging must take place in a relatively short portion of the operating cycle; the duration of the process differs in 2- and 4-stroke cycle engines. In a 2-stroke cycle engine, the process takes place during the latter part of the downstroke (expansion) and the early part of the upstroke (compression). In a 4-stroke cycle engine, scavenging takes place when the piston is nearing and passing TDC during the latter part of an upstroke (exhaust) and the early part of a downstroke (intake). The intake and exhaust openings are both open during this interval of time. The overlap of intake and exhaust permits the air from the blower to pass through the cylinder into the exhaust manifold, cleaning out the exhaust gases from the cylinder and, at the same time, cooling the hot engine parts.

When scavenging air enters the cylinder of an engine, it must be so directed that the waste gases are removed from the remote parts of the cylinder. The two

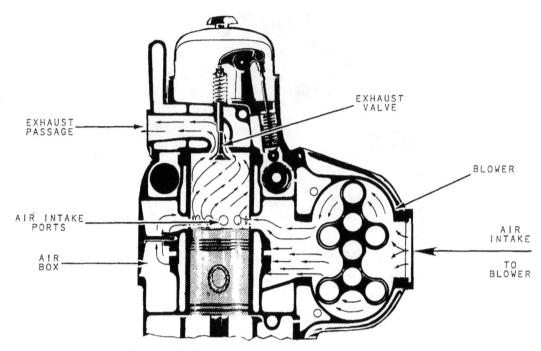

Figure 17-23.–Valve uniflow system in a 2-stroke diesel engine.

principal methods by which this is accomplished are sometimes referred to as port uniflow scavenging and valve uniflow scavenging. In the uniflow method of scavenging, both the air and the burned gases flow in the same direction. This action causes a minimum of turbulence and improves the effectiveness of the scavenging action. An example of a valve uniflow system is shown in figure 17-23. Methods of scavenging in a diesel engine are shown in figure 17-24.

An increase in airflow into the cylinders of an engine can serve to increase power output, in addition to being used for scavenging. Since the power of an engine comes from the burning of fuel, an increase in power requires more fuel since each pound of fuel requires a certain amount of air for combustion. Supplying more air to the combustion spaces that can be supplied through the action of atmospheric pressure and

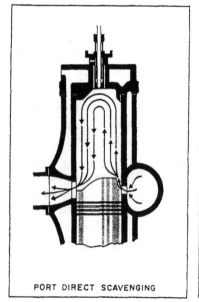
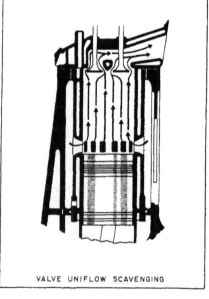
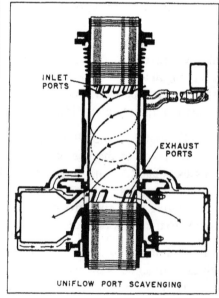

Figure 17-24.–Methods of scavenging in diesel engines.

piston action (in 4-stroke cycle engines) or scavenging air (in 2-stroke cycle engines) is called supercharging.

In some 2-stroke cycle diesel engines, the cylinders are supercharged during the air intake simply by an increase in the pressure of scavenging air. The same blower is used for supercharging and scavenging. Scavenging is done when air is admitted under low pressure into the cylinder while the exhaust ports or valves are open. Supercharging is done with the exhaust ports or valves closed, a condition that enables the blower to force air under pressure into the cylinder and thereby increase the amount of air available for combustion. The increase in pressure, resulting from the compression action of the blower, will depend on the type of installation. With this increase in pressure and the amount of air available for combustion, there is a corresponding increase in the air-fuel ratio and in combustion efficiency within the cylinder. In other words, an engine of a given size that is supercharged can develop more power than an engine of the same size that is not supercharged.

For a 4-stroke cycle diesel engine to be supercharged, a blower must be added to the intake system since the exhaust and intake in an unsupercharged engine are performed by the action of the piston. The timing of the valves in a supercharged 4-stroke cycle engine is also different from that in a similar engine that is not supercharged. In the supercharged engine, the closing of the intake valve is slowed down so that the intake valves or ports are open for a longer time after the exhaust valves close. The increased time that the intake valves are open (after the exhaust valves close) allows more air to be forced into the cylinder before the start of the compression event. Study figure 17-25 so that you will understand how the opening and closing of the intake and exhaust valves or ports affect both scavenging and supercharging. Also note the differences in these processes as they occur in supercharged 2- and 4-stroke cycle engines. In figure 17-25, the circular pattern represents crankshaft rotation. Some of the events occurring in the cycles are shown in degrees of shaft rotation for purposes of illustration and comparison only. In studying figure 17-25, keep in mind that the crankshaft of a 4-stroke cycle engine makes two complete revolutions in one cycle of operation, while the shaft in a 2-stroke cycle engine makes only one revolution per cycle. Also, keep in mind that the exhaust and intake events in a 2-stroke cycle engine do not involve complete piston strokes as they do in a 4-stroke cycle engine.

EXHAUST SYSTEM

The system that functions primarily to carry gases away from the cylinders of an engine is called the exhaust system. In addition to this principal function, an exhaust system may be designed to (1) muffle exhaust noise, (2) quench sparks, (3) remove solid material from exhaust gases, and (4) furnish energy to a turbine-driven supercharger. The principal parts that may be used in combination to accomplish the functions of an engine exhaust system are shown in figure 17-26.

ENGINE FUEL SYSTEMS

The method of getting fuel into the cylinder is one of the major differences between gasoline and diesel engines. As pointed out earlier, fuel for gasoline engines is mixed with air outside the cylinder and the mixture is

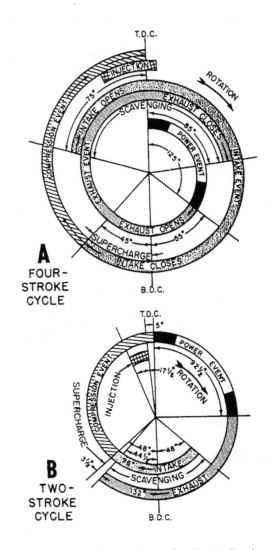

Figure 17-25.—Scavenging and supercharging in diesel engines.

then drawn into the cylinder and compressed. On the other hand, fuel for diesel engines is injected or sprayed into the combustion space after the air is already compressed. The equipment that supplies fuel to the cylinders of a gasoline engine differs from that of a diesel engine.

There are several types of fuel-injection systems in use. The function of each type is, however, the same. The primary function of a fuel-injection system is to deliver fuel to the cylinders under specified conditions. The conditions must be according to the power requirements of the engine.

The first condition to be met is that of the injection equipment. The quantity of fuel injected determines the amount of energy available through combustion to the engine. Smooth engine operation and even distribution of the load between the cylinders depend upon the same volume of fuel being admitted to a particular cylinder each time it fires and upon equal volumes of fuel being delivered to all cylinders of the engine. The measuring device of a fuel injection system must also be designed to vary the amount of fuel being delivered as changes in load and speed vary.

In addition to measuring the amount of fuel injected, the system must properly time injection to ensure efficient combustion so maximum energy can be obtained from the fuel. Early injection tends to develop excessive cylinder pressures and extremely early injection will cause knocking. Late injection tends to decrease power output and, if extremely late, it will cause incomplete combustion. In many engines, fuel-injection equipment is designed to vary the time of injection as speed or load varies.

A fuel system must also control the rate of injection. The rate at which fuel is injected determines the rate of combustion. The rate of injection at the start should be low enough that excessive fuel does not accumulate in the cylinder during the initial ignition delay (before combustion begins). Injection should proceed at such a rate that the rise in combustion pressure is not excessive, yet the rate of injection must be such that fuel is introduced as rapidly as is permissible to obtain complete combustion. An incorrect rate of injection will affect engine operation in the same way as improper timing. If the rate of injection is too high, the results will be similar to those caused by an excessively early

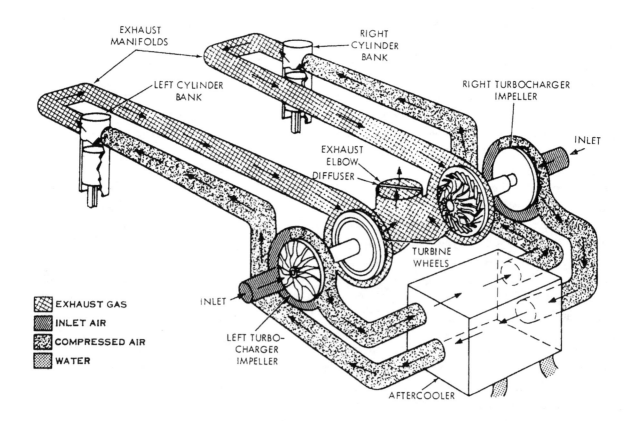

Figure 17-26.–Intake and exhaust system.

17-28

injection; if the rate is too low, the results will be similar to those caused by an excessively late injection.

A fuel-injection system must increase the pressure of the fuel sufficiently to overcome compression pressures and to ensure proper distribution of the fuel injected into the combustion space. Proper distribution is essential if the fuel is to mix thoroughly with the air and burn efficiently. While pressure is a prime contributing factor, the distribution of the fuel is influenced in part by atomization and penetration of the fuel. As used in connection with fuel injection, atomization means the breaking up of the fuel as it enters the cylinder into small particles that form a mistlike spray. Penetration is the distance through which the fuel particles are carried by the kinetic energy imparted to them as they leave the injector or nozzle.

Atomization is obtained when the liquid fuel under high pressure passes through the small opening or openings in the injector or nozzle. As the fuel enters the combustion space, high velocity is developed because the pressure in the cylinder is lower than the fuel pressure. The friction created as the fuel passes through the air at high velocity causes the fuel to break up into small particles. Penetration of the fuel particles depends chiefly upon the viscosity of the fuel, the fuel-injection pressure, and the size of the opening through which the fuel enters the cylinder.

Fuel must be atomized into particles sufficiently small so as to produce a satisfactory ignition delay period. However, if the atomization process reduces the size of the fuel particles too much, they will lack penetration–the smaller the particles the less the penetration. Lack of sufficient penetration results in the small particles of fuel igniting before they have been properly distributed. Since penetration and atomization tend to oppose each other, a compromise in the degree of each is necessary in the design of fuel-injection equipment if uniform fuel distribution is to be obtained. The pressure required for efficient injection and, in turn, proper distribution is dependent upon the compression pressure in the cylinder, the size of the opening through which the fuel enters the combustion space, the shape of the combustion space, and the amount of turbulence created in the combustion space.

The fuel system of a gasoline engine is basically similar to that of a diesel engine except that a carburetor is used instead of injection equipment. While injection equipment handles fuel only, the carburetor handles both air and fuel. The carburetor must meet requirements similar to those of an injection system, except that in the carburetor air is also involved. The carburetor must accurately meter fuel and air in varying percentages according to engine requirements. The carburetor also vaporizes the fuel charge and then mixes it with the air in the proper ratio. The amount of fuel mixed with the air must be carefully regulated and must change with the different speeds and loads of an engine. The amount of fuel required by an engine that is warming up is different from the amount required by an engine that has reached operating temperature. Special fuel adjustment is needed for rapid acceleration. All of these varying requirements are met automatically by the modern carburetor.

ENGINE IGNITION SYSTEMS

The methods by which the fuel mixture is ignited in the cylinders of diesel and gasoline engines differ as much as the methods of obtaining a combustible mixture in the cylinders of the two engines. An ignition system as such is not commonly associated with diesel engine. There is no one group of parts in a diesel engine that functions only to cause ignition as there is in a gasoline engine. However, a diesel engine does have an ignition system; otherwise, combustion would not take place in the cylinders.

In a diesel engine, the parts that may be considered as forming the ignition system are the piston, the cylinder liner, and the cylinder head. These parts are not commonly thought of as forming an ignition system since they are generally associated with other functions such as forming the combustion space and transmitting power. Nevertheless, ignition in a diesel engine depends upon the piston, the cylinder, and the head. These parts not only form the space where combustion takes place but also provide the means by which the air is compressed to generate the heat necessary for self-ignition of the combustible mixture. In other words, both the source (air) of ignition heat and its generation (compression) are wholly within a diesel engine.

This is not true of a gasoline engine because the combustion cycles of the two types of engines are different. In a gasoline engine, even though the piston, the cylinder, and the head form the combustion space, as in a diesel engine, the heat necessary for ignition is caused by energy from a source external to the combustion space. The completion of the ignition process involving the transformation of mechanical energy into electrical energy and then into heat energy requires several parts–each performing a specific function. The parts that make the transformation of energy and the system that they form are commonly thought of when reference is made to an ignition system.

The spark that causes the ignition of the explosive mixture in the cylinders of a gasoline engine is produced when electricity is forced across a gap formed by two electrodes in the combustion chamber. The electrical ignition system furnishes the spark periodically to each cylinder at a predetermined position of piston travel. To accomplish this function, an electrical ignition system must have either a source of electrical energy or a means of developing electrical energy. In some cases, a storage battery is used as the source of energy; in other cases, a magneto generates electricity for the ignition system. The voltage from either a battery or a magneto is not sufficiently high enough to overcome the resistance created by pressure in the combustion chamber and to cause the proper spark in the gap formed by the two electrodes in the combustive chamber. Therefore, an ignition system must include a device that increases the voltage of the electricity supplied to the system sufficiently to cause a "hot" spark in the gap of the spark plug. The device that performs this function is generally called an ignition coil or induction coil.

Since a spark must occur momentarily in each cylinder at a specific time, an ignition system must include a device that controls the timing of the flow of electricity to each cylinder. This control is accomplished by interrupting the flow of electricity from the source to the voltage increasing device (ignition coil). The interruption of the flow of electricity also plays an important part in the process of increasing voltage. The interrupting device is generally called the breaker assembly. A device that will distribute electricity to the different cylinders in the proper firing order is also necessary. The part that performs this function is called the distributing mechanism. Spark plugs to provide the gaps and wiring and switches to connect the parts of the system are essential to complete an ignition system.

All ignition systems are basically the same except for the source of electrical energy. The source of energy is frequently used as a basis for classifying ignition systems–thus, the battery ignition system and the magneto-ignition system.

ENGINE COOLING SYSTEMS

A great amount of heat is generated within an engine during operation. The combustion process produces the greater portion of this heat; however, compression of gases within the cylinders and friction between moving parts add to the total amount of heat developed within an engine. Since the temperature of combustion alone is about twice that at which iron melts, it is apparent that without some means of dissipating heat, an engine could

operate for only a very limited time. Without proper temperature control, the lubricating oil film between moving parts would be destroyed, proper clearance between parts could not be maintained, and metals would tend to fail.

When fuel burns in the cylinder of an engine, only about one-third of the heat from the fuel changes into mechanical energy and then leaves the engine in the form of brake horsepower. The rest of the heat shows up as unwanted heat in the form of (1) hot exhaust gases, (2) frictional heat from the rubbing surfaces, and (3) heating of the metal walls that form the combustion chamber, cylinder head, cylinder, and piston. The job of the cooling system is to remove the unwanted heat from these parts so that the following problems can be prevented:

— Overheating and the resulting breakdown of the lubricating oil film that separates the rubbing surfaces of the engine

— Overheating and the resulting loss of strength of the metal itself

— Excessive stresses in or between the engine parts resulting from unequal temperatures

Unwanted heat is transferred through the mediums of water, lubricating oil, air, and fuel. The water of the closed cooling system removes the greatest portion (approximately one-fourth) of the unwanted heat generated by combustion. The balance of the heat is carried away from the engine by the lubricating oil, the fuel, and the air. If the heat lost through cooling could be turned into work by the engine, the output of the engine would be almost doubled. However, the loss of valuable heat is necessary for an engine to operate.

The cooling system of a marine engine may be of the open or closed type. In the open system, the engine is cooled directly by seawater. In the closed system, fresh water (or an antifreeze solution) is circulated through the engine. The fresh (jacket) water is then cooled as it passes through a cooling device where heat is carried away by a constant flow of seawater. The closed system is the design that is most commonly used on marine internal-combustion engines. However, some older marine installations use a system of the open type. The cooling systems of diesel and gasoline engines are similar mechanically and in the function performed.

The cooling system of an engine may include such parts as pumps, coolers, radiators, water manifolds, valves, expansion tank, piping, strainers, connections, and instruments.

Even though there are many types and models of engines used by the Navy, the cooling systems of most of these engines include the same basic parts. The design and location of parts, however, may differ considerably from one engine to another.

Figures 17-27 and 17-28 show the freshwater and seawater systems of an engine that employs the heat exchanger type of cooling system.

ENGINE LUBRICATING SYSTEMS

For proper operation of an engine, the contacting surfaces of all moving parts must be prevented from touching each other so that friction and wear can be reduced to a minimum. Sliding contact between two dry metal surfaces under load will cause excessive friction, heat, and wear. Friction, heat, and wear can greatly be reduced if metal-to-metal contact is prevented. When a clean film of lubricant is used between the metal surfaces, contact is automatically reduced.

Lubrication and the system that supplies lubricating oil to engine parts that involve sliding or rolling contact are as important to successful engine operation as air,

fuel, and heat are to combustion. It is important not only that the proper type of lubricant be used, but also that the lubricant be supplied to the engine parts at the specific flow rate and temperature and that provisions be made to remove any impurities that enter the system. The engine lubricating oil system is designed to fulfill these requirements.

The lubricating system of an internal-combustion engine consists of two main divisions: (1) one that is inside the engine (internal), and (2) one that is outside the engine (external). The internal system consists mainly of passages and piping. The external system includes several components that aid in supplying the oil in the proper quantity at the proper temperature and free of impurities. In the majority of lubricating systems for internal-combustion engines, the external system includes such parts as tanks and sumps, pumps, coolers, strainers, and filters and purifiers. A schematic diagram of a typical lubricating oil system is shown in figure 17-29.

The engine system that supplies the oil required to perform the functions of lubrication is of the pressure type in practically all modern internal-combustion

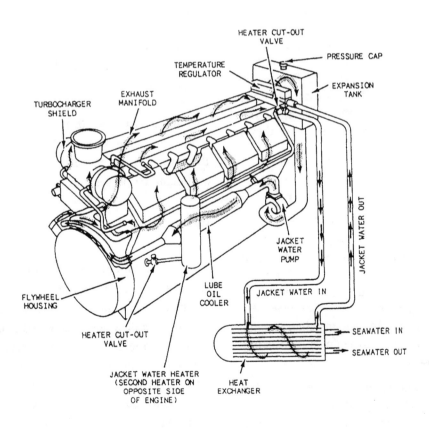

Figure 17-27.–Jacket-water (heat exchanger) cooling system.

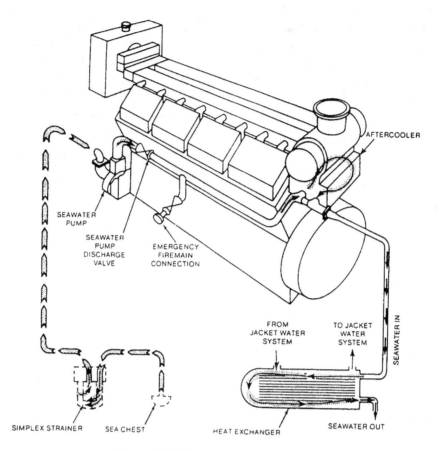

Figure 17-28.–Seawater cooling system.

engines. Although there are many variations in lubricating systems, the components and method of operation are basically the same whether the system is in a diesel or a gasoline engine. Any variance is generally due to differences in engine design and in opinions of manufacturers as to the best location of the component parts of the system. In many cases, similar types of components are used in the systems of diesel and gasoline engines.

TRANSMISSION OF ENGINE POWER

The basic characteristics of an internal-combustion engine make it necessary, in many cases, for the drive mechanism to change both the speed and the direction of shaft rotation in the driven mechanism. There are various methods for making changes of speed and direction during the transmission of power from the driving unit to the driven unit. In most installations, the job is accomplished by a drive mechanism consisting principally of gears and shafts.

The process of transmitting engine power to a point where it can be used in performing useful work involves a number of factors. Two of these factors are torque and speed.

TORQUE

Torque is the force that tends to cause a rotational movement of an object. The crankshaft of an engine supplies a turning force to the gears and shafts that transmit power to the driven unit. Gears are used to increase or decrease torque. If the right combination of gears is installed between the engine and the driven unit, the torque is increased, and the turning force is then sufficient to operate the driven unit.

SPEED

If maximum efficiency is to be obtained, an engine must operate at a certain speed. To obtain efficient engine operation in some installations, the engine may need to operate at a higher speed than that required for efficient operation of the driven unit. In other installations, the speed of the engine may need to be lower than the speed of the driven unit. Through a combination of gears, the speed of the driven unit can be increased or decreased so that the proper speed ratio exists between the units.

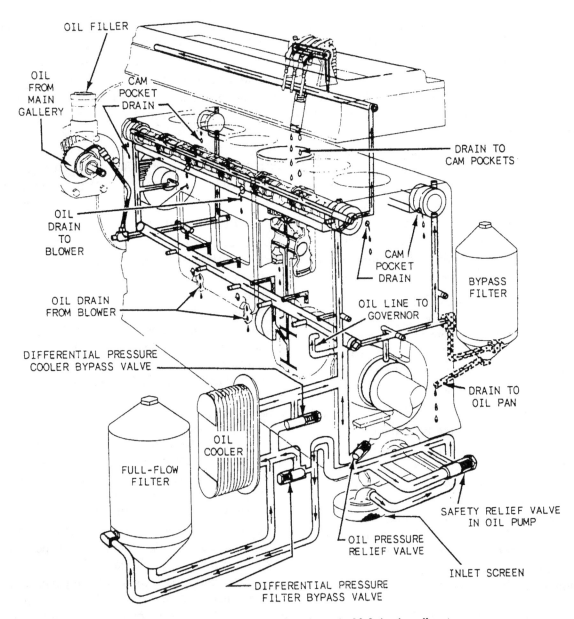

OIL FILLER

OIL FROM MAIN GALLERY

CAM POCKET DRAIN

DRAIN TO CAM POCKETS

OIL DRAIN TO BLOWER

CAM POCKET DRAIN

BYPASS FILTER

OIL DRAIN FROM BLOWER

OIL LINE TO GOVERNOR

DIFFERENTIAL PRESSURE COOLER BYPASS VALVE

DRAIN TO OIL PAN

OIL COOLER

FULL-FLOW FILTER

SAFETY RELIEF VALVE IN OIL PUMP

OIL PRESSURE RELIEF VALVE

INLET SCREEN

DIFFERENTIAL PRESSURE FILTER BYPASS VALVE

Figure 17-29.–Schematic diagram of a typical lubricating oil system.

TYPES OF DRIVE MECHANISMS

The term *indirect drive* describes a drive mechanism that changes both speed and torque. Drives of this type are common to many marine engine installations. Where the speed and the torque of an engine do not need to be changed to drive a machine satisfactorily, the mechanism used is a direct drive. Drives of this type are commonly used when the engine furnishes power for the operation of auxiliaries, such as generators and pumps.

Indirect Drives

The drive mechanism of most engine-powered ships and of many boats is the indirect type. With indirect drive, the power developed by the engine is transmitted to the propeller indirectly, through an intermediate mechanism that reduces the shaft speed. Speed may be reduced mechanically, by a combination of gears or by electrical means.

MECHANICAL DRIVES.–Mechanical drives include devices that reduce the shaft speed of the driven unit, provide a means for reversing the direction of shaft rotation in the driven unit, and permit quick-disconnect of the driving unit from the driven unit.

Propellers operate most efficiently in a relatively low rpm range. The most efficient designs of diesel engines, however, operate in a relatively high rpm range. So that both the engine and propeller may operate efficiently, the drive mechanisms in many installations include a device that permits a speed reduction from the engine crankshaft to the propeller shaft. The combination of gears that brings about the speed reduction is called a reduction gear. In most diesel engine installations, the reduction ratio does not exceed 3 to 1. There are some units, however, that have reductions as high as 6 to 1.

The propelling equipment of a boat or a ship must provide backing down power as well as forward motive power. In some ships, backing down is accomplished by reversing the pitch of the propeller; in other ships, however, backing down is accomplished by reversing the direction of rotation of the propeller shaft. In mechanical drives, the direction of rotation of the propeller shaft is reversed by use of reverse gears.

The drive mechanism of a ship or boat must do more than reduce speed and change the direction of shaft rotation. Most drive mechanisms have a clutch. The clutch disconnects the drive mechanism from the propeller shaft and permits the engine to be operated without turning the propeller shaft.

The arrangement of the components in an indirect drive varies, depending upon the type and size of the installation. In some small installations, the clutch, the reverse gear, and the reduction gear may be combined in a single unit. In other installations, the clutch and the reverse gear may be in one housing and the reduction gear in a separate housing attached to the reverse-gear housing. Drive mechanisms arranged in either manner are usually called transmissions. The arrangement of the components in two different types of transmissions are shown in figures 17-30 and 17-31.

In the transmission shown in figure 17-30, the housing is divided into two sections by the bearing carrier. The clutch assembly is in the forward section and the gear assembly is in the after section of the housing. In the transmission shown in figure 17-31, note that the clutch assembly and the reverse gear assembly are in one housing, while the reduction gear unit is in a separate housing (attached to the clutch and the reverse gear housing).

In large engine installations, the clutch and the reverse gear may be combined or they may be separate units located between the engine and a separate reduction gear or the clutch may be separate and the reverse gear and the reduction gear may be combined. An assembly of the last type is shown in figure 17-32.

In most geared-drive, multiple-propeller ships, the propulsion units and their drive mechanisms are independent of each other (fig. 17-33). In others, two or

more engines can drive a single propeller. In one type of installation, the CODOG (combination diesel or gas turbine) system, each propeller is driven by a diesel engine or both propellers are driven by one gas turbine. The diesel engines are used for normal cruising and maneuvering in confined waters. Each diesel drives a propeller shaft independently of the other. The gas turbine is used for high-speed operation. The single gas turbine drives both propeller shafts through the drive mechanism. This combination permits a large cruising range along with high speed whenever it is needed.

ELECTRIC DRIVES.–Electric drives are used in the propulsion plants of some diesel-driven ships. With electric drive, there is no mechanical connection between the engine(s) and the propeller(s). In such plants, the diesel engines are connected directly to generators. The electricity produced by such an engine-driven generator is transmitted through cables to a motor, which is connected to the propeller shaft directly, or indirectly, through a reduction gear. When a speed reduction gear is included in a diesel-electric drive, the gear is located between the motor and the propeller.

The generator and the motor of a diesel-electric drive may be of the alternating current (ac) or the direct current (dc) type. Almost all diesel-electric drives in the Navy, however, are of the dc type. Since the speed of a dc motor varies directly with the voltage furnished by the generator, the control system of an electric drive is arranged so that the generator voltage can be changed at any time. An increase or decrease in generator voltage is used to control the speed of the propeller. Generator voltage may be changed electrically, by changes in engine speed, and by a combination of these methods. The controls of an electric drive may be remotely located from the engine, such as the in the pilot house.

In an electric drive, the direction of rotation of the propeller is not reversed by a reverse gear. The electrical system is arranged so that the flow of current through the motor can be reversed. This reversal of current flow causes the motor to revolve in the opposite direction. Thus, the direction of rotation of the motor and of the propeller can be controlled by manipulating the electrical controls.

Direct Drives

In some marine engine installations, power from the engine is transmitted to the drive unit without a change in shaft speed; that is, by a direct drive. In a direct-drive, the connection between the engine and the driven unit

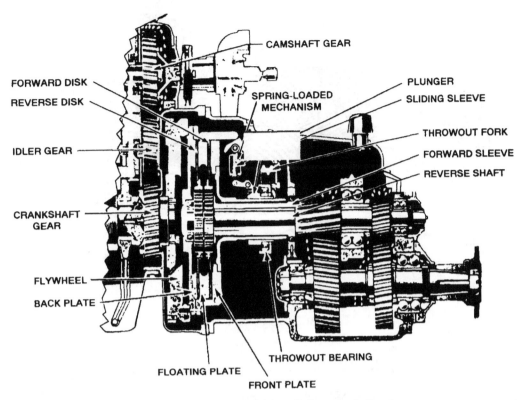

Figure 17-30.–Transmission with an independent oil system.

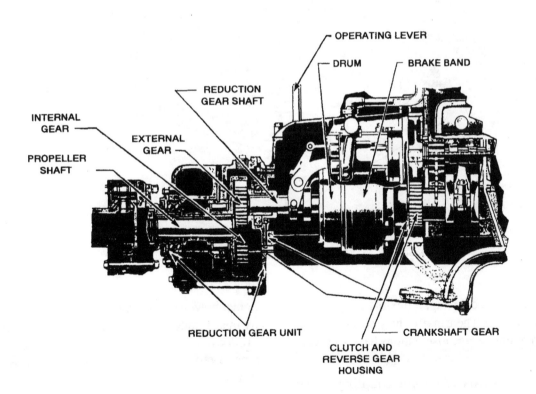

Figure 17-31.–Clutch and reverse gear assembly with an attached reduction gear unit.

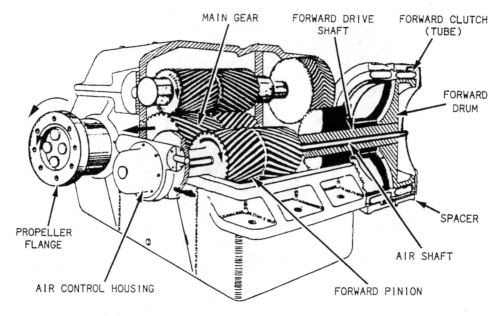

A. FORWARD SHAFT ROTATION.

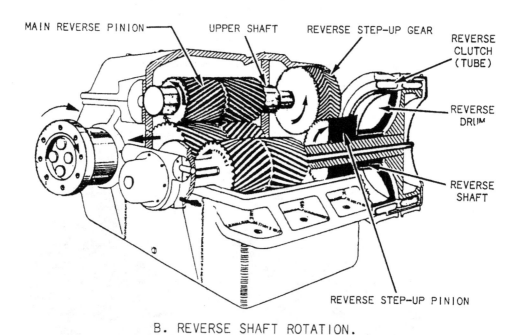

B. REVERSE SHAFT ROTATION.

Figure 17-32.–Clutch and reverse-reduction gear assembly.

may consist of a solid coupling, a flexible coupling, or a combination of both. There may or may not be a clutch in a direct-drive unit. Some installations have a reverse gear.

SOLID COUPLINGS.–Solid couplings vary considerably in design. Some solid couplings consist of two flanges bolted solidly together. In other direct drives, the driven unit is attached directly to the engine crankshaft by a nut. Solid couplings offer a positive means of transmitting torque from the crankshaft of an engine; however, a solid connection does not allow for any misalignment between the input and output shafts, nor does it absorb any of the torsional vibrations transmitted from the engine crankshaft or shaft vibrations.

FLEXIBLE COUPLINGS.–Since solid coupling do not absorb vibration and do not permit any misalignment, most direct drives consist of a flexible coupling, which uses a flexible member or element to connect two flanges or hubs together. Connections of

17-36

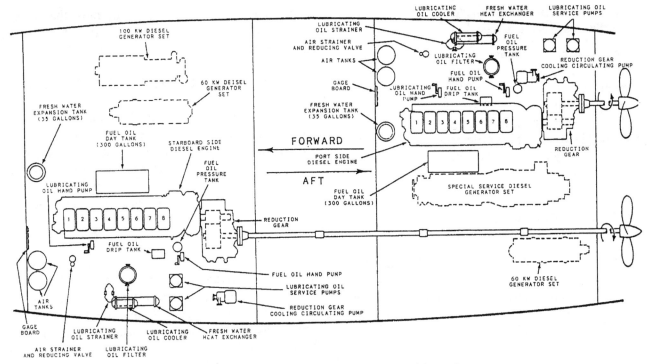

Figure 17-33.–Example of independent propulsion units.

the flexible type are common to the drives of many auxiliaries, such as engine-generator sets. Flexible couplings are also used in indirect drives to connect the engine to the drive mechanism.

The two solid halves of a coupling are joined by a flexible element. The flexible element may be made of rubber, neoprene, steel springs, or gears. Figure 17-34 shows an example of a grid-type flexible coupling that uses a steel spring (view A) and of a gear-type coupling (view B).

CLUTCHES, REVERSE GEARS, AND REDUCTION GEARS

Clutches may be used on direct-driven propulsion Navy engines to disconnect the engine from the propeller shaft. In small engines, clutches are usually combined with reverse gears and used for maneuvering the ship. In large engines, special types of clutches are used to obtain special coupling or control characteristics and to prevent torsional vibration.

Reverse gears are used on marine engines to reverse the direction of rotation of the propeller shaft when maneuvering the ship, without changing the direction of rotation of the engine. They are used principally on relatively small engines. If a high output engine has a reverse gear, the gear is used for low-speed operation only and does not have full load and full speed capacity.

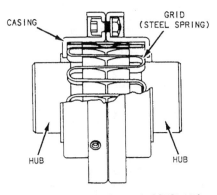

A. GRID-TYPE FLEXIBLE COUPLING.

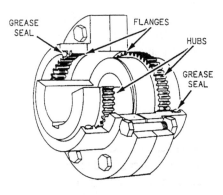

B. GEAR-TYPE FLEXIBLE COUPLING.

Figure 17-34.–Flexible couplings.

For maneuvering ships with large direct-propulsion engines, the engines are reversed.

Reduction gears are used to obtain low propeller-shaft speed with a high engine speed. Speed reduction gears resolve two conflicting requirements: (1) for minimum weight and size for a given power output, engines must have a relatively high rotative speed; and (2) for maximum efficiency, propellers must rotate at a relatively low speed particularly where high thrust capacity is desired.

FRICTION CLUTCHES AND GEAR ASSEMBLIES

Friction clutches are most commonly used with smaller high-speed engines up to 500 hp. However, certain friction clutches in combination with a jaw-type clutch are used with engines up to 1,400 hp and pneumatic clutches with a cylindrical friction surface with engines up to 2,000 hp.

There are two general types of friction clutches: disk and band. Disk-type clutches are classified as either mechanical or hydraulic.

Friction clutches are engaged when two friction surfaces are mechanically forced into contact with each other by toggle-action linkage through stiff springs or through the use of hydraulic or pneumatic pressure.

TWIN-DISK CLUTCH AND GEAR MECHANISM

One of the several types of transmissions used by the Navy is the twin-disk transmission mechanism, shown in figure 17-35. The twin-disk transmission is equipped with a duplex clutch and a reverse and reduction gear unit, all contained in a housing at the afterend of the engine. (See fig. 17-35.) Parts A, B, and C of the clutch assembly are bolted to the flywheel on the crankshaft of the engine and rotate at the same speed as the engine.

The clutch assembly is contained in the part of the housing nearest the engine. It is a dry-type, twin-disk clutch with two friction disks or clutch plates. Each disk is connected to a separate reduction gear train in the afterpart of the housing. The disk (D) and the gear train for reverse rotation are connected by a hollow shaft (F). The disk (E) and the gear train for reverse rotation are connected by a shaft (G), which runs through the center of a shaft (F). Since the gears for forward and reverse rotation of the twin-disk clutch and gear mechanism remain in mesh at all times, there is no shifting of gears.

When the mechanism is shifted, only the floating pressure plate, located between the forward and reverse disks, is moved.

AIRFLEX CLUTCH AND GEAR ASSEMBLY

On large diesel-propelled ships, the clutch, reverse gear, and reduction gear unit have to transmit an enormous amount of power. To maintain the weight and size of the mechanism as low as possible, special clutches have been designed for large diesel installations. One of these is the airflex clutch and gear assembly. A typical airflex clutch and gear assembly for ahead and astern rotation is shown in figure 17-32.

The airflex clutch and gear assembly, shown in figure 17-32, consists of two clutches–one for forward rotation and one for reverse rotation. The clutches are bolted to the engine flywheel by a steel spacer so that they both rotate with the engine at all times and at engine speed. Each clutch has a flexible tube (gland) on the inner side of a steel shell. Before the tubes are inflated, they will rotate out of contact with the drums, which are keyed to the forward and reverse drive shafts. When air under pressure (100 psi) is sent into one of the tubes, the inside diameter of the clutch decreases. This causes the friction blocks on the inner tube surface to come in contact with the clutch drum, locking the drive shaft with the engine.

Forward Rotation

The parts of the airflex clutch that give the propeller ahead rotation are illustrated in view A of figure 17-32. The clutch tube nearest the engine (forward clutch) is inflated to contact and drive the forward drum with the engine. The forward drum is keyed to the forward drive shaft, which carries the double helical forward pinion at the afterend of the gear box. The forward pinion is in constant mesh with the double helical main gear, which is keyed on the propeller shaft. By following through the gear train, you can see that for ahead motion the propeller rotates in a direction opposite to the rotation of the engine.

Reverse Rotation

The parts of the airflex clutch that give the propeller astern rotation are shown in view B of figure 17-32. The reverse clutch is inflated to engage the reverse drum, which is then driven by the engine. The reverse drum is keyed to the short reverse shaft that surrounds the forward drive shaft. A large reverse step-up pinion transmits the motion to the large reverse step-up gear on

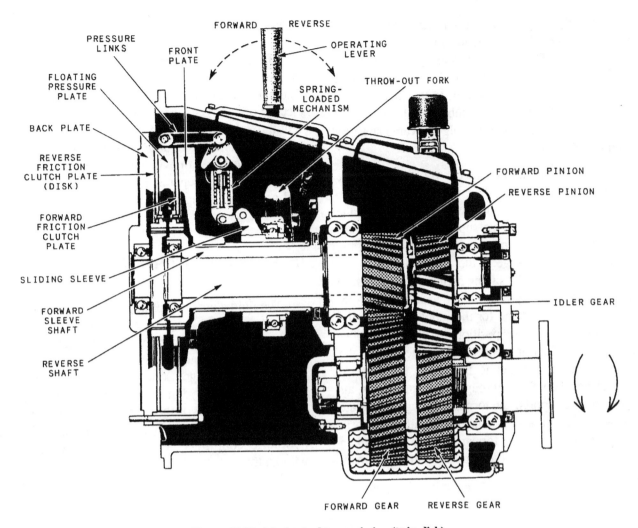

Figure 17-35.–Mechanical transmission (twin-disk).

the upper shaft. The upper shaft rotation is opposite to the engine's rotation. The main reverse pinion on the upper shaft is in constant mesh with the main gear. By tracing through the gear train, you can see that for reverse rotation the propeller rotates in the same direction as the engine.

The diameter of the main gear of the airflex clutch is approximately 2 1/2 times as great as that of the forward and reverse pinions. Thus, there is a speed reduction of 2 1/2 to 1 from either pinion to the propeller shaft.

Since the forward and main reverse pinions are in constant mesh with the main gear, the set that is not clutched in will rotate as idlers driven from the main gear. The idling gears rotate in a direction opposite to their rotation when carrying the load. For example, with the forward clutch engaged, the main reverse pinion rotates in a direction opposite to its rotation for astern motion (note the dotted arrow in view A of fig. 17-32). Since the drums rotate in opposite directions, a control

mechanism is installed to prevent the engagement of both clutches simultaneously.

Airflex Clutch Control Mechanism

The airflex clutch is controlled by an operating lever that works the air control housing, located at the afterend of the forward pinion shaft. The control mechanism, shown with the airflex clutches in figure 17-36, directs the air into the proper paths to inflate the clutch glands (tubes). The air shaft, which connects the control mechanism to the clutches, passes through the forward drive shaft. The supply air enters the control housing through the air check valve and must pass through the small air orifice. The restricted orifice delays the inflation of the clutch to be engaged during shifting from one direction of rotation to the other. The delay is necessary to allow the other clutch to be fully deflated and out of contact with its drum before the inflating clutch can make contact with its drum.

The supply air goes to the rotary air joint in which a hollow carbon cylinder is held to the valve shaft by spring tension. This prevents leakage between the stationary carbon seal and the rotating air valve shaft. The air goes from the rotary joint to the four-way air valve. The sliding-sleeve assembly of the four-way valve can be shifted endwise along the valve shaft by operating the control lever.

When the shifter arm on the control lever slides the valve assembly away from the engine, air is directed to the forward clutch. The four-way valve makes the connection between the air supply and the forward clutch. There are eight neutral ports that connect the central air supply passage in the valve shaft with the sealed air chamber in the sliding member. In the neutral position of the four-way valve, as shown in figure 17-36, the chamber is a dead end for the supply air. With the shifter arm in the forward position, the sliding member uncovers eight forward ports that connect with the forward passages conducting the air to the forward clutch. The air now flows through the neutral ports, air chamber, forward ports, and forward passages to inflate the forward clutch gland. As long as the shifter arm is in the forward position, the forward clutch will remain inflated, and the entire forward air system will remain at a pressure of 100 psi.

Lubrication

On most large gear units, a separate lubrication system is used. One lubrication system is shown in figure 17-37. Oil is picked up from the gearbox by an electric-driven, gear-type lubricating oil pump and is sent through a strainer and cooler. After being cleaned and cooled, the oil is returned to the gearbox to cool and lubricate the gears. In twin installations, such as shown in figure 17-37, a separate pump is used for each unit, and a standby pump is interconnected for emergency use.

HYDRAULIC CLUTCHES (COUPLINGS)

The fluid clutch (coupling) is widely used on Navy ships. A hydraulic coupling eliminates the need for a mechanical connection between the engine and the reduction gears. Couplings of this type operate with a minimum of slippage. Some slippage is necessary for operation of the hydraulic coupling, since torque is transmitted because of the principle of relative motion between the two rotors. Power is transmitted through hydraulic couplings very efficiently (97 percent) without the transmission of torsional vibrations or load shocks from the engine to the reduction gear. The power

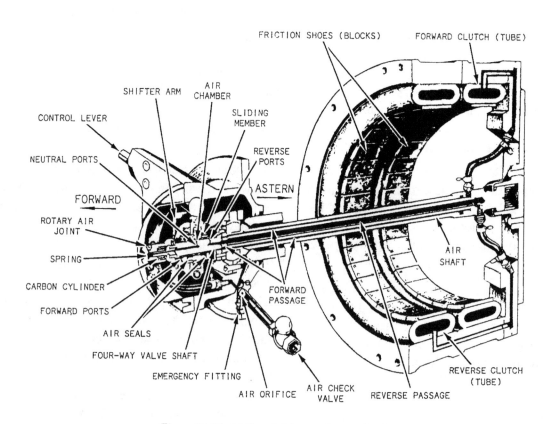

Figure 17-36.–Airflex clutches and control valves.

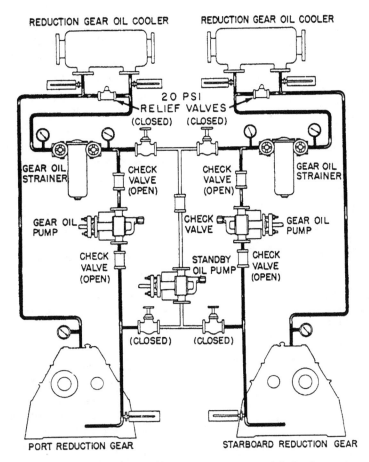

Figure 17-37.--Schematic diagram of a reverse gear lubrication system.

loss from the small amount of slippage is transformed into heat that is absorbed by the oil in the system.

Hydraulic Coupling Assemblies

The two rotors and the oil sealing cover of a typical hydraulic coupling are shown in figure 17-38. The primary rotor (impeller) is attached to the engine crankshaft. The cover is bolted to the secondary rotor and surrounds the primary rotor.

Each rotor is shaped like a half doughnut with radial partitions. A shallow trough is welded into the partitions around the inner surface of the rotor. The radial passages tunnel under this trough (as indicated by the white arrows in fig. 17-38).

When the coupling halves are assembled, the two rotors are placed facing each other to form a series of circular chambers. (See fig. 17-39.) The rotors do not quite touch each other; the clearance between them is 1/4 to 5/8 inch, depending on the size of the coupling. The curved radial passages of the two rotors are opposite each other so that the outer passages combine to make a circular passage, except for the small gaps between the rotors.

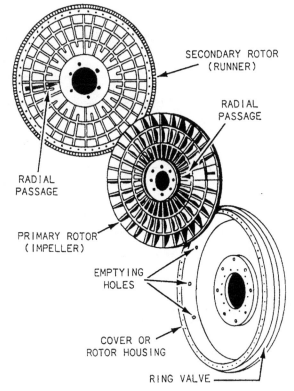

Figure 17-38.--Runner, impeller, and cover of a hydraulic coupling.

17-41

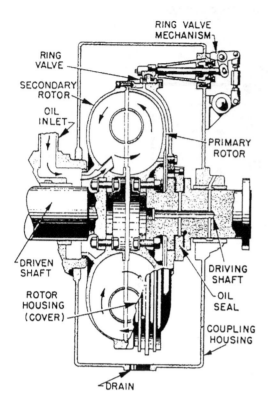

Figure 17-39.–Hydraulic coupling assembly.

In the hydraulic coupling assembly, shown in figure 17-39, the driving shaft is secured to the engine crankshaft and the driven shaft goes to the reduction gear box. The oil inlet admits oil directly to the rotor cavities, which become completely filled. The rotor housing is bolted to the secondary rotor and has an oil-sealed joint with the driving shaft. A ring valve, going entirely around the rotor housing, can be operated by the ring valve mechanism to open or close a series of emptying holes in the rotor housing. When the ring valve is opened, the oil will fly out from the rotor housing into the coupling housing, draining the coupling completely in 2 or 3 seconds. Even when the ring valve is closed, some oil leaks out into the coupling housing, and additional oil enters through the inlet. From the coupling housing, the oil is drawn by a pump to a cooler, then sent back to the coupling.

Another coupling assembly used on several classes of Navy ships is the hydraulic coupling with piston-type, quick-dumping valves. The operation of this coupling is similar to the one previously described. A series of piston valves are located around the periphery of the rotor housing. The piston valves are normally held in the closed position by springs. When air or oil pressure is admitted to the valves, the pistons are moved axially to uncover drain ports, allowing the coupling to empty. A hydraulic coupling with piston-type, quick-dumping valves is used when extremely rapid declutching is not required. It offers greater simplicity and lower cost than a hydraulic coupling with a ring valve mechanism.

Another type of self-contained unit for certain diesel engine drives is the scoop control coupling, shown in figure 17-40. In couplings of this type, the oil is picked

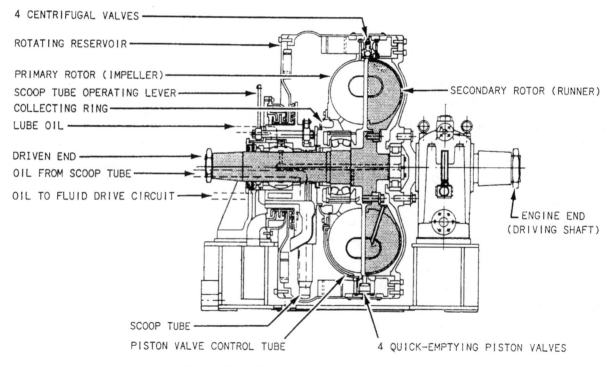

Figure 17-40.–Scoop control hydraulic system.

up by one of two scoop tubes (one tube for each direction of rotation), mounted on the external manifold. Each scoop tube contains two passages: a smaller one (outermost), which handles the normal flow of oil for cooling and lubrication, and a larger one, which rapidly transfers oil from the reservoir directly to the working circuit

The scoop tubes are mechanically operated from the control stand through a system of linkages. As one tube moves outward from the shaft centerline and into the oil annulus, the other is being retracted.

Four spring-loaded centrifugal valves are mounted on the primary rotor. These valves are arranged to open progressively as the speed of the primary rotor decreases. The arrangement provides the necessary oil flow for cooling as it is required. Quick-emptying piston valves are provided to give rapid emptying of the circuit when the scoop tube is withdrawn from contact with the rotating oil annulus.

Under normal circulating conditions, oil fed into the collector ring passes into the piston valve control tubes. These tubes and connecting passages conduct oil to the outer end of the pistons. The centrifugal force of the oil in the control tube holds the piston against the valve port, thus sealing off the circuit. When the scoop tube is withdrawn from the oil annulus in the reservoir, the circulation of oil will be interrupted, and the oil in the control tubes will be discharged through the orifice in the outer end of the piston housing. This releases the pressure on the piston and allows it to move outward, thus opening the port for rapid discharge of oil. Resumption of oil flow from the scoop tube will fill the control tubes, and the pressure will move the piston to the closed position.

Principles of Operation

When the engine is started and the coupling is filled with oil, the primary rotor turns with the engine crankshaft. As the primary rotor turns the oil in, its radial passages flow outward under centrifugal force. (See arrows in fig. 17-39.) This forces oil across the gap at the outer edge of the rotor and into the radial passages of the secondary rotor, where the oil flows inward. The oil in the primary rotor is not only flowing outward but is also rotating. As the oil flows over and into the secondary rotor, it strikes the radial blades in the rotor.

The secondary rotor soon begins to rotate and pick up speed, but it will always rotate more slowly than the primary rotor because of drag on the secondary shaft. Therefore, the centrifugal force of the oil in the primary rotor will always be greater than that of the oil in the secondary rotor. This causes a constant flow from the primary rotor to the secondary rotor at the outer ends of the radial passages and from the secondary rotor to the primary rotor at the inner ends.

The power loss in the hydraulic clutch is small (3 percent) and is caused by friction in the fluid itself. This means that approximately 97 percent of the power delivered to the primary rotor is transmitted to the reduction gear. The loss power is transformed into heat that is absorbed by the oil—which is the reason for sending part of the oil through a cooler at all times.

MAINTENANCE

Keeping an internal-combustion engine (diesel or gasoline) in good operating condition demands a well-planned procedure of periodic inspection, maintenance, and repair. If inspections are made regularly, many maladjustments can be detected and corrected before a serious casualty results. A planned maintenance program will help to prevent major casualties and the occurrence of many operating troubles. There may be times when service requirements interfere with a planned maintenance program. In this event, routine maintenance must be performed as soon as possible after the specified interval of time has elapsed. Necessary corrective measures should be accomplished as soon as possible; if repair jobs are allowed to accumulate, the result may be hurried and incomplete work. Since the Navy uses so many models of internal-combustion engines, it is impossible to specify any detailed overhaul procedure that is adaptable to all models. However, there are several general rules that apply to all engines. They are as follows:

- Study the appropriate manufacturer's instruction manuals and maintenance pamphlets, which contain detailed repair procedures, before attempting any repair work. Pay particular attention to tolerances, limits, and adjustments.

- Observe the highest degree of cleanliness when you handle engine parts during overhaul.

- Before you start repair work, be sure that all required tools and replacements for known defective parts are available.

- Keep detailed records of repairs. Such records should include the measurements of parts, hours in use, and new parts installed. An analysis of such records will indicate the hours of operation that may be expected from the various engine parts. This knowledge is helpful as an aid in determining when a part should be renewed to avoid a failure.

• Detailed information on preventive maintenance is contained in the PMS Manual for the engineering department. All preventive maintenance should be accomplished according to the Planned Maintenance Subsystem (3-M Systems) that is based upon the proper use of the PMS manuals, maintenance requirement cards (MRCs), and schedules for the accomplishment of planned maintenance actions. An MRC is shown in figure 17-41

PMS does not cover certain operating checks and inspections that are required as a normal part of the regular watchstanding routine. For example, you will not find such things as hourly pressure and temperature checks or routine oil level checks listed as maintenance requirements under the PMS. Even though these routine operating checks are not listed as PMS requirements, you must of course still perform them according to all applicable watchstander's instructions.

REFERENCES

Engineman 1 & C, NAVEDTRA 10543, Naval Education and Training Program Management Support Activity, Pensacola, Florida, 1987.

Engineman 3, NAVEDTRA 10539, Naval Education and Training Program Management Support Activity, Pensacola, Florida, 1989.

Figure 17-41.–Maintenance requirement card.

Fireman, NAVEDTRA 10520, Naval Education and Training Program Management Support Activity, Pensacola, Florida, 1987.

CHAPTER 18

PIPING, FITTINGS, AND VALVES

LEARNING OBJECTIVES

Upon completion of this chapter, you should be able to do the following:

1. Identify the factors used in determining the types of materials found in piping and fittings.

2. Identify the various types of steam traps used in the Navy.

3. Discuss the different types of valves used to control the flow of liquids and gases.

4. Describe the common types of pressure-control valves used to relieve pressure, reduce pressure, and control or regulate pressure.

5. Recognize the symbol/color coding used to identify valves, fittings, flanges, and unions.

6. Discuss the symbol numbers used to identify types of packing and gaskets.

7. Become familiar with the instruction/technical manuals used to observe safety precautions when performing repairs to piping systems.

INTRODUCTION

This chapter deals with pipe, tubing, fittings, valves, and related components that make up the shipboard piping systems used for the transfer of fluids. The general arrangement and layout of the major engineering piping systems is discussed in chapter 9 of this text. In this chapter, we are concerned with the practical aspects of piping system design and with the actual piping system components—pipe, tubing, fittings, and valves.

DESIGN CONSIDERATIONS

Each piping system and all its components must be designed to meet the particular condition of service that will be encountered in actual use. The nature of the contained fluid, the operating pressures and temperatures of the system, the amount of fluid that must be delivered, and the required rate of delivery are some of the factors that determine the materials used, the types of valves and fittings used, the thickness of the pipe or tubing, and many other details. Piping systems that must be subjected to temperature changes are designed to allow for expansion and contraction. Special problems that might arise, such as water hammer, turbulence, vibration, erosion, corrosion, and creep, are also considered in the design of piping systems.

The requirements governing the design and arrangement of components for shipboard piping systems are covered in detail by contract specifications and by a number of plans and drawings. The information given in this chapter is not intended as a detailed listing but merely as a general guide to the design requirements of shipboard piping systems.

All shipboard piping is installed in such a way that it will not interfere with the operation of the ship's machinery or with the operation of doors, hatches, scuttles, or openings covered by removable plates. As far as possible, piping is installed so that it will not interfere with the maintenance and repair of machinery or of the ship's structure. If piping must be installed in the way of machinery or equipment that requires periodic dismantling or overhaul, or if it must be installed in the way of other piping systems or electrical systems, the piping should be designed for easy removal. Piping that is vital to the propulsion of the ship is not installed where it would have to be dismantled to permit routine maintenance on machinery or other systems. Piping is not normally installed in such a way as to pass through voids, fuel-oil tanks, ballast tanks, feed tanks, and similar spaces,

Valves, unions, and flanges are carefully located to permit isolation of sections of piping with the least possible interference to the continued operation of the rest of the system. The type of valve used in any particular location is specified on the basis of the service conditions to be encountered. For example, gate valves are widely used in locations where turbulent flow characteristics of other types of valves might be detrimental to the components of the system.

Unnecessary high points and low points are avoided in piping systems. Where high points and low points are unavoidable, vents, drains, or other devices are installed to ensure proper functioning of the system and equipment served by the system.

Various joints are used in shipboard piping systems. The joints used in any system depend upon the piping service, the pipe size, and the construction period of the ship. Older naval ships have threaded flanges in low-pressure piping, rolled-in joints for steel piping that is too large for the threaded flanges, and spelter-brazed flanges for copper and brass piping. Most of the ships in service are built using welded joints to the maximum practicable extent in systems that are fabricated of carbon steel, alloy steel, or other weldable material. Flanged joints made up with special gaskets are also used.

Components welded in piping system must be accessible for repair, reseating, and overhaul while in place. They are located so that they can be removed, preheated, rewelded, and stress relieved when major repairs or replacements are necessary. Complex assemblies, such as assemblies of valves, strainers, and traps in high-pressure drain systems, are designed to be removed as a group if they cannot be repaired while in place and if they require frequent overhaul.

Flanged and union joints are placed where they will be least affected by piping system stresses. In general, this means that joints are not located at bends or offsets in the piping.

Valves are designed so that they can be operated with the maximum practicable amount of force and with the maximum practicable convenience. If personnel must stand on slippery deck plates to turn a valve handwheel, reach over their heads or around a corner, they cannot apply the same amount of torque that they could apply to a more conveniently located handwheel. Therefore, the location of the handwheels is an important design consideration. Toggle mechanisms or other mechanical advantage devices are used where the amount of torque required to turn a handwheel is more than could normally be applied by one person. If mechanical advantage devices are not sufficient to produce easy operation of the valve, power operation is used.

If accidental opening or closing of a valve could endanger personnel or jeopardize the safety of the ship, locking devices are used. Any locking device installed on a valve must be designed so that it can be easily operated by authorized personnel, but it must be complex enough to discourage casual or indiscriminate operation by other persons.

Supports used in shipboard piping systems must be strong enough to support the weight of the piping, its contained fluid, and its insulation and lagging. Supports must carry the loads imposed by the expansion and contraction of the piping and by the working of the ship, and they must be able to support the piping with complete safety. Supports are designed to permit the movement of the piping necessary for flexibility of the system. A sufficient number of supports are used to prevent excessive vibration of the system under all conditions of operation, but the support must not cause excessive constraint of the piping. Supports are used for heavy valves and fittings,so that the weight of the valves and fittings will not be entirely supported by the pipe.

PIPE AND TUBING

The Naval Sea Systems Command (NAVSEA) defines piping as an assembly of pipe or tubing, valves, fittings, and related components. These form a whole or a part of a system used to transfer fluids (liquids and gases).

IDENTIFICATION

In commercial usage, there is no clear distinction between pipe and tubing. The correct designation for each tubular product is established by the manufacturer. If the manufacturer calls a product pipe, it is a pipe; if they call it tubing, it is tubing. In the Navy, however, a distinction is made between pipe and tubing based on its dimensions.

There are three important dimensions of any tubular product: outside diameter (OD), inside diameter (ID), and wall thickness. A tubular product is called tubing if its size is identified by actual measured OD and by actual measured wall thickness. A tubular product is called pipe if its size is identified by a nominal dimension called nominal pipe size (NPS) and by reference to a wall thickness schedule designation.

The size identification of tubing is simple enough, since it consists of actual measured dimensions. However, the terms used to identify pipe sizes require some explanation. A nominal dimension, such as NPS, is close to but not necessarily identical with an actual measured dimension. For example, a pipe with an NPS of 3 inches has an actual measured OD of 3.50 inches. A pipe with an NPS of 2 inches has an actual measured OD of 2.375 inches. In the larger sizes (about 12 inches), the NPS and the actual measured OD are the same. For example, a pipe with an NPS of 14 inches has an actual measured OD of 14 inches. Nominal dimensions are used to simplify the standardization of pipe fittings, pipe taps, and threading dies.

The wall thickness of pipe is identified by reference to wall thickness schedules established by the American Standards Association. For example, a reference to schedule 40 for a steel pipe with a nominal pipe size of 3 inches indicates that the wall thickness of the pipe is 0.216 inch. A reference to schedule 80 for a steel pipe for the same NPS (3 inches) indicates that the wall thickness of the pipe is 0.300 inch.

A schedule designation does not identify any one particular wall thickness unless the NPS is also specified. For example, we have said that a schedule 40 steel pipe with an NPS of 3 inches has an actual wall thickness of 0.216 inch. But if we look at schedule 40 for a steel pipe with an NPS of 4 inches, we will find that the wall thickness is 0.237 inch. These examples are used merely to illustrate the meaning of wall thickness schedule designations. Many other values can be found in pipe tables given in engineering handbooks.

You have probably seen pipe identified as standard (Std), extra strong (XS) and double extra strong (XXS). These designations, which are still used to some extent, also refer to wall thickness. However, pipe is manufactured in a number of different wall thicknesses. Some pipe do not fit into the standard, extra strong, and double extra strong classifications. The wall thickness schedules are being used increasingly to identify the wall thickness of pipe. They provide identification of more wall thicknesses than can be identified under the strong, extra strong, and double extra strong classifications.

We have briefly described the standard ways of identifying the size and wall thickness of pipe and tubing. However, you will sometimes see pipe and tubing identified in other ways. For example, you may see some tubing identified by ID rather than by OD.

Many different kinds of pipe and tubing are used in shipboard piping systems. A few shipboard applications that may be of particular interest are noted in the following paragraphs.

Seamless chromium-molybdenum alloy steel pipe is used for some high-pressure, high-temperature systems. The upper limits for the piping is 1500 psig and 1050°F.

Seamless carbon steel tubing is used in oil, steam, and feedwater lines operating at 775°F and below. Different types of this tubing are available. The type used in any particular system depends upon the working pressure of the system.

Seamless carbon-molybdenum alloy tubing is used for feedwater discharge piping, boiler pressure superheated steam lines, and boiler pressure saturated steam lines. Several types of this tubing are available. The type used in any particular case depends upon the boiler operating pressure and the superheater outlet temperature. The upper limits for any class of this tubing is 1500 psig and 875°F.

Seamless chromium-molybdenum alloy steel tubing is used for high-pressure, high-temperature steam service. This type of alloy tubing is available with different percentages of chromium and molybdenum. The upper limits are 1500 psig and 1050°F.

Welded carbon steel tubing is used in some water, steam, and oil lines where the temperature does not exceed 450°F. There are several types of this tubing. Each type is specified for the certain services and certain service conditions.

Nonferrous pipe and nonferrous tubing are used for many shipboard systems. Nonferrous metals are used chiefly where their special properties of corrosion resistance and high heat conductivity are required. Various types of seamless copper tubing are used for refrigeration lines, plumbing and heating systems, lubrication systems, and other shipboard systems. Copper nickel alloy tubing is widely used aboard ship. Seamless brass tubing is used in systems that must resist the corrosive action of salt water and other fluids. It is available in types and sizes suitable for operating pressures up to 4000 psig. Seamless aluminum tubing is used for dry lines in the sprinkling systems and for some bilge and sanitary drain systems.

Many other kinds of pipe and tubing besides the kinds mentioned in this chapter are used in shipboard piping systems. You should remember that design considerations govern the selection of any particular pipe or tubing for the particular system. Although many kinds of pipe and tubing look almost exactly alike from the outside, they may respond very differently to pressures, temperatures, and other service conditions. Therefore, each kind of pipe and tubing can be used only for the specified applications.

PIPE FITTINGS

Pipe or tubing alone does not constitute a piping system. To make the pipe or tubing into a system, you must have a variety of fittings, connections, and accessories by which the sections of pipe or tubing can be properly joined and the flow of the transferred fluid may be controlled. This section deals with some of the pipe fittings most commonly used in shipboard piping systems. These fittings include unions, flanges, expansion joints, and flareless fluid connections.

UNIONS

Union fittings are provided in piping systems to allow the pipe to be taken down for repairs or alterations. Unions are available in many different materials and designs to withstand a wide range of pressures and temperatures. Figure 18-1 shows some commonly used types of unions.

FLANGES

Flanges are used in piping stems to allow easy removal of piping and other equipment. The materials used and the design of the flanges are governed by the requirements of service. Flanges in steel piping systems are usually welded to the pipe or tubing. Flanges in nonferrous systems are usually brazed to the pipe or tubing.

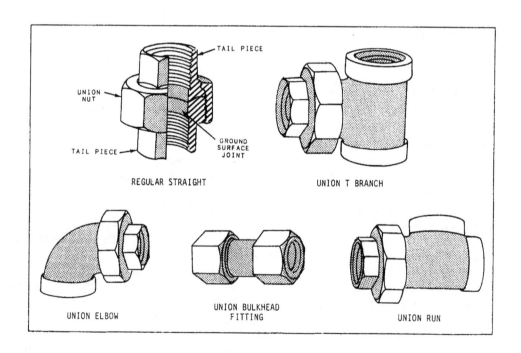

11.313

Figure 18-1.—Unions.

EXPANSION JOINTS

Expansion joints are used in some piping systems to allow the piping to expand and contract with temperature changes, without damage to the piping. Two basic types of expansion joints are used in shipboard systems: sliding-type joints and flexing-type joints.

Sliding-type expansion joints include sleeve joints, rotary joints, ball and socket joints, and joints made up of some combination of these types. The amount of axial and rotary motion that can be absorbed by any particular type of sliding expansion joint depends upon the specific design of the joint.

Flexing-type expansion joints are those in which motion is absorbed by the flexing action of a bellows or some similar device. There are various kinds of flexing-type expansion joints, each kind designed to suit the requirements of the particular system in which it is installed. Figure 18-2 illustrates the general principle of a bellows-type expansion joint.

Expansion joints are not always used in piping systems, even when allowance must be made for expansion and contraction of the piping. The same effect can be achieved by using directional changes and expansion bends or loops.

FLARELESS FITTINGS

A flareless fitting is used for connecting sections of tubing in some high-pressure shipboard systems. This fitting, which is generally known as the bite-type fitting, is very useful for certain applications because it is smaller and lighter in weight than the conventional fittings previously used to join tubing. The bite-type fitting is used on selected systems where the tubing is between 1/8 and 2 inches in OD.

The bite-type fitting, shown in figure 18-3, consists of a body, a ferrule or sleeve that grips the tubing, and a nut. The fitting is not used in places where there is insufficient space for proper tightening of the nut, where piping or equipment would have to be removed to gain access to the fitting, or where the tubing cannot be easily deflected for ready assembly or breakdown of the joint. The fitting is sometimes used on gauge board or instrument panel tubing, provided the gauge board is designed to be removed as a unit when repairs are required.

STEAM TRAPS

Steam traps in steam lines drain condensate from the lines without allowing steam to escape. There are many different kinds of steam traps. They all consist essentially of a valve and some device or arrangement that will cause the valve to open and close as necessary to drain the condensate without allowing the escape of steam. Some designs are suitable for low pressures and temperatures, others for high pressures and temperatures. A few common types of steam traps are the mechanical, thermostatic, and orifice.

MECHANICAL STEAM TRAPS

Mechanical steam traps may be of the ball-float type or the bucket type.

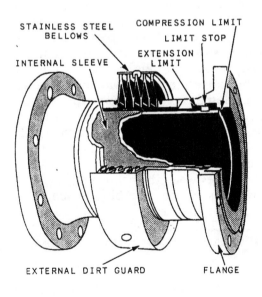

38.127

Figure 18-2.—Bellows-type expansion joint.

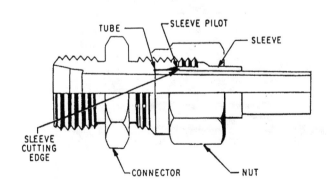

Figure 18-3.—Double-male flareless fitting (bite type).

Ball-Float Trap

In a ball-float steam trap, such as the one shown in figure 18-4, the valve of the trap is connected to the float in such a way that the valve opens when the float rises. The operating principle of the trap is quite simple. When steam cools, it condenses (changes state) back to water. The liquid water, called condensate, flows by gravity into the chamber around the ball valve. As the water level rises, the float is lifted, thereby lifting the valve plug and opening the valve. The condensate drains out and the float settles to a lower position, closing the valve. The condensate that passes out of the trap is returned to the feed system. In the figure, the two white circles directly above and below the ball float are connections (holes) for a gauge glass, which the operator checks to set the desired liquid level in the steam trap.

Bucket-Type Trap

Figure 18-5 shows a common bucket-type steam trap that is suitable for high pressures and temperatures and has a large capacity. Its operation may be described as follows: As soon as sufficient water enters the trap, the bucket, being buoyant, floats and closes the valve. As condensation increases, the body of the trap fills and water enters the bucket, causing it to sink. The bucket being attached to the discharge valve, opens the discharge valve and the trap begins to discharge, continuing to do so until the

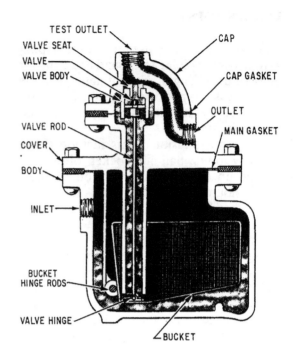

Figure 18-5.—Bucket-type steam trap.

condensation is blown out of the body to the edge of the bucket. At this point, the water in the bucket continues to be forced out until the bucket again becomes buoyant and rises, closing the valve.

THERMOSTATIC STEAM TRAPS

There are several types of thermostatic steam traps. In general, these traps are more compact and have fewer moving parts than most mechanical steam traps. The operation of a bellows-type thermostatic steam trap is controlled by expansion of the vapor of a volatile liquid enclosed in a bellows-type element. Steam enters the trap body and heats the volatile liquid in the sealed bellows, thus causing an expansion of the bellows. The valve is attached to the bellows in such a way that the valve closes when the bellows expands. The valve remains closed, trapping steam in the trap body. As the steam cools and condenses, the bellows cools and contracts, thereby opening the valve and allowing the condensate to drain.

The impulse and bimetallic steam traps are two examples of those that use the thermostatic principle.

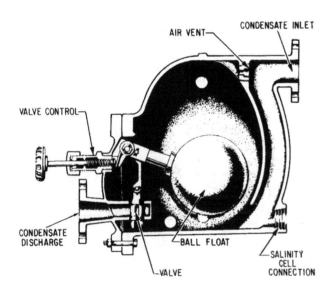

Figure 18-4.—Ball-float steam trap.

Impulse Steam Traps

Impulse steam traps (fig. 18-6) are used in some steam drain collecting systems aboard ship. Steam and condensate pass through a strainer before entering the trap. A circular baffle keeps the entering steam and condensate from impinging in the cylinder or on the disk.

The impulse steam trap operates on the principle that hot water under pressure tends to flash into steam when the pressure is reduced. To understand how the principle is used, we will consider the arrangement of parts shown in the figure and see what happens to the flow of condensate under various conditions.

The only moving part in the steam trap is the disk. This disk is rather unusual in design. Near the top of the disk is a flange that acts as a piston. As may be seen in the figure, the working surface above the flange is larger than the working surface below the flange. The importance of having this larger effective area above the flange will become apparent.

A control orifice runs through the disk from top to bottom, being considerably smaller at the top than at the bottom. The bottom part of the disk extends through and beyond the orifice in the seat. The upper part of the disk (including the flange) is inside the cylinder. The cylinder tapers inward, so the amount of clearance between the flange and the cylinder varies according to the position of the valve. When the valve is open, the clearance is greater than when the valve is closed.

When the trap is first cut in, pressure from the inlet (chamber A) acts against the underside of the flange and lifts the disk off the valve seat. Condensate is thus allowed to pass out through the orifice in the seat, and, at the same time, a small amount of condensate (called control flow) flows up past the flange into chamber B. The control flow discharges through the control orifice, into the outlet side of the trap, and the pressure in chamber B remains lower than the pressure in chamber A.

As the line warms up, the temperature of the condensate flowing through the trap increases. The reverse taper of the cylinder varies the amount of flow around the flange until a balanced position is reached in which the total force exerted above the flange is equal to the total force exerted below the flange. It is important to note that there is still a pressure difference between chamber A and chamber B. The force is equalized because the effective area above the flange is larger than the effective area below the flange. The difference in working area is such that the valve maintains an open, balanced position when the pressure in

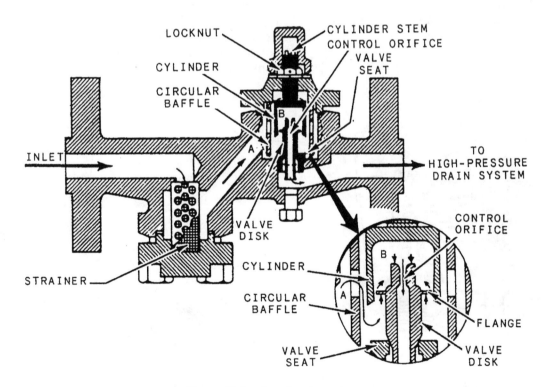

Figure 18-6.—Impulse steam trap.

chamber B is 86 percent of the pressure in chamber A.

As the temperature of the condensate approaches its boiling point, some of the control flow going to chamber B flashes into steam as it enters the low-pressure area. Since the steam has a much greater volume than the water from which it is generated, pressure builds up in the space above the flange (chamber B). When the pressure in the space is 86 percent of the inlet pressure (chamber A), the force exerted on the top of the flange pushes the entire disk downward and thus closes the valve.

With the valve closed, the only flow through the trap is past the flange and through the control orifice. When the temperature of the condensate entering the trap drops slightly, condensate enters chamber B without flashing into steam. Pressure in chamber B is thus reduced to the point where the valve opens and allows condensate to flow through the orifice in the valve seat. Thus, the entire cycle is repeated continuously.

With a normal condensate load, the valve opens and closes at frequent intervals, discharging a small amount of condensate at each opening. With a heavy condensate load, the valve remains wide open and allows a continuous discharge of condensate.

Bimetallic Steam Traps

Bimetallic steam traps (fig. 18-7) are used in many ships to drain condensate from main steam lines, auxiliary steam lines, and other steam components. The main working parts of this steam trap are a segmented bimetallic element and a ball-type check valve.

The bimetallic element has several bimetallic strips fastened together in a segmented fashion as shown in the figure. One end of the bimetallic element is fastened rigidly to a part of the valve body. The other end, which is free to move, is fastened to the top of the stem of the ball-type check valve.

Line pressure acting on the check valve keeps the valve open. When the steam enters the trap body, the bimetallic element expands unequally because of the different response to the temperature of the two metals. The bimetallic element deflects upward at its free end, thus moving the valve stem upward and closing the valve. As the steam cools and condenses, the bimetallic element moves downward, toward the horizontal position, thus opening the valve and allowing some condensate to flow out through the valve. As the flow of condensate begins, an unbalance of line pressure across the valve is created. Since the line pressure is greater on the upper side of the ball of the check valve, the valve now opens wide and allows a full capacity flow of condensate.

ORIFICE STEAM TRAPS

Figure 18-8 shows the assembly of an orifice-type steam trap. Constant-flow orifices may be used in systems of 150 psi and above where condensate load and pressure remain near constant.

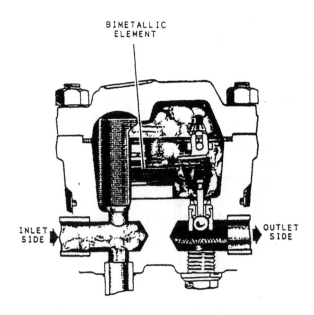

Figure 18-7.—Bimetallic steam trap.

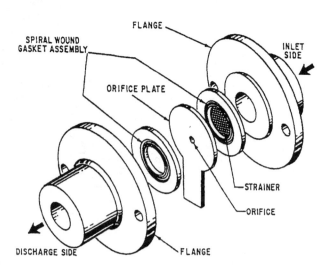

Figure 18-8.—Orifice-type steam trap.

Several variations of the orifice-type steam trap exist, but all have one thing in common—they have no moving parts. One or more restricted passageways or orifices allow condensate to trickle through but does not allow steam to flow through. Besides orifices, some orifice-type steam traps have baffles.

A constant-flow drain orifice operates on a thermodynamic principle. The variable density of condensate helps its operation. Density changes with temperature. As the temperature of the condensate decreases, the density of the condensate increases, as well as the flow of condensate through the orifice. The reverse is also true. As the temperature of the condensate increases, the density of the condensate decreases, as well as the flow of the condensate through the orifice.

Because of the difference in densities between the steam and condensate, the condensate will flow through the orifice at a faster rate.

Other operating aspects of the orifice are size, pressure, and condensate load.

By calculating the condensate flow based on the condensing rate of the equipment and by knowing the pressure of the system, you can select an orifice of the proper size. Flow rate through the orifice is expressed in pounds per hour (lb/hr).

The advantages of the orifice-type steam trap over other types warrant their use in all systems of 150 psi and above.

STRAINERS AND FILTERS

Strainers are fitted in practically all piping lines to prevent the passage of grit, scale, dirt, and other foreign matter, which could obstruct pump suction valves, throttle valves, or other machinery parts.

Figure 18-9 illustrates three common types of strainers. View A shows a bilge suction strainer located in the bilge pump suction line between the suction manifold and the pump. Any debris that enters the piping is collected in the strainer basket. The basket can be removed for cleaning by loosening the strongback screws, removing the cover, and lifting the basket out by its handle. View B shows a duplex oil strainer that is commonly used in fuel-oil and lubricating-oil lines, where it is essential to maintain an uninterrupted flow of oil. The flow may be diverted from one basket to the other, while one is being cleaned. View C shows a manifold steam strainer. This type of strainer is desirable where

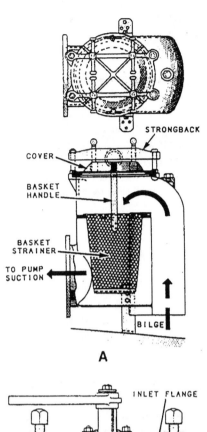

A

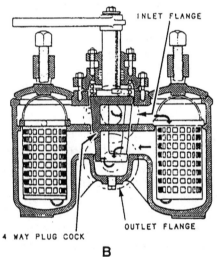

B

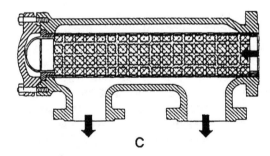

C

A. Bilge suction strainer   B. Duplex oil strainer
C. Manifold steam strainer

11.329

Figure 18-9.—Strainers.

space is limited, since it eliminates the use of separate strainers and their fittings. The cover is located so that the strainer basket can be removed for cleaning.

Metal-edge filters are used in the lubricating systems of many auxiliary units. A metal-edge filter consists of a series of metal plates or disks. Turning a handle moves the plates or disks across each other in such a way as to remove any particles that have collected on the metal surfaces. Some metal-edge filters have magnets to aid in removing fine particles of magnetic materials.

VALVES

Every piping system must have some means of controlling the amount and direction of the flow of the contained fluid through the lines. The control of fluid flow is accomplished by the installation of valves.

Valves are usually made of bronze, brass, iron, or steel. Steel valves are either cast or forged and are made of either plain steel or alloy steel. Alloy steel valves are used in high-pressure, high-temperature systems. The disks and seats of these valves are usually surfaced with stellite, an extremely hard chromium-cobalt alloy.

Bronze and brass valves are not used in high-temperature systems. Also, they are not used in systems in which they would be exposed to severe conditions of pressure, vibration, or shock. Bronze valves are widely used in saltwater systems. The seats and disk of bronze valves used for seawater service are often made of Monel, a metal that is highly resistant to corrosion and erosion.

Many different types of valves are used to control the flow of liquids and gases. The basic valve types can be divided into two groups, stop valves and check valves. Stop valves are those that are used to shut off, or partially shut off, the flow of fluid. Stop valves are controlled by the movement of the valve stem. Check valves are those that are used to permit the flow of fluid in only one direction. Check valves are designed to be controlled by the movement of the fluid itself.

Stop valves include globe valves, gate valves, plug valves, needle valves, and butterfly valves.

Check valves include ball-check valves, swing-check valves, and lift-check valves.

Combination stop-check valves are valves that function either as stop valves or check valves, depending upon the position of the valve stem.

In addition to the basic types of valves, many special valves, which cannot really be classified either as stop valves or as check valves, are found in the engineering spaces. Many of these special valves serve to control the pressure of the fluid and are therefore generally called pressure-control valves. Others are identified by names that indicate their general function, for example, thermostatic recirculating valves. The following sections deal first with the basic types of stop valves and check valves and then with some of the more complex special kinds of valves.

GLOBE VALVES

Globe valves are one of the most common types of stop valves. Globe valves get their name from the globular shape of their bodies. It is important to note, however, that other types of valves may also have globe-shaped bodies; hence, it is not always possible to identify a globe valve merely by external appearance. The internal structure of the valve, rather than the external shape, is what distinguishes one type of valve from the other.

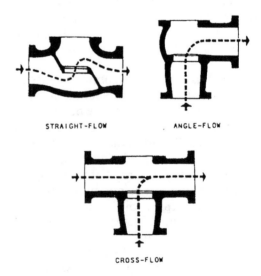

STRAIGHT-FLOW   ANGLE-FLOW

CROSS-FLOW

Figure 18-10.—Types of globe valve bodies.

The disk of a globe valve is attached to the valve stem. The disk seats against a seating ring or a seating surface and thus shuts off the flow of fluid. When the disk is moved off the seating surface, fluid can pass through the valve. Globe valves may be used partially open as well as fully open or fully closed. Globe valve inlet and outlet openings are arranged in several ways, to suit varying requirements of flow. Figure 18-10 shows three common types of globe valve bodies. In the straight type, the fluid inlet and outlet openings are in line with each other. In the angle type, the inlet and outlet openings are at an angle to each other. An angle-type globe valve is used where a stop valve is needed at a 90-degree turn in a line. The cross-type globe valve has three openings rather than two. This type of valve is often used in connection with bypass piping. Figure 18-11 shows a cutaway view of a globe valve.

Globe valves are commonly used in steam, air, oil, and water lines. On many ships, the surface blow valves, the bottom blow valves, the boiler stops, the feed stop valves, and many guarding valves are of this type. Globe valves are also used as stop valves on the suction side of many pumps, as recirculating valves, and as throttle valves.

GATE VALVES

Gate valves are used when a straight-line flow of fluid and minimum flow restriction are needed. Gate valves are so named because the part that either stops or allows flow through the valve acts somewhat like the opening or closing of a gate and is called, appropriately, the gate. The gate is usually wedge shaped. When the valve is wide open, the gate is fully drawn up into the bonnet. This leaves an opening for flow through the valve the same size as the pipe in which the valve is installed. Therefore, there is little pressure drop of the flow restriction through the valve. Gate valves are not suitable for throttling purposes. The control of flow would be difficult because of valve design, and the flow of the fluid slapping against a partially open gate can cause extensive damage to the valve. Except as specifically authorized, gate valves should not be used for throttling.

Gate valves are classified as either rising stem or nonrising stem valves. The nonrising stem valve is shown in figure 18-12. The stem is threaded on the lower end into the gate. As the handwheel on the stem is rotated, the gate travels up and down the stem on the threads while the stem remains vertically stationary. This type of valve will almost always have a pointer type of indicator threaded onto the upper end of the stem to indicate valve position.

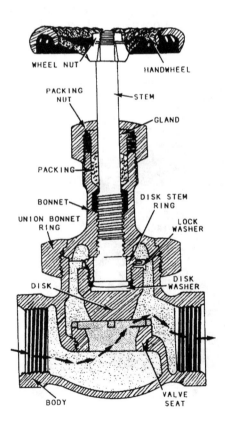

Figure 18-11.—Cutaway view of globe stop valve.

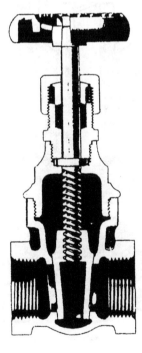

Figure 18-12.—Cutaway view of a gate valve (nonrising stem type).

The rising-stem gate valve is shown in figure 18-13. It has the stem attached to the gate. The gate and stem rise and lower together as the valve is operated.

Gate valves used in steam systems have flexible gates. This type of gate prevents binding of the gate within the valve when the valve is in the closed position. When steam lines are heated, they will expand, causing some distortion of valve bodies. A solid gate fits snugly between the seat of a valve in a cold steam system. When the system is heated and pipes elongate, the seats will compress against the gate. The gate will be wedged between the seats and will clamp the valve shut. This problem is overcome by using a flexible gate. This type of gate is best described as two circular plates that are flexible around the hub. These two plates are attached to each other. The gate can then flex as the valve seat compresses it. This prevents clamping.

The major problem with the flexible gates (if installed with the stem below the horizontal) is that water tends to collect in the body neck. Then under certain conditions, the admission of steam may cause the valve body neck to rupture, the bonnet to lift off, or the seating ring to collapse. To prevent this, correct warming-up procedures must be followed. Also, some very large gate valves have a three-position vent and bypass valve. This valve allows venting from the bonnet either upstream or downstream of the valve and has a position for bypassing the valve.

BUTTERFLY VALVES

Butterfly valves (fig. 18-14) may be used in a variety of systems aboard ship. These valves can be used effectively in freshwater, saltwater, JP-5, F-76 (naval distillate), lube-oil, and chill-water systems. The butterfly valve is light in weight, relatively small, and quick-acting. It provides positive shut-off and can be used for throttling.

The butterfly valve has a body, a resilient seat, a butterfly disk, a stem, packing, a notched position plate, and a handle. The resilient seat is under compression when it is mounted in the valve body, thus making a seal around the periphery of the disk and both upper and lower points where the stem passes through the seat. Packing is provided to form a positive seal around the stem

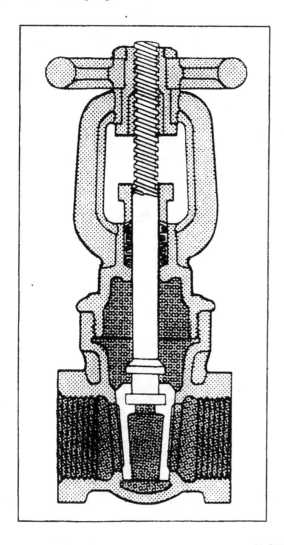

11.317

Figure 18-13.—Cutaway view of gate valve (rising stem type).

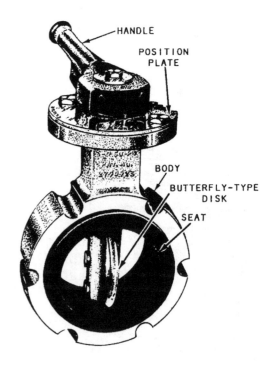

Figure 18-14.—Butterfly valve.

18-12

for added protection in case the seal formed by the seat should become damaged.

To close or open a butterfly valve, turn the handle only one-quarter turn to rotate the disk 90°. Some larger butterfly valves may have a handwheel that operates through a gearing arrangement to operate the valve. This method is used especially where space limitation prevents use of a long handle.

Butterfly valves are relatively easy to maintain. The resilient seat is held in place by mechanical means, and neither bonding nor cementing is

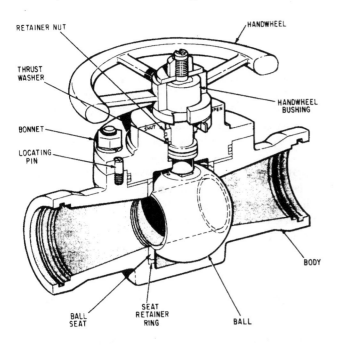

Figure 18-15.—Ball valve.

necessary. Because the seat is replaceable, the valve seat does not require lapping, grinding, or machine work.

BALL VALVES

Ball valves, as the name implies, are stop valves that use a ball to stop or start the flow of fluid. The ball valve, shown in figure 18-15, performs the same function as the disk in the globe valve. When the valve handle is operated to open the valve, the ball rotates to a point where the hole through the ball is in line with the valve body inlet and outlet. The valve is shut by a 90-degree rotation of the handwheel for most valves. The ball is rotated so that the hole is perpendicular to the flow openings of the body.

Most ball valves are of the quick-acting type. They require only a 90-degree turn to operate the valve either completely open or closed. However, many are planetary gear operated. This type of gearing allows the use of a relatively small handwheel and operating force to operate a fairly large valve. The gearing does, however, increase the operating time for the valve. Some ball valves contain a swing check located within the ball to give the valve a check valve feature. Ball valves are normally found in seawater, sanitary, trim and drain, air, hydraulic, and oil transfer systems.

PLUG VALVES

Plug valves (sometimes referred to as plug cocks) are frequently used in gasoline and oil feed pipes as well as water drain lines.

The body of a plug valve, as illustrated in figure 18-16, is shaped like a cylinder with holes

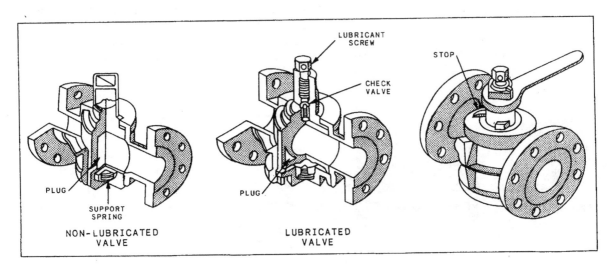

139.39

Figure 18-16.—Plug valves.

or ports in the cylinder wall in line with the pipes in which the valve is mounted. Either a cone-shaped or cylindrical plug, attached to the handle, fits snugly into the valve body. A hole bored through the plug is in line with the ports in the valve body. Turning the plug valve handle (which is in line with the hole in the plug) lines up the hole in the plug with the ports in the valve body so that fluid can pass through the valve. The flow can be stopped by turning the plug 90° (one-quarter turn) from the open position.

Some plug valves are designed as three- or four-way selector valves. Three or more pipes are connected to a single valve in line with the same numbers of ports in the cylinder wall. Two or more holes drilled in the plug provide a variety of passages through the valve. When a valve of this kind is located in a fuel line, the liquid may be drawn from any one of two or three tanks by setting the handle in different positions.

NEEDLE VALVES

Needle valves are used to make relatively fine adjustments in the flow of fluid. A needle valve has a long tapered point at the end of the valve stem. This needle acts as a disk. Because of the long taper, part of the needle passes through the opening in the valve seat before the needle actually seats. This arrangement permits a very gradual increase or decrease in the size of the opening. This allows more precise control of flow than can be obtained with an ordinary globe valve.

CHECK VALVES

Check valves allow fluid to flow in a system in only one direction. They are operated by the

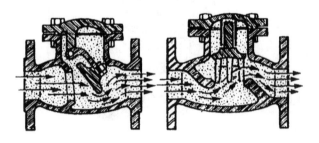

A. SWING CHECK   B. LIFT CHECK

Figure 18-17.—Check valves.

flow of fluid in the piping. A check valve may be of the swing type, lift type, or ball type. Figure 18-17 shows a swing-check valve and a lift-check valve. Check valves may also be built into globe valves or ball valves.

STOP-CHECK VALVES

Most valves can be classified as being either stop valves or check valves. Some valves, however, function either as stop valves or as check valves, depending upon the position of the valve stem. These valves are known as stop-check valves.

Stop-check valves are shown in cross section on figure 18-18. This type of valve looks very much like a lift-check valve. However, the valve stem is long enough so that when it is screwed all the way down, it holds the disk firmly against the seat. This prevents any flow of fluid. In this position, the valve acts as a stop valve. When the stem is raised, the disk can be opened by pressure on the inlet side. In this position, the valve acts as a check valve, allowing the flow of fluid in only one direction. The maximum lift of the disk is controlled by the position of the valve stem. Therefore, the position of the valve stem can limit the amount of fluid passing through the valve even when the valve is operating as a check valve.

Stop-check valves are used in various locations throughout the engineering plant. Perhaps the most familiar example is the boiler feed-check valve, which is actually a stop-check valve rather than a true check valve. Stop-check valves are used in many drain lines, on the discharge side of many pumps, and as exhaust steam valves on auxiliary machinery.

SPECIAL-PURPOSE VALVES

There are many types of automatic pressure-control valves. Some of them merely provide an escape for excessive pressure; others reduce or regulate pressure.

Relief Valves

Relief valves are installed in piping systems to protect them from excessive pressure. These valves have an adjusting screw, a spring, and a disk. The

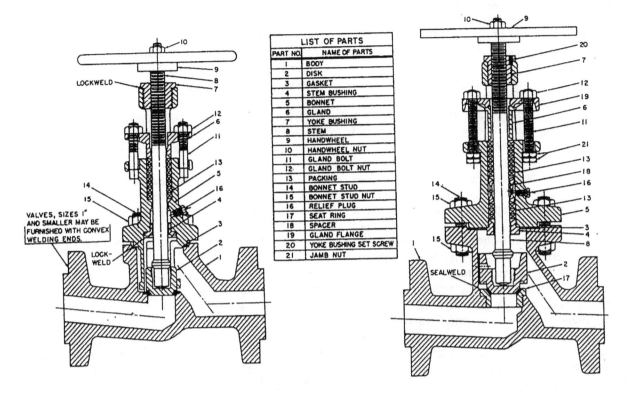

LIST OF PARTS	
PART NO.	NAME OF PARTS
1	BODY
2	DISK
3	GASKET
4	STEM BUSHING
5	BONNET
6	GLAND
7	YOKE BUSHING
8	STEM
9	HANDWHEEL
10	HANDWHEEL NUT
11	GLAND BOLT
12	GLAND BOLT NUT
13	PACKING
14	BONNET STUD
15	BONNET STUD NUT
16	RELIEF PLUG
17	SEAT RING
18	SPACER
19	GLAND FLANGE
20	YOKE BUSHING SET SCREW
21	JAMB NUT

Figure 18-18.—Stop-check valve.

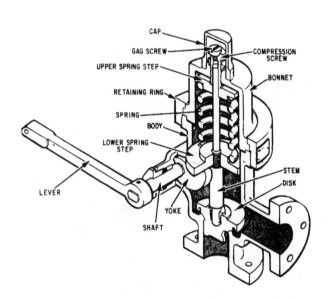

Figure 18-19.—Relief valve.

force exerted on the disk by the spring sets the relieving pressure. Most relief valves simply open when the preset pressure is reached and close when the pressure drops slightly below the lifting pressure. Many relief valves will also have a lever so the valve can be opened by hand for test purposes. Figure 18-19 shows a relief valve of this type.

Sentinel Valves

Sentinel valves are simply small relief valves installed in some systems to warn of impending overpressurization. Sentinel valves relieve the pressure of the system. If the situation causing the sentinel valve to lift is not corrected, a relief valve (if installed) will lift to protect the system or component. If a relief valve is not installed, action must be taken quickly to secure the piece of equipment or system to reduce the pressure.

Pressure-Reducing Valves

Reducing valves are automatic valves that provide a steady pressure into a system that is at a lower pressure than the supply system. Reducing valves of one type or another are found in steam, air, lube-oil, seawater, and other systems. A reducing valve can normally be set for any desired downstream pressure within the design limits of

18-15

the valve. Once the valve is set, the reduced pressure will be maintained. This is true regardless of changes in supply pressure but must be at least as high as the reduced pressure desired. It is also true regardless of the amount of reduced pressure fluid that is used.

Pressure-reducing valves for piping systems are usually installed in reducing stations, like the one shown in figure 18-20. In addition to a pressure-reducing valve, a reducing station should contain at least four other valves. Two of these are stop valves located in the inlet piping and outlet piping for the reducing valve. These valves (V1 and V2) are shut to isolate the pressure-reducing valve from the piping system in the event the valve needs repair. Some reducing valves may also have a stop valve in the downstream sensing line (V3). There should be a bypass valve (V4) used for throttling service or to manually control downstream pressure when the reducing valve is inoperative. The bypass valve is normally shut. Finally, there should be a relief valve (V) to prevent overpressurization of the piping system downstream of the reducing station in the event the reducing valve fails open or the manual bypass valve is misadjusted.

There are three basic designs of pressure-reducing valves in use. They are spring-loaded reducing valves, pneumatic-pressure-controlled (gas loaded) reducing valves, and air-pilot-operated diaphragm-type reducing valves. There are many different styles within these three types. These are discussed in the following paragraphs.

SPRING-LOADED REDUCING VALVES.— One type of spring-loaded reducing valve is shown in figure 18-21. These valves are used in a wide variety of applications. Low-pressure air reducers, auxiliary machinery cooling-water reducing stations, and some reduced-steam system reducers are of this type. The valve simply uses spring pressure against a diaphragm to open the valve. On the bottom of the diaphragm, the outlet

pressure (the pressure in the reduced-pressure system) of the valve forces the disk upward to shut the valve. When the outlet pressure drops below the set point of the valve, spring pressure overcomes the outlet pressure and forces the valve stem downward, opening the valve. As the outlet pressure increases, approaching the desired value, the pressure under the diaphragm begins to overcome the spring pressure. This forces the valve stem upward, shutting the valve. Downstream pressure can be adjusted by removing the valve cap and turning the adjusting screw, which varies the spring pressure against the diaphragm. This particular spring-loaded valve will fail in the open position in case of a diaphragm rupture.

INTERNAL PILOT-ACTUATED PRESSURE-REDUCING VALVES.— The internal pilot-actuated pressure-reducing valve shown in figure 18-22 uses a pilot valve to control the main valve. The pilot valve controls the flow of upstream fluid, which is ported to the pilot valve, to the operating piston, which operates the main valve. The main valve is opened by the operating piston and closed by the main valve spring. The pilot valve opens when the adjusting spring pushes downward on the pilot diaphragm. It closes when downstream pressure exerts a force that exceeds

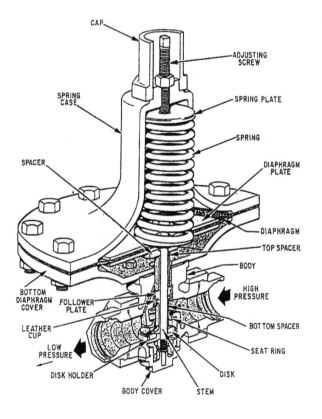

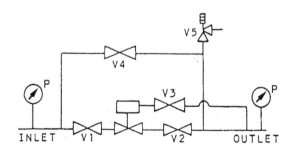

Figure 18-20.—Pressure-reducing station.

Figure 18-21.—Pressure-reducing (spring-loaded) valve.

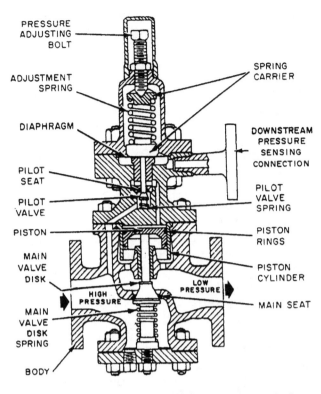

Figure 18-22.—Internal pilot-actuated pressure-reducing valve.

the force of the adjusting spring. When the pilot valve shuts off or throttles the flow of upstream fluid to the operating piston, the main valve then pushes the valve and stem upward to throttle or close the main valve. When downstream pressure falls below the set point, the adjusting spring force acts downward on the diaphragm. This action overcomes the force of the downstream system pressure, which is acting upward on the diaphragm. This opens the pilot valve, allowing upstream pressure to the top of operating piston to open the main valve.

PNEUMATIC-PRESSURE-CONTROLLED REDUCING VALVES.—For engines that use compressed air as a power source, starting air comes directly from the ship's medium or high-pressure air service line or from starting air flasks, which are included in some systems for the purpose of storing starting air. From either source, the air, on its way to the engine, must pass through a pressure-reducing valve, which reduces the higher pressure to the operating pressure required to start a particular engine.

One type of pressure-reducing valve is the regulator shown in figure 18-23, in which

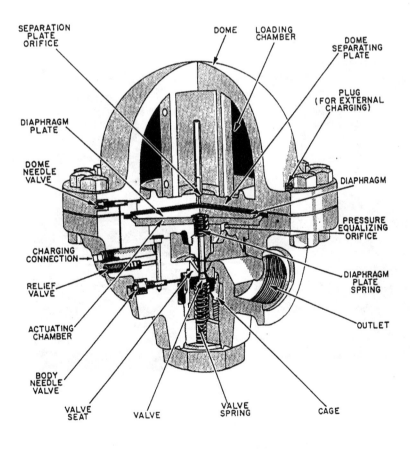

Figure 18-23.—Pneumatic-pressure-controlled reducing valve.

compressed air, sealed in a dome, furnishes the regulating pressure that actuates the valve. The compressed air in the dome performs the same function as a spring used in a more common type of regulating valve.

The dome is tightly secured to the valve body, which is separated into an upper (low-pressure outlet) and a lower (high-pressure inlet) chamber by the main valve. At the top of the valve stem is another chamber, which contains a rubber diaphragm and a metal diaphragm plate. This chamber has an opening leading to the low-pressure outlet chamber. When the outlet pressure drops below the pressure in the dome, air in the dome forces the diaphragm and the diaphragm plate down on the valve stem. This partially opens the valve and permits high-pressure air to pass the valve seat into the low-pressure outlet and into the space under the diaphragm. As soon as the pressure under the diaphragm is equal to that in the dome, the diaphragm returns to its normal position, and the valve is forced shut by the high-pressure air acting on the valve head.

When the dome-type regulator is used in the air start system for a diesel engine during the starting event, the regulator valve continuously and rapidly adjusts for changes in air pressure by partially opening and closing to maintain a safe, constant starting air pressure. When the engine starts and there is no longer a demand for air, pressure builds up in a low-pressure chamber to equal the pressure on the dome, and the valve closes completely.

AIR-PILOT-OPERATED DIAPHRAGM CONTROL VALVES.—These valves are extensively used on naval ships. The valves and their control pilots are available in several designs to meet different requirements. They may be used as unloading valves to reduce pressure or to provide continuous regulation of pressure and temperature. They may also be used for the control of liquid levels.

The air-operated control pilot may be either direct-acting or reverse-acting. A direct-acting pilot is shown in figure 18-24. In this type of pilot, the controlled pressure (the pressure from the discharge side of the diaphragm control valve) acts on top of a diaphragm in the control pilot. This pressure is balanced by the pressure exerted by

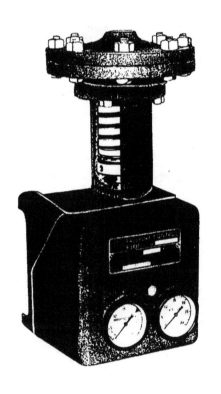

Figure 18-24.—Air-operated control pilot.

the pilot adjusting spring. When the controlled pressure increases and overcomes the pressure exerted by the pilot adjusting spring, the pilot valve stem is forced downward. This action opens the pilot valve to increase the amount of operating air pressure going from the pilot to the diaphragm control valve. In a reverse-acting pilot, therefore, an increase in controlled pressure produces a decrease in operating air pressure.

In the diaphragm control valve shown in figure 18-25, operating air from the pilot acts on a diaphragm contained in the superstructure of the valve operator or positioner. It is direct-acting in some valves and reverse-acting in others. If the valve operator is direct-acting, the operating air pressure from the control pilot is applied to the top of the valve diaphragm. When the valve operator is reverse-acting, the operating air pressure from the pilot is applied to the underside of the valve diaphragm.

Figure 18-25, view A, shows a very simple type of direct-acting diaphragm control valve. The operating air pressure from the control pilot is applied to the top of the valve diaphragm. The valve in view A is a downward-seating valve. Therefore, any increase in operating air pressure pushes the valve stem downward. This tends to close the valve.

Figure 18-25, view B, is also a direct-acting diaphragm control valve. The operating air pressure from the control pilot is applied to the top of the valve diaphragm. The valve shown in view B is more complicated than the valve shown in view A. View B is an upward-seating valve rather than a downward-seating valve. Any increase in operating pressure from the control pilot tends to open this valve rather than to close it.

The type of air-operated control pilot, the type of positioner of the diaphragm control valve, as well as the purpose of the installation determine how the diaphragm control valve and its air-operated control pilot are installed in relative position to each other.

To see how these factors are related, consider the following installation: A diaphragm control valve and its air-operated control are used to

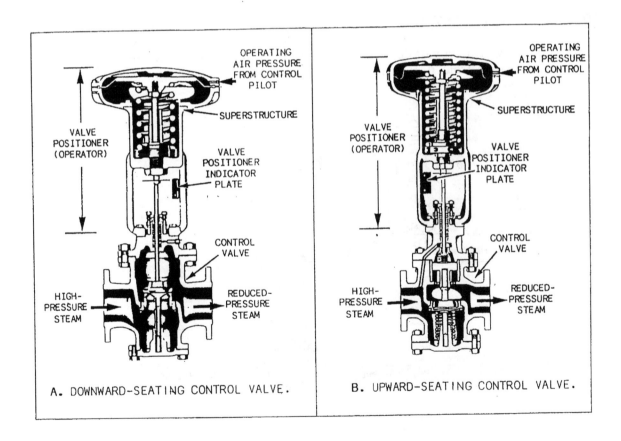

Figure 18-25.—Diaphragm control valves.

supply reduced steam pressure. Figure 18-26 shows one kind of arrangement that could be used. We will assume that the service requirements indicate the need for a direct-acting, upward-seating, diaphragm control valve. What kind of control pilot (direct-acting or reverse-acting) should be used in this installation?

In a direct-acting control pilot, when the controlled pressure (discharge pressure from the diaphragm control valve) increases, increased pressure is applied to the diaphragm. The valve stem is pushed downward and the valve in the control pilot is opened. This sends an increased amount of operating air pressure from the control pilot to the top of the diaphragm control valve. The increased air operating air pressure acting on the diaphragm of the valve pushes the stem downward. Since this is an upward-seating valve, this action opens the diaphragm control valve still wider. Obviously, this type of installation will not work. In this application, an increase in controlled pressure must result in a decrease in operating air pressure. A reverse-acting control pilot must be used for this particular installation.

REMOTE-OPERATED VALVES

Remote-operating gear provides a means of operating certain valves from distant stations.

Remote-operating gear may be mechanical, hydraulic, pneumatic, or electrical. A reach rod or series of reach rods and gears may be used to operate engine-room valves in cases where valves are difficult to reach.

Other remote-operating gear is installed as emergency equipment. Some split-plant valves, main drainage system valves, and overboard valves are equipped with remote operating gear. These valves can be operated normally or, in an emergency, may be operated from the remote stations. Remote-operating gear also includes a valve position indicator to show whether the valve is open or closed.

UNLOADING VALVES

An automatic unloading valve (also called a dumping valve) is installed at each main and auxiliary condenser. The function of the unloading valve is to discharge steam from the auxiliary exhaust line to condensers whenever the auxiliary exhaust line pressure exceeds the design operating pressure.

An automatic unloading valve is shown in figure 18-27. Auxiliary exhaust steam is led through valve A to the top of the actuating valve diaphragm. The actuating valve is double seated,

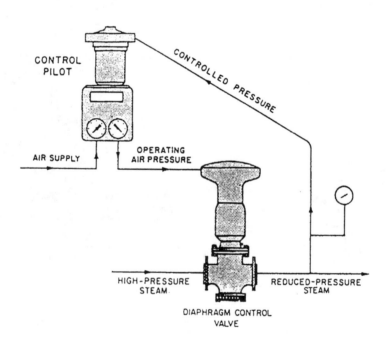

Figure 18-26.—Arrangement of control pilot and diaphragm control valve for supplying reduced steam pressure.

and one side is open when the other is closed. When the auxiliary exhaust line pressure is less than the pressure for which the unloading valve is set, the upper seat is closed and the lower seat is open. The valve is held by the diaphragm spring. Steam passes into the line through valve B and goes under the unloading valve diaphragm. The pressure acting on this diaphragm holds the unloading valve up and closed. If the auxiliary exhaust pressure exceeds the pressure of the actuating valve diaphragm spring, the diaphragm is forced downward and the lower seat closes while the upper seat opens. This makes a direct connection between the top and the bottom of the unloading valve diaphragm through the actuating valve. The equalized

pressure on the diaphragm allows the auxiliary exhaust pressure to force the unloading valve down and steam is unloaded to the condenser. The unloading pressure can be adjusted by turning an adjusting screw, thereby changing the force exerted on the actuating valve diaphragm.

THERMOSTATIC RECIRCULATING VALVES

Thermostatic recirculating valves are used in systems where it is necessary to recirculate a fluid to maintain the temperature within certain limits. Thermostatic recirculating valves are designed to operate automatically.

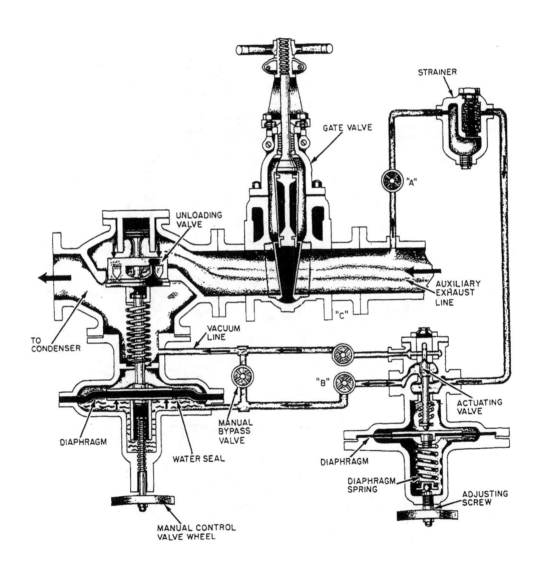

47.61

Figure 18-27.—Automatic unloading valve.

The thermostatic recirculating valve shown in figure 18-28 is used to recirculate condensate from the discharge side of the main air-ejector condenser to the main condenser. The valve is actuated by the temperature of the condensate. When the condensate temperature becomes higher than the temperature for which the valve is set, the thermostatic bellows expands and automatically opens the valve, allowing condensate to be sent back to the condenser.

VALVE MANIFOLDS

A valve manifold is used when it is necessary to take suction from one of several sources and to discharge to another unit or several units of the same or separate group. One example of a manifold is shown in figure 18-29. This manifold is used in the fuel-oil filling and transfer system, where provision must be made for the transfer of oil from any tank to any other tank, to the fuel-oil service system, or to another ship. The manifold valves are frequently of the stop-check type.

IDENTIFICATION OF VALVES, FITTINGS, FLANGES, AND UNIONS

Most valves, fittings, flanges, and unions used on naval ships are marked with identification symbols of various kinds. The few valves and

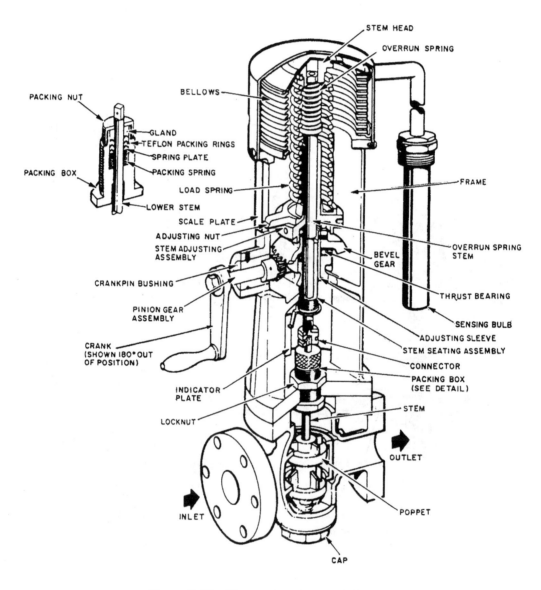

Figure 18-28.—Thermostatic recirculating valve.

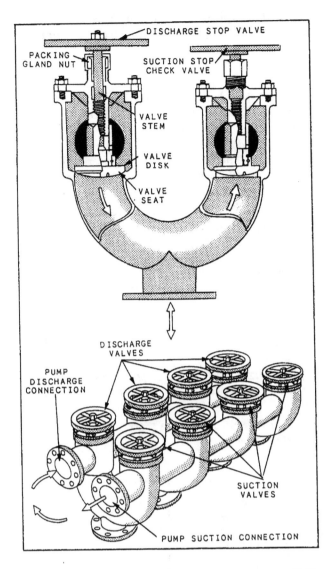

PACKING GLAND NUT
DISCHARGE STOP VALVE
SUCTION STOP CHECK VALVE
VALVE STEM
VALVE DISK
VALVE SEAT

DISCHARGE VALVES
PUMP DISCHARGE CONNECTION
SUCTION VALVES
PUMP SUCTION CONNECTION

47.63

Figure 18-29.—Valve manifold.

fittings that are made on board repair ships or tenders or at naval shipyards are usually marked with symbols indicating the manufacturing activity, the size, the melt or casting number, and the material. They may also be marked with an arrow to indicate the direction of flow.

Commercially manufactured valves, fittings, flanges, and unions may be identified according to the requirements of the applicable specifications. However, many valves, fittings, flanges, and unions are now identified according to a standard marking system developed by the Manufacturers Standardization Society (MSS) of the valves and fittings industry. Identification markings in this system usually include the manufacturer's name or trademark, the pressure

and service for which the product is intended, and the size (in inches). When appropriate, the material identification, limiting pressure, and other identifying data are included.

The MSS standard identification markings are generally cast, forged, stamped, or etched on the exterior surface of the product. In some cases, however, the markings are applied to an identification plate rather than to the actual surface of the product.

The service designation in the MSS system of marking usually includes a letter to indicate the type of service and numerals to indicate the pressure rating in psi. The letters used in the service designations are the following:

A	Air
G	Gas
L	Liquid
O	Oil
W	Water
D-W-V	Drainage, waste, and vent

When the primary service rating is for steam, and when no other service is indicated, the service designation may consist of numerals only. For example, the number 600 marked on the body of a valve would indicate that the valve is suitable for steam service at 600 psi. If the valve is designed for water at 600 psi, the service designation would be 600 W. Service designations are also used in combination; for example, 3000 WOG indicates a product suitable for water, oil, or gas service at 3000 psi.

Some abbreviations that are commonly used for material identification in the MSS system include the following:

AL	Aluminum
B	Bronze
CS	Carbon steel
CI	Cast iron
HF	Cobalt-chromium-tungsten alloy (hard facing)
CU NI	Copper-nickel alloy

NI CU	Nickel-copper alloy
SM	Soft metal (such as lead, Babbitt, copper)
CR 13	13-percent chromium steel
18 8	18-8 stainless steel
18 8SMO	18-8 stainless steel with molybdenum
SH	Surface-hardened steel (such as Nitralloy)

Some examples of the MSS standard identification marking symbols are given in figure 18-30.

VALVE HANDWHEEL IDENTIFICATION AND COLOR CODING

Valves are identified by markings inscribed on the rims of handwheels, by a circular label plate secured by the handwheel nut, or by label plates

---

3-INCH CAST STEEL SCREWED FITTING SUITABLE FOR WATER, OIL, OR GAS SERVICE AT 1000 PSI:

Manufacturer's identification	A B CO
Service designation	1000 WOG
Material designation	STEEL
Size	3

2-INCH CAST IRON FLANGED FITTING FOR USE IN REFRIGERATION SYSTEM:

Manufacturer's identification	A B CO
Service designation	300 GL
Temperature designation	300 F
Size	2

CAST BRASS FITTING FOR DRAINAGE, WASTE, AND VENT SERVICE:

Manufacturer's identification	A B CO
Service designation	D-W-V

2-INCH BRONZE VALVE RECOMMENDED BY THE MANUFACTURER FOR 200 PSI STEAM SERVICE:

Manufacturer's identification	A B CO
Service designation	200
Size	2

4-INCH STEEL VALVE WITH 13 PERCENT CHROMIUM STEEL VALVE STEM, DISK, AND SEAT, SUITABLE FOR 1500 PSI STEAM SERVICE AT TEMPERATURE OF NO MORE THAN 850° F:

VALVE BODY MARKING:

Manufacturer's identification	A B CO
Service designation	1500
Material designation	STEEL
Size	4

IDENTIFICATION PLATE MARKING:

Manufacturer's identification	A B CO
Service designation	1500
Limiting temperature	MAX 850 F
Body material designation	STEEL
Valve stem material designation	STEM CR 13
Valve disk material designation	DISC CR 13
Valve seat material designation	SEAT CR 13
Size	4

Figure 18-30.—Examples of MSS standard identification markings for valves, fittings, and unions.

attached to the ships structure or to adjacent piping.

Piping system valve handwheels and operating levers are marked for training and casualty control purposes with a standardized color code. Color code identification should be in conformance with the color scheme of table 18-1. Implementation of this color scheme provides uniformity among all naval surface ships and shore-based training facilities.

Table 18-1.—Valve Handwheel Color Code

FLUID	VALVE HANDWHEEL & OPERATING LEVER
STEAM	WHITE
POTABLE-WATER	DARK BLUE
NITROGEN	LIGHT GRAY
HP AIR	DARK GRAY
LP AIR	TAN
OXYGEN	LIGHT GREEN
SALT WATER	DARK GREEN
JP-5	PURPLE
FUEL OIL	YELLOW
LUBE OIL	STRIPED YELLOW/BLACK
FIRE PLUGS	RED
FOAM DISCHARGE	STRIPED RED/GREEN
GASOLINE	YELLOW
FEEDWATER	LIGHT BLUE
HYDRAULIC	ORANGE
HYDROGEN	CHARTREUSE
HELIUM	BUFF
HELIUM/OXYGEN	STRIPED BUFF/GREEN
SEWAGE	GOLD

PACKING AND GASKET MATERIALS

Gasket materials are used to seal fixed joints in steam, water, fuel, air, lube-oil, and other piping systems. Packing materials are used to seal joints that slide or rotate under operating conditions (moving joints). There are many commercial types and forms of packing and gasket materials. The Navy has simplified the selection of packing and gasket materials commonly used in naval service. NAVSEA has prepared a packing and gasket chart (Mechanical Standard Drawing B0153). This chart shows the symbol numbers and the recommended applications of all types and kinds of packing and gasket materials. A copy of the chart should be located in all engineering spaces.

A four-digit symbol number identifies each type of packing and gasket. The first digit indicates the class service with respect to fixed and moving joints. For example, if the first digit is 1, it indicates a moving joint (moving rods, shafts, valve stems, and so forth). If the first digit is 2, it indicates a fixed joint (such as flanged or bonnet). The second digit indicates the material of which the packing or gasket is primarily composed. This may be vegetable fiber, rubber, metal, and so on. The third and fourth digits indicate the different styles or forms of the packing or the gaskets made from the material. To find the right packing material, check the maintenance requirement card (MRC) or the NAVSEA packing and gasket chart. The MRC lists the symbol numbers and the size and number of rings required. The NAVSEA packing and gasket chart lists symbol numbers and includes a list of materials. For additional information concerning packing and gasket materials, refer to *NSTM*, chapter 078.

INSULATION AND LAGGING

The purpose of insulation is to retard the transfer of heat from piping that is hotter than the surrounding atmosphere or to piping that is cooler than the surrounding atmosphere. Insulation helps to maintain the desired temperature in all systems. In addition, it prevents sweating of piping that carries cool or cold fluids. Insulation also serves to protect personnel from being burned when they come in contact with hot surfaces.

Insulation is actually composite covering that includes (1) the insulating material itself, (2) the lagging or covering, and (3) the fastening that is used to hold the insulation and lagging in place. In some instances, the insulation is covered by material that serves both as lagging and as a fastening device. Cold piping additionally requires a vapor barrier coating to prevent condensation of water vapor within the insulation.

Insulating materials commonly used by the Navy include magnesia, calcium silicate, diatomaceous silica, mineral wool, fibrous glass, and high-temperature insulating cement.

Lagging, which may consist of cloth, tape, or sheet metal, serves to protect the relatively soft insulating material from damage. It also gives added support to insulation that may be subjected to heavy or continuous vibration, and it provides a smooth surface that may be painted.

Lagging is secured in place by sewing or by fire-resistant adhesives, insulating cement, or sealing compounds. The method used to fasten the lagging in place depends on the type of insulation used, the type of lagging used, and the service requirements of the piping surfaces to be insulated. *NSTM*, chapter 635 (old chapter 9390), gives detailed information on the selection of insulation, lagging, and fastening of insulation, lagging, and fastenings for various applications.

FLANGE SAFETY SHIELDS

A fuel fire in a main engine room or an auxiliary machinery room can be caused by a leak at a fuel-oil or lube-oil pipe flange connection. Even the smallest leak can spray fine droplets of oil on nearby hot surfaces. To reduce this possibility, flange safety shields are provided around the piping flanges of inflammable liquid systems, especially in areas where the fire hazard is apparent. The spray shields are usually made of aluminized glass cloth and are simply wrapped and wired around the flange.

PIPE HANGERS

Pipe hangers and supports are designed and located to support the combined weight of the piping, fluid, and insulation. They absorb the movements imposed by the thermal expansion of the pipe and the motion of the ship. The pipe hangers and supports prevent excessive vibration of the piping and resilient mounts or other materials. They are used in the hanger arrangement to break metal-to-metal contact to lessen unwanted sound transmissions.

One type of pipe hanger is the variable spring hanger. It provides support by directly compressing a spring or springs. The loads carried by the hangers are equalized by adjustment of the hangers when they are hot. These hangers have load scales attached to them with a traveling arm or pointer that moves in a slot alongside the scale. This shows the degree of pipe movement from cold to hot. The cold and hot positions are marked on the load scale.

INSPECTIONS, MAINTENANCE, AND SAFETY

Reasonable care must be given to the various piping assemblies as well as to the units connected to the piping systems. Unless the piping system is in good condition, the connected units of machinery cannot operate efficiently and safely.

The most important factor in maintaining piping systems in satisfactory condition is keeping joints, valves, and fittings tight. To ensure this condition, frequent tests and inspections must be conducted.

Maintenance of valves, pipe fittings, and piping are performed using the applicable MRC.

Many packing and insulation materials contain asbestos fibers, which is a recognized health hazard. *NSTM*, chapter .635, gives detailed information on asbestos handling.

Safety precautions dictated in MRCs and OPNAVINST 3140.32B must be observed when working on piping systems.

REFERENCES

Boiler Technician 3 & 2, NAVEDTRA 10535-H, Naval Education and Training Program Manangement Support Activity, Pensacola, Florida, 1985.

Engineman 3, NAVEDTRA 10539, Naval Education and Training Program Management Support Activity, Pensacola, Florida, 1989.

Fireman, NAVEDTRA 10520-H, Naval Education and Training Program Management Support Activity, Pensacola, Florida, 1987.

Gas Turbine Systems Technician (Electrical) 3/Gas Turbine Systems Technician (Mechanical) 3, Volume 1, NAVEDTRA 10563, Naval Education and Training Program Management Support Activity Center, Pensacola, Florida, 1989.

Machinist's Mate 3 & 2, NAVEDTRA 10524-F1, Naval Education and Training Program Management Support Activity, Pensacola, Florida, 1987.

CHAPTER 19

PUMPS

LEARNING OBJECTIVES

Upon completion of this chapter, you should be able to do the following:

1. Define some of the terms commonly used in connection with pumps.

2. Recognize fundamental principles of operation of the various types of pumps used aboard ships.

INTRODUCTION

This chapter deals with shipboard pumps used aboard many surface ships. In general, we are concerned here with the driven end of the units rather than with the driving end. The auxiliary steam turbines used to drive many pumps were discussed in chapter 14 of this text. The electric motors used to drive other units will be discussed in chapter 20.

As discussed is chapter 8, the pump is one of the five basic elements in any thermodynamic cycle. The function of the pump is to move the working substance from the low-pressure side of the system to the high-pressure side. In the conventional steam turbine propulsion plant, the pump of the thermodynamic cycle is actually three pumps; the condensate pump, the feed booster pump, and the main feed pump.

In addition to these three pumps, a large number of pumps are used for other purposes aboard ship. Pumps supply seawater to the firemains, circulate cooling water for condensers and coolers, empty the bilges, transfer fuel oil, discharge fuel oil to the burners, supply lubricating oil to main and auxiliary machinery, supply seawater to the distilling plant, pump the distillate into storage tanks, supply liquid under pressure for use in hydraulically operated equipment, and provide a variety of other vital services.

Pumps are used to move any substance that flows or that can be made to flow. Most commonly, pumps are used to move water, oil, and other liquids. However, air, steam, and other gases are also fluid and can be moved with pumps, as can substances such as molten metal, sludge, and mud.

A pump is essentially a device that uses an external source of power to apply a force to a fluid to move the fluid from one place to another. A pump develops no energy of its own; it merely transforms energy from the external source (steam turbine, electric motor, and so forth) into mechanical kinetic energy, which is manifested by the motion of the fluid. This kinetic energy is then used to do work—for example, to raise a liquid from one level to another, as when water is raised from a well; to transport a liquid through a pipe, as when oil is carried through a pipeline; to move a liquid against some resistance, as when water is pumped to a boiler under pressure; or to force a liquid through a hydraulic system, against various resistances, for the purpose of doing work at some point.

PRINCIPLES AND DEFINITIONS

Before considering specific designs of shipboard pumps, it may be helpful to examine briefly certain basic concepts and to define some of the terms commonly used in connection with pumps.

FORCE-PRESSURE-AREA RELATIONSHIPS

When we strike the end of a bar, the main force of the blow is carried straight through the other end. This happens because the bar is rigid. The direction of the blow almost entirely determines the direction of the transmitted force. The more rigid the bar, the less force is lost inside the bar or transmitted outward at right angles to the direction of the blow.

When we apply pressure to the end of a column of confined liquid, the pressure is transmitted not only straight through to the other end but also equally and undiminished in every direction. Figure 19-1 illustrates the difference between pressure applied to a rigid bar and pressure applied to a column of contained liquid.

The principle that pressure is transmitted equally and undiminished in all directions through a contained liquid is known as Pascal's principle. This principle may be regarded as the basic law or foundation of the science of hydraulics.

An important corollary of Pascal's principle is that the transmission of pressure through a liquid is not altered by the shape of the container. This idea is illustrated in figure 19-2. If the pressure due to the weight of the liquid is 8 psi at any one point on the horizontal line, H, it is 8 psi at every point along line H. The pressure due to the weight of the liquid at any level thus depends upon the vertical distance from the chosen level to the surface of the liquid. The vertical distance between the horizontal levels in a liquid is known as the head of the liquid. (Since various kinds of head enter into pump calculations, the term *head* is more fully discussed later.)

Pressure is defined as force per unit area. Alternatively, we may say that force is equal to pressure times area. Figure 19-3 shows how a force of 20 pounds acting on a piston with an area of 2 square inches can produce a force of 200 pounds on a piston with an area of 20 square inches. The system will also work the same in reverse. If we consider piston 2 as the input piston and piston 1 as the output piston, the output force would be 1/10 the input force.

We are now in a position to state the first basic rule: If two pistons are used in a hydraulic system, the force acting on each will be directly proportional to its area, and the magnitude of each force will be the product of the pressure and the area.

The second basic rule for two pistons in a hydraulic system such as the one shown in

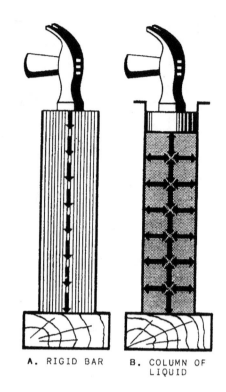

Figure 19-1.—Results of applied pressure.

A. RIGID BAR   B. COLUMN OF LIQUID

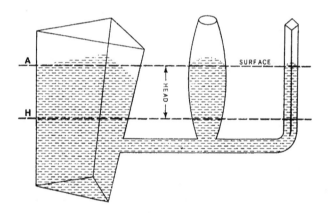

Figure 19-2.—Pressure in liquid is not affected by the shape of the vessel.

figure 19-3 may be stated as follows: The distance moved by each piston is inversely proportional to the area of the piston. Thus, if piston 1 in figure 19-3 is pushed down 1 inch, piston 2 will be raised 1/10 inch.

Consideration of the two basic rules just stated leads us to another basic rule: The input force multiplied by the distance through which it moves is exactly equal to the output force multiplied by the distance through which it moves (disregarding energy losses due to friction). In essence, this rule

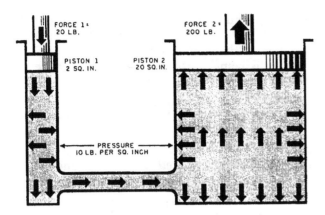

Figure 19-3.—Relationship of force, pressure, and area in a simple hydraulic system.

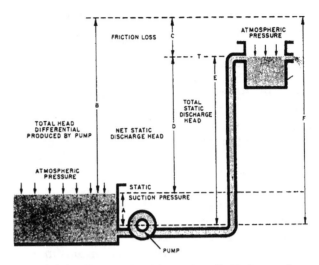

Figure 19-4.—Pressure head (pump installed below surface of supply liquid).

is merely another statement of the general energy equation—energy in equals energy out.

PUMP CAPACITY

The capacity of a pump is the amount of liquid the pump can handle in a given period of time. For marine applications, the capacity of a pump is usually stated in gallons per minute (gpm).

PRESSURE HEAD

The power required to drive a pump is a function of pump capacity and the total head against which the pump operates. Previously, we defined head quite simply as the vertical distance between two horizontal levels in a liquid. Since a pump may be installed above, at, or below the surface of the source of supply, it is obvious that other factors must enter into the discussion of pressure head as applied to pumps.

When the pump is installed at the same level as the free surface of the source of supply, no new considerations need apply since the pump merely acts on the liquid like any other applied force.

When the pump is installed below this level, as shown in figure 19-4, a certain amount of energy in the form of gravity head will already be available when the liquid enters the pump. In other words, there is a static pressure head, A, on the suction side of the pump. This head is part of the total input head necessary to produce the output head, F, that is required to raise the liquid to the top, T, of the discharge reservoir. The action of the pump produces the total head differential, B, which can be broken down into friction loss, C, and net static discharge head, D. Since D is the vertical distance from the surface of the supply liquid in the discharge reservoir, it is clear that our previous definition of head would

apply only to D. The total static discharge head, E, is the vertical distance from the center of the pump to the surface of the liquid in the discharge reservoir; thus, E is equal to D plus A.

As may be seen in figure 19-4, atmospheric pressure is acting upon the free surface of the supply liquid and upon the free surface of the liquid in the discharge reservoir. Since atmospheric pressure is exerted equally on both sides of the pump, in this system, the two heads created by the atmospheric pressure cancel out.

Now consider the case of a pump that is installed in a vertical distance, A, above the free surface, S, of the supply liquid (fig. 19-5). In this

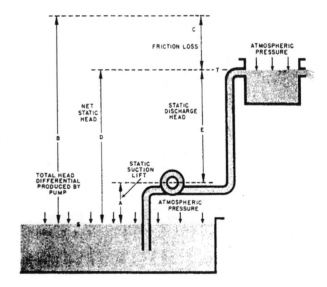

Figure 19-5.—Pressure head (pump installed above surface of supply liquid).

case, energy must be supplied merely to get the liquid into the pump (static suction lift, A). In addition, energy must be supplied to produce the static discharge head, E, if the liquid is to be raised to the top of the discharge reservoir, T. Here, the total head differential produced by the pump action, B, is the total energy input. It is divided into A on the suction side—the head required to raise the liquid to the pump—and C (friction loss) plus E (static discharge head) on the discharge side. Atmospheric pressure cancels out here, as before, except that in this case the atmospheric pressure at S is required to lift the liquid to the pump.

In both cases just described, we have dealt with systems which were open to the atmosphere. Now, let us examine the head relationships in a closed system, such as the one shown in figure 19-6.

In this system, a pump is being used to drive a work piston back and forth inside a cylinder. We will assume that the pump must develop a pressure equivalent to the head, H, to drive the piston back and forth against the resistance offered. Under this assumption, the total head differential, B, must be produced by the pump after the system has begun to operate. B is the sum of the friction head (or friction losses), C, and the static discharge head, D. Since D is equal to H, D therefore produces the pressure required to do the work.

Since the liquid returns to its original level and the system is closed throughout, a siphon effect

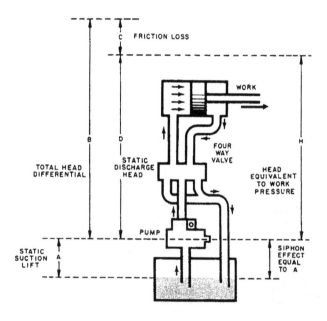

Figure 19-6.—Pressure head (closed system).

in the return pipe will exactly balance the static suction lift, A. Therefore, atmospheric pressure plays a part in the operational system only when the system is being started up, before the entire system has been filled with liquid.

At this point, let us consider the various ways in which the term *head* has been used, and attempt to formulate a definition. From previous discussion, we may infer that head is (1) measured in feet, (2) somehow related to pressure, and (3) taken as some kind of a measure of energy. But what is it?

Basically, head is a measure of the pressure exerted by a column or body of liquid because of the weight of the liquid. In the case of water, we find that a column of fresh water 2.309 feet high exerts a pressure of 1 pound per square inch (psi). When we refer to a head of water of 2.309 feet, we know that the water is exerting a pressure of 1 psi because of its own weight. Thus, a reference to a head of so many feet of water does imply a reference to the pressure exerted by that water.

The situation is somewhat different when we have a horizontal pipe through which water is being pumped. In this case, the head is calculated as the vertical distance that would correspond to the pressure. If the pressure in the horizontal pipe is 1 psi, then the head on the liquid in the pipe is 2.309 feet. Further calculations show that a head of 1 foot corresponds to a pressure of 0.433 psi.

The relationship between head and energy can be clarified by considering that (1) work is a form of energy—mechanical energy in transition; (2) work is the product of a force times the distance through which it acts; and (3) for liquids, the work performed is equal to the volume of liquid moved times the head against which it is moved. Thus, the head relationships actually indicate some of the energy relationships for a given quantity of liquid.

VELOCITY HEAD

The head required to impart velocity to a liquid is known as velocity head. It is equivalent to the distance through which the liquid would have to fall to acquire the same velocity. If we know the velocity of the liquid, we can compute the velocity head by the formula

$$H_v = \frac{V^2}{2g}$$

where

H_v = velocity head (in feet)
V = velocity of liquid (in feet per second)
g = acceleration due to gravity (32.2 ft per second per second)

In a sense, velocity head is obtained at the expense of pressure head. Whenever a liquid is given a velocity, some part of the original static pressure head must be used to impart this velocity. However, velocity head does not represent a total loss, since at least a portion of the velocity head can always be reconverted to static pressure head.

FRICTION HEAD

The force or pressure required to overcome friction is also obtained at the expense of the static pressure head. Unlike velocity head, however, friction head cannot be "recovered" or reconverted to static pressure head, since fluid friction results in the conversion of mechanical kinetic energy to thermal energy. Since the thermal energy is usually wasted, friction head must be considered as a total loss from the system.

BERNOULLI'S THEOREM

At any point in a system, the static pressure head will always be the original static pressure head minus the velocity head and minus the friction head. Since both velocity head and friction head represent energy that comes from the original static pressure head, the sum of the static pressure head, the velocity head, and the friction head at any point in a system must add up to the original static pressure head. This general principle, which is also known as Bernoulli's theorem, may also be expressed as

$$Z_1 + \frac{P_1}{D} + \frac{V_1^2}{Zg} = Z_2 + \frac{P_2}{D} + \frac{V_2^2}{Zg} +$$
$$[J(U_2 - U_1) - Wk - JQ)$$

where

Z = elevation (in feet)

P = absolute pressure (in pounds per square foot)

D = density of liquid (in pounds per cubic foot)

V = velocity (in feet per second)

g = acceleration due to gravity (32.2 ft per second per second)

J = the mechanical equivalent of heat (778 foot-pounds per Btu)

U = internal energy (in Btu)

Wk = work (in foot-pounds)

Q = heat transferred (in Btu)

When written in this form, Bernoulli's theorem may be readily recognized as a special statement of the general energy equation. The bracketed term represents energy in transition as work, energy in transition as heat, and the increase in internal energy of the fluid arising from friction and turbulence. In some cases of fluid flow, all elements in the bracketed term are of such small magnitude that they may be safely disregarded.

Consideration of Bernoulli's theorem indicates that the term *pressure head*, as used in connection with pumps and other hydraulic equipment, is actually a measure of mechanical potential energy; that the term *velocity head* is a measure of mechanical kinetic energy; and that the term *friction head* is a measure of the energy that departs from the system as thermal energy, which remains in the liquid, generally unusable, in the form of internal energy.

TYPES OF PUMPS

Pumps are so widely used for such varied services that the number of different designs is almost overwhelming. As a general rule, however, it may be stated that all pumps are so designed to move fluid substances from one point to another by pushing, pulling, or throwing, or by some combination of these three methods.

Every pump has a power end and a fluid end. The power end may be a steam turbine, a reciprocating steam engine, a steam jet, or an electric motor. In steam-driven pumps, the power end is often called the steam end. The fluid end is usually called the pump end. However, it may also be called the liquid end, the water end, the oil end, or some other term to indicate the nature of the fluid substance being pumped.

Pumps are classified in a number of different ways according to various design and operational features. Perhaps the basic distinction is between positive-displacement pumps and continuous-flow pumps. Pumps may also be classified according to the type of movement that causes the pumping action—reciprocating, rotary, centrifugal, propeller, and jet pumps. Another classification may be made according to speed; some pumps run at variable speed, others at constant speed. Some pumps have a variable capacity, others discharge at a constant rate. Some pumps are self-priming, others require a positive pressure on the suction side before they can begin to operate. These and other distinctions are noted as appropriate in the following discussion of specific types of pumps.

CENTRIFUGAL PUMPS

Centrifugal pumps are widely used aboard ship for pumping water and other nonviscous liquids. The centrifugal pump uses the throwing force of a rapidly revolving impeller. The liquid is pulled in at the center or eye of the impeller and is discharged at the outer rim of this impeller. By the time the liquid reaches the outer rim of the impeller, it has acquired a considerable velocity. The liquid is then slowed down by being led through a volute or through a series of diffusing passages. As the velocity of the liquid decreases, its pressure increases—or, in other words, some of the mechanical kinetic energy of the liquid is transformed into mechanical potential energy. In the terminology commonly used in discussions of pumps, the velocity head of the liquid is partially converted to pressure head.

Centrifugal pumps are not positive-displacement pumps. When a centrifugal pump is operating at a constant speed, the amount of liquid discharged (capacity) varies with the discharge pressure according to the relationship inherent in the particular pump design. The relationships among capacity, total head (pressure), and power are usually expressed by a characteristics curve (generally given in the manufacturers' technical manuals or in the outline assembly drawings of pumps).

Capacity and discharge pressure can be varied by changing the pump speed. However, centrifugal pumps should be operated at or near their rated capacity and discharge pressure whenever possible. Impeller vane angles and the sizes of the pump waterways can be designed for maximum efficiency at only one combination of speed and discharge pressure; under other operating conditions, the impeller vane angles and the sizes of the waterways will be too large or too small

for efficient operation. Therefore, a centrifugal pump cannot operate satisfactorily over long periods of time at excess capacity and low discharge pressure or at reduced capacity and high discharge pressure.

Centrifugal pumps are not self-priming. The casing must be flooded before a pump of this type will function. For this reason, most centrifugal pumps are located below the level from which suction is to be taken. Priming can also be effected by using another pump to supply liquid to the pump suction—for example, the feed booster pump supplies suction pressure for the main feed pump. Some centrifugal pumps have special priming pumps, air ejectors, or other devices for priming.

When two or more centrifugal pumps are installed to operate in parallel, it is particularly important to avoid operating the pumps at low capacity, since it is possible that a unit having a slightly lower discharge pressure might be pushed off the line and, thus, forced into a shutoff position.

Because of the danger of overheating, centrifugal pumps can operate at zero capacity for only short periods of time. The length of time varies. For example, a fire pump might be able to operate for as long as 15 to 30 minutes before losing suction, but a main feed pump would overheat in a matter of a few seconds if operated at zero capacity.

Most centrifugal pumps and particularly boiler feed pumps, fire pumps, and others which may be required to operate at low capacity or in shutoff condition for any length of time are fitted with recirculation lines from the discharge side of the pump back to the source of suction supply. The main feed pump, for example, has a recirculating line going back to the deaerating feed tank. An orifice allows the recirculation of the minimum amount of water required to prevent overheating of the pump. On boiler feed pumps, the recirculating lines must be kept open whenever the pumps are in operation.

On centrifugal pumps, there must always be a slight leakoff through the packing in the stuffing boxes to keep the packing lubricated and cooled. Stuffing boxes are used either to prevent the gross leakage of liquid from the pump or to prevent the entrance of air into the pump; the purpose served depends upon whether the pump is operating with a positive suction head or is taking suction from a vacuum.

If a centrifugal pump is operating with a positive suction head, the pressure inside the pump is sufficient to force a small amount of liquid through the packing when the packing gland is properly set upon. On multistage pumps, it is sometimes necessary to reduce the pressure on one or both of the stuffing boxes. This is accomplished by using a bleedoff line which is tapped into the stuffing box between the throat bushing and the packing.

If a pump is taking suction at or below atmospheric pressure, a supply of sealing water must be furnished to the packing glands to ensure the exclusion of air. Some of this water must be allowed to leak off through the packing. Most centrifugal pumps use the pumped liquid as the lubricating, cooling, and sealing medium. On some pumps, an external sealing liquid is used.

Types of Centrifugal Pumps

There are many different types of centrifugal pumps, but the two most commonly used on board ship are the volute pump and the diffuser pump.

VOLUTE PUMP.—In the volute pump, shown in figure 19-7, the impeller discharges into a volute (a gradually widening spiral channel in the pump casing). As the liquid passes through the volute and into the discharge nozzle, a great part of its kinetic energy (velocity head) is converted into potential energy (pressure head).

DIFFUSER PUMP.—In the diffuser pump, shown in figure 19-8, the liquid leaving the impeller is first slowed down by the stationary diffuser vanes that surround the impeller. The liquid is forced through gradually widening passages in the diffuser ring into the volute (casing). Since both the diffuser vanes and the volute reduce the velocity of the liquid, there is an almost complete conversion of kinetic energy to potential energy.

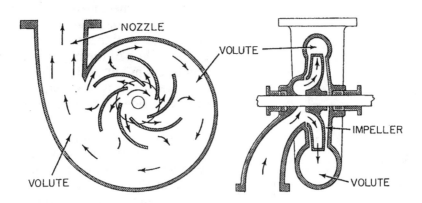

Figure 19-7.—A simple volute pump.

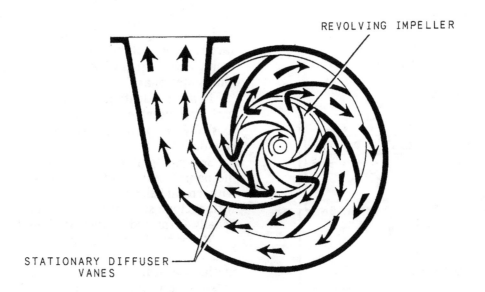

Figure 19-8.—A diffuser pump.

19-7

Classification of Centrifugal Pumps

Centrifugal pumps may be classified in several ways. For example, they must be either single stage or multistage. A single-stage pump has only one impeller. A multistage pump has two or more impellers housed together in one casing. As a rule, each impeller acts separately, discharging to the suction of the next stage impeller. This arrangement is called series staging. Centrifugal pumps are also classified as horizontal or vertical, depending upon the position of the pump shaft.

The impellers used on centrifugal pumps may be classified as single suction or double suction. The single-suction impeller (fig. 19-9, view A) allows liquid to enter the eye from one side only. The double-suction impeller (fig. 19-9, view B)

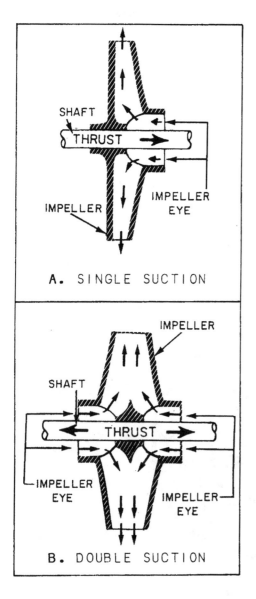

Figure 19-9.—Centrifugal pumps.

allows liquid to enter the eye from two directions.

Impellers are also classified as closed or open. Closed impellers have side walls that extend from the eye to the outer edge of the vane tips. Open impellers do not have these side walls. Most centrifugal pumps used in the Navy have closed impellers.

Construction of Centrifugal Pumps

The casing for the liquid end of a centrifugal pump with a single-suction impeller is made with an end plate which can be removed for inspection and repair of the pump. A pump with a double-suction impeller is generally made so that one-half of the casing may be lifted without disturbing the pump. Figure 19-10 is an exploded view of a typical small centrifugal water pump.

Since an impeller rotates at high speed, it must be carefully machined to minimize friction. An impeller must be balanced to avoid vibration. A close radial clearance must be maintained between the outer hub of the impeller and the part of the pump casing in which the hub rotates. The purpose of this is to minimize leakage from the discharge side of the pump casing to the suction side.

Because of the high rotational speed of the impeller and the necessary close clearance, the rubbing surfaces of both the impeller hub and the casing at that point are subject to stress, which causes rapid wear. To eliminate the need for replacing an entire impeller and pump casing solely because of wear in this location, most centrifugal pumps are designed with replaceable casing wearing rings. The impeller hub wearing face is 0.050 inch over size to accommodate the casing wearing ring which is attached to the casing and is stationary.

In most centrifugal pumps, the shaft is fitted with a replaceable sleeve. The advantage of using a sleeve is that it can be replaced more economically than the entire shaft.

Seal piping (liquid seal) is installed to cool the mechanical seal. Most pumps in saltwater service with a total head of 30 psi or more are also fitted with cyclone separators. A cyclone separator is usually furnished in each stuffing box.

Mechanical seals instead of packing are used in a variety of centrifugal pumps. One type of mechanical seal is shown in figure 19-11. Spring pressure keeps the rotating seal face snug against the stationary seal face. The rotating seal and all of the assembly below it are affixed to the pump

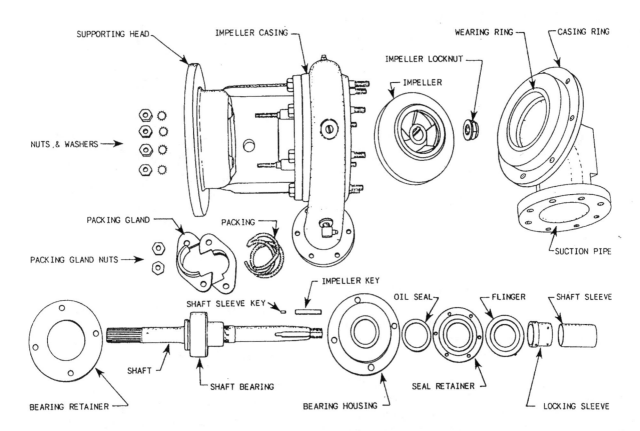

Figure 19-10.—Exploded view of a centrifugal water pump.

shaft. The stationary seal face is held stationary by the seal gland and packing ring. A static seal is formed between the two seal faces and the sleeve. System pressure within the pump assists the spring in keeping the rotating seal face tight against the stationary seal face. The type of material used for the seal face depends upon the service of the pump. Most water service pumps use a carbon material for the seal faces. When a seal wears out, it is simply replaced. New seals should not be touched on the sealing face because body acid and grease cause the seal face to prematurely pit and fail.

Bearings support the weight of the impeller and shaft and maintain the position of the impeller—both radially and axially. Some bearings are grease-lubricated with grease cups and vent plugs constructed on the housing to allow for periodic relubrication.

ROTARY PUMPS

Rotary pumps are positive-displacement pumps. The theoretical displacement of a rotary pump is the volume of liquid displaced by the rotating elements on each revolution of the shaft. The capacity of a rotary pump is defined as the quantity of liquid (in gpm) actually delivered under specified conditions. Thus, the capacity is equal to the displacement times the speed (rpm) minus whatever losses may be caused by slippage, suction lift, viscosity of the pumped liquid, amount of entrained or dissolved gases in the liquid, and so forth.

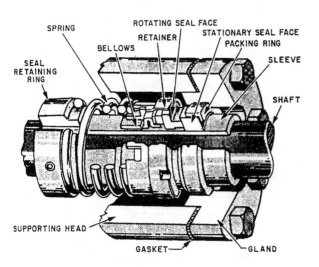

Figure 19-11.—Type 1 mechanical seal.

All rotary pumps work by rotating parts which trap the liquid at the suction side and force it through the discharge outlet. Gears, screws, lobes, vanes, and cam-and-plunger arrangements are commonly used as the rotating elements in rotary pumps.

Rotary pumps are particulary useful for pumping oil and other heavy, viscous liquids. This type of pump is used for fuel-oil service, fuel-oil transfer, lubricating-oil service, and other similar services. Rotary pumps are also used for pumping nonviscous liquids, such as water or gasoline, particulary where the pumping problem involves a high-suction lift.

The power end of a rotary pump is usually an electric motor or an auxiliary steam turbine. Some lubricating-oil pumps that supply oil to the propulsion turbine bearings and to the reduction gears are attached to and driven by either the propulsion shaft or the quill shaft of the reduction gear.

Rotary pumps are designed with very small clearances between rotating parts and stationary parts. The small clearances are necessary to minimize slippage from the discharge side back to the suction side. Rotary pumps are designed to operate at relatively slow speeds to maintain these clearances. Operation at high speeds would cause erosion and excessive wear which, in turn, would result in increased clearances.

Classification of rotary pumps is generally made on the basis of the type of rotating element. The main features of some common types of rotary pumps are discussed in the following paragraphs.

Simple Gear Pump

The simple gear pump (fig. 19-12) has two spur gears that mesh together and revolve in opposite directions. One gear is the driving gear, the other is the driven gear. Clearances between the gear teeth and the casing and between the gear faces and the casing are only a few thousandth of an inch. The action of the unmeshing gears draw the liquid into the suction side of the pump. The liquid is then trapped in the pockets formed by the gear teeth and the casing so that it must follow along the teeth. On the discharge side, the liquid is forced out by the meshing of the gears. Simple gear pumps of this type are frequently used as lubricating pumps on pumps and other auxiliary machinery.

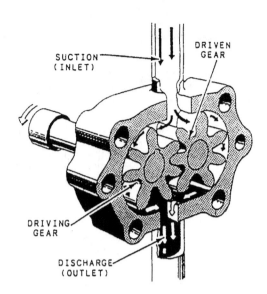

Figure 19-12.—A simple gear rotary pump.

Herringbone Gear Pump

The herringbone gear pump (fig. 19-13) is a modification of the simple gear pump. In this type of pump, one discharge phase begins before the previous discharge phase is entirely complete. This overlapping tends to give a steadier discharge than that obtained with a simple gear pump. Herringbone gear pumps are sometimes used for low-pressure, fuel-oil service and lubricating-oil service.

Helical Gear Pump

The helical gear pump (fig. 19-14) is still another modification of the simple gear pump. Because of the helical gear design, the overlapping of successive discharges from the spaces between the teeth is even greater than it is in the herringbone gear pump. The discharge flow is even smoother. Since the discharge flow is smooth in the helical gear pump, the gears can be designed with a small number of large teeth. This design allows for increased capacity without sacrificing smoothness of flow.

The pumping gears in this type of pump are driven by a set of timing and driving gears which also function to maintain the required close clearances while preventing actual metal-to-metal contact between the pumping gears. Metallic contact between the pumping gears would provide a tighter seal against leakage. However, it would cause rapid wear of the teeth because foreign matter in the pumped liquid would act like an abrasive on the contact surfaces.

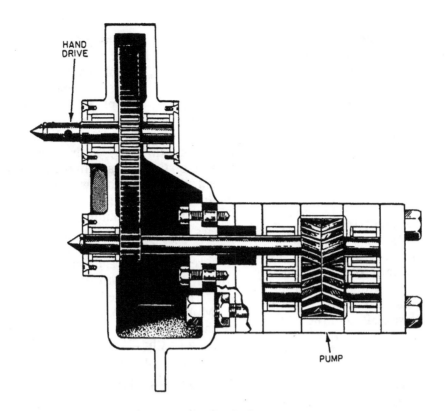

Figure 19-13.—Herringbone gear pump.

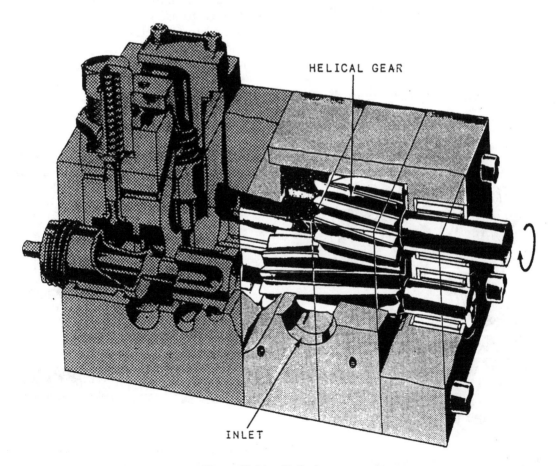

Figure 19-14.—Helical gear pump.

19-11

Roller bearings at both ends of the gear shafts maintain proper alignment and, thus, minimize friction losses in the transmission of power. Stuffing boxes are used to prevent leakage at the shafts.

The helical gear pump is used to pump nonviscous liquids and light oils at high speeds and to pump viscous liquids at lower speeds.

Lobe Pump

The lobe pump is still another variation of the simple gear pump. A lobe pump (heliquad type) is illustrated in figure 19-15. The lobes are considerably larger than gear teeth, but there are only two or three lobes on each rotor. The rotors are driven by external spur gears on the rotor shafts. Some lobe pumps are made with replaceable inserts (gibs) at the extremities of the lobes. These inserts take up the wear that would otherwise be sustained by the ends of the lobes. In addition, they maintain a tight seal between the lobe ends and the casing. The inserts are usually seated on a spring. In this way, they automatically compensate for considerable wear of both the gibs and the casing. Replaceable cover plates (liner plates) are fitted at each end of the casing where the lobe faces cause heavy wear.

Screw Pumps

Several different types of screw pumps exist. The differences between the various types are the number of intermeshing screws and the pitch of the screws. Figure 19-16 shows a positive-displacement, double-screw, low-pitch pump. Figure 19-17 shows a triple-screw, high-pitch pump. Screw pumps are used aboard ship to pump fuel and lube oil and to supply pressure to the hydraulic system. In the double-screw pump, one rotor is driven by the drive shaft and the other by a set on timing gears. In the triple-screw pump, a central rotor meshes with two idler rotors.

In the screw pump, liquid is trapped and forced through the pump by the action of rotating screws. As the rotor turns, the liquid flows in between the threads at the outer end of each pair of screws. The threads carry the liquid along within the housing to the center of the pump, where it is discharged.

Most screw pumps are equipped with mechanical seals. If the mechanical seal fails, the stuffing box has the capability of accepting two rings of conventional packing for emergency use.

PROPELLER PUMPS

Propeller pumps are used primarily where there is a large volume of liquid with a relatively low total head requirement. These pumps are usually limited to use where the total head does not exceed 40 to 60 feet.

The propeller pump is used chiefly for the main condenser circulating pump. In most ships,

Figure 19-15.—Lobe pump (heliquad type).

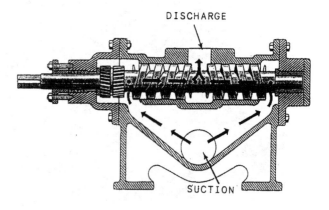

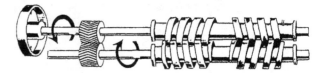

38.105

Figure 19-16.—Positive-displacement, double-screw, low-pitch pump.

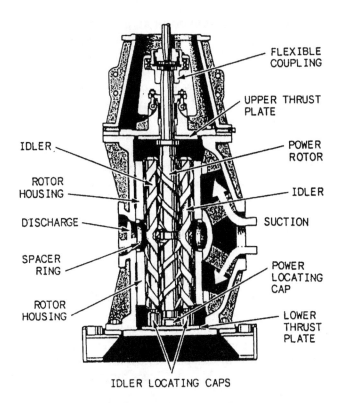

IDLER

ROTOR HOUSING

DISCHARGE

SPACER RING

ROTOR HOUSING

FLEXIBLE COUPLING

UPPER THRUST PLATE

POWER ROTOR

IDLER

SUCTION

POWER LOCATING CAP

LOWER THRUST PLATE

IDLER LOCATING CAPS

Figure 19-17.—Triple-screw, high-pitch pump.

this pump has an emergency suction for pumping out the engine room.

The main condenser circulating pump is of the vertical propeller type. The pump unit consists of three major parts: the propeller, together with its bearings and shaft; the pump casing; and the driving unit, which may be an auxiliary steam turbine or electric motor.

The propeller is a multibladed screw propeller having a large pitch. The blades are thick at the roots and flare out toward the tips. The blades and hubs are cast or forged in one piece and are then machined and balanced. The lower shaft bearing is a water-lubricated, sleeve bearing. The shaft packing gland prevents excessive leakage of water between the casing and the shaft.

FANS

"Fan" is the name commonly given to a nonpositive-displacement compressor. Fans operate on the same principle as nonpositive-displacement pumps. There are two types of fans: centrifugal and axial.

In a centrifugal fan, the flow of liquid through the fan element is in a radial direction, just as in a centrifugal pump. This type of fan uses an impeller (very similar to a pump impeller) to impart a high velocity to the fluid by the use of

centrifugal force. Centrifugal compressors (fans) are commonly used as refrigeration compressors and gas turbine compressors in small gas turbines.

In an axial fan, the flow of the fluid through the fan element is in an axial direction, just as in a propeller pump. Several shapes of fan elements are used. One type uses a propeller very similar to an airplane propeller. This type of fan is commonly called a propeller fan and is used almost exclusively as the forced draft blower on conventional steam-powered ships. Another type uses blading very similar to turbine blading. This type of compressor is commonly used on larger gas turbines.

VARIABLE-STROKE PUMPS

Variable-stroke (also called variable-displacement) pumps are most commonly used on naval ships as part of an electrohydraulic transmission for anchor windlasses, cranes, winches, steering gear, and other equipment. In these applications, the variable-stroke pump is sometimes referred to as the A end, and the hydraulic motor which is driven by the A end is then called the B end.

Although variable-stroke pumps are often classified as rotary pumps, they are actually reciprocating pumps of special design. A rotary motion is imparted to a cylinder barrel or cylinder block in the pump by a constant-speed electric motor, but the actual pumping is done by a set of pistons reciprocating inside cylindrical openings in the cylinder barrel or cylinder block.

There are two general types of variable-stroke pumps in common use: the axial-piston pump and the radial-piston pump. In the axial-piston pump, the pistons are arranged parallel to each other and to the pump shaft. In the radial-piston pump, the piston are arranged radially from the shaft.

Axial-Piston Varible-Stroke Pump

This type of pump usually has either seven or nine single-acting pistons that are evenly spaced around a cylinder barrel. An uneven number of pistons is always used to avoid pulsations in the discharge flow. (Note that the term *cylinder barrel* as used here actually refers to a cylinder block which holds all the cylinders.) The piston rods make a ball-and-socket connection with a socket ring. The socket ring rides on a thrust bearing carried by a casting called the tilting box or tilting block.

The socket ring that actually revolves, is actually fitted into the tilting box, which does not revolve. Figure 19-18 shows diagrammatically the arrangement of the cylinder barrel, the socket ring, and the tilting box. Although only one piston is shown in this illustration, the others fit similarly into the cylinder barrel and into the socket ring.

Figure 19-19 illustrates diagrammatically the manner in which the position of the tilting box affects the position of the pistons. (Note that this is not a continuous cross-sectional view, since for illustrative purposes two pistons are shown.) To understand how the pumping action takes place, let us follow one piston as the cylinder barrel and socket ring makes one complete revolution. When the tilting box is set perpendicular to the shaft, as in figure 19-19, view A, the piston does not move back and forth within its cylindrical opening as the cylinder barrel and socket ring revolve. Thus, the piston is in the same position with respect to its own cylindrical opening when it is at the top position as it is when the cylinder barrel has completed half a revolution and has carried

the piston to the bottom position. Since the piston does not reciprocate, there is no pumping action when the tilting box is in this position even though the cylinder barrel and socket ring are revolving.

In figure 19-19, view B, the tilting box is set at an angle so that it is farther away from the top of the cylinder barrel and closer to the bottom of the cylinder barrel. As the cylinder barrel and the socket ring revolve, the piston is pulled outward as it is carried from the bottom position to the top position, and it is pushed inward as it is carried from the top position to the bottom position. Thus, the piston makes one suction stroke (from the bottom position to the top position) and one discharge stroke (from the top position to the bottom position) for each complete revolution of the cylinder barrel.

In figure 19-19, view C, we see the tilting box set at a somewhat larger angle. Because there is more distance between the cylinder barrel and the socket ring at the top, and less distance between them at the bottom, the piston now moves further in each stroke and, thus, displaces more liquid on the discharge stroke.

Although we have considered the position of only one piston, it is obvious that the others are being similarly positioned as the cylinder barrel and socket ring revolve. At any given moment, some pistons are making suction strokes and some pistons are making discharge strokes. In a nine-piston pump, for example, four pistons will be making suction strokes, four will be making discharge strokes, and one will be at the end of its stroke and will therefore be momentarily motionless.

Each cylindrical opening in the cylinder barrel has a port in the face of the cylinder barrel. Each port except one will be either a suction port or a discharge port, depending on the position of the

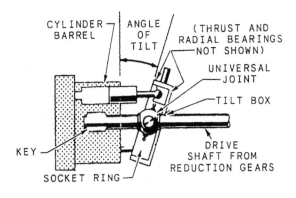

Figure 19-18.—Diagram showing the cylinder barrel, socket ring, and tilting box in an axial-piston, variable-stroke pump.

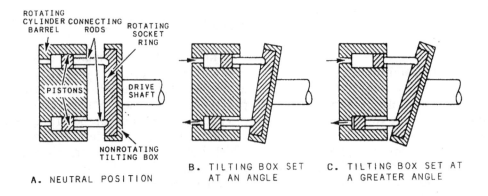

Figure 19-19.—Diagram showing how the tilting box position affects the position of the pistons.

piston in the cylindrical opening. The face of the cylinder barrel bears against the valve plate, a nonrotating piece which has two semicircular ports, one for suction and one for discharge. When the piston is at the top position at the end of the suction stroke, the port for that piston is over the top land (the term *land* refers to the space between ports) on the valve plate. When a piston is at the bottom position at the end of the discharge stroke, the port is over the bottom land on the valve plate. Figure 19-20 shows the ports on the face of the cylinder barrel and in the valve plate.

When the A end is used alone as a constant-speed, variable-capacity pump, the tilting box is often so designed that it can be tilted in one direction only. In this case, the flow of the pumped liquid is always in the same direction. When the A end is used as part of an electro-hydraulic system, the tilting box is most commonly designed to be tilted in either direction. In this case, the flow of the pumped liquid is in either direction. It is therefore clear that the position of the tilting box controls both the direction of flow and the amount of flow.

Radial-Piston Variable-Stroke Pump

The radial-piston variable-stroke pump is similar in general principle to the axial-piston pump just described, but the arrangement of the component parts is somewhat different. In the radial-piston pump, the cylinders are arranged radially in a cylinder body that rotates around a nonrotating central cylindrical valve. Each cylinder communicates with horizontal ports in the central cylindrical valve. Plungers or pistons that extend outward from each cylinder are pinned at their outer ends to slippers that slide around the inside of a rotating floating ring or housing.

The floating ring is constructed so that it can be shifted offcenter from the pump shaft. When it is centered or in the neutral position, the pistons do not reciprocate and the pump does not function, even though the electric motor is still causing the pump to rotate. If the floating ring is forced offcenter to one side, the pistons reciprocate and the pump operates. If the floating ring is forced offcenter to the other side of the pump shaft, the pump also operates but the direction of flow is reversed. Thus, both the direction of flow and the amount of flow are determined by the position of the cylinder body and the relative position of the floating ring.

RECIPROCATING PUMPS

A reciprocating pump moves water or other liquid by a plunger or piston that reciprocates (travels back and forth) inside a cylinder. Reciprocating pumps are positive-displacement pumps; each stroke displaces a definite quantity of liquid, regardless of the resistance against which the pump is operating.

Reciprocating pumps in naval service are usually classified as follows:

—Direct-acting or indirect-acting

—Simplex (single) or duplex (double)

—Single-acting or double-acting

—High-pressure or low-pressure

—Vertical or horizontal

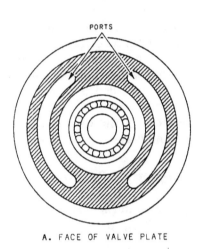

A. FACE OF VALVE PLATE

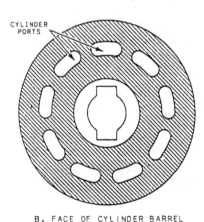

B. FACE OF CYLINDER BARREL

Figure 19-20.—Suction and discharge ports in the face of the cylinder barrel and in the valve plate.

The reciprocating pump shown in figure 19-21 is a direct-acting, simplex, double-acting, high-pressure, vertical pump. Now, let us see what all these terms mean, with reference to the pump shown in the illustration.

Direct- and Indirect-acting Pumps

The pump shown in the figure is direct-acting because the pump rod is a direct extension of the piston rod. Therefore, the piston in the power end is directly connected to the plunger in the liquid end. Most reciprocating pumps used in the Navy are direct-acting. In an indirect-acting pump, there is some intermediate mechanism between the piston and the pump plunger. The intermediate mechanism may be a lever or a cam. This arrangement can be used to change the relative length of the strokes of piston and plunger or to vary the relative speed between piston and plunger. If there is no intermediate mechanism,

the pump may use a rotating crankshaft, such as a chemical proportioning pump in a distilling unit.

Simplex and Duplex Pumps

The pump shown in the figure is called a single or simplex pump because it has only one liquid cylinder. Simplex pumps may be either direct-acting or indirect-acting. A double or duplex pump is an assembly of two single pumps placed side by side on the same foundation; the two steam cylinders are cast in a single block and the two liquid cylinders are cast in another block. Duplex reciprocating pumps are seldom found in modern combatant ships but were once commonly used in the Navy.

Single-acting and Double-acting Pumps

In a single-acting pump, the liquid is drawn into the liquid cylinder on the first, or suction, stroke and is forced out of the cylinder on the return, or discharge, stroke. In a double-acting pump, each stroke serves both to draw in liquid and to discharge liquid. As one end of the cylinder is filled, the other end is emptied; on the return stroke, the end that was just emptied is filled and the end that was just filled is emptied.

The pump shown in the figure is double-acting, as are most of the of the reciprocating pumps used in the Navy.

High- and Low-pressure Pumps

The pump shown in the figure is designed to operate with a discharge pressure which is higher than the pressure of the steam operating the piston in the steam cylinder; in other words, it is a high-pressure pump. In a high-pressure pump, the steam piston is larger in diameter than the plunger in the liquid cylinder. Since the area of the steam piston is greater than the area of the plunger in the liquid cylinder, the total force exerted by the steam against the steam piston is concentrated on the smaller working area of the plunger in the liquid cylinder. Therefore, the pressure per square inch is greater in the liquid cylinder than in the steam cylinder. A high-pressure pump discharges a comparatively small volume of liquid against a high pressure. A low-pressure pump, on the other hand, has a comparatively low discharge pressure but a large volume of discharge. In a low-pressure pump, the steam piston is smaller than the plunger in the liquid cylinder.

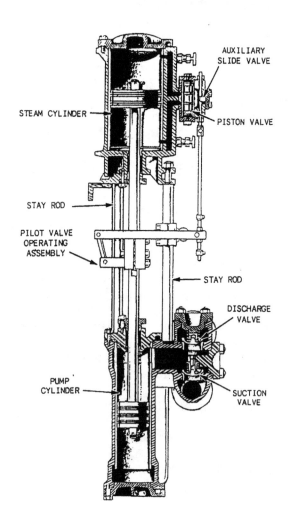

Figure 19-21.—Reciprocating pump.

The standard way of designating the size of a reciprocating pump is by giving three dimensions, in the following order:

1. The diameter of the steam piston
2. The diameter of the pump plunger
3. The length of the stroke

For example, a 12- by 11- by 18-inch reciprocating pump has a steam piston that is 12 inches in diameter, a pump plunger that is 11 inches in diameter and a stroke that is 18 inches in length. The designation enables you to tell immediately whether the pump is a high-pressure or a low-pressure pump.

Vertical and Horizontal Pump

Finally, the pump shown in the figure is classified as vertical because the steam piston and the pump plunger move up and down. Most reciprocating pumps in naval use are vertical; however, you may occasionally encounter a horizontal pump, in which the piston moves back and forth rather than up and down.

The following discussion of reciprocating pumps is generally concerned with direct-acting, simplex, double-acting, vertical pumps. Most reciprocating pumps used in the Navy are of this type.

The power end of a reciprocating pump consists of a bored cylinder in which the steam piston reciprocates. The steam cylinder is fitted with heads at each end; one head has an opening to accommodate the piston rod. Steam inlet and exhaust ports connect each end of the steam cylinder with the steam chest. Drain valves are installed in the steam cylinder so that water resulting from condensation may be drained off.

Automatic timing of the admission and release of steam to and from each end of the steam cylinder is accomplished by various types of valve arrangements. Figure 19-22 shows the piston-type valve gear. It consists of a main piston-type slide valve and a pilot-slide valve. Since the rod from the pilot valve is connected to the pump rod by a valve-operating assembly, the position of the pilot valve is controlled by the position of the piston in the steam cylinder. The pilot valve furnishes actuating steam to the main piston-type valve, which, in turn, admits steam to the top or to the bottom of the steam cylinder at the proper time.

The valve-operating assembly that connects the pilot-valve-operating rod and the pump rod is shown in figure 19-23. As the crosshead arm

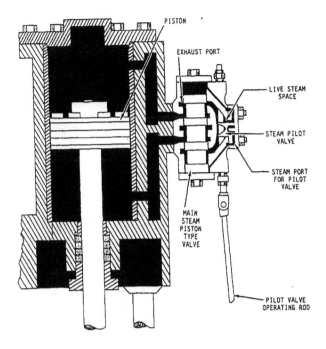

Figure 19-22.—Piston-type valve gear for the steam end of a reciprocating pump.

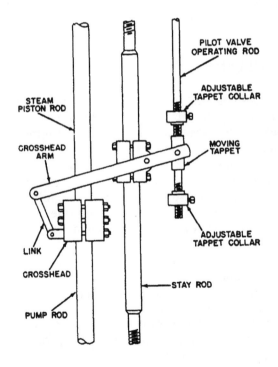

Figure 19-23.—Valve-operating gear of a reciprocating pump.

(sometimes called the rocker arm) is moved up and down by the movement of the pump rod, the moving tappet slides up and down on the pilot-valve-operating rod. The tappet collars are adjusted so that the pump will make the full designed stroke.

The liquid end of a reciprocating pump has a piston and a cylinder assembly similar to that of the power or steam end. The piston in the liquid end is often called a plunger. A valve chest, sometimes called a water chest, is attached to the liquid cylinder. The valve chest contains two sets of suction and discharge valves, one set to serve the upper end of the liquid cylinder and one to serve the lower end. The valves are arranged so that the pump takes suction from the suction chamber and discharges through the discharge chamber on both up and down strokes.

An adjustable relief valve is fitted to the discharge chamber to protect the pump and the piping against excessive pressure.

Although reciprocating pumps were once widely used aboard ship for a variety of services, their use on Navy ships is now generally limited to emergency feed pumps and fuel-oil tank stripping and bilge pumps.

JET PUMPS

Devices that use the rapid flow of a fluid to entrain another fluid and thereby move it from one place to another are called jet pumps. Jet pumps are sometimes not considered to be pumps because they have no moving parts. In view of our previous definition of a pump as a device that uses an external source of power to apply force to a fluid to move the fluid from one place to another, it will be apparent that a jet pump is indeed a pump.

Jet pumps are generally considered in two classes: ejectors, which use a jet of steam to entrain air, water, or other fluid; and eductors, which use a flow of water to entrain and thereby pump water. The basic principles of operation of these two devices are identical.

A simple jet pump is shown in figure 19-24. In this pump, steam under pressure enters chamber C through pipe A, which is fitted with a nozzle, B. As the steam flows through the

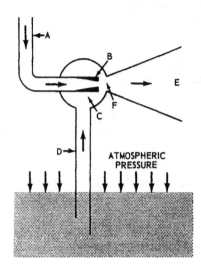

Figure 19-24.—Simple jet pump (ejector type).

nozzle, the velocity of the steam is increased. The fluid in the chamber at point F, in front of the nozzle, is driven out of the pump through the discharge line, E, by the force of the steam jet. The size of the discharge line increases gradually beyond the chamber to decrease the velocity of the discharge and thereby transform some of the velocity head to pressure head. As the steam jet forces some of the fluid from the chamber into the discharge line, pressure in the chamber is lowered and the pressure on the surface of the supply fluid forces fluid up through the inlet, D, into the chamber, and out through the discharge line. Thus, the pumping action is established.

Jet pumps of the ejector type are occasionally used aboard ship to pump small quantities of drains overboard. Their primary use on naval ships is not in the pumping of water but in the removal of air and other noncondensable gases from the main and auxiliary condensers.

An eductor is shown in figure 19-25. The principle of operation is the same as that just described for the ejector-type jet pump; however, water is used instead of steam. On Navy ships, eductors are used to pump water from bilges, to dewater compartments, and to supply positive pressure head for pumps used in firefighting.

PUMP MAINTENANCE

Pumps require a certain amount of routine maintenance and, upon occasion, some repair

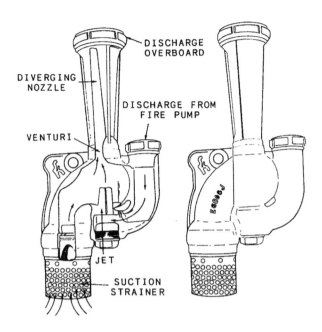

Figure 19-25.—Eductor-type jet pump.

DISCHARGE
OVERBOARD

DIVERGING
NOZZLE

DISCHARGE FROM
FIRE PUMP

VENTURI

JET

SUCTION
STRAINER

work. Because pumps are so widely used for various services in the Navy, you should consult the manufacturer's technical manual for details concerning the repair of a specific unit. Routine maintenance should be performed according to the Planned Maintenance Subsystem requirements.

REFERENCES

Engineman 3, NAVEDTRA 10539, Naval Education and Training Program Management Support Activity, Pensacola, Florida, 1989.

Fireman, NAVEDTRA 10520-H, Naval Education and Training Program Management Support Activity, Pensacola, Florida, 1987.

Gas Turbine Systems Technician (Electrical)/Gas Turbine Systems Technician (Mechanical) 3, Volume 1, NAVEDTRA 10563, Naval Education and Training Program Management Support Activity, Pensacola, Florida, 1989.

Machinist's Mate 3 & 2, NAVEDTRA 10524-F1, Naval Education and Training Program Management Support Activity, Pensacola, Florida, 1987.

Naval Ships' Technical Manual, Chapter 503, "Pumps," NAVSEA S9086-RH-STM-000/CH-503, Change 3, Naval Sea Systems Command, Washington, D.C., 1981.

CHAPTER 20

SHIPBOARD ELECTRICAL SYSTEMS

LEARNING OBJECTIVES

Upon completion of this chapter, you should be able to do the following:

1. Describe basic electrical and electron theory.

2. Describe the relationship between magnetism and electricity.

3. Identify the most commonly used methods of producing a voltage aboard ship.

4. Describe the three conditions necessary to produce a voltage by magnetism.

5. State Ohm's law.

6. Describe the relationship of resistance in a series circuit.

7. Describe the relationship of current in a parallel circuit.

8. State the practical use of a Wheatstone bridge.

9. Describe the makeup of a dc generator and a three-phase ac generator.

10. Describe the types of dc and ac generators and motors.

11. Describe the effects of inductance and capacitance in an ac circuit.

12. Describe the function and parts of a transformer.

13. State the purpose of the ship's power distribution system and describe its components.

14. Describe the control and safety devices used in the distribution electrical power.

15. Describe the purpose and basic operation of a ship's degaussing system.

16. Describe the two types of cathodic protection systems.

17. Describe the operation principles of gyroscopes and their function on board ships.

INTRODUCTION

Shipboard electrical systems include a variety of equipment that provides many services vital to the operation of naval ships. These systems distribute electrical power throughout the ship for offensive and defensive weapons, the ship's movement, and shipboard habitability. Since the systems and equipment using electric power are often under the cognizance of a division other than the electrical division, a joint responsibility frequently exists for the operation, maintenance, and repair of electrical systems and equipment.

This chapter provides some information on basic electrical theory and gives a brief description of shipboard electrical systems and equipment.

BASIC ELECTRICAL THEORY

The word *electric* is derived from the Greek word meaning amber. The ancient Greeks used the word to describe the strong forces of attraction and repulsion that were exhibited by amber after it had been rubbed with a cloth. Since scientists are still unable to define electricity clearly, and since many of the phenomena which occur cannot be completely explained, theories can only be postulated from the reactions observed.

Through research and experiment, scientists have observed and described many predictable characteristics of electricity and have postulated certain rules that are often called laws. These laws of electricity, together with the electron theory, are the bases for our present concepts of electricity.

ELECTRON THEORY

Every atom is primarily an electrical system with high-speed planetary electrons orbiting around its nucleus. The electron, whose negative charge forms a natural unit of electricity, is bound to the atom by the positive charge within the nucleus.

The electrons in outer orbits of certain elements are easily separated from the positive nuclei of their parent atoms. Should an outside force be applied, one of these loosely bound electrons will be released from the parent atom, becoming a free electron, and will travel to another atom. It is on this ability of an electron to move about from one atom to another that the electron theory is based.

Elements such as silver, copper, gold, and aluminum have many loosely bound electrons and are considered to be good conductors of electricity. In materials used as insulators, electron flow from one atom to another is relatively nonexistent, since the planetary electrons in the outer orbital shells are more tightly bound to their parent nuclei.

Ordinarily an atom is most likely to be in that state in which the internal energy is at a minimum, having a neutral electrical charge. However, if an atom absorbs enough energy from an outside source, loosely bound electrons in the outer orbital shells will leave the atom. An atom that has lost or gained one or more electrons is said to be ionized. If an atom loses electrons, it becomes positively charged and is referred to as a positive ion; if an atom gains electrons, it is referred to as a negative ion and is said to have a negative charge. A positive ion will attract any free electron in its surroundings to reach a neutral state.

STATIC ELECTRICITY

When two bodies have unlike charges, one positive and the other negative, an electrical force is exerted between the two. This force is called a static charge or an electrostatic force.

One of the easiest ways to create a static charge is by friction. When two pieces of material (matter) are rubbed together, electrons can be "wiped off" one material onto the other. If the materials used are both good conductors, it is difficult to get a detectable charge on either, since equalizing currents can flow easily between the conducting materials. These currents equalize the charges almost as fast as they are created. A static charge is more easily created between nonconducting materials. Since nonconducting

materials are poor conductors (insulators), very little equalizing current can flow and an electrostatic charge builds up. When the charge becomes great enough, current will flow regardless of the poor conductivity of the materials. These currents will cause visible sparks and produce a crackling sound.

Charged Bodies

One of the fundamental laws of electricity is that like charges repel each other and unlike charges attract each other. A positive charge and a negative charge, being unlike, tend to move toward each other. In the atom, the negative electrons are drawn toward the positive protons in the nucleus. This attractive force is caused by its rotation about the nucleus. As a result, the electrons remain in orbit and are not drawn into the nucleus. Electrons repel each other because of their like negative charges, and protons repel each other because of their like positive charges.

The law of charged bodies may be demonstrated by a simple experiment using two pith (paper pulp) balls suspended near one another by threads, as shown in figure 20-1, and a hard rubber rod. If the hard rubber rod is rubbed to give it a negative charge and then held against the right-hand ball in view A, the rod will give off a negative charge to the ball. The right-hand ball will have a negative charge with respect to the left-hand ball. When released, the two balls will be drawn together (fig. 20-1, view A). They will touch and remain in contact with each other until the left-hand ball gains a portion of the negative charge of the right-hand ball,

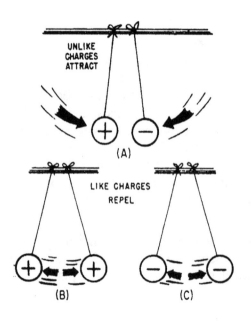

Figure 20-1.–Reaction between charged bodies.

then they will swing apart (fig. 20-1, view C). If a positive or a negative charge is placed on both balls (fig. 20-1, view B), the balls will repel each other.

Coulomb's Law of Charges

The amount of attracting or repelling force that acts between two electrically charged bodies in free space depends on two things–(1) their charges and (2) the distance between them. This relationship between attracting or repelling charged bodies was first discovered by a French scientist named Charles A. Coulomb. Coulomb's law of charge states that charged bodies attract or repel each other with a force that is directly proportional to the product of their charges and inversely proportional to the square of the distance between them.

The practical unit of charge a body has is expressed in coulombs. One coulomb is the charge carried by approximately 6×10^{18} electrons.

Electric Current Flow

A difference of potential exists between two bodies having opposite electrostatic charges. If a path is provided between the two bodies, electrons will flow from the negatively charged body to the positively charged body until the charges have equalized and the difference of potential no longer exists. This movement of electrons is called electric current. The rate of flow is measured in amperes. One ampere may be defined as the flow of 1 coulomb per second past a fixed point in a conductor.

The force or difference in potential that causes electrons to flow from one charged body to another is called electromotive force (emf). Electromotive force is measured in volts. One volt may be defined as the potential difference between two points when 1 joule of work is required to move a 1-coulomb charge between these points.

MAGNETISM

Magnetism is generally defined as that property of a material that enables it to attract pieces of iron. A material possessing this property is known as a magnet. The word originated with the ancient Greeks, who found stones possessing this characteristic. Materials that are attracted by a magnet, such as iron, steel, nickel, and cobalt, have the ability to become magnetized. These are called magnetic materials. Materials, such as paper, wood, glass, or tin, that are not attracted by magnets, are considered nonmagnetic. Nonmagnetic materials cannot become magnetized.

Magnets produced from magnetic materials are called artificial magnets. They can be made in a variety of shapes and sizes and are used extensively in electrical apparatus. Artificial magnets are generally made from special iron or steel alloys that are usually magnetized electrically. The material to be magnetized is inserted into a coil of insulated wire and a heavy flow of electrons is passed through the wire. Magnets can also be produced by stroking a magnetic material with magnetite or with another artificial magnet. The forces causing magnetization are represented by magnetic lines of force, very similar in nature to electrostatic lines of force.

Artificial magnets are usually classified as permanent or temporary, depending on their ability to keep their magnetic properties after the magnetizing force has been removed. Magnets made from substances that retain a great deal of their magnetism, such as hardened steel and certain alloys, are called permanent magnets. These materials are relatively difficult to magnetize because of the opposition offered to the magnetic lines of force as the lines of force try to distribute themselves throughout the material. The opposition that a material offers to the magnetic lines of force is called reluctance. All permanent magnets are produced from materials having a high reluctance.

A material with a low reluctance, such as soft iron or annealed silicon steel, is relatively easy to magnetize but will keep only a small part of its magnetism once the magnetizing force is removed. Materials of this type that easily lose most of their magnetic strength are called temporary magnets. The amount of magnetism that remains in a temporary magnet is called its residual magnetism. The anility of a material to keep an amount of residual magnetism is called the retentivity of the material.

One difference between a permanent and a temporary magnet is reluctance–a permanent magnet having a high reluctance and a temporary magnet having a low reluctance. Another difference is the permeability of their materials, or the ease with which magnetic lines of force distribute themselves throughout the material. A permanent magnet, produced from material with a high reluctance, has low permeability. A temporary magnet, produced from material with a low reluctance, has a high permeability.

The space surrounding a magnet where magnetic forces act is known as the magnetic field.

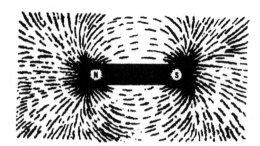

Figure 20-2.–Pattern formed by iron filings.

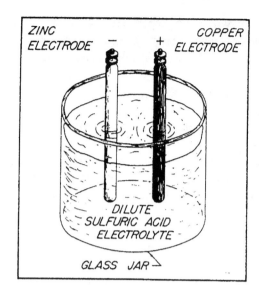

Figure 20-3.–Voltaic cell.

A pattern of this directional force can be obtained by performing an experiment with iron filings. Place a piece of glass over a bar magnet and sprinkle the iron filings on the surface of the glass. The magnetizing force of the magnet will be felt through the glass and each iron filing becomes a temporary magnet. If the glass is now tapped gently, the iron particles will align themselves with the magnetic field surrounding the magnet as the compass needle did previously. The filings form a definite pattern, a visible representation of the forces comprising the magnetic field. Examination of the arrangements of iron filings in figure 20-2 will indicate that the magnetic field is very strong at the poles and weakens as the distance from the poles increases. It is also apparent that the magnetic field extends from one pole to the other, constituting a loop about the magnet.

PRODUCING A VOLTAGE

Presently, there are six commonly used methods of producing a voltage. Some of these methods are more widely used than others, and some are used mostly for specific applications. Following is a list of the six known methods of producing a voltage:

• Friction–Voltage produced by rubbing certain materials together

• Pressure–Voltage produced by squeezing crystals of certain substances

• Heat–Voltage produced by heating the joint (junction) where two unlike metals are joined

• Light–Voltage produced by light striking photosensitive (light sensitive) substances

• Chemical action–Voltage produced by chemical reaction in a battery cell

• Magnetism–Voltage produced in a conductor when the conductor moves through a magnetic field, or

a magnetic field moves through the conductor in such a manner as to cut the magnetic lines of force of the field

Magnetism and chemical action are the two methods most commonly used aboard ship; therefore, our discussion will be limited to these two methods.

Voltage Produced By Chemical Action

Voltage may be produced chemically when certain substances are exposed to chemical action.

If two dissimilar substances are immersed in a solution that produces a greater chemical action on one substance than on the other, a difference of potential will exist between the two. If a conductor is then connected between them, electrons will flow through the conductor to equalize the charge. This arrangement is called a primary cell. The two metallic pieces are called electrodes and the solution is called electrolyte. The voltaic cell shown in figure 20-3 is a simple example of a primary cell. The difference of potential results because material from one or both of the electrodes goes into solution in the electrolyte and, in the process, ions form in the vicinity of the electrodes. Due to the electric field associated with the charged ions, the electrodes acquire charges.

The amount of difference in potential between the electrodes depends principally on the metals used. The type of electrolyte and the size of the cell have little or no effect on the potential difference produced.

There are two types of primary cells: the wet cell and the dry cell. In a wet cell, the electrolyte is a liquid. A cell with a liquid electrolyte must remain in an upright position and is not readily transportable. An automotive battery is an example of this type of cell. The dry cell, much more commonly used than the wet cell, is not actually dry, but contains an electrolyte mixed with other materials to form a paste. Flashlights and portable radios are commonly powered by dry cells.

Batteries are formed when several cells are connected together to increase electrical output.

Voltage Produced By Magnetism

Magnets or magnetic devices are used for thousands of different jobs. A most useful and widely employed application of magnets is the production of vast quantities of electric power from mechanical sources. The mechanical power may be provided by a number of different sources, such as gasoline or diesel engines, water turbines, steam turbines, or gas turbines. However, the final conversion of these energies to electricity is done by generators employing the principle of electromagnetic induction.

There are three fundamental conditions that must exist before a voltage can be produced by magnetism.

1. There must be a magnetic field in the conductor's vicinity.

2. There must be a conductor in which the voltage will be produced.

3. There must be relative motion between the field and the conductor.

With these conditions, when a conductor or conductors move across a magnetic field so as to cut the lines of force, electrons within the conductor are propelled in one direction or another. Thus, an electric force, or voltage, is produced.

The production of a voltage by magnetic induction is illustrated in figure 20-4. If the ends of a conductor are connected to a low-reading voltmeter or galvanometer and the conductor is moved rapidly down through a magnetic field, there is a momentary reading on the meter. When the conductor is moved up through the field, the meter deflects in the opposite direction. If the conductor is held stationary and the magnet is moved so the field cuts across the conductor, the meter is deflected in the same manner as when the conductor was moved and the field was stationary.

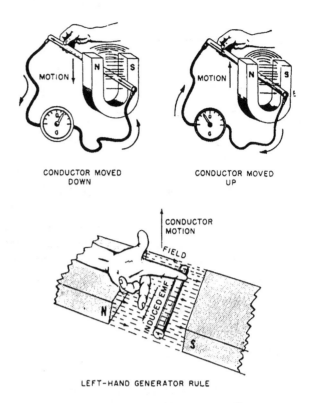

CONDUCTOR MOVED DOWN

CONDUCTOR MOVED UP

LEFT-HAND GENERATOR RULE

Figure 20-4.–Left-hand generator rule.

The voltage developed across the conductor terminals by electromagnetic induction is known as an induced emf, and the resulting current that flows is called induced current. The induced emf exists only so long as relative motion occurs between the conductor and the field.

There is a definite relationship between the direction of flux, the direction of motion of the conductor, and the direction of the induced emf. When two of these directions are known, the third can be found by applying the left-hand rule for generators. To find the direction of the emf induced in a conductor, extend the thumb, the index finger, and the second finger of the left hand at right angles to each other, as shown in figure 20-4. Point the index finger in the direction of the flux (toward the south pole) and the thumb in the direction in which the conductor is moving in respect to the fields. The second finger then points in the direction in which the induced emf will cause the electrons to flow.

DIRECT-CURRENT CIRCUITS

An electric circuit is a complete path through which electrons can flow from the negative terminal of the voltage source, through the connecting wires (conductors), through the load, and back to the positive

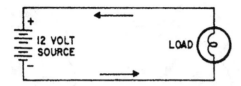

Figure 20-5.–Simple electric current.

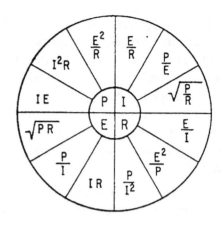

Figure 20-6.–Summary of basic Ohm's law formulas.

terminal of the voltage source (see fig. 20-5). The resistance[1] of a circuit (opposition to current flow) controls the amount of current flow through the circuit. The unit of electrical resistance, the ohm (Ω), is named after the German physicist George Simon Ohm, who in the nineteenth century proved by experiment the constant proportionality between current and voltage in the simple electric circuit.

OHM'S LAW

In the early part of the nineteenth century, George Simon Ohm proved by experiment that a precise relationship exists between current, voltage, and resistance. This relationship is called Ohm's law and is stated as follows:

The current in a circuit is directly proportional to the applied voltage and inversely proportional to the circuit resistance. Ohm's Law may be expressed as an equation:

$$I = \frac{E}{R}$$

where:

I = intensity of current, in amperes

E = difference in potential, in volts

R = resistance, in ohms

As stated in Ohm's law, current is inversely proportional to resistance. This means, as the resistance in a circuit increases, the current decreases proportionally.

In the equation, if any two quantities are known, the third can be determined.

In addition to the volt, the ampere, and the ohm, the unit of power frequently appears in electric circuit calculations. Power, whether electrical or mechanical, pertains to the rate at which work is being done. Work is done whenever a force causes motion. When a mechanical force is used to lift or move a weight, work is done. However, force exerted without causing motion, such as the force of a compressed spring acting between two fixed objects, is not work.

The basic unit of power is the watt. Power in watts is equal to the voltage across a circuit multiplied by current through the circuit. This represents the rate at any given instant at which work is being done. The symbol P indicates electrical power. Thus, the basic power formula is $P = E \times I$. Where E is voltage and I is current in the circuit. The amount of power changes when either voltage or current, or both voltage and current, are caused to change.

The various implications of Ohm's law may be derived from the algebraic transposition of the units I, E, R, and P. A summary of the 12 basic formulas that may be derived from transposing these units is given in figure 20-6. The unit in each quadrant of the smaller circle is equivalent to the quantities in the same quadrant of the larger circle.

SERIES CIRCUITS

The analysis of a series circuit to determine values for voltage, current, resistance, and power is relatively simple. It is necessary only to draw or to visualize the

---

1 All conductors have some resistance, and therefore a circuit made up of nothing but conductors would have some resistance, however small it might be. In circuits containing long conductors, through which an appreciable amount of current is drawn, the resistance of the conductors becomes important. For the purposes of this chapter, however, the resistance of the conducting wires is neglected.

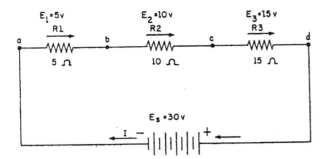

Figure 20-7.–Series circuit for demonstrating Kirchhoff's law of voltages.

circuit, to list the known values and to determine the unknown values by Ohm's law and Kirchhoff's law of voltages.

Kirchhoff's law of voltages states that the algebraic sum of all the voltages in any complete electric circuit is equal to zero. In other words, the sum of all positive voltages must be equal to the sum of all negative voltages. For any given voltage rise, there must be an equal voltage drop somewhere in the circuit. The voltage rise (potential source) is usually regarded as the power supply, such as a battery. The voltage drop is usually regarded as the load, such as a resistor. The voltage drop may be distributed across a number of resistive elements, such as a string of lamps or several resistors. However, according to Kirchhoff's law, the sum of their individual voltage drops must always equal the voltage rise supplied by the power source.

The statement of Kirchhoff's law can be translated into an equation, from which many unknown circuit factors may be determined. (See fig. 20-7.) Note that the source voltage E_s is equal to the sum of the three load voltages E_1, E_2, and E_3. In equation form,

$$E_s = E_1 + E_3$$

The following procedure may be used to solve problems applicable to figure 20-7:

1. Note the polarity of the source emf (E_s) and indicate the electron flow around the circuit. Electron flow is out from the negative terminal of the source, through the load, and back to the positive terminal of the source. In the example, the arrows indicate electron flow in a clockwise direction around the circuit.

2. To apply Kirchhoff's law, it is necessary to establish a voltage equation. The equation is developed by tracing around the circuit and noting the voltage absorbed (that is, the voltage drop) across each part of the circuit, and expressing the sum of these voltages according to the voltage law. It is important that the trace

be made around a closed circuit, and that it encircle the circuit only once. Thus, a point is arbitrarily selected at which to start the trace. The trace is then made and, upon completion, the terminal point coincides with the starting point.

3. Sources of emf are preceded by a plus sign if, in tracing through the source, the first terminal encountered is positive; if the first terminal is negative the emf is preceded by a minus sign.

4. Voltage drops along wires and across resistors (loads) are preceded by a minus sign if the trace is in the assumed direction of electron flow; if in the opposite direction, the sign is plus.

5. If the assumed direction of electron flow is incorrect, the error is indicated by a minus sign preceding the current as obtained in solving for circuit current. The magnitude of the current is not affected.

The preceding rules may be applied to the example of figure 20-7 as follows:

1. The left terminal of the battery is negative, the right terminal is positive, and electron flow is clockwise around the circuit.

2. The trace may arbitrarily be started at the positive terminal of the source and continued clockwise through the source to its negative terminal. From this point, the trace is continued around the circuit to a, b, c, d, and back to the positive terminal, thus completing the trace once around the entire closed circuit.

3. The first term of the voltage equation is $+E_s$.

4. The second, third, and fourth terms are $-E_1$, $-E_2$, and $-E_3$, respectively. Their algebraic sum is equated to zero, as follows:

$$E_s - E_1 - E_2 - E_3 = 0$$

Transposing the voltage equation and solving for E_s,

$$E_s = E_1 + E_2 + E_3$$

Since $E = IR$ from Ohm's law, the voltage drop across each resistor may be expressed in terms of the current and resistance of the individual resistor, as follows:

$$E_s = IR_2 + IR_3$$

where R_1, R_2, and R_3 are the resistances of resistors R1, R2, and R3, respectively. E_s is the source voltage and I is the circuit current.

E_s may be expressed in terms of the circuit current and total resistance as IR_t. Substituting IR_t for E_s, the voltage equation becomes

$$IR_t = IR_1 + IR_2 + IR_3$$

Since there is only one path for current in the series circuit, the total current is the same in all parts of the circuit. Dividing both sides of the voltage equation by the common factor I, an expression is derived for the total resistance of the circuit in terms of the resistances of the individual devices:

$$R_t = R_1 + R_3$$

Therefore, in series circuits, the total resistance is the sum of the resistances of the individual parts of the circuit. In the example of figure 20-7, the total resistance is $5 + 10 + 15 = 30$ ohms. The total current may be found by applying the equation

$$I_t = \frac{E_t}{R_t} = \frac{30}{30} = 1 \text{ ampere}$$

The power absorbed by resistor R_1 is I^2R_1, or $1_2 \times 5 = 5$ watts. Similarly, the power absorbed by R_2 is $1^2 \times 10 = 10$ watts, and the power absorbed by R_3 is $1^2 \times 15 = 15$ watts. The total power absorbed is the arithmetic sum of the power of each resistor, or $5 + 10 + 15 = 30$ watts. The value is also calculated by $P_t = E_t I_t = 30 \times 1 = 30$ watts.

PARALLEL CIRCUITS

The parallel circuit differs from the simple series circuit in that two or more resistors, or loads, are connected directly to the same source of voltage. The parallel circuit has more than one path for current. The more paths (or resistors) that are added in parallel, the less opposition there is to the flow of electrons from the source. This condition is opposite to the effect that is produced in the series circuit where added resistors increase the opposition to the electron flow.

As may be seen from figure 20-8, the same voltage is applied across each of the parallel resistors. In this case the voltage applied across the resistors is the same as the source voltage, E_s.

Current flows from the negative terminal of the source to point a where it divides and passes through the three resistors to point b and back to the positive terminal of the voltage source. The amount of current flowing through each individual branch depends on the source voltage and on the resistance of that branch, the lower the resistance of the branch, the higher will be the

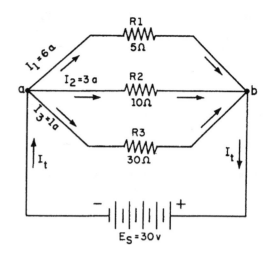

Figure 20-8.—Parallel electric circuit.

current through that branch. The individual currents can be found by the application of Ohm's law to the individual resistors. Thus,

$$I_1 = \frac{E_s}{R_1} = \frac{30}{5} = 6 \text{ amperes}$$

and

$$I_2 = \frac{E_s}{R_2} = \frac{30}{10} = 3 \text{ amperes}$$

and

$$I_3 = \frac{E_s}{R_3} = \frac{30}{30} = 1 \text{ ampere}$$

The total current, I_t, of the parallel circuit is equal to the sum of the currents through the individual branches. This, in slightly different words, is Kirchhoff's law. In this case, the total current is

$$I_t = I_1 + I_2 + I_3 = 6 + 3 + 1 = 10 \text{ amperes}$$

To find the equivalent, or total resistance (R_t) of the combination shown in figure 20-8, Ohm's law is used to find each of the currents $(I_t, I_1, I_2,$ and $I_3)$ in the preceding formula. The total current is equal to the sum of the branch currents. Thus,

$$\frac{E_s}{R_t} = \frac{E_s}{R_1} + \frac{E_s}{R_2} + \frac{E_s}{R_3}$$

or

$$\frac{E_s}{R_t} = E_s \left(\frac{1}{R_1} + \frac{1}{R_2} + \frac{1}{R_3} \right)$$

Both sides of this equation may be divided by E_s without changing the value of the equation; therefore,

$$\frac{1}{R_t} = \frac{1}{R_1} + \frac{1}{R_2} + \frac{1}{R_3}$$

The total resistance of the circuit shown in figure 20-8 may be determined by the preceding equation. Thus,

$$\frac{1}{R_t} = \frac{1}{5} + \frac{1}{10} + \frac{1}{30}$$

and

$$\frac{1}{R_t} = \frac{10}{30}$$

Taking the reciprocals of both sides,

$$R_t = \frac{30}{10} = 3 \ ohms$$

A useful rule to remember in computing the equivalent resistance of a dc parallel circuit is that the total resistance is always less than the smallest resistance in any of the branches.

In addition to adding the individual branch currents to obtain the total current in a parallel circuit, the total current may be found directly by dividing the applied voltage by the equivalent resistance, R_t. For example, in figure 20-8:

$$I_t = \frac{E_s}{R_t} = \frac{30}{3} = 10 \ amperes$$

Three or more resistors may be connected in series and parallel combinations to form a compound circuit. One basic series-parallel circuit composed of three resistors is shown in figure 20-9.

The total resistance, R_t, of figure 20-9 is determined in two steps. First, the resistance, $R_{2,3}$, of the parallel combination of R_2 and R_3 is determined as

$$R_{2,3} = \frac{R_2 R_3}{R_2 + R_3} = \frac{3 \times 6}{3 + 6} = \frac{18}{9} = 2 \ ohms$$

The sum of $R_{2,3}$ and R_1 (that is, R_t) is

$$R_t = R_{2,3} + R_1 = 2 + 2 = 4 \ ohms$$

If the total resistance, R_t, and the source voltage, E_s, are known, the total current, I_t, may be determined by Ohm's law. Thus, in figure 20-9,

$$E_{ab} = I_t R_1 = 5 \times 2 = 10 \ volts$$

and

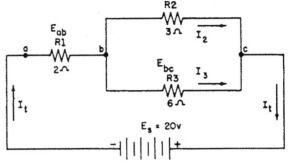

R1 IN SERIES WITH PARALLEL COMBINATION OF R2 AND R3

Figure 20-9.–Compound electric circuit.

$$E_{bc} = I_t R_{2,3} = 5 \times 2 = 10 \ volts$$

According to Kirchhoff's voltage law, the sum of the voltage drops around the closed circuit is equal to the source voltage. Thus,

$$E_{ab} + E_{bc} = E_s$$

or

$$10 + 10 = 20 \ volts$$

If the voltage drop, E_{bc}, across $R_{2,3}$–that is, the drop between points b and c–is known, the current through the individual branches may be determined as

$$I_2 = \frac{E_{bc}}{R_2} = \frac{10}{3} = 3.333 \ amperes$$

and

$$I_3 = \frac{E_{bc}}{R_3} = \frac{10}{6} = 1.666 \ amperes$$

According to Kirchhoff's current law, the sum of the currents flowing in the individual parallel branches is equal to the total current. Thus,

$$I_2 + I_3 = I_t$$

or

$$3.333 + 1.666 = 5 \ amperes \ (approx.)$$

The total current flows through R_1; at point b it divides between the two branches in inverse proportion to the resistance of the branches. Twice as much goes through R_2 as through R_3 because R_2 has one-half the resistance of R_3. Thus, 3.333 (or two-thirds of 5) amperes flow through R_2; and 1.666 (or one-third of 5) amperes flow through R_3.

WHEATSTONE BRIDGE

The Wheatstone bridge is widely used for precision measurements of resistance. The circuit diagram of a

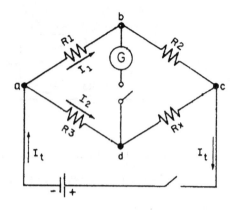

Figure 20-10.—Wheatstone bridge circuit diagram.

Wheatstone bridge is shown in figure 20-10. Resistors R1, R2, and R3 are precision variable resistors. The value of R_x is an unknown value of resistance that must be determined. After the bridge has been properly balanced (galvanometer [G] reads zero [0]), the unknown resistance may be determined by a simple formula. The galvanometer is inserted across terminals b and d to indicate the condition of balance. When the bridge is properly balanced, no difference in potential

exists across terminals b and d; when switch S2 is closed, the galvanometer reading is zero. Should the bridge become unbalanced due to a change in resistance of R_x, the difference of potential between terminals b and d will cause a deflection in the galvanometer.

When this type of circuit is used as a component of a resistance thermometer, R_x is the temperature-sensing element. The resistance of R_x varies directly with the temperature; a change in temperature results in an unbalanced bridge and a deflection of the galvanometer.

DIRECT-CURRENT GENERATORS

A dc generator is a rotating machine that converts mechanical energy into electrical energy using the principle of magnetic induction. This conversion is accomplished by rotating an armature, which carries conductors, in a magnetic field, inducing an emf in the conductors.

A dc generator (fig. 20-11) consists essentially of a steel frame or yoke containing the pole pieces and field windings; an armature consisting of a group of copper conductors mounted in a slotted cylindrical core; a

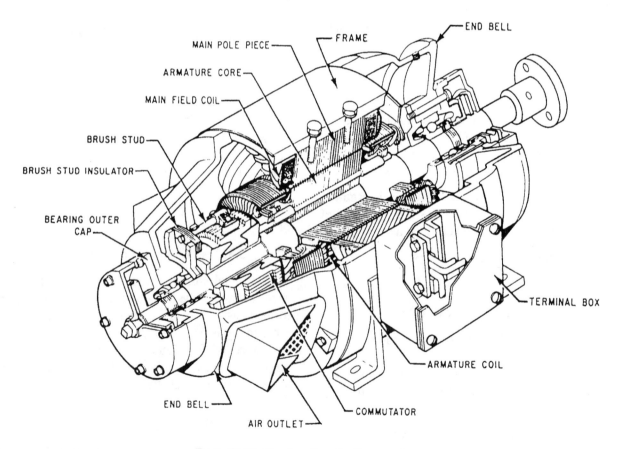

Figure 20-11.—Construction of a dc generator.

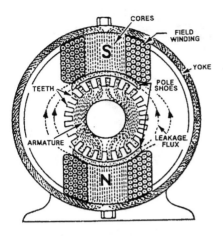

Figure 20-12.–Magnetic circuit of a 2-pole generator.

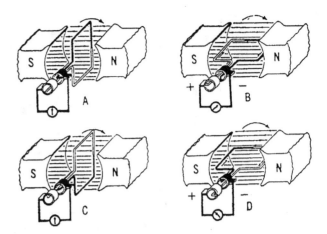

Figure 20-13.–Single-coil generator with commutator.

commutator for maintaining the current in one direction through the external circuit; and brushes with brush holders to carry the current from the commutator to the external load circuit.

The frame, in addition to providing mechanical support for the pole pieces, serves as a portion of the magnetic circuit in that it provides a path for the magnetic flux between the poles.

Figure 20-11 shows the entire generator with the components installed. The cutaway drawing helps you to see the physical relationship of the components to each other.

GENERATING A VOLTAGE

The field windings of a dc generator receive current either from an external dc source or directly across the armature, thus becoming electromagnets. They are connected so that they produce alternate north and south poles, and when energized, they establish magnetic flux in the field yoke, pole pieces, air gap, and armature core, as shown in figure 20-12.

The armature is mounted on a shaft and is rotated through the field by an outside energy source (prime mover). Thus, we have a magnetic field, a conductor, and relative motion between the two–which are the three essentials for producing a voltage by magnetism. If the output of the armature is connected across the field windings, the voltage and the field current at start will be small because of the small residual flux in the field poles. However, as the generator continues to run, the small voltage across the armature will circulate a small current through the field coils and the field will become stronger. In a self-excited generator, this action causes the generator voltage to rise quickly to the proper value and the machine is said to "build up" its voltage.

The simplest generator armature winding is a loop or single coil. Rotating this loop in a magnetic field will induce an emf whose strength is dependent upon the strength of the magnetic field and the speed of rotation of the conductor.

A single-coil generator with each coil terminal connected to a bar of a two-segment metal ring is shown in figure 20-13. The two segments of the split ring are insulated from each other and the shaft, thus forming a simple commutator that mechanically reverses the armature coil connections to the external circuit at the same instant that the direction of generated voltage reverses in the armature coil.

The emf developed across the brushes is pulsating and unidirectional. Figure 20-14 is a graph of the pulsating emf for one revolution of a single-loop armature in a 2-pole generator. A pulsating direct voltage of this characteristic (called ripple) is unsuitable for most applications. In practical generators, more coils and more commutator bars are used to produce an output

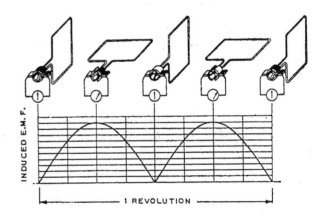

Figure 20-14.–Pulsating voltage from a single-coil armature.

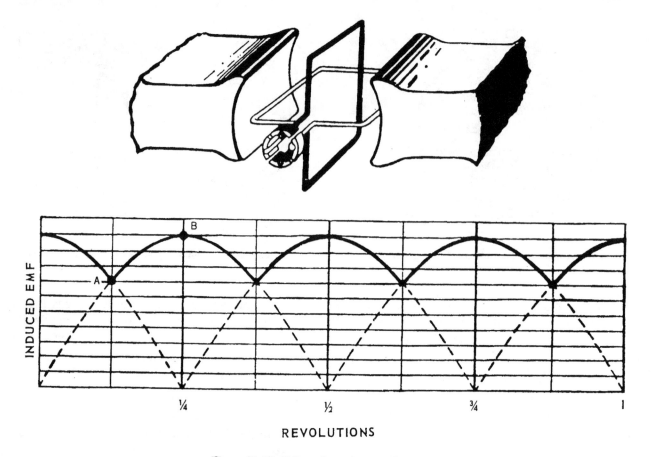

Figure 20-15.–Voltage from a two-coil armature.

voltage waveform with less ripple. Figure 20-15 shows the reduction in ripple obtained by the use of two coils instead of one. Since there are now four commutator segments and only two brushes, the voltage cannot fall any lower than point A; therefore, the ripple is limited by the rise and fall between points A and B. By adding still more armature coils, the ripple can be reduced still more.

TYPES OF DC GENERATORS

Dc generators are usually classified according to the manner in which the field windings are connected to the armature circuit (fig. 20-16).

A separately excited dc generator is indicated in view A of figure 20-16. In this machine, the field

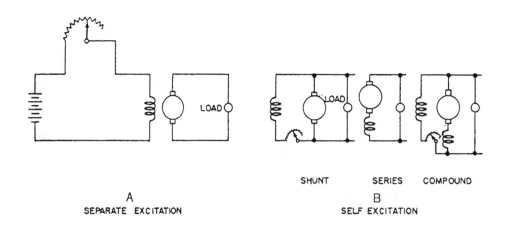

A
SEPARATE EXCITATION

SHUNT   SERIES   COMPOUND

B
SELF EXCITATION

Figure 20-16.–Types of dc generators.

windings are energized from a dc source other than its own armature.

Self-excited dc generators are classed according to the type of field connection they use. There are three general types of field connections–series wound, shunt-wound (parallel), and compound-wound (fig. 20-16, view B). A shunt-wound generator has its field windings connected in parallel with the armature; whereas, the field windings of a series-wound generator are connected in series with the armature. The compound-wound generator employs both shunt-wound and series-wound field windings.

The dc generator most widely used in the Navy is the stabilized shunt generator, which employs a light series field winding on the same poles with the shunt field windings. This type of generator has good voltage regulation characteristics and at the same time ensures good parallel operation.

VOLTAGE CONTROL

Voltage control is an imposed action, usually through an external adjustment, for increasing or decreasing terminal voltage. Voltage control is either manual or automatic. In most cases, the process involves changing the resistance of the field circuit. By changing the field circuit resistance, the field current is controlled. Controlling the field current permits control of the terminal voltage. The major difference between the various voltage regulator systems is merely the method by which the field circuit resistance is controlled.

DIRECT-CURRENT MOTORS

The construction of a dc motor is essentially the same as that of a dc generator. The dc generator converts mechanical energy into electrical energy and the dc motor converts the electrical energy into mechanical energy. A dc generator may be made to function as a motor by applying a suitable source of direct voltage across the normal output electrical terminals.

Dc motors are classed according to the way in which the field coils are connected. There are three general types of dc motors–shunt, series, and compound. Each type has characteristics that are advantageous under given load conditions.

Shunt motors have the field coils connected in parallel with the armature circuit. This type of motor, with constant potential applied, develops variable torque at an essentially constant speed, even under changing load conditions. Such loads are found in drives

for such machine shop equipment as lathes, milling machines, drills, planers, and shapers.

Series motors have the field coils connected in series with the armature circuit. This type of motor, with constant potential applied, develops variable torque, but its speed varies widely under changing load conditions. The speed of a series motor is low under heavy loads, but it becomes excessively high under light loads. Series motors are commonly used to drive electric cranes, hoists, and winches.

A compound motor has two field windings. One is a shunt field connected in parallel with the armature; the other is a series field that is connected in series with the armature. The shunt field gives this type of motor the constant speed advantage of a regular shunt motor. The series field gives it the advantage of being able to develop a large torque when the motor is started under a heavy load. The compound motor has both shunt- and series-motor characteristics. The compound motor develops an increased starting torque over the shunt motor and has less variation in speed than the series motor.

The operation of a dc motor is based on the principle that a current-carrying conductor placed in, and at right angles to, a magnetic field tends to move in a direction perpendicular to the magnetic lines of force.

There is a definite relationship between the direction of the magnetic field, the direction of current in the conductor, and the direction in which the conductor tends to move. This relationship is called the right-hand rule for motors (fig. 20-17).

To find the direction of motion of a conductor, extend the thumb, index finger, and middle finger of your right hand so they are at right angles to each other. If the forefinger is pointed in the direction of magnetic flux (north to south), and the middle finger is pointed in the direction of current flow in the conductor, then the thumb will point in the direction the conductor will move.

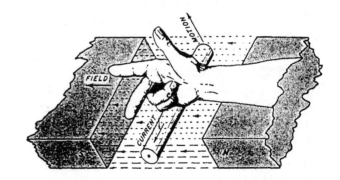

Figure 20-17.–Right-hand motor rule for electron flow.

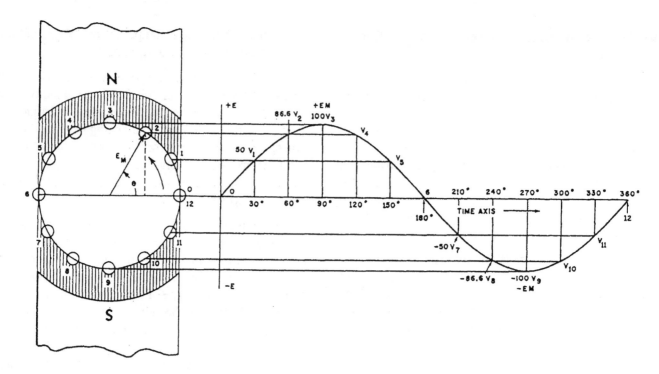

Figure 20-18.–Generation of sine-wave voltage.

ALTERNATING-CURRENT THEORY

Just as a current flowing in a conductor produces a magnetic field around the conductor, the reverse of this process is true. A voltage can be generated in a circuit by moving a conductor so it cuts across lines of magnetic force or, conversely, by moving the lines of force so they cut across the conductor. An ac generator uses this principle of electromagnetic induction to convert mechanical energy into electrical energy.

In alternating current, electrons move first in one direction and then in the other. The direction of the current reverses periodically and the magnitude of the voltage is constantly changing. This variation in current is represented graphically in sine-wave form in figure 20-18.

The vertical projection (dotted line in fig. 20-18) of a rotating vector may be used to represent the voltage at any instant. Vector E_M represents the maximum voltage induced in a conductor rotating at uniform speed in a 2-pole field (points 3 and 9). The vector is rotated counterclockwise through one complete revolution (360°). The point of the vector describes a circle. A line drawn from the point of the vector perpendicular to the horizontal diameter of the circle is the vertical projection of the vector.

The circle also describes the path of the conductor rotating in the bipolar field. The vertical projection of the vector represents the voltage generated in the conductor at any instant corresponding to the position of the rotating vector as indicated by angle θ. Angle θ represents selected instants at which the generated voltage is plotted. The sine curve plotted at the right of the figure represents successive values of the ac voltage induced in the conductor as it moves at uniform speed through the 2-pole field, because the instantaneous values of rotationally induced voltage are proportional to the sine of angle θ that the rotating vector makes with the horizontal.

The sine wave in figure 20-18 represents one complete revolution of the armature or one voltage cycle. The frequency of ac voltage is measured in cycles per second (cps) and may be determined by the following formula:

$$f = \frac{P \times rpm}{120}$$

where

f = frequency, in cps (according to the National Bureau of Standards Special Publication 304, frequency in cycles per second in the International Systems of Units is expressed as Hertz [Hz]. One hertz equals 1 cycle per second.)

rpm = revolutions per minute

P = number of poles in the generator

A generator made to deliver 60 cps, and having two field poles, would need an armature designed to rotate at 3,600 rpm.

PROPERTIES OF AC CIRCUITS

Resistance, the opposition to current flow, has the same effect in an ac circuit as it does in a dc circuit. However, in the application of Ohm's law to ac circuits, other properties must be taken into consideration. Inductance is that property of an electrical circuit that opposes any change in the current flow and capacitance is that property that opposes any change in voltage. Since ac current is constantly changing in magnitude and direction, the properties of inductance and capacitance are always present. The amount of opposition to current flow in an inductive circuit is called its inductive reactance. Inductive reactance is measured in ohms and its symbol is X_L. The value of inductive reactance depends on the inductance of the circuit and the frequency of the applied voltage. The formula for inductive reactance is as follows:

$$X_L = 2\pi fL$$

where:

X_L = inductive reactance, in ohms

π = 3.1416

f = frequency of the alternating current, in Hz

L = inductance, in henrys

The current flowing in a capacitive circuit is directly proportional to the capacitance and to the rate at which the applied voltage is changing. The rate at which the voltage changes is determined by the frequency. The opposition that a capacitor offers to ac is inversely proportional to frequency and to capacitance. This opposition is called capacitive reactance. The symbol for capacitive reactance is XC. The value of the capacitive reactance is inversely proportional to the capacitance of the circuit and the frequency of the applied voltage. The formula for capacitive reactance is as follows:

$$X_C = \frac{1}{2\pi fC}$$

where:

X_C = capacitive reactance, in ohms

π = 3.1416

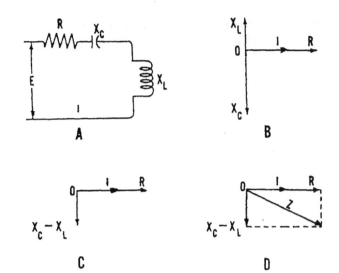

Figure 20-19.–Vector solution of an ac circuit.

f = frequency, in Hz

C = capacitance, in farads

The effects of capacitance and inductance in an ac circuit are exactly opposite. Inductive reactance causes the current to lag the applied voltage, and capacitive reactance causes the current to lead the applied voltage. These effects tend to neutralize each other, and the combined reactance is the difference between the individual reactances.

The total opposition offered to the flow of current in an ac circuit is the impedance, Z. The impedance of a circuit, expressed in ohms, is composed of the capacitive reactance, the inductive reactance, and the resistance.

The effects of capacitive reactance, inductive reactance, and resistance in an ac circuit can be shown graphically with vectors. For example, consider the series circuit shown in figure 20-19, view A. The vector representation of the reactances is shown in figure 20-19, view B. Because the inductive reactance and the capacitive reactance are exactly opposite, they are subtracted directly and the difference is shown in figure 20-18, view C, as capacitive reactance. The resultant is found vectorially by constructing a parallelogram, as shown in figure 20-19, view D. The resultant vector is also the hypotenuse of a right triangle; therefore,

$$Z = \sqrt{R^2 + (X_C - X_L)^2}$$

According to Ohm's law for ac circuits, the effective current through a circuit is directly proportional to the effective voltage and inversely proportional to the impedance. Thus,

$$I = \frac{E}{Z}$$

where:

I = current, in amperes

E = emf, in volts

Z = impedance, in ohms

ALTERNATING-CURRENT GENERATORS

Most electric power used today is generated by ac generators. The ac generators are also used in aircraft and automobiles.

The ac generators come in many different sizes, depending on their intended use. For example, any one of the huge generators at Boulder Dam can produce millions of volt-amperes, while the relatively small generators used on aircraft produce only a few thousand volt-amperes.

Regardless of their size, all generators operate on the same basic principle–a magnetic field cutting through conductors, or conductors passing through a magnetic field. All generators have at least two distinct sets of conductors. They are (1) a group of conductors in which the output voltage is generated and (2) a second group of conductors through which direct current is passed to obtain an electromagnetic field of fixed polarity. The conductors that generate the output voltage are always referred to as the armature windings. The conductors that originate the electromagnetic field are always referred to as the field windings.

Since motion is required between the armature and field, ac generators are built in two major assemblies–the stator and the rotor. The rotor rotates inside the stator. It may be driven by any one of a number of commonly used power sources, such as gas or steam turbines, electric motors, and internal-combustion engines.

There are various types of ac generators used today. They all perform the same basic function. The types that we will discuss in the following paragraphs are typical of the most common ones encountered in shipboard electrical systems.

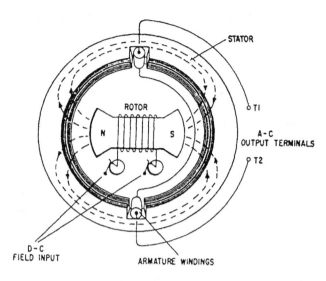

Figure 20-20.–Revolving-field ac generator.

REVOLVING ARMATURE

In the revolving-armature ac generator, the stator provides a stationary electromagnetic field. The rotor, acting as the armature, revolves in the field, cutting the lines of force, thereby producing the desired output voltage. In this generator, the armature output is taken from slip rings, retaining its alternating characteristic.

The use of the revolving-armature ac generator is limited to low-power, low-voltage applications. The primary reason for this limitation is its output power is conducted through sliding contacts (slip rings and brushes). These contacts are subject to frictional wear and sparking. In addition, they are exposed and liable to arc-over at high voltages.

REVOLVING FIELD

The revolving-field ac generator (fig. 20-20) is the most widely used type. In this type of generator, where brushes are installed, direct current from a separate source is passed through windings on the rotor by slip rings and brushes. This maintains a rotating electromagnetic field of fixed polarity (similar to a rotating bar magnet). The rotating magnetic field produced by the rotor extends outward and cuts through the armature windings imbedded in the surrounding stator. As the rotor turns, alternating voltages are induced in the windings since magnetic fields of first one polarity and then the other cut through them. Since the output power is taken from stationary windings, the output may be connected through fixed terminals (T1 and T2 in fig. 20-20). This is advantageous because there are no sliding contacts, and the whole output circuit is continuously insulated, thus minimizing the danger of arc-over.

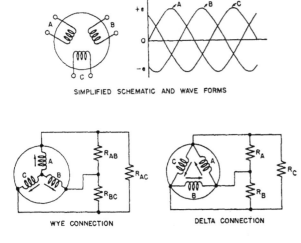

SIMPLIFIED SCHEMATIC AND WAVE FORMS

WYE CONNECTION   DELTA CONNECTION

Figure 20-21.–Three-phase ac generator.

Slip rings and brushes are adequate for the dc field supply because the power level in the field is much smaller than in the armature circuit.

THREE-PHASE GENERATORS

A three-phase ac generator, as the name implies, has three single-phase windings spaced so the voltage induced in each winding is 120° out of phase with the voltages in the other two windings. A schematic diagram of a three-phase stator showing all the coils becomes complex, and it is difficult to see what is actually happening. A simplified schematic diagram showing all the windings of a single phase lumped together as one winding is shown in figure 20-21, view A. The rotor is omitted for simplicity. The waveforms of voltage are shown to the right of the schematic. The three voltages are 120° apart and are similar to the voltages that would be generated by three single-phase ac generators whose voltages are out of phase by angles of 120°. The three phases are independent of each other.

Wye Connection

Rather than have six leads come out of the three-phase ac generator, one of the leads from each phase may be connected to form a common junction. The stator is then said to be wye, or star, connected. The common lead may or may not be brought out of the machine. If it is brought out, it is called the neutral. The simplified schematic (fig. 20-21, view B) shows a wye-connected stator with the common lead not brought out. Each load is connected across two phases in series. R_{AB} is connected across phases A and B in series; R_{AC}

is connected across phases A and C in series; and R_{BC} is connected across phases B and C in series. Thus, the voltage across each load is larger than the voltage across a single phase. In a wye-connected ac generator, the three start ends of each single-phase winding are connected to a common neutral point and the opposite, or finish, ends are connected to the line terminals, A, B, and C. These letters are always used to designate the three phases of a three-phase system, or the three line wires to which the ac generator phases connect. When unbalanced loads are used, a neutral may be added, as shown in the figure by the broken line between the common neutral point and the loads. The neutral wire serves as a common return circuit for all three phases and maintains a voltage balance across the loads. No current flows in the neutral wire when the loads are balanced.

Delta Connection

A three-phase stator may also be connected, as shown in figure 20-21, view C. This is called the delta connection. In a delta-connected ac generator, the start end of one phase winding is connected to the finish end of the third; the start of the third phase winding is connected to the finish of the second phase winding; and the start of the second phase winding is connected to the finish of the first phase winding. The three junction points are connected to the line wires leading to the load. The generator is connected to a three-phase, three-wire circuit, which supplies a three-phase, delta-connected load at the right-hand end of the three-phase line. Because the phases are connected directly across the line wires, phase voltage is equal to line voltage. When the generator phases are properly connected in delta, no appreciable current flows within the delta loop when there is no external load connected to the generator. If any of the phases is reversed with respect to its correct connection, a short-circuit current flows within the windings of no load, causing damage to the windings.

VOLTAGE REGULATION

When the load on an ac generator is changed, the terminal voltage varies with the load. The amount of variation depends on the design of the generator and on the amount of reactance from the inductive or capacitive loads. Under practical shipboard operating conditions, the load varies widely with the starting and stopping of motors.

The only practical way to regulate the voltage output of an ac generator is to control the strength of the

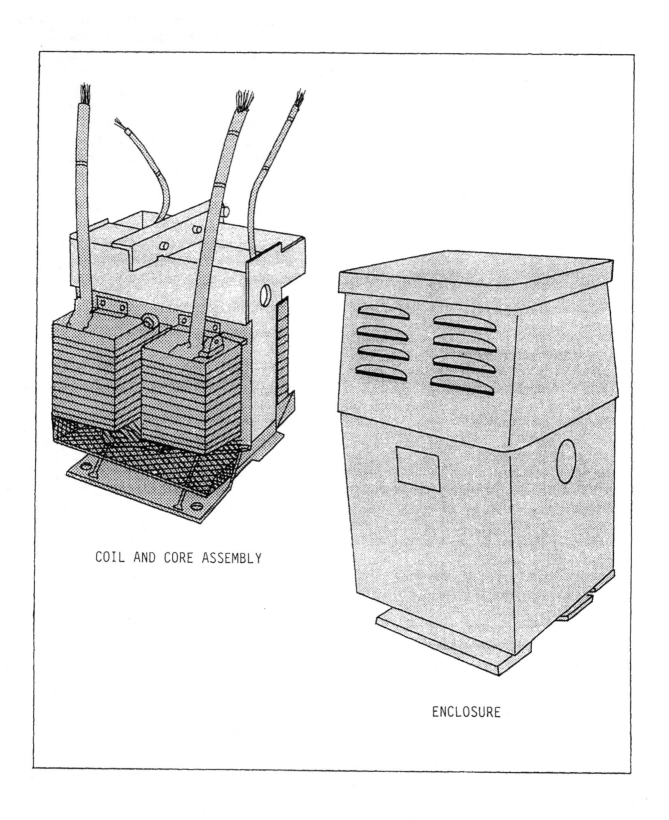

COIL AND CORE ASSEMBLY

ENCLOSURE

Figure 20-22.–A single-phase transformer.

rotating magnetic field. The strength of the electromagnetic field may be varied by changing the amount of current flowing through the coil, which is done by connecting a rheostat in series with the coil. Thus, voltage regulation in an ac generator is accomplished by varying the field current. This allows a relatively large ac voltage to be controlled by a much smaller dc voltage and current.

Since manual adjustment of ac voltage is not practical when the load fluctuates rapidly, automatic voltage regulators are used. The construction and operating principles of voltage regulators varies; however, the essential function of any voltage regulator is to use the ac output voltage, which the regulator is designed to control, as a sensing influence to control the amount of current the exciter supplies to its own control field.

TRANSFORMERS

A transformer (fig. 20-22) is a device that has no moving parts and that transfers energy from one circuit to another by electromagnetic induction. The energy is always transferred without a change in frequency, but usually with changes in voltage and current. A step-up transformer receives electrical energy at one voltage and delivers it at a higher voltage. Conversely, a step-down transformer receives energy at one voltage and delivers it at a lower voltage. Transformers require little care and maintenance because of their simple, rugged, and durable construction. The efficiency of transformers is high. Because of this, transformers are responsible for the more extensive use of alternating current than direct current. The conventional constant-potential transformer is designed to operate with the primary connected across a constant-potential source and to provide a secondary voltage that is substantially constant from no load to full load.

Various types of small, single-phase transformers are used in electrical equipment. In many installations, transformers are used on switchboards to step down the voltage for indicating lights. Low-voltage transformers are included in some motor control panels to supply control circuits or to operate overload relays.

Instrument transformers include potential, or voltage, transformers and current transformers. Instrument transformers are commonly used with ac instruments when high voltages or large currents are to be measured.

Electronic circuits and devices employ many types of transformers to provide the necessary voltages for proper electron-tube operation, interstage coupling, signal amplification, and so forth. The physical construction of these transformers differs widely.

Power-supply transformers, used in electronic circuits, are single-phase, constant-potential transformers with either one or more secondary windings, or a single secondary with several tap connections. These transformers have a low volt-ampere capacity and are less efficient than large constant-potential power transformers. Most power-supply transformers for electronic equipment are designed to operate at a frequency of 50 to 60 Hz. Aircraft power-supply transformers are designed for a frequency of 400 Hz. The higher frequencies permit a savings in size and weight of transformers and associated equipment.

CONSTRUCTION

The typical transformer has two windings insulated electrically from each other. These windings are wound on a common magnetic core made of laminated sheet steel. The principal parts of a transformer are (1) the core, which provides a circuit of low reluctance for the magnetic flux; (2) the primary winding, which receives the energy from the ac source; (3) the secondary winding, which receives the energy by mutual induction from the primary and delivers it to the load; and (4) the enclosure.

When a transformer is used to step up the voltage, the low-voltage winding is the primary. Conversely, when a transformer is used to step down the voltage, the high-voltage winding is the primary. The primary is always connected to the source of the power; the secondary is always connected to the load. It is common practice to refer to the windings as the primary and secondary rather than the high-voltage and low-voltage windings.

VOLTAGE AND CURRENT RELATIONS

The operation of the transformer is based on the principle that electrical energy can be transferred efficiently by mutual induction from one winding to another. When the primary winding is energized from an ac source, an alternating magnetic flux is established in the transformer core. This flux links the turns of both primary and secondary, thereby inducing voltages in them. Because the same flux cuts both windings, the same voltage is induced in each turn of both windings.

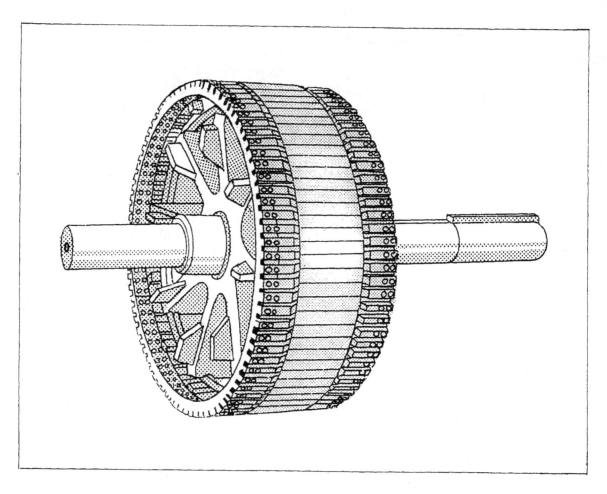

Figure 20-23.–Cage-type induction motor rotor.

Hence, the total induced voltage in each winding is proportional to the number of turns in that winding; that is,

$$\frac{E_1}{E_2} = \frac{N_1}{N_2}$$

where E_1 and E_2 are the induced voltage in the primary and secondary windings, and N_1 and N_2 are the number of turns in the primary and secondary windings. In ordinary transformers, the induced primary voltage is almost equal to the applied primary voltage; hence, the applied primary voltage and the secondary induced voltage are approximately proportional to the respective number of turns in the two windings.

ALTERNATING-CURRENT MOTORS

An ac motor is particularly well suited for constant speed applications. This is because its speed is determined by the frequency of the ac voltage applied to the motor terminals. Ac motors are manufactured in many different sizes, shapes, and ratings for use in a wide variety of applications. These motors are designed for use with either polyphase or single-phase power systems. Since this discussion cannot possibly cover all aspects of all kinds of ac motors, only the principles of polyphase induction motor will be covered. Information on other types of motors may be found in the Navy Electricity and Electronics Training Series (NEETS), module 5, and in various manufacturers' technical manuals.

The induction motor is the most commonly used ac motor. It is simple and rugged in construction and costs very little to manufacture. It consists essentially of a stator and a rotor; it can be designed to suit most applications requiring constant speed and variable torque. Examples are found in washing machines, refrigeration compressors, benchgrinders, and table saws.

The stator of a polyphase induction motor consists of a laminated steel ring with slots on the inside circumference. The stator winding is similar to the ac generator stator winding and is generally of the two-layer distributed preformed type. Stator phase windings are symmetrically placed on the stator and may be either wye connected or delta connected.

Most induction motors used by the Navy have a cage-type rotor consisting of a laminated cylindrical core with parallel slots in the outside circumference to hold the windings in place. The rotor winding is constructed of individual short circuited bars connected to end rings (fig. 20-23).

In induction motors, the rotor currents are supplied by electromagnetic induction. The stator windings contain two or more out-of-time-phase currents that produce corresponding magnetomotive forces that establish a rotating magnetic field across the air gap. This magnetic field rotates continuously at constant speed, regardless of the load on the motor. The stator winding corresponds to the primary winding of a transformer. The induction motor derives its name from the mutual induction (transformer action) that takes place between the stator and the rotor under operating conditions. The magnetic revolving field produced by the stator cuts across the rotor conductors, thus inducing a voltage in the conductors, which causes rotor current to flow. Hence, motor torque is developed by the interaction of the rotor current and the magnetic revolving field.

AC POWER DISTRIBUTION SYSTEM

The ac power distribution system aboard a ship consists of the ac power plant, switchboards that distribute the power, and the equipment that consumes the power. The power plant is either the ship's service electric plant or the emergency electric plant. The power distribution system is made up of the ship's service power distribution system, the emergency power distribution system, and the casualty power distribution system.

Most ac power distribution systems in naval ships are 450-volt, three-phase, 60-Hz, three-wire systems. The lighting distribution systems are 115-volt, three-phase, 60-Hz, three-wire systems supplied from the power circuits through transformer banks. On some ships, the weapons systems, some I.C. circuits, and aircraft starting circuits receive electrical power from a 400-cps system.

SHIP'S SERVICE POWER

The ship's service power distribution system is the electrical system that normally supplies electric power to the ship's equipment and machinery. The switchboards and associated generators are located in separate engineering spaces to minimize the possibility that a single hit will damage more than one switchboard.

The ship's service generators and distribution switchboards are interconnected by bus ties so any switchboard can be connected to feed power from the generators to one or more of the other switchboards. The bus ties also connect two or more switchboards so the generator plants can be operated in parallel (or the switchboards can be isolated for split-plant operation). In large installations, power distribution to loads is from the generator and distribution switchboards or switchgear groups to the load centers, to distribution panels, and to the loads. Distribution may also be direct from the load centers to some loads.

On some ships, such as large aircraft carriers, a system of zone control of the ship's service and emergency distribution is provided. Essentially, the system establishes a number of vertical zones that contain one or more load center switchboards supplied through bus feeders from the ship's service switchgear group. A load center switchboard supplies power to the electrical loads within the electrical zone in which it is located. Thus, zone control is provided for all power within the electrical zone. An emergency switchboard may supply more than one zone, depending on the number of emergency generators installed. Figure 20-24 shows the ship's service and emergency power distribution system in a large aircraft carrier.

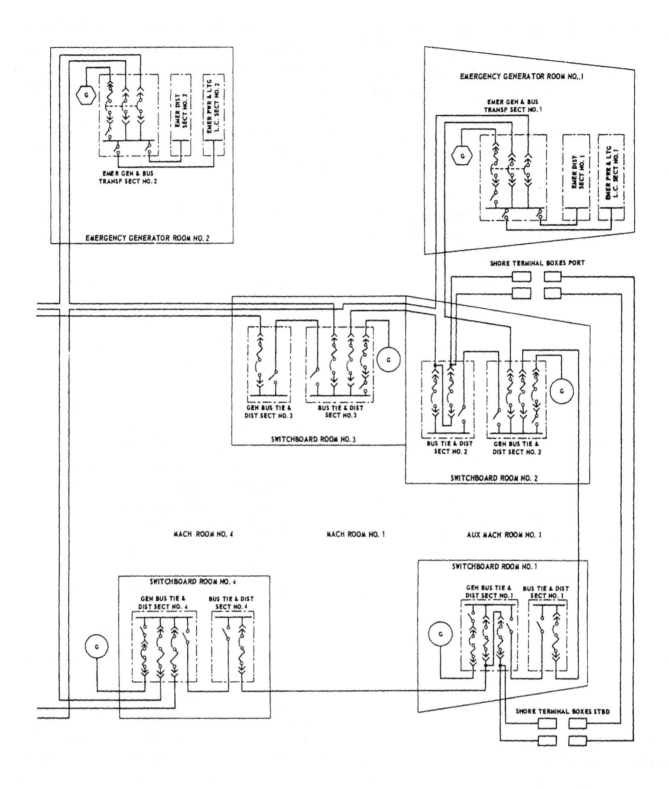

Figure 20-24.–Ship's service and emergency power distribution system in a large carrier.

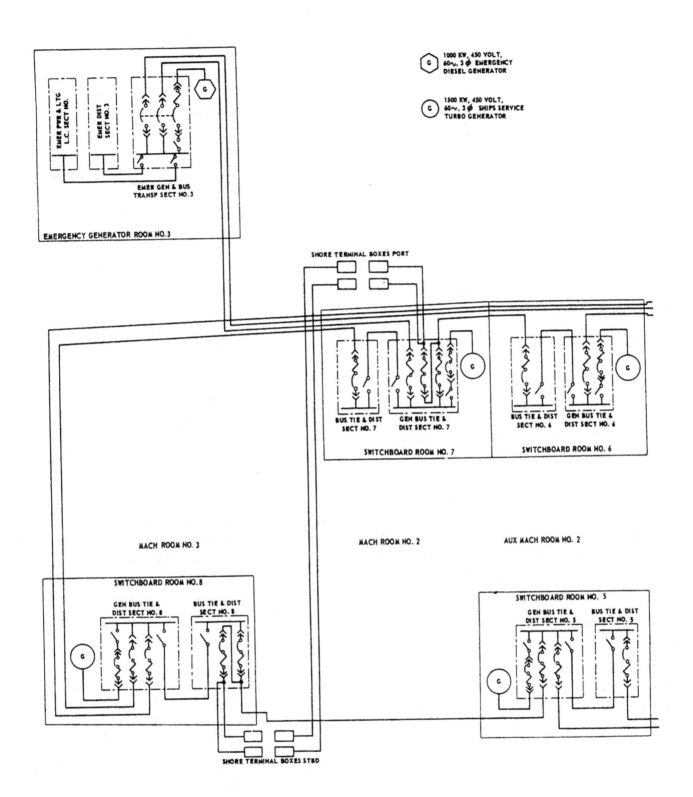

Figure 20-24.–Ship's service and emergency power distribution system in a large carrier–Continued.

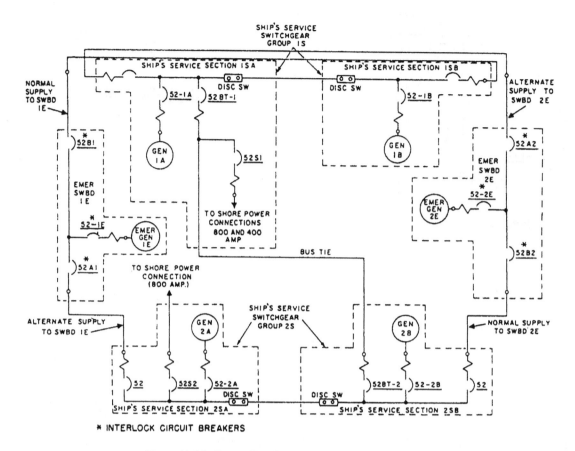

Figure 20-25.--Power distribution system on the CG-23.

In small installations, the distribution panels are fed directly from the generator and distribution switchboards. The distribution panels and load centers, if installed, are located centrally with respect to the loads they feed. This arrangement simplifies the installation and requires less weight, space, and equipment than if each load were connected to a switchboard. Figure 20-25 shows the power distribution system on the CG-23.

At least two independent sources of power are provided for selected vital loads. The distribution of this dual supply is accomplished by (1) a normal and an alternate ship's service feeder; (2) a normal ship's service feeder and an emergency feeder; or (3) a normal and an alternate ship's service feeder and an emergency feeder.

The normal and alternate feeders to a common load run from different ship's service switchboards and are located below the waterline on opposite sides of the ship to minimize the possibility that both will be damaged by a single hit. The lighting circuits are supplied from the secondaries of 450/115-volt transformer banks connected to the ship's service power system. In large ships, the transformer banks are installed near the lighting distribution panels, at some distance from the

generator and distribution switchboards. In small ships, the transformer banks are located near the generator and distribution switchboards and energize the switchboard buses that supply the lighting circuits.

EMERGENCY POWER DISTRIBUTION SYSTEMS

The emergency power distribution system supplies an immediate and automatic source of electric power to a limited number of selected vital loads in the event of failure of the normal ship's service power distribution system. The system, which is separate and distinct from the ship's service power distribution system, includes one or more emergency distribution switchboards. Each emergency switchboard is supplied by its associated emergency generator. The emergency feeders run from the emergency switchboard and terminate in manual or automatic bus transfer equipment at the distribution panels or loads for which emergency power is provided. The emergency power distribution system is a 450-volt, three-phase, 60-Hz system with transformer banks at the emergency distribution switchboards to provide 120-volt, three-phase power for the emergency lighting system.

The emergency generators and switchboards are located in separate spaces from those containing the ship's service generators and distribution switchboards. The emergency feeders are located near the centerline and higher in the ship (above the water line) than the normal and alternate ship's service feeders. This arrangement provides for horizontal separation between the normal and alternate ship's service feeders and vertical separation between these feeders and the emergency feeders, thereby minimizing the possibility of damaging all three types of feeders simultaneously.

The emergency switchboard is connected by feeders to at least one and usually to two different ship's service switchboards. One of these switchboards is the preferred source of ship's service power for the emergency switchboard, and the other is the alternate source. The emergency switchboard and distribution system are normally energized from the preferred source of ship's service power. If this source of power should fail, bus transfer equipment automatically transfers the emergency switchboard to the alternate source of the ship's service power. If both the preferred and alternate sources of the ship's service power fails, the prime mover of the emergency generator starts automatically (within 10 seconds after power failure), and the emergency switchboard is automatically transferred to the emergency generator.

When the voltage is restored on either the preferred or alternate source of the ship's service power, the emergency switchboard is automatically retransferred to the source that is available, or to the preferred source if voltage is restored on both the preferred and alternate sources. However, this is only on certain types of ships. Some systems require the emergency generator to be tripped manually before the ABT will transfer back to the available preferred or alternate source. The emergency generator must be manually shut down. The emergency switchboard and distribution system are always energized by a ship's service generator or by the emergency generator. Therefore, the emergency distribution system can always supply power to a vital load if both the normal and alternate sources of the ship's service power to this load fails. The emergency generator will not start automatically if the emergency switchboard is receiving power from a ship's service generator.

A feedback tie from the emergency switchboard to the ship's service switchboard (fig. 20-25) is provided on most ships. The feedback tie permits a selected portion of the ship's service switchboard load to be supplied from the emergency generator. This feature facilitates starting up the machinery after major alterations and repairs and provides power to operate necessary auxiliaries and lighting during repair periods when shore power and ship's service power are not available.

CASUALTY POWER DISTRIBUTION SYSTEMS

Damage to ship's service and emergency distribution systems in wartime led to the development of the casualty power system. This system provides the means for making temporary connections to vital circuits and equipment. The casualty power distribution system is limited to those facilities that are necessary to keep the ship afloat and permit it to get out of the danger area. It also provides a limited amount of armament, such as weapons systems and their directors, to protect the ship when in a damaged condition.

Optimum continuity of service is ensured in ships provided with ship's service, emergency, and casualty power distribution systems. If one generating plant should fail, a remote switchboard can be connected by the bus tie to supply power from the generator or generators that have not failed.

If a circuit or switchboard fails, the vital loads can be transferred to an alternate feeder and source of ship's service power by a transfer switch near the load.

If both the normal and alternate sources of the ship's service power fail because of a generator, switchboard, or feeder casualty, the vital auxiliaries can be shifted to an emergency feeder that receives power from the emergency switchboard.

If the ship's service and emergency circuits fail, temporary circuits can be rigged with the casualty power distribution system and used to supply power to vital auxiliaries if any of the ship's service or emergency generators can be operated.

The casualty power system includes suitable lengths of portable cable stowed on racks throughout the ship. Permanently installed casualty power bulkhead terminals form an important part of the casualty power system. They are used for connecting the portable cables on opposite sides of bulkheads so power may be transmitted through compartments without loss of watertight integrity; also included are permanently installed riser terminals between decks. The vital equipment selected to receive casualty power will have a terminal box mounted on or near the equipment or

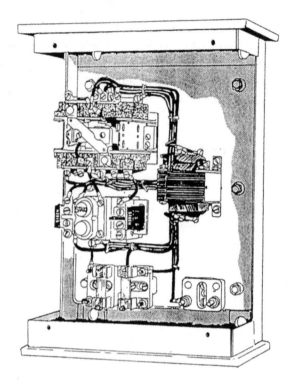

Figure 20-26.–Across-line, three-phase controller.

panel concerned and connected in parallel with the normal feeder for the equipment.

Sources of supply for the casualty power system are provided at each ship's service and emergency generator switchboard. A casualty power riser terminal is installed on the back of the switchboard or switchgear group and connected to the busses through a 225- or 250-ampere AQB circuit breaker. This circuit breaker is connected between the generator circuit breaker and the generator disconnect links. By opening the disconnect links, you will isolate the generator from the switchboard. Then, it can be used exclusively for casualty power purposes.

CONTROL AND SAFETY DEVICES

The distribution of electric power requires the use of many devices to control the current and to protect the circuits and equipment. Control devices are those electrical accessories that govern, in some predetermined way, the power delivered to any electrical load. In its simplest form, a control device applies voltage to or removes it from a single load. In more complex control systems, the initial switch may set into action other control devices that govern the motor speeds, the compartment temperatures, the depth of liquid in a tank, the aiming and firing of guns, or the direction of guided missiles. In fact, all electrical systems and equipment are controlled in some manner by one or more controllers.

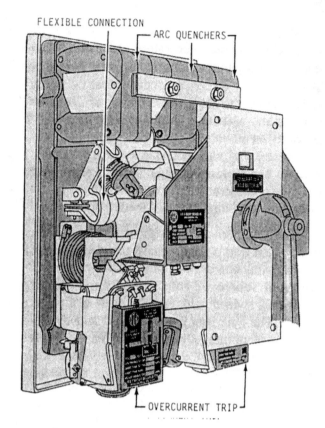

Figure 20-27.–Type ACB circuit breaker.

Switchboards make use of hand-operated (manual) switches as well as electrically operated controls. Manually operated switches are those familiar electrical items that can be operated by motions of the hand, such as with a pushing, pulling, or twisting motion. The type of action required to operate the manually operated switch is indicated by the names of the controls–push-button switch, pull-chain switch, or rotary switch.

Automatic switches are devices that perform their function of control through the repeated closing and opening of their contacts, without requiring a human operator. Limit switches and float switches are representative automatic switches.

The Navy uses many different types of switches and controllers, which range from the very simple to the very complex. A typical ac across-the-line magnetic controller is shown in figure 20-26.

The simplest protective device is a fuse, consisting of a metal alloy strip or wire and terminals for electrically connecting the fuse into the circuit. The most important characteristic of a fuse is its current-versus-time or "blowing" ability. Three time ranges for existence of overloads can be broadly defined as fast (5 microseconds through 1/2 second), medium (1/2 second to 5 seconds), and delayed (5 to 25 seconds).

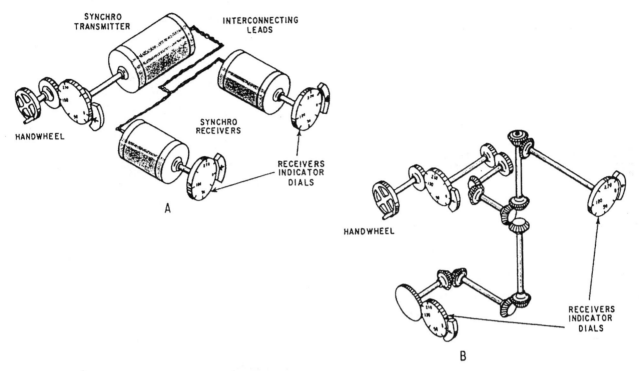

Figure 20-28.–Simple synchro system.

Normally, when a circuit is overloaded or when a fault develops, the fuse element melts and opens the circuit that it is protecting. However, all fuse openings are not the result of current overload or circuit faults. Abnormal production of heat, aging of the fuse element, poor contact due to loose connections, oxides or other corrosion products forming within the fuse holder, and unusually high ambient temperatures will alter the heating conditions and the time required for the element to melt.

A more complex type of protective device is the circuit breaker. In addition to acting as protective devices, circuit breakers perform the function of normal switching and are used to isolate a defective circuit while repairs are being made.

Circuit breakers are available in manually or electrically operated types. Some types may be operated both ways, while others are restricted to one mode. Figure 20-27 shows a circuit breaker that may be operated either manually or electrically. When the breaker is operated electrically, the operation is usually in conjunction with a pilot device such as relay or switch. Electrically operated circuit breakers employ an electromagnet, used as a solenoid, to trip a release mechanism that causes the breaker contacts to open. The energy to open the breaker is derived from a coiled spring, and the electromagnet is controlled by the contacts in a pilot device.

Circuit breakers designed for high currents have a double-contact arrangement, consisting of the main bridging contacts and the arcing contacts. When the circuit opens, the main contacts open first, allowing the current to flow through the arc contacts and thus preventing burning of the main contacts. When the arc contacts are open, they pass under the front of the arc runner, causing a magnetic field to be set up that blows the arc up into the arc quencher and quickly opens the circuit.

SYNCHROS AND SERVOMECHANISMS

Synchros, as identified by the Armed Forces, are ac electromagnetic devices that are used primarily for the rapid and accurate transmission of information between equipment and stations. Synchros are, in effect, single-phase transformers in which the primary-to-secondary coupling may be varied by physically changing the relative orientation of these two windings.

Synchro systems are used throughout the Navy to provide a means of transmitting the position of a remotely located device to one or more indicators located away from the transmitting area.

View A of figure 20-28 shows a simple synchro system that can be used to transmit different types of data. When the handwheel is turned, an electrical signal

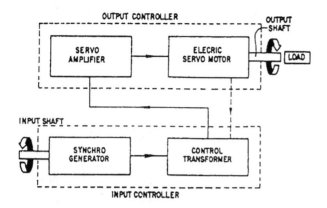

Figure 20-29.–Simplified block diagram of a servomechanism.

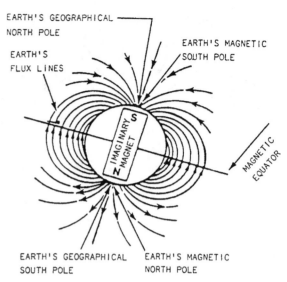

Figure 20-30.–Earth's magnetic field.

is generated by the synchro transmitter and is transmitted through interconnecting leads to the synchro receivers. The synchro receivers will always turn the same amount and direction and at the same speed as the synchro transmitter.

View B figure 20-28 shows the same type of system using mechanical linkage. As may be readily seen, mechanical systems are impracticable because of the need for associated belts, pulleys, gears, and rotating shafts.

Synchro systems are widely used for input control of electromechanical devices (servomechanisms) that position an object according to a variable signal. The essential components of a servomechanism system are the input controller and the output controller.

The input controller provides the means, either mechanical or electrical, where the human operator may actuate or operate a remotely located load.

The output controller of a servomechanism system is the component (or components) in which power amplification and conversion occur. This power is usually amplified by vacuum tube or magnetic amplifiers and then converted by the servomotor into mechanical motion of the direction required to produce the desired function.

Figure 20-29 shows a simplified block diagram of a servomechanism. When the shaft of the input controller is rotated in either direction, a voltage is induced in the rotor of the control transformer. This voltage is fed to the amplifier, where it is sent through the necessary stages of amplification and drives the servomotor.

DEGAUSSING

A steel-hulled ship is like a huge floating magnet with a large magnetic field surrounding it. As the ship moves through the water, this field also moves and adds to or subtracts from the earth's magnetic field. Because of its magnetic field, the ship can act as a triggering device for magnetic sensitive ordnance or devices. Degaussing is the method used to reduce a ship's magnetic field to minimize the distortion of the earth's magnetic field. This, in turn, reduces the possibility of detection by these magnetic sensitive ordnances or devices.

EARTH'S MAGNETIC FIELD

The magnetic field of the earth is larger than the magnetic field of a ship. The earth's magnetic field acts upon all metal objects on or near the earth's surface.

Figure 20-30 shows the earth as a huge permanent magnet, 6,000 miles long, extending from the Arctic to the Antarctic polar region. Lines of force from this magnet extend all over the earth's surface, interacting with all ferrous materials on or near the surface. Since many of these ferrous materials themselves become magnetized, they distort the background field into areas of increased or decreased magnetic strength. Thus, the lines of magnetic force at the earth's surface do not run in straight, converging lines like the meridians on a globe, but appear more like the isobar lines on a weather map.

By convention, the positive external direction of the magnetic field of a bar magnet is from the north pole to the south pole. Lines of force for the earth's field,

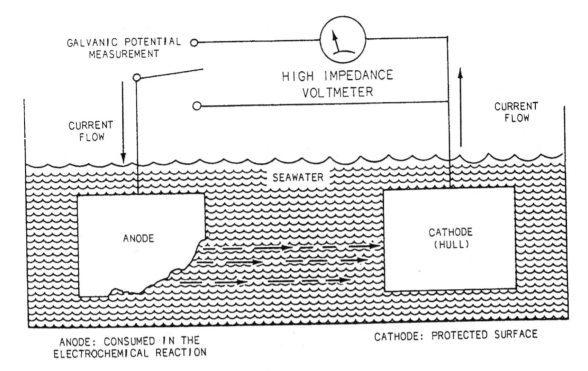

Figure 20-31.–Dry-cell battery circuit.

however, leave the earth in the Southern Hemisphere and reenter in the Northern Hemisphere. For this reason, think of the polar region in the Arctic as the north-geographic, south-magnetic pole. Note in figure 20-30 that the magnetic lines of force form closed loops, arching from the earth's magnetic core to outer space, and then reentering the earth in the opposite hemisphere. Since all lines of magnetic force return to their points of origin, they form closed magnetic circuits. It is impossible to eliminate the earth's field; however, the effect a ship has on the earth's magnetic field may be lessened. If a ship is close to a magnetic mine or torpedo, the distortion caused by the ship's field will activate the firing mechanism to detonate the mine or torpedo. The purpose of degaussing is to prevent the ship from distorting the earth's magnetic field.

SHIPBOARD DEGAUSSING INSTALLATION

A shipboard degaussing installation consists of one or more coils of electric cable in specific locations inside the ship's hull, a dc power source to energize these coils, and a means of controlling the magnitude and polarity of the current through the coils. Compass-compensating equipment, consisting of compensating coils and control boxes, is also installed as a part of the degaussing system. This equipment is used to compensate for the deviation effect of the degaussing coils on the ship's magnetic compasses. Naval ships are tested periodically at magnetic range stations to determine the configuration of the ship's magnetic field. Sensitive measuring coils, located at or near the bottom of the channel, and recording equipment respond to the signals induced in the coils as the ship passes over them. These measurements indicate the distortion of the earth's magnetic field caused by the ship and are used to determine the values of current needed in the ship's degaussing coils to neutralize this distortion.

CATHODIC PROTECTION

Cathodic protection reduces the corrosion or deterioration of metal caused by a reaction with its environment (ship's hull and seawater). The chemical action that is created is similar to the electrochemical action of a battery or cell. Figure 20-31 shows a dry-cell battery circuit. So you may understand the electrochemical theory shown in figure 20-31, you need to use the conventional theory where current flows from positive to negative. The positive current is indicated by a positive deflection of the voltmeter needle when the positive terminal of the meter is connected to the cathode (positive terminal) of the cell. As the electrochemical action continues, the process will eventually corrode, or consume, the anode that is providing the current to light the lamp. This process is called electrochemical action.

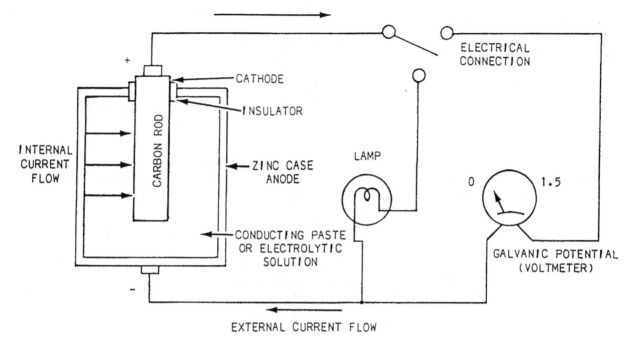

Figure 20-32.–Electrochemical corrosion cell.

In a marine environment, corrosion is an electrochemical process caused when two dissimilar metals are immersed in seawater, with the seawater acting as the electrolyte. This process is shown in the electrochemical corrosion cell (fig. 20-32). You must understand that in an electrochemical cell a metal that is more corrosion prone always has a higher driving voltage than the metal that is less corrosion prone. In cathodic protection, the more corrosion-prone metal is the anode (zinc) and the less corrosion-prone metal is the cathode (steel hull). The rate of corrosion is directly related to the magnitude of the potential difference and is called the open- or half-cell potential of metals. Some of the factors affecting the amount of corrosion are stray currents, resistivity, and the temperature of the seawater.

Stray-current corrosion is caused by an external current leaving the hull of a vessel and entering the seawater. If the connection between the ship and welding machine is not correctly made (fig. 20-33), or no return lead to the welder is connected, you could have current flow between the ship's hull and the pier, causing corrosion to form on the hull.

Seawater resistivity is the concentration of ions in seawater, which acts as a resistance to current flow between two dissimilar metals. Normal seawater generally has a nominal resistivity of 20 to 22 ohms/cm at a temperature of 20°C (68°F). In brackish (riverfed) or fresh water this resistivity may vary.

There are two types of cathodic protection systems, the sacrificial anode and the impressed current.

SACRIFICIAL ANODE SYSTEM

The sacrificial anode system is based on the principle that a more reactive metal, when installed near a less reactive metal and submerged in an electrolyte such as seawater, will generate a potential of a sufficient magnitude to protect the less reactive metal. In this process, the more reactive metal is sacrificed. Sacrificial anodes attached to a ship's hull slowly oxidize and generate a current. This system does not have an onboard control of protecting current, and it depends on the limited current output of the anode. This type of system requires anode replacement on a fixed schedule (usually every 3 years on naval ships). The system is rugged and simple, requires little or no maintenance, and always protects the ship.

The following is a list of sacrificial anodes:

- Zinc
- Aluminum
- Magnesium
- Iron
- Steel waster pieces

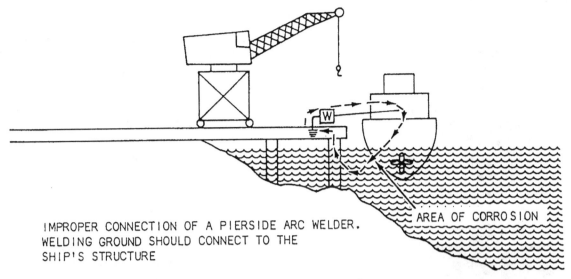

IMPROPER CONNECTION OF A PIERSIDE ARC WELDER. WELDING GROUND SHOULD CONNECT TO THE SHIP'S STRUCTURE

AREA OF CORROSION

Figure 20-33.–Stray-current corrosion.

IMPRESSED CURRENT CATHODIC PROTECTION SYSTEM

The impressed current cathodic protection (ICCP) system uses an external source of electrical power provided by a regulated dc power supply to provide the current necessary to polarize the hull. The protective current is distributed by specially designed inert anodes of platinum-coated tantalum. The principal advantage of an ICCP system is its automatic control feature, which continuously monitors and varies the current required for corrosion protection. If the system is secured, no corrosion protection is provided.

The following is a list of components of the ICCP system:

- Power supply
- Controller
- Anodes
- Reference electrode
- Stuffing tube
- Shaft grounding assembly
- Rudder ground (including stabilizer if installed)
- Dielectric shield

GYROCOMPASSES

Gyrocompass systems provide information that is used for remote indicators and various navigational, radar, sonar, and fire control systems throughout a ship.

A free gyroscope is a universal-mounted, spinning mass. In its simplest form, the universal mounting is a system that allows three degrees of freedom of movement. The spinning mass is provided by a heavy rotor. Figure 20-34 illustrates a free gyroscope. As you can see in the figure, the rotor axle is supported by two bearings in the horizontal ring. This ring is supported by two studs mounted in two bearings in the larger vertical

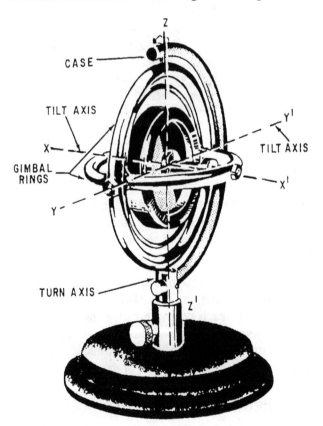

Figure 20-34.–The gyroscope.

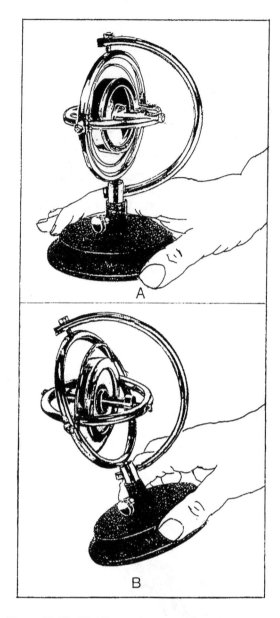

Figure 20-35.–Rigidity of plane of a spinning gyroscope.

or three degrees of freedom: (1) freedom to spin, (2) freedom to tilt, and (3) freedom to turn. The three degrees of freedom allow the rotor to assume any position within the case. The rotor is free to spin on its own axle, or the X axle, the first degree of freedom. The inner gimbal is free to tilt about the horizontal or Y axle, the second degree of freedom. The outer gimbal ring is free to turn about the vertical or Z axis, the third degree of freedom.

GYROSCOPIC PROPERTIES

When a gyroscope rotor is spinning, it develops two characteristics, or properties, that it does not possess when at rest: rigidity of plane and precession. These two properties make it possible to convert a free gyroscope into a gyrocompass.

Rigidity of Plane

When the rotor of the gyroscope is set spinning with its axle pointed in one direction (fig. 20-35, view A), it will continue to spin with its axle pointed in that direction, no matter how the case of the gyroscope is positioned (fig. 20-35, view B). As long as the bearings are frictionless and the rotor is spinning, the rotor axle will maintain its plane of spin with respect to a point in space. This property of a free gyroscope is termed *rigidity of plane.*

Newton's first law of motion states that a body in motion continues to move in a straight line at a constant speed unless acted on by an outside force. Any point in a spinning wheel tries to move in a straight line, but being a part of the wheel, must travel in an orbit around its axle. Although each part of the wheel is forced to travel in a circle, it still resists change. Any attempt to change the alignment or angle of the wheel is resisted by both the mass of the wheel and the velocity of that mass. This combination of mass and velocity is the kinetic energy of the wheel, and kinetic energy gives the rotor rigidity of plane. *Gyroscopic inertia* is another term that is frequently used interchangeably with rigidity of plane.

A gyroscope can be made more rigid by making its rotor heavier, by causing the rotor to spin faster, and by concentrating most of the rotor weight near its circumference. If two rotors with cross sections like those shown in figure 20-36 are of equal weight and rotate at the same speed, the rotor in figure 20-36, view B, will have more rigidity than the rotor in figure 20-36, view A. This condition exists because the weight of the rotor in figure 20-36, view B, is concentrated near the

ring. These two rings are called the inner gimbal and outer gimbal, respectively. The outer gimbal is then mounted with two studs and bearings to a larger frame called the case.

The rotor and both gimbals are pivoted and balanced about their axes. The axes (marked X, Y, and Z) are perpendicular to each other, and they intersect at the center of gravity of the rotor. The bearings of the rotor and two gimbals are essentially frictionless and have negligible effect on the operation of the gyroscope.

THREE DEGREES OF FREEDOM

As you can see in figure 20-34, the mounting of the gimbals allows movement in three separate directions,

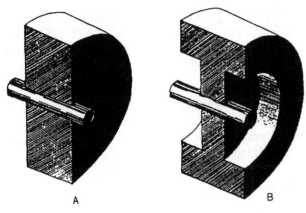

Figure 20-36.—Weight distribution in rotors.

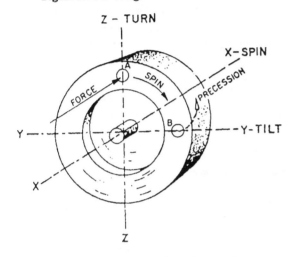

Figure 20-37.—Direction of precession.

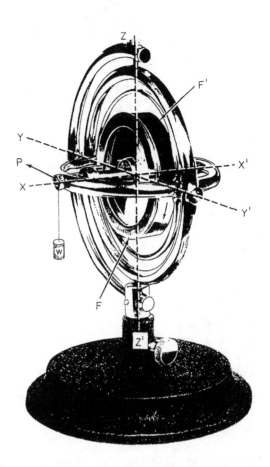

Figure 20-38.—Continuous precession.

circumference. Both gyroscope and gyrocompass rotors are shaped like the rotors shown in figure 20-36, view B.

Precession

Precession describes how a gyro reacts to any force that attempts to tilt or turn it. Though vector diagrams can help explain why precession occurs, it is more important to know how precession affects gyro performance.

The rotor of a gyro has one plane of rotation as long as its axle is aligned with, or pointed at, one point in space. When the axle tilts, turns, or wobbles, the plane of rotation of the rotor changes. Plane of rotation means the direction that the axle is aligned or pointed.

Torque is a force that tends to produce rotation. Force acts in a straight line, at or on a point. Torque occurs within a plane and about an axle or axis of rotation. If the force acts directly on the point of an axis, no torque is produced.

Because of precession, a gyro will react to the application of torque by moving at right angles to the direction of the torque. If the torque is applied

downward against the end of the axle of a gyro that is horizontal, the gyro will swing to the right or left in response. The direction in which it will swing depends on the direction the rotor is turning.

A simple way to predict the direction of precession is illustrated in figure 20-37. The force that tends to change the plane of rotation of the rotor is applied to point A at the top of the wheel. This point does not move in the direction of the applied force, but a point displaced 90° in the direction of rotation moves in the direction of the applied force. This results in the rotor turning left about the Z axis and is the direction of precession.

Any force that tends to change the plane of rotation causes a gyroscope to precess. Precession continues as long as there is a force acting to change the plane of rotation, and precession ceases immediately when the force is removed. When a force (torque) is applied, the gyroscope precesses until it is in the plane of the force. When this position is reached, the force is about the spinning axis and can cause no further precession.

If the plane in which the force acts or moves at the same rate and in the same direction as the precession that it causes, the precession will be continuous. This is illustrated by figure 20-38, in which the force attempting

BLACK LETTERING ON YELLOW TAG

Figure 20-39.–A CAUTION tag.

to change the plane of rotation is provided by a weight, W, suspended from the end of the spin axle, X. Although the weight is exerting a downward force, the torque is felt 90° away in the direction of rotation. If the wheel rotates clockwise, as seen from the weighted end, precession will occur in the direction of arrow P. As the gyroscope precesses, it carries the weight around with it so forces F and F^1 continuously act at right angles to the plane of rotation, and precession continues indefinitely. In other words, the rotor will turn to the right and continue turning until the weight is removed.

SAFETY PRECAUTIONS

Because of the possibility of injury to personnel, the danger of fire, and the chance of damage to material, all repair and maintenance work on electrical equipment should be performed only by duly authorized and assigned persons. Equipment that you are intending to repair or overhaul must be de-energized and tagged out by use of either a CAUTION or DANGER tag.

CAUTION TAG

A CAUTION tag (fig. 20-39) is a yellow tag used as a precautionary measure to provide temporary special instructions or to indicate that unusual caution must be exercised to operate the equipment. These instructions must state the specific reason that the tag is installed. Use of phrases such as DO NOT OPERATE WITHOUT EOOW PERMISSION is not appropriate since equipment or systems are not operated unless permission from the responsible supervisor has been obtained. A CAUTION tag cannot be used if personnel or equipment could be endangered while performing evolutions using normal operating procedures; a DANGER tag is used in this case.

DANGER TAG

Safety must always be practiced by persons working around electric circuits and equipment to prevent injury from electric shock and from short circuits caused by accidentally placing or dropping a conductor of electricity across an energized line. The arc and fire started by these short circuits, even where the voltage is relatively low, may cause extensive damage to equipment and serious injury to personnel.

NO work will be done on electrical circuits or equipment without permission from the proper authority, and until all safety precautions are taken. One of the most important precautions is the proper use of DANGER tags, commonly called RED tags (fig. 20-40).

DANGER tags are used to prevent the operation of equipment that could jeopardize your safety or endanger the equipment systems or components. When equipment is red tagged, under no circumstances will it be operated. After the work has been completed the tag or tags should be removed by the same person who put the tags in place.

The covers of fuse boxes and junction boxes should be kept securely closed except when work is being done. Safety devices such as interlocks, overload relays, and fuses should never be altered or disconnected except to be replaced. Fuses should be removed and replaced only after the circuit has been deenergized. When a fuse blows, it should be replaced only with a fuse of the correct current and voltage ratings. When possible, a circuit should be carefully checked before the replacement is made, since the burned-out fuse is often the result of a circuit fault. Safety or protective devices must never be changed or modified in any way without specific authorization.

REFERENCES

Electrician's Mate 3 & 2, NAVEDTRA 10546-F, Naval Education and Training Program Management Support Activity, Pensacola, Florida, 1988.

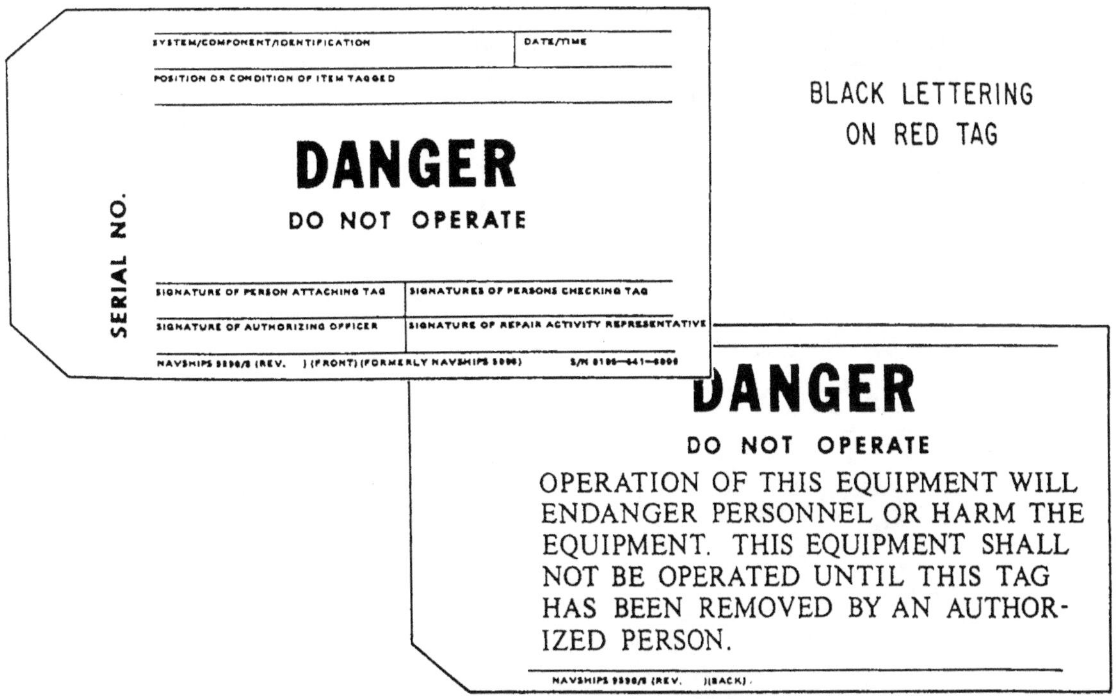

SYSTEM/COMPONENT/IDENTIFICATION   **DATE/TIME**

POSITION OR CONDITION OF ITEM TAGGED

DANGER

DO NOT OPERATE

SIGNATURE OF PERSON ATTACHING TAG   **SIGNATURES OF PERSONS CHECKING TAG**

SIGNATURE OF AUTHORIZING OFFICER   **SIGNATURE OF REPAIR ACTIVITY REPRESENTATIVE**

NAVSHIPS 9890/8 (REV.) (FRONT) (FORMERLY NAVSHIPS 5090)   S/N 0105—641—0800

SERIAL NO.

BLACK LETTERING
ON RED TAG

DANGER

DO NOT OPERATE

OPERATION OF THIS EQUIPMENT WILL
ENDANGER PERSONNEL OR HARM THE
EQUIPMENT. THIS EQUIPMENT SHALL
NOT BE OPERATED UNTIL THIS TAG
HAS BEEN REMOVED BY AN AUTHOR-
IZED PERSON.

NAVSHIPS 9890/8 (REV.)(BACK)

Figure 20-40.–A DANGER tag.

IC Electrician 3, NAVEDTRA 10559-A. Naval Education and Training Program Management Support Activity, Pensacola, Florida, 1989.

Navy Electricity and Electronics Training Series, NAVEDTRA 172-01-00-88, Module 1, *Introduction to Matter, Energy, and Direct Current*, Naval Education and Training Program Management Support Activity, Pensacola, Florida, 1988.

Navy Electricity and Electronics Training Series, NAVEDTRA 172-02-00-88, Module 2, *Introduction to Alternating Current and Transformers*, Naval Education and Training Program Management Support Activity, Pensacola, Florida, 1988.

Navy Electricity and Electronics Training Series, NAVEDTRA 172-05-00-79, Module 5, *Introduction to Generators and Motors*, Naval Education and Training Program Management Support Activity, Pensacola, Florida, 1979.

Navy Electricity and Electronics Training Series, NAVEDTRA 172-15-00-85, Module 15, *Principles of Synchros, Servos, and Gyros*, Naval Education and Training Program Management Support Activity, Pensacola, Florida, 1985.

CHAPTER 21

DISTILLING PLANTS

LEARNING OBJECTIVES

Upon completion of this chapter, you should be able to do the following:

1. Identify the basic principles of distillation and terms relating to the distillation process.

2. Identify the various types of distilling units used aboard Navy ships.

3. Discuss the operating principles of distilling units.

INTRODUCTION

Naval ships must be self-sustaining as far as production of fresh water is concerned. The large quantities of fresh water required aboard ship for boiler feed, drinking, cooking, bathing, and washing make it impracticable to provide storage tanks large enough for more than a few days' supply. Therefore, all naval ships depend on distilling plants to produce fresh water of extremely high purity from seawater.

PRINCIPLES OF DISTILLATION

The principle by which distilling plants produce fresh water from seawater is simple. The distillation process consists of heating seawater to the boiling point and condensing the vapor (steam) to obtain fresh water (distillate). This leaves behind the impurities of the seawater. The distillation process for a shipboard plant is illustrated in figure 21-1. Notice that the seawater after boiling is identified as brine.

At a given pressure, the rate at which seawater is evaporated in a distilling plant is dependent upon the rate at which heat is transmitted to the water. The rate of heat transfer to the water is dependent upon several factors; of major importance are the temperature difference between the substance giving up heat and the substance receiving heat, the available surface area through which heat may flow, and the coefficient of heat transfer of the substances and materials involved in the various heat exchangers that constitute the distilling plant. Additional factors such as the velocity of flow of the fluids and the cleanliness of the heat transfer surfaces also have a marked effect upon heat transfer in a distilling plant.

Since a shipboard distilling plant consists of several heat exchangers, each serving one or more specified purposes, the plant as a whole provides an excellent illustration of many thermodynamic processes and concepts. Practical manifestations of heat transfer–including heating, cooling, and change of phase–abound in the distilling plant, and the significance of the pressure-temperature relationships of liquids and their vapors is evident.

The seawater that is the raw material of the distilling plant is a solution of water and various minerals and salts. Seawater also contains suspended matter such as vegetable and animal growths and bacteria and other microorganisms. When properly operated, naval distilling plants produce fresh water that contains only slight traces of chemical salts and no biological contaminants.

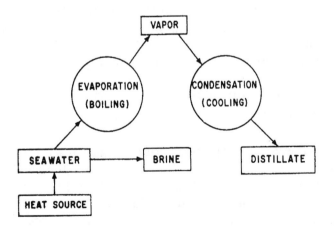

Figure 21-1.–Simplified diagram of the shipboard distillation process.

Distilling plants are not effective, however, in removing volatile gases or liquids that have a lower boiling point than water, nor are they effective in killing all microorganisms. These dissolved gases and liquids will simply boil into the vapor and be combined with the fresh water (distillate).

One of the problems that arises in the distillation of seawater occurs because some of the salts present in seawater are negatively soluble–that is, they are less soluble in hot water than they are in cold water. A negatively soluble salt remains in solution at low temperatures but precipitates out of solution at higher temperatures. The crystalline precipitation of various sea salts forms scale on heat transfer surfaces and thereby interferes with heat transfer. In naval distilling plants, this problem is partially avoided by designing the plants to operate under vacuum or at approximately atmospheric pressure.

The use of low pressures (and therefore low boiling temperatures) has the additional advantage of greater thermodynamic efficiency than can be achieved when higher pressures and temperatures are used. With low pressures and temperatures, less heat is required to raise the temperature of the feedwater (seawater) to make it boil; therefore, less heat is wasted by the plant.

DEFINITION OF TERMS

Before getting into the discussion of the process of distillation, you should learn the meaning of the terms in the following paragraphs. These terms apply basically to all types of distilling plants now in naval use. Additional terms that apply specifically to a particular type of distilling unit are defined as necessary in subsequent discussion.

DISTILLATION: The process of boiling seawater and cooling and condensing the resulting vapor to produce fresh water.

EVAPORATION: The process of boiling seawater to separate it into freshwater vapor and brine. Evaporation is the first half of the process of distillation.

CONDENSATION: The process of cooling the freshwater vapor produced by evaporation to produce usable fresh water. Condensation is the second half of the process of distillation.

FEED: The seawater that is the raw material of the distilling unit; also called seawater feed or evaporator feed.

VAPOR: The product of the evaporation of seawater. The terms *vapor* and *freshwater vapor* are used interchangeably.

DISTILLATE: The product resulting from the condensation of the freshwater vapor produced by the evaporation of seawater. Distillate is also called condensate, fresh water, freshwater condensate, and seawater distillate. However, the use of the term *condensate* should be avoided whenever there is any possibility of confusion between the condensate of the distilling plant and the condensate that results from the condensation of steam in the main and auxiliary condensers. In general, it is best to use the term *distillate* when referring to the product resulting from the condensation of vapor in the distilling plant.

SALINITY: The concentration of chemical salts in water. Salinity is measured by electrical devices, called salinity cells, of either equivalents per million (epm) or parts per million (ppm).

BRINE: Water in which the concentration of chemical salts is higher than it is in seawater.

TYPES OF DISTILLING UNITS

Naval ships use two general types of distilling plants: the vapor compression type and the low-pressure steam distilling unit. The major differences between the two are the kinds of energy used to operate the units and the pressure under which distillation takes place. Vapor compression units use electrical energy (for heaters and a compressor). Steam distilling units use low-pressure steam from either the auxiliary exhaust systems or the auxiliary steam system. In addition, vapor compression units boil the feedwater at a pressure slightly above atmospheric while the low-pressure steam units depend on a relatively high vacuum for operation.

VAPOR COMPRESSION DISTILLING UNITS

The vapor compression distilling plant is used in submarines and small diesel-driven surface craft where the daily requirements do not exceed 4,000 gallons per day (gpd). Nuclear submarines use the vapor compression plant as a backup for the steam-operated unit and for in-port operation.

A vapor compression distilling unit is shown in cutaway view in figure 21-2 and schematically in

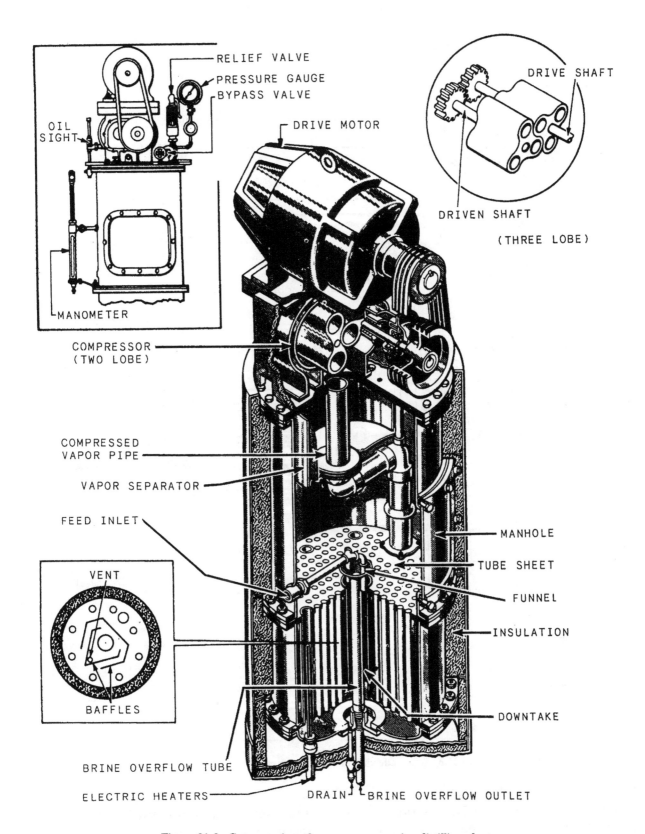

RELIEF VALVE
PRESSURE GAUGE
BYPASS VALVE
OIL SIGHT
MANOMETER

DRIVE SHAFT
DRIVEN SHAFT
(THREE LOBE)

DRIVE MOTOR

COMPRESSOR
(TWO LOBE)

COMPRESSED
VAPOR PIPE

VAPOR SEPARATOR

FEED INLET

VENT

BAFFLES

MANHOLE

TUBE SHEET

FUNNEL

INSULATION

DOWNTAKE

BRINE OVERFLOW TUBE

ELECTRIC HEATERS

DRAIN

BRINE OVERFLOW OUTLET

Figure 21-2.–Cutaway view of a vapor compression distilling plant.

21-3

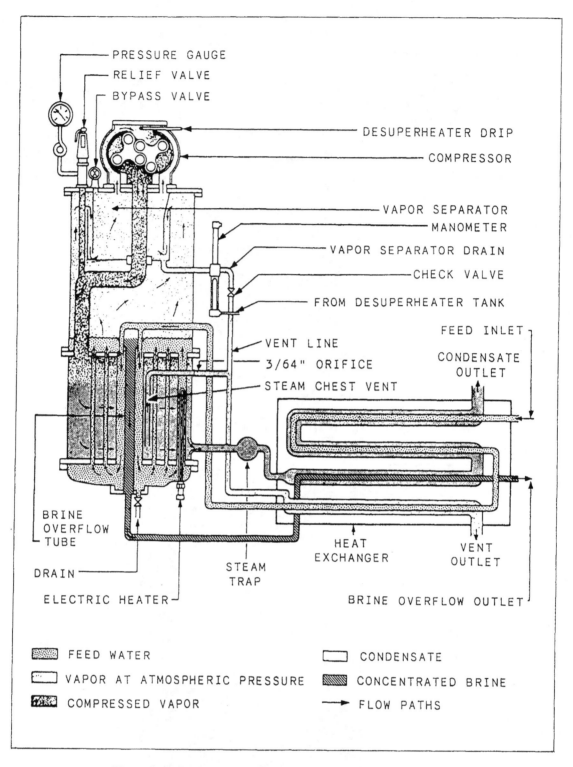

Figure 21-3.–Schematic view of a vapor compression distilling plant.

figure 21-3. The unit consists of three main components–the evaporator, compressor, and heat exchanger–and several accessories and auxiliaries.

Evaporator

The cylindrical shell in which vaporization and condensation occur is commonly called the evaporator.

The evaporator consists of two principal elements: the steam chest and the vapor separator.

The steam chest includes all space within the evaporator shell except the space that is occupied by the vapor separator. The steam chest (vapor chest) is considered to have an evaporating side and a condensing side. The evaporating side includes the space within the tubes of the tube bundle (which is located in the lower part of the evaporator shell) and the space that communicates with the inside of the tubes. The condensing side includes the space that surrounds the external surfaces of the tubes; this space communicates with the discharge side of the compressor by a pipe, as shown in figure 21-3.

The tube bundle is enclosed in a shell. At the top and bottom of the bundle, the tube ends are expanded into tube sheets. Most of the tubes are small; but a few, set near the periphery, are larger. Each of the larger tubes contains an electric heater. As the seawater feed flows through these larger tubes, it is heated to the boiling point by the heaters.

The feed inlet pipe extends horizontally to the center of the evaporator, where it branches into a Y. The two ends of the feed pipe turn downward into the downtake, as shown in figure 21-2. Seawater feed enters the evaporator through the horizontal inlet pipe, pours into the downtake, and passes down to the bottom head of the evaporator shell; from there, the feed flows upward through the tubes.

A funnel is installed inside the downtake at the top. The top of the funnel is about 2 inches above the top of the evaporator tubes, and the brine level in the evaporator shell is thus maintained at this height. About one-half to two-thirds of the feed is vaporized; the remaining brine overflows continuously into the funnel and then into the brine overflow tube that is installed inside the downtake. The overflow tube leads the brine out through the bottom of the evaporator shell to the heat exchanger. In the heat exchanger, the brine gives up its heat and raises the temperature of the incoming feed.

The vapor separator is an internal compartment located at the top of the evaporator shell. The separator consists of two cylindrical baffles. One cylinder extends downward from the upper head plate of the evaporator; the other extends upward and is fitted around the upper cylinder to form a baffle. The floor of the separator is formed by the bottom of the outer cylinder. The space between the two cylinders provides a passage for the vapor flowing from the evaporating side of the steam chest to the suction side of the compressor.

The vapor from the boiling seawater rises up through the space between the shell wall and the outer cylinder of the separator; it then flows downward through the space between the cylinders of the separator and enters the separator chamber. From the separator

chamber, the vapor travels upward to the intake side of the compressor.

During this roundabout passage through the vapor separator, the vapor is separated from any entrained particles of water. The water drops to the floor of the separator and is continuously drained away. This water has a high salt concentration and must be continuously drained to keep it from entering the compressor and thus getting into the condensing side of the evaporator, where it would contaminate the distillate.

Vapor Compressor

The vapor that flows upward from the separator is compressed by a positive-displacement compressor. The type of compressor discussed here has two three-lobe rotors of the type shown in the insert on figure 21-2 and in figure 21-3. Two-lobe compressors of the type shown in the main part of figure 21-2 were an earlier design.

The two rotors are enclosed in a compact housing that is mounted on the evaporator. The three lobes on each rotor are designed to produce a continuous and uniform flow of vapor. The vapor enters the compressor housing at the bottom and then passes upward between the inner and outer walls of the housing to the rotor chamber, where it fills the space between the rotor lobes. The vapor is then carried around the cylindrical sides of the housing, and a pressure is developed at the bottom as the lobes roll together. Clearances are provided so the rotor lobes do not actually touch each other and do not touch the housing.

The shaft of one rotor is fitted with a drive pulley on one end and a gear on the other end. This gear meshes with a gear on the shaft of the other rotor to provide the necessary drive for the second rotor.

Heat Exchanger

The heat exchanger preheats the incoming seawater feed by two heat exchange processes. In one process, the seawater feed is heated by the distillate that is being discharged from the distilling unit to the ship's tanks. In the other process, seawater feed is heated by the brine overflow that is being discharged overboard or to the brine collecting tank.

The heat exchanger is a horizontal double-tube unit. Either seawater or brine flows through the inner tubes, while distillate flows through the space between the inner and the outer tubes.

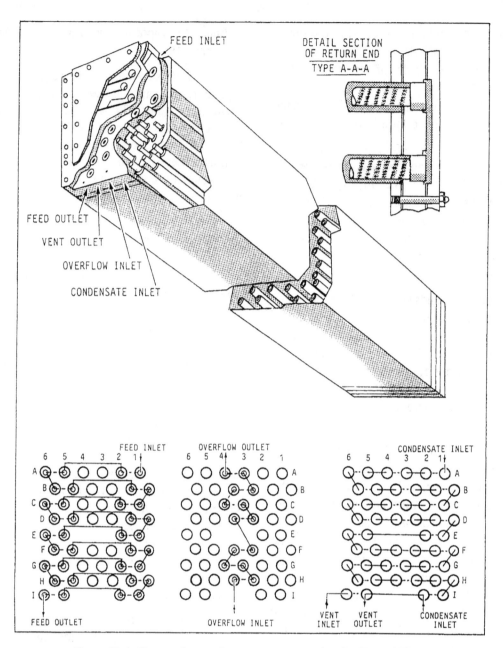

Figure 21-4.—Heat exchanger for a vapor compression distilling plant.

Figure 21-4 shows the construction of the heat exchanger and also illustrates the flow paths. There are four distinct flow paths: feed, brine overflow, condensate (distillate), and vent.

Accessories and Auxiliaries

Several accessories and auxiliaries are required for the operation of the vapor compression distilling unit. These include feed, distillate, and brine overflow pumps; feed regulating and flow control valves; relief valves; compressor bypass valves; rotameters; and a variety of pressure and temperature gauges.

The Vapor Compression Process

Now that the principal parts of a vapor compression unit have been described, let us summarize briefly the sequence of events within the unit and consider some of the factors that are important in the vapor compression process of distillation.

The cold seawater feed enters the heat exchanger and is heated there to about 190° or 200°F. From the heat exchanger, the feed goes into the evaporator. Here it flows down the downtake and into the bottom of the evaporator shell, then upward in the tubes. Boiling and evaporation take place in the tubes at atmospheric

21-6

pressure. About one-half to two-thirds of the incoming feed is evaporated; the remainder flows out through the brine overflow, thus maintaining a constant water level within the evaporator.

The vapor thus generated rises and enters the vapor separator, where any particles of moisture that may be present are separated from the vapor and drained out of the separator. The vapor goes to the suction side of the compressor. In the compressor, distilled water drips onto the rotors and thus desuperheats the vapor as it is compressed. The vapor is compressed to a pressure of about 3 to 5 pounds above atmospheric pressure and is discharged to the space surrounding the tubes in the steam chest. As the vapor condenses on the outside of the smaller tubes, the distillate drops down and collects on the bottom tube plate. Every time a pound of compressed vapor condenses, approximately a pound of vapor is formed in the evaporator section; the compressor suction is kept supplied with the right amount of vapor.

The distillate is drawn off through a steam trap and flows into the heat exchanger at a temperature of about 220°F. As it flows through the heat exchanger, the distillate gives up heat to the incoming feed and is cooled to within about 18°F of the cold feedwater temperature. Noncondensable gases, together with a small amount of vapor, flow into the vent line and then to the heat exchanger.

Meanwhile, the seawater that is not vaporized in the evaporator is flowing continuously into the funnel down the brine overflow tube and into the heat exchanger. The temperature of this brine is about 214°F. In passing through the heat exchanger, the hot brine raises the temperature of the seawater feed that is entering through the heat exchanger.

The entire distillation cycle is started by using the electric heaters to bring the seawater feed temperature up to the boiling point and to generate enough vapor for compressor operation. After the cycle has been started and the compressor is adequately supplied with vapor, the normal operating cycle begins and the electric heaters are used from then on only to provide the heat necessary to make up for heat losses. After the unit has become fully operational, the heat input from the heaters is only a small part of the total heat input.

The major part of the heat input comes from the compression work that is done on the vapor by the compressor. The major energy transformations involved in normal operation are thus from electrical energy (put in at the compressor motor) to mechanical energy (work done by the compressor on the vapor) to thermal energy. The thermal energy thus supplied is used to boil the seawater feed and keep the process going.

The compression process serves another vital function in the vapor compression distilling unit. Since the boiling point of seawater is several degrees higher than the boiling point of fresh water at any given pressure, the boiling seawater in the evaporator is actually above 212°F and would therefore be too hot to condense the freshwater vapor if the vapor were at the same pressure as the boiling seawater. By compressing the vapor, the boiling point of the vapor is raised above the boiling point of the seawater at atmospheric pressure. Therefore, the compressed vapor can be condensed on the outside of the tubes in which seawater feed is being boiled. This process would not be possible without the pressure difference between the evaporating side and the condensing side of the unit, and this pressure difference is created by the compression of the vapor.

LOW-PRESSURE STEAM DISTILLING UNITS

The low-pressure steam distilling unit is used in all steam-driven and gas turbine-driven surface ships and nuclear submarines. Low-pressure steam distilling units are "low pressure" from two points of view: First, they use low-pressure steam as the source of energy; and second, their operating shell pressure is less than atmospheric pressure.

There are three major types of low-pressure steam distilling units: (1) submerged-tube, (2) flash-type, and (3) vertical-basket.

Submerged-Tube Distilling Plants

There are three classifications, or arrangements, of submerged-tube distilling plants: (1) two-shell double-effect, (2) Soloshell double-effect, and (3) triple-effect. The principal difference between the double-effect type and the triple-effect type is the number of stages of evaporation.

In a submerged-tube distilling plant, feed floods into the bottom of the unit and surrounds the tubes that contain circulating low-pressure steam. The steam in the tubes causes the surrounding feed to boil and produce steam (vapor). The vapor passes up into the moisture separators, where any entrained (drawn in) seawater droplets are removed. The clean vapor then passes on and is condensed into distillate. Submerged-tube distilling plants are found on older ships.

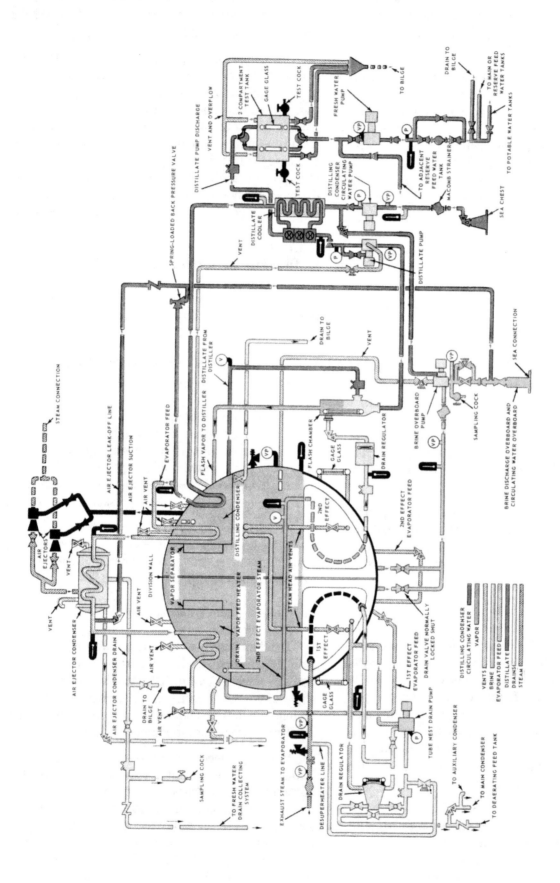

C47.114

Figure 21-5.—Schematic diagram of a Soloshell double-effect distilling plant.

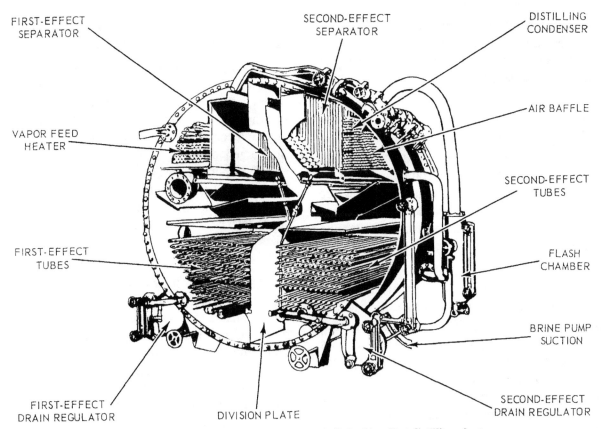

FIRST-EFFECT SEPARATOR

SECOND-EFFECT SEPARATOR

DISTILLING CONDENSER

VAPOR FEED HEATER

AIR BAFFLE

SECOND-EFFECT TUBES

FIRST-EFFECT TUBES

FLASH CHAMBER

BRINE PUMP SUCTION

FIRST-EFFECT DRAIN REGULATOR

DIVISION PLATE

SECOND-EFFECT DRAIN REGULATOR

Figure 21-6.–Cutaway view of a two-shell double-effect distilling plant.

SOLOSHELL DOUBLE-EFFECT PLANTS.– Most Soloshell double-effect units have capacities of 12,000 gpd or less. However, some Soloshell units of 20,000 gpd capacity are in use.

A Soloshell double-effect unit is shown schematically in figure 21-5 and in cutaway view in figure 21-6. The unit consists of a single cylindrical shell that is mounted with the long axis in a horizontal position. A longitudinal vertical partition plate divides the shell into a first-effect shell and a second-effect shell. The first-effect shell contains the first-effect tube bundle, a vapor separator, and the vapor feed heater. The second-effect shell contains the second-effect tube bundle, a vapor separator, and the distilling condenser. A distillate cooler, not a part of the main cylindrical shell, is mounted at any convenient location as piping arrangements permit. Another separate unit, the air ejector condenser, is mounted on brackets on the outside of the evaporator shell. The air ejector takes suction on the second-effect part of the shell, maintaining it under a vacuum of approximately 26 in.Hg. A lesser vacuum–about 16 in.Hg–is maintained in the first-effect shell.

Steam for the distilling unit is obtained from the auxiliary exhaust line through a regulating valve. This valve is adjusted to maintain a constant steam pressure of 1 to 5 psig in the line between the regulating valve and a control orifice. The size of the opening in the control orifice determines the amount of steam admitted to the distilling unit and hence controls the output of distilled water.

When the steam pressure is reduced by the regulating valve, the steam becomes superheated. Since superheat increases the rate of scale formation, provision is made for desuperheating the steam. This is done by spraying hot water into the steam line between the control orifice and the point where the steam enters the first-effect shell. The hot water for desuperheating the steam is taken from the first-effect drain pump discharge.

After being desuperheated, the steam passes into the first-effect tube nest, where it heats the seawater feed that surrounds the first-effect tubes. The seawater boils, generating steam that is called vapor to distinguish it from the steam that is the external source of energy for the unit. The condensate that results from the condensation of the supply steam is discharged by the first-effect drain pump to the low-pressure drain system or to the condensate system and is thus eventually used again in the boiler feed system.

Although the vapor generated in the first-effect shell is pure water vapor, it does contain small particles of liquid feed. As the vapor rises, a series of baffles above the surface of the water begins the process of separating the vapor and the water particles.

After passing through the baffles, the vapor enters the vapor separator. As the vapor passes around the hooked edges of the baffles and vanes in the separator, it is forced to change direction several times; and with each change of direction, some water particles are separated from the vapor. The hooked edges trap particles of water and drain them away, discharging them back into the feed at a distance from the vapor separator.

After passing through the first-effect vapor separator, the vapor goes to the vapor feed heater. Seawater feed passes through the tubes of the vapor feed heater, and part of the vapor is condensed as it flows over the tubes of the heater. This distillate, together with the remaining uncondensed vapor, goes through an external crossover pipe and enters the tube nest of the second-effect shell. The remaining vapor is now condensed as it gives up the rest of its latent heat to the seawater feed in the second-effect shell.

Since the pressure in the second-effect shell is considerably less than the pressure in the first-effect shell, the introduction of the vapor and the distillate from the first-effect shell causes the seawater feed in the second-effect shell to boil and vaporize.

The vapor thus generated in the second-effect shell passes through baffles just above the surface of the water and then goes to the second-effect vapor separator. From the vapor separator, it passes to the distilling condenser. The condensing tubes nearest the incoming vapor are used as a feed heating section; the vapor condenses on the outside of the tubes and thus heats the incoming seawater feed that is circulating through the tubes. The remainder of the vapor is condensed in the condensing section and is discharged to the test tanks as distillate.

The first-effect distillate that was used in the second-effect tubes to boil and vaporize the feed in the second-effect shell is discharged through the second-effect tube nest drain regulator and is led to the distilling condenser by way of a flash chamber. The flash chamber is essentially a receptacle within which the vapor, liberated when the second-effect drains are reduced to a pressure and temperature corresponding to the distilling condenser vacuum, is separated from the condensate and directed to the distilling condenser. As

may be seen in figure 21-6, the flash chamber is located just outside of the second-effect shell.

The distilling condenser circulating water pump takes suction from the sea and discharges the seawater through the shell of the distillate cooler (which is external to the unit) and then through the tubes of the distilling condenser. Some of the cooling water is then discharged overboard; but a portion (which is now called evaporator feed) goes through the feed heating section of the distilling condenser, through the air ejector condenser, and through the first-effect vapor feed heater before it is discharged to the first-effect shell. These paths of the distilling condenser circulating water and the evaporator feed may be traced in figure 21-5.

As previously described, some of the seawater feed in the first-effect shell is boiled and vaporized by the supply steam. The remaining portion becomes more dense and has a higher salinity than the original seawater feed; this denser, saltier water is called brine to distinguish it from seawater. After a certain amount of seawater feed has been vaporized in the first-effect shell, the remaining brine is led to the second-effect shell through a pipe that has a manually controlled feed regulating valve installed in it. When the feed regulating valve is open, the higher pressure in the first-effect shell causes the brine to flow from the first-effect shell to the second-effect shell. After the brine has been used as feed to generate vapor in the second-effect shell, the remaining brine is discharged overboard by the brine overboard discharge pump.

TWO-SHELL DOUBLE-EFFECT PLANTS.– Two-shell double-effect units of 20,000 gpd capacity are used on some ships. A typical unit of this kind is shown in figure 21-7. The unit consists of two cylindrical evaporator shells, mounted horizontally with the long axes of the shells parallel. The first-effect vapor feed heater is built into the upper part of the first-effect shell. The distilling condenser and the distillate cooler are built into separate shells, which are usually mounted between the two evaporator shells. The air ejector condenser is also a separate unit, though it is mounted on one of the shells. The operation of the two-shell double-effect unit is almost precisely the same as the operation of the Soloshell double-effect unit. The flow paths of steam, condensate, seawater, brine, vapor, and distillate may be traced out on figure 21-7.

THREE-SHELL TRIPLE-EFFECT PLANTS.– Three-shell triple-effect distilling units are similar to the double-effect units previously discussed, except the triple-effect units have an intermediate evaporating stage.

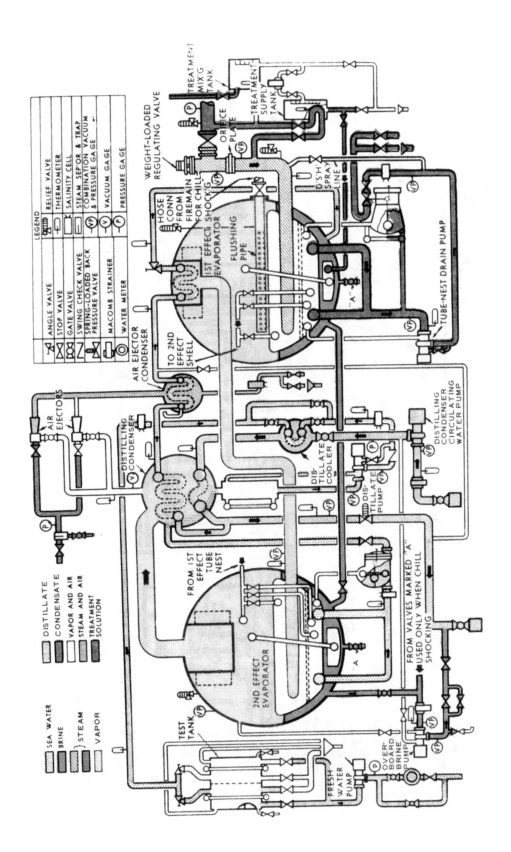

C47.116

Figure 21-7.—Schematic diagram of a two-shell double-effect distilling plant.

21-11

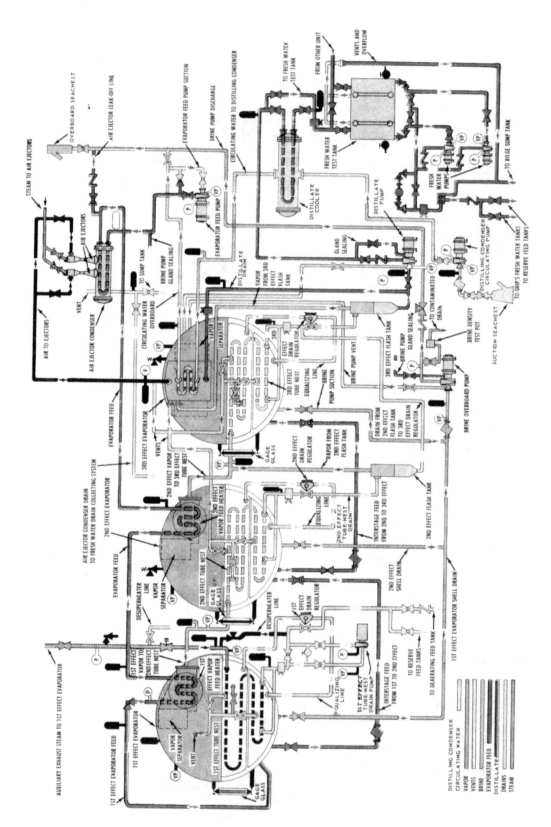

C47.115

Figure 21-8.—Schematic diagram of a triple-effect distilling plant.

A triple-effect distilling unit is shown schematically in figure 21-8. Although there are several kinds of triple-effect units, the general relationships shown in this illustration hold for any triple-effect plant.

A standard 20,000 gpd triple-effect unit consists of three horizontal cylindrical shells, set side by side with their axes parallel. The first- and second-effect vapor feed heaters are built into the front end of the second- and third-effect evaporator shells. The distilling condenser is contained within the third-effect shell. The air ejector condenser and the distillate cooler are in separate shells and are mounted on the third-effect shell.

Another 20,000 gpd triple-effect design consists essentially of three horizontal shells bolted together end to end, with vertical partition plates between each shell to separate the effects. Vapor separators in independent shells are installed in the vapor piping between effects and between the third effect and the distilling condenser. The first- and second-effect vapor feed heaters are in separate shells and are mounted in the piping at the inlet to the second-effect and third-effect tube nests, respectively. The two sections of the distilling condenser and the distillate cooler are built into a single shell and independently mounted as space and piping arrangements may permit. The air ejector condenser is also a separately mounted unit.

A standard 30,000 gpd triple-effect unit is also in use. This is similar to the standard 20,000 gpd unit, except the 30,000 gpd unit is larger.

There are two types of 40,000 gpd triple-effect units that may be regarded as standard, since both are widely used in naval ships. The first type uses the same arrangement as the standard 20,000 gpd triple-effect unit but has the larger components needed for the increased capacity. The second type consists of three horizontal shells, usually mounted side by side with axes parallel. In this design, both vapor feed heaters and distilling condensers are built as three independent units, each mounted separately outside the evaporator shells. The air ejector condenser and the distillate cooler are also in independent shells and are separately mounted outside the evaporator shells.

Triple-effect units operate virtually the same as the Soloshell and the two-shell double-effect units previously described, except the comparable actions in a triple-effect unit are spread out through more equipment and through one more effect. In a triple-effect unit, the seawater feed is piped to the first-effect shell, then to the second-effect shell, and then to the third-effect shell. Steam from the auxiliary exhaust line is used to vaporize the feed in the first-effect shell; in the second-effect and third-effect shells, the vapor is generated by the heat given up by vapor generated in the previous shell. In the triple-effect units, as in the double-effect units, this sequence of events is possible because the vacuum is greatest in the shell of the final effect and least in the shell of the first effect.

Flash-Type Distilling Plants

The flash-type distilling plant is widely used throughout the Navy. Flash-type plants have some distinct advantages over the submerged-tube plant. One is that the flash type "flashes" the feed into vapor (steam) rather than boiling it inside the evaporator shell. The flashing process involves heating the feed before it enters the evaporator shell. The shell is under a relatively high vacuum. The feed is heated to a temperature at which it will flash into vapor when it enters the vacuum. This design has no submerged heat transfer surfaces within the evaporator shell, such as the steam tubes in the submerged-tube unit. The elimination of these surfaces greatly reduces the scale formation problem of evaporators and allows prolonged operation at maximum efficiency. Any scale that may form is composed mainly of soft calcium carbonate compounds that are relatively easy to remove.

Flash-type plants consist of two or more stages. Two-stage plants of 12,000 gpd capacity are installed on destroyer-type ships. Five-stage plants of 50,000 gpd capacity are installed on larger ships.

Each stage of a flash-type plant has a flash chamber, a feed box, a vapor separator, and a distilling condenser. A two-stage or three-stage air ejector, a distillate cooler, and a feedwater heater are also provided. Feedwater passes through the tubes of the distillate cooler, the stage distilling condenser, and the air ejector condenser. In each of these heat exchangers, the feed picks up heat. The final heating is done by low-pressure steam admitted to the shell of the feedwater heater. From this heater, the feedwater enters the first-stage feedbox and comes out through orifices into the flash chamber. As the heated feedwater enters the chamber, a portion flashes or vaporizes because the pressure in the chamber is lower than the saturation pressure corresponding to the temperature of the hot feed. The vapor condenses on the tubes of the first-stage distilling condenser. The feed that does not vaporize in the first chamber passes to the second chamber. The process is repeated in each stage and the brine remaining in the last stage is removed by the brine overboard pump. Vapor formed in each stage passes through a vapor separator and into the stage

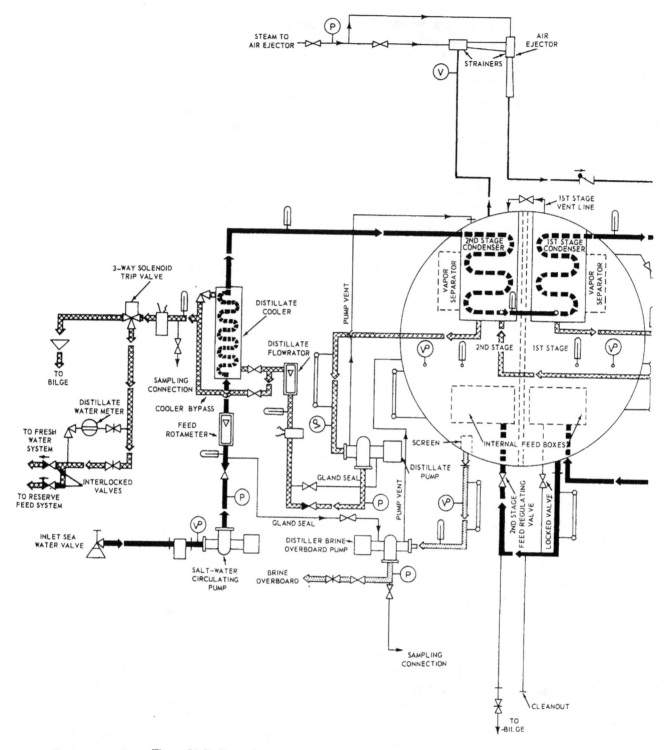

Figure 21-9.—General arrangement of a two-stage flash-type distilling plant.

distilling condenser, where it is condensed into distillate. The distillate passes through a loop seal on its way to the distilling condenser of the next stage. The distillate pump removes the distillate from the last stage and discharges it through the distillate cooler and the solenoid-operated dump valve to the ship's tanks.

The general arrangement of a two-stage flash-type plant is shown in figure 21-9; a five-stage plant is shown in figure 21-10. Both figures are foldouts at the end of this chapter. The major circuits are shown in each illustration.

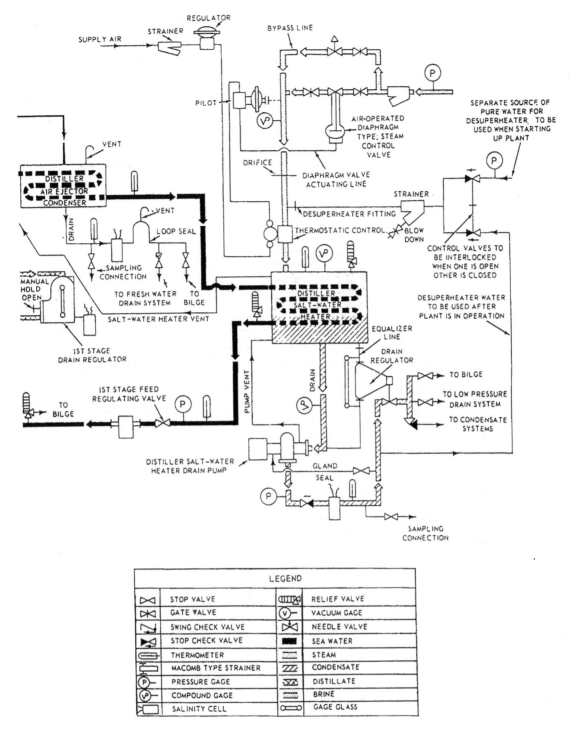

LEGEND			
⋈	STOP VALVE	⊞	RELIEF VALVE
⋈	GATE VALVE	Ⓥ	VACUUM GAGE
⋈	SWING CHECK VALVE	⋈	NEEDLE VALVE
⋈	STOP CHECK VALVE	▬	SEA WATER
⊏⊐	THERMOMETER	⋯	STEAM
⊏⊐	MACOMB TYPE STRAINER	⫻	CONDENSATE
Ⓟ	PRESSURE GAGE	⋈	DISTILLATE
Ⓥ	COMPOUND GAGE	⋯	BRINE
⊐	SALINITY CELL	⊂⊃	GAGE GLASS

Figure 21-9.–General arrangement of a two-stage flash-type distilling plant–Continued.

Vertical-Basket Distilling Plants

Single-stage, vertical-basket distilling plants are used extensively in our nuclear submarines. These plants have a capacity of approximately 8,000 gpd and are sometimes called the 8K evaporators.

Vertical-basket distilling plants use 150 psig auxiliary steam, reduced to 20 psi for the heat source. The steam is fed into the basket of the evaporator. The basket is a cylindrical, corrugated shell, located within the bottom of the evaporator shell. The corrugation provides a greater surface area for heat

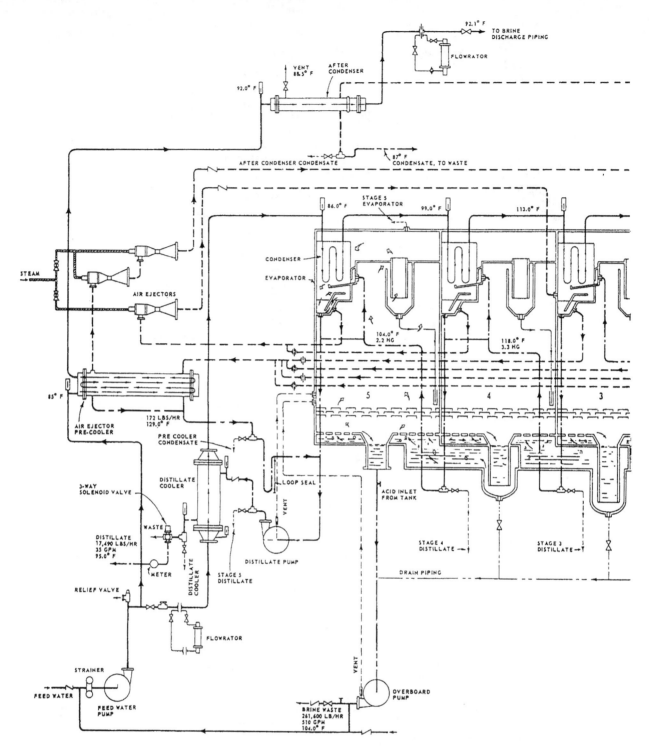

Figure 21-10.—General arrangement of a five-stage flash-type distilling plant.

transfer between the basket and the surrounding feedwater.

Some ships are equipped with vertical-basket distilling units. A unit of this type is shown in figure 21-11. The unit shown has two effects; however, some units of this type have more than two effects.

The vertical-basket unit consists of two or more evaporators, a distiller condenser, vapor feed heaters, a distillate cooler, and air ejectors. The major difference between a vertical-basket unit and a submerged-tube unit is in the design of the evaporators. In the vertical-basket unit, each evaporator consists of a vertical shell in which a deeply corrugated vertical

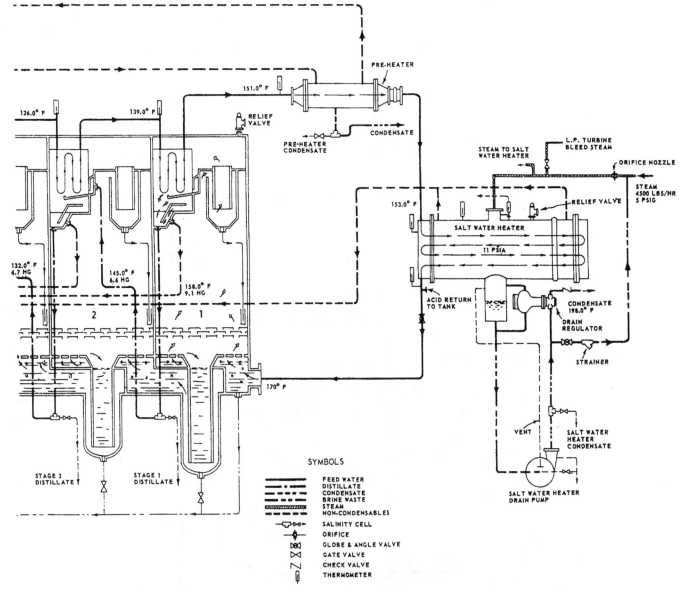

SYMBOLS

————	FEED WATER
— · — · —	DISTILLATE
— — — —	CONDENSATE
·········	BRINE WASTE
▨▨▨▨▨	STEAM
– – – –	NON-CONDENSABLES
⊸▷◦◁	SALINITY CELL
⊸◊⊸	ORIFICE
⋈	GLOBE & ANGLE VALVE
⋈	GATE VALVE
⊿	CHECK VALVE
⊟	THERMOMETER

Figure 21-10.–General arrangement of a five-stage flash-type distilling plant–Continued.

basket is installed. Figure 21-12 shows a sectional view of the evaporator and basket.

Low-pressure steam is admitted to the inside of the first-effect basket. This steam boils the feedwater in the space between the outside of the basket and the shell of the evaporator. The condensate resulting from the condensation of steam drains downward and is returned to the boiler feed system. The vapor generated from the boiling seawater feed passes through the cyclonic separator above the evaporation section, where most of the entrained liquid particles are removed from the vapor by centrifugal force. The vapor continues through the second vapor separator (called the snail), where the remaining water droplets are separated from the vapor.

The liquid particles from both separators drain downward and become part of the brine drains.

The vapor generated in the first-effect shell passes from the steam dome of the first-effect shell. It goes through the vapor feedwater heater and then enters the steam chest and evaporator basket of the second-effect shell. The first-effect vapor boils the second-effect feed and thus causes the generation of the second-effect vapor. The second-effect vapor goes through the cyclonic separator and the snail in the second-effect shell. From the steam dome, this vapor then goes to the distilling condenser, where the vapor is condensed on the outside of the tubes. The second-effect distillate drains down and collects in the flash tank.

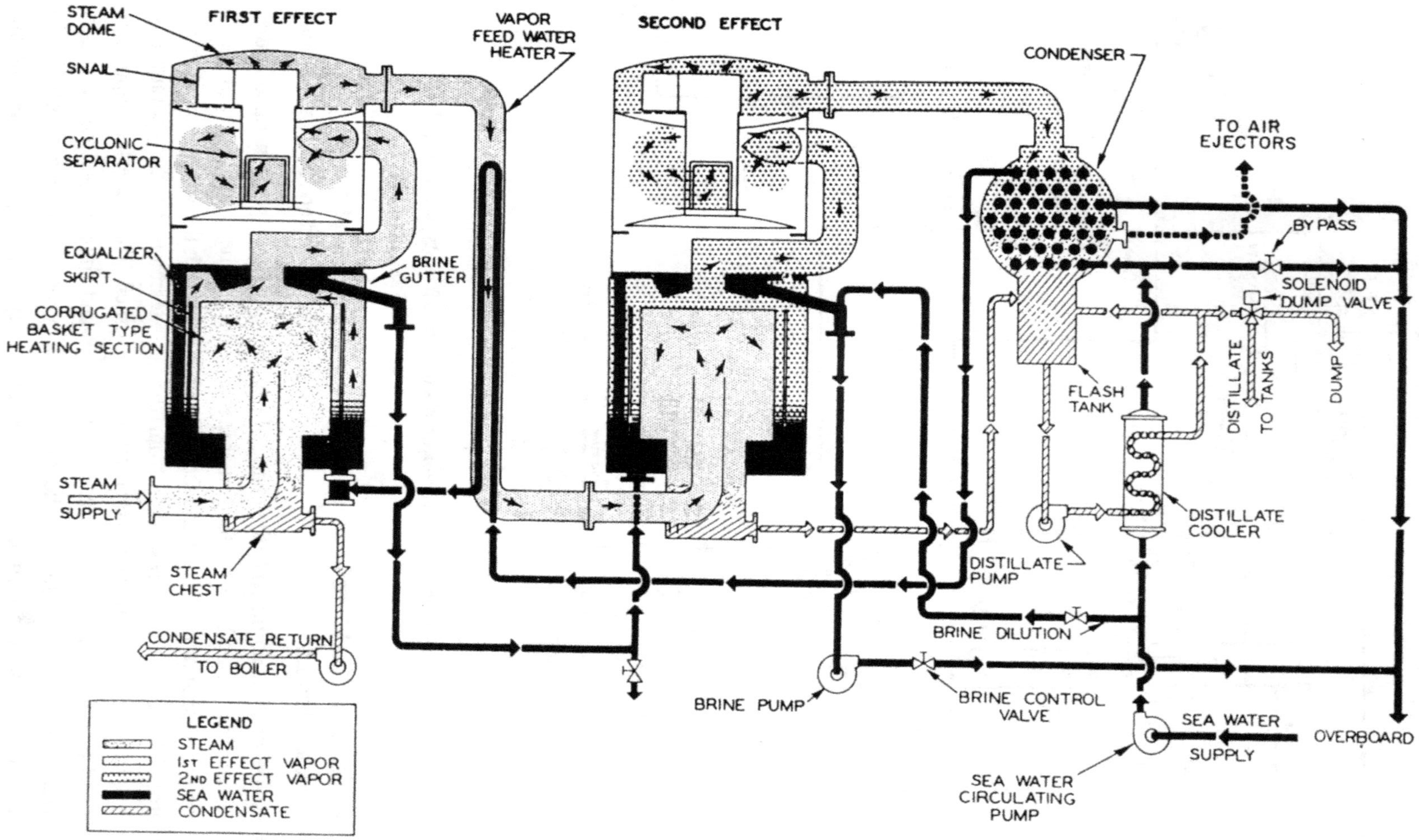

Figure 21-11.—Vertical-basket double-effect distilling plant.

21-18

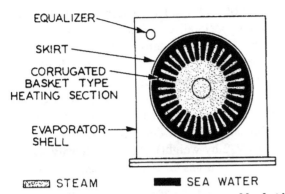

EQUALIZER
SKIRT
CORRUGATED BASKET TYPE HEATING SECTION
EVAPORATOR SHELL

▨ STEAM   ■ SEA WATER

Figure 21-12.–Sectional view of the evaporator and basket in a vertical-basket distilling plant.

As the first-effect vapor is being used to boil the second-effect feed, some of the vapor condenses. This distillate drains downward into the second-effect steam chest and is discharged to the flash tank at the bottom of the distilling condenser, where it mixes with the distillate formed from the second-effect vapor. The distillate is removed from the flash tank by the distillate pump and is discharged through the distillate cooler and the solenoid-operated dump valve to the ship's tanks. Should the salinity of the distillate exceed 0.065 epm, the dump valve automatically dumps the distillate to the bilges.

Seawater flows through the tubes of the distillate cooler and the distilling condenser, creating a suction for the brine pump and maintaining a back pressure for the feed system. About 25 percent of the seawater passes through supplementary heating sections in the distilling condenser to the air ejector condenser and feeds the evaporator shells in parallel. As the seawater passes through the air ejector condenser, it condenses the air ejector steam; the resulting condensate drains to an atmospheric drain tank.

DISTILLING PLANT OPERATION

Although a detailed discussion of distilling plant operation is beyond the scope of this text, certain operational considerations should be noted. The factors mentioned here apply primarily (although not exclusively) to low-pressure steam distilling units.

Naval distilling plants are designed to produce distillate of very high quality. The distillate must meet specific standards of chloride content. The chloride content of distillate discharged to the ship's tanks must not exceed 0.065 epm. Any distilling unit that cannot produce distillate of this quality is not considered to be operating properly.

Steady operating conditions are essential to the satisfactory operation of a distilling unit. Fluctuations in the pressure and temperature of the first-effect generating steam will cause fluctuations of pressure and temperature throughout the unit. Such fluctuations may cause priming, with increased salinity of the distillate, and also may cause erratic operation of the feed and brine pump. Rapid fluctuations of pressure in the last effect tend to cause priming.

To achieve satisfactory operation of a distilling unit, the designed vacuum in all effects must be maintained. When the unit is operated at less than the designed vacuum, the heat level rises throughout the unit and there is an increased tendency toward scale formation. Scale formation is highly undesirable, since scale interferes with heat transfer and thus reduces the capacity of the unit. Excessive scale formation may also impair the quality of the distillate.

Various methods have been used to retard scale formation in distilling units. To retard the formation of scale on evaporator tubes and to minimize priming, solutions are continuously injected into the evaporator.

A chemical compound, either Ameroyal or PD-8, is used in all Navy evaporators that are feed-treated. Ameroyal increases the production of distilled water by decreasing downtime of plants for scale removal.

The removal of scale is accomplished by a procedure called chill shocking (cold shocking). For chill shocking, the unit is secured and pumped dry while it is still hot. Then, cold seawater is introduced and the resulting thermal shock causes scale to flake off and fall to the bottom of the tube nest. The unit is then pumped dry, the loose scale is removed, and the unit is filled with water and started up again. Chill shocking is an effective way of removing scale, but it is somewhat laborious and time-consuming. A particular disadvantage of the chill shocking process is that it requires each operating distilling unit to be out of production for an hour or more each day. This can lead to serious water shortages under some circumstances.

Another method that can be used to remove scale formation is mechanical cleaning. Mechanical cleaning should be used only as a last resort. Clean with chemicals if possible. The evaporator tube nest must be withdrawn from the shell for mechanical cleaning. Lifting gear suitable to the type of installation is usually provided to help remove the tube nest. Clean the tubes with a light scaling tool operated by a light air hammer.

Chemical cleaning is faster, more economical, more effective, and less damaging to evaporator parts than

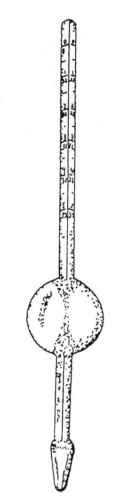

Figure 21-13.—Salinometer.

in the rate of scaling of the evaporator tube surfaces. This causes poor quality distillate due to excessive vapor formation. A high brine density indicates that too much of the feedwater is being converted into distillate. The ideal value of brine density is just under 1.5 thirty-seconds. The density should never exceed this level. Since the average seawater contains approximately 1 part of dissolved sea salts to 32 parts of water (1/32 by weight), the brine density should be just under 1 1/2 times the density of seawater or 1.5/32.

Brine density is measured with a salinometer (fig. 21-13), which works on the same principle as a hydrometer. Salinometers are graduated in thirty-seconds, from 0/32 to 5/32, and calibrated on four different temperatures: 110°, 115°, 120° and 125°F.

Special restrictions are placed upon the operation of distilling units when the ship is operating in contaminated waters. Because most distilling plants operate at low pressures (and therefore low temperatures), the distillate is not sterilized by the boiling process in the evaporators and may contain dangerous microorganisms or other matter harmful to health. All water in harbors, rivers, inlets, and bays, land-locked water, and the open sea within 10 miles of the entrance to such water must be considered contaminated unless a specific determination to the contrary is made. In other areas, contamination may be declared to exist by the fleet surgeon or representatives, as local conditions may warrant. When the ship is operating in contaminated waters, the distilling units must be operated in strict accordance with special procedures established by the Naval Ship Systems Command.

Additional information on distilling plants can be found in *Machinist's Mate 3 & 2*, NAVEDTRA 12144, chapter 9.

REFERENCES

Machinist's Mate 3 & 2, NAVEDTRA 12144, Naval Education and Training Program Management Support Activity, Pensacola, Florida, 1991.

mechanical cleaning. In chemical cleaning, a heated diluted acid solution circulates through the saltwater circuits of the system. The three acids used are hydrochloric, sulfamic, and citric.

The concentration of brine (or brine density, as it is called) has a direct bearing on the quality of the distillate. If the brine concentration is too low, there will be a loss in capacity and economy. A low brine density indicates that not enough of the feedwater is being converted to distillate. It means poor efficiency in the plant, which results in a reduced output capacity. If the brine concentration is too high, there will be an increase

CHAPTER 22

REFRIGERATION AND AIR CONDITIONING

LEARNING OBJECTIVES

Upon completion of this chapter, you should be able to do the following:

1. Identify the most common types of refrigeration plants used aboard Navy ships.

2. Identify the major components of refrigeration plants and their functions.

3. Explain the purpose of air conditioning and identify the various principles affecting its efficiency.

4. Identify the safety precautions to observe when handling refrigerants.

INTRODUCTION

Refrigeration equipment is used aboard ship for several purposes, including the refrigeration of ship's stores and cargo, the cooling of water, and the conditioning of air for certain spaces. The distinction between refrigeration and air conditioning is that refrigeration is only a cooling process, while air conditioning is a process of treating air to control simultaneously its temperature, humidity, cleanliness, and distribution to meet the requirements of the conditioned spaces.

REFRIGERATION

Refrigeration is a general term. It describes the process of removing heat from spaces, objects, or materials and maintaining them at temperatures below that of the surrounding atmosphere. To produce a refrigeration effect, the material to be cooled needs only to be exposed to a colder object or environment. The heat will flow in its "natural" direction–that is, from the warmer material to the colder material. For example, a pan of hot water placed on a cake of ice will be cooled by the flow of heat from the hot water to the ice. We can maintain this refrigeration effect as long as the ice lasts. But no matter how much ice we have, we cannot produce a refrigeration effect any greater than the cooling of the water to 32°F. We cannot, for example, cause the water to freeze by this method, since freezing would require the removal of the latent heat of fusion from the water after it had been cooled to 32°F. For this process, we would need a temperature difference that does not exist when both the water and the ice are at

32°F. When the purpose of refrigeration is the production of ice or the maintenance of temperatures lower than 32°F at atmospheric pressure, it is obvious that ice is not a suitable refrigerant. Refrigeration, then, usually means an artificial way of lowering the temperature. Mechanical refrigeration is a mechanical system or apparatus that transfers heat from one substance to another.

Refrigeration is a process involving the flow of heat and is therefore a thermodynamic process. (Thermodynamic cycles are discussed in chapter 8 of this text.) In this text, we have been concerned primarily with a closed cycle in which thermal energy (in the form of heat) is converted into mechanical energy (in the form of work). A closed cycle is one in which the working fluid never leaves the system except through accidental leakage. Instead, the working fluid undergoes a series of processes that are of such a nature that the fluid is returned periodically to its initial state and is then used again. Now, instead of wanting to convert heat into work, we want to remove heat from a body and we want to continue to remove heat from this body even after its temperature has been lowered below that of its surroundings to maintain the body at its lowered temperature. In other words, we want to extract heat from a cold body and discharge it to a warm area. The question is How can this be done, since we know from the second law of thermodynamics that heat cannot, of itself, flow from a colder body or region to a warmer one? It is entirely possible to extract heat from a body at a low temperature and discharge it to a body or region at a higher temperature, provided a suitable expenditure of energy is made to accomplish this. The energy supplied to the refrigeration cycle for this purpose is in the form of work (mechanical energy) done on the working fluid (refrigerant) by a compressor. (A compressor provides the required energy in a vapor-compression refrigeration cycle, the cycle most commonly used in naval refrigeration plants. Other kinds of refrigeration cycles use other forms of energy to accomplish the same purpose–namely, to raise the temperature of the refrigerant after it has absorbed heat from the space or object to be cooled.) In a refrigeration cycle, the refrigerant must alternate between low temperatures and high temperatures. When the refrigerant is at a low temperature, heat flows from the space or object to be cooled to the refrigerant. When the refrigerant is at a high temperature, heat flows from the refrigerant to a condenser. The energy supplied as work is used to raise the temperature of the refrigerant to a high enough value so the refrigerant will be able to reject heat to the condenser.

Because the energy transformations in a refrigeration cycle occur in an order that is precisely the reverse of the sequence in a power cycle, the refrigeration cycle is sometimes said to be one in which heat is pumped uphill. This view of a refrigeration cycle is entirely legitimate, provided the reverse order of energy transformations does not imply actual thermodynamic reversibility. True thermodynamic reversibility was considered an impossibility. A refrigeration cycle does not give us something for nothing. Instead we must put energy into the cycle to extract heat at a low temperature and discharge it at a higher temperature.

DEFINITIONS OF REFRIGERATION TERMS

Some of the standard terms used in the discussion of refrigeration are defined in this section. A few of these terms have been defined in chapter 8 of this manual but are briefly noted here because of their importance in the study of refrigeration.

UNIT OF HEAT: The British thermal unit (Btu) is the standard unit of heat measurement used in refrigeration, as in most other engineering applications. By definition, 1 Btu is equal to 778.26 ft-lb.

SPECIFIC HEAT: The amount of heat required to raise the temperature of 1 pound of a substance 1°F. All substances are compared to water, which has a heat of 1 Btu per pound per degree Fahrenheit (1 Btu/lb/°F).

SENSIBLE HEAT: Heat that is given off or absorbed by a substance without changing its state.

LATENT HEAT OF VAPORIZATION: The amount of heat required to change the state of a substance from a liquid to a vapor without a change in temperature.

LATENT HEAT OF FUSION: The heat that must be removed from a liquid to change it into a solid (or, on the other hand, the amount of heat that must be added to a solid to change it to a liquid) without any change in temperature.

REFRIGERATING EFFECT: Since the heat removed from an object that is being refrigerated is absorbed by the refrigerant, the refrigerating effect of a refrigeration cycle is defined as the heat gain per pound of refrigerant.

REFRIGERATION TON: The unit of measure for the amount of heat removed. The capacity of a refrigeration unit is usually stated in tons. The refrigeration ton is based on the cooling effect of 1 ton (2,000 pounds) of ice at 32°F melting in 24 hours. The latent heat of fusion of ice (or water) is approximately 144 Btu. Therefore, the number of Btu required to melt 1 ton of ice is 144 × 2,000 or 288,000. The standard refrigeration ton is defined as the transfer of 288,000 Btu in 24 hours. On an hourly basis, the refrigeration ton is 12,000 Btu per hour (288,000 divided by 24 equals 12,000). The refrigeration ton is the standard unit of measure used to designate the heat removal capacity of a refrigeration unit. It is not a measure of the ice-making capacity of a machine, since the amount of ice that can be made depends upon the initial temperature of the water and other factors.

COEFFICIENT OF PERFORMANCE: The coefficient of performance of a refrigeration cycle is comparable to the thermal efficiency of a power cycle. The thermal efficiency of a power cycle is given by the equation

$$\textit{Thermal efficiency} = \frac{work\ output}{heat\ input}$$

Since thermal efficiency is a function of absolute temperature alone in the Carnot cycle, the equation may also be given as

$$\textit{Thermal efficiency} = \frac{T_s - T_r}{T_s}$$

where T_s is the absolute temperature at the heat source and T_r is the absolute temperature at the heat receiver.

For the refrigeration cycle, the coefficient of performance is given by the equation

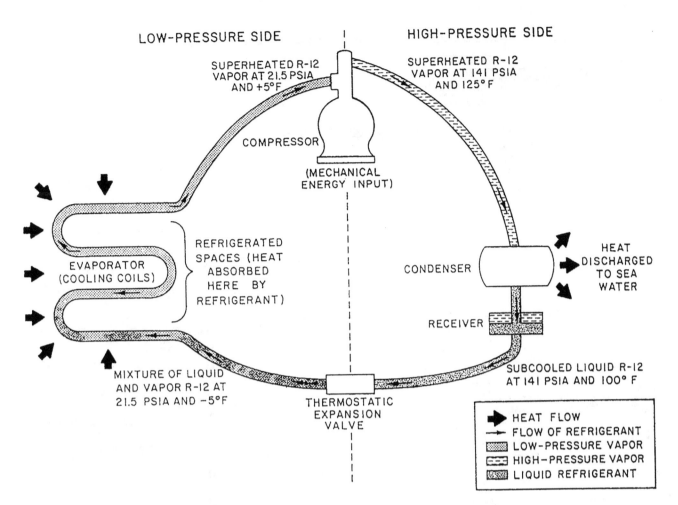

LOW-PRESSURE SIDE | HIGH-PRESSURE SIDE

SUPERHEATED R-12 VAPOR AT 21.5 PSIA AND +5°F

SUPERHEATED R-12 VAPOR AT 141 PSIA AND 125°F

COMPRESSOR

(MECHANICAL ENERGY INPUT)

EVAPORATOR (COOLING COILS)

REFRIGERATED SPACES (HEAT ABSORBED HERE BY REFRIGERANT)

CONDENSER

HEAT DISCHARGED TO SEA WATER

RECEIVER

MIXTURE OF LIQUID AND VAPOR R-12 AT 21.5 PSIA AND −5°F

SUBCOOLED LIQUID R-12 AT 141 PSIA AND 100° F

THERMOSTATIC EXPANSION VALVE

HEAT FLOW
FLOW OF REFRIGERANT
LOW-PRESSURE VAPOR
HIGH-PRESSURE VAPOR
LIQUID REFRIGERANT

Figure 22-1.–Schematic representation of an R-12 refrigeration cycle.

$$Coefficient\ of\ performance = \frac{refrigerating\ effect}{work\ output}$$

which, as in the power cycle, can be shown to be a function of absolute temperature alone.

R-12 PLANT

Most refrigeration systems used by the Navy use R-12 as the refrigerant. Chemically, R-12 is dichlorodifluoromethane (CCl_2F_2). R-12 has such a low boiling point that it cannot exist as a liquid unless it is confined in a container under pressure. For example, R-12 boils at −21°F at atmospheric pressure, at 0°F at 9.17 psig, at 50°F at 46.69 psig, and at 100°F at 116.9 psig. Because of its low boiling point, R-12 is well suited for use in refrigeration systems designed for only moderate pressures. It also has the advantage of being practically nontoxic, nonflammable, nonexplosive, and noncorrosive. It also does not poison or contaminate foods.

The R-12 refrigeration system is classified as a mechanical system of the vapor-compression type. It is a mechanical system because the energy input is in the form of mechanical energy (work). It is a vapor-compression system because compression of the vaporized refrigerant is the process that allows the refrigerant to discharge heat at a high temperature.

R-12 CYCLE

The basic cycle of an R-12 refrigeration cycle is shown schematically in figure 22-1. As an introduction to the system, you should trace the refrigerant through the entire cycle, noting especially the points at which the refrigerant changes from liquid to vapor and from vapor to liquid. You also should note the concomitant flow of heat in one direction or another.

As shown in figure 22-1, the cycle has two pressure sides. The low-pressure side extends from the orifice of the thermostatic expansion valve (TXV) up to and including the intake side of the compressor cylinders. The high-pressure side extends from the discharge side of the compressor to the TXV. The condensing and evaporating pressures and temperatures indicated in figure 22-1 are not standard for all refrigeration plants, since pressures and temperatures are established as part

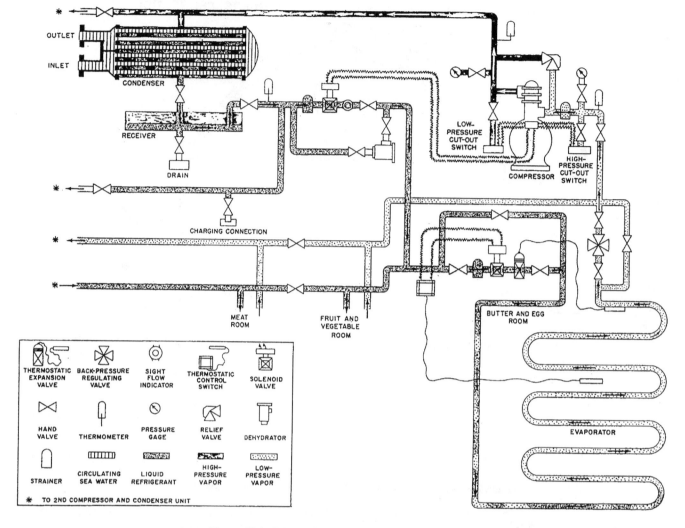

Figure 22-2.–Diagram of an R-12 refrigeration system.

of the design of any refrigeration system. Note also that the pressures and temperatures shown in figure 22-1 are theoretical rather than actual values, even for this system. If the system were in actual operation, the pressures and temperatures would vary slightly because they are dependent upon the temperature of the cooling water entering the condenser, the amount of heat absorbed by the refrigerant in the evaporator, and other factors.

Liquid R-12 enters the TXV at high pressure from the high-pressure side of the system. The refrigerant leaves the outlet of the expansion valve at a much lower pressure and enters the low-pressure side of the system. Because of the low pressure, the liquid refrigerant begins to boil and some of it flashes to a vapor.

From the TXV the refrigerant passes into the cooling coil (or evaporator). The boiling point of the refrigerant under the low pressure in the evaporator is about 20°F lower than the temperature of the spaces in which the cooling coil is installed. As the liquid boils

and vaporizes, it picks up its latent heat of vaporization from the space being cooled. The refrigerant continues to absorb heat until all the liquid has been vaporized. By the time the refrigerant leaves the cooling coil, it has not only absorbed this latent heat of vaporization but also picked up some additional heat–that is the vapor has become slightly superheated. As a rule, the amount of superheat is about 4° to 12°F.

The refrigerant leaves the evaporator as low-pressure superheated vapor, having absorbed heat and thus cooled the space. The remainder of the cycle is used to dispose of this heat and convert the refrigerant back into a liquid state so it can again vaporize in the evaporator and absorb the heat again.

The low-pressure superheated vapor is drawn out of the evaporator to the suction side of the compressor. The compressor is the unit that keeps the refrigerant circulating through the system. In the compressor cylinders, the refrigerant is compressed from a

Figure 22-3.–High-pressure side of an R-12 installation aboard ship.

low-pressure, low-temperature vapor to a high-pressure vapor and its temperature rises accordingly.

The high-pressure R-12 vapor is discharged from the compressor into the condenser. Here the refrigerant condenses, giving up its superheat and its latent heat of condensation. The condenser may be air or water cooled. The refrigerant still at high pressure, is now a liquid again.

From the condenser, the refrigerant flows into a receiver that serves as a storage place for the liquid refrigerant. From the receiver, the refrigerant goes to the TXV and the cycle begins again.

From this brief summary of an R-12 vapor compression refrigeration system, you can see that the cycle is indeed one in which heat is pumped uphill as a result of the arrangements that cause the refrigerant to

go through successive phases of expansion, evaporation, compression, and condensation.

MAJOR COMPONENTS

The major components of a shipboard R-12 refrigeration plant are shown diagrammatically in figure 22-2. The primary parts of the system are the TXV, the evaporator, the compressor, the condenser, and the receiver. Additional equipment required to complete the plant includes piping, pressure gauges, thermometers, various types of control switches and control valves, strainers, relief valves, sight-flow indicators, dehydrators, and charging connections. Figure 22-3 shows most of the components on the high-pressure side of an R-12 system as actually installed aboard ship.

In the following discussion of the major components of an R-12 system, we will deal with the

22-5

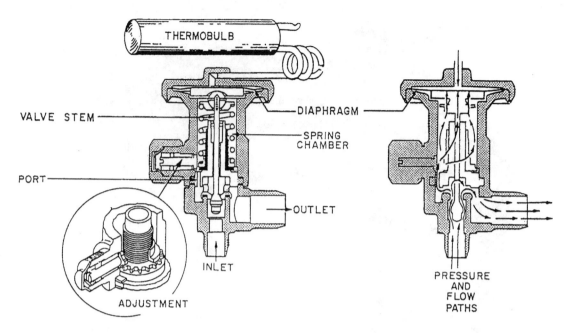

Figure 22-4.–Thermostatic expansion valve.

R-12 system as though it had only one evaporator, one compressor, and one condenser. As you can see from figure 22-2, however, a shipboard refrigeration system may (and usually does) include more than one evaporator, and it may include an additional compressor and condenser units to provide operational flexibility and to protect against loss of refrigerating capacity.

Thermostatic Expansion Valve

The TXV, shown in figure 22-4, is a reducing valve between the high-pressure side and the low-pressure side of the system. The valve regulates the amount of refrigerant to the cooling coil. The amount of refrigerant needed in the coil depends, of course, on the temperature of the space being cooled.

The thermo bulb, which controls the opening and closing of the TXV, is clamped to the cooling coil, near the outlet. The substance in the thermo bulb will vary depending on the refrigerant used. Control tubing connects the bulb with the area above the diaphragm in the TXV. When the temperature at the bulb rises, the R-12 expands and transmits a pressure to the diaphragm. This causes the diaphragm to be moved downward, thus opening the valve and allowing more refrigerant to enter the cooling coil. When the temperature at the bulb falls, the pressure above the diaphragm decreases and the valve closes. Thus, the temperature near the evaporator outlet controls the operation of the TXV.

Evaporator

The evaporator consists of a coil of copper, aluminum, or aluminum alloy tubing installed in the space to be refrigerated. Figure 22-5 shows some of this tubing. The liquid R-12 enters the tubing at a reduced pressure and, therefore, with a lower boiling point. In passing through the expansion valve, going from the high-pressure side of the system to the low-pressure side, some of the refrigerant boils and vaporizes because of the reduced pressure and some of the remaining liquid refrigerant is thereby cooled to its boiling point. Then, as the refrigerant passes through the evaporator, the heat flowing to the coil from the surrounding air causes the rest of the liquid refrigerant to boil and vaporize. After the refrigerant has absorbed its latent heat of

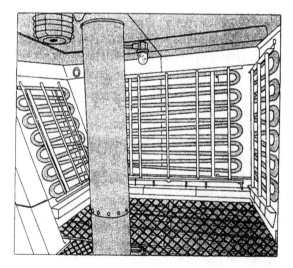

Figure 22-5.–Evaporator tubing.

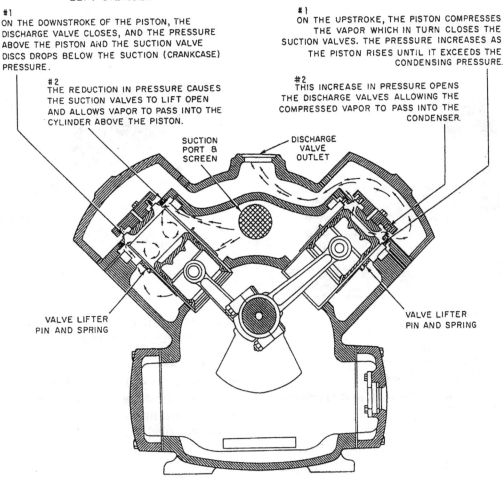

LEFT CYLINDER

#1
ON THE DOWNSTROKE OF THE PISTON, THE
DISCHARGE VALVE CLOSES, AND THE PRESSURE
ABOVE THE PISTON AND THE SUCTION VALVE
DISCS DROPS BELOW THE SUCTION (CRANKCASE)
PRESSURE.

#2
THE REDUCTION IN PRESSURE CAUSES
THE SUCTION VALVES TO LIFT OPEN
AND ALLOWS VAPOR TO PASS INTO THE
CYLINDER ABOVE THE PISTON.

RIGHT CYLINDER

#1
ON THE UPSTROKE, THE PISTON COMPRESSES
THE VAPOR WHICH IN TURN CLOSES THE
SUCTION VALVES. THE PRESSURE INCREASES AS
THE PISTON RISES UNTIL IT EXCEEDS THE
CONDENSING PRESSURE.

#2
THIS INCREASE IN PRESSURE OPENS
THE DISCHARGE VALVES ALLOWING THE
COMPRESSED VAPOR TO PASS INTO THE
CONDENSER.

SUCTION
PORT &
SCREEN

DISCHARGE
VALVE
OUTLET

VALVE LIFTER
PIN AND SPRING

VALVE LIFTER
PIN AND SPRING

Figure 22-6.–Reciprocating compressor.

vaporization (that is, after it is entirely vaporized), the refrigerant continues to absorb heat until it becomes superheated by approximately 10°F. The amount of superheat is determined by the amount of liquid refrigerant admitted to the evaporator. This, in turn, is controlled by the spring adjustment of the TXV. A temperature range of 4° to 12°F of superheat is considered desirable. It increases the efficiency of the plant and evaporates all of the liquid. This prevents liquid carry-over into the compressor.

Compressor

The compressor in a refrigeration system is a pump. It is used to pump heat uphill from the cold side to the hot side of the system.

The heat absorbed by the refrigerant in the evaporator must be removed before the refrigerant can again absorb latent heat. The only way the vaporized refrigerant can be made to give up the latent heat of vaporization that it absorbed in the evaporator is by cooling and condensing it. Because of the high temperature of the available cooling medium (seawater), the only way to make the vapor condense is by first compressing it.

The vapor drawn into the compressor is at very low pressure and very low temperature. In the compressor, both the pressure and the temperature are raised. When we raise the pressure, we also raise the temperature. Therefore, we have raised its condensing temperature, which allows us to use seawater as a cooling medium in the condenser. In other words, the compressor raises the pressure of the vaporized refrigerant sufficiently high to permit heat transfer and condensation to take place in the condenser.

In addition to this primary function, the compressor also keeps the refrigerant circulating and maintains the required pressure difference between the high-pressure and the low-pressure sides of the system.

Many different types of compressors are used in refrigeration systems. The designs of compressors vary depending on the application of the refrigerants used in the system. Figure 22-6 shows a motor-driven,

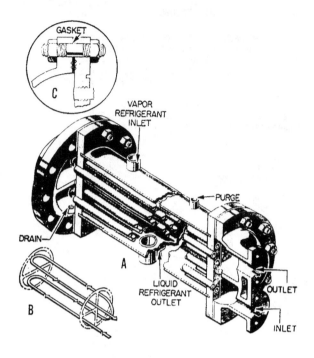

A. Cutaway view.
B. Water-flow diagram.
C. Arrangement of heat joint.

Figure 22-7.–Water-cooled condenser for an R-12 refrigeration system.

single-acting, two-cylinder, reciprocating compressor, such as those commonly used in naval shipboard refrigeration plants.

Condenser

The compressor discharges the high-pressure, high-temperature refrigerant vapor to the condenser, where it flows around the tubes through which seawater is being pumped. As the vapor gives up its superheat to the circulating seawater, the temperature of the vapor drops to the condensing point. As soon as the temperature of the vapor drops to its condensing point at the existing pressure, the vapor condenses and in the process gives up the latent heat of vaporization that it picked up in the evaporator. The refrigerant, now in liquid form, is subcooled slightly below its condensing point. This is done at the existing pressure to make sure it will not flash into vapor.

A water-cooled condenser for an R-12 refrigeration system is shown in figure 22-7. Circulating water is obtained through a branch connection from the firemain or by an individual pump taking suction from the sea. A water regulating valve is usually installed to control the flow of cooling water through the condenser. The purge connection shown in figure 22-6 is on the refrigerant

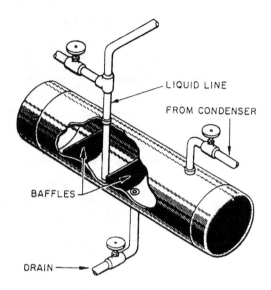

Figure 22-8.–Receiver.

side. It is used to remove air and other noncondensable gases that are lighter than the R-12 vapor.

Most condensers used in naval refrigeration plants are of the water-cooled type. However, some small units have air-cooled condensers. These consist of tubing with external fins to increase the heat transfer surface. Most air-cooled condensers have fans to ensure positive circulation of air around the condenser tubes.

Receiver

The receiver, shown in figure 22-8, acts as a temporary storage space and surge tank for the liquid refrigerant. The receiver also serves as a vapor seal to keep vapor out of the liquid line to the TXV. Receivers may be constructed for either horizontal or vertical installment.

Accessories and Controls

In addition to the five major components just described, a refrigeration system requires several controls and accessories. The most important of these are discussed briefly in the following paragraphs.

DEHYDRATOR.–A dehydrator, or dryer, containing silica-gel or activated alumina, is placed in the liquid refrigerant line between the receiver and the TXV. In older installations, such as the one shown in figure 22-2, bypass valves allow the dehydrator to be cut in or out of the system. In newer installations, the dehydrator is installed in the liquid refrigerant line without any bypass arrangement. A refrigerant dehydrator is shown in figure 22-9.

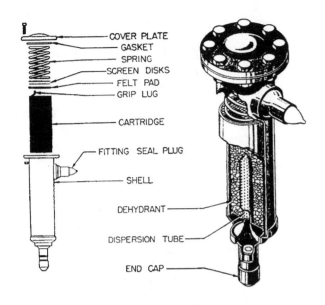

Figure 22-9.–Refrigerant dehydrator.

COVER PLATE
GASKET
SPRING
SCREEN DISKS
FELT PAD
GRIP LUG
CARTRIDGE
FITTING SEAL PLUG
SHELL
DEHYDRANT
DISPERSION TUBE
END CAP

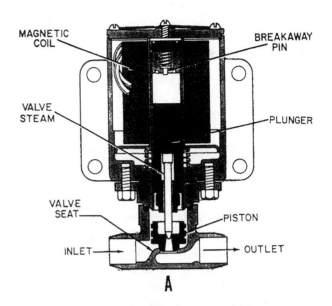

MAGNETIC COIL
BREAKAWAY PIN
VALVE STEAM
PLUNGER
VALVE SEAT
PISTON
INLET
OUTLET

A

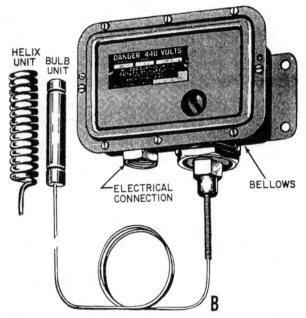

HELIX UNIT
BULB UNIT
DANGER 440 VOLTS
ELECTRICAL CONNECTION
BELLOWS

B

Figure 22-10.–Solenoid valve and thermostatic control switch.

SOLENOID VALVE AND THERMOSTATIC CONTROL SWITCH.–A solenoid valve is installed in the liquid line leading to each evaporator. Figure 22-10 shows a solenoid valve and the thermostatic control switch that operates it. The thermostatic control switch is connected by long flexible capillary tubing to a thermal bulb that is located in the refrigerated space. When the temperature in the refrigerated space drops to the desired point, the thermal bulb causes the thermostatic control switch to open. This action closes the solenoid valve and shuts off all flow of liquid refrigerant to the TXV. When the temperature in the refrigerated space rises above the desired point, the thermostatic control switch closes, the solenoid valve opens, and liquid refrigerant once again flows to the TXV.

The solenoid valve and its related thermostatic control switch maintain the proper temperature in the refrigerated space. You may wonder why the solenoid valve is necessary if the TXV controls the amount of refrigerant admitted to the evaporator. Actually, the solenoid valve is not necessary in systems that have only one evaporator. In systems that have more than one evaporator, where there is wide variation in load, the solenoid valve provides additional control to prevent the spaces from becoming too cold at light loads.

In addition to the solenoid valve installed in the line to each evaporator, a large refrigeration plant usually has a main liquid line solenoid valve installed just after the receiver. If the compressor stops for any reason except normal suction pressure control, the main liquid line solenoid valve closes. This prevents liquid refrigerant

from flooding the evaporator and flowing to the compressor suction. Extensive damage to the compressor can result if liquid is allowed to enter the compressor suction.

EVAPORATING PRESSURE REGULATING VALVE.–In some ships, several refrigerated spaces of varying temperatures are maintained by one compressor. In these cases, an evaporator pressure regulating valve is installed at the outlet of each evaporator except the evaporator in the space in which the lowest temperature is to be maintained. The evaporator pressure regulating valve is set to keep the pressure in the coil from falling below the pressure corresponding to the lowest evaporator temperature desired in that space.

LOW-PRESSURE CUTOUT SWITCH.–The low-pressure cutout switch is the control that causes the compressor to go on or off as required for the normal operation of the refrigeration plant. This switch is located on the suction side of the compressor and is actuated by pressure changes in the suction line. When the solenoid valves in the lines to the various evaporators are closed so the flow of refrigerant to the evaporators is stopped, the pressure of the vapor in the compressor suction line drops quickly. When the suction pressure has dropped to the desired pressure, the low-pressure cutout switch causes the compressor motor to stop. When the temperature in the refrigerated space has risen enough to operate one or more of the solenoid valves, refrigerant is again admitted to the cooling coils, and the compressor suction pressure builds up again. At the desired pressure, the low-pressure cutout switch closes, starting the compressor again and repeating the cycle.

HIGH-PRESSURE CUTOUT SWITCH.–A high-pressure cutout switch is connected to the compressor discharge line to protect the high-pressure side of the system against excessive pressures. The design of this switch is very similar to the low-pressure cutout switch. However, the low-pressure cutout switch is designed to CLOSE when the suction pressure reaches its upper normal limit. The high-pressure cutout switch is designed to OPEN when the discharge pressure is too high. The high-pressure cutout switch is normally set to stop the compressor when the pressure reaches 160 psi and to start it again when the pressure drops to 140 psi. As previously noted, the low-pressure cutout switch is the compressor control for normal operation of the plant. The high-pressure cutout switch, on the other hand, is a safety device only and does not have control of compressor operation under normal conditions.

SPRING-LOADED RELIEF VALVE.–A spring-loaded relief valve is installed in the compressor discharge line as an additional precaution against excessive pressures. The relief valve is set to open at about 225 psig. Therefore, it functions only in case of failure or improper setting of the high-pressure cutout switch. If the relief valve opens, it discharges high-pressure vapor to the suction side of the compressor.

WATER REGULATING VALVE.–A water regulating valve, as shown in figure 22-11, controls the quantity of circulating water flowing through the

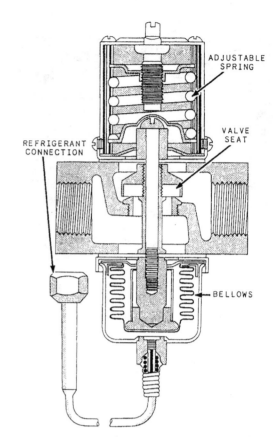

Figure 22-11.–Water regulating valve (cross section).

refrigerant condenser. The valve is located either at the inlet to the condenser or at the outlet from the condenser. The valve is actuated by the refrigerant pressure in the compressor discharge line. This pressure acts upon a diaphragm (or in some valves, a bellows arrangement) that transmits motion to the valve stem. As temperature of the circulating water increases, the temperature of the refrigerant vapor increases. This causes the pressure of the refrigerant to increase and thereby raises the condensation point. When this occurs, the increased pressure of the refrigerant causes the water regulating valve to open wider, thus automatically permitting more circulating water to flow through the condenser. When the condenser is cooler than necessary, the water regulating valve allows less water to flow through the condenser. Thus, the flow of cooling water through the condenser is automatically maintained at the rate actually required to condense the refrigerant under varying conditions of load and temperature.

WATER FAILURE SWITCH.–A water failure switch stops the compressor in the event of failure of the circulating water supply. This is a pressure-actuated switch. It is similar to the low-pressure cutout switch and the high-pressure cutout switch previously described. If the water failure switch fails to function, the refrigerant pressure in the condenser will quickly

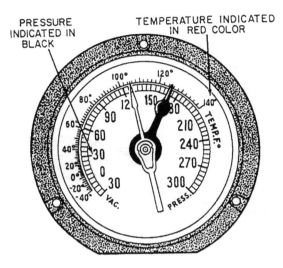

PRESSURE INDICATED IN BLACK

TEMPERATURE INDICATED IN RED COLOR

Figure 22-12.–Compound R-12 pressure gauge.

build up to the point where the high-pressure cutout switch stops the compressor.

STRAINER.–Because of the solvent action of R-12, any particles of grit, scale, dirt, and metal that the system may contain are very readily circulated through the refrigerant lines. To avoid damage to the compressor from such foreign matter, a strainer is installed in the compressor suction connection. In addition, a liquid strainer is installed in the liquid line leading to each evaporator; these strainers protect the solenoid valves and the TXVs.

PRESSURE GAUGES AND THERMOMETERS.–Several pressure gauges and thermometers are used in refrigeration systems. Figure 22-12 shows a compound refrigerant gauge. The temperature markings on this gauge show the boiling point (or condensing point) of the refrigerant at each pressure; the gauge cannot measure temperature directly. The dark pointer (which is actually red in color) is a stationary marker that can be set manually to indicate the maximum working pressure.

A water pressure gauge is installed in the circulating water line to the condenser to indicate failure of the circulating water supply.

Standard thermometers of appropriate range are provided for the refrigerant systems.

REFRIGERANT PIPING.–Refrigerant piping is normally made of copper. Copper is good for this purpose because (1) it does not become corroded by the refrigerants; (2) the internal surface of the tubing is smooth enough to minimize friction; and (3) copper tubing is easily shaped to meet installation requirements.

AIR CONDITIONING

Air conditioning is a field that deals with the design, construction, and operation of equipment used to establish and maintain desirable indoor air conditions. It is used to maintain the environment of an enclosure at any required temperature, humidity, and purity. Simply stated, air conditioning involves the cooling, heating, dehumidifying, ventilating, and purifying of air.

Aboard ship, air conditioning keeps the ship's crew comfortable, alert, and physically fit. The temperature, humidity, cleanliness, quantity, and distribution of the conditioned air supply is a matter of vital concern.

The comfort and efficiency of the crew is not the only immediate reason for shipboard air conditioning. Mechanical cooling, heating, or ventilating must be provided for several spaces for a variety of reasons. Ammunition spaces must be kept below a certain temperature to prevent deterioration of the ammunition; gas storage spaces must be kept cool to prevent the buildup of excessive pressures in containers; electrical and electronic equipment must be maintained at certain temperatures, with controlled humidity, in air that is relatively free of dust and dirt.

PRINCIPLES OF AIR CONDITIONING

To achieve the objectives of air conditioning, you must take into account several factors. The principal factors that are important about air conditioning are discussed in the following sections.

Humidity

Humidity is the vapor content of the atmosphere; it has great influence on human comfort. Excessive humidity and too little humidity both lead to discomfort and impaired efficiency; hence, the measurement and control of the moisture content of the air is an important phase of air conditioning.

SATURATED AIR.–Air can hold varying amounts of water vapor. It depends on the temperature of the air at a given atmospheric pressure. As the temperature rises, the amount of moisture the air can hold increases. But, for every temperature there is a definite limit to the amount of moisture the air can hold. When air contains the maximum amount of moisture it can hold at a specific temperature and pressure, the air is said to be saturated.

The saturation point is usually called the dewpoint. If the temperature of saturated air falls below its

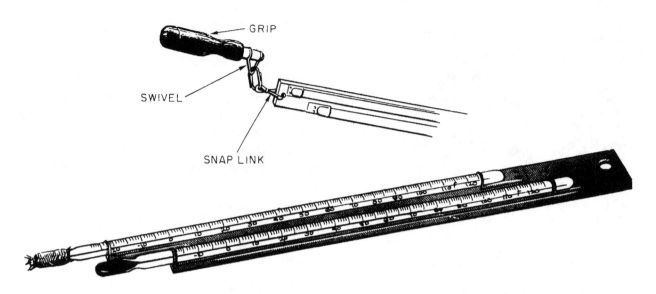

Figure 22-13.–A standard sling psychrometer.

dewpoint, some of the water vapor in the air must condense to water. An example is the dew that is visible in early morning after a drop in temperature. Another is the "sweating" of cold water pipes as water vapor from the warm air condenses on the cold surfaces of the pipes.

ABSOLUTE AND SPECIFIC HUMIDITY.–The amount of water vapor in the air is expressed in terms of the weight of the water vapor. This weight is usually given in grains (7,000 grains = 1 pound). Absolute humidity is the weight of water vapor (in grains) per cubic foot of air. Specific humidity is the weight of water vapor (in grains) per pound of air. The weight of water vapor refers only to the weight of the moisture that may be present in the vapor state, such as rain or dew.

RELATIVE HUMIDITY.–Relative humidity is the ratio of the weight of water vapor in a sample of air to the weight of water vapor the same sample of air would contain if saturated at the existing temperature. This ratio is usually stated as a percentage. For example, when air is fully saturated, the relative humidity is 100 percent. When air contains no moisture at all, its relative humidity is 0 percent. If air is half saturated–that is, holding half as much moisture as it is capable of holding at the existing temperature–its relative humidity is 50 percent.

The deciding factor in human comfort is the relative humidity–not the absolute or specific humidity. This is true because it is the relative humidity that affects evaporation. Moisture travels from regions of greater wetness to regions of lesser wetness. If the air above a liquid is saturated, the two are in balance and no moisture can travel from the liquid to the air; that is, the liquid cannot evaporate. If the air above the liquid is only partly saturated, some moisture can travel to the air; that is, some evaporation can take place.

An example may illustrate the difference between absolute or specific humidity and relative humidity. When the temperature of the air is 76°F and with the specific humidity of the air remaining constant at 120 grains per pound, the relative humidity is nearly 90 percent, the air is nearly saturated. At such a relative humidity, the body may perspire freely but the perspiration does not evaporate rapidly; thus, a general feeling of discomfort results. However, when the temperature of the air is 86°F, with the specific humidity remaining constant at 120 grains per pound, the relative humidity would then be only 64 percent. Although the absolute amount of moisture in the air is the same, the relative humidity is lower, because at 86°F the air can hold more water vapor than it can hold at 76°F. The body is now able to evaporate its excess moisture and the general feeling of comfort is much greater, though the temperature is 10° higher.

Temperature

To test the effectiveness of air-conditioning equipment and to check the humidity of a space, two different temperatures are usually considered. These are the dry-bulb and the wet-bulb temperatures.

The dry-bulb temperature is the temperature sensible heat of the air, as measured by an ordinary dry-bulb thermometer. The dry-bulb temperature reflects the sensible heat of the air.

The wet-bulb temperature is the temperature of the air as measured by a wet-bulb thermometer. A wet-bulb

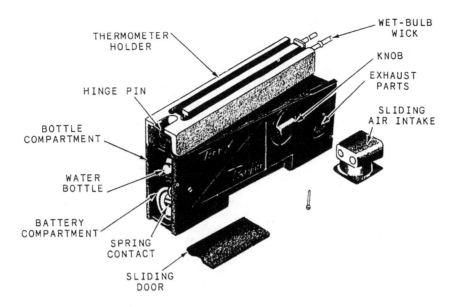

Figure 22-14.--Exposed view of a hand electric psychrometer.

thermometer is an ordinary thermometer with a loosely woven cloth sleeve or wick placed around the bulb that is then wet with water. The water in the sleeve or wick is made to evaporate by a current of air at high velocity. The evaporation lowers the temperature of the wet-bulb thermometer. The difference between the dry-bulb and the wet-bulb temperatures is called the wet-bulb depression. When the dry-bulb temperature is the same as the wet-bulb temperature (that is, evaporation cannot take place), the air is saturated. The condition of saturation is unusual, however, and a wet-bulb depression is normally expected.

The wet-bulb and the dry-bulb thermometers are usually mounted side by side on a frame that has a handle or a short chain attached. This allows the thermometers to be whirled in the air, thus providing a high velocity air current to promote evaporation. Such a device is known as a sling psychrometer (fig. 22-13). Motorized psychrometers are provided with a small motor-driven fan and dry-cell batteries. An exposed view of a hand electric psychrometer is shown in figure 22-14. Motorized psychrometers are generally preferred and are gradually replacing sling psychrometers. With either type of psychrometer, you should observe the wet-bulb temperature at intervals as the water is being evaporated. The point at which there is no further drop in temperature on the wet-bulb thermometer is the wet-bulb temperature for that space.

As you can see from this discussion, the wet-bulb depression is an indication that latent heat of vaporization has been used to vaporize the water in the sleeve or wick around the wet-bulb thermometer.

When the air contains some moisture but is not saturated, the dewpoint temperature is lower than the dry-bulb temperature; the wet-bulb temperature is between the dewpoint and the dry-bulb temperatures. As the amount of moisture in the air increases, the difference between the dry-bulb temperature and the wet-bulb temperature becomes less. When the air is saturated, the dewpoint temperature, the dry-bulb temperature, and the wet-bulb temperature are the same.

Effect of Air Motion

In perfectly still air, the layer of air around a body absorbs the sensible heat given off by the body and increases in temperature. This layer of air also absorbs some of the water vapor given off by the body, thus increasing in relative humidity. This means the body is surrounded by an envelope of moist air that is at a higher temperature and relative humidity than the ambient air. Therefore, the amount of heat that the body can lose to this envelope of motionless air is considerably less than the amount it can lose to the ambient air. When the air is set in motion past the body, the motionless envelope of air is broken up and replaced by ambient air, thereby increasing the heat loss from the body. When the increased heat loss improves the heat balance of the body, the sensation of a "breeze" is felt; when the increase is excessive, the rate of heat loss makes the body feel cool and the sensation of a "draft" is felt.

Sensation of Comfort

From the previous discussion, it is evident that the three factors–temperature, relative humidity, and air

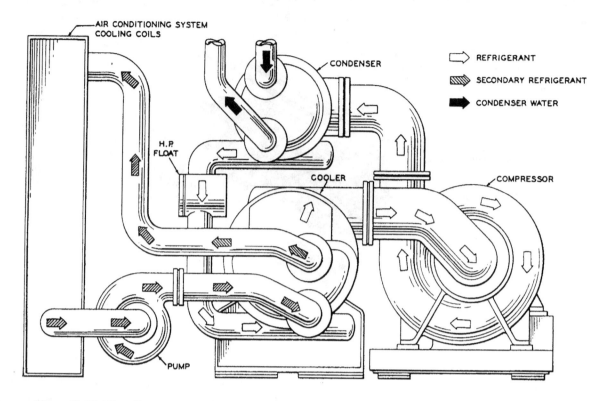

Figure 22-15.--Flow diagram of a chilled water circulating system with a single-stage centrifugal compressor.

motion--are closely interrelated in their effects upon the comfort and health of personnel aboard ship. In fact, a given combination of temperature, relative humidity, and air motion produces the same feeling of warmth or coolness as a higher or lower temperature in conjunction with a compensating relative humidity and air motion.

The term used to identify the net effect of these three factors is *effective temperature*. Effective temperature cannot be measured by an instrument, but it can be found on a special psychometric chart when the dry-bulb temperature and the air velocity are known.

Although all of the combinations of temperature, relative humidity, and air motion of a particular effective temperature may produce the same feeling of warmth or coolness, they are not all equally comfortable or healthful. For best health and comfort conditions, you need a relative humidity of 40 to 50 percent for cold weather and from 50 to 60 percent in warm weather. An overall range of 30 to 70 percent is acceptable.

MECHANICAL COOLING EQUIPMENT

Most working and living spaces on newer ships are air conditioned. The equipment used on these ships was carefully tested to see which types would best cool and dehumidify ship compartments. Two basic types of equipment have been found most effective and are now in general use. They are chilled water circulating systems and self-contained air conditioners.

Chilled Water Circulating Systems

Two types of chilled water circulating systems are used for mechanical cooling aboard ship. Both systems use chilled water as the secondary refrigerant, but one type uses R-12 as the primary refrigerant and the other uses R-11. R-12 systems use reciprocating compressors; R-11 systems use centrifugal compressors.

Both types of chilled water circulating systems operate on the same general principle. The secondary refrigerant (chilled water) is circulated to the various cooling coils. Heat from the spaces being cooled is absorbed by the chilled water and is removed from the water by the primary refrigerant in a water chiller.

Figure 22-15 illustrates the flow of primary refrigerant, secondary refrigerant, and condenser water in an R-11 chilled water circulating system that has a single-stage centrifugal compressor. The primary refrigerant vapor goes from the evaporator to the compressor, where it is compressed. It is then discharged to the condenser. In the condenser, the primary refrigerant vapor condenses, giving up its superheat, its latent heat of vaporization, and its heat of compression to the cooling water that flows through the

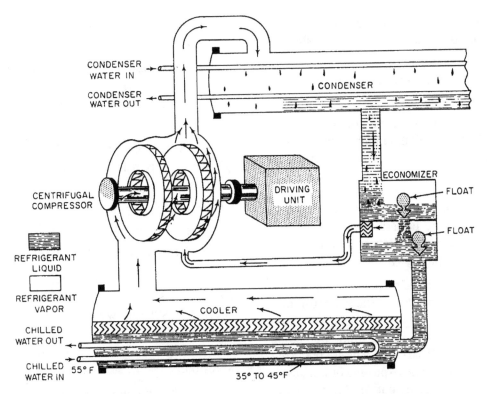

Figure 22-16.–Chilled water circulating system with a two-stage centrifugal compressor.

condenser tubes. The liquid primary refrigerant then passes through a high-pressure float valve to the cooler.

The secondary refrigerant picks up heat in the coils of the air-conditioned space and carries this heat to the cooler. The function of the cooler is to transfer the heat from the secondary refrigerant to the primary refrigerant that surrounds the tubes of the cooler. As this heat is transferred, the liquid primary refrigerant absorbs its latent heat of vaporization, boils, and vaporizes. The quantity of liquid refrigerant thus evaporated varies directly with the amount of heat picked up by the secondary refrigerant. The vaporized primary refrigerant goes to the compressor and the same sequence of events is repeated in cyclical manner.

Figure 22-16 illustrates an R-11 chilled water circulating system with a two-stage centrifugal compressor. The refrigerant vapor coming from the cooler goes into an opening around the hub of the first wheel of the centrifugal compressor. The blades in the rapidly rotating wheel impart velocity to the vapor. The vapor is then directed to the hub of the second wheel, where it is compressed and discharged to the condenser.

Between the condenser and the cooler, the liquid refrigerant passes through an economizer. A float in the upper chamber of the economizer allows the passage of refrigerant into the lower chamber. By connecting the

economizer to the second stage of the compressor, the pressure in the lower chamber is greatly reduced. The reduced pressure causes some of the liquid refrigerant to flash into vapor, thus cooling the remainder of the refrigerant. Thus, the economizer acts as an interstage flash cooler and increases the efficiency of the plant. A float in the lower chamber of the economizer allows the passage of the refrigerant into the cooler. In the cooler, the liquid refrigerant absorbs heat from the water and changes from liquid to a vapor.

Self-Contained Air Conditioners

Ships without central-type air conditioning may use self-contained air-conditioning units. NAVSEA approval is required.

A self-contained air-conditioning unit is simply the type of air conditioner you see installed in the windows of many homes. All that is required for installation is to mount the proper brackets for the unit case and provide electrical power.

These units use nonaccessible hermetically sealed compressors (motor and compressors are contained in a welded steel shell). For this reason, shipboard maintenance or the motor-compressor unit is impractical. The TXV used in these units is preset and nonadjustable. However, a thermostat and fan speed control are normally provided for comfort adjustment.

HEATING AND VENTILATION

Aboard ship, heating is accomplished by steam heaters installed in the ventilation ducts and by space heaters. On steam-driven ships, the steam for the heaters is supplied at reduced pressure from an auxiliary steam system. On diesel-driven ships, the steam is supplied by an auxiliary boiler. Some electric heaters are also used aboard ship. These electric heaters are used primarily for heating spaces that are located at a considerable distance from the steam piping system.

Ventilation is accomplished by fans that supply and exhaust through ventilation duct systems. Most fans used in duct systems are of the axial-flow type because they require less space for installation, but some centrifugal fans are used. Centrifugal fans are preferred for exhaust systems that handle explosive or hot gases. The motors of these fans, being outside the airstream, cannot ignite the explosive gases.

Bracket fans are used in hot weather to provide local circulation. These fans are normally installed in living, hospital, office, commissary, supply, and berthing spaces. Where air-conditioning systems are used, bracket fans are sometimes used to facilitate proper circulation and direction of cold air.

Portable-axial fans with flexible air hoses are used aboard ship for ventilating holds and cofferdams. They are also used in unventilated spaces to clear out stale air or gases before personnel enter, and for emergency cooling of machinery. Most portable fans are of the axial-flow type, driven by electric, "explosive-proof" motors. On ships carrying gasoline, a few air turbine-driven centrifugal fans are normally provided. You can place greater confidence in the explosive-proof characteristics of these fans. Portable fans are used for such purposes as temporary ventilation of compartments after painting, exhausting toxic gases from closed spaces and tanks, and cooling hot areas around machinery while repairs are being made.

SAFETY PRECAUTIONS

Refrigerants are furnished in cylinders for use in shipboard refrigeration and air-conditioning systems. The following precautions must be observed by personnel when handling, using, and storing these cylinders:

1. Never drop cylinders nor permit them to strike each other violently.

2. Never use a lifting magnet or a sling (rope or chain) when handling cylinders. A crane may be used if a safe cradle or platform is provided to hold the cylinders.

3. Keep caps provided for valve protection on cylinders except when the cylinders are being used.

4. Whenever refrigerant is discharged from a cylinder, weigh it immediately and record the weight of the refrigerant remaining in the cylinder.

5. Never attempt to mix gases in a cylinder.

6. Never put the wrong refrigerant into a refrigeration system! No refrigerant except the one for which the system was designed should ever be introduced into the system. In some cases, putting the wrong refrigerant into a system may cause a violent explosion.

7. When a cylinder has been emptied, close the cylinder valve immediately to prevent the entrance of air, moisture, or dirt. Also, be sure to replace the valve protection cap.

8. Never use cylinders for any purpose other than their intended purpose. DO NOT use them as rollers, supports, and so on.

9. DO NOT tamper with the safety devices, the valves, or the cylinders.

10. Open cylinder valves slowly. Never use wrenches or other tools except those provided by the manufacturer.

11. Make sure the threads on regulators or other connections are the same as those on the cylinder valve outlets. Never force connections that do not fit.

12. DO NOT use regulators and pressure gauges provided for use with a particular gas on cylinders containing other gases.

13. Never attempt to repair or alter cylinders or valves.

14. Never fill R-12 cylinders beyond 80 percent of capacity.

15. Whenever possible, store cylinders in a cool, dry place in an upright position. If the cylinders are exposed to excessive heat, a dangerous increase in pressure will occur. If cylinders must be stored in the open, take care that they are protected against extremes

of weather. NEVER allow a cylinder to be subjected to a temperature above 125°F.

16. Never allow R-12 to come in contact with a flame or red-hot metal! When exposed to excessively high temperatures, R-12 breaks down into phosgene gas, an extremely poisonous substance. Because R-12 is such a powerful freezing agent that even a very small amount can freeze the delicate tissues of the eyes, causing permanent damage, goggles must be worn by all personnel who may be exposed to a refrigerant, particularly in its liquid form. If refrigerant does get in the eyes, the person suffering the injury should receive medical treatment immediately to avoid permanent damage to the eyes. In the meantime, IMMEDIATELY FLUSH EYES WITH FRESH WATER FOR A MINIMUM OF 15 MINUTES. Make sure the person does not rub his or her eyes.

CAUTION

Do not use anything except clean water for this type of eye injury.

(**NOTE**: If large leaks are indicated, the soap method should be used to detect leaks; for minute leaks, the halide torch should be employed.)

If R-12 comes in contact with the skin, it may cause frostbite. This injury should be treated as any other case of frostbite. Immerse the affected part in a warm bath for about 10 minutes, then dry carefully. DO NOT rub or massage the affected area. R-12 is considered a fluid of low toxicity. However, in closed spaces, high concentrations displace the oxygen in the air and thus do not sustain life. If a person should be overcome by R-12, remove the person IMMEDIATELY to a well-ventilated place and get medical attention at the earliest opportunity. Watch the person's breathing. If the person is not breathing, give CPR.

REFERENCES

Machinist's Mate 3 & 2, NAVEDTRA 12144, Naval Education and Training Program Management Support Activity, Pensacola, Florida, 1991.

Naval Ships' Technical Manual, Chapter 516, "Refrigeration Systems," NAVSEA S9086-RW-STM-010, Naval Sea Systems Command, Washington, D.C., 1986.

CHAPTER 23

COMPRESSED AIR PLANTS

LEARNING OBJECTIVES

Upon completion of this chapter, you should be able to do the following:

1. Identify the classifications of air compressors used aboard Navy ships.

2. Identify the operating principles and functions of air compressors.

3. Identify the purpose of compressed air receivers.

4. Identify the shipboard uses of compressed air.

5. Explain the purpose behind the prairie-masker system.

6. Explain the reason moisture should be removed from compressed air.

7. State the purpose and need for oil-free compressed air compressors and how they operate.

8. Explain why compressors must run unloaded and how it is accomplished.

9. Recognize the procedures for maintaining air compressors.

10. Identify the safety precautions to be observed to minimize the hazards inherent in the process of air compression.

INTRODUCTION

Compressed air serves many purposes aboard ship. These purposes include, but are not limited to, the operation of pneumatic tools and equipment, diesel engine starting and speed control, torpedo charging, aircraft starting and cooling, air deballasting, and the operation of pneumatic boiler and propulsion control systems. Compressed air is supplied to the various systems by high-, medium-, or low-pressure air compressors, depending on the need of the ship. Reducing valves may be used to reduce high-pressure air to a lower pressure for a specific use. This chapter will discuss equipment used to compress the air and supply it to the compressed air systems.

Compressed air represents a storage of energy. Work is done on the working fluid (air) so work can later be done by the working fluid. Air compression may be either an adiabatic or an isothermal process. Adiabatic compression results in a high internal energy level of the air being discharged from the compressor. However, much of the extra energy provided by adiabatic compression may be dissipated by heat losses since compressed air is usually held in an uninsulated receiver until it is used. Isothermal compression is, in theory, the most economical method of compressing air because it requires the least work to be done on the working fluid.

However, the isothermal compression of air requires a cooling medium to remove heat from the compressor and its contained air during the compression process. The more closely isothermal compression is approached, the greater the cooling effect required; in a compressor of finite size, then, we reach a point at which it is no longer practicable to continue to strive for isothermal compression.

In actual practice, the process of air compression is approximately isothermal when considered from start to finish, and approximately adiabatic when considered within any one stage of the compression process. To achieve some benefits from each type of process (isothermal and adiabatic), most air compressors are designed with more than one stage and with a cooling arrangement after each stage. Multistaging and after-stage cooling have the further advantages of preventing the development of excessively high temperatures in the compressor and in the accumulator, reducing the horsepower requirements, condensing some of the entrained moisture, and increasing volumetric efficiency.

The accumulator is found in all compressed air plants, although the size of the unit varies according to the needs of the system. The accumulator (also called a receiver) helps to eliminate pulsations in the discharge line of the air compressor, acts as a storage tank during intervals when the demand for air exceeds the capacity of the compressor, and allows the compressor to shut down during periods of light load. Overall, the accumulator retards increases and decreases in the pressure of the system, thereby lengthening the start-stop-start cycle of the compressor.

AIR COMPRESSORS

The compressor is the heart of any compressed air system. It takes in atmospheric air, compresses it to the pressure desired, and pumps the air into supply lines or into storage for later use. It comes in different designs, construction, and methods of compression.

COMPRESSOR CLASSIFICATIONS

An air compressor is generally classified according to capacity (high, medium, or low), the type of compressing element, the type of driving unit, how it is connected to the driving unit, the pressure developed, and whether the discharged air is oil free. Because of our need for oil-free air aboard ship, the oil-free air compressor is replacing most of the standard low-pressure air compressors.

Types of Compressing Elements

Shipboard air compressors may be centrifugal, rotary, or reciprocating. The reciprocating type is selected for capacities from 200 to 800 cubic feet per minute (cfm) and for pressures of 100 to 5,000 psi. The rotary lobe type is selected for capacities up to 8,800 cfm and for pressures of no more than 20 psi. The centrifugal type is selected for 800 cfm or greater capacities (up to 2,100 cfm in a single unit) and for pressures up to 125 psi.

Most general-service-use air compressors aboard ship are reciprocators (fig. 23-1). In this type of compressor the air is compressed in one or more

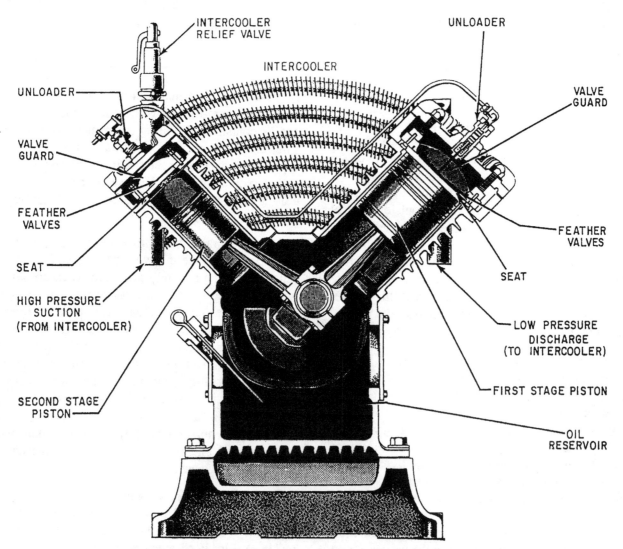

Figure 23-1.—A simple two-stage reciprocating low-pressure air compressor.

cylinders. This is very much like the compression that takes place in an internal-combustion engine.

Sources of Power

Compressors may be driven by electric motors or steam turbines. Aboard ship, most low- and high-pressure air compressors are driven by electric motors.

Drive Connections

The driving unit may be connected to the compressor by one of several methods. When the compressor and the driving unit are mounted on the same shaft, they are close coupled. This method is usually restricted to small capacity compressors driven by electric motors. However, the high-speed, single-stage, centrifugal, turbine-driven units serving prairie-masker systems in FF 1052 and similar ships are close coupled. Flexible couplings join the driving unit to the compressor when the speed of the compressor and the speed of the driving unit are the same. This is called a direct-coupled drive.

V-belt drives are commonly used with small, low-pressure, motor-driven compressors and with some medium-pressure compressors. In a few installations, a rigid coupling is used between the compressor and the electric motor of a motor-driven compressor. In a steam-turbine drive, compressors are usually (not always) driven through reduction gears. In centrifugal (high-speed) compressors, they are usually driven through speed increasing gears.

Pressure Classifications

According to the *Naval Ships' Technical Manual (NSTM)*, compressors are classified as low, medium, or high pressure. Low-pressure compressors have a discharge pressure of 150 psi or less. Medium-pressure compressors have a discharge pressure of 151 psi to 1,000 psi. Compressors that have a discharge pressure above 1,000 psi are classified as high pressure.

Most low-pressure reciprocating air compressors are of the two-stage type. They have either a vertical V (fig. 23-1) or a vertical W (fig. 23-2) arrangement of cylinders. The V-type compressors have one cylinder for the first (lower pressure) stage of compression and one cylinder for the second (higher pressure) stage of compression. The W-type compressors have two

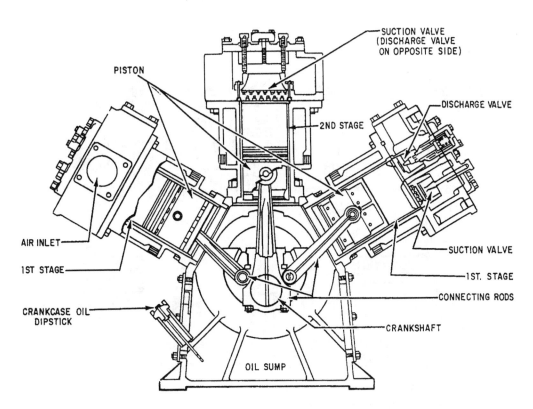

Figure 23-2.–Low-pressure reciprocating air compressor, vertical W configuration.

cylinders for the first stage of compression and one cylinder for the second stage. This arrangement is also shown in view A of figure 23-3. Notice that the pistons in the lower-pressure stage (1) have a larger diameter than the pistons in the higher-pressure stage (2).

Medium-pressure air compressors are of the two-stage, vertical, duplex, single-acting type. Many medium-pressure compressors have differential pistons. This type of piston has more than one stage of compression during each stroke of the piston (fig. 23-3, view A).

Most high-pressure compressors are motor-driven, liquid-cooled, four-stage, single-acting units with vertical cylinders. Figure 23-3, view B, shows the cylinder arrangements for high-pressure air compressors installed in Navy ships. Small capacity, high-pressure air systems may have three-stage compressors. Large capacity, high-pressure air systems may be equipped with four-, five-, or six-stage compressors.

OPERATING CYCLE OF RECIPROCATING AIR COMPRESSORS

Reciprocating air compressors are similar in design and operation. The following discussion describes the operating cycle during one stage of compression in a single-stage, single-acting compressor.

The cycle of operation, or compression cycle, within an air compressor cylinder (shown in fig. 23-4) includes two strokes of the piston: a suction stroke and a compression stroke. The suction stroke begins when the piston moves away from top dead center (TDC). The air under pressure in the clearance space (above the piston) expands rapidly until the pressure falls below the pressure on the opposite side of the inlet valve. At this point, the difference in pressure causes the inlet valve to open, and air is admitted to the cylinder. Air continues to flow into the cylinder until the piston reaches bottom dead center(BDC).

The compression stroke starts as the piston moves away from BDC and continues until the piston reaches TDC again. When the pressure in the cylinder equals the pressure on the opposite side of the air inlet valve, the inlet valve closes. Air continues to be compressed as the piston moves toward TDC. The pressure in the cylinder becomes great enough to force the discharge valve open against the discharge line pressure and the pressure of the valve springs. (The discharge valve opens shortly before the piston reaches TDC.) During the remainder of the compression stroke, the air that has been compressed in the cylinder is discharged at almost constant pressure through the open discharge valve. This

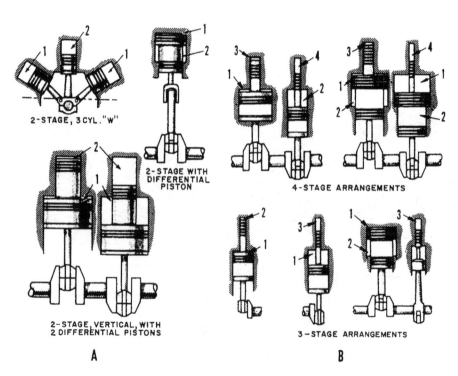

Figure 23-3.–Air compressor cylinder arrangements: (A) Low- and medium-pressure compressors; (b) High-pressure compressor.

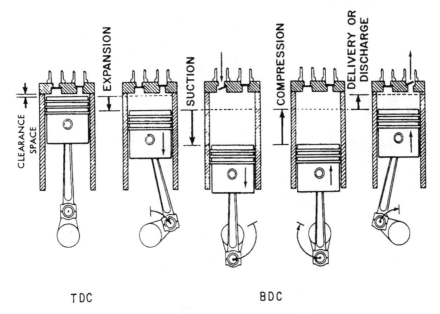

Figure 23-4.–The compression cycle.

cycle is completed twice for each revolution of the crankshaft in double-acting compressors, once on the down stroke and once on the upstroke.

COMPONENT PARTS OF RECIPROCATING AIR COMPRESSORS

Reciprocating air compressors consist of a system of connecting rods, a crankshaft, and a flywheel. These parts transmit power developed by the driving unit to the pistons as well as lubrication systems, cooling systems, control systems, and unloading systems.

Compressing Element

The compressing element of a reciprocating compressor consists of the air valves, cylinders, and pistons.

VALVES.–The valves are made of special steel and come in several different types. The opening and closing of the valves is caused by the difference between (1) the pressure of the air in the cylinder and (2) the pressure of the external air on the intake valve or the pressure of the discharged air on the discharge valve.

Two types of valves commonly used in high-pressure air compressors are shown in figure 23-5. The strip, or feather, valve is shown in view A. It is used for the suction and discharge valves of the lower-pressure stages, that is, 1 and 2. The valve shown in the figure is a suction valve; the discharge valve assembly (not shown) is identical except that the pistons

of the valve seat and the guard are reversed. At rest, the thin strips lie flat against the seat. They cover the slots and seal any pressure applied to the guard side of the valve. The following action works in either a suction or discharge operation (depending on the valve service). As soon as pressure on the seat side of the valve exceeds the pressure on the guard side, the strips flex against the contoured recesses in the guard. As soon as the pressure equalizes or reverses, the strips unflex and return to their original position flat against the seat.

The disk-type valve in figure 23-5, view B, is used for the suction and discharge valves of the higher-pressure stages, that is, 3 and 4. The fourth stage assembly is shown. These valves are of the spring-loaded, dished-disk type. At rest, the disk is held against the seat by the spring. It is sealed by pressure applied to the keeper side of the valve. The following action works in either a suction or discharge operation (depending on the valve service). As soon as the pressure on the seat side of the valve exceeds the pressure on the keeper side, the disk lifts against the stop in the keeper. This compresses the spring and permits air to pass through the seat, around the disk, and through the openings in the sides of the keeper. As soon as the pressure equalizes or reverses, the spring returns the disk to the seat.

CYLINDERS.–The designs of cylinders depend mostly upon the number of stages of compression required to produce the maximum discharge pressure. Several common cylinder arrangements for low-, and medium-pressure air compressors are shown in figure

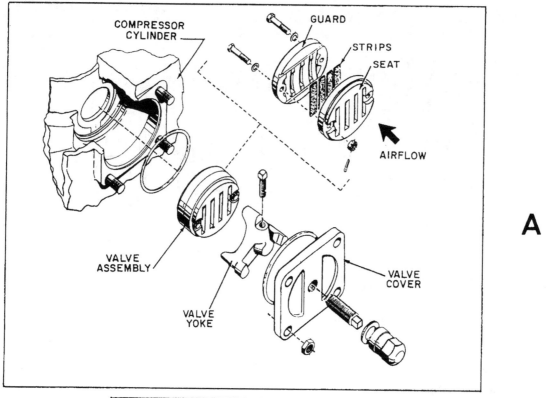

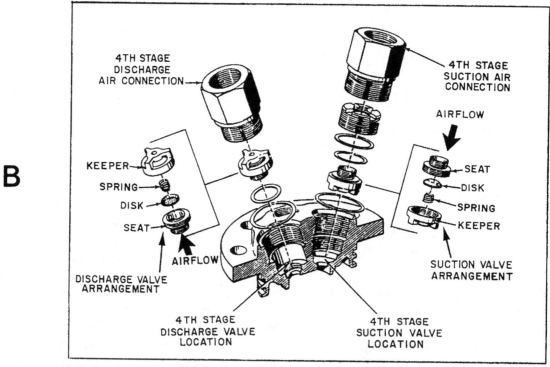

Figure 23-5.–High-pressure air compressor valves: (A) Valve arrangement for lower-pressure stages (suction shown); (B) Valve arrangement for higher-pressure stages (both suction and discharge are shown).

23-3, view A. Several arrangements for cylinders and pistons of high-pressure compressors are shown in figure 23-3, view B. The stages are numbered 1 through 4, and a three- and a four-stage arrangement are shown.

In five- and six-stage compressors, the same basic stage arrangement is followed.

PISTONS.–The pistons may be of two types: trunk pistons or differential pistons. Trunk pistons (fig. 23-6, view A) are driven directly by the connecting rods. The upper end of a connecting rod is fitted directly to the piston by a wrist pin. This causes a tendency for the piston to develop a side pressure against the cylinder walls. To distribute the side pressure over a wide area of the cylinder walls or liners, pistons with long skirts are used. This type of piston minimizes cylinder wall wear. Differential pistons (fig. 23-6, view B) are modified trunk pistons with two or more different diameters. These pistons are fitted into special cylinders. They are arranged so more than one stage of compression is achieved by one piston. The compression for one stage takes place over the piston crown; compression for the other stage(s) takes place in the annular space between the large and small diameters of the piston.

Lubrication Systems

There are generally three types of lubrication systems in reciprocating compressors. They are for high-pressure, low-pressure, and nonlubricated systems. High-pressure air compressor cylinders are generally lubricated by an adjustable mechanical force-feed lubricator (except nonlubricated compressors). This unit is driven from a reciprocating or rotary part of the compressor. Oil is fed from the cylinder lubricator by separate lines to each cylinder. A check valve at the end of each feed line keeps the compressed air from forcing the oil back into the lubricator. Each feed line has a sight-glass oil flow indicator. Lubrication begins automatically as the compressor starts up. The amount of oil that must be fed to the cylinder depends on the cylinder diameter, the cylinder wall temperature, and the viscosity of the oil. Figure 23-7 shows the lubrication connections for the cylinders. The type and grade of oil used in compressors is specified in the equipment technical manual. The correct type is vital to the operation and reliability of the compressor.

The running gear is lubricated by an oil pump that is attached to the compressor and driven from the compressor shaft. This pump is usually a gear type. It draws oil from the reservoir (oil sump) in the compressor base and delivers it through a filter to an oil cooler (if installed). From the cooler, the oil is distributed to the top of each main bearing, to spray nozzles for the reduction gears, and to the outboard bearings. The crankshaft is drilled so oil fed to the main bearings is picked up at the main bearing journals and carried to the crank journals. The connecting rods contain passages that conduct lubricating oil from the crank bearing up to the piston pin bushings. As oil leaks out from the various bearings, it drips back into the oil sump (in the base of the compressor) and is recirculated.

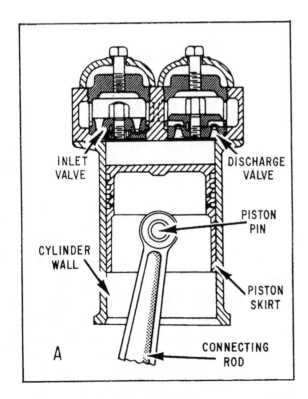
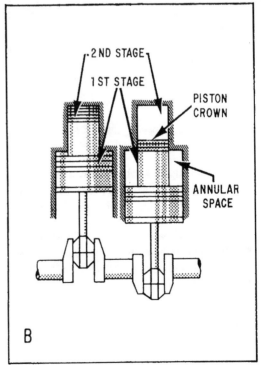

Figure 23-6.–Air compressor pistons: (A) Trunk type; (B) Differential type.

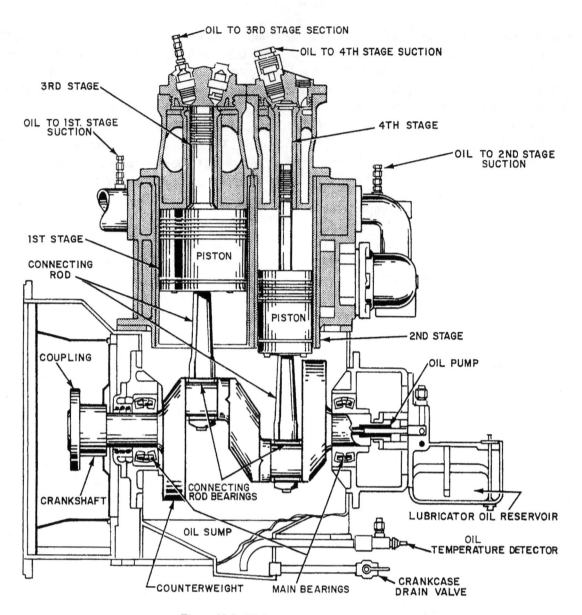

Figure 23-7.–High-pressure air compressor.

Oil from the outboard bearings is carried back to the sump by drain lines.

Low-pressure air compressor lubrication is shown in figure 23-8. This system is similar to the running gear lubrication system for the high-pressure air compressor.

Nonlubricated reciprocating compressors have lubricated running gear (shaft and bearings) but no lubrication for the piston and valves. This design produces oil-free air.

Cooling Systems

Most high-, and medium-pressure reciprocating compressors are cooled by the ship's auxiliary fresh water or by seawater supplied from the ship's fire main or machinery cooling water system. Compressors located outside the larger machinery spaces may have an attached circulating water pump as a standby source of cooling water. Small low-pressure compressors are air-cooled by a fan driven by the compressors. These compressors supply ship's service air, diesel engine starting air, and some small capacity high-pressure air compressors.

The path of water in the cooling water system of a typical four-stage compressor is illustrated in figure 23-9. Not all cooling water systems have identical path-of-water flow. However, in systems equipped with oil coolers, it is important that the coldest water be available for circulation through the cooler. Valves usually control the water to the cooler independently of

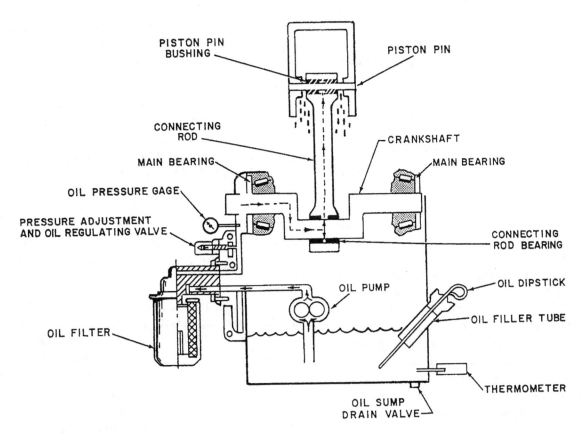

Figure 23-8.–Lubricating oil system of a low-pressure air compressor.

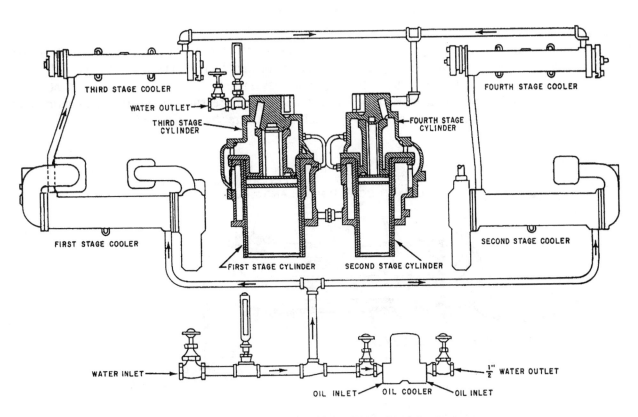

Figure 23-9.–Cooling water system in a typical multistage air compressor.

the rest of the system. Thus, oil temperature can be controlled without harmful effects to the other parts of the compressor. It is very important to cool the air in the intercoolers and aftercoolers as well as the cylinder jackets and heads. The amount of cooling water required depends on the capacity (cfm) and pressure. High-pressure air compressors require more cooling water (for the same cfm) than the low-pressure units.

When seawater is used as the cooling agent, all parts of the circulating system must be of corrosion-resistant materials. The cylinders and heads are therefore composed of a bronze alloy with water jackets cast integral with the cylinders. Each cylinder is generally fitted with a liner of special cast iron or steel to withstand the wear of the piston. Wherever practicable, cylinder jackets are fitted with handholes and covers so the water spaces can be inspected and cleaned. Jumpers are usually used to make water connections between the cylinders and heads because they prevent possible leakage into the compression spaces. In some compressors, however, the water passes directly through the joint between the cylinder and the head. With this latter type, the joint must be properly gasketed to prevent leakage that, if allowed to continue, will damage the compressor.

The intercoolers and aftercoolers remove heat generated during compression and promote condensation of any vapor that may be present. Figure 23-10 is a diagram of a basic cooler and separator unit. It shows the collected condensate in the separator section. The condensate must be drained at regular intervals to prevent carryover into the next stage. If condensate accumulates at low points, it may cause water hammer or freezing and bursting of pipes in exposed locations. It could also cause faulty operation of pneumatic tools and possible damage to electrical apparatus when air is used for cleaning.

The interstage cooling reduces the maximum temperature in each cylinder. This reduces the amount of heat that must be removed by the water jacket at the cylinder. Also, the resulting lower temperature in the cylinder ensures better lubrication of the piston and the valves. Figure 23-11 illustrates the pressures and temperatures through a four-stage compressor. The intercoolers and the aftercoolers (on the output of the final stage) are of the same general construction. The exception is that the aftercoolers are designed to withstand a higher working pressure than the intercoolers.

Water-cooled intercoolers may be the straight tube and shell type (fig. 23-10) or, if size dictates, the coil type. In coolers with an air discharge pressure below 150 psi, the air may flow either through the tubes or over and around them. In coolers with an air discharge pressure above 250 psi, the air generally flows through the tubes. In tubular coolers, baffles deflect the air or water in its course through the cooler. In coil-type coolers, the air passes through the coil and the water flows around the coils.

Control System

The control system of a reciprocating air compressor may include one or more control devices. These may include start-stop control, constant-speed control, speed-pressure governing, and automatic high-temperature shut-down devices.

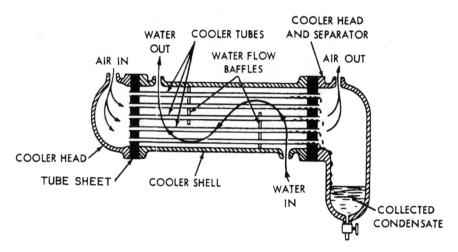

Figure 23-10.–Basic cooler and separator.

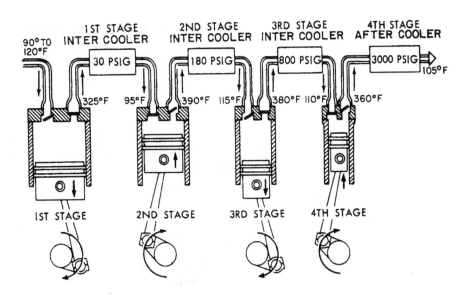

Figure 23-11.–Pressure and temperature resulting from multistaging and interstage cooling.

Control or regulating systems for air compressors in use by the Navy are largely of the start-stop type. In these, the compressor starts and stops automatically as the receiver pressure falls or rises within predetermined set points. On electrically driven compressors, the system is very simple. The receiver pressure operates against a pressure switch that opens when the pressure upon it reaches a given limit. It closes when the pressure drops a predetermined amount. Centrifugal compressors do not have automatic start-stop controls mainly because of their high horsepower. An automatic load/unload control system is used instead.

Some electrically driven units, such as the medium-pressure system, are required to start at either of two pressures. In these, one of two pressure switches is selected with a three-way valve or a pet cock that admits pressure from the air accumulator to the selected pressure switch. Another method is to direct the air from the receiver through a three-way valve to either of two control valves set for the respective range of pressures. A line is run from each control valve to a single pressure switch that may be set at any convenient pressure; the setting of the control valve selected will determine the operation of the switch.

The constant-speed control regulates the pressure in the air receiver by controlling the output of the compressor. This is done without stopping or changing the speed of the unit. This control prevents frequent starting and stopping of compressors when there is a fairly constant but low demand for air. Control is provided by directing air to unloading devices through a control valve set to operate at a predetermined pressure.

Automatic high-temperature shutdown devices are fitted on all recent designs of high-pressure air compressors. If the cooling water temperature rises above a safe limit, the compressor will stop and will not restart automatically. Some compressors have a device that will shut down the compressor if the temperature of the air leaving any stage exceeds a preset value.

Unloading Systems

Reciprocating air compressor unloading systems remove all but the friction loads on the compressor. In other words, they automatically remove the compression load from the compressor while the unit is starting and automatically apply the compression load after the unit is up to operating speed. Units with start-stop control have the unloading system separate from the control system. Compressors with constant-speed control have the unloading and control systems as integral parts of each other.

We cannot give a detailed explanation for every type of unloading device used to unload air compressor cylinders. Still, you should know something about several of the unloading methods. These include closing or throttling the compressor intake, holding intake valves off their seats, relieving intercoolers to the atmosphere, relieving the final discharge to the atmosphere (or opening a bypass from the discharge to the intake), opening cylinder clearance pockets, using

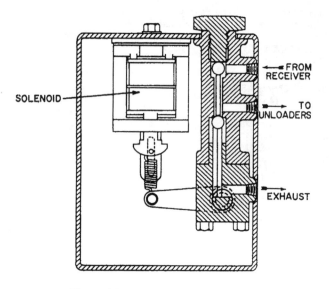

Figure 23-12.—Magnetic-type unloader.

a solenoid-operated valve connected with the motor starter.

ROTARY-CENTRIFUGAL AIR COMPRESSORS

A nonreciprocating type of air compressor aboard ship is variously called a rotary compressor, a centrifugal compressor, or a liquid piston compressor. Actually, the unit is something of a mixture. It operates partly on rotary principles and partly on centrifugal principles. It may, more accurately, be called a rotary-centrifugal compressor.

The rotary-centrifugal compressor supplies low-pressure compressed air. It can supply air that is completely free of oil. Therefore, it is often used as the compressor for pneumatic control systems and for other applications where oil-free air is required.

The rotary-centrifugal compressor, shown in figure 23-13, consists of a round, multibladed rotor that revolves freely in an elliptical casing. The elliptical casing is partially filled with high-purity water. The curved rotor blades project radially from the hub. The blades, together with the side shrouds, form a series of pockets or buckets around the periphery. The rotor is

miscellaneous constant-speed unloading devices, and various combinations of these methods.

The following paragraphs will discuss one example of a typical compressor unloading device–the magnetic-type unloader. Figure 23-12 illustrates the unloader valve arrangement. This unloader consists of

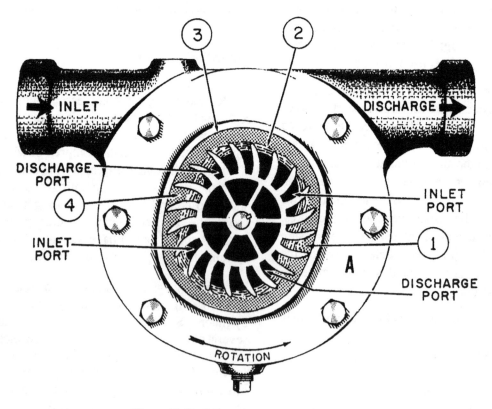

Figure 23-13.—Rotary-centrifugal compressor.

keyed to the shaft of an electric motor. It revolves at a speed high enough to throw liquid out from the center by centrifugal force. This causes a solid ring of liquid to revolve in the casing at the same speed as the rotor but following the elliptical shape of the casing. This action alternately forces the liquid to enter and recede from the buckets in the rotor at high velocity.

To follow through a complete cycle of operation, look at figure 23-13. Start at point A (located at right center). The chamber (1) is full of liquid. The liquid, because of centrifugal force, follows the casing, withdraws from the rotor, and pulls air in through the inlet port. At (2) the liquid has been thrown outward from the chamber in the rotor and has been replaced with atmospheric air. As the rotation continues, the converging wall (3) of the casing forces the liquid back into the rotor chamber. This compresses the trapped air and forces it out through the discharge port. The rotor chamber (4) is now full of liquid and ready to repeat the cycle, which takes place twice in each revolution.

A small amount of water must be constantly supplied to the compressor to make up for the water that is carried over with the compressed air. The water that is carried over with the compressed air is removed in a refrigeration-type dehydrator.

COMPRESSED AIR RECEIVERS

An air receiver is installed in each space that houses air compressors (except centrifugal and rotary lobe types). The receiver is an air storage tank. If demand is greater than the compressor capacity, some of the stored air is supplied to the system. If demand is less than the compressor capacity, the excess is stored in the receiver or accumulator until the pressure is raised to its maximum setting. At that time, the compressor unloads or stops. Thus, in a compressed air system, the receiver minimizes pressure variations in the system and supplies air during peak demand. This will minimize start-stop cycling of air compressors. Air receivers may be horizontal or vertical. Vertically mounted receivers have convex bottoms. These permit proper draining of accumulated moisture, oil, and foreign matter.

All receivers have fittings such as inlet and outlet connections and drain connections and valves. They have connections for an operating line to compressor regulators, pressure gauges, relief valves (set at approximately 12 percent above normal working pressure of the receiver). They also have manhole plates (depending on the size of the receiver). The discharge line between the compressor and the receiver is as short

and straight as possible. This eliminates vibration caused by pulsations of air and reduces pressure losses caused by friction.

In high-pressure (HP) air systems, air receivers are called air flasks. Air flasks are usually cylindrical in shape, with belled ends and female-threaded necks. The flasks are constructed in shapes to conform to the hull curvature for installation between hull frames.

One or more air flasks connected together constitute an air bank.

COMPRESSED AIR SUPPLY SYSTEMS

The remainder of the compressed air system is the piping and valves that distribute the compressed air to the points of use.

HIGH-PRESSURE AIR

Figure 23-14, view A, shows the first part of an HP air system aboard a surface ship. The 3,000/150 psi reducing station is used for emergencies or abnormal situations to provide air to the LP air system.

LOW-PRESSURE AIR

Low-pressure (LP) air (sometimes called LP ship's service air) is the most widely used air system aboard the ship. Figure 23-14, view B, shows the first part of an LP air system. Many of the LP air systems are divided into subsystems: vital and nonvital air.

Vital air is used primarily for engineering purposes, such as automatic boiler controls, water level controls, and air pilot-operated control valves. Vital air is also supplied to electronics systems. Vital air systems are split between all main machinery groups with cross-connect capability.

Nonvital air has many different purposes, such as laundry equipment, tank-level indicating systems, and airhose connection. Air for a nonvital air system is supplied through a priority valve. This valve will shut automatically to secure air to nonvital components when the pressure in the air system drops to a specified set point. It will reopen to restore nonvital air when pressure in the system returns to normal. This system gives the vital air first priority on all the air in the LP system.

PRAIRIE-MASKER AIR SYSTEMS

A special-purpose air system installed in many surface ships is the prairie-masker air system. This

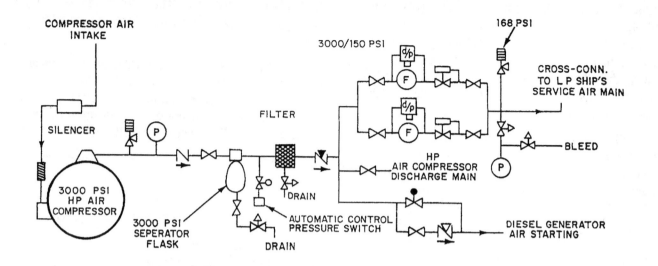

(A) HP AIR COMPRESSOR AND CROSS CONNECTION REDUCING STATION

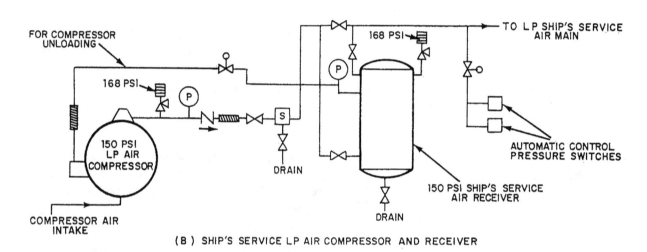

(B) SHIP'S SERVICE LP AIR COMPRESSOR AND RECEIVER

LEGEND					
	PRESSURE REDUCING VALVE		GLOBE VALVE	d/p	DIFFERENTIAL PRESSURE INDICATOR
	PRESSURE SWITCH		NEEDLE VALVE	S	LP AIR PURIFIER
	GLOBE VALVE LOCKED SHUT		CHECK VALVE SWING	F	FILTER
	GLOBE VALVE LOCKED OPEN		CHECK VALVE LIFT		ANGLE RELIEF VALVE
P	PRESSURE GAGE		FLEXIBLE CONNECTOR		

A. HP air compressor and cross-connection reducing station.   B. Ship's service LP air compressor and receiver.

Figure 23-14.–HP and LP air compressor and receiver.

system supplies "disguise" air to a system of emitter rings or belts surrounding the hull and to the propeller blades through the propulsion shafts.

The emitter rings contain small holes that release the masker air into the sea, coating the hull with air bubbles. These bubbles disguise the shape of the ship so it cannot be seen accurately by enemy sonar.

The prairie air passing through the propulsion shafts is emitted to sea by small holes in the propeller blades.

The air supply for the prairie-masker system is provided by a turbocompressor. The turbocompressor is composed of five major parts contained in one compact unit. They are the turbine-driven compressor, lube water tanks, air inlet silencer, lube water system, and control system.

The turbine-driven compressor consists of a single-stage centrifugal compressor driven by a single-stage impulse turbine. The compressor impeller and the turbine wheel are mounted at opposite ends of the same shaft. Two water-lubricated bearings support the rotor assembly. The compressor runs at speeds approaching 40,000 rpm. A control system for the unit provides constant steam admission, overspeed trip, overspeed alarm, low lube pressure trip and alarm, and a high lube water temperature alarm.

MOISTURE REMOVAL

The removal of moisture from compressed air is an important part of compressed air systems. If air at atmospheric pressure, with even a very low relative humidity, is compressed to 3,000 psi or 4,500 psi, it becomes saturated with water vapor. Some moisture is removed by the intercoolers and aftercoolers, as seen earlier in this chapter. Also, air flasks, receivers, and banks have low point drains to periodically drain any collected moisture. However, many shipboard uses of air require air with an even smaller moisture content than is obtained through these methods. In addition, moisture in air lines can create other problems that are potentially hazardous, such as the freezing of valves and controls. This can occur, for example, if very-high-pressure air is throttled to a very low pressure at a high flow rate. The venturi effect of the throttled air produces very low temperatures that will cause any moisture in the air to freeze into ice. This makes the valve (especially an automatic valve) either very difficult or impossible to operate. Also, droplets of water in an air system with a high pressure and high flow rate can cause serious water hammer within the system. For these reasons, air dryers or dehydrators are used to dry the compressed air. Two basic types of air dehydrators are in use: the desiccant type and the refrigerated type.

Desiccant-Type Dehydrators

A desiccant is a drying agent. More practically, a desiccant is a substance with a high capacity to remove (absorb) water or moisture. It also has a high capacity to give off that moisture so the desiccant can be reused.

Compressed air system dehydrators use a pair of desiccant towers (flasks full of desiccant). One is on service dehydrating the compressed air while the other is being reactivated. A desiccant tower is normally reactivated by passing dry, heated air through the tower being reactivated in the direction opposite to normal dehydration airflow. The hot air evaporates the collected moisture and carries it out of the tower to the atmosphere. The purge air is heated by electrical heaters. Once the tower that is reactivating has completed the reactivation cycle, it is placed on service to dehydrate air and the other tower is reactivated.

Another type of desiccant dehydrator in use is the Heat-Les Dryer. These units require no electrical heaters or external source of purge air. Figure 23-15, view A, shows the compressed air entering at the bottom of the left tower. It then passes upward through the desiccant where it is dried to a very low moisture content. The dry air passes through the check valve to the dry air outlet. Simultaneously, a small percentage of the dry air passes through the orifice between the towers and flows down through the right tower. This dry air reactivates the desiccant and passes out through the purge exhaust. At the end of the cycle, the towers are automatically reversed, as shown in view B of figure 23-15.

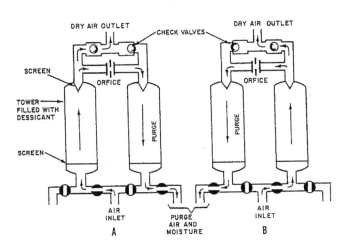

Figure 23-15.–Heat-Les dehydrator.

Refrigerated-Type Dehydrators

Refrigeration is another method of removing moisture from compressed air. The compressed air passes over a set of refrigerated cooling coils. Oil and moisture vapors will condense from the air and can be collected and removed via a low point drain.

Some installations may use a combination of a refrigerated dehydrator and desiccant dehydrators to purify the compressed air.

OIL-FREE, LOW-PRESSURE AIR COMPRESSOR

The oil-free air compressors, like all reciprocating air compressors, have a compression and running gear chamber. The oil-free air compressors (fig. 23-16) have

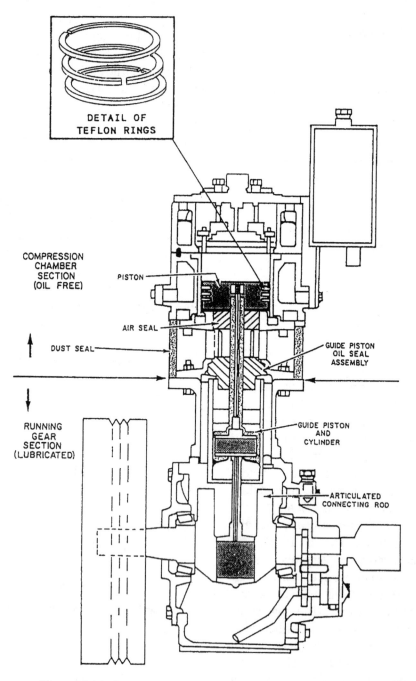

Figure 23-16.--Cutaway view of an oil-free, low-pressure compressor.

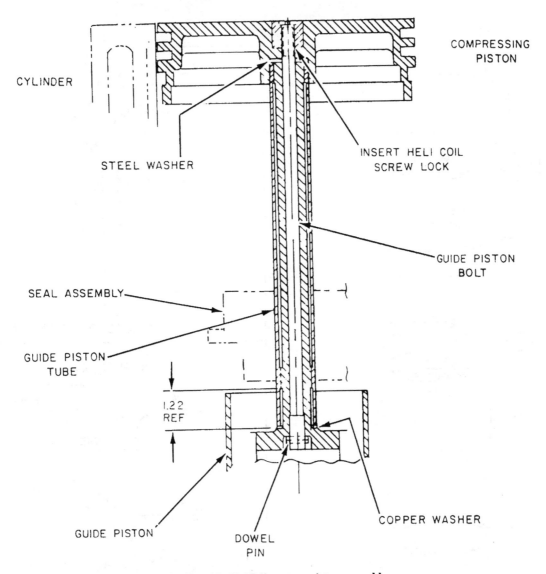

Figure 23-17.–Hollow-type piston assembly.

CYLINDER

COMPRESSING PISTON

STEEL WASHER

INSERT HELI COIL SCREW LOCK

GUIDE PISTON BOLT

SEAL ASSEMBLY

GUIDE PISTON TUBE

1.22 REF

GUIDE PISTON

DOWEL PIN

COPPER WASHER

two major components that provide oil-free discharged air. They are the guide piston and the guide piston seal assembly.

The compression piston (fig. 23-17) is hollow, and is connected to the guide piston via the guide piston seal assembly. The piston rings are made from a Teflon-bronze material that will become damaged if contacted with lube oil.

The guide piston seal assembly (fig. 23-18) contains oil control rings, retainer rings, cup, and cover. This seal prevents oil from entering the compression chamber by scraping the piston connecting rod of oil from the running gear chamber.

The running gear chamber consists of a system of connecting rods, a crankshaft, and a flywheel. (**NOTE:** The connecting rods are connected to the guide piston in the oil-free air compressor.)

The valve assemblies are of the strip/feather type previously discussed in this chapter.

UNLOADER SYSTEM

The Worthington oil-free, low-pressure air compressor unloader system consists of three suction valve unloaders. There is one for each cylinder head located directly above the suction valve assemblies (fig. 23-19). The major components of the unloader assembly are the piston (108), springs (105, 106, and 107), and

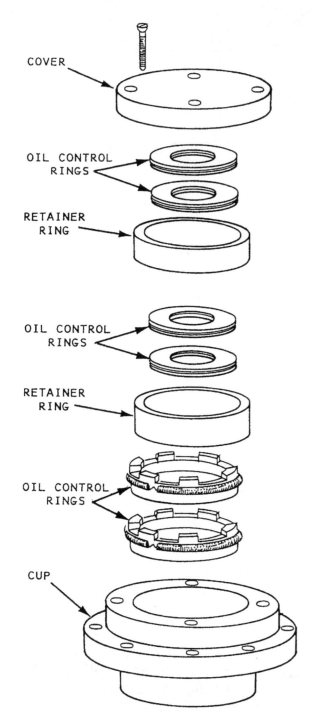

Figure 23-18.—Guide piston seal assembly.

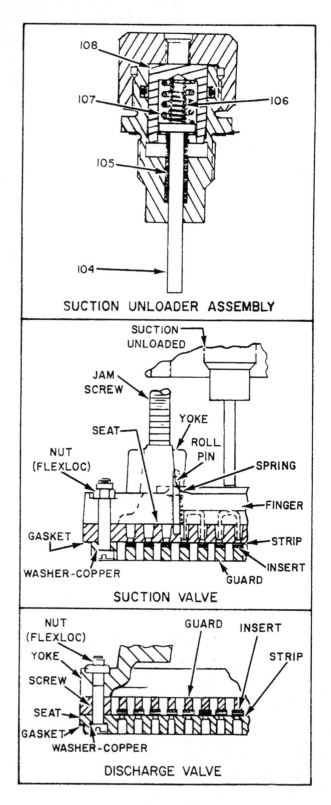

Figure 23-19.—Valve assembly.

finger (104). The unloader assemblies (fig. 23-20) are actuated by a solenoid-operated valve that admits air to the top of the piston, forcing the piston and finger downward, unseating the suction valves, and causing the air compressor to be in an unloaded condition. The compressor becomes loaded when the solenoid-operated valve bleeds the air off the top of the piston in the unloader assemblies. The pistons and fingers are forced upward by a spring, causing the

suction valves to seat. A flow control valve is provided to prevent instant full loading of the compressor. This is done by controlling the amount of airflow from the unloader assemblies and the solenoid-operated valve.

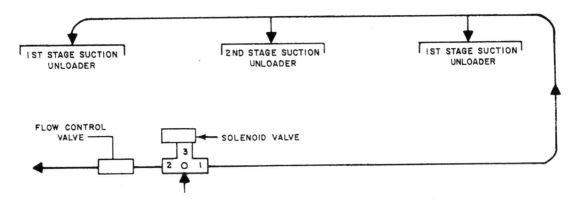

Figure 23-20.–Unloaded system from the Worthington S-100 cfm 125 psi oil-free, low-pressure air compressor.

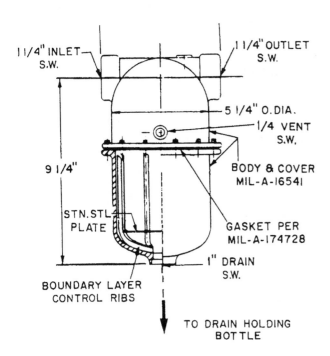

Figure 23-21.–Separator.

The oil-free air compressor has a moisture separator (fig.23-21) that receives air from an air cooler. It drains the collected moisture to a drain collecting holding bottle where it is removed by either the automatic drain system or the manual drain valves. The separator has a level probe that shuts off the air compressor if the drains are backed up because of a malfunction of the automatic drain system.

The automatic drain system operates using a solenoid-operated valve and a three-way air-operated valve (fig. 23-22). The solenoid-operated valve is energized by a time relay device in the motor controller. The solenoid-operated valve admits air to the three-way air-operated valve, which allows drainage of the moisture separator and the holding bottle.

Worthington oil-free air compressors have a freshwater cooling system that provides cooling of the cylinder walls. The fresh water is mixed with an antifreeze solution that allows continuous cleaning of the water passages and piping. The freshwater cooling system (fig. 23-23) consists of a pump, expansion tank, heat exchanger, and thermostatic valve. The thermostatic valve maintains freshwater temperature at a prescribed setting by recirculating the fresh water through the heat exchanger.

COMPRESSED AIR PLANT OPERATION AND MAINTENANCE

Any air compressor or air system must be operated in strict compliance with approved operating procedures. Compressed air is potentially very dangerous. Cleanliness is of greatest importance in all maintenance that requires the opening of compressed air systems.

SAFETY PRECAUTIONS

There are many hazards associated with pressurized air, particularly air under high pressure. Dangerous explosions have occurred in high-pressure air systems because of diesel effect. If a portion of an unpressurized system or component is suddenly and rapidly pressurized with high-pressure air, a large amount of heat is produced. If the heat is excessive, the air may reach the ignition temperature of the impurities present in the air and piping (oil, dust, and so forth). When the ignition temperature is reached, a violent explosion will occur as these impurities ignite. Ignition temperatures also may result from other causes. Some are rapid pressurization of a low-pressure dead end portion of the piping system, malfunctioning of compressor aftercoolers, and leaky or dirty valves. Use every precaution to have only clean, dry air at the compressor inlet.

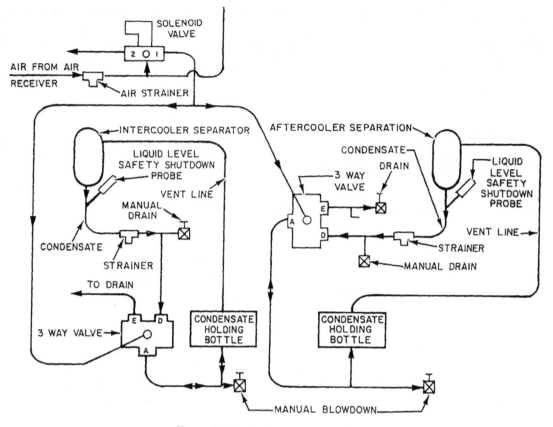

Figure 23-22.–Condensate drain system.

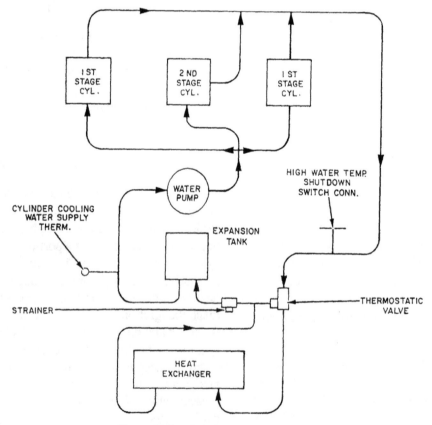

Figure 23-23.–Cylinder cooling water flow diagram.

Air compressor accidents have also been caused by improper maintenance procedures. These accidents can happen when you disconnect parts under pressure, replace parts with units designed for lower pressures, and install stop valves or check valves in improper locations. Improper operating procedures have resulted in air compressor accidents with serious injury to personnel and damage to equipment.

You should try to minimize the hazards inherent in the process of compression and in the use of compressed air. Strictly follow all safety precautions outlined in the NAVSEA technical manuals and in *NSTM*, chapter 9590 (551). Some of these hazards and precautions are as follows:

–Explosions can be caused by dust-laden air or by oil vapor in the compressor or receiver. The explosions are triggered by abnormally high temperatures, which may be caused by leaky or dirty valves, excessive pressurization rates, and faulty cooling systems.

–NEVER use distillate fuel or gasoline as a degreaser to clean compressor intake filters, cylinders, or air passages. These oils vaporize easily and will form a highly explosive mixture with the air under compression.

–Secure a compressor immediately if you observe that the temperature of the air discharged from any stage exceeds the maximum temperature recommended.

–NEVER leave the compressor station after starting the compressor unless you are sure that the control, unloading, and governing devices are operating properly.

–To prevent damage due to overheating, do NOT run compressors at excessively high speeds. Maintain proper cooling water circulation.

–If the compressor is to remain idle for any length of time and is in an exposed position in freezing weather, thoroughly drain the compressor circulating water system.

–Before working on a compressor, be sure the compressor is secured and cannot start automatically or accidentally. Completely blow down the compressor, then secure all valves (including control and unloading valves) between the compressor and the receiver. Follow appropriate tag-out procedures for the compressor control and the isolation valves. (Always leave the pressure gauge open when the gauges are in place.)

–When cutting air into the whistle, siren, or a piece of machinery, be sure the supply line to the equipment has been properly drained of moisture. When securing the supply of air to the affected equipment, be sure all drains are left open.

–Before disconnecting any part of an air system, be sure the part is not under pressure. Always leave the pressure gauge cutout valves open to the sections to which they are attached.

–Avoid rapid operation of manual valves. The heat of compression caused by a sudden flow of high pressure into an empty line or vessel can cause an explosion if oil or other impurities are present. Slowly crack open the valves until flow is noted, and keep the valves in this position until pressure on both sides has equalized. Keep the rate of pressure rise under 200 psi per second.

REFERENCES

Machinist's Mate 3 & 2, NAVEDTRA 10524-F, Naval Education and Training Program Management Support Activity, Pensacola Florida, 1987.

CHAPTER 24

OTHER AUXILIARY MACHINERY

LEARNING OBJECTIVES

Upon completion of this chapter, you should be able to do the following:

1. Identify the characteristics of electrohydraulic transmission that make it advantageous for use on board naval ships.

2. Identify the purposes and principles of operation of the different types of winches and capstans.

3. Identify the function of and the different types of anchor windlasses.

4. Identify distinctive features of electrohydraulic elevators.

5. Identify the purpose of and the types of conveyors installed on naval ships.

6. Identify the major components of and explain the operating cycle of a steam catapult.

INTRODUCTION

In addition to the shipboard auxiliary machinery described in previous chapters of this text, there are several other units of machinery that are essential to the operation of a ship that are directly or indirectly of concern to engineering department personnel. Such auxiliary machinery includes steering gears and their remote control equipment, elevators, winches, capstans, windlasses, and catapults. Some of this machinery may be located within the engineering spaces of the ship, but many of the units are located outside the engineering spaces and are sometimes called outside machinery.

ELECTROHYDRAULIC DRIVE MACHINERY

Hydraulic units drive or control steering gears, windlasses, winches, capstans, airplane cranes, ammunition hoists, and distant control valves. This chapter contains information on some hydraulic units that will concern you.

The electrohydraulic type of drive operates several kinds of machinery. Here are some of the advantages of electrohydraulic machinery.

- Tubing, which can readily transmit fluids around corners, conducts the liquid that transmits the force. Tubing requires very little space.

- The machinery operates at variable speeds.

- Operating speed can be closely controlled from minimum to maximum limits.

- The controls can be shifted from no load to full load rapidly without damage to machinery.

ELECTROHYDRAULIC SPEED GEAR

An electrohydraulic speed gear is most frequently used in electrohydraulic applications. Different variations of the basic design are used for specific applications, but the principles remain the same. Basically, the unit consists of an electric motor-driven hydraulic pump (A-end) and a hydraulic motor (B-end).

The B-end (fig. 24-1) is already on stroke and will be made to rotate by the hydraulic force of the oil acting on the pistons. Movement of the pistons' A-end is controlled by a tilt box (also called a swash plate) in which the socket ring is mounted, as shown in part A of figure 24-1.

The length of piston movement, one way or the other, is controlled by movement of the tilt box and by the amount of angle at which the tilt box is placed. The length of the piston movement controls the amount of fluid flow. When the drive motor is energized, the A-end is always in motion. However, with the tilt box in a neutral or vertical position, there is no reciprocating

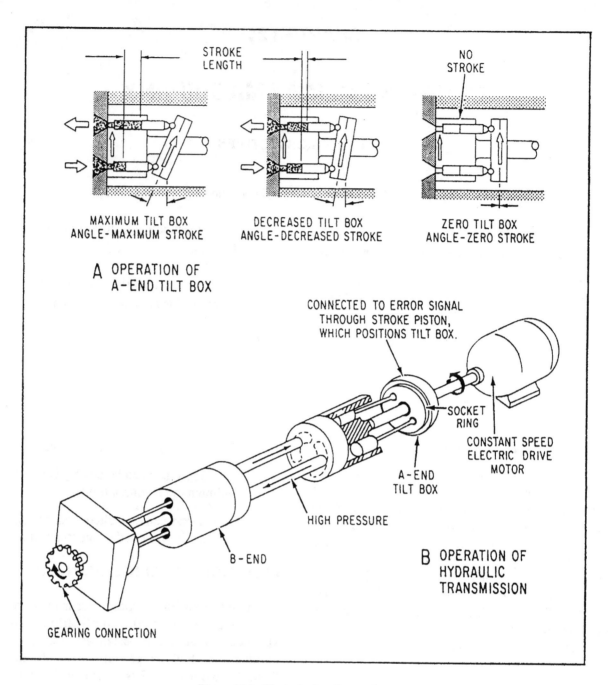

STROKE
LENGTH

NO
STROKE

MAXIMUM TILT BOX
ANGLE-MAXIMUM STROKE

DECREASED TILT BOX
ANGLE-DECREASED STROKE

ZERO TILT BOX
ANGLE-ZERO STROKE

A OPERATION OF
A-END TILT BOX

CONNECTED TO ERROR SIGNAL
THROUGH STROKE PISTON,
WHICH POSITIONS TILT BOX.

SOCKET
RING

CONSTANT SPEED
ELECTRIC DRIVE
MOTOR

A-END
TILT BOX

HIGH PRESSURE

B-END

B OPERATION OF
HYDRAULIC
TRANSMISSION

GEARING CONNECTION

Figure 24-1.–Electrohydraulic speed gear.

motion of the pistons. Therefore, no oil is pumped to the B-end. Any movement of the tilt box, no matter how slight, causes pumping action to start. This causes immediate action in the B-end because of the transmission of force by the hydraulic fluid.

When you need reciprocating motion, such as in a steering gear, the B-end is replaced by a piston or ram. The force of the hydraulic fluid causes the movement of the piston or ram. The tilt box in the A-end can be controlled locally (as on the anchor windlass) or by remote control (as on the steering gear).

ELECTROHYDRAULIC STEERING GEAR

The steering gear transmits power from the steering engine to the rudder stock. The term *steering gear* frequently includes the driving engine and the transmitting mechanism.

Many different designs of steering gear are in use, but the principle of operation for all of them is similar. One type of electrohydraulic steering gear is shown in figure 24-2. It consists essentially of (1) a ram unit and (2) a power unit.

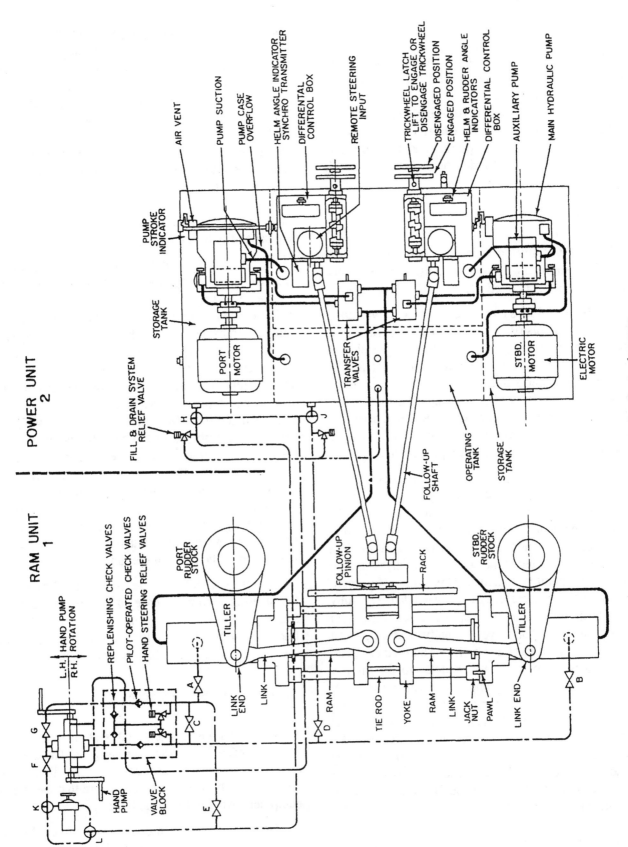

Figure 24-2.—Electrohydraulic steering gear.

24-3

Ram Unit

The ram unit is mounted athwartship and consists of a single ram operated by opposed cylinders. The ram is connected by links to the tillers of the twin rudders. When oil pressure is applied to one end of the operating cylinder, the ram will move, causing each rudder to move with it. Oil from the opposite end of the cylinder is returned to the suction side of the main hydraulic pump in the power unit.

Power Unit

The power unit consists of two independent pumping systems. Two systems are used for reliability. One pump can be operated while the other is on standby.

Each pumping system consists of a variable-delivery, axial-piston main pump and a vane-type auxiliary pump. Both are driven by a single electric motor through a flexible coupling. Each system also includes a transfer valve with operating gear, relief valves, a differential control box, and trick wheels. The whole unit is mounted on a bedplate, which serves as the top of an oil reservoir. Steering power is taken from either of the two independent pumping systems.

The pumps of the power unit are connected to the ram cylinders by high-pressure piping. The two transfer valves are placed in the piping system to allow for the lineup of one pump to the ram cylinders with the other pump isolated. A hand lever and mechanical linkage (not shown) are connected to the two transfer valves so both valves are operated together. This allows for rapid shifting from the on-service pumping unit to the standby unit; it also prevents lining up both pumps to the ram at the same time. The hand lever is usually located between the trick wheels. It has three positions marked P, N, and S. P denotes the port pump connected to the ram; N denotes neutral (neither pump connected to the ram); and S denotes the starboard pump connected to the ram. Also, the hand lever is usually connected to motor switches. This permits the operator to connect the selected pump to the ram and start the pump drive motor in one quick operation. In most modern ships, this valve is electrically controlled by the motor controller and by pressure switches.

Principles of Operation

The on-service hydraulic pump is running at all times and is a constant-speed pump. Unless steering is actually taking place, the tilt box of the main hydraulic pump is at zero stroke, and no oil is being moved within the main system. The auxiliary pump provides control oil and supercharge flows for the system. Assume that a steering order signal comes into the differential control box. It can come from either the remote steering system in the ship's wheelhouse or the trick wheel. The control box mechanically positions the tilt box of the main hydraulic pump to the required angle and position. Remember that direction of fluid and flow may be in either direction in a hydraulic speed gear. It depends on which way the tilt box is angled. For this reason, the constant speed, unidirectional motor can be used to drive the main hydraulic pump. The pump will still have the capability to drive the ram in either direction.

With the main hydraulic pump now pumping fluid into one of the ram cylinders, the ram will move, moving the rudders. A rack and gear are attached to the rudder yoke between the rudder links. As the ram and the rudder move, the rack gear moves, driving the follow-up pinion gear. The pinions drive follow-up shafts, which feed into the differential box. This feedback or servo system tells the differential control box when the steering operation has been completed. As the ordered rudder angle is approached, the differential control box will begin realigning the tilt box of the main hydraulic pump. By the time the desired rudder angle is reached, the tilt box is at zero stroke. This means that the ordered signal (from the pilothouse or trick wheel) and the actual signal (from the follow-up shafts) are the same. If either of these change, the differential control box will react accordingly to cause the main hydraulic unit to pump oil to one end or the other of the ram.

The trick wheels provide local-hydraulic control of the steering system in case of failure of the remote steering system. A hand pump and associated service lines are also provided for local-manual operation of the ram in case of failure of both hydraulic pump units.

Operation and Maintenance

Operating instructions and system diagrams are normally posted near the steering gear. The diagrams describe the various procedures and lineups for operation of the steering gear. Be sure that the standby equipment is ready for instant use.

General maintenance of the steering gear requires that you clean, inspect, and lubricate the mechanical parts and maintain the hydraulic oil at the proper level and purity. The Planned Maintenance System (PMS) lists the individual requirements for the equipment. The electricians maintain the electrical portion of the steering system, including the control system.

WEIGHT-HANDLING EQUIPMENT

You should be familiar with the construction, operation, and maintenance of anchor windlasses, cranes, and winches. The following paragraphs will discuss such machinery and other weight-handling equipment. Most steam-powered, weight-handling equipment is obsolete. If your ship has steam-powered equipment, refer to the technical manual for information.

Anchor Windlasses

In a typical electrohydraulic mechanism, one constant-speed electric motor drives two variable-stroke pumps through a coupling and reduction gear. Other installations include two motors, one for driving each pump. Each pump normally drives one wildcat. However, if you use a three-way plug, cock-type valve, either pump may drive either of the two wildcats. The hydraulic motors drive the wildcat shafts with a multiple-spur gearing and a locking head. The locking head allows you to disconnect the wildcat shaft and permits free operation of the wildcat, as when dropping anchor.

Each windlass pump is controlled either from the weather deck or locally. The controls are handwheels on shafting leading to the pump control. The hydraulic system will require your attention. Be certain the hydraulic system is always serviced with the specified type of clean oil.

There are three types of anchor windlasses: the electric, electrohydraulic, and hand-driven. Hand-driven windlasses are used only on small ships where the anchor gear can be handled without excessive effort by operating personnel.

The major work on a hand-driven windlass is to keep the linkage, friction shoes, locking head, and brake in proper adjustment and in satisfactory operating condition at all times. In an electrohydraulic windlass, your principal concern is the hydraulic system.

A windlass is used intermittently and for short periods of time. However, it must handle the required load under severe conditions. This means that you must maintain and adjust the machinery when it is not in use. This practice will prevent deterioration and ensure dependable use.

Windlass brakes must be kept in satisfactory condition if they are to function properly. Wear and compression of brake linings will increase the clearance between the brake drum and band after a windlass has been in operation. Brake linings and clearances should be inspected frequently. Adjustments should be made according to the manufacturer's instructions.

Follow the lubrication instructions furnished by the manufacturer. If a windlass has been idle for some time, lubricate it. This protects finished surfaces from corrosion and prevents seizure of moving parts.

The hydraulic transmissions of electrohydraulic windlasses and other auxiliaries are manufactured with close tolerances between moving and stationary parts. Use every precaution to keep dirt and other abrasive material out of the system. When the system is replenished or refilled, use only clean oil. Strain it as it is poured into the tank. If a hydraulic transmission has been disassembled, clean it thoroughly before reassembly. Before installing piping or valves, clean their interiors to remove any scale, dirt, preservatives, or other foreign matter.

Winches

Winches are used to heave in on mooring lines, to hoist boats, as top lifts on jumbo booms of large auxiliary ships, and to handle cargo. Power for operating shipboard winches is usually furnished by electricity and, on some older ships, by steam. Sometimes delicate control and high acceleration without jerking are required, such as for handling aircraft. Electrohydraulic winches are usually installed for this purpose. Most auxiliary ships are equipped with either electrohydraulic or electric winches.

CARGO WINCHES.–Some of the most common winches used for general cargo handling are the double-drum, double-gypsy and the single-drum, single-gypsy units. Four-drum, two-gypsy machines are generally used for minesweeping.

ELECTROHYDRAULIC WINCHES.–Electrohydraulic winches (fig. 24-3) are always the drum type. The drive equipment is like most hydraulic systems. A constant-speed electric motor drives the A-end (variable-speed hydraulic pump), which is connected to the B-end (hydraulic motor) by suitable piping. The drum shaft is driven by the hydraulic motor through reduction gearing.

Winches normally have one horizontally mounted drum and one or two gypsy heads. If only one gypsy is required, it may be easily removed from or assembled on either end of the drum shaft. When a drum is to be used, it is connected to the shaft by a clutch.

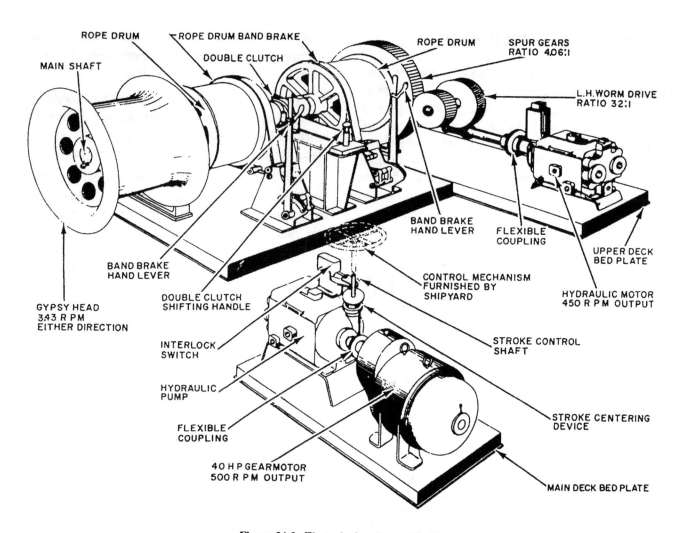

Figure 24-3.–Electrohydraulic winch unit.

ELECTRIC WINCHES.–An electrically driven winch is shown in figure 24-4. This winch is a single-drum, single-gypsy type. The electric motor drives the unit through a set of reduction gears. A clutch engages or disengages the drum from the drum shaft. Additional features include an electric brake and a speed control switch.

Capstans

The terms *capstan* and *winch* should not be confused. A winch has a horizontal shaft, and a capstan has a vertical shaft. The type of capstan installed aboard ship depends on the load requirements and the type of power available. In general, a capstan consists of a single head mounted on a vertical shaft, reduction gearing, and a power source. The types, classified according to power source, are electric and steam.

Electric capstans are usually of the reversible type. They develop the same speed and power in either direction. Capstans driven by ac motors run at either full,

one-half, or one-third speed. Capstans driven by dc motors usually have from three to five speeds in either direction of rotation.

Maintenance of Winches and Capstans

In several respects, the maintenance of a winch or a capstan is similar to that for a windlass. Where band brakes are used on the drums, inspect the friction linings regularly and replace when necessary. Take steps to prevent oil or grease from accumulating on the brake drums. Check the operation of brake-actuating mechanisms, latches, and pawls periodically.

Frequently inspect winch drums driven by friction clutches for deterioration in the friction material. Check also to see if oil and grease are preventing proper operation. Lubricate the sliding parts of positive clutches properly. Check the locking device on the shifting gear to see if it will hold under load.

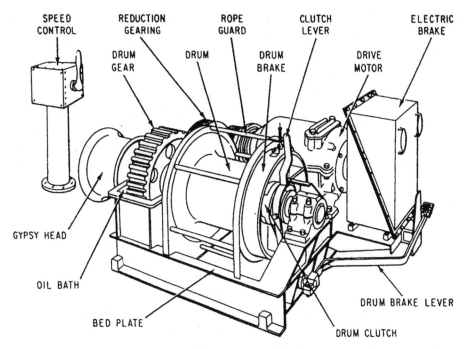

Figure 24-4.–Electric winch.

Labels in figure: SPEED CONTROL, REDUCTION GEARING, ROPE GUARD, CLUTCH LEVER, ELECTRIC BRAKE, DRUM GEAR, DRUM, DRUM BRAKE, DRIVE MOTOR, GYPSY HEAD, OIL BATH, BED PLATE, DRUM BRAKE LEVER, DRUM CLUTCH

Cranes

Cranes are designed to meet the following conditions:

- Hoist, lower, top, and rotate a rated load at the specified speed and against a specified list of the ship.

- Handle 150 percent of the rated load at no specified speed.

- Withstand a static, suspended load of 200 percent of the rated load without damage or distortion to any part of the crane or structure.

The types of cranes installed on ships vary according to the equipment handled.

The crane equipment generally includes the boom, king post, king post bearings, sheaves, hook, rope, machinery platforms, rotating gear, drums, hoisting, topping and rotating drives, and controls. The important components are described in the paragraphs that follow.

BOOMS.–A boom, used as a mechanical shipboard appliance, is a structural unit used to lift, transfer, or support heavy weights. A boom is used in conjunction with other structures or structural members that support it, and various ropes and pulleys, called blocks, that control it.

KING POST BEARINGS.–Bearings on stationary king posts take both vertical load and horizontal strain at the collar, located at the top of the king post. On rotating king posts, bearings take both vertical and horizontal loads at the base and horizontal reactions at a higher deck level.

SHEAVES AND ROPES.–The hoisting and topping ropes are led from the drums over sheaves to the head of the boom. The sheaves and ropes are designed according to recommendations by the Naval Sea Systems Command. This command sets the criteria for selection of sheave diameter, size, and flexibility of the rope. Sufficient fair-lead sheaves are fitted to prevent fouling of the rope. A shock absorber is installed in the line, hoisting block, or sheave at the head of the boom to take care of shock stresses.

MACHINERY PLATFORMS.–Machinery platforms carry the power equipment and operator's station. These platforms are mounted on the king post above the deck.

ROTATING GEAR AND PINIONS.–Rotation of the crane is accomplished by vertical shafts with pinions engaging a large rotating gear.

DRUMS.–The drums of the hoisting and topping winches are generally grooved for the proper size wire rope. The hoisting system uses single or multiple part lines as required. The topping system uses a multiple purchase as required.

OPERATION AND MAINTENANCE OF CRANES.–The hoisting whips and topping lifts of cranes are usually driven by hydraulic variable-speed gears through gearing of various types. This provides the wide range of speed and delicate control required for load handling. The cranes are usually rotated by an electric motor connected to worm and spur gearing. They also may be rotated by an electric motor and hydraulic variable-speed gear connected to reduction gearing.

Some electrohydraulic cranes have automatic slack line take-up equipment. This consists of an electric torque motor geared to the drum. These cranes are used to lift boats, aircraft, or other loads from the water. The torque motor assists the hydraulic motor drive to reel in the cable in case the load is lifted faster by the water than it is being hoisted by the crane.

Electrohydraulic equipment for the crane consists of one or more electric motors running at constant speed. Each motor drives one or more A-end variable-displacement hydraulic pumps. The pump strokes are controlled through operating handwheels. START, STOP, and EMERGENCY RUN push buttons at the operator's station control the electric motors. Interlocks prevent starting the electric motors when the hydraulic pumps are on stroke. B-end hydraulic motors are connected to the A-end pumps by piping. They drive the drums of the hoisting and topping units or the rotating machinery.

Reduction gears are located between the electric motor and the A-end pump and between the B-end hydraulic motor and the rotating pinion. Each hoisting, topping, and rotating drive has an electric brake on the hydraulic motor output shaft. This brake is interlocked with the hydraulic pump control. It will set when the hydraulic control is on neutral or when electric power is lost. A centering device is used to find and retain the neutral position of the hydraulic pump.

Relief valves protect the hydraulic system. These valves are set according to the requirements of *NSTM*, chapter 556.

Cranes usually have a rapid slack take-up device consisting of an electric torque motor. This motor is connected to the hoist drum through reduction gearing. This device works in conjunction with the pressure stroke control on the hydraulic pump. It provides for fast acceleration of the hook in the hoisting direction under light hook conditions. Thus, slack in the cable is prevented when hoisting is started.

Some cranes have a light-hook paying-out device mounted on the end of the boom. It pays out the hoisting cables when the weight of the hook and cable beyond the boom-head sheave is insufficient to overhaul the cable as fast as it is unreeled from the hoisting drum.

When the mechanical hoist control is in neutral, the torque motor is not energized and the cable is gripped lightly by the action of a spring. Moving the hoist control to LOWER energizes the torque motor. The sheaves clamp and pay out the cable as it is unreeled from the hoist drum. When the hoist control is moved to HOIST, the torque motor is reversed and unclamps the sheaves. A limit switch opens and automatically de-energizes the paying-out device.

Maintain cranes according to the PMS requirements or the manufacturers' instructions. Keep the oil in the replenishing tanks at the prescribed levels. Keep the system clean and free of air. Check the limit stop and other mechanical safety devices regularly for proper operation. When cranes are not in use, secure them in their stowed positions. Secure all electric power to the controllers.

Elevators

Some of the hydraulic equipment that you maintain will be found in electrohydraulic elevator installations. Modern carriers use elevators of this type. The elevators described in this chapter are now in service in some of the ships of the CV class. These ships are equipped with four deck-edge airplane elevators that have a maximum lifting capacity of 79,000 to 105,000 pounds. The cable lift platform of each elevator projects over the side of the ship and is operated by an electrohydraulic plant.

ELECTROHYDRAULIC POWER PLANT.— The electrohydraulic power plant for the elevators consists of the following components:

- A horizontal plunger-type hydraulic engine

- Multiple variable-delivery parallel piston-type pumps

- Two high-pressure tanks

- One low-pressure tank

- A sump tank system

- Two constant-delivery vane-type pumps (sump pumps)

- An oil storage tank

- A piping system and valves

- A nitrogen supply

The hydraulic engine is operated by pressure developed in a closed hydraulic system. Oil is supplied to the system in sufficient quantity to cover the baffle plates in the high-pressure tanks and allow for piston displacement. Nitrogen is used because air and oil in contact under high pressure form an explosive mixture. Air should not be used except in an emergency. Nitrogen, when used, should be kept at 97 percent purity.

The hydraulic engine has a balanced piston-type valve with control orifices and a differential control unit. This control assembly is actuated by an electric motor and can be operated by hand. To raise the elevator, move the valve off-center to allow high-pressure oil to enter the cylinder. High-pressure oil entering the cylinder moves the ram. The ram works through a system of cables and sheaves to move the platform upward. The speed of the elevator is controlled by the amount of pressure in the high-pressure tank and the control valve.

When the elevator starts upward, the pressure in the high-pressure tank drops. The pressure drop automatically starts the main pumps. These pumps transfer oil from the low-pressure tank to the high-pressure system until the pressure is restored. An electrical stopping device automatically limits the stroke of the ram and stops the platform at the proper position at the flight deck level.

To lower the elevator, move the control valve in the opposite direction. This permits the oil in the cylinder to flow into the exhaust tank. As the platform descends, oil is discharged to the low-pressure tank (exhaust tank). The original oil levels and pressures, except leakages, are reestablished. The speed of lowering is controlled by the control valve and the cushioning effect of the pressure in the exhaust tank. Leakage is drained to the sump tanks. It is then automatically transferred to the pressure system by the sump pumps. An electrically operated stopping device automatically slows down the ram and stops the platform at its lower level (hangar deck).

SAFETY FEATURES.–Some of the major safety features incorporated into modern deck-edge elevators are as follows:

– If the electrical power fails while the platform is at the hangar deck, there will be enough pressure in the system to move the platform to the flight deck one time without the pumps running.

– Some platforms have serrated safety shoes. If the hoisting cable should break on one side, the shoes will wedge the platform between the guide rails. This will stop the platform with minimum damage.

– A main pump may have a pressure-actuated switch to stop the pump motors when the discharge pressure is excessive. They also may have to relieve the pressure when the pressure switch fails to operate.

– The sump pump system has enough capacity to return the unloaded platform from the hangar deck to the flight deck.

– The oil filter system may be used continuously while the engine is running. This allows part of the oil to be cleaned with each operation of the elevator.

CONVEYORS

Two types of conveyors are used for shipboard handling: gravity and powered.

GRAVITY CONVEYORS

Gravity conveyors may be wheel, roller, or ball type with straight, curved, or angular sections provided in the quantities and lengths required to facilitate stores strikedown. Folding conveyor stands are furnished with each conveyor section. Conveyors are usually 18 inches wide, but they may be 12 inches wide for use in restricted passageways. Most conveyors are of the roller type and are used to handle dry provisions and chilled and frozen food (fig. 24-5).

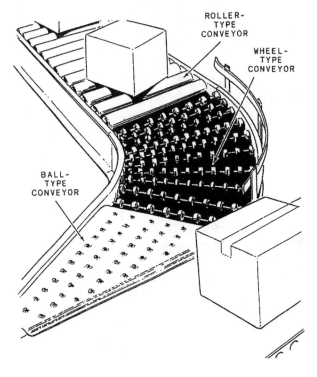

Figure 24-5.–Gravity conveyor.

POWERED CONVEYORS

Powered conveyors are configured either vertically or horizontally.

Vertical Conveyors

Vertical conveyors (figs. 24-6 and 24-7), for shipboard use, consist of the following components:

- Structural frame (head, tail, and intermediate sections)

- Drive system

- Conveyor system

- Operating controls

- Safety devices

The structural frame may be designed as a truss frame installed in a trunk closure. Shields running the length of the conveyor provide a smooth unbroken surface in the area of the moving tray loads and isolate

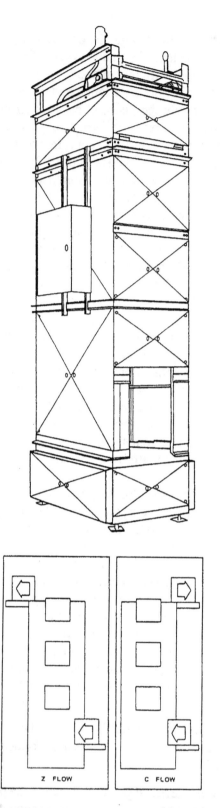

Figure 24-7.–Vertical conveyor, pallet, tray type.

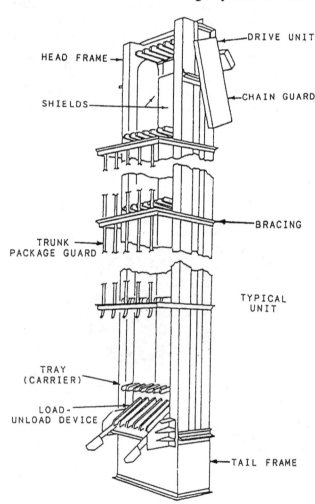

Figure 24-6.–Vertical conveyor, package, tray type.

the load side from the idle return side of the conveyor tray cycle.

The drive system components are the friction clutch (package conveyor), magnetic clutch (pallet conveyor), speed reducer, motor brake, drive shafts with chain sprockets, and connecting roller chain. The drive units are located at the head section of the conveyor frame and can be positioned at the side, back, or top of the conveyor.

The conveying system consists of the chain sprockets mounted in the head and tail sections of the conveyor frame. The carrier (tray) chain is driven by the head chain sprockets. Each tray is supported on two sides by the carrier chain, and each tray is guided on two sides by cam guide arms with rollers that ride in guide tracks mounted to the conveyor frame.

The operating controls consist of a motor controller that provides electrical power for the conveyor electrical components, a switching network for operation on electrical circuits, and a control station that provides operating switches for directional control: STOP, stops the conveyor; UP-DOWN, controls the direction of the conveyor; and RUN, starts the conveyor. An EMERGENCY RUN push button permits operation of the conveyor when the thermal overload relay in the motor controller has tripped. A communication system is provided at control stations for operating personnel to control conveyor operation.

Safety Features

Safety devices are installed for the safety of personnel and to increase the reliability of the conveyor.

The lockout device, located at each control station, secures the operating controls from unauthorized operation. When secured, the lockout device permits operation of the STOP push button from each control station to stop the conveyor motion.

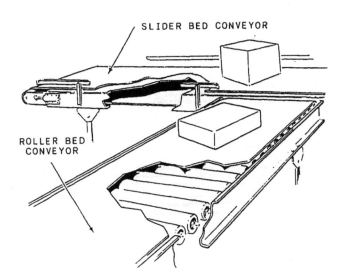

Figure 24-8.–Powered belt conveyor.

The package conveyor has a load-unload device capable of loading and unloading the conveyor at each load station, and it can be placed in three positions:

1. Load position (horizontal) for UP direction loading

2. Unload position (30-degree incline) for DOWN direction unloading

3. Stowed position (vertical)

An interlock switch is placed at each load-unload device to prevent downward operation of the conveyor when the load-unload device is in the load position.

A door block device is provided at each package conveyor load station equipped with a load-unload device so the trunk door will not close unless the load-unload device is in the stowed position.

Two-way communication should continuously be maintained between operating levels to prevent injury to personnel or damage to equipment.

Horizontal Conveyors

Horizontal conveyors are similar to vertical conveyors except that belts or driven rollers (figs. 24-8 and 24-9) are used in place of chains and trays to support the loads. Powered conveyors can bridge a span and operate at an incline.

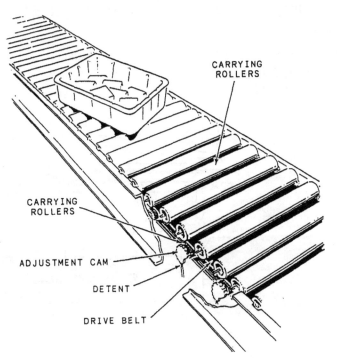

Figure 24-9.–Powered roller conveyor.

To ensure accident-free conveyor operation, use the following procedures:

1. Inspect all interlocks and safety devices to make sure they are operational before further conveyor operations, according to PMS requirements.

2. Verify all warning plates are in place.

3. Establish positive communications between all operating control stations, using sound-powered telephones or intercom systems.

4. Do not use the conveyor trunk as a voice tube.

5. Use the two-man rule at all times while operating the conveyor.

STEAM CATAPULTS

Steam catapults are the steam-powered, direct-drive, flush-deck type. The most significant differences between the various types of steam catapults are the length and capacity.

Steam is the principal source of energy and is supplied to the catapults by the ship's main engineering spaces. The steam is drawn from the main spaces to the catapult receivers where it is stored at the desired pressure. From the receivers, it is directed to the launching valves and provides the energy to launch aircraft.

The major components of a steam catapult are shown in figure 24-10.

PRINCIPLES OF OPERATION

Steam is admitted to the steam receivers through the flow control valve (steam charging valves for catapults equipped with the wet receiver system). The number of steam receivers used varies from 1 to 4. For catapults with wet or dry receivers, the steam pressure is selected for each launch. Ships that have the capacity selector valve (CSV) installed use the constant-pressure system. The launching cylinders are preheated by using an internal and external preheating system to prevent thermal shock and to minimize possible damage to the launching engine when superheated steam is admitted through the launching valves into the launching cylinders.

After making sure all prelaunching steps and precautions have been taken, the catapult officer signals the deck-edge operator to fire the catapult. The deck-edge operator then operates the controls at that station that causes steam to be admitted from the steam receivers into the launching cylinders.

The launching pistons (one in each of the two launching cylinders) are connected via mating dogs on the shuttle and connector of the piston assemblies. The steam admitted to the launching cylinders drives the pistons through the cylinders. The power generated thereby is transmitted to the shuttle and the shuttle is accelerated forward.

At the end of the power run, the forward motion of the shuttle and launching pistons are halted by a water brake assembly (fig. 24-11). When the shuttle and the launching pistons have been halted, the grab, driven by the retraction engine, advances along the deck track and engages the shuttle. The retraction engine is then

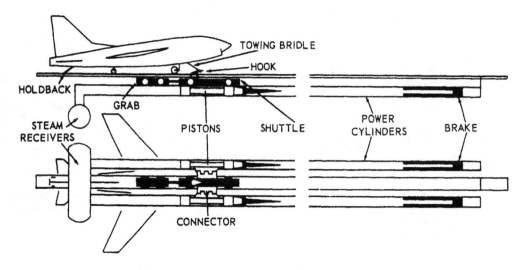

Figure 24-10.–Major components of a steam catapult.

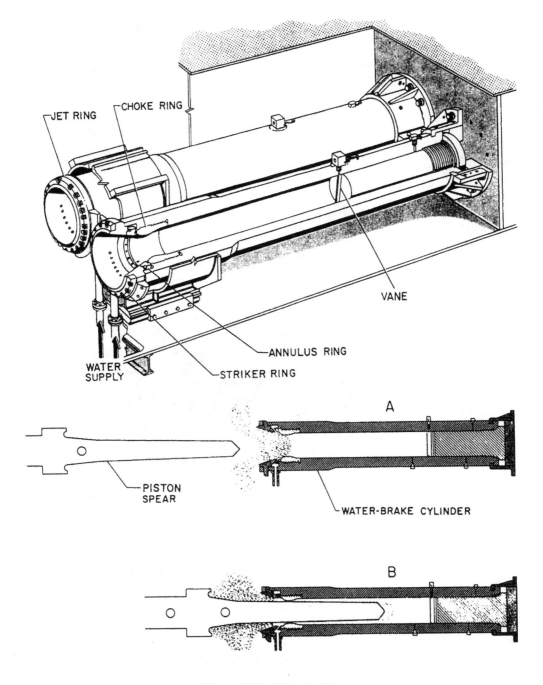

JET RING
CHOKE RING
VANE
ANNULUS RING
WATER SUPPLY
STRIKER RING
PISTON SPEAR
WATER-BRAKE CYLINDER
A
B

Figure 24-11.–Water brakes.

reversed and returns the grab, shuttle, and launching pistons to the battery position.

The control console and the deck-edge control panel are used in conjuction to direct and integrate the functions of the electrical and hydraulic systems and control the sequence of operation of the catapult through a normal launching cycle.

CATAPULT SYSTEMS

Each steam catapult consists basically of eight major systems:

1. Launching system. This system may be defined as those components to which access can be gained at the flight deck level.

2. Steam system. There are two basic steam systems associated with steam catapults. They are the dry receiver system and the wet receiver (constant-pressure) sytem.

3. Retraction system. The two retraction engine systems in use are the linear and the rotary. The majority of catapults installed aboard carriers are equipped with the rotary retraction

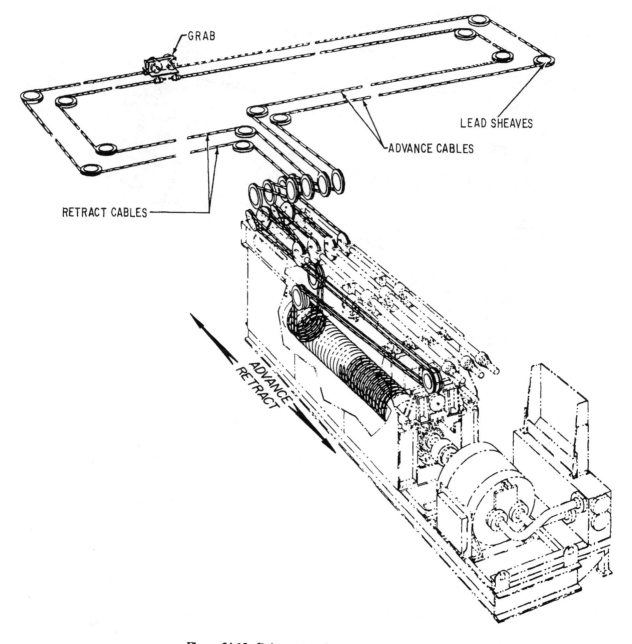

GRAB

LEAD SHEAVES

ADVANCE CABLES

RETRACT CABLES

ADVANCE

RETRACT

Figure 24-12.–Drive system (rotary retraction system).

(fig. 24-12) engines. Rotary retraction engines are installed aboard CV-67 and subsequent carriers. They also are being installed on all new carriers. The retraction engine provides the power to retract the shuttle and the launching engine pistons after the catapult has been fired. It is also used to advance and maneuver the grab forward and aft.

4. Drive system. This system provides the means of transferring the motion of the retraction engine to the grab for advance and retraction of the shuttle and piston assemblies.

5. Hydraulic system. This system supplies pressurized fluid to the hydraulic components of the catapult.

6. Bridle tension system. This system exerts a force on the shuttle to tension the aircraft before launching.

7. Lubrication system. This system provides lubricating oil for launching engine cylinder walls by injecting lubricating oil through the cylinder covers with a spray pattern that ensures even lubrication of the cylinder walls before passage of the launching engine pistons.

8. Control system. The control system provides for the control of the catapult during all phases of operation.

REFERENCES

Aviation Boatswain's Mate E 3 & 2, NAVEDTRA 10302-D1, Naval Education and Training Program Management Support Activity, Pensacola, Florida, 1983.

Machinist's Mate 3 & 2, NAVEDTRA 12144, Naval Education and Training Program Management Support Activity, Pensacola, Florida, 1991.

CHAPTER 25

ENGINEERING ADMINISTRATION

LEARNING OBJECTIVES

Upon completion of this chapter, you should be able to do the following:

1. Identify the purposes of various logs, records, and operating orders.

2. Describe the objective of the Navy's 3-M program and the policy toward maintaining equipment readiness.

3. Describe the most common records of the 3-M Systems.

4. Describe the Quality Assurance Program, its components, goals, and organization.

5. Identify the purpose of the tag-out program and the proper sequence in tagging out a piece of equipment.

6. Describe the precautions to be taken to avoid heat-related casualties.

7. Describe the purpose of the Navy's Hearing Protection and Noise Abatement Program.

8. Identify the purpose of the Personnel Qualification Standards (PQS) Program.

INTRODUCTION

There are many things to be done to ensure the proper and safe operation of the engineering plant. There are logs and records to be maintained, reports to be made to higher authority, and operating orders to be carried out.

For several years, significant losses of dollars and man-hours have resulted from injuries, illnesses, and property damage attributed to workplace hazards. All naval personnel must strive to eliminate or control all identified hazards as best they can within their capabilities.

We will discuss some of the logs, records, and operating orders. We also will discuss some of the systems and programs that reduce equipment downtime and protect personnel during operation and maintenance.

RECORD SYSTEMS

Accurate, legible, and up-to-date engineering records plus timely, accurate, and legible reports reflect efficient administration. Records maintained by the engineering department and reports submitted to the engineer officer provide the data for engineering reports to higher authority. The engineer officer uses reliable records and reports to keep up with the state of material and performance in the department.

The type commander (TYCOM) is coordinator of administrative matters for ships assigned. The TYCOM issues the necessary directives to regulate administrative records and reports required of ships within the command. The directives generally include a list of the records and recurring reports required by other commands and agencies of the Navy according to OPNAVINST 5214.2B. With this information, the commanding officer can establish a system to maintain current and accurate records and to forward reports in a correct and timely fashion. Instructions for the disposal of shipboard records are published in part III of *Disposal of Navy and Marine Corps Records*, SECNAVINST P5212.5C.

The engineer officer must be familiar with the engineering records and reports system. The engineer officer is responsible for (1) keeping record reference files containing complete information on the methods used to maintain required records, (2) keeping a report tickler file, (3) separating record reference files alphabetically and filing record reference cards alphabetically by subject, and (4) arranging report tickler file cards in order of the occurrence of the report—daily, weekly, monthly, and so forth. The engineer officer can use the Recurring Reports Records form, NAVEXOS 4179, for both files.

There is no simple way to assure the accuracy of records and reports. The first step is to establish the responsibility for keeping the records and preparing the reports within the department. The next step is to assign the duty of checking and verifying the data contained in the reports. The engineering department and division organization manuals provide excellent means of fixing departmental record-keeping responsibilities. The department training program should train personnel to obtain data and maintain records.

Some engineering records are mandatory (required by law), while others are necessary for efficient operation of the engineering plant. This chapter covers the records and reports that are basic to a well-administered engineering department of any large ship.

The engineer officer prepares the ship's operational reports that deal principally with engineering matters. The engineer officer should refer to directives of appropriate fleet and other operational commanders for requirement frequency, format, and timely submission of specific operational reports.

Logs should be properly maintained and kept as a record. They can warn you of upcoming equipment problems or help you troubleshoot an existing problem. They also can give you the information you need to make reports to higher authority. There are also legal requirements for certain logs in the engineering department.

In most cases each major piece of machinery in the engineering plant will be covered by an operating log. This is a daily log that runs from midnight to midnight. Readings are normally taken every hour when the equipment is in operation. The log will have a place to record important data about the piece of equipment covered by that log. Some examples are steam pressure, bearing temperatures, and oil pressure. Also, the log will have the minimum and maximum readings allowed for normal operation. These readings give you a quick reference point for detecting abnormal operation. If a reading is below the minimum or above the maximum, circle it in red and take the necessary action to correct the problem. Note in the remarks section what you did or are doing to correct it. For reasons previously mentioned, you should make sure you enter the correct reading in the appropriate block on the log. Most operating logs are retained onboard ship for 2 years. After this time they can be destroyed according to part

III of the *Disposal of Navy and Marine Corps Records*, SECNAVINST P5212.5, current edition.

LEGAL RECORDS

The Engineering Log, NAVSHIPS 3120/2, and the Engineer's Bell Book, NAVSHIPS 3120/1, are legal records of the engineering department. Completed Engineering Log and Engineer's Bell Book sheets are preserved on board as permanent records. They will be given up only in obedience to a demand from a Navy court or board or from the Navy Department. Sometimes it is necessary for these records or portions thereof to be removed from the ship. If so, a photostatic copy of the material to be removed is prepared for the ship's files and certified as a true copy by the engineer officer. Completed Engineering Log and Bell Book sheets may be destroyed 3 years after the date of the last entries. When a ship is stricken from the list of naval ships, its current Engineering Log and Bell Book sheets are forwarded to the nearest Naval Records Management Center. Sheets less than 3 years old (at time of inactivation) are kept on board when a ship is placed in an inactive status.

Engineering Log

The Engineering Log is in three parts: NAVSEA 3120/2A, Title Page; NAVSEA 31210/2B, Engineering Log; and NAVSEA 3120/2C, Continuation Sheet. The instructions for filling out these forms are in NAVSEA 3120/2D.

The Engineering Log is a complete daily record, by watches. It is a record of important events and data pertaining to the engineering department and the ship's propulsion plant. The log must show the average hourly rpm (to the nearest tenth) for all shafts; the speed in knots; the total engine miles steamed for the day; all major speed changes; draft and displacement upon getting underway and anchoring; fuel, water, and lubricating oil on hand, received, and expended; the disposition of the engines, boilers, and principal auxiliaries and any changes in their disposition; any injuries to engineering department personnel; any casualties to machinery, equipment, or material; and such other matters specified by competent authority.

Entries in the Engineering Log must be made according to instructions given in (1) 3120/2D; (2) the *Naval Ships' Technical Manual*, chapter 090; and (3) directives issued by the TYCOM. Each entry must be a complete statement using standard phraseology. The TYCOM's directives contain other specific requirements pertaining to the remarks section of Engineering Logs for ships of the type; the engineer officer must ensure compliance with these directives.

NOTE: Do not keep a rough log. Keep the Engineering Log current. Enter each event into the Engineering Log as it happens.

The original Engineering Log, prepared neatly and legibly in ink or pencil, is a legal record. The remarks should be prepared by–and signed by–the engineering officer of the watch (EOOW) (underway) or the engineering department duty officer (in port), whichever applies. The log may NOT contain any erasures. When a correction is necessary, draw a single line through the original entry so the entry remains legible. The correct entry must be clear and legible. Corrections, additions, or changes are made only by the person required to sign the log for the watch. This person then initials the margin of the page.

The engineer officer verifies the accuracy and completeness of all entries and signs the log daily. The commanding officer approves the log and signs it on the last calendar day of each month and on the date he or she relinquishes command. The log sheets must be submitted to the engineer officer in time to allow him or her to check and sign them before noon of the first day following the date of the log sheet(s). Completed pages of the log are filed in a post-type binder and are numbered consecutively. They begin with the first day of each month and run through the last day of the month.

When the commanding officer (or engineer officer) directs a change or addition to the Engineering Log, the person concerned must comply unless he or she believes the proposed change or addition is incorrect. In that event, the commanding officer (or engineer officer) enters his or her comments and signs the log. After the log has been signed by the commanding officer, it may not be changed without the CO's permission or direction.

Engineer's Bell Book

The Engineer's Bell Book is a record of all bells, signals, and other orders received by the throttleman regarding movement of the ship's propellers. Entries are made in the Bell Book by the throttleman as soon as an order is received. The assistant usually makes the entries when the ship is entering or leaving port, or engaging in any maneuver that may involve frequent speed changes. This allows the throttleman to devote his or her attention to answering the signals.

The Bell Book is maintained in the following manner:

1. A separate bell sheet is used for each shaft each day, except where more than one shaft is controlled by the same throttle station. In that case, the same bell sheet is used to record the orders for all shafts controlled by the station. All sheets for the same date are filed together as a single record.

2. The time of receipt of the order is recorded in column 1.

3. The order received is recorded in column 2. Minor speed changes are recorded by entering the number of rpm ordered. Major speed changes are recorded using the following symbols:

- 1/3–ahead 1/3 speed

- 2/3–ahead 2/3 speed

- I–ahead standard speed

- II–ahead full speed

- III–ahead flank speed

- Z–Stop

- B1/3–back 1/3 speed

- B2/3–back 2/3 speed

- BF–back full speed

- BEM–back emergency speed

4. The number of revolutions corresponding to the major speed change ordered is entered in column 3. When the order received is recorded as rpm in column 2 (minor speed changes), do not make an entry in column 3.

5. The shaft revolution counter reading (total revolutions) at the time of the speed change is recorded in column 4. The shaft revolution counter reading is taken hourly on the hour while underway.

Ships and crafts with controllable reversible pitch propellers also use column 4 to record responses to speed change orders. However, they record changes in the propeller pitch in feet and fractions of feet. Entries for astern pitch are preceded by the letter *B*. Entries are made of counter readings each hour on the hour. This information helps in the calculation of miles steamed during those hours when the propeller pitch remains constant at the last value set in response to a signaled order.

On ships with gas turbine propulsion plants, a bell logger provides an automatic printout each hour. This printout is also provided when propeller rpm or pitch is changed by more than 5 percent, when the engine order telegraph is changed, or when the controlling station is shifted. Provisions must be made for manual logging of data if the bell logger is out of commission.

Before going off watch, the EOOW signs the Bell Book on the line following the last entry for his or her watch. The next officer of the watch continues the record on the following line. In machinery spaces where an EOOW is not stationed, the watch supervisor signs the bell sheet.

NOTE: A common practice is to have the throttleman also sign the Bell Book before it is signed by the EOOW or his or her relief.

In ships or crafts with controllable pitch propellers, bridge personnel control the engines and maintain the Bell Book.

Some smaller ships with controllable pitch propellers sometimes need to switch control of the engines between the engine room and the bridge. For that purpose they maintain two Bell Books, and the personnel in control of the engines at any one time make entries in the Bell Book. When control shifts from one to the other, say from the bridge to the engine room, bridge personnel enter the time they gave control to the engine room. At the same time, engine-room personnel enter the time they assumed control. When the Bell Book is maintained by bridge personnel, the officer of the deck (OOD) signs it. When it is maintained by engine-room personnel, the EOOW signs it. At the end of the day, the two sets of Bell Sheets are consolidated and approved so there is only one official set for the day.

Alterations or erasures are not permitted in the Bell Book. An incorrect entry should be corrected by drawing a single line through the entry and recording the correct entry on the following line. The EOOW, the OOD, or the watch supervisor should initial changes.

OPERATING ORDERS

Engineering operating records help ensure regular inspection of operating machinery and provide data for performance analysis. They should be reviewed daily at the level specified by appropriate directives. Operating records are not intended to replace frequent inspections of operating machinery by supervisory personnel. Also, they are not to be trusted to warn of impending casualties. Personnel who maintain operating records must be properly indoctrinated. They must be trained to obtain, interpret, and record data correctly and to report any abnormal conditions. Acceptable high and low readings and abnormal readings must be permanently recorded on operating logs for each machinery type. Abnormal readings should be circled in red and reported to the watch supervisor.

The TYCOM's directives specify which engineering operating records will be maintained and prescribe the forms to be used when no standard record forms are provided. The engineer officer may require additional operating records if he or she finds them necessary.

The operating records discussed in this chapter are generally retained on board for 2 years. They may then be destroyed according to current disposal regulations. Complete records must be stowed where they will be properly preserved and easily located in case of need.

Propulsion Steam Turbine and Reduction Gear Operating Record

The Propulsion Steam Turbine and Reduction Gear Operating Record, NAVSEA 9231/1, is a daily record maintained for each main engine in operation. In ships with more than one main engine in the same engine room, a separate sheet is maintained for each engine, but common entries are omitted from the record for the port engine.

The watch supervisor enters the remarks and signs the record for his or her watch. The petty officer in charge of the engine room checks the accuracy of the record and signs in the space provided on the back of the record. The main propulsion assistant notes the contents and signs the record. Any unusual conditions noted in the record should be reported to the engineer officer immediately.

Gas Turbine Operating Record

On ships with gas turbines, bells and engine operating parameters are logged automatically by computer. The system can produce printouts at regular intervals or on demand. The data comes from two line printers; one for the bell logger and one for the data logger. The bell logger prints only bell signals and replies to those signals. The data logger prints all information other than bell signals. Examples of data logger printouts are logs for data on engine parameters, alarms, status changes, trends in operating parameters, and demand prints of any of the logs.

Diesel Engine Operating Record

The Diesel Engine Operating Record, NAVSEA 9231/2, is a complete daily record for each operating propulsion and auxiliary diesel engine in the ship. The watch supervisor writes and signs the remarks for his or her watch. The petty officer in charge of the ship's diesel engines checks the accuracy of the entries and signs the record in the space provided. The engineer officer notes the contents and approves the record daily by signing it.

AC/DC Electric Propulsion Operating Record

The AC/DC Electric Propulsion Operating Record, NAVSEA 9235/1, is a daily record for each operating propulsion generator and motor in ships (except submarines) equipped with ac or dc electric propulsion machinery. A separate record sheet is used for each shaft. Exceptions are ships with more than two generators or two motors per shaft, where as many sheets as required are used.

Data is entered on the record and the remarks are written and signed by the Electrician's Mate (EM) of the watch. The accuracy of the entries is checked by the EM in charge of the electric propulsion equipment and the electrical officer. Space is provided on the record for the daily approval and signature of the engineer officer.

Boiler Room Operating Record

The Boiler Room Operating Record, NAVSEA 9221/6, is a complete record for each steaming fireroom. Space is provided on the back of the record for the operating data of all fireroom auxiliary machinery. Entries are checked for accuracy by the fireroom supervisor. The B division officer also checks and initials the record. The engineer officer checks the entries and approves the record daily by signing it in the space provided for his or her signature.

Electrical Log

The Electrical Log, NAVSEA 9600/1, is a complete daily record for each operating ship's service generator. Entries for the prime movers are generally recorded by the generator watch (MM). Electrical data are recorded by the switchboard watch (EM). Each signs the remarks made for his or her watch.

The accuracy of the entries is checked by the EM in charge of the ship's service generators. Both the M and

E division officers check the record for accuracy and any evidence of impending casualties. Each officer initials the record to indicate he or she has checked it. The engineer officer notes the content and signs the record daily in the space provided.

Distilling Plant Operating Record

There is a distilling plant operating record for each of the three principal types of distilling plants in use aboard naval ships. The records are (1) the Low Pressure Distilling Plant Operating Record, NAVSEA 9530/3, (2) the Flash Type Distilling Plant Operating Record, NAVSEA 9530/1, and (3) the Vapor Compression Distilling Plant Operating Record, NAVSEA 9530/2. Each is a complete daily record maintained for each distilling plant in operation. Personnel of the watch record data and remarks in the record. The watch supervisor signs the remarks for his or her watch, and the petty officer in charge of the ship's distilling plants checks all entries for accuracy and signs the record. The division officer (M or A, as applicable) reviews and initials the record. Space is provided on the back of the record for the daily signature of the engineer officer.

Refrigeration/Air Conditioning Equipment Operating Record

The Refrigeration/Air Conditioning Equipment Operating Record, NAVSEA 9516/1, is a complete daily record for each operating refrigeration plant and air conditioning plant (except package units). Spaces on the front of the record are for entries applicable to both refrigeration and air conditioning plants. Data are recorded in 2-hour intervals in this record. The A division officer reviews the contents and initials the record daily.

Gyrocompass Operating Record

The Gyrocompass Operating Record is a locally prepared, complete daily record for each operating master gyrocompass. The form for the log is prepared according to the TYCOM's directives. Columns in the log should provide space for recording the times of starting and stopping the gyrocompass, total hours of operation since delivery of the gyrocompass, and important operating data pertaining to the gyrocompass installation. The petty officer in charge of the interior communications (IC) equipment checks the accuracy of the log, and the electrical officer notes its contents daily.

IC Room Operating Record

The IC Room Operating Record is a daily record of major electrical equipment in operation in the IC room and is maintained by the IC watch. The form for the record is prepared locally according to the TYCOM's directives. On small ships the gyrocompass log and the IC room record may be maintained on the same form. Important data, such as voltages and currents of major units of IC equipment (IC switchboard, telephone switchboard, and motor generator sets), should be recorded on the form. The IC Room Operating Record is checked and approved in the manner described for the Gyrocompass Operating Record.

Air Compressor Operating Record

Some large ships maintain an Air Compressor Operating Record that contains important data such as temperatures and pressures pertaining to air compressors in operation. When required by the TYCOM, the Air Compressor Operating Record is prepared locally according to the TYCOM's directives. Contents of the record should be checked by the petty officer in charge of the air compressors and the appropriate division officer.

SHIPS' MAINTENANCE AND MATERIAL MANAGEMENT (3-M) SYSTEMS

The primary objective of the Navy Ships' Maintenance and Material Management (3-M) Systems is to manage maintenance and maintenance support in a manner that will ensure maximum equipment operational readiness. OPNAVINST 4790.4B, volumes I, II, and III, contain all of the detailed procedures and instructions for the effective operation of the 3-M Systems. That includes examples of the forms discussed in this chapter. Other instructions on the 3-M Systems are found in the TYCOM's maintenance manuals.

The following paragraphs will discuss the most common records of the 3-M Systems that must be kept current in the engineering department.

PLANNED MAINTENANCE SCHEDULES

In an effective Planned Maintenance System (PMS), PMS schedules must be accurately filled out and posted in a timely manner. PMS schedules are categorized as cycle, quarterly, and weekly.

Cycle Schedule

The Cycle PMS Schedule displays the planned maintenance requirements to be performed between major overhauls of the ship. The following information must be filled in on the cycle schedules: ship's name and hull number, work center designator code, maintenance index page (MIP) number, component's or system's name, and maintenance scheduled in each quarter after overhaul.

The engineer officer must supervise all cycle scheduling of engineering department maintenance, and then sign and date the Cycle PMS Schedule before it is posted.

If there is a need to rewrite the Cycle PMS Schedule, the old schedules should be filed with the last quarterly schedule with which it was used.

Quarterly Schedule

The Quarterly PMS Schedule is a visual display of the work center's PMS requirements to be performed during a specific 3-month period. Spaces are provided to enter the work center, quarter after overhaul, department head's signature, date prepared, and the months covered. The schedule has 13 columns, one for each week in the quarter. These permit scheduling of maintenance requirements on a weekly basis throughout the quarter. There are also columns to enter the MIP number and PMS requirements that may require rescheduling. There are "tic" marks across the top of the scheduling columns for use in showing the in-port/underway time of the ship for the quarter.

The engineer officer must supervise scheduling of PMS on the quarterly schedule for his or her department. The engineer officer must then sign and date the schedule before it is posted. At the end of each quarter, the engineer officer must review the quarterly schedule, check the reasons for PMS actions not accomplished, and sign the form in the space provided on its reverse side. The division officer is responsible for updating the quarterly schedule every week. Completed quarterly schedules should be kept on file for 1 year.

Weekly Schedule

The Weekly PMS Schedule is a visual display of the planned maintenance scheduled for a given work center during a specific week. The work center supervisor uses weekly schedules to assign and monitor work on the PMS tasks by work center personnel.

The Weekly PMS Schedule contains blank spaces to be filled in for work center code, date of current week, division officer's signature, MIP number minus the date code, component names, names of personnel responsible for specific maintenance requirements, outstanding major repairs, and situation requirements.

The work center supervisor is responsible for completing the Weekly PMS Schedule and for updating it every day.

FEEDBACK FORM

The PMS Feedback Report Form, OPNAV Form 4790/7B, provides maintenance personnel with the means to report discrepancies and problems and to request PMS coverage. All PMS Feedback Reports are sent to NAVSEACANs or TYCOMs, based on the category of the feedback report.

Feedback reports are originated in the work center and must be signed by the originator. They are then screened and signed by the division officer and the engineer officer before being forwarded to the 3-M coordinator. The 3-M coordinator will date and sign the feedback report, serialize it, and return the green copy to the originating work center. The originating work center will file the green copy until an answer to the feedback report is received.

SHIP'S MAINTENANCE ACTION FORM

The Ship's Maintenance Action Form, OPNAV 4790/2K, is used by maintenance personnel to report deferred maintenance and completed maintenance. This form also allows the entry of screening and planning information for management and control of intermediate maintenance activity (IMA) workloads.

The OPNAV 4790/2K is originated in the work center. It is screened for accuracy and legibility, and initialed by the division officer and engineer officer before being forwarded to the 3-M coordinator. When the form is used to defer maintenance, the 3-M coordinator will send two copies back to the originating work center to hold on file. When the deferred maintenance is completed, one of the copies is used to document the completion of the maintenance.

CURRENT SHIP'S MAINTENANCE PROJECT

The standard Current Ship's Maintenance Project (CSMP) is a computer-produced report listing deferred maintenance and alterations that have been identified through Maintenance Data Collection System (MDCS) reporting. Copies of the CSMP should be received monthly. The engineer officer is provided with a copy for each of the engineering department work centers, and each work center is provided a copy that shows only its own deferred maintenance.

The purpose of the CSMP is to give shipboard maintenance managers a consolidated list of deferred corrective maintenance. They can use the list to manage and control maintenance in the deferred items. The work center supervisor is responsible for ensuring the CSMP accurately describes the material condition of the work center.

Each month when a new CSMP is received, verified, and updated, the old CSMP may be destroyed.

OPNAVINST 4790.4B contains the instructions and procedures needed to complete and route all 3-M Systems forms.

ADDITIONAL RECORDS

The engineering department records and reports discussed in this section inform responsible personnel of coming events (including impending casualties). They supply data for the analysis of equipment performance, provide a basis for design comparison and improvement, or provide information for the improvement of maintenance techniques and the development of new work methods. The records are those papers that must be compiled and retained on board (in original or duplicate form) for prescribed periods. They are primarily used for reference in administrative and operational matters. The reports are of either a one-time or recurring nature. Recurring reports are required at prescribed or set intervals, while one-time reports need to be made on the occurrence of a given situation.

ENGINEER OFFICER'S NIGHT ORDER BOOK

The engineer officer keeps a Night Order Book as part of the engineering records. This book includes (1) orders concerning the operation of the engineering plant, (2) any special orders or precautions concerning the speed and operation of the main engines, and (3) all other orders for the night for the EOOW. The Night Order Book is prepared and maintained according to instructions issued by the TYCOM. Some TYCOMs require that the Night Order Book have a specific format that is standard for ships of the type. Others allow use of a locally prepared form but specify certain contents of the book.

The Night Order Book must contain orders covering routine recurring situations (engineering department standing orders) as well as orders for the night for the EOOW. Standing orders are issued by the engineer officer as a letter-type directive (instruction), according to the ship's directives systems. A copy of the instruction is posted in the front of the Night Order Book. Orders for the night for the EOOW generally specify the boilers and other major items of machinery to be used during the night watches. A form similar to the one illustrated in figure 25-1 is in use in some ships for the issuance of the engineer officer's night orders.

The Night Order Book is maintained in port and at sea. In the temporary absence of the engineer officer in port, the engineering department duty officer maintains it. Underway, the Night Order Book is delivered to the EOOW before 2000 and is returned to the log room before 0800 of the following day. In addition to the EOOW, principal engineering watch supervisors and the oil king should read and initial the night orders for the watch. In port, the leading duty petty officer of each engineering division and the principal watch supervisors should read and initial the night orders.

STEAMING ORDERS

Steaming orders are written orders issued by the engineer officer. They list the major machinery units and readiness requirements of the engineering department based upon the time set to get the ship underway. Generally, a locally prepared form similar to the one illustrated in figure 25-2 is used to issue the steaming orders. The orders normally specify the (1) engine combinations to be used, (2) times to light fires and cut in boilers, (3) times to warm up and test main engines, (4) times to start and parallel ship's service generators, (5) standard speed, and (6) EOOW and principal watch supervisors. Early posting of steaming orders is essential to get a ship with a large engineering plant underway.

GYROCOMPASS SERVICE RECORD

A Gyrocompass Service Record Book is furnished to the ship for each gyrocompass installed. The book is a complete record of inspections, tests, and repairs to the gyrocompass and must always remain with its associated gyrocompass. The front of the book contains complete instructions for maintaining the record–they must be followed carefully. If the Gyrocompass Service

Record Book is lost or damaged, use the *Navy Stock List of Publications and Forms*, NAVSUP 2002, to get a replacement. Include the mark, modification, and serial number of the gyrocompass for which the book is intended.

QUALITY ASSURANCE PROGRAM

The Quality Assurance (QA) program was established to provide personnel with information and guidance necessary to administer a uniform policy of maintenance and repair of ships and submarines. The QA program is intended to impart discipline into the repair of equipment, safety of personnel, and configuration control, thereby enhancing ship's readiness.

The various QA manuals set forth minimum QA requirements for both the surface fleet and the submarine force. If more stringent requirements are imposed by higher authority, such requirements take precedence. If conflicts exist between the QA manual and previously issued letters and transmittals by the appropriate force commanders, the QA manual takes precedence. Such conflicts should be reported to the appropriate officials.

The instructions contained in the QA manual apply to every ship and activity of the force. Although the requirements are primarily applicable to the repair and maintenance done by the force IMAs, they also apply to maintenance done aboard ship by ship's force. In all cases, when specifications cannot be met, a departure from specifications request must be completed and reported.

Because of the wide range of ship types and equipment and the varied resources available for maintenance and repair, the instructions set forth in the QA manual are general in nature. Each activity must implement a QA program to meet the intent of the QA manual. The goal should be to have all repairs conform to QA specifications.

PROGRAM COMPONENTS

The basic thrust of the QA program is to make sure you comply with technical specifications during all work on ships of both the surface fleet and the submarine force. The key elements of the program are as follows:

* Administrative. This includes training and qualifying personnel, monitoring and auditing programs, and completing the QA forms and records.

ENGINEER OFFICER'S NIGHT ORDERS   Date 6 June 19--

USS SAMPLE (DDG-34)

At or enroute from __Naples, Italy__ to __Genoa, Italy__

Standard speed is ____15____ knots __141__ rpm, or as ordered.

Anticipated speed changes: _____ knots at ___: _____ at _____

Be prepared for _____ knots with Boilers _____ at _____

Boilers in use __1A & 2B__ ; Standby Boilers __1B & 2A__

Sprayer plates in use: __1712__

Operate with engineering plant __Split__

Operate ship's service generators __1A & 2A__ ; generators __1B & 2B__ in

standby condition.

Evaporators __1 & 2__ distill to ship's tanks until __0200__ and

then shift to __RFT__

Carry out standing night orders published in the front of this book.

REMARKS:

Carry out normal steaming watch routine &
keep bilges dry. Call me at 0600 if not
needed theretofore.

In case of trouble or doubt, call me in room __112__, telephone __222__ ;

and __Lt. Corley__ in room __118__, telephone __236__

INITIALS

20 - 24   _____   _____   _____

00 - 04   _____   _____   _____

04 - 08   _____   _____   _____

                              L. S. Savage, LCDR, USN
                              ENGINEER OFFICER

Figure 25-1.—Engineer officer's night orders (sample).

```
                    U.S.S. SPEEDWILL CV-3333
                    ENGINEERING DEPARTMENT
                         STEAMING ORDERS

                                        Date ___17 August___ , 19___

The ship is scheduled to get underway at 1400, _____18 August_____ , 19___

1.  The Engineering Department shall report ready to get underway at 1345 (zero time)

2.  Boilers Nos. 1, 2, 3, 4, 6, 8 shall be used.

3.  Steaming watches ( 3rd  steaming section) below as follows:

        SPACE                       BOILERS            MACHINERY

No. 1 Machinery Space
No. 2 Machinery Space
No. 3 Machinery Space
No. 4 Machinery Space

4.  Light Boiler Nos.  1A, 2B  at  0800 ; on line   1030
        Boiler Nos.   3A, 4B  at  1000 ; on line   1230
        Boiler Nos.  _____ at _____; on line  _____

5.  Commence warming-up main engines at   1130  , and follow warming-up schedule.

6.  Warm up Nos.   2B  1250-K.W. generators at  0800 ; on line (idle)  0900  .
        Nos.   4A  1250-K.W. generators at  1200 ; on line (idle)  1245  .
        Nos. _____ 1250-K.W. generators at _____; on line (idle) _____.

7.  Test main engines at  1330 . Report the Engineering Department ready to get
    underway to Engineer Officer.

8.  Have all burners made up with  3208  sprayer plates.

9.  Operate  2  set(s) evaporators, and distill as directed by  E.O.O.W.  .

10. Calls                               On Station:
        Engineer Officer        0700        Duty Officer      R.E. Caldwell
        Assistant Engineer Officer  0630    Junior Duty Officer G.E. Holt
        Main Propulsion                     Duty MMC          Miller, MMC
        Assistant               0600        Duty MMCM         Smith, MMC
                                            Duty BTC          Walters, BTC
                                            Duty EMC          Land, EMC

11. Watch           E.O.O.W.                        J.O.O.W.
    0000-0400   D. D. Harper, Lt(jg)        R. S. Smith, Ensign
    0400-0800   R. C. Johnson, Lt.          E. E. Robertson, Ensign
    0800-1200   C. C. Smart, Ch. Mach.      D. F. Edwards, Ensign
    1200-1600   R. E. Caldwell, Lt(jg)      G. E. Holt, Mach.
    1600-1800   D. D. Harper                R. S. Smith
    1800-2000   R. C. Johnson               E. E. Robertson
    2000-2400   C. C. Smart                 D. F. Edwards

Unless notified otherwise, standard speed will be  15  knots,  115  rpm.


                                    Glenn Taylor, LCDR, USN.
                                         (Engineer Officer)



Copies: Duty Officer and C.P.O.'s, Steaming M.S.,
            Div. B.B.'s, File.
```

Figure 25-2.—Steaming orders (sample).

• Job Execution. This includes preparing work procedures, meeting controlled material requirements, requisitioning material, conducting in-process control of fabrication and repairs, testing and recertifying, and documenting any departure from specifications.

CONCEPTS OF QUALITY ASSURANCE

The ever-increasing technical complexity of present-day surface ships and submarines has spawned the need for special administrative and technical procedures known collectively as the QA program. The QA concept is fundamentally the prevention of defects. This encompasses all events from the start of maintenance operations until their completion. It is the responsibility of all maintenance personnel. Achievement of QA depends on prevention of maintenance problems through your knowledge and special skills. As a supervisor, you must consider QA requirements whenever you plan maintenance. The fundamental rule for you to follow for all maintenance is that TECHNICAL SPECIFICATIONS MUST BE MET AT ALL TIMES.

Prevention is concerned with regulating events rather than being regulated by them. It relies on eliminating maintenance failures before they happen. This extends to safety of personnel, maintenance of equipment, and virtually every aspect of the total maintenance effort.

Knowledge is obtained from factual information. This knowledge is acquired through the proper use of data collection and analysis programs. The MDCS provides maintenance managers unlimited quantities of factual information. The experienced maintenance manager provides management with a pool of knowledge. Correct use of this knowledge provides the chain of command with the tools necessary to achieve maximum shipboard readiness.

Special skills, normally not possessed by production personnel, are provided by a staff of trained personnel for analyzing data and supervising QA programs.

The QA program provides an efficient method for gathering and maintaining information on the quality characteristics of products and on the source and nature of defects and their impact on current operations. It permits decisions to be based on facts rather than intuition or memory. It provides comparative data that will be useful long after details of particular times or events have been forgotten. QA requires both authority and assumption of responsibility for action.

A properly functioning QA program points out problem areas to maintenance managers so they can take appropriate action to accomplish the following:

–Improve the quality, uniformity, and reliability of the total maintenance effort.

–Improve the work environment, tools, and equipment used in the performance of maintenance.

–Eliminate unnecessary man-hour and dollar expenses.

–Improve the training, work habits, and procedures of maintenance personnel.

–Increase the excellence and value of reports and correspondence originated by the maintenance activity.

–Distribute required technical information more effectively.

–Establish realistic material and equipment requirements in support of the maintenance effort.

To obtain full benefits from a QA program, teamwork must be achieved first. Blend QA functions in with the interest of the total organization and you produce a more effective program. Allow each worker and supervisor to use an optimum degree of judgment in the course of the assigned daily work; a person's judgment plays an important part in the quality of the work. QA techniques supply each person with the information on actual quality. This information provides a challenge to the person to improve the quality of the work. The resulting knowledge encourages the best efforts of all your maintenance personnel.

QA is designed to serve both management and production equally. Management is served when QA monitors the complete maintenance effort of the department, furnishes factual feedback of discrepancies and deficiencies, and provides the action necessary to improve the quality, reliability, and safety of maintenance. Production is served by having the benefit of collateral duty inspectors formally trained in inspection procedures; it is also served in receiving technical assistance in resolving production problems. Production personnel are not relieved of their basic responsibility for quality work when you introduce QA to the maintenance function. Instead, you increase their responsibility by adding accountability. This accountability is the essence of QA.

GOALS

The goals of the QA program are to protect personnel from hazardous conditions, increase the time between equipment failure, and ensure proper repair of failed equipment. The goals of the QA program are intended to improve equipment reliability, safety of personnel, and configuration control. Achievement of these goals will ultimately enhance the readiness of ship and shore installations. There is a wide range of ship types and classes in the fleet, and there are equipment differences within ship classes. This complicates maintenance support and increases the need for a formalized program that will provide a high degree of confidence that overhaul, installations, repairs, and material will consistently meet conformance standards.

THE QUALITY ASSURANCE ORGANIZATION

The QA program for naval forces is organized into different levels of responsibility. For example, the QA program for the Naval Surface Force for the Pacific Fleet is organized into the following levels of responsibility: TYCOM, readiness support group/area maintenance coordinator, and the IMAs. The QA program for the submarine force is organized into four levels of responsibility: TYCOM, group and squadron commanders, IMA commanding officers, and ship commanding officer/officers in charge. The QA program for the Naval Surface Force for the Atlantic Fleet is organized into five levels of responsibility: force commander, audits, squadron commanders, IMAs, and force ships.

The QA program organization (Navy) begins with the commanders in chief of the fleets, who provide the basic QA program organization responsibilities and guidelines.

The TYCOMs provide instruction, policy, and overall direction for implementation and operation of the force QA program. TYCOMs have a force QA officer assigned to administer the force QA program.

The commanding officers (COs) are responsible to the force commander for QA in the maintenance and repair of the ships. The CO is responsible for organizing and implementing a QA program within the ship to carry out the provisions of the TYCOMs QA manual.

The CO makes sure all repair actions performed by ship's force conform to provisions of the QA manual as well as pertinent technical requirements.

The CO makes sure all work requests requiring special controls are properly identified and applicable supporting documentation is provided to the maintenance or repair activity using the applicable QA form.

The CO also makes sure departures from specifications are reported, required audits are conducted, and adequate maintenance is performed for the material condition necessary to support continued unrestricted operations.

The quality assurance officer (QAO) is responsible to the CO for the organization, administration, and execution of the ship's QA program according to the QA manual. On most surface ships other than IMAs, the QAO is the chief engineer, with a senior chief petty officer assigned as the QA coordinator. The QAO is responsible for the following:

–Coordinating the ship's QA training program

–Maintaining ship's QA records and inspection reports according to the QA manual

–Maintaining auditable departure from specification records

–Reviewing procedures and controlled work packages prepared by the ship before submission to the engineer

–Conducting QA audits as required by the QA manual and following up on corrective actions to ensure compliance with the QA program

–Maintaining liaison with the IMA office for all work requiring QA controls

–Providing QA guidance to the supply department when required

–Preparing quality assurance/quality control (QA/QC) reports (as required) by higher authority

–Maintaining liaison with the ship engineer in all matters pertaining to QA to ensure compliance with the QA manual

The ship quality control inspectors (SQCIs), usually the work center supervisor and two others from the work center, must have a thorough understanding of the QA program. Some of the other responsibilities an SQCI will have are as follows:

–Maintain ship records to support the QA program.

–Inspect all work for conformance to specifications.

–Ensure that only calibrated equipment is used in acceptance testing and inspection of work.

–Witness and document all tests.

–Make sure all materials or test results that fail to meet specifications are recorded and reported.

–Train personnel in QC.

–Initiate departure from specification reports (discussed later) when required.

–Ensure all inspections beyond the capabilities of the shop's QA inspector are performed and accepted by IMA before final acceptance and installation of the product by the ship.

–Report all deficiencies and discrepancies to the ship's QA coordinator (keeping the division officer informed).

–Develop controlled work packages for all ship repair work requiring QA controls.

LEVELS OF ESSENTIALITY

A number of early failures in certain submarine and surface ship systems were due to the use of the wrong material. This led to a system for prevention involving levels of essentiality. A level of essentiality is simply a range of controls in two broad categories representing a certain high degree of confidence that procurement specifications have been met. These categories are

- verification of material, and

- confirmation of satisfactory completion of tests and inspections required by the ordering data.

Levels of essentiality are codes, assigned by the ship according to the QA manual, that indicate the degree to which the ship's system, subsystem, or components are necessary or indispensable in the performance of the ship's mission. Levels of essentiality also indicate the impact that catastrophic failure of the associated part or equipment would have on ship's mission capability and personnel safety.

LEVELS OF ASSURANCE

QA is divided into three levels: A, B, and C. Each level reflects certain quality verification requirements of individual fabrication in process or repair items. Here, verification refers to the total of quality of controls, tests, and/or inspections. The levels of assurance are as follows:

Level A: Provides for the most stringent or restrictive verification techniques. This normally will require both QC and test or inspection methods.

Level B: Provides for adequate verification techniques. This normally will require limited QC and may or may not require tests or inspections.

Level C: Provides for minimum or "as necessary" verification techniques. This normally will require very little QC or tests and inspections.

LEVELS OF CONTROL

QC also may be assigned generally to any of the three levels–A, B, or C. Levels of control are the degrees of control measures required to assure reliability of repairs made to a system, subsystem, or component. Furthermore, levels of control (QC techniques) are the means by which we achieve levels of assurance.

An additional category that you will see is level I. This is reserved for systems that require maximum confidence that the composition of installed material is correct.

CONTROLLED MATERIAL

Some material, as part of a product destined for fleet use, has to be systematically controlled from procurement, receipt, stowage, issue, fabrication, repair, and installation to ensure both quality and material traceability. Controlled material is any material you use that must be accounted for (controlled) and identified throughout the manufacturing and repair process, including installation, to meet the specifications required of the end product. Controlled material must be inspected by your controlled material petty officer (CMPO) for required attributes before you can use it in a system or component and must have inspection documentation maintained on record. You must retain traceability through the repair and installation process. These records of traceability must be maintained for 7 years (3 years aboard ship and 4 years in record storage). Controlled material requires special marking and tagging for identification and separate storage to preclude loss of control. The repair officer (RO) may designate as controlled material any material that requires material traceability.

Under this definition, controlled material has two meanings. The first meaning applies to items considered critical enough to warrant the label of controlled material. Your CMPOs will be responsible for inspecting the material when it is received, stowing it separately from other material, providing custody, and seeing that controlled assembly procedures are used during its installation.

NAVAL OCCUPATIONAL SAFETY AND HEALTH PROGRAM

It is Navy policy to provide a safe and healthy workplace for all personnel. These conditions can be ensured through an aggressive and comprehensive occupational safety and health program fully endorsed by the Secretary of the Navy and implemented through the appropriate chain of command.

The material discussed in this section stresses the importance of electrical and general safety precautions. All electrical equipment is hazardous; hence, all safety precautions must be strictly observed. The main purposes of safety are to protect personnel and material and to ensure that unsafe equipment operations do not occur. You have the responsibility to recognize unsafe conditions and to take appropriate actions to correct any discrepancies. You must always observe safety precautions. You also should know and be able to perform the proper action when a mishap occurs.

Remember: Mishaps seldom just happen; they are caused. Another point to remember is never to let familiarity breed contempt. Hundreds of people have been injured by mishaps and many have died from injuries. Most of those mishaps could have been prevented had the individuals involved heeded the appropriate safety precautions. Preventing mishaps that are avoidable will help you in the Navy and possibly determine whether or not you survive.

The following paragraphs will discuss some of the safety and health programs that you will be involved with in the day-to-day operations of the ship.

SAFETY RESPONSIBILITIES

All individuals have the responsibility to understand and observe safety standards and regulations that are established for the prevention of injury to themselves and other people and damage to property and equipment. As an individual, you have a responsibility to yourself and to your shipmates to do your part in preventing mishaps. You also have the responsibility of setting a good example; you cannot ignore safety regulations and expect others to follow them.

Personnel should always observe the following safety practices:

- Observe all posted operating instructions and safety precautions.

- Report any unsafe condition or any equipment or material you think might be unsafe.

- Warn others of hazards or their failure to observe safety precautions.

- Wear or use approved protective clothing or protective equipment.

- Report any injury or evidence of impaired health that occurs during your work or duty to your supervisor.

- Exercise reasonable caution as appropriate to the situation in the event of an emergency or other unforseen hazardous condition.

- Inspect equipment and associated attachments for damage before using the equipment. Be sure the equipment is suited for the job.

Safety must always be practiced by people working around electrical circuits and equipment to prevent injury from electrical shock and from short circuits caused by accidentally placing or dropping a conductor of electricity across an energized line. The arc and fire started by these short circuits, even where the voltage is relatively low, may cause extensive damage to equipment and serious injury to personnel.

No work will be done on electrical circuits or equipment without permission from the proper authority, and until all safety precautions are taken.

EQUIPMENT AND INSTRUMENT TAG-OUT

For several years significant losses of dollars and man-hours have resulted from injuries, illnesses, and property damage attributed to workplace hazards. All naval personnel must strive to eliminate or control all identified hazards as best they can within their capabilities.

Whenever repairs are to be made to equipment or piping systems, that section of the system or the equipment must be isolated and tagged out. The tag-out program provides a procedure to be used when a component, a piece of equipment, a system, or a portion of a system must be isolated because of some abnormal condition. The tag-out program also provides a procedure to be used when an instrument becomes unreliable or is not operating properly. The major difference between equipment tag-out and instrument tag-out is that labels are used for instrument tag-out and tags are used for equipment tag-out.

Tag-out procedures are described in OPNAVINST 3120.32B and represent the minimum requirements for tag-out. These procedures are mandatory and are standardized aboard ships and repair activities. The following definitions are used in the tag-out bill:

• Authorizing officer–This officer has the authority to sign tags and labels and to cause tags and labels to be issued or cleared. The authorizing officer is always the officer responsible for supervising the tag-out log. The commanding officer designates authorizing officers by billet or watch station. The authorizing officer for engineering is normally the EOOW underway and the engineering duty officer (EDO) in port.

• Department duty officer (DDO) (repair activities only)– This officer is designated as DDO on the approved watch bill or plan of the day.

• Engineering officer of the watch (EOOW)–This officer may be either the EOOW or the EDO, depending on engineering plant conditions.

Tags

The purpose of using tags is to prevent the improper operation of a component, piece of equipment, system, or portion of a system when isolated or in an abnormal condition.

Equipment that you are intending to repair or perform PMS on must be de-energized and tagged out by use of either a CAUTION or a DANGER tag.

CAUTION TAGS.–A CAUTION tag, NAVSHIPS 9890/5 (fig. 25-3), is a yellow tag used as a precautionary measure to provide temporary special instructions or to indicate that unusual caution must be exercised to operate equipment. These instructions must state the specific reason that the tag is installed. Use of phrases such as DO NOT OPERATE WITHOUT EOOW PERMISSION is not appropriate since equipment or systems are not operated unless permission from the responsible supervisor has been obtained. A CAUTION tag cannot be used if personnel or equipment could be endangered while performing evolutions using normal operating procedures. A DANGER tag must be used in this case.

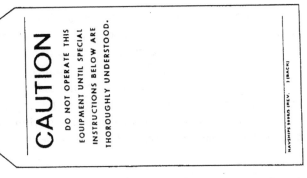

Figure 25-3.–CAUTION tag (colored YELLOW).

DANGER TAG.–The DANGER tag, NAVSHIPS 9890/8 (fig. 25-4), commonly called the red tag, is used to prevent the operation of equipment that could jeopardize the safety of personnel or endanger the

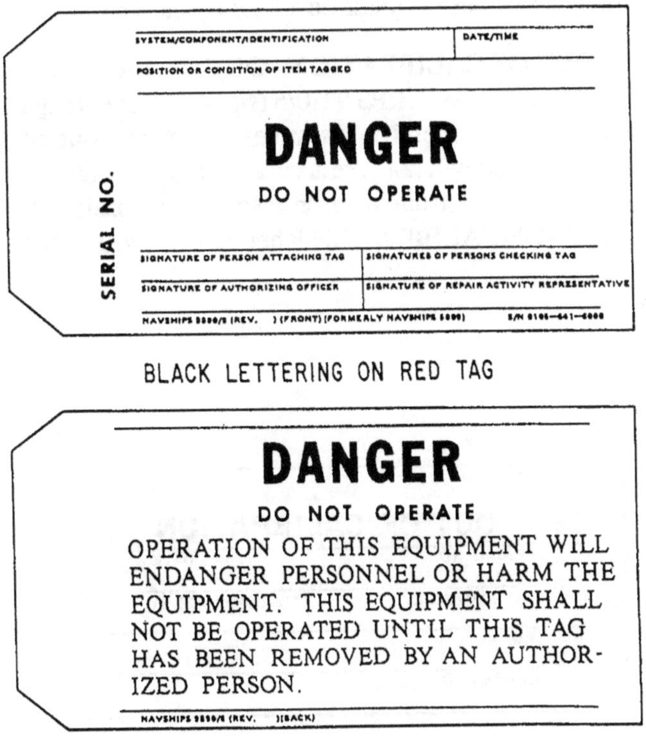

BLACK LETTERING ON RED TAG

DANGER

DO NOT OPERATE

OPERATION OF THIS EQUIPMENT WILL ENDANGER PERSONNEL OR HARM THE EQUIPMENT. THIS EQUIPMENT SHALL NOT BE OPERATED UNTIL THIS TAG HAS BEEN REMOVED BY AN AUTHORIZED PERSON.

Figure 25-4.–DANGER tag (colored RED).

equipment systems or components. When equipment is red tagged, under no circumstances will it be operated. When a major system is being repaired or when PMS is being performed by two or more repair groups, both parties will hang their own tags. This prevents one group from operating or testing circuits that could jeopardize the safety of personnel from the other group.

No work should be done on energized or de-energized switchboards without approval of the commanding officer, the engineer officer, and the electrical officer.

All supply switches or cutout switches from which power could be fed should be secured in the off or open (safety) position and red tagged. Circuit breakers should have a handle locking device installed. The proper use of red tags cannot be overstressed. When possible, double red tags should be used, such as tagging open the main power supply breaker and removing and tagging the removal of fuses of the same power supply.

Labels

Labels are used to warn operating or maintenance personnel that an instrument is unreliable or is not in normal operating condition. There are two types of labels used on instruments, OUT-OF-CALIBRATION and OUT-OF-COMMISSION. The decision as to which label to use is made on a case-by-case basis.

OUT OF CALIBRATION.–OUT-OF-CALIBRATION labels, NAVSHIPS 9210/6 (fig. 25-5), are orange labels used to identify instruments that are out of calibration and will not give accurate measurements. In general, if the instrument error is small and consistent, an OUT-OF-CALIBRATION label may be used. This

Figure 25-6.–Out-of-commission label.

label indicates that the instrument may be used, but only with extreme caution.

OUT OF COMMISSION.–OUT-OF-COMMISSION labels, NAVSHIPS 9890/7 (fig. 25-6), are red labels used to identify instruments that will not indicate correct measurements because they are defective or isolated from the system. This label indicates that the instrument cannot be relied on and must be repaired and recalibrated or be reconnected to the system before it can be used properly.

Tag-out Logs

The number of tag-out logs on a ship depends on the ship's size. For example, a minesweeper or nonnuclear-powered submarine may need only one tag-out log; a major surface combatant may need a separate log for each major department. Individual force commanders specify the number of logs needed and their location.

A tag-out log is a record of authorization of each effective tag-out action. It is the control document used to administer the entire tag-out procedure. Supervisory personnel must review the logs during watch relief. The tag-out log includes the following documents:

- A copy of OPNAVINST 3120.32B and any amplifying directives needed to administer the system.

- The DANGER/CAUTION Tag-out Index and Record of Audit (Index/Audit Record). This is a sequential list of all tag-outs issued. It provides a ready reference of existing tag-outs, ensures that serial numbers are sequentially issued, and is useful in conducting audits of the log. A sample of this index is shown in figure 25-7. Index pages with all tag-outs listed as cleared may be removed by the department head.

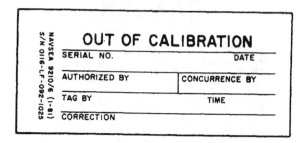

Figure 25-5.–Out-of-calibration label.

DANGER/CAUTION TAG-OUT INDEX AND RECORD OF AUDIT (INDEX/AUDIT RECORD)				
LOG SERIAL	DATE ISSUED	TYPE (DANGER/CAUTION)	DESCRIPTION (SYSTEM, COMPONENT, WORK PERMIT OR TEST DOCUMENT REFERENCE)	DATE CLEARED

Figure 25-7.–DANGER/CAUTION tag-out index and record of audit.

• DANGER/CAUTION Tag-out Record Sheet (fig. 25-8, views A and B). All tags that have been used in the tag-out of a system are logged on one DANGER/CAUTION tag-out record sheet along with the reason for the tag-out. All effective sheets are kept in one section of the log.

• Instrument Log (fig. 25-9). OUT-OF-CALIBRATION and OUT-OF-COMMISSION instruments are logged in the instrument log.

• Cleared DANGER/CAUTION Tag-out Record Sheets. The sheets that have been cleared and completed are transferred to this section of the log until they are reviewed and removed by the department head.

Tag-out Procedures

Assume that a requirement for tags has been identified, and the affected system will be out of commission as a result of the tag-out action. At that time, the authorizing officer must ask the commanding officer and the responsible department head for permission to begin the tag-out. The authorizing officer also must notify the responsible division officer of the requirement for the tag-out. On ships having damage control central (DCC), the authorizing officer must notify the DCC if the affected system or component will be out of commission. The authorizing officer should have approval from either the OOD or the EOOW if the tag-out will affect systems under their responsibility. After permission has been obtained, the authorizing officer should direct the preparation of the tag-out record sheet and tags. The following procedures should be used. These procedures may be modified during overhaul periods at the discretion of the commanding officer.

1. PREPARATION OF TAGS AND RECORD SHEET. DANGER and CAUTION tags and the associated tag-out record sheet should be prepared as follows:

a. The person designated to prepare the tag-out is normally the ship's force petty officer in charge of the work. This person fills out and signs the record sheet and prepares the tags. Ditto marks, arrows, or similar devices may not be used on the tag-out record sheet.

b. A tag-out record sheet is prepared for a stated purpose. All tags used for that purpose are listed on one record sheet and continued on additional sheets as necessary. The stated purpose may include several work items. Each record sheet is assigned a log serial number in sequence. The index/audit record is used to assign log serial numbers. Log serial numbers are also used to identify all tags associated with a given purpose. Each tag is given its own sequential number as it is entered in the record sheet. For example, tag 7-16 would be the sixteenth tag issued on a single record sheet with the log serial number seven. To differentiate among tag-out logs, a prefix system, approved by the commanding officer, is used with the log serial numbers.

c. The tag-out record sheet includes references to other documents that apply. Some examples are work permits, work procedures, repair directives, reentry control forms, test forms, and rip-out forms. Certain information should be obtained either from reference documents or from the personnel requesting the work. Some examples are the reasons for the tag-out, the hazards involved, the amplifying instructions, and the

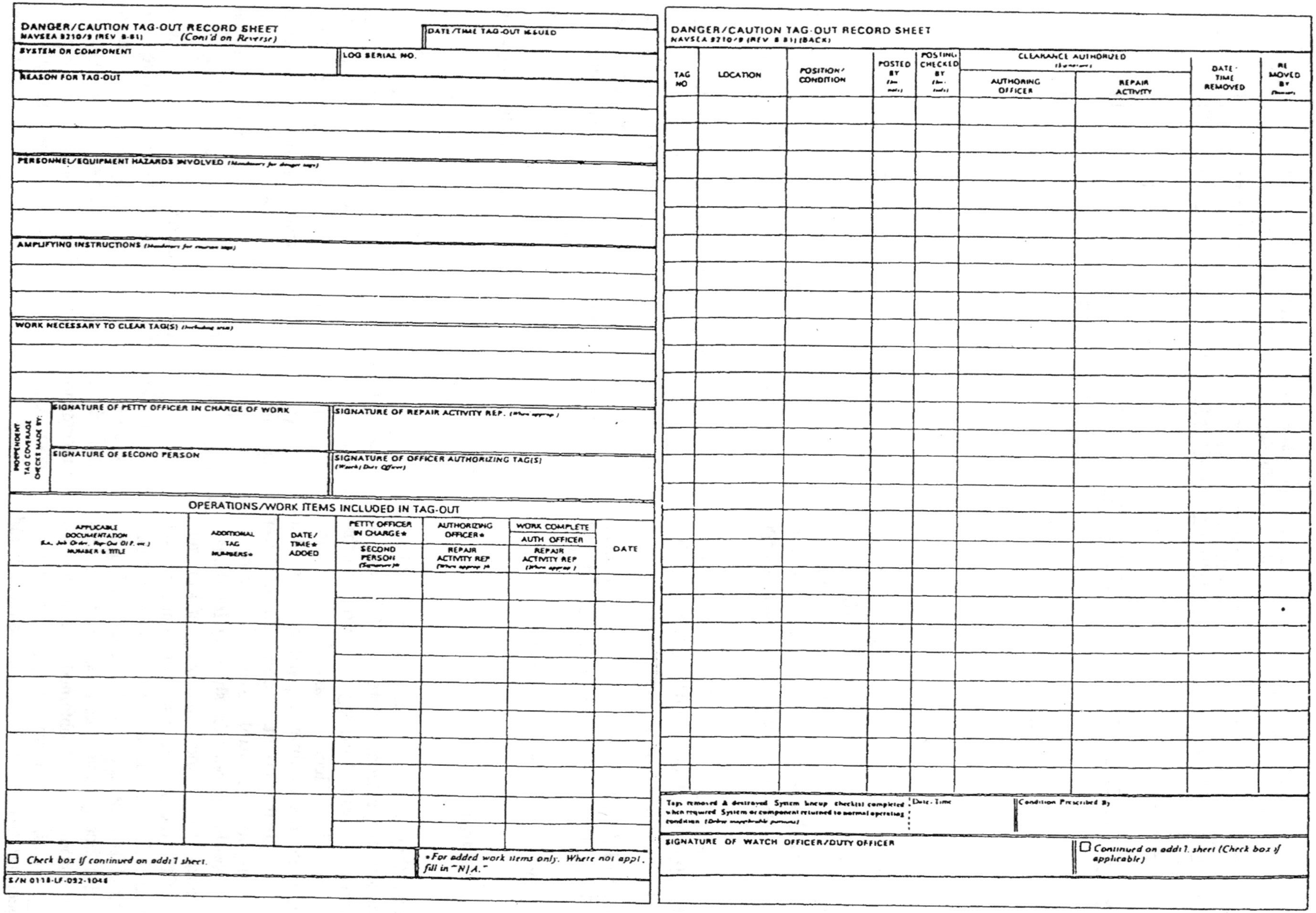

DANGER/CAUTION TAG-OUT RECORD SHEET
NAVSEA 9210/9 (REV 8-81) (Cont'd on Reverse)

DATE/TIME TAG-OUT ISSUED

SYSTEM OR COMPONENT | LOG SERIAL NO.

REASON FOR TAG-OUT

PERSONNEL/EQUIPMENT HAZARDS INVOLVED

AMPLIFYING INSTRUCTIONS

WORK NECESSARY TO CLEAR TAG(S)

INDEPENDENT TAG COVERAGE CHECKS MADE BY:

SIGNATURE OF PETTY OFFICER IN CHARGE OF WORK | SIGNATURE OF REPAIR ACTIVITY REP.

SIGNATURE OF SECOND PERSON | SIGNATURE OF OFFICER AUTHORIZING TAG(S)

OPERATIONS/WORK ITEMS INCLUDED IN TAG-OUT

APPLICABLE DOCUMENTATION NUMBER & TITLE	ADDITIONAL TAG NUMBERS*	DATE/TIME* ADDED	PETTY OFFICER IN CHARGE* / SECOND PERSON	AUTHORIZING OFFICER* / REPAIR ACTIVITY REP	WORK COMPLETE AUTH OFFICER / REPAIR ACTIVITY REP	DATE

☐ Check box if continued on add'l sheet.

*For added work items only. Where not appl. fill in "N/A."

S/N 0116-LF-092-1044

DANGER/CAUTION TAG-OUT RECORD SHEET
NAVSEA 9210/9 (REV 8-81) (BACK)

TAG NO	LOCATION	POSITION/CONDITION	POSTED BY	POSTING CHECKED BY	CLEARANCE AUTHORIZED: AUTHORIZING OFFICER	CLEARANCE AUTHORIZED: REPAIR ACTIVITY	DATE-TIME REMOVED	REMOVED BY

Tags removed & destroyed. System lineup checklist completed when required. System or component returned to normal operating condition. | Date-Time | Condition Prescribed By

SIGNATURE OF WATCH OFFICER/DUTY OFFICER

☐ Continued on add'l. sheet (Check box if applicable)

Figure 25-8.–DANGER/CAUTION tag-out record sheet (front/back).

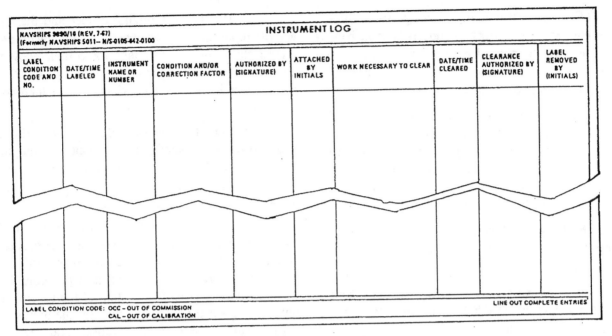

LABEL CONDITION CODE AND NO.	DATE/TIME LABELED	INSTRUMENT NAME OR NUMBER	CONDITION AND/OR CORRECTION FACTOR	AUTHORIZED BY (SIGNATURE)	ATTACHED BY INITIALS	WORK NECESSARY TO CLEAR	DATE/TIME CLEARED	CLEARANCE AUTHORIZED BY (SIGNATURE)	LABEL REMOVED BY (INITIALS)

NAVSHIPS 9690/16 (REV. 7-67) (Formerly NAVSHIPS 5011— N/S-0105-842-0100)

INSTRUMENT LOG

LABEL CONDITION CODE: OCC – OUT OF COMMISSION
CAL – OUT OF CALIBRATION

LINE OUT COMPLETE ENTRIES

Figure 25-9.—Instrument log.

work necessary to clear the tags. This information should be detailed enough to give watch standers a clear understanding of the purpose of, and necessity for, each tag-out action.

d. Use enough tags to completely isolate the system, piping, or circuit being worked on. Be sure you use tags to prevent the operation of a system or component from all the stations that could exercise control. Use system diagrams or circuit schematics to determine the number of tags needed. Indicate the location and position/condition of each tagged item by an easily identifiable means. Some examples are MS-1, STBD TG BKR, OPEN, SHUT, BLANK FLANGE INSTALLED.

e. After the tags and the tag-out record sheet have been filled out, have a second person make an independent check of the tag-out coverage and usage. That person should use the appropriate circuit schematics and system diagrams. The second person verifies the completeness of the tag-out action by signing the record sheet.

f. The authorizing officer then reviews the record sheet and tags for adequacy and accuracy. When satisfied, the officer signs the record sheet and the tags.

(1) If a tag-out is requested by a repair activity, the repair activity representative (shop supervisor or equivalent) must sign the tag-out record sheet. This shows that the repair activity is satisfied with the completeness of the tag-out. Verified tags alert all personnel that the repair activity must approve the removal of the tags.

(2) If the repair activity representative's concurrence is not required, this space on the record sheet need not be filled in.

(3) On ships with DCC, the authorizing officer annotates the tag-out record sheet in the upper right-hand corner with the words *DCC notified*, and then initials it. This ensures that DCC knows the extent of the tag-out and the status of the material condition of the unit.

(4) The authorizing officer then authorizes installation of the tags.

g. The person attaching the tag must make sure the item tagged is in the prescribed position/condition. If not, get permission from the authorizing officer to change an item to the prescribed condition or position. As each tag is attached and the position/condition verified, the person attaching the tag must sign the tag and initial the record sheet.

NOTE: Only qualified ship's force personnel may position equipment and affix tags and labels. Attach the tags so they will be noticed by anyone who wants to operate the component. Do not attach tags to breaker covers or valve caps that may be removed later.

h. After all tags have been attached, a second person must independently verify proper positioning and tag attachment, sign each tag, and initial the record sheet. If repair activity concurrence is required, a repair activity representative must witness the verification, sign the tags, and initial the tag-out record sheet.

NOTE: Only qualified ship's force personnel may perform the second check of tag installation.

i. Sometimes additional tags are required because of added work on an existing tag-out record sheet. In that case, the DANGER and CAUTION tags and tag-out record sheet must be handled as follows:

(1) The person preparing the change to the existing tag-out record sheet must make sure the purpose of the existing record sheet remains unchanged by the new work and associated tags.

(2) Will out the tag-out record sheet to reflect the added work. Prepare whatever additional tags are required. Review the reason for the tag-out, the hazards involved, the amplifying instructions, and the work necessary to clear tags. Do this on the existing tag-out record sheet to make sure it reflects the old work and the new work being added to the record sheet. After this review of the record sheet is completed, the petty officer in charge of the work signs the first coverage check block next to the added work item.

(3) Number each tag added to the existing tag-out sequentially in the tag series for the original tag-out. Annotate the serial numbers of these tags next to the associated new work item on the record sheet. Enter the updated number of effective tags at the top of the record sheet by crossing through the previous number and adding the new number.

(4) After the new tags and the tag-out record sheet have been filled out and signed by the petty officer in charge of the work, a second review is made. The second person makes an independent check of the tag coverage and usage by referring to appropriate schematics and diagrams. This person signs the record sheet in the block for the new work item to show satisfaction with the completeness of the tag-out actions. This includes both the additional and the previously issued tags.

(5) The authorizing officer and the repair activity representative, when required, review the entire record sheet and the new tags for completeness and accuracy. They then sign their respective blocks for the added work item. The authorizing officer then issues the tags.

j. Do not start work until all the DANGER tags required for the protection of personnel or equipment have been attached according to established procedures.

2. REMOVING DANGER AND CAUTION TAGS. Remove these tags immediately after the situation requiring the tag-out has been corrected. As each work item identified on the tag-out record sheet is completed, delete it from the tag-out record sheet. Completed work items listed in the Operations/Work Items Included in Tag-Out section of the record sheet must be signed off. This is done by the authorizing officer (and repair activity representative, when required) in the designated signature block. All DANGER tags must be properly cleared and removed before a system or portion of a system can be operationally tested and restored to service. To remove individual tags, the authorizing officer must make sure the remaining tags provide adequate protection for work, testing, or operations that still remain to be performed. Tags may only be removed following the signed authorization of the authorizing officer. When a tag-out action was initiated by a repair activity, an authorized representative of that repair activity must agree that the job is complete. A shop supervisor or equivalent must sign the tag-out record sheet before the tags may be removed. As the tags are removed, the date/time of removal must be initialed. Ditto marks are not allowed. All tags must be returned immediately to the authorizing officer. This officer then requires a system lineup or a lineup check. Tags that have been removed must be destroyed after they have been delivered to the authorizing officer. All tags associated with one tag-out action must be destroyed and the system or component returned to normal operating (shutdown) condition. The authorizing officer then certifies these actions by entering the date and the time when the system lineup or lineup check was completed. In a case where a system or component restoration was performed according to a specific document, reference to that document is made in the Condition Prescribed By block. Inapplicable portions of the statements on the record sheet are lined out and initialed when a valve lineup check is not required or the system is not returned to a normal condition. The authorizing officer also must enter the date and time cleared on the appropriate line of the tag-out index/audit record. The completed record

sheets must be removed from the effective section of the log and placed in the completed section. They will be reviewed and removed by a designated officer. On ships having a DCC, the authorizing officer must notify DCC that the tag-out has been cleared. The authorizing officer annotates the completed tag-out record sheet in the lower right-hand corner on the reverse side with the words *DCC notified*, and then initials it.

a. When any component is tagged more than once, the DANGER tag takes precedence over all other tags. All DANGER tags must be removed and cleared according to this procedure before the equipment may be operationally tested or operated.

b. A missing or damaged tag is reissued by indicating on the tag-out record sheet, on the line corresponding to the tag, that the tag was missing or damaged and that a replacement was issued. The new tag is issued using the next number in the tag-out record sheet. The authorizing officer should sign the tag-out record sheet to authorize the clearing of the damaged or missing tags and to authorize their replacement. Agreeing repair activity signatures must be obtained as appropriate.

3. ISSUANCE AND REMOVAL OF LABELS. Labels are issued and removed in a manner similar to that required for tags.

a. The authorizing officer authorizes the use of labels by signing the label and the instrument log. When labels are required for reactor plant systems and reactor plant support systems, the repair activity representative agrees by signing on the label and in the instrument log next to that of the authorizing officer.

b. Second check signatures are not required on the label and the instrument log.

c. When the out-of-commission or the out-of-calibration labels are signed, they must be affixed to the exterior surfaces of the affected instrument. This must be done so operators can easily determine the instrument's status.

d. A different procedure is used for installed instruments not associated with propulsion plants on nuclear-powered ships and for portable test and radiac equipment. In these cases, the labels may be replaced by those affixed by a qualified instrument repair or calibration facility.

Tag-out Information

A tag-out procedure is necessary because of the complexity of modern ships and the cost, delays, and hazards to personnel that can result from the improper operation of equipment. Learn and use the following guidelines:

–Enforce the tag-out procedure at all times. It is necessary during normal operations as well as during construction, testing, repair, or maintenance.

–Do not use tags or labels as substitutes for other safety measures. Examples are chaining or locking valves, removing fuses, or racking out circuit breakers. However, you must attach tags to the fuse panel, the racked-out circuit breaker cabinet, or a locked valve to show a need for action. You do not need to use tags where a device will be locked during normal operations.

–Use tags to show the presence of, and the requirement for, freeze seals, blank flanges, or similar safety devices. When equipment or components are placed out of commission, use the tag-out procedures to control the status of the affected equipment. Examples are disconnecting electrical leads, providing jumpers, or pulling fuses for testing or maintenance.

–Never use tag-outs to identify valves, to mark leaks, or for any purpose not specified in the tag-out procedure.

–Do not laminate tags or labels for reuse. The reuse of tags or labels is not allowed.

–The absence of a tag or label may not be taken as permission for unauthorized operation of equipment.

–Whenever a tag or label is issued, correct the situation requiring the tag or label so it can be removed as soon as possible.

–The tag-out procedure is for use by the ship's personnel on the equipment and systems for which they are responsible. However, repair activity personnel should use the procedure to the maximum extent practicable with systems and equipment that are still under construction.

–OPNAVINST 3120.32B is also required when work is being done by an intermediate level maintenance activity on equipment or systems that are the responsibility of the ship's force. Sometimes a ship is under construction or assigned to a repair activity not under the control of the TYCOM. When that happens, the ship's force and the repair activity may have to agree on the use of tags and labels. In this case, the tag-out system should be formal in nature and familiar to both the repair activity and the ship's force.

–Any person who knows of a situation requiring tags or labels should request that they be issued and applied.

–When using labels, you should list on the log any associated requirements specified for installation procedures, test procedures, work permits (ripouts or reentries), or system turnover agreements.

–Decide on a case-by-case basis whether an OUT-OF-COMMISSION or an OUT-OF-CALIBRATION instrument label is to be used. In general, if the instrument error is small and consistent, you can use an OUT-OF-CALIBRATION label and the operator may continue to use the instrument. When an OUT-OF-CALIBRATION label is used, mark on the label the magnitude and units of the required correction. However, when you use an OUT-OF-COMMISSION label, the instrument should not be used.

–Use enough tags to completely isolate a section of piping or circuit being worked on, or to prevent the operation of a system or component from all stations that could exercise control. Use system diagrams or circuit schematics to determine the adequacy of all tag-out actions.

–Careful planning of tag-outs can significantly reduce the number of record sheets and tags. Planning also can reduce the effort required to perform audits particularly during periods of overhaul or repair.

HEAT STRESS PROGRAM

Heat stress may occur in many work spaces throughout the Navy. Heat stress is any combination of air temperature, thermal radiation, humidity, airflow, and workload that may stress the human body as it attempts to regulate its temperature. Heat stress becomes excessive when your body's capability to adjust is exceeded. This results in an increase in body temperature. This condition can readily produce fatigue, severe headaches, nausea, and poor physical and/or mental performance. Prolonged exposure to heat stress could cause heatstroke or heat exhaustion and severe impairment of the body's temperature-regulating ability. These conditions can be life-threatening if not immediately and properly treated. Recognizing personnel with heat stress symptoms and getting them prompt medical attention is an all-hands responsibility.

Primary causes that increase heat stress conditions are as follows:

• Excessive steam and water leaks

• Boiler air casing leaks

• Missing or deteriorated lagging on steam piping, valves, and machinery

• Clogged ventilation systems or an inoperative fan motor

• Ships operating in hot or humid climates

Heat stress can be controlled somewhat by making sure that all lagging and insulation are in their proper place, that steam and hot water leaks are corrected as soon as possible, and that all the ventilation systems are operating as designed. There are other measures you can take to help reduce heat stress. You can make sure readings are taken and recorded at each watch or work station every hour and at any other time the temperature exceeds 100°F dry-bulb temperature. You can make sure the readings are reported to the EOOW so a heat survey can be conducted and corrective action can be taken.

HEARING CONSERVATION AND NOISE ABATEMENT PROGRAM

Historically, hearing loss has been recognized as an occupational hazard related to certain rates. Exposure to high-intensity noise occurs as a result of either impulse or blast noise, such as gunfire or rocket fire, or from continuous or intermittent sounds, such as jet or propeller aircraft, marine engines, and a myriad of noise sources from machinery and equipment.

Hearing loss has been and continues to be a source within the Navy. Hearing loss attributed to such occupational exposure to hazardous noise, the high cost of related compensation claims, and the resulting drop in productivity and efficiency have highlighted a significant problem that requires considerable attention. The goal of the Navy Hearing Conservation Program is to prevent occupational noise-related hearing loss among Navy personnel.

Hazardous noise areas and equipment must be so designated and appropriately labeled. The loud, high-pitched noise produced by an operating propulsion plant can cause hearing loss. A hearing loss can seldom be restored. For this reason, ear protection must be worn in all areas where the sound level is 85 dB or greater. In these places, warning signs must be posted cautioning about noise hazard that may cause loss of hearing.

Personnel working in or entering designated hazardous noise areas must have hearing protection devices available at all times. When noise sources are operating, personnel must wear their hearing protection devices regardless of exposure time.

PERSONNEL QUALIFICATION STANDARDS PROGRAM

Personnel Qualification Standards (PQS) are a list of knowledges and skills required to qualify for a specific watch station, maintain specific equipment, or perform as a team member within a unit. The purpose of the PQS program is to assist in qualifying personnel to perform their assigned duties. The PQS program is not a training program, but it does provide objectives to be met through training. The PQS program is most effective when it is used as a key element of a well-structured and dynamic unit training program. This program also helps personnel prepare for advancement. The program places most of the responsibility for learning on the individual and encourages self-achievement. It also provides a means for you to monitor the progress of your personnel.

Each PQS guides trainees toward a specific qualification goal by telling them exactly what they must learn to achieve that goal. Each qualification standard is divided into three sections: Fundamentals, Systems, and Watchstations. The Fundamentals section covers basic knowledge needed to understand the specific equipment or duties and provides an analysis of those fundamentals that broadly apply. The Systems section deals with the major working parts of the installation, organization, or equipment the PQS is concerned with. The Watchstations section defines the actual duties, assignments, and responsibilities you will be performing to obtain your qualification.

ENGINEERING OPERATIONAL SEQUENCING SYSTEM (EOSS)

The many types of engineering plants that exist in today's modern Navy require an ever-increasing range and depth of operational knowledge by engineering personnel at all levels of shipboard operations. The Engineering Operational Sequencing System (EOSS) provides each of these levels with the required information to enable the engineering plant to respond to any demands placed upon it that are within its design capability.

The EOSS is a set of systematic and detailed written procedures, using charges, instructions, and diagrams that provide the information required for the operation and casualty control functions of a specific shipboard propulsion plant.

The system will improve the operational readiness of the ship's engineering plant by increasing its operational efficiency, providing better engineering plant control, reducing operational casualties, and extending equipment life.

The EOSS defines the levels of control and operation within the engineering plant and provides each supervisor and operator with the necessary information, in easily understood words, about each watch station. The EOSS is composed of three basic parts: the user's guide, engineering operational procedures (EOP), and engineering operational casualty control (EOCC).

• The USER'S GUIDE, installed with each system, is a booklet that explains the EOSS package and how it is used to the ship's best advantage. It contains samples of the various system documents and explains how they are used. Recommendations on how to introduce the EOSS and how to train the ship's personnel in the use of these procedures for the system installed aboard ship are also included.

• The EOP section of EOSS contains all the information necessary to operate the ship's engineering plant properly. They also aid in scheduling, controlling, and directing plant evolutions from receiving shore service to underway and back down.

• The EOCC section of EOSS contains information about recognizing certain symptoms of a casualty and the probable causes and effects of each casualty. It also contains information on preventive action that may be taken to preclude a casualty as well as procedures for controlling single source and multiple (cascade effect) casualties.

REFERENCES

Engineering Administration, NAVEDTRA 10858-F1, Naval Education and Training Program Management Support Activity, Pensacola, Florida, 1988.

IC Electrician 3, NAVEDTRA 10559-A, Naval Education and Training Program Management Support Activity, Pensacola, Florida, 1989.

Interior Communications Electrician, Volume 3, NAVEDTRA 12162, Naval Education and Training Program Management Support Activity, Pensacola, Florida, 1992.

Machinist's Mate 3 & 2, NAVEDTRA 12144, Naval Education and Training Program Management Support Activity, Pensacola, Florida, 1991.

Naval Ships' Technical Manual, Chapter 090, "Inspections, Tests, Records, and Reports," NAVSEA S9086-CZ-STM-000, Naval Sea Systems Command, Washington, D. C., 1988.

Quality Assurance Manual, COMNAVSURFLANT-INST 9090.2, Commander Naval Surface Force, Atlantic, Norfolk, Va., November 1985.

Ships' Maintenance and Material Management (3-M) Manual, OPNAVINST 4790.4B, Chief of Naval Operations, Washington, D. C., August 1987.

APPENDIX I

GLOSSARY

ABSOLUTE PRESSURE–Actual pressure (includes atmospheric pressure).

AFTERCOOLER–A terminal heat-transfer unit after the last stage.

AIR EJECTOR–A jet pump that removes air and noncondensable gases.

AIR LOCK SYSTEM–A system of control devices combined to hold all final operating elements in the position existing before loss of air supply pressure.

AIR REGISTER–A device in the casing of a boiler that regulates the amount of air for combustion and provides a circular motion to the air.

ALLOY–A mixture of two or more metals.

ALTERNATING CURRENT (ac)–Current that is constantly changing in value and direction at regularly recurring intervals.

AMBIENT TEMPERATURE–The air temperature of the room or shipboard space.

ATMOSPHERE ESCAPE PIPING–Piping that leads from safety or large volume relief valves in machinery spaces, up the outer stack to the atmosphere.

ATMOSPHERIC PRESSURE–The pressure exerted by the atmosphere in all directions as indicated by a barometer. Standard atmospheric pressure is considered to be 14.7 pounds per square inch (psi), which is equivalent to 29.92 inches of mercury (in.Hg).

ATOMIZATION–The spraying of a liquid through a nozzle so the liquid is broken into tiny droplets or particles.

AUTOMATIC CONTROLLER–An instrument or device that operates automatically to regulate a controlled variable in response to a setpoint and/or input signal.

AUTOMATIC CONTROL SYSTEMS–A combination of instruments or devices arranged systematically to control a process or operation at a setpoint without assistance from operating personnel.

AUTOMATIC OPERATION–Operation of a control system and the process under control without assistance from the operator.

AUXILIARIES–Propulsion plant equipment not powered by main steam.

AUXILIARY–Systems or components functioning in a secondary capacity to the main boilers and propulsion turbines, such as pumps, air ejectors, and blowers.

AXIAL–In a direction parallel to the axis. Axial movement is movement parallel to the axis.

BABBITTED–Lined with a babbitt metal (containing tin, copper, and antimony).

BACK PRESSURE–(1) Refers to the resistance to the flow of exhaust fluids through the exhaust system. (2) The pressure exerted on the exhaust side of a pump or engine.

BAFFLE–A plate installed to disperse (scatter) motion and/or change the direction of flow of fluids.

BALLASTING–The process of filling empty tanks with salt water to protect the ship from underwater damage and to increase its stability.

BDC (bottom dead center)–The position of a reciprocating piston at its lowest point of travel.

BEARING–A mechanical component that supports and guides the location of another rotating or sliding member.

BELL BOOK–An official record of engine orders received and answered.

BIAS OR BIASING–The act of adding to or subtracting from a control system signal.

BIMETALLIC–Two dissimilar metals with different rates of expansion when subjected to temperature changes.

BLEEDER–A small cock valve, or plug, that drains off small quantities of air or fluids from a container or system.

BLOCK DIAGRAM–A drawing of a system using blocks for components to show the relationship of the components.

BLOWING OF TUBES–A procedure that uses steam to remove soot and carbon from the tubes of steaming boilers.

BLUEPRINT–A reproduced copy or drawing (usually having white lines on a blue background).

BOILER–A strong metal tank or vessel composed of tubes, drums, and headers, in which water is heated by the gases of combustion to form steam.

BOILER BLOW PIPING–Piping from the individual boiler blow valves to the overboard connection at the skin of the ship.

BOILER CENTRAL CONTROL STATION–A centrally located station for directing the control of all boilers in the fireroom.

BOILER DESIGN PRESSURE–Pressure specified by the manufacturer, usually about 103 percent of normal steam drum operating pressure.

BOILER EFFICIENCY–The efficiency of a boiler is the ratio of the Btu per pound of fuel absorbed by the water and steam to the Btu per pound of fuel fired. In other words, boiler efficiency is output divided by input, or heat use divided by heat available. Boiler efficiency is expressed as a percentage.

BOILER FEEDWATER–Deaerated water in the piping system between the deaerating feed tank and the boiler.

BOILER FULL-POWER CAPACITY–The total quantity of design steam flow required to develop the specified horsepower of the ship, divided by the number of boilers installed in the ship. Also expressed as the number of pounds of steam generated per hour at a specified pressure and temperature.

BOILER INTERNAL FITTINGS–All parts inside the boiler that control the flow of steam and water.

BOILER LOAD–The steam output demanded from a boiler, generally expressed in pounds per hour (lb/hr).

BOILER REFRACTORY–Materials used in the boiler furnace to protect the boiler from the heat of combustion.

BOILER WATER–The water actually contained in the boiler.

BONNET–A cover used to guide and enclose the tail end of a valve spindle.

BOTTOM BLOW–A procedure used to remove suspended solids and sludge from a boiler.

BOURDON TUBE–A C-shaped hollow metal tube that is used in a gauge for measuring pressures of 15 psi and above. One end of the C is welded or silver-brazed to a stationary base. Pressure on the hollow section forces the tube to try to straighten. The free end moves a needle on the gauge face.

BRAZING–A method of joining two metals at high temperature with a molten alloy.

BRINE–Any water in which the concentration of chemical salts is higher than seawater.

BRITISH THERMAL UNIT (Btu)–A unit of heat used to measure the efficiency of combustion. It is equal to the quantity of heat required to raise 1 pound of water 1°F.

BRITTLENESS–That property of a material that causes it to break or snap suddenly with little or no prior sign of deformation.

BUCKET WHEEL–The steel wheel or disc, fitted to a turbine shaft, to which the blading is attached.

BULL GEAR–The largest gear in a reduction gear train–the main gear, as in a geared turbine drive.

BURNERMAN–Person in the fireroom who tends the burners in the boilers.

BUS–The common connection between a group of line cutout switches. The bus may be in one single piece or it may be divided; may be free or directly connected to a jack outlet.

BUSHING–A renewable lining for a hole through which a moving part passes.

BUS TRANSFER–A device for selecting either of two available sources of electrical power. It may be accomplished either manually or automatically.

BUTTERFLY VALVE–A lightweight, relatively quick-acting, positive shutoff valve.

BYPASS–To divert the flow of gas or liquid. Also, the line that diverts the flow.

CALIBRATION–The procedure required to adjust an instrument or device to produce a standardized output with a given input. The amount of deviation from the standard must first be determined to ascertain the proper correction requirements.

CAPILLARY TUBE–A slender, thin-walled, small-bored tube used with remote-reading indicators.

CARBON MONOXIDE–A deadly, colorless, odorless, and tasteless gas formed by incomplete burning of hydrocarbons.

CARBON PACKING–Pressed segments of graphite used to prevent steam leakage around shafts.

CARRYOVER–(1) Boiler water entrained with the steam (by foaming or priming). (2) Particles of seawater trapped in vapor in a distilling plant and carried into the condensate.

CASING–A housing that encloses the rotating element (rotor) of a pump or turbine.

CASING THROAT–An opening in a turbine or pump casing through which the shaft protrudes.

CASUALTY–An event or series of events in progress during which equipment damage and/or personnel injury has already occurred. The nature and speed of these events are such that proper and correct procedural steps will serve only to limit personnel injury.

CASUALTY POWER SYSTEM–Portable cables that are rigged to transmit power to vital equipment in an emergency.

CELSIUS–Thermometer scale on which the boiling point of water is 100° and the freezing point is 0°.

CENTRIFUGAL FORCE–The outward force on a rotating body.

CHARACTERIZER–A control system component that acts to alter a signal in a predetermined manner to match a nonlinear parameter in the process under control.

CHECK VALVE–A valve that permits a flow of liquid in one direction only.

CHEMICAL ENERGY–Energy stored in chemicals (fuel) and released during combustion of the chemicals.

CHILL SHOCKING–A method that uses steam and cold water to remove scale from the tubes of a distilling plant.

CHLORIDE–A compound of the chemical element chlorine with another element or radical.

CHLORINE–A heavy, greenish-yellow gas used in water purification, sewage disposal, and the preparation of bleaching solutions. It is poisonous in concentrated form.

CIRCUIT BREAKER–An electrical switching device that provides circuit overload protection.

CIRCULATING WATER–Water circulated through a heat exchanger (condenser or cooler) to transmit heat away from an operating component.

CLARIFIER–A water tank containing baffles that slow the rate of water flow sufficiently to allow heavy particles to settle to the bottom and light particles to rise to the surface. This separation permits easy removal, thus leaving the clarified water. The clarifier is sometimes referred to as a settling tank or sedimentation basin.

CLASSIFICATION AND/OR TYPE–A method of identifying and sorting various equipment and materials. For example: (1) check valves–swing, stop, and so forth; (2) valve–solenoid, manual, and so forth.

CLUTCH–A form of coupling that is designed to connect or disconnect a driving or driven member.

COLD IRON CONDITION–An idle plant, as in a destroyer when all port services are received from an external source such as shore or tender.

COMBINING TUBES–Short open-ended tubes in which the inner surfaces are paralleled, or nearly so, and used to combine two inlets into a single outlet.

COMBUSTIBLE–A material that can burn.

COMBUSTION–The burning of fuel in a chemical process accompanied by the evolution of light and heat.

COMBUSTION CONTROL SYSTEM–A system that regulates fuel rate and combustion airflow to a boiler so that steam is produced at a constant pressure and fuel is burned with optimum combustion efficiency.

COMBUSTION EFFICIENCY–The ratio of the energy in the combustion gases, theoretically available for absorption by the boiler under actual operating conditions, to the energy available had the fuel been burned with the minimum theoretical combustion air.

COMPONENT–Individual unit, or part of a system; also, the major units that, when suitably connected, comprise a system.

CONDENSATE–Water produced in the cooling system of the steam cycle from steam that has returned from the turbine or from steam that has returned from various heat exchangers. The water is used over again to generate steam in the boiler for an endless repetitive cycle.

CONDENSATE DEPRESSION–The difference between the temperature of condensate in the condenser hotwell and the saturation temperature corresponding to the vacuum maintained in the condenser.

CONDUCTIVITY (OF WATER)–The ability of water to conduct an electric current. It is expressed in micromhos/cm. Generally, the amount of dissolved solids is directly proportional to the conductivity.

CONTROLLER (AUTOMATIC)–A device or group of devices arranged to automatically regulate a controlled variable according to a command or setpoint signal.

CONTROL POWER–Power that controls or operates a component or component part.

CONVECTION–The transmission of heat by the circulation of a liquid or a gas such as air. Convection may be forced by use of a pump or fan, or it may occur naturally due to heated air or liquid or liquid rising and forcing the colder air or liquid downward.

COOLANT–Liquid in the cooling system.

COOLER–Any device that removes heat. Some devices, such as oil coolers, remove heat to waste in overboard seawater discharge; other devices, such as ejector coolers, conserve heat by heating condensate for boiler feedwater.

COOLING SYSTEM–Heat removal process that uses mechanical means to remove heat to maintain the desired air temperature. The process may also result in dehumidification.

CORROSION–A gradual wearing away or alteration of metal by a chemical or electrochemical process. Essentially, it is an oxidizing process, such as the rustling of iron by the atmosphere.

COUPLING–A device for securing together adjoining ends of piping, shafting, and so forth in a manner to permit disassembly when necessary.

CRITICAL SPEED–The speed at which the centrifugal force of a rotating element tends to overcome the natural weight of the element, causing distortion and vibration.

CROSS-CONNECT–To align piping of systems to provide flow between machinery groups.

CROSS-CONNECTED PLANT–A method of operating two or more plants as one unit from a common steam supply.

CURTIS STAGE–A velocity-compounded impulse turbine stage having one pressure drop in the nozzles and two velocity drops in the blading.

DAMPING–A characteristic of a system that results in dissipation of energy and causes decay in oscillations. The negative feedback of an output rate of change.

DEAERATE–The process of removing dissolved oxygen.

DEAERATING FEED TANK (DFT)–A unit in the steam-water cycle used to (1) free the condensate of dissolved oxygen, (2) heat the feedwater, and (3) act as a reservoir for feedwater.

DEBALLASTING–The process by which salt water is emptied from tanks.

DEGREE OF SUPERHEAT–The amount by which the temperature of steam exceeds the saturation temperature.

DESIGN PRESSURE (BOILER)–The pressure specified by a manufacturer as a criterion in design. (In a boiler, it is approximately 103 percent operating pressure.)

DESIGN TEMPERATURE (BOILER)–The intended operating temperature at the superheater outlet, at some specified rate of operation. The specified rate of operation is normally full-power capacity.

DIFFUSER–A device that spreads a fluid out in all directions and increases fluid pressure while decreasing fluid velocity.

DIRECT CURRENT (dc)–Current that moves in one direction only.

DIRECT DRIVE–One in which the drive mechanism is coupled directly to the driven member.

DIRECT-DRIVEN–Driven at the same speed as the driver (not having reduction gears).

DISTILLATE–The product (fresh water) resulting from the condensation of vapors produced by the evaporation of seawater.

DISTILLATION–The process of evaporating seawater, then cooling and condensing the resulting vapors. Produces fresh water from seawater by separating the salt from the water.

DOUBLE REDUCTION–A reduction gear assembly that reduces the high input rpm to a lower output rpm in two stages.

DUPLEX PUMP–A pump that has two liquid cylinders and is referred to as double-acting.

DUPLEX STRAINER–A strainer containing two separate elements, independent of each other.

ECONOMIZER–A heat transfer device that uses the gases of combustion to preheat the feedwater in the boiler before it enters the steam drum.

EFFICIENCY–The ratio or the output to the input. Also, the degree of conversion of heat of steam to usable mechanical power output.

ELASTICITY–The ability of a material to return to its original size and shape.

ELECTRICAL ENERGY–Energy derived from the forced induction of electrodes from one atom to another.

ELECTROLYSIS–A chemical action that takes place between unlike metals in systems using salt water.

EMULSIFIED OIL–A chemical condition of oil in which the molecules of the oil have been broken up and suspended in a foreign substance (usually water).

ENGINEERING LOG–A legal record of important events and data concerning the machinery of a ship.

ENGINE ORDER TELEGRAPH–A device on the ship's bridge to give orders to the engine room. Also called annunciator.

EVAPORATION–The action that takes place when a liquid changes to a vapor or gas.

FAIL POSITION–The operating or physical position to which a device will go upon loss of its actuating electrical, electronic, pneumatic, or hydraulic control signal.

FEEDBACK–Information about a process output that is communicated to the process input.

FEEDER–An electrical conductor or group of conductors between different generating or distributing units of a power system.

FEEDWATER–Water that meets the requirements of *NSTM*, chapter 220 (9560), for use in a boiler.

FERROUS METAL–Metal with a high iron content.

FLASH POINT OF OIL–The temperature at which oil vapor will flash into fire although the main body of the oil will not ignite.

FLEXIBLE COUPLING–A coupling that transmits rotary motion from one shaft to another while compensating for minor misalignment between the two units.

FORCE–Anything that tends to produce or modify motion.

FREQUENCY–The number of vibrations, cycles, or changes in direction in a unit of time.

FRESHWATER DRAINS–A collective term that refers to drainage from steam heating systems and warming-up drainage from other higher pressure steam systems. These drains are of feedwater quality and are returned to the boiler condensate system.

FRICTION–Resistance to relative motion between two bodies in contact with each other.

FUSE–A protective device that will open a circuit if the current flow exceeds a predetermined value.

GAUGE PRESSURE–Pressure above atmospheric pressure.

GAIN–The ratio of the signal change, which occurs at the output of a device, to the change at the input.

GAS FREE–A term that describes a space that has been tested and found safe for entry.

GENERATOR–A machine that converts mechanical energy into electrical energy.

GOVERNOR–A speed-sensitive device designed to control or limit the speed of a turbine or engine.

HARDENING–The heating and rapid cooling (quenching) of metal to induce hardness.

HARDNESS–A quality exhibited by water containing various dissolved salts, principally calcium and magnesium. Can result in a heat transfer resistant scale on the steam-generating surfaces.

HEAT EXCHANGER–Any device that is designed to allow the transfer of heat from one fluid (liquid or gas) to another.

HERTZ (Hz): Frequency per second of alternating current. Formerly referred to as cycles per second.

HORSEPOWER (hp)–A unit to indicate the time rate of doing work equal to 550 foot-pounds per second or 33,000 foot-pounds per minute.

HOTWELL–Reservoir attached to the bottom of a condenser for collecting condensate.

HUMIDITY–The vapor content of the atmosphere. Humidity can vary depending on air temperature;

the higher the temperature, the more vapor the air can hold.

HYDRAULICS–The study of liquid in motion.

HYDROCARBON–Chemical compound of hydrogen and carbon; all petroleum fuels are composed of hydrocarbons.

HYDROGEN–A highly explosive, light, invisible, nonpoisonous gas used in underwater welding and cutting operations.

HYDROMETER–An instrument used to determine the specific gravity of liquids.

HYDROSTATIC–Static (nonmoving) pressure generated by pressurizing liquid.

IGNITION COMPRESSION–When the heat generated by compression in an internal-combustion engine ignites the fuel (as in a diesel engine).

IGNITION TEMPERATURE–The minimum temperature to which the substance (solid, liquid, or gas) must be heated to cause self-sustained combustion.

IMPELLER–An encased, rotating element provided with vanes that draw in fluid at the center and expel it at a high velocity at the outer edge.

INERTIA–The tendency of a stationary object to remain stationary and of moving objects to remain in motion.

INLET PLENUM–That section of the GTE inlet air passage that is contained within the engine enclosure. Applies to GTM and GTGS engines.

INJECTOR–A device used in a diesel engine to force fuel into the cylinders.

INSULATION–A material that retards heat transfer.

INTERCOOLER–An intermediate heat transfer unit between two successive stages, as in an air compressor.

INTERFACE–Surface or area between two abutting parts usually of different materials or systems.

INTEGRATED THROTTLE CONTROL (ITC)–Has control electronics located in the PACC that allows single lever control of throttle and pitch of one shaft. Two levers (one per shaft) are located at the SCC and at the PACC.

INTERLOCK–A feature or device in one system or component that affects the operation of another system or component. Generally, a safety device, but it may be used to control the operating sequence of components.

KEY–A parallel-sided piece inserted into a groove cut part way into each of two parts, which prevents slippage between the two parts.

KEYWAY–A slot cut in a shaft, pulley hub, wheel hub, and so forth. A square key is placed in the slot and engages a similar keyway in the mating piece. The key prevents slippage between the two parts.

KINETIC ENERGY–Energy in motion producing work.

LATENT HEAT–Heat that is given off or absorbed by a substance while it is changing its state.

LATENT HEAT OF CONDENSATION–The amount of heat (energy) required to change the state of a substance from a vapor to a liquid without a change in temperature.

LOCAL MANUAL OPERATION–Direct manual positioning of a control valve or power operator by a handwheel or lever.

LOOP SEAL–A vertical U-bend in drain piping in which a water level is maintained to create an airtight seal.

LUBRICATING OIL PURIFIER–A unit that removes water and sediment from lubricating oil by centrifugal force.

MAIN CONDENSER–A heat exchanger that converts exhaust steam to feedwater.

MAIN INJECTION (SCOOP INJECTION)–An opening in the skin of a ship designed to deliver cooling water to the main condenser and main lubricating oil cooler by the forward motion of the ship.

MANIFOLD–A fitting with numerous branches used to convey fluids between a large pipe and several smaller pipes.

MAXIMUM OPERATING PRESSURE–The highest pressure that can exist in a system or subsystem under normal operating conditions. This pressure is determined by such influences as pump or compressor shutoff pressures, pressure-regulating valve lockup (no-flow) pressure, and maximum chosen pressure at the system source.

MAXIMUM SYSTEM PRESSURE–The highest pressure that can exist in a system or subsystem during any condition. Normal, abnormal, and emergency operation and casualty conditions must be considered in determining the maximum system pressure. In any system or subsystem with relief valve protection, the nominal setting of the relief valve must be taken as maximum system pressure (relief valve accumulation may be ignored).

MICROMHO–Electrical unit used with salinity indicators for measuring the conductivity of water. It is equivalent to the quantity of one divided by the resistance of the water to electrical conductivity.

MONITORING POINT–The physical location at which any indicating device displays the value of a parameter at some control station.

MORPHOLINE–A chemical that prevents sludge in boilers by neutralizing the acidic quality of condensate, thereby reducing corrosion in condensate and feedwater piping, which forms as sludge in boilers.

MOTIVE STEAM–Steam that performs work in the turbine steam path.

NAVY DISTILLATE (ND)–A fuel used in steam-powered ships of the Navy. ND is a fuel of the middle to higher distillation range. Military specification MIL-F-24397 (ships), NATO symbol F85 covers the requirements for Navy distillate fuel.

NAVY SPECIAL FUEL OIL (NSFO)–A fuel oil used in steam-powered ships of the Navy. NSFO is a blend of heavy residuum and heavy distillates (both cracked and virgin) with cutter stock distillate, used to adjust viscosity. The blend is made at the refinery in proper proportions to comply with Military specification MIL-F-859, NATO symbol F77.

NONFERROUS METAL–Metals that are composed primarily of some element or elements other than iron.

NOZZLE–That portion of a turbine that converts the heat energy of steam into a directed steam and sets the amount of steam flow.

NOZZLE-BLOCK–The turbine part that takes steam from the turbine chest and directs it into the first stage of the turbine.

NOZZLE DIAPHRAGM–A removable metal ring inserted in the casing between the stages of the turbine. This ring contains the nozzles by which steam flows from one stage to another.

OIL POLLUTION ACTS–The Oil Pollution Act of 1924, the Oil Pollution Act of 1961, and the Water Quality Improvement Act of 1970 prohibit the overboard discharge of oil or water that contains oil in port in any sea area within 12 miles of land and in special prohibited zones.

OPERATING CHARACTERISTICS–The combination of a parameter and its setpoints.

OPERATING PRESSURE–The constant pressure at which a component is designed to operate in service.

OPERATING TEMPERATURE–The actual temperature of a component during operation.

ORIFICE PLATE–A place with an opening fitted between flanges in piping systems to reduce velocity and pressure in steam traps and the steam supply to distilling plants.

OVERLOAD RELAY–An electrical protective device that automatically trips when a circuit draws excessive current.

OXIDATION–The process of various elements and compounds combining with oxygen. The corrosion of metals is generally a form of oxidation; on iron, for example, it is iron oxide, or oxidation.

PANTING–A series of pulsations caused by minor, recurrent explosions in the firebox of a ship's boiler. It is usually caused by a shortage of air.

PARALLEL CIRCUIT–An electrical circuit with two or more resistance units wired to permit current flow through both units at the same time. Unlike the series circuit, the current in the parallel circuit does not have to pass through one unit to reach the other.

PARAMETER–A variable, such as temperature, pressure, flow rate, voltage, current, and frequency, that may be indicated, monitored, checked, or sensed in any way during operation or testing.

PERIPHERY–The curved line that forms the boundary of a circle (circumference), ellipse, or similar figure. Also, the outer bounds of something as distinguished from the center or internal regions.

pH–A chemistry term that denotes the degree of acidity or alkalinity of a solution. The pH of a water solution may have any value between 0 and 14. A solution with a pH of 7 is neutral; above 7, it is alkaline; below 7, it is acidic.

PINION–A gear that meshes with a larger mating gear.

PIPING MAIN–The larger or primary piping, extending throughout the boundaries of a system, to which components or subsystems are interconnected by smaller branch lines.

PITOMETER LOG–A device that indicates the speed of a ship and the distance traveled by measuring water pressure on a tube projected outside the ship's hull.

PLUG-COCK–A valve that has a rotating plug that is drilled for passage of flushes.

PNEUMATIC–Driven or operated by air pressure.

PNEUMERCATOR–A type of manometer that measures the volume of liquid in tanks.

POUNDS PER SQUARE INCH (psi)–Unit of pressure.

POSITIONER–That part of a control drive loaded by a control signal, which supplies energy to an actuator in such a manner that the final control element is positioned according to the control signal.

POTENTIAL ENERGY–Energy at rest; stored energy.

POWER–The time rate of doing work.

POWER LEVER ANGLE (PLA)–A rotary actuator mounted on the side of the GTM fuel pump and its output shaft lever. It is mechanically connected to the MFC power lever. The PLA actuator supplies the torque to position the MFC power lever at the commanded rate.

POWER TAKEOFF (PTO)–The drive shaft between the GTGS gas turbine engine and the reduction gear. It transfers power from the gas turbine to the reduction gear to drive the generator.

POWER TURBINE (PT)–The GTM turbine that converts the GG exhaust into energy and transmits the resulting rotational force via the attached output shaft.

PPM (PARTS PER MILLION)–Concentration of the number of parts of a substance dissolved in a million parts of another substance. It is used to measure the salt content of water. If 1 pound of sea salt were dissolved in 1,000,000 pounds of water, the sea salt concentration would be 1.00 ppm.

PRAIRIE AIR SYSTEM–Disguises the sonar signature of the ship's propellers by emitting cooled bleed air from small holes along the leading edges of the propeller blades. The resulting air bubbles disturb the thrashing sound so identification of the type of ship through sonar detection becomes unreliable.

PRIMARY ELEMENT–That part of a measuring device that affects, or is affected by, the quantity being measured to produce a signal capable of being sensed by a transmitter or indicator.

PRIME MOVER–The source of motion, such as a turbine or an automobile engine.

PROTECTIVE FEATURE–A feature of a component or component part designed to protect a component or system from damage.

PUMP CAPACITY–The amount of fluid a pump can move in a given period of time. It is usually stated in gallons per minute (gpm).

PUNCHING TUBES–A process for cleaning the interiors of boiler tubes.

PURGE–To make free of an unwanted substance; as to bleed air out of a fuel system.

PURPLE-K-POWDER (PKP)–A purple powder composed of potassium bicarbonate that is used on class B fires. It can be used on class C fires; however, CO_2 is a better agent for such electrical fires because it leaves no residue.

QUILL SHAFT–A reduction gear shaft that connects the first reduction gear to the second reduction pinion.

RACE (BEARING)–The inner or outer ring that provides a contact surface for the balls or rollers in a bearing.

RADIAL BEARINGS–Bearings designed to carry loads applied in a plane perpendicular to the axis of the shaft and used to prevent movement in a radial direction.

RADIAL THRUST BEARINGS–Bearings designed to carry a combination of radial and thrust loads. The loads are applied both radially and axially with a resultant angular component.

RADIATION–Transfer of heat in the form of waves similar to light and radio waves, without physical contact between the emitting and the receiving regions.

REACTION TURBINE–A turbine in which the major part of the driving force is received from the reactive force of steam as it leaves the blading.

RECEIVER INDICATORS–Pressure-sensitive instruments indicating the loading pressure signals in percentage.

RECIRCULATION SYSTEM–The process of removing heat and moisture with cooled air by mechanical or natural distribution ductwork. The process may include filtering, heating, and dehumidifying.

RECTIFIER–A device for converting alternating current into direct current.

RECTIFY–To make an alternating current flow in one direction only.

REFRACTORY–Various types of heat resistant, insulating material used to line the insides of boiler furnaces.

RELATIVE HUMIDITY (RH)–The ratio of the weight of water vapor in a quantity of air to the weight of water vapor that quantity of air would hold if saturated at the existing temperature. Usually expressed as a percentage; for example, if air is holding half the moisture it is capable of holding at the existing temperature, the RH is 50 percent.

RELAY–A magnetically operated switch that makes and breaks the flow of current in a circuit. Also called the cutout and circuit breaker.

REMOTE MANUAL OPERATION–Human operation of a process by manual manipulation of loading signals to the final control elements.

REMOTE OPERATING GEAR–Rods or flexible cables attached to valve wheels so the valves can be operated from another compartment or level.

RESERVE FEEDWATER–Water stored in tanks for use in the boiler feedwater system as needed.

ROOT VALVE–A valve located where a branch line comes off the main line.

SAFETY VALVE–An automatic, quick opening and closing valve that has a reset pressure lower than the lift pressure.

SALINE/SALINITY–(1) Constituting or characteristic of salt. (2) Relative salt content of water.

SALIENT POLE GENERATOR–A generator whose field poles are bolted to the rotor, as opposed to a generator whose field poles are formed by imbedding field windings in slot in a solid rotor.

SALINOMETER–A hydrometer that measures the concentration of salt in a solution.

SATURATION TEMPERATURE–The temperature at which a liquid boils under a given pressure. For a given pressure there is a corresponding saturation temperature.

SAYBOLT VISCOMETER–An instrument that determines the fluidity or viscosity (resistance to flow) of an oil.

SCALE–An undesirable deposit, mostly calcium sulfate, that forms in the tubes of boilers.

SEA CHEST–An arrangement for supplying seawater to engines, condensers, and pumps and for discharging waste from the ship to the sea. It is a cast fitting or a built-up structure located below the waterline of the vessel and having means for attachment of the piping. Suction sea chests are fitted with strainers or gratings.

SECURE–(1) To make fast or safe. (2) The order given on completion of a drill or exercise. (3) The procedure followed with any piece of equipment that is to be shut down.

SECURED PLANT CONFIGURATION–The condition in which engines of a set (GTM) are disengaged from the reduction gear/propulsion shaft.

SEDIMENT–An accumulation of matter that settles to the bottom of a liquid.

SENSIBLE HEAT–Heat that is given off or absorbed by a substance without changing its state.

SENSING POINT–The physical and/or functional point in a system at which a signal may be detected and monitored or may cause some automatic operation to result.

SEPARATOR–A trap for removing oil and water from compressed gas before it can collect in the lines or interfere with the efficient operation of pneumatic systems.

SERIAL DATA BUS–Major communication link between ship control equipment, PAMCE, and PLOE. The bus is time-shared between the consoles. Control and status information is exchanged in the form of serial data words.

SERVICE TANKS–Tanks in which fluids for use in the service systems are stored.

SHIP CONTROL CONSOLE (SCC)–Located on the bridge and is part of the SCE. It has equipment for operator control of ship's speed and direction.

SHIP CONTROL EQUIPMENT–Bridge located equipment of the ECSS and includes SCC, SCEEE, PSCU, BWDU, and RPIU.

SHIP CONTROL EQUIPMENT ELECTRONIC ENCLOSURE (SCEEE)–The SCEEE (located in

CIC) is part of ship control equipment. It has power supplies that provide the various operating voltages required for the SCC, BWDU, and RPIU.

SHAFT ALLEY–The long compartment of a ship in which the propeller shafts revolve.

SIMPLEX PUMP–A pump that has only one liquid cylinder.

SLIDING FEET–A mounting for turbines and boilers to allow for expansion and contraction.

SLUDGE–(1) The sediment in the lower portion of a secured boiler resulting from the settling of suspended solids in the boiler water. The sediment may include, besides the suspended solids, oil and other contaminants. (2) The sediment left in fuel oil tanks.

SOLID COUPLING–A device that joins two shafts rigidly.

SPECIFIC GRAVITY–The relative weight of a given volume of a specific material as compared to the weight of an equal volume of water.

SPECIFIC HEAT–The amount of heat required to raise the temperature of 1 pound of a substance 1°F. All substances are compared to water, which has a specific heat of 1 Btu/lb/°F.

SPLIT PLANT–A method of operating propulsion plants so they are divided into two or more separate and complete units.

STANDBY EQUIPMENT–Two identical auxiliaries that perform one function. When one auxiliary is running, the standby is so connected that it may be started if the first fails.

STATIC–A force exerted by reason of weight alone as related to bodies at rest or in balance.

START AIR COMPRESSOR (SAC)–An LP air compressor, driven by the SSDGs, that provides air to start the GTMs on FFG-7 class ships.

TANK STRIPPING–Stripping is the process of removing normally small amounts of water that collect in the bottom of fuel or other tanks.

VARIABLE STATOR VANE (VSV)–Compressor stator vanes that are mechanically varied to provide optimum, stall-free compressor performance over a wide operating range. The inlet guide vanes (IGVs) and stage 1 through 6 stator vanes of the main propulsion gas turbine compressors are variable.

WASTE HEAT BOILER–Each waste heat boiler is associated with a GTGS and uses the hot GT exhaust to convert feedwater to steam for various ship's services.

INDEX

CPSIA information can be obtained
at www.ICGtesting.com
Printed in the USA
FSHW02n0032060918
51848FS